SUPERSTRONG FIELDS IN PLASMAS

SUPERSTRONG FIELDS IN PLASMAS

First International Conference

Varenna, Italy   August-September 1997

EDITORS
M. Lontano
Istituto di Fisica del Plasma "Piero Caldirola," C.N.R., Milano, Italy

G. Mourou
Center for Ultrafast Optical Science, University of Michigan

F. Pegoraro
Dipartimento di Fisica, Università di Pisa, Pisa, Italy

E. Sindoni
International School of Plasma Physics "Piero Caldirola," Milano, Italy

AIP CONFERENCE PROCEEDINGS 426

American Institute of Physics   **Woodbury, New York**

Editor:

Maurizio Lontano
Istituto di Fisica del Plasma "Piero Caldirola"
Consiglio Nazionale delle Ricerche
EURATOM-ENEA-CNR Association
via R. Cozzi 53
20125 Milano
ITALY

Email: lontano@ifp.mi.cnr.it

Authorization to photocopy items for internal or personal use, beyond the free copying permitted under the 1978 U.S. Copyright Law (see statement below), is granted by the American Institute of Physics for users registered with the Copyright Clearance Center (CCC) Transactional Reporting Service, provided that the base fee of $15.00 per copy is paid directly to CCC, 222 Rosewood Drive, Danvers, MA 01923. For those organizations that have been granted a photocopy license by CCC, a separate system of payment has been arranged. The fee code for users of the Transactional Reporting Service is: 1-56396-748-0/ 98 /$15.00.

© 1998 American Institute of Physics

Individual readers of this volume and nonprofit libraries, acting for them, are permitted to make fair use of the material in it, such as copying an article for use in teaching or research. Permission is granted to quote from this volume in scientific work with the customary acknowledgment of the source. To reprint a figure, table, or other excerpt requires the consent of one of the original authors and notification to AIP. Republication or systematic or multiple reproduction of any material in this volume is permitted only under license from AIP. Address inquiries to Office of Rights and Permissions, 500 Sunnyside Boulevard, Woodbury, NY 11797-2999; phone: 516-576-2268; fax: 516-576-2499; e-mail: rights@aip.org.

L.C. Catalog Card No. 98–70235
ISBN 1-56396-748-0
ISSN 0094-243X
DOE CONF- 9708740

Printed in the United States of America

CONTENTS

Preface xi
Conference Staff xiii
Acknowledgments xv
Welcome address: Prof. G. Lampis xvii

1. FUNDAMENTAL ATOMIC AND PLASMA PROCESSES AND NONLINEAR PHENOMENA IN THE PRESENCE OF ULTRAINTENSE FIELDS

Atoms in Intense Laser Fields 3
    C. J. Joachain
Few Optical-Cycle Pulse Interactions with Plasmas:
Models and Nonlinear Effects 15
    A. M. Sergeev, V. B. Gildenburg, A. V. Kim, M. Lontano,
    and M. Quiroga-Teixeiro
Parametric Instabilities and Electron Heating in Ultra-Intense
Laser-Plasma Interaction 32
    P. Mora, J. C. Adam, A. Héron, S. Guérin, G. Laval, and B. Quesnel
Radiation at $2\omega_p$ from Inverse Two-Plasmon Decay in Overdense Plasma
Driven by Ultrashort Laser Pulses 41
    R. Lichters, J. Meyer-ter-Vehn, and A. M. Pukhov
Emission of Electromagnetic Waves by Accelerated Short
Laser Pulses in a Plasma 49
    N. L. Tsintsadze and J. T. Mendonca
Ionization of Atoms in Intense Fields: Semiclassical Approach,
Fluid Equations, and Effects on Laser Propagation 55
    F. Cornolti, A. Macchi, and E. C. Jarque
Nonlinear Reflection of High Intensity Picosecond Laser Pulse
from Overdense Plasma 61
    A. A. Andreev, V. I. Bayanov, A. B. Vankov, A. A. Kozlov, I. V. Kurnin,
    K. Y. Platonov, N. A. Solovyev, S. A. Chizhov, and V. E. Yashin
Study of a Plasma Diffraction Grating Induced by Superstrong
Crossed Laser Beams 67
    L. Plaja and L. Roso
Surface Electromagnetic Waves at the Interface between Vacuum
and Plasma Produced by Ultrashort Laser Pulses 73
    S. A. Magnitskii, V. T. Platonenko, and A. V. Tarasishin
Self-Confinement Plasma Effect in Intense Laser Interaction
with a Cluster Gas 79
    A. V. Kim, D. Anderson, M. D. Chernobrovtseva, and M. Lisak
Ionization Induced Self-Guiding of Femtosecond Laser Pulses in Gases 85
    M. Lontano, M. D. Chernobrovtseva, A. V. Kim, and A. M. Sergeev

2. RELATIVISTIC NONLINEAR OPTICS OF ULTRASHORT LASER PULSES IN PLASMAS

Relativistic Nonlinear Optics in Plasmas by 3D PIC Simulations............. 93
    A. M. Pukhov and J. Meyer-ter-Vehn
Nonlinear Optics in Relativistic Plasmas 103
    D. Umstadter, S.-Y. Chen, G. S. Sarkisov, A. Maksimchuk,
    and R. Wagner
Ultraintense Magnetic Fields in Laser Plasma Interaction:
Their Generation and Influence on Light Propagation...................... 113
    F. Pegoraro, S. V. Bulanov, F. Califano, T. Zh. Esirkepov, M. Lontano,
    N. M. Naumova, A. M. Pukhov, and V. A. Vshivkov
Nonlinear Weibel Instability in the Interaction of an Underdense Plasma
with a "Relativistic" Laser Pulse 123
    F. Califano, R. Prandi, F. Pegoraro, and S. V. Bulanov
Relativistic Effects in Laser-Atom Interactions at Ultra-High Intensities 129
    N. J. Kylstra, A. E. Ermolaev, and C. J. Joachain
Formation of Ionization Fronts by Intense Short Laser Pulses
Propagating in a Neutral Gas.. 135
    C. C. Rosa, L. Oliveira e Silva, N. Lopes, and J. T. Mendonça
LPIC++ a Parallel One-Dimensional Relativistic Electromagnetic
Particle-in-Cell Code for Simulating Laser-Plasma-Interaction 141
    R. E. W. Pfund, R. Lichters, and J. Meyer-ter-Vehn
Enhancement of Ponderomotively Generated Wakefields in 1D.............. 147
    R. J. Kingham and A. R. Bell
Relativistic Electron Dynamics in Intersecting Laser Pulses 153
    Z.-M. Sheng and J. Meyer-ter-Vehn
Magnetic Field Generation in a Low Density Plasma Wake of a
Short Laser Pulse.. 159
    Z.-M. Sheng, J. Meyer-ter-Vehn, and A. M. Pukhov
Magnetic Fields in Highly Compressed Plasma Formed at the Front
of a Laser-Driven Channel 164
    A. V. Tatarinov and T. J. M. Boyd
A New Type of Longitudinal Plasma Waves Induced in an
Isotropic Plasma with a Superstrong Short Laser Pulse 170
    L. N. Tsintsadze and N. L. Tsintsadze
On the Theory of Relativistically Intense EM Wave Propagation
in Overdense Plasma and the Boundary-Value Problem 176
    L. N. Tsintsadze, K. Mima, and K. Nishikawa
Breaking of the Laser Excited Wakefield Due to Electron Beam............. 182
    L. N. Tsintsadze, K. Mima, and K. Nishikawa
On Self-Consistent Stationary Propagation of Relativistically Coupled
Electromagnetic and Electrostatic Waves in Three Component Plasma 186
    L. N. Tsintsadze and K. Nishikawa
Propagation of Relativistic Ultrashort Laser Pulse in Near-Critical
Density Plasma 192
    K. Nagashima, Y. Kishimoto, and H. Takuma

3. SOLID DENSITY PLASMAS AND OTHER ULTRAINTENSE LASER DRIVEN EXOTIC STATES OF MATTER

3.1 LASER-SOLID

Ultraintense Laser-Solid Interaction Phenomena 201
    P. Mulser, D. Bauer, S. Hain, and F. Cornolti

Measurement of Acceleration in Femtosecond Laser-Plasmas 213
    R. Häßner, W. Theobald, S. Niedermeier, K. Michelmann, T. Feurer,
    H. Schillinger, and R. Sauerbrey

Generation of High Order Optical Harmonics in Steep Plasma Density Gradients 221
    D. von der Linde

Isochoric Heating of Solid Density Matter with Ultra-High Intensity Femtosecond Laser 231
    Z. Jiang, P. Gallant, J. C. Kieffer, H. Pépin, O. Peyrusse, and J. L. Miquel

X-Ray Production and SHG from Femtosecond Plasmas Induced in Modified Solid Targets 241
    V. M. Gordienko, M. S. Dzhidzhoev, M. A. Joukov, A. B. Savel'ev,
    A. A. Shashkov, and R. V. Volkov

High Intensity 30 Femtosecond Laser Pulse Interaction with Thin Foils 253
    A. Giulietti, A. Barbini, P. Chessa, D. Giulietti, L. A. Gizzi,
    and D. Teychenne'

Optimizing Harmonics from Solid Targets 264
    M. Zepf, G. Pretzler, U. Andiel, D. M. Chambers, A. E. Dangor,
    P. A. Norreys, J. S. Wark, I. F. Watts, and G. D. Tsakiris

Soft X-Ray Spectroscopy and Heat Transport in Plasmas Created by High-Contrast Fs-Laser Pulses 270
    A. Saemann and K. Eidmann

Generation of a Train of Attosecond Pulses in the Reflected Field from a Laser-Plasma Interaction 276
    L. Plaja, L. Roso, K. Rzążewski, and M. Lewenstein

Kinetic Approach to Superintense Laser-Solid Interaction 282
    H. Ruhl, F. Cornolti, F. Califano, and A. Macchi

High Quality Shocks Produced by Lasers: Application to Equations of State Measurements 288
    S. Bossi, D. Batani, A. Bernardinello, L. Muller, A. Benuzzi, M. Koenig,
    B. Faral, T. A. Hall, and Th. Löwer

3.2 LASER-GAS

Coherent X-Ray Generation at 2.7 nm using 25 fs Laser Pulses 296
    A. Rundquist, Z. Chang, H. Wang, I. Christov, H. C. Kapteyn,
    and M. M. Murnane

Compression of High Energy Femtosecond Pulses by the
Hollow Fiber Technique: Generation of sub-5-fs Multigigawatt Pulses........ 304
    M. Nisoli, S. Stagira, S. De Silvestri, O. Svelto, S. Sartania, Z. Cheng,
    M. Lenzner, Ch. Spielmann, and F. Krausz

The Interaction of Atomic Clusters with Intense Laser Fields 314
    M. H. R. Hutchinson, T. Ditmire, J. W. G. Tisch, E. Springate,
    J. P. Marangos, M. B. Mason, and R. A. Smith

Theoretical/Experimental Studies of Ultraviolet High-Power-Density
Self-Trapped Channels 322
    A. B. Borisov, B. D. Thompson, A. McPherson, F. G. Omenetto,
    T. Nelson, W. A. Schroeder, K. Boyer, and C. K. Rhodes

Phase-Matched Optical Parametric Conversion of Ultrashort Pulses
in a Hollow Waveguide 331
    C. G. Durfee, S. Backus, M. M. Murnane, and H. C. Kapteyn

Anomalous Emission of Soft X-Rays from OFI-Plasmas 336
    G. Pretzler and E. E. Fill

Laser-Plasma Harmonics with High-Contrast Pulses and
Designed Prepulses.. 342
    R. S. Marjoribanks, L. Zhao, F. W. Budnik, G. Kulcsár, A. Vitcu,
    H. Higaki, R. Wagner, A. Maksimchuk, D. Umstadter, S. P. Le Blanc,
    and M. C. Downer

Interaction of TW Laser Pulses with High Density Gas Jet Targets
Near the Threshold for Relativistic Self-Focusing.......................... 348
    G. D. Tsakiris, R. Fedosejevs, and X. F. Wang

Energy Deposition and Transport Dynamics in Plasmas Produced
by Intense, Short Pulse Irradiation of Atomic Clusters 354
    T. Ditmire, R. A. Smith, E. T. Gumbrell, J. W. G. Tisch,
    and M. H. R. Hutchinson

Harmonic Generation During the Ionization of a Plasma by a
Short Light Pulse 360
    E. Conejero Jarque and L. Plaja

Time-Resolved Imaging of High Harmonic Radiation Using
Chirped Laser Pulses... 366
    J. W. G. Tisch, T. Ditmire, D. D. Meyerhofer, N. Hay, M. B. Mason,
    and M. H. R. Hutchinson

3.3 FAST ELECTRON PRODUCTION

Propagation in Compressed Matter of Hot Electrons Created
by Short Intense Lasers... 372
    D. Batani, A. Bernardinello, V. Masella, F. Pisani, M. Koenig,
    J. Krishnan, A. Benuzzi, S. Ellwi, T. A. Hall, P. A. Norreys, A. Djaoui,
    D. Neely, S. Rose, P. Fews, and M. H. Key

Generation of Superhot Electrons by Intense Field Structures............... 377
    R. R. E. Salomaa, S. J. Karttunen, P. Mulser, T. J. H. Pättikangas,
    and W. Schneider

Hot Particle Generation in Ultrahigh Intensity Laser-Plasma Interaction 383
    C. Toupin, E. Lefebvre, and G. Bonnaud

Fast Electron Transport in Solid Targets 389
    J. R. Davies and A. R. Bell

4. LASERS FOR ULTRAHIGH INTENSITY PHYSICS

Ultrahigh Intensity Laser: Present and Future 397
    J. Nees, S. Biswal, F. Druon, J. Faure, M. Nantel, G. Mourou,
    A. Nishimura, H. Takuma, J. Itatani, J. C. Chanteloup, and C. Hönninger

Terawatt Picosecond CO_2 Laser Technology for Strong Field Physics Applications 415
    I. V. Pogorelsky

How to Measure Femtosecond Pulses 423
    M. A. Franco, J.-F. Ripoche, H. R. Lange, J. P. Chambaret, P. Rousseau,
    B. S. Prade, and A. Mysyrowicz

Methods for the Shaping High-Power Picosecond Laser Pulses with a High-Contrast Ratio 439
    V. A. Malinov, A. V. Charukchev, V. N. Chernov, N. V. Nikitin,
    S. L. Potapov, V. M. Efanov, and P. M. Yarin

30 TW Laser Facility "Progress-P" 445
    V. G. Borodin, A. V. Charukchev, V. N. Chernov, V. M. Komarov,
    S. V. Krasov, V. A. Malinov, V. M. Migel, N. V. Nikitin, V. S. Popov,
    and S. L. Potapov

Methods and Means of Superstrong Field Formation on Multiterawatt Laser Facility "Progress-P" 450
    V. N. Chernov, V. G. Borodin, A. V. Charukchev, V. A. Malinov,
    V. M. Migel, N. V. Nikitin, R. R. Gerke, and I. Yu. Yusupov

High Intensity Ultraviolet Laser and Next Generation Sources 454
    F. G. Omenetto, K. Boyer, J. W. Longworth, T. Nelson, W. A. Schroeder,
    and C. K. Rhodes

Development of a High-Peak and High-Average Power Ti:Sapphire Laser System 461
    K. Yamakawa, M. Aoyama, S. Matsuoka, Y. Akahane, H. Takuma,
    D. Fittinghoff, and C. P. J. Barty

Design Considerations for the Stretcher of a 35-fs Chirped Pulse Amplification Laser System 467
    M. B. Mason and M. H. R. Hutchinson

Implementation of a CPA Line-Focus Travelling-Wave for Highly Efficient Saturated Lasing of Ne-like Ti and Ge 473
    C. N. Danson, P. V. Nickles, R. Allott, A. Behjat, J. Collier, A. Demir,
    M. P. Kalachnikov, M. H. Key, C. L. S. Lewis, D. Neely, D. A. Pepler,
    G. J. Pert, M. Schnürer, W. Sandner, V. N. Shlyaptsev, G. J. Tallents,
    P. J. Warwick, E. Wolfrum, and J. Zhang

A Multi-Channel Soft X-Ray Flat-Field Spectrometer 479
    D. Neely, D. M. Chambers, C. N. Danson, P. A. Norreys, S. G. Preston,
    F. Quinn, M. Roper, J. S. Wark, and M. Zepf

200 TW Upgrade of the Vulcan Nd:Glass Laser Facility 485
    C. B. Edwards, C. N. Danson, M. H. R. Hutchinson, D. Neely,
    and B. Wyborn

Saturation in Transient Gain Scheme of Collisionally Pumped Germanium X-Ray Laser 491
    K. A. Janulewicz, P. V. Nickles, M. P. Kalachnikov, M. Schnürer,
    W. Sandner, S. B. Healy, G. J. Pert, P. J. Warwick, C. L. S. Lewis,
    C. N. Danson, D. Neely, E. Wolfrum, A. Behjat, A. Demir, and G. J. Tallents

5. APPLICATIONS OF ULTRA-INTENSE FIELDS

Recent Progress in Coherent XUV Generation at RAL 499
    M. Zepf, J. Zhang, D. M. Chambers, A. E. Dangor, A. G. MacPhee, J. Lin,
    E. Wolfrum, J. Nilsen, T. W. Barbee, Jr., C. N. Danson, M. H. Key, C. L. S. Lewis,
    D. Neely, P. A. Norreys, S. G. Preston, R. M. N. O'Rourke, G. J. Pert, R. Smith,
    G. J. Tallents, I. F. Watts, and J. S. Wark

The Scope and Present Status of JAERI "Advanced Photon Research" Program 509
    H. Takuma

High Energy Electron Acceleration by Laser Wakefields 516
    K. Nakajima, H. Nakanishi, A. Ogata, M. Kando, H. Ahn, H. Dewa,
    S. Kondo, H. Kotaki, H. Sakai, T. Ueda, M. Uesaka, T. Watanabe,
    and K. Yoshii

Measurement of Ultra High Field Propagation and the Excited Wakefield 526
    H. Dewa, H. Ahn, M. Kando, H. Kotaki, K. Nakajima, and A. Ogata

Mega-Ampere Ion Currents and Nuclear Reactions in the Focal Laser Spot 532
    V. V. Goloviznin and T. J. Schep

6. ASTROPHYSICS APPLICATIONS OF ULTRAINTENSE LASER PULSES

Ultradense Hydrogen in Astrophysics, High-Pressure Metal Physics and Fusion Studies 541
    S. Ichimaru and H. Kitamura

Supernova Hydrodynamics Experiments on Nova 551
    B. A. Remington, S. G. Glendinning, K. Estabrook, R. J. Wallace,
    R. London, R. A. Managan, A. Rubenchik, D. Ryutov, K. S. Budil, J. Kane,
    D. Arnett, R. P. Drake, R. McCray, and E. Liang

Laboratory Astrophysics with Intense and Ultraintense Lasers 560
    H. Takabe

Author Index 571

Preface

Over the past ten years the technique of Chirped Pulse Amplification, along with important progress in short pulse generation, has revolutionized our ability to produce high-peak-power pulses. It is now possible to produce with tabletop systems pulses with peak power exceeding 10 TW and intensities approaching 10^{20} W/cm^2, that is, four to five orders of magnitude greater than was previously possible. In the near future we expect that this trend will continue, to reach the theoretical intensity limit of the best amplifying materials, i.e., 10^{23} W/cm^2. This revolution in laser intensities gives access to a fundamentally new physical regime governed by extreme electromagnetic fields, laser pressure, density, temperature, and acceleration.

Entirely new types of phenomena have already been observed, and more are predicted. For instance, the generation of very high harmonics of the laser up to the water window has been demonstrated and will lead to the development of bright, coherent x-ray sources with attosecond duration. These ultrashort bursts of coherent radiation will make possible the investigation of ultrafast phenomena in the kilo-electron-volt range. The electron acceleration in the wake of a plasma wave produced by an ultraintense laser pulse has been shown to experience acceleration gradients of the order of 200 GeV/m or three to four orders of magnitude greater than conventional technology. Let's point out that this quantum leap in laser intensity is very similar to the one that occurred in electronics in the 1960s with the introduction of the integrated circuit. The enormous light pressure, far exceeding the thermal pressure, combined with large lateral ponderomotive force will be used in novel Inertial Confinement Fusion to bore a hole in a pre-imploded DT target to ignite it. Extreme conditions of pressure, temperature, magnetic field, and acceleration — found only in stellar interiors or close to the horizon of a black hole — can be re-created for a very short amount of time by this university-type, compact, ultraintense laser. Finally, the interaction of these ultraintense pulses with superrelativistic electrons makes nonlinear quantum dynamical effects observable.

The first International Conference on Superstrong Fields in Plasmas was held August 27 - September 2, 1997, at the superb site of the Villa Monatero, Varenna, Italy. It was intended to focus on the most advanced theoretical and experimental results of laser-matter interaction in the different intensity regimes ranging from 10^{14} to 10^{21} W/cm^2. The great success of this workshop, with more than 100 participants coming from fourteen Countries, was a testimony to the excitement that this field experiences.

The present volume is divided into six main sections dealing with particular aspects of laser-matter interaction in the high- and ultrahigh-intensity regime:

1. Fundamental atomic and plasma processes
2. Relativistic nonlinear optics
3. Physics of solid density plasmas
4. Lasers for ultrahigh-intensity physics
5. Applications of ultrastrong fields
6. Application of ultraintense pulses to astrophysics

Finally, from an educational point of view, the variety of scientific topics discussed at this conference demonstrates clearly that compact, ultraintense lasers may bring some of the "Big Science" done on large instruments back to university settings. The large number and variety of topics is bound to attract large numbers of students to scientific and engineering disciplines.

Maurizio Lontano
ISTITUTO DI FISICA DEL PLASMA
"PIERO CALDIROLA"
Consiglio Nazionale delle Ricerche
EURATOM-ENEA-CNR Association
Milan, Italy

Gerard Mourou
CENTER FOR ULTRAFAST
OPTICAL SCIENCE
University of Michigan
Ann Arbor, U.S.A.

Francesco Pegoraro
DIPARTIMENTO DI FISICA
Universita' di Pisa
Pisa, Italy

Elio Sindoni
INTERNATIONAL SCHOOL
OF PLASMA PHYSICS
"PIERO CALDIROLA"
Milan, Italy

Conference staff

Conference Chairmen
G. Mourou   F. Pegoraro

International Scientific Committee

P. Corkum	M.H.R. Hutchinson	G. Lampis
M. Lontano (Secretary)	G. Mourou	A. Mysyrowicz
F. Pegoraro	M. Perry	A.M. Sergeev
T. Tajima	H. Takuma	D. von der Linde

Local Organizing Committee
International School of Plasma Physics "Piero Caldirola"

D. Palumbo (President)
E. Sindoni (Director)

Conference Secretariat
D. Pifferetti

Acknowledgments

This meeting was supported by

International School of Plasma Physics "Piero Caldirola"
Istituto di Fisica del Plasma "Piero Caldirola", C.N.R.,
 EURATOM-ENEA-CNR Association, Milano
Dipartimento di Fisica, Universita' di Milano
Amministrazione Provinciale di Lecco
Ente per le Nuove Tecnologie, l'Energia e l'Ambiente (ENEA)
Consiglio Nazionale delle Ricerche
Camera di Commercio di Lecco
Centro Innovazione Lecco S.p.A

Welcome address: Prof. G. Lampis
Director of the Istituto di Fisica del Plasma "Piero Caldirola" del C.N.R., Milano

Ladies and Gentlemen,

It is an honour for me to welcome the Participants to the First International Conference on "Superstrong Fields in Plasmas" on behalf of the Organizers of the Conference:
1) The "International School of Plasma Physics" and
2) The Institute for Plasma Physics of the National Council for Research of Italy,
both dedicated to the Name of their Founder: Prof. Piero Caldirola.

This formal act has been requested to me by Prof. Donato Palumbo, President of the School, and by Prof. Elio Sindoni, Director of the School, and obedience is a pleasure.

I would like to express my gratitude to the Persons involved in the organization of this Conference:
- The Director Prof. Elio Sindoni and the Staff he has with him.
- The International Scientific Committee.
- Dr. Maurizio Lontano who is the Scientific Secretary of the Conference.
- The Secretariat of Villa Monastero which represents the Local Organizing Committee, and in particular Mrs. Donatella Pifferetti.

Finally, I wish a fruitful and pleasant stay in Varenna to all the Participants and to their accompanying families.

Thank you

1. FUNDAMENTAL ATOMIC AND PLASMA PROCESSES AND NONLINEAR PHENOMENA IN THE PRESENCE OF ULTRAINTENSE FIELDS

ATOMS IN INTENSE LASER FIELDS

C.J. Joachain

*Physique Théorique, Faculté des Sciences CP 227,
Université Libre de Bruxelles, B-1050 Bruxelles, Belgium*

Abstract. The recent development of lasers capable of delivering short pulses of very intense radiation, over a frequency range extending from the infrared to the ultraviolet, has led to the discovery of new phenomena in laser interactions with atomic systems. In the first part of this article, a survey is given of the basic properties found in the study of multiphoton processes such as the multiphoton ionization of atoms, the emission by atoms of high order harmonics of the exciting laser light and laser-assisted electron-atom collisions. The second part contains a review of the main non-perturbative methods, based on the Floquet approach or the numerical solution of the time-dependent equations for the wave function, which have been used to perform theoretical studies of these processes.

INTRODUCTION

Recently, super-intense laser fields have become available over a wide frequency range extending from the infrared to the ultraviolet, in the form of short pulses yielding intensities well beyond the value $I_a = 3.5 \times 10^{16}$ W cm^{-2} corresponding to the atomic unit of electric field strength $\mathcal{E}_a = 5 \times 10^9$ V cm^{-1}. Such laser fields are strong enough that they compete with the Coulomb forces in controlling the electron dynamics in atomic systems. As a result of this interplay, atoms and molecules in intense laser fields exhibit novel, fascinating properties which have been discovered essentially via the study of multiphoton processes. These modified properties generate in turn new behaviour of bulk matter in intense laser fields, with wide ranging applications such as the interaction of ultrastrong laser fields with plasmas.

In the next section, a brief account will be given of the new physics discovered by studying multiphoton processes in atoms. The last section is devoted to the main non-perturbative theoretical methods which have been used to analyze these processes. Detailed reviews of high-intensity laser-atom physics can be found in the volume edited by Gavrila (1992), in the articles of Burnett et al. (1993), Joachain (1994), Di Mauro and Agostini (1995) and Protopapas et al. (1997) and in the Proceedings of the 7th International Conference on Multiphoton Processes, edited by Lambropoulos and Walther (1997).

MULTIPHOTON PROCESSES

In this section, we shall discuss three important multiphoton processes occuring in atoms (ions): multiphoton ionization, harmonic generation and laser-assisted electron-atom collisions.

1. Multiphoton ionization of atoms and ions.

We begin by considering the multiphoton (single) ionization (MPI) reaction

$$n\hbar\omega + A^q \to A^{q+1} + e^- \tag{1}$$

This process was first observed by Voronov and Delone (1965), who used a ruby laser to induce seven-photon ionization of xenon, and by Hall et al.(1965), who recorded two-photon ionization from the negative ion I^-. In the following years, important results were obtained by several experimental groups, in particular at Saclay concerning the dependence of the ionization yields on the intensity, absolute measurements of MPI cross sections, and the resonance-enhancement of MPI. A crucial step was made in our understanding of MPI when experiments detecting the photo-electrons were performed. In this way Agostini et al. (1979) discovered that the ejected electron in the reaction (1) could absorb photons in excess of the minimum required for ionization to occur. The study of this excess-photon ionization, known as "above threshold ionization" (ATI) has been one of the central themes of multiphoton physics in recent years.

The ATI photo-electron energy spectra consist of several peaks, separated by the photon energy $\hbar\omega$. As the intensity increases, peaks at higher energies appear, whose intensity dependence does not follow the (non-vanishing) lowest order perturbation theory (LOPT) prediction according to which the ionization rate for an n-photon process is proportional to I^n. Another remarkable feature is that as the intensity increases the low-energy ATI peaks are reduced in magnitude. The reason for this peak suppression is that the energies of the atomic states are Stark-shifted in the presence of a laser field. For low laser frequencies (e.g. a Nd-YAG laser with $\hbar\omega = 1.165$ eV), the AC Stark shifts for the lowest bound states are small in magnitude. On the other hand, the induced Stark shifts of the Rydberg and continuum states are essentially given by the electron ponderomotive energy U_p, which is the cycle-averaged kinetic energy of the electron and is given in atomic units (a.u) by

$$U_p = \frac{\mathcal{E}_0^2}{4\omega^2} \tag{2}$$

where $\mathcal{E}_0$ is the electric field strength. It is worth stressing that the ponderomotive energy U_p is proportional to I/ω^2 and may become quite large. For example, in the case of the Nd-YAG laser, $U_p = \hbar\omega = 1.165$ eV at the intensity $I \simeq 10^{13}$ W cm^{-2}. Since the energies of the Rydberg and continuum states are shifted upwards relative to the lower bound states by about U_p, there is a corresponding increase in the ionization threshold. If this threshold increase is such that $E_i + n\hbar\omega - U_p < 0$, where E_i denotes the unperturbed (field-free) energy of the initial state, then the peak in the ATI spectrum corresponding to ionization by n photons will be suppressed. We also remark that for short (sub-picosecond) pulses, the ATI peaks exhibit a sub-structure (Freeman et al., 1987), due to the fact that the time varying pulse brings different states of the atom into multiphoton resonance because of the intensity-dependent Stark shifts.

For increasing laser field strengths approaching the Coulomb field binding of the electron, ($I \simeq 10^{14}$ W cm^{-2}) and for low laser frequencies, the sharp ATI peaks of the photo-electron spectrum gradually disappear, giving rise to a continuous distribution (Augst et al., 1989; Mevel et al., 1993). This phenomenon can be interpreted qualitatively by using a quasi-static model in which the bound electron experiences an effective potential formed by adding to the atomic potential the contribution due to the laser electric field. This quasi-static approach was used by Keldysh (1965) to study tunneling ionization in the low frequency limit, and pursued by several authors (Faisal, 1973; Reiss, 1980; Ammosov et al., 1986). An important quantity in these studies is the Keldysh adiabaticity parameter γ, defined as the ratio of the laser and tunneling frequencies, which is given by

$$\gamma = \left(\frac{2\omega^2 I_p}{I}\right)^{1/2} \qquad (3)$$

where I_p is the ionization potential. If $\gamma \ll 1$ then tunneling dynamics will dominate. Above a critical intensity I_c (which is about 1.4×10^{14} W cm^{-2} for atomic hydrogen in the ground state), the electron can "flow over the top" of the barrier, so that field ionization occurs and the atom ionizes in about one orbital period.

The semi-classical, "recollision picture" developed by Kulander et al. (1993), Corkum (1993) and others is based on the idea that strong field ionization occurs in two steps. In the first (bound-free) step, an electron is liberated from its parent atom by tunneling ionization. In the second (free-free) step, the interaction with the laser field dominates, a fact which was used earlier in the "simpleman" (classical) picture of a quivering electron (van Linden van den Heuvell and Muller, 1988). As the phase of the field reverses, the electron can be accelerated back towards the atomic core. Scattering of the electron by the core then leads to single or multiple ionization, while radiative recombination can lead to harmonic generation.

This semi-classical two step model has been very useful in explaining a number of features found in recent experiments, such as the existence of a plateau in the photo-electron energy spectrum of noble gases (Paulus et al. 1994 a) and "rings" in angular distributions of photo-electrons (Yang et al. 1993; Paulus et al. 1994 b). It has also played an important role in our understanding of harmonic generation, a subject to which we now turn our attention.

2. Harmonic generation.

Atoms irradiated by an intense laser field can emit radiation at high-order multiples, or harmonics of the pump laser field (see e.g. Mc Pherson et al., 1987; L'Huillier and Balcou, 1993). This phenomenon, which can be enhanced in a suitable gaseous medium, has attracted considerable interest as a source of short pulse, high frequency coherent radiation.

The theoretical treatment of harmonic generation by an intense laser pulse focused into a gaseous medium has two main aspects. First, the microscopic, single atom response to the laser field must be analyzed. Then, the single atom spectra must be combined to obtain the macroscopic harmonic fields generated from the coherent emission of all the atoms in the laser focus; this is done by using the single-atom polarization fields as source terms in the propagation (Maxwell) equations.

We shall only discuss here the microscopic aspect of the problem. In response to a laser field, the electrons in an atom oscillate, the source of harmonic generation being the polarization of the medium induced by the laser field,

$$\mathbf{P}(t) = \mathcal{N}\mathbf{d}(t) \tag{4}$$

where $\mathcal{N}$ is the atomic density and $\mathbf{d}(t)$ is the laser-induced dipole moment. If $|\Psi(t)>$ denotes the atomic state vector in the presence of the laser field, then $\mathbf{d}(t)$ is defined as the expectation value of the electric dipole operator in the state $|\Psi(t)>$. In weak fields, the electrons oscillate essentially at the fundamental frequency ω. As the field intensity increases, the electron oscillations cause the emission of radiation at additional frequencies ω_N which, for an initial state of given parity, are odd multiples of the laser frequency, i.e. $\omega_N = N\omega$ with $N = 3, 5, ...$

The harmonic intensity distribution consists of a rapid decrease over the first few harmonics, followed by a plateau of approximately constant intensities, and then a cut - off. It is important to note that the existence of a plateau is a non-perturbative feature. It was discovered in the framework of time-dependent Schrödinger equation (TDSE) calculations (Krause et al., 1992) that the cut-off frequency in the harmonic spectrum is given approximately (in a.u.) by

$$\omega_c = I_p + 3\, U_p \tag{5}$$

where I_p is the ionization potential and U_p is the ponderomotive energy. In the two-step "recollision model", the maximum returning kinetic energy of a classical electron recolliding with the atomic one is given by 3.2 U_p, so that the highest energy which can be radiated is $I_p + 3.2\, U_p$ (Kulander et al., 1993; Corkum 1993). This result has also been recovered in semi-classical strong field calculations (L'Huillier et al. 1993; Lewenstein et al. 1994; Antoine et al., 1996) and is in good agreement with experiment.

Finally, we mention that high-order harmonics can be used to generate pulses of extremely short duration, in the attosecond range (1 as = 10^{-18}s). This topic is discussed by Lewenstein in this volume.

3. Laser-assisted electron-atom collisions.

An electron scattered by an atom (ion) in the presence of a laser field can absorb or emit radiation. Since these radiative collisions involve continuum states of the electron-atom (ion) system, they are often called "free-free transitions" (FFT). In weak fields, only one photon processes have a large enough probability to be observed. However, as the field strength is increased, multi-photon processes become important. Examples of laser-assisted electron-atom collisions are "elastic" collisions

$$e^- + A(i) + n\hbar\omega \to e^- + A(i) , \qquad (6)$$

inelastic collisions

$$e^- + A(i) + n\hbar\omega \to e^- + A(f) \qquad (7)$$

and single ionization (e, 2e) collisions

$$e^- + A(i) + n\hbar\omega \to A^+(f) + 2e^- \qquad (8)$$

where $A(i)$ and $A(f)$ denote an atom A in the initial state i and the final state f, respectively, and $A^+(f)$ means the ion A^+ in the final state f. We remark that positive values of n correspond to photon absorption (inverse bremsstrahlung), negative ones to photon emission (stimulated bremsstrahlung) and $n = 0$ to a collision process without net absorption or emission of photons, but in the presence of the laser field.

Direct information on laser-assisted electron-atom collisions is obtained by performing three-beam experiments, in which an atomic beam is crossed in coincidence by a laser and an electron beam, and the scattered electrons, having undergone FFT, are recorded. Several experiments of this kind have been performed, in which the exchange of photons between the electron-atom system and the laser field has been observed in laser-assisted "elastic" (Weingartshofer et al., 1977, 1983; Wallbank and Holmes, 1994) and inelastic (Mason and Newell, 1987; Wallbank et al., 1988, 1990; Luan et al., 1991) processes. It is worth noting that even at modest intensities ($I \simeq 10^8$ W cm^{-2}), multiphoton processes have been observed (see e.g. Weingartshofer et al., 1983) and must be analyzed by using non-perturbative methods.

NON-PERTURBATIVE METHODS

In this section we shall give a survey of the main non-perturbative methods which have been used to study laser-atom interactions at high intensities. We shall use a semi-classical approach in which the laser field is treated classically, while the atomic system is treated by using quantum mechanics; this approach is entirely justified for the intense fields considered here.

1. Floquet theory

Let us consider an atomic system in a laser field treated classically as a spatially homogeneous electric field $\mathcal{E}(t)$ of arbitrary polarization. Although more general laser fields can be considered (for example two-colour fields), we shall assume that we are dealing with a monochromatic field. Neglecting relativiste effects, we start from the time-dependent Schrödinger equation (in a.u.)

$$i\frac{\partial}{\partial t}|\Psi(t)>= [H_{at} + H_{int}(t)]|\Psi(t)> \qquad (9)$$

where H_{at} is the field-free atomic Hamiltonian and the laser-atom interaction term H_{int} can be written in both the length or the velocity gauge in the form

$$H_{int}(t) = H_+ e^{-i\omega t} + H_- e^{-i\omega t} . \qquad (10)$$

Since the Hamiltonian $H = H_{at} + H_{int}$ is periodic, the Floquet method (see e.g. Shirley, 1965; Chu, 1985) can be used the write the state vector in the Floquet-Fourier form

$$|\Psi(t)>= e^{-iEt} \sum_{n=-\infty}^{+\infty} e^{-in\omega t} |\phi_n> \qquad (11)$$

where the quasi-energy E is complex and the harmonic components $|\phi_n>$ satisfy the time-independent system of coupled equations

$$(E + n\omega - H_{at})|\phi_n>= H_+|\phi_{n-1}> + H_-|\phi_{n+1}> \qquad (12)$$

with $n = 0, \pm 1, \pm 2, ...$ These equations, together with appropriate boundary conditions, form an eigenvalue problem for the quasi-energies. These can be expressed as

$$E = E_i + \Delta - i\frac{\Gamma}{2} \qquad (13)$$

where E_i is the energy of the initial unperturbed (field free) state, Δ is the AC Stark shift and Γ is the total ionization rate.

Potvliege, Shakeshaft, Dörr and co-workers have performed detailed Floquet calculations for atomic hydrogen. Following Maquet et al. (1983), the system of coupled equations (12) is solved by expanding each harmonic component in position space, $\phi_n(\mathbf{r}) \equiv <\mathbf{r}|\phi_n>$, on a discrete basis set of Sturmian functions. This Sturmian-Floquet approach has been applied to calculate quasi-energy spectra, total and partial ionization rates, angular distributions and harmonic generation yields in atomic hydrogen. A review of this work has been given by Potvliege and Shakeshaft (1992).

In experiments performed at Bielefeld, Rottke et al. (1990) have measured photo-electron spectra for multiphoton ionization of H(1s) by linearly polarized sub-picosecond pulses having a maximum intensity of about 10^{14} W cm^{-2}, and a wavelength in the range 596-616 nm. The spectra exhibit ATI peaks, each peak containing sub-peaks due to Rydberg levels moving in and out of resonance because of the intensity-dependent Stark shifts, as explained above. The theoretical photo-electron yields, calculated by Dörr et al (1990) by using the Sturmian-Floquet ionization rates, are in good agreement with the experimental data of the Bielefeld group.

The Floquet theory has also been used by Gavrila and co-workers to study the behaviour of atoms irradiated by very intense, high frequency laser fields. Using a version of the theory initially formulated by Gavrila and Kaminski (1984) in the Kramers (acceleration) frame, they showed that at superintensities and when the frequency of the laser field is substantially larger than the threshold frequency for one-photon ionization, the ionization rate decreases as the intensity increases. This phenomenon, called "adiabatic stabilization", has been discussed in detail by Gavrila (1992). The experimental verification of the stabilization phenomenon is very difficult, because atoms subjected to a laser pulse will ionize during the rising edge of the pulse (i.e. at lower intensities), before reaching the stabilization regime. For this reason, experiments investigating stabilization have been carried out on Rydberg states (de Boer et al., 1993). Such experiments have given indications of stabilization in the case of the circular 5g state of neon.

2. The R-matrix-Floquet theory.

We shall now describe a non-perturbative method - the R-matrix-Floquet (RMF) theory - which, since its formulation a few years ago by Burke, Francken and Joachain (1990, 1991), has become a useful tool for studying multiphoton processes in intense laser fields. The RMF theory treats multiphoton ionization, harmonic generation and laser-assisted electron-atom collisions in a unified way. It is completely ab-initio and is applicable to an arbitrary atom (ion), allowing an accurate description of electron correlation effects. In its present form, however, it is limited to processes involving at most one unbound electron, thus excluding multiphoton multiple ionization and laser-assisted electron impact ionization.

Our starting point is the time-dependent Schrödinger equation (9) for an atomic system in a laser field. According to the R-matrix method (Wigner, 1946) we subdivide configuration space into two regions. The internal region is defined by the condition that the radial coordinates r_i of all N electrons of the atom (ion) are such that $r_i \leq a$ ($i = 1, 2, ...N$), where the sphere of radius a envelops the charge distribution of the target atom states retained in the calculation. In this region exchange effects involving all N electrons are important. The external region is defined so that one of the N electrons lies on or outside the sphere of radius a, while the remaining $N - 1$ electrons are confined within this sphere. Hence in this region exchange effects between the "external" electron and the remaining $N - 1$ electrons can be neglected.

Having divided configuration space into an internal and an external region, the time-dependent Schrödinger equation (9) is solved in these two regions separately, using the Floquet method. On the boundary ($r = a$), the solutions are matched, using the R-matrix, which relates the radial wave functions to their derivatives. The boundary conditions for $r \to \infty$ are formulated in the acceleration frame, because in this frame the channels decouple asymptotically.

The RMF theory has been applied to study multiphoton processes for a number of atoms and ions. These include the multiphoton ionization of atomic hydrogen (Dörr et al., 1993), of two-electron systems (H^-, He) (Purvis et al., 1993; Dörr et al., 1995 a) and of complex atoms and ions such as Ne, Ar, F^- and Cl^-. A wide variety of interesting resonance effects have been studied, including quantum interference between two resonant pathways (Kylstra et al., 1995) and the occurrence of laser-induced degeneracies (LIDS) involving autoionizing states (Latinne et al., 1995). The RMF theory has also been applied to two-colour processes (van der Hart, 1996), harmonic generation (Gebarowski et al., 1997) and laser-assisted electron-proton scattering (Dörr et al., 1995 b). A recent review of the RMF theory and its applications has been given by Joachain (1997).

3. Numerical integration of the time-dependent Schrödinger equation.

Advances in computer technology over the past ten years have made possible the direct numerical integration of the time-dependent Schrödinger equation (TDSE) for single electron atoms in laser fields. The first calculations of this kind were carried out by Kulander (1987). These single electron calculations are "exact" for hydrogenic systems, but only approximate for atoms or ions with more than one electron.

Among the numerous "exact" TDSE calculations performed for atomic hydrogen, we mention the analysis of the time evolution of H atoms in super-intense, high frequency laser fields (Latinne et al., 1994; Dörr et al., 1995 c) the calculation of high-order ATI spectron (Paulus et al., 1996; Cormier and Lambropoulos, 1997) and the two-colour calculations of Véniard et al. (1995), in which laser-assisted single photon ionization was studied in the simultaneous presence of a low frequency laser and one of its high harmonics.

In the case of complex atoms, nearly all the TDSE calculations have been performed by using the single active electron (SAE) approximation, in which only one atomic electron interacts with the laser field, and effective potentials replace the core electrons (see e.g. Kulander et al., 1992; Schafer et al., 1993; Yang et al., 1993). TDSE calculations in which electron correlation effects are included are extremely difficult. Using B-splines to construct basis functions, Zhang and Lambropoulos (1995) have calculated ATI spectra for helium in which configuration-interaction effects are taken into account. More recently, this approach has been used by Zhang and Lambropoulos (1996) to calculate photo-electron spectra in Mg involving mutiple continua. Parker et al. (1996) have studied multiphoton processes in helium by direct numerical integration of the TDSE on a Cray T3D, but by restricting the Coulomb interaction between the two electrons to a few multipoles in order to reduce interprocessor communication. Results on ionization and double excitation of helium by short intense laser pulses, obtained by direct numerical integration of the TDSE, have been reported recently by Scrinzi and Piraux (1997).

A stringent test for theories of multiphoton processes including electron correlation effects is to calculate accurately double ionization rates for two-electron systems in intense laser fields. Promising results in this direction have been obtained recently by Faisal and Becker (1997), using S-matrix theory.

Let us now consider briefly relativistic effects. A relativistic treatment of laser atom interactions becomes necessary when the magnitude of the "quiver" velocity of an electron in the laser field is of the order of the velocity of light, i.e. when the electron ponderomotive energy $U_p \simeq mc^2$. For a low frequency laser field ($\hbar\omega \simeq 1eV$) this is the case when the intensity $I > 10^{18}$ W cm^{-2}. While the three-dimensional relativistic laser-atom interaction problem has been solved classically by numerical methods (Keitel and Knight, 1995), quantum mechanical calculations have been restricted to lower dimensional treatments. Thus, Protopapas et al. (1996) have studied the time-dependent relativistic Schrödinger equation in the acceleration frame for a one-dimensional model atom in the high frequency, ultra-high intensity (stabilizaion) regime, while Kylstra et al. (1997) have solved numerically the corresponding time-dependent Dirac equation. The calculations of Kylstra et al. are discussed in this volume. In both calculations, it was found that relativistic effects lead to a reduction of ionization, i.e. an increase of atomic stabilization. Retardation and spin effects as well as the influence of the magnetic field component are clearly not included in these one-dimensional calculations. The future investigation of these effects will require the solution of the time-dependent Dirac equation in two and three dimensions.

REFERENCES

1. Agostini P, Fabre F, Mainfray G, Petite G and Rahman N K 1979, *Phys. Rev. Lett.* **42**, 1127.
2. Ammosov M V, Delone N B and Krainov V P 1986, *Sov. Phys.-JETP* **64** 1191.
3. Antoine P, L'Huillier A and Lewenstein M 1996, *Phys. Rev. Lett.* **77** 1234.
4. Augst S, Strickland D, Meyerhofer D D, Chin S L and Eberly J H 1989, *Phys. Rev. Lett.* **63** 2212.
5. Burke P G, Francken P and Joachain C J 1990, *Europhys. Lett.* **13**, 617.
6. Burke P G, Francken P and Joachain C J 1991, *J. Phys. B* **24**, 761.
7. Burnett K, Reed V C and Knight P L 1993, *J. Phys. B* **26**, 561.
8. Chu S I 1985, *Adv. At. Phys.* **21**, 197.
9. Corkum P B 1993, *Phys. Rev. Lett.* **71**, 1994.
10. Cormier E and Lambropoulos P 1997, *J. Phys. B* **30**, 77.
11. de Boer M P, Hoogenraad J H, Vrijen R B, Noordam L D and Muller H G 1993, *Phys. Rev. Lett.* **71**, 3263.
12. Di Mauro L F and Agostini P, 1995, *Adv. At. Mol. Phys.* **35**, 79.
13. Dörr M, Potvliege R M and Shakeshaft R, 1990, *Phys. Rev. Lett.* **64**, 2003.
14. Dörr M, Burke P G, Joachain C J, Noble C J, Purvis J and Terao-Dunseath M 1993, *J. Phys. B* **26**, L 275.
15. Dörr M, Purvis J, Terao-Dunseath M, Burke P G, Joachain C J and Noble C J 1995 a, *J. Phys. B* **28**, 4481.
16. Dörr M, Terao-Dunseath M, Burke P G, Joachain C J, Noble C J and Purvis J 1995 b, *J. Phys. B* **28**, 3545.
17. Dörr M, Latinne O and Joachain C J 1995 c, *Phys. Rev. A* **52**, 4289.
18. Faisal F H M, 1973, *J. Phys. B* **6**, L 89.
19. Faisal F H M and Becker A 1997, in "Multiphoton Processes 1996", ed. by Lambropoulos P and Walther H, (Inst. of Phys. Publ., Bristol), p. 118.
20. Freeman R R, Bucksbaum P H, Milchberg H, Darack S, Schumacher D and Geusic M E 1987, *Phys. Rev. Lett.* **59**, 1092.
21. Gavrila M 1992, ed. "Atoms in Intense Laser Fields", *Adv. At. Mol. Opt. Phys. Suppl.* **1**.
22. Gavrila M and Kaminski J Z 1984, *Phys. Rev. Lett.* **52**, 613.
23. Gebarowski R, Burke P G, Taylor K T, Dörr M, Bensaid M and Joachain C J 1997, *J. Phys. B* **30**, 1837.
24. Hall J L, Robinson E J and Branscomb L M 1965, *Phys. Rev. Lett.* **14**, 1013.
25. Joachain C J 1994, in "Laser Interactions with Atoms, Solids and Plasmas", ed. by More R M (Plenum Press, New York), p. 39.
26. Joachain C J 1997, in "Multiphoton Processes 1996", ed. by Lambropoulos P and Walther H (Institute of Physics, Publ., Bristol), p. 46.
27. Keitel C H and Knight P L 1995, *Phys. Rev. A* **51**, 1420.
28. Keldysh L V 1965, *Sov. Phys.-JETP* **20**, 1307.
29. Krause J L, Schafer K J and Kulander K C 1992, *Phys. Rev. Lett.* **68**, 3535.
30. Kulander K C 1987, *Phys. Rev. A* **36**, 2726.

31. Kulander K C, Schafer K J and Krause J C 1992, in "Atoms in Intense Laser Fields", ed. by Gavrila M, *Adv. At. Mol. Phys. Suppl.* **1**, 247.
32. Kulander K C, Schafer K J and Krause J C 1993, in "Super-Intense Laser-Atom Physics", ed. by Piraux B, L'Huillier A and Rzazewski K (Plenum Press, New York), p. 95.
33. Kylstra N J, Dörr M, Joachain C J and Burke P G 1995, *J. Phys. B* **28**, L 685.
34. Kylstra N J, Ermolaev A M and Joachain C J 1997, *J. Phys. B* **30**, L 449.
35. Lambropoulos P and Walther H, eds. 1997, "Multiphoton Processes 1996" (Inst. of Physics Publ., Bristol).
36. Latinne O, Joachain C J and Dörr M 1994, *Europhys. Lett.* **26**, 333.
37. Latinne O, Kylstra N J, Dörr M, Purvis J, Terao-Dunseath M, Joachain C J, Burke P G and Noble C J 1995, *Phys. Rev. Lett.* **74**, 46.
38. Lewenstein M, Balcou Ph., Ivanov M Yu, L'Huillier A and Corkum P B 1994, *Phys. Rev. A* **49**, 2117.
39. L'Huillier A and Balcou Ph. 1993, *Phys. Rev. Lett.* **70**, 774.
40. L'Huillier A, Lewenstein M, Salières P, Balcou Ph, Ivanov M Yu, Larsson J and Wahlström C G 1993, *Phys. Rev. A* **48**, R 3433.
41. Luan S, Hippler R and Lutz H O 1991, *J. Phys. B* **24**, 3241.
42. Mason H J, and Newell W R 1987, *J. Phys. V* **20**, L 323.
43. Maquet A, Chu S I and Reinhardt W P 1983, *Phys. Rev. A* **27**, 2946.
44. Mc Pherson A, Gibson G, Jara H, Johann U, Luk T S, Mc Intyre I, Boyer K and Rhodes C K 1987, *J. Opt. Soc. Am.B* **4**, 595.
45. Mevel E, Breger P, Trainham R, Petite G, Agostini P, Migus A, Chambaret J P and Antonetti A 1993, *Phys. Rev. Lett.* **70**, 406.
46. Parker J, Taylor K T, Clark C W and Blodgett-Ford S. 1996, *J. Phys. B* **29**, L 33.
47. Paulus G G, Nicklich W, Xu H, Lambropoulos P and Walther H 1994 a, *Phys. Rev. Lett.* **72**, 2851.
48. Paulus G G, Nicklich W and Walther H 1994 b, *Europhys. Lett.* **27**, 267.
49. Paulus G G, Nicklich W, Zacher F, Lambropoulos P and Walther H 1996, *J. Phys. B* **29**, L249.
50. Potvliege R M and Shakeshaft R, 1992, in "Atoms in Intense Laser Fields", ed. by Gavrila M, Adv. At. Mol. Phys. Suppl. **1**, 373.
51. Protopapas M, Keitel C H and Knight P L 1996, *J. Phys. B* **29**, L 595.
52. Protopapas M, Keitel C H and Knight P L 1997, *Rep. Progr. Phys.* **60**, 389.
53. Purvis J, Dörr M, Terao-Dunseath M, Joachain C J, Burke P G and Noble C J 1993, *Phys. Rev. Lett.* **71**, 3943.
54. Reiss H R 1980, *Phys. Rev. A* **22**, 1786.
55. Rottke H, Wolff B, Brickwedde M, Feldmann D and Welge K H 1990, *Phys. Rev. Lett.* **64**, 404.
56. Schafer K J et al. 1993, *Phys. Rev. Lett.* **70**, 1599.
57. Scrinzi A and Piraux B 1997, *Phys. Rev. A* **56**, R 13.
58. Shirley J H 1965, *Phys. Rev. B* **138**, 979.
59. van der Hart H W 1996, *J. Phys. B* **29**, 2217.

60. van Linden van den Heuwell H B and Muller H G 1988, in "Multiphoton Processes" ed. by Smith S J and Knight P L , (Cambridge Univ. Press, London).
61. Véniard V, Taieb R and Maquet A 1995, *Phys. Rev. Lett.* **74**, 4161.
62. Voronov G S and Delone N B 1965, *JETP Letters* **1**, 66.
63. Wallbank B and Holmes J K 1994, *J. Phys. B* **27**, 1221.
64. Wallbank B, Holmes J K, Le Blanc L and Weingarshofer A 1988, *Z. Phys. D* **10**, 467.
65. Wallbank B, Holmes J K and Weingartshofer A 1990, *J. Phys. B* **22**, 777.
66. Weingartshofer A, Holmes J, Caudle G, Clarke E and Kruger H 1977, *Phys. Rev. Lett.* **39**, 269.
67. Weingartshofer A, Holmes J, Sabbagh J and Chiu S 1983, *J. Phys. B* **16**, 1808.
68. Wigner E 1946, *Phys. Rev.* **70**, 15.
69. Yang B, Schafer K J, Walker B, Kulander K C, Agostini P and Di Mauro L F 1993, *Phys. Rev. Lett.* **71**, 3770.
70. Zhang J and Lambropoulos P 1995, *J. Phys. B* **28**, L101.
71. Zhang J and Lambropoulos P 1996 *Phys. Rev. Lett.* **77**, 2186.

Few-Optical-Cycle Pulse Interactions with Plasmas: Models and Nonlinear Effects

A.M.Sergeev, V.B.Gildenburg, A.V.Kim,
M.Lontano[†], M.Quiroga-Teixeiro[‡]

*Institute of Applied Physics, Russian Academy of Sciences,
Nizhny Novgorod 603600, Russia*

[†]*Institute of Plasma Physics, National Counsil of Research,
Milano, Italy*

[‡]*Institute for Electromagnetic Field Theory,
Chalmers University of Technology, S-412 96 Göteborg, Sweden*

Abstract. We present an analysis of nonlinear optical effects when a high intense ultrashort laser pulse comprising only few field cycles propagates through an ionized medium. We introduce a model for the optical pulse and plasma based on the real electric field representation and the quantum mechanical description of electron dynamics in the ionization process. In the first part, we study the nonlinear atom response to such short driving pulses and the accompanied process of attosecond continuum production. In the second part, we analyze some collective plasma phenomena with ultrashort driving pulses, for which transverse effects are important. Among them, the processes of leaking mode self-channeling at saturable ionization, induced ionization scattering and large blueshifting with pulse shortening to a single-optical-cycle burst are of particular interest.

INTRODUCTION

Recent progress in developing techniques for generation and compression of ultrashort laser pulses has resulted in mastering the range of duration shorter than 5 fs.[1,2] The use of Ti:Sa laser crystals and broad-band optical elements with programmable dispersion, from the one side, and application of new ideas of nonlinear pulse spectrum manipulation, from the other side, have provided the optical community with a remarkable instrument, a source of coherent optical radiation with the pulse duration comparable with a single period of electromagnetic field. It has turned out that the technological and experimental progress in this area has to some degree left behind the theoretical comprehension of what may happen in nonlinear interactions of such short light bursts with matter. Most of the concepts of ultrashort pulse interactions rely on the validity of slowly varying field amplitude approximations, which in general is obviously not the case for few or single optical cycle pulses. In this report we make an attempt to put forward and analyze several problems that will arise in investigation of few-optical-cycle pulses propagation in ionized media. Our main goal is to draw the attention to the specificity of ionization nonlinear wave processes and outline new physical effect rather than to give a general approach for description phenomena beyond the slowly varying amplitude approximation.

In the following section we introduce a model for the optical pulse and plasma response based on the real electric field representation and the quantum mechanical description of electron dynamics in the ionization process. We will give an example when even in the case of propagation in vacuum, at decreasing the number of optical cycle in a pulse, one can observe a nontrivial spatial dynamics of the light burst, which is untypical for the usual situation with a longer pulse duration. In the next section we will describe the ionization plasma response in the frames of the introduced model. We will demonstrate when and how the application of few-optical-cycle pulses can advance the process of XUV production at plasma ionization and discuss the dependence of this phenomenon on the absolute phase of the optical field. In this interaction the XUV radiation is produced by the plasma in the form of the of attosecond continuum that is quite sensitive to the phase matching conditions. That is why we will pay a special attention to analyze the limiting values of the frequency conversion efficiency and XUV pulse duration. In the final section we will discuss collective nonlinear ionization phenomena that can strongly affect the spatial and temporal properties of few-optical-cycle pulses. Among them, the effects of leaking mode self-channeling at saturable ionization, induced ionization scattering and pulse shortening to a single-optical-cycle burst will be especially interesting and unusual.

MODELS OF FEW-OPTICAL-CYCLE PULSE PROPAGATION AND IONIZATION PROCESSES

When constructing a model for an optical pulse with a small number of field cycles, that propagates through and ionizes a medium, one has to forget about the traditional approximation of the slowly varying complex amplitude and to employ the real electric field and the real polarization of the medium. This is caused both by the necessity to correctly describe linear diffraction and dispersion effects in the supershort pulse limit and by the impossibility to represent the nonlinear response of the medium as a function (or a functional) of only the optical field amplitude. The main assumptions we put to outline a range of phenomena under discussion and to facilitate their description are as follows.

We will mainly deal with propagation effects, which means that a characteristic distance of evolution (in z-direction) is larger than the longitudinal pulse scale L_z. Another geometrical factor is that in our consideration typically the transverse size of the pulse $L_\perp$ will exceed essentially both L_z and the wavelength λ_o, which is quite reasonable for not so tight focusing and the pulse duration range of interest. We will focus our attention on the intensity range 10^{14}-10^{16} W/cm^2, where the ionization nonlinearity plays a dominant role in ultrashort pulse interactions with gaseous media and relativistic corrections to the free electron motion are yet negligible. Since the characteristic time intervals are well within few tens of femtosecond, the medium response is supposed to be that of mostly mobile particles, i.e. electrons occupying outer atom orbits or being freed from parent atoms due to field-induced ionization. Finally, except for one specific case in the last section, we assume the gas pressure not so high, so that the produced plasma density is undercritical with respect to the current radiation frequency. Strictly speaking, only under this assumption, propagation effects with long-distance evolution and the absence of reflected electromagnetic waves can be analyzed as a separate class of nonlinear wave phenomena in a plasma.

The general description of an arbitrary electromagnetic field in a medium is given by the equation

$$\nabla \times \nabla \times \vec{E} + \frac{1}{c^2}\frac{\partial^2 \vec{E}}{\partial t^2} = -\frac{4\pi}{c^2}\frac{\partial^2 \vec{P}}{\partial t^2}. \tag{1}$$

To derive a simplified equation for description of propagation effects in a plasma, let us assume for a moment that we have a quasi plane wave structure (i.e. omit the dependence on the transverse coordinates x and y) in a rather rarified medium ($\vec{P}$ is negligibly small). Then we obviously arrive at a one-dimensional wave equation governing the propagation of any wave form along the direction z in the vacuum without any change. In the presence of the omitted factors, having in mind their contribution accumulated over a long distance of propagation through the plasma, it is convenient to introduce a new time variable $\tau = t - z/c$ and then assume that $\left|\frac{\partial}{\partial z}\right| \ll \frac{1}{c}\left|\frac{\partial}{\partial \tau}\right|$. It results in the following simplified equation for the real values of the optical field strength and polarization

$$-\frac{2}{c}\frac{\partial^2 \vec{E}}{\partial \tau \partial z} + \Delta_\perp \vec{E} = \frac{4\pi}{c^2}\frac{\partial^2 \vec{P}}{\partial \tau^2}. \tag{2}$$

This equation, being a first-order equation in the direction of propagation z, preserves the main advantage of the slowly varying amplitude approximation. At the same time, it does not impose limitations on the pulse duration and is valid for both a cw harmonic wave and an optical monopulse. To demonstrate its potentialities, let us derive from (2) an equation for a field with well defined envelope and carrier structures, $\vec{E} = \frac{1}{2}(\vec{A}e^{i\omega_o t} + c.c.)$. In the case of propagation in the vacuum, we obtain

$$2i\frac{\omega_o}{c}\frac{\partial \vec{A}}{\partial z} + \frac{1}{1+i\dfrac{\partial}{\omega_o \partial \tau}}\Delta_\perp \vec{A} = 0. \tag{3}$$

Analyzing Eq. (3) it is easy to see that for a slowly varying amplitude, i.e. $\left|\dfrac{\partial}{\omega_o \partial \tau}\right| \ll 1$, we indeed arrive at a common equation governing the diffraction pattern of the opical field in vacuum. On the other hand, for a not so long envelope we are not able to neglect the corresponding term comprising the time derivative. As a result the diffraction of optical field at each pulse cross-section exhibits different evolution and the pulse does not preserve in general self-similarity in the spatial structure at propagation in vacuum. This is an example to demonstrate that dynamics of ultrashort pulses with a small number of field periods becomes nontrivial even in the simplest case of propagation in a linear system, not to say of interaction processes in nonlinear media. Eq. (2) will be used below to study the effect of attosecond continuum production at atom ionization at rapidly growing fronts of ultrashort laser radiation.

In the case we can not neglect a reflected wave or radiation scattering at rather large angles to the propagation direction, we have to keep the second-order derivative in z. Here a simplification of the initial equation (1) can be obtained only by reducing the number of spatial variables and investigating a pulse with a given field polarization. So, for a two-dimensional field structure ($\partial/\partial y=0$) with the TE wave polarization we have the following equation for the y-components of the electric field and plasma polarization

$$\frac{\partial^2 E}{\partial z^2} + \frac{\partial^2 E}{\partial x^2} - \frac{1}{c^2}\frac{\partial^2 E}{\partial t^2} = \frac{4\pi}{c^2}\frac{\partial^2 P}{\partial t^2}. \tag{4}$$

Eq. (4) will be used in the following study to describe the phenomena of induced ionization scattering and leaking mode self-channeling at saturable ionization nonlinearity.

The formulation of an adequate description for the medium response to a few-optical-cycle pulse is a key issue of our modeling. We will represent a model of the medium in the form of an equation for quantum particles characterizing by a rather arbitrary set of quantum levels. We assume the gas of these particles to be quite rarified, so that any particle-particle interaction can be neglected during the time of the pulse propagation through a given point of the medium. This assumption allows us to formulate a Schroedinger equation for the Ψ-function of electrons in single-electron noninteracting atoms, considering the electron moving in the superposition of an intraatomic steady-state potential V and an external potential produced by the oscillating laser field in the dipole approximation

$$i\hbar\frac{\partial \Psi}{\partial t} = -\frac{\hbar^2}{2m}\Delta\Psi + V(\vec{r})\Psi + e\vec{r}\vec{E}\Psi. \tag{5}$$

To close the set of equations for the optical pulse and the medium response we represent the polarization of matter as the total dipole momentum of N particles per unit volume

$$\vec{P} = -eN\int |\Psi|^2 \vec{r}\, d\vec{r}. \tag{6}$$

By this the right-hand side of the wave equation may be rewritten in the form

$$\frac{\partial^2 \vec{P}}{\partial t^2} = \frac{eN}{m}\int |\Psi|^2 \nabla V\, d\vec{r} + \frac{e^2 N \vec{E}}{m}. \tag{7}$$

In spite of briefness of the suggested formulation, this model of an optical medium has quite a universal character. The diversity of optical nonlinearities is contained in the form of intraatomic potential $V(\vec{r})$. By choosing it properly, one can simulate at the microscopic level a nonlinear response of a wide range of optical systems.

PLASMA RESPONSE AT FIELD-INDUCED IONIZATION AND GENERATION OF ATTOSECOND CONTINUUM

The application of laser pulses with supershort duration of the order of 10 fs has recently been proposed for enhancement of high order harmonic emission and subfemtosecond pulse production in the XUV range.[3-5] Such short driving pulses contain only a small number of optical cycles and therefore high-energy photon bursts are generated due to atom ionization at rapidly increasing field amplitudes. It is obvious that the efficiency of this process becomes dependent on the concrete field distribution over the pulse or, in other words, on the absolute phase of the optical field. It should be also noted that several other schemes have been proposed for attosecond pulse generation. Corkum, Burnett, and Ivanov[6] have proposed to use the elliptically polarized light to take advantage of the strong polarization dependence of the harmonic emission process. To produce a train of attosecond pulses, one pulse each half cycle of the driving laser field, Antoine, L'Huillier and Lewenstein have suggested filtering and combining several plateau harmonics.[7] Schafer and Kulander[8] have suggested to generate single subfemtosecond pulse by using ultrashort pump laser and compression techniques for harmonics at the end of the plateau.

The idea of an enhancement of energetic x-ray burst production with a decrease of the pulse duration consists in the following.[4] At a sharply increasing laser field amplitude, atoms may be ionized with high probability (or the main part of the electron Ψ-function may be detached from the intraatomic potential) for a fraction of an optical cycle when the field strength passes the ionization threshold value. For the subsequent laser period the electron wave packet including almost all particles will be accelerated by the laser field and may collide with the parent ion core at a velocity higher than that for the case of slowly varying field amplitude. By employing a 1D quantum model, consideration of this effect in detail was presented in the work,[9] where it was treated as nonadiabatic effect in high-harmonic generation with ultrashort pulses. It was observed that for shorter pulses, a redistribution of the energy between the different harmonics orders occurs, where the lower orders are weaker, while the higher ones are stronger, than for longer pulses.

The efficiency of this mechanism may be easily illustrated by considering the classical picture of the relevant trajectories of freed electrons that are responsible for the high-energy photon emission. Let us assume that the leading front of the pulse is characterized by the exponential growth of the field amplitude with an increment β measured in the laser field frequencies. We will be interested in the phase φ_{col} and the energy U_{col} of return collisions of classical electrons with parent ions depending on the electron release phase φ. In the case $\beta=0$, considering the one-dimensional electron trajectory, precisely half of the electrons will collide with ions at least once. These are the electrons released from intraatomic potential during quarters of the field cycles after achieving the local maximum values. The electrons released within quarters of the increasing field cycles do not return to parent ions at all. The picture changes for $\beta > 0$. It is seen in figure 1a that with increasing β the boundary of the "non-return phase" shifts to the negative domain, i.e. the electrons released before the field maximum may participate in the bremsstrahlung. In this case the maximum of the return collision energy (Fig. 1b) broadens and shifts from $\approx 18°$ at $\beta = 0$ to the earlier phases, so that it arrives at $\varphi = 0$ when $\beta \approx 0.2$. In general, this fact indicates improved conditions for the soft X-ray burst generation and an increase in the maximum energy of emitted photons. However, it is clear that the process of soft X-ray burst generation is rather sensitive to a specific temporal profile of the pulse and one can imagine a situation when the non-return effect of the most released electrons will lead to a dramatic reduction of the bremsstrahlung efficiency.

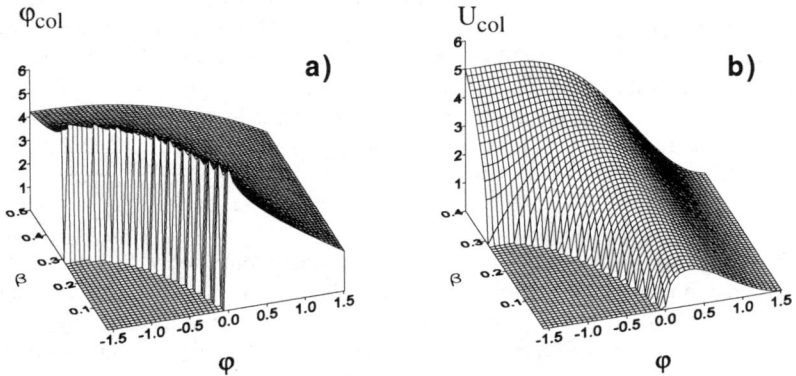

Figure 1. The phase φ_{col} (a) and the energy U_{col} (b) of electron return collisions with parent atoms versus the phase of electron production φ and the rate β of the field amplitude increase.

This can happen for example if the field passes the range of strengths that are critical for the atom decay in the growing phase but close to the pulse center when the amplitude reaches the maximum. Then a freed electron bunch is mainly generated for $\varphi < 0$, while the value β is close to 0. It is interesting to note that if the non-return condition is fulfilled near the axis of the 3D pulse cross-section, at the periphery, vice versa, the excitation of short wavelength bursts does occur, since electrons are released for lower field envelopes near the field maximum $\varphi = 0$. This may cause hollow distributions in the spatial domain where subfemtosecond pulses are generated.

From the above simple consideration we conclude that the high-energy photon burst emission may be controlled by shaping the driving laser pulse. So, the idea to produce ultrashort bursts needs exploration in the frames of more realistic models. A principal question that arises is how collective processes in the ionized gas affect the characteristics of the burst. In other words, is there an optimal propagation length for producing ultrashort XUV radiation? How is its conversion efficiency affected by the temporal electric field distribution? To investigate this problem quantitatively we have employed a 1+1 simulation model developed in our previous work[10] to consider the self-consistent dynamics of linearly polarized electromagnetic wave propagating in a gas of single-type atoms. It should be noted that propagation model was also developed by Rae, Burnett and Cooper[11] and applied to the high-harmonic emission with long femtosecond laser pulses. The set of equations are the Maxwell equations for a field $E(z,t)$ in the medium written in the simplest form of a one-dimensional scalar wave equation taken in a reduced form (2) and the Schroedinger equation (5) with the intraatomic potential $V(x)$ (x-direction along electric field) for the single electron Ψ-function.

$$\frac{\partial^2 E}{\partial z \partial t} = -\frac{1}{2}(E+R), \tag{8}$$

$$i\frac{\partial \Psi}{\partial t} = -\frac{1}{2}\frac{\partial^2 \Psi}{\partial x^2} - V(x)\Psi + xE(z,t)\Psi. \tag{9}$$

Here $R(t) = \int |\Psi|^2 \frac{\partial V}{\partial x} dx$ is the source of harmonic emission in the reduced wave equation, z is the dimensionless coordinate along the laser pulse propagation direction, measured in $me^4 c/\omega_p \hbar^3$ units, $t \to t - z/c$ is the local time, and $\omega_p^2 = 4\pi e^2 N/m$ is proportional to the gas density N. The latter scaling ratio demonstrates that the result of the consideration will depend on the product of the gas density and the propagation path, but not on these factors separately. We note that this propagation model (reduced form of the wave equation) neglects the reflection light effect assuming that plasma density appearing during ionization process is much less than the critical one. Another variables of time, intraatomic coordinate, intraatomic potential and electric field in a pulse are normalized by the corresponding atomic values $t_a = \hbar^3/me^4$, $x_a = \hbar^2/me^2$, $V_a = me^4/\hbar^2$, $E_a = m^2 e^5/\hbar^4$. We have chosen the potential in the form of $V(x) = (1+x^2)^{-0.5}$ with a binding energy of 0.67 a.u. and considered laser-gas interaction for the few-optical-cycle pulses.

Figure 2 shows the time dependence of the driving field that has the Gaussian envelope with carrier frequency of 0.2 a.u. and the response function $R(t)$, for the field amplitude of 2.1 a.u. Since the R function is responsible for the high-energy photon emission due to the electron-core interaction, we see that it is possible to ionize atom almost completely for one half period. During this time interval the electromagnetic radiation emitted by atoms mainly consists of comparatively low frequency components (as follows from the curve for the response function) and is produced at electron tunneling[12] through the barrier formed by the atomic potential and the driving field.

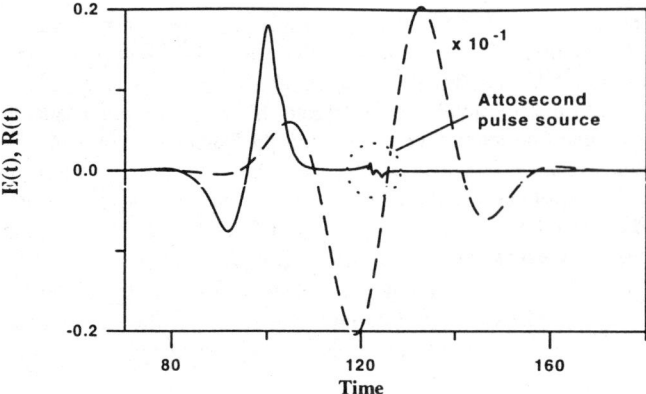

Figure 2. The time dependence of the driving field (dotted line) and the response function R caused by electron-core interaction (solid line).

For the subsequent half-period, the electron wave packet is free ($R \sim 0$), and then, returning back and colliding with parent ion ($t \sim 125$), it generates radiation with higher frequencies. As it is seen in figure 3, where the frequency spectrum of electric field is shown at various distances of pulse propagation, the plateau region is extended over the harmonic order up to $n \sim 80$-90. We also see that spectral intensities in several frequency domains have different dependencies on the propagation distance. In figure 4, we present the value of energy contained in definite intervals of spectrum versus the propagation path. We see that, at the beginning, energies of the generated radiation are increased with the interaction distance, then saturated at a level depending on harmonic frequencies and further they have oscillation character along propagation path.

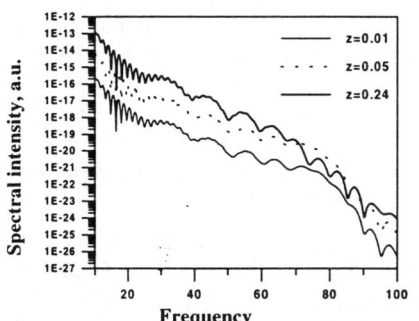

Figure 3. Field spectrum at the various propagation distances.

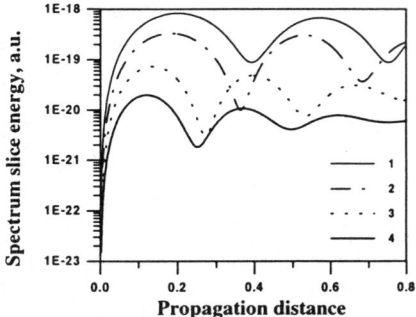

Figure 4. Energy of high-frequency components integrated over definite spectral intervals versus the propagation distance:
1 – from 55 ω_0 to 60 ω_0; 2 – from 60 ω_0 to 65 ω_0; 3 – from 65 ω_0 to 70 ω_0; 4 – from 75 ω_0 to 80 ω_0.

Here we only note that the saturation effect in the energy amplification is mainly caused by wave dispersion in the emerging plasma and it is more essential at higher frequencies (see Fig. 3 and 4).

As we have seen in figure 2, the source of high-energy photon emission is localized in a very narrow time interval in optical cycle. This means that the pulse duration of high-energy photon bursts generated by sub-10 fs pump lasers must be much shorter than the laser period, i.e. it may lie in few hundred attosecond range. As we are specifically interested in the harmonic emission on a time scale less than an optical cycle, it is necessary to use a time-frequency or wavelet analysis to analyze the time profile of high-energy photon bursts. Such analysis has recently been used to study the temporal behavior of the harmonic emission process in a single atom response.[13] The form of the analyzing wavelet, which we used here, is the Morlet's wavelet:[12] $W(t,t_o) = \exp(it)\exp[-(t-t_o)^2/\sigma^2]$ where σ is a fixed parameter and must be choosen as long as $\sigma > 2\pi$.

As an example, we focus on the spectral region expanding from $55\omega_o$ to $60\omega_o$ (ω_o is the fundamental frequency). Figure 5 shows the nonlinear dynamics of electric field envelope composed by spectral components in this region along the propagation distance. The pulse duration of the high energy photon bursts is more than 20 times less than pump laser period, i.e. we can expect that by using sub-10 fs, 800 nm laser pulses, single XUV pulses with duration of about 100 attosecond will be directly generated when a gas medium is ionized by such ultrafast pump laser.

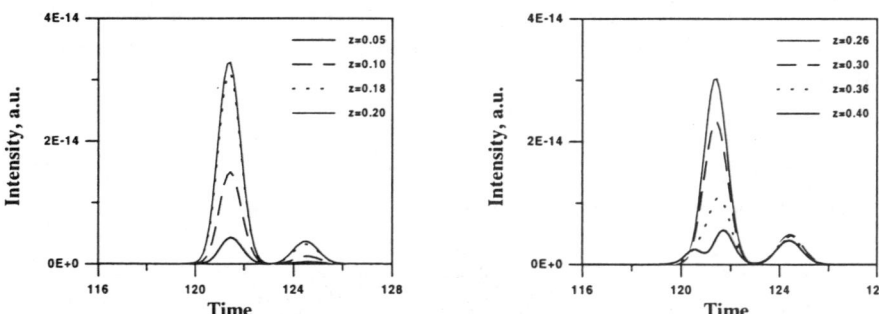

Figure 5. Electric field envelope of the attosecond pulse under the same conditions as Fig. 4 at the various propagation distances. The pulse is composed by spectral components in the interval from $55\omega_o$ to $60\omega_o$.

An important feature of the attosecond pulse generation is oscillation dynamics (nonmonotonic dependence) along the propagation path, as it is also clearly seen in figure 4. So, there is an optimal nonlinear interaction length for efficient attosecond pulse production in a given spectral interval. For higher photon energies this optimal length becomes shorter. Oscillating character of attosecond pulse evolution along pump laser propagation points to the crucial role of the phase mismatching effects. As it is shown in previous works,[4,11] for the case of long femtosecond pump pulses the main role in efficient high-order harmonics generation is played by phase mismatching due to blueshifting of the fundamental frequency which arises self-consistently due to dynamic change in the refractive index. Ionization process is responsible for both plasma-induced blueshifting and high harmonic emission effects. In the "single-cycle" regime, the attosecond pulse source is localized in a narrow temporal interval within one optical cycle (Fig. 2) and, therefore, its position is very sensitive to the temporal distribution of the laser field. As follows from our simulations, plasma dispersion gives the main contribution to the modification of the electric field profile and, thus, defines the efficiency of attosecond pulse production. So, the physical mechanisms of phase mismatching in these two cases, for long femtosecond and few-optical-cycle pulses are quantitatively different.

COLLECTIVE NONLINEAR IONIZATION EFFECTS

Consideration of the wave equation consistently with the Schroedinger equation in 3D or even in 2D geometry encounters severe computational difficulties. However we can separate two distinct aspects of this problem: the process of high-order harmonic generation for which the Schroedinger equation is necessary, and spatio-temporal dynamics of a driving laser pulse and produced plasma. In the latter case, instead of the microscopic quantum model of ionized atoms we can employ a simpler macroscopic model. In this section, we will concentrate on the ionization-induced propagation effects which can be describe in the frames of the wave equation (4) (or the simplified version (2)) and an equation for the current of freed electrons $J = (\partial P/\partial t)$ in a plasma with variable density N_e

$$\frac{\partial J}{\partial t} = \frac{e^2 N_e}{m} E. \tag{10}$$

The electron density N_e will be governed by a simple dynamical equation

$$\frac{\partial N_e}{\partial t} = W(E) = 4\Omega(N - N_e)\frac{E_a}{|E|}\exp\left(-\frac{2}{3}\frac{E_a}{|E|}\right) \tag{11}$$

that describes the ionization by the well-known static expression for the field-dependent rate at each instant of time.[15] Here $E_a = m^2 e^5/\hbar^4 = 5.14 \cdot 10^9$ V/cm is a characteristic atomic field, $\Omega = me^2/\hbar^3 = 4.16 \cdot 10^{16}$ s^{-1} is the atomic frequency unit, N is the gas atom density.

We start our consideration with *the ionization-induced blue shifting effect*, which was previously studied in detail for long femtosecond pulses[16-19]. Although the possibility of a large (~100%) frequency upshift was predicted by using spatiotemporal geometric optics and direct integration of the 1D wave equation,[20,21] in the experiments with intense laser pulses focused in a dense gas large spectral blueshifting was not observed; only several percents of fundamental frequency was measured.[22] One of the main reason is that the pulse length was too large, i.e. exceeded the Rayleigh length. In this case, only a small forward part of the pulse (near the leading edge) produces ionization, propagates in the time-varying plasma, and hence undergoes frequency upshift. Most part of the pulse determining its spectrum as a whole does not ionize the gas (even though it is far from full ionization) because its field amplitude is decreased due to transverse defocusing in the plasma produced by the leading part. Thus, for ultrashort laser pulses, especially, in the "few-optical-cycle" regime this effect may be of great interest due to large spectral conversion. Now we present the simulation results demonstrating large spectral blueshifting of an ultrashort (consisting of several wavelength) intense pulse focused in a gas and formation of the half-wave ionizing "leader".[23]

The following initial conditions were set

$$N_e = 0, E = F(x,z), \frac{\partial F}{\partial t} = G(x,z). \tag{12}$$

The functions $F(x, z)$, $G(x, z)$ were chosen so that the wave packet, in the absence of ionization ($W = 0$), moves in the $+z$ direction, forming at some time $t = t_o$ the focused pulse of finite length and wavelength with the center at the point $x = z = 0$ and with a Gaussian transverse profile,

$$E(x,z,t_o) = A(x,z) = E_o f(z)\sin(kz)\exp\left(-\frac{x^2}{2a^2}\right) \tag{13}$$

Here E_o is the maximum field amplitude, $f(z)$ is the pulse envelope function, a is the minimal effective transverse dimension of the pulse, $k = 2\pi/\lambda_o = \omega_o/c$, ω_o is the fundamental laser frequency. The functions $F(x, z)$, $G(x, z)$ were found on the basis of time reversibility of the field evolution in vacuum by the numerical integration of Eq.(4) with $N_e = 0$ at the initial conditions

$$E(x,z,-t_o) = 2A(x,z),$$
$$\frac{\partial E(x,z,-t_o)}{\partial t} \equiv 0. \tag{14}$$

In Figs.6 and 7, the simulation results are presented for rectangular envelope function $f(z)$ and the following parameter values: $kL = 6\pi$ (pulse length $L = 3\lambda_o$), $ka = 10$ (focal region length $L_f = ka^2 = 16\lambda_o$), $\Omega/\omega_o = 22$, $N/N_{cr} = 0.135$, $3E_o/2E_a = 0.25$. This case corresponds to the 10 fs laser pulse with the wavelength $\lambda_o = 1\mu m$ and the power of 80 MW, focused with f/7 optics to the peak intensity of 10^{15} W/cm^2 (radius of the focal spot $a = 1.6$ μm) into a 5 atm hydrogen gas. Time dependencies $E(t)$ and spectra of the field E_ω at some points of the z-axis are shown in dimensionless variables: $t \to \omega_o t$, $E \to 3E/2E_a$. Fig. 6 distinctly demonstrates the upshift of the time-spectrum maximum of the incident pulse as a whole (at $x = 0$) after passing the ionizing region, which is here of up to $\Delta\omega/\omega_0 = 42\,\%$. We also note that in this case the maximum value of electron density is determined by the ionization of an ultrashort high-amplitude "leader" that is formed at the leading edge of the pulse (see Fig. 7). The leader consisting of only one or two half-waves, propagates continuously in the region of both comparatively small average plasma density N_e and high value of its growth rate. As a result the leader is only slightly subjected to defocusing in the plasma but undergoes a sharp decrease in the wavelength. As we can see from the plots of $E(t)$ at $x = 0$ (Fig. 7), during and after passing the focal region, the first wave period of the pulse is $T \approx (1/3 \div 1/4)T_0$, where $T_0 = 2\pi/\omega_0$.

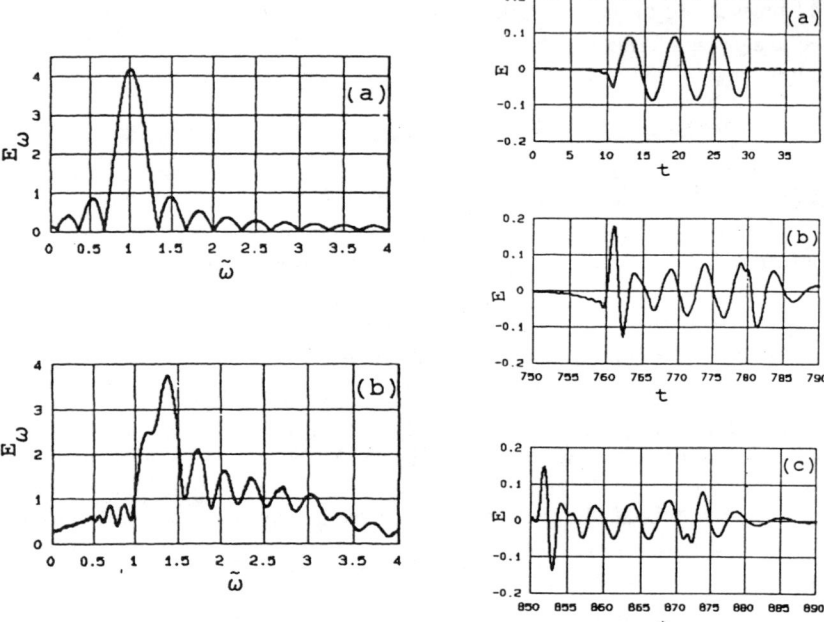

FIGURE 6. Time spectra of the electric field E_ω (in arbitrary units) versus $\tilde{\omega}=\omega/\omega_0$ at the points $x = 0$ and (a) $z = -780$, (b) $z = 60$.

FIGURE 7. Plots of E versus time t at the points $x = 0$ and (a) $z = -780$, (b) $z = -30$, (c) $z = 60$.

The next collective nonlinear phenomenon that will be discussed in this section is *the structural instability* of ultrashort laser pulses focused in an ionized gas of high density. It is well known that electromagnetic waves propagating in a medium with ionization nonlinearity undergo ionization instabilities,[24] in particular, for intense short laser pulse, when optical field-induced ionization is dominant, the main role is played by the so-called induced ionization scattering.[25,26] This instability manifests itself in splitting of the ionized region into smaller plasma bunches with different ionization degrees and causes increased scattering of the incident pulse. The development of the instability may completely change the structural and temporal characteristics of the laser field and produced plasma. Here we will demonstrate that the structural instability takes place even with few-optical-cycle laser pulses. In Figs.8 and 9 results of simulation of Eqs. (4), (10), (11) for a pulse with the Gaussian shape focused on a gas layer with density of $N/N_{cr} = 4$ are presented. Incident pulse has the form $E=\exp(-0.1\ t^2)\sin(\pi t)\exp(-0.02\ x^2)$. Here the field is measured in atomic units, time in π/ω_o, and the spatial coordinates in π/k. Development of the structural instability in this case is distinctly seen in Fig.8 where the plasma density is near periodically laminated in the transverse direction. In this case it leads to increased scattering of the incident pulse at the angles more than 30° to the direction of pulse propagation.

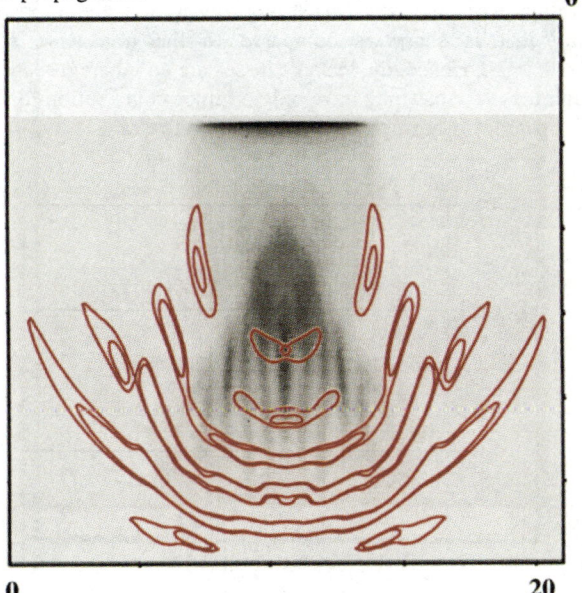

FIGURE 8. Plasma density distribution and electric field contours.
Incident pulse propagates from the top to down.

Fig.9 shows the temporal profiles of the electric field of the transmitted pulse on the axis ($x=0$) and at the angles of 15° and 30°. We want to emphasize here the effect of shortening of the pulse length, which takes place for the scattered radiation (Fig. 9c and d). An effective pulse length at these angles is so short that such process may be used for single-cycle pulse production. The effect of shortening can qualitatively be explained by the fact that the ionization scattering as a parametric process needs some time for development (the growth rate of the short-scale density perturbations) and its efficiency has nonlinear dependence on intensity of the driving pulse.

Finally we address another nonlinear ionization effect, *the leaking mode self-guiding* with intense few-optical-cycle laser pulses. As has been recently shown, the saturable ionization nonlinearity alone, without any focusing nonlinearities, is a sufficient mechanism for self-channeling of an intense laser pulse.[27]

At first sight, this statement looks absurd since in accord with a common concept nonlinearity with a growing dependence of the refractive index on the field intensity is needed for the self-guiding effect. The idea of self-guiding at defocusing ionization nonlinearity consists in the following. Owing to a strong dependence of the field ionization rate on the field intensity a laser pulse can produce a plasma distribution that is smooth near the axis and sharply cut at the periphery of the cross-section (a plasma filament with sharp boundaries). This distribution, inspite of negative variation of the refractive index at the axis, can guide an electromagnetic wave in the form of a leaking mode with exponentially small losses over the distances of many free-space Rayleigh lengths. As distinct from the common self-guiding effect where the field localization is achieved due to the total internal reflection at the periphery of a waveguide, in this case the quasi-localization is obtained due to a strong reflection of the trapped wave from the plasma boundary that is sharp as compared to the transverse scale (transverse wavelength) of this wave. Hence, the leakage losses are an inherent feature of the plasma waveguide though these factors may have only a minor contribution to the overall wave dissipation as compared, for example, to the ionization losses.

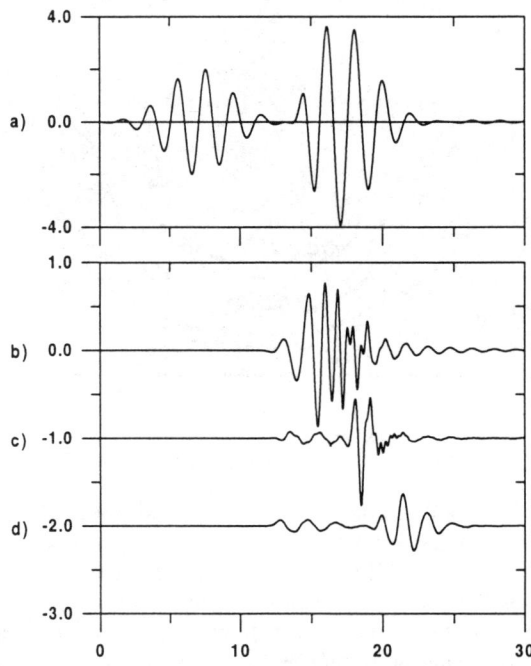

FIGURE 9. Electric field in units $E_a/6$ versus time in units π/ω in: a) incident and reflecting pulses, b) directly transmitted pulse, c) and d) scattered pulses.

Formation of a channel with a quasi-rectangular radial plasma profile can be facilitated in the case of saturation of the ionization, which is typical for this nonlinearity and corresponds to the complete depletion of one or several electronic states in atoms. The result of saturation is flattening of the plasma profile near the axis and a corresponding decrease of refraction in the center of the channel where the main part of the laser energy is propagated. This regime seems easier to be implemented in single-species gases at not so high pressure when the saturation can be reached before the free electron density becomes too large and leaves no chance to balance the strong refraction. This idea was checked for the relatively long pulses by employing paraxial approximation for a slowly varying field amplitude.[27] Here we show that the idea of self-guiding may be extended to the few-optical-cycle regime with the same results. Figs. 10 and 11 demonstrate the effect of the long plasma channel creation effect by an ultrashort focused pulse. Representative space distributions of the wave electric field $E(x, z)$ and the plasma density $N(x, z)$ at some time t are shown in dimensionless variables: $x \to kx$, $z \to kz$, $t \to \omega_0 t$, $E \to 3E/2E_a$, $N_e \to N_e/N_{cr}$. We note that in these figures the focal point and the focal region length are 0, 18, respectively.

The parameter values correspond to the rectangular laser pulse with duration of 20 fs, $\lambda_0 = 1$ μm, power of 60 MW, focused at $f/3$ to a peak intensity of $4 \cdot 10^{15}$ W/cm^2 into 3 atm hydrogen gas (pulse length $L = 6\lambda_0$, radius of the focal sport $a = 0.7$ μm, focal region length $L_F = ka^2 = 3$ μm). We can see that a completely ionized plasma channel is created in this case (Fig. 10). The full channel length is 54 μm, that is 19 times more that the focal length L_F ($13L_F$ before focus point and $6L_F$ behind one). Figure 11 presents space distributions of the electric field $E(x, z)$ at three time moments. Note that to the moment $t = 240$, when the pulse passes the focal region, its transverse structure is well localized and trapped by the quasi-rectangular plasma channel. The field amplitude on the axis ($x=0$) is much higher than one out of channel. This indicates that radiation losses of leaking mode in our case are quite small and, therefore, long self-guiding of the laser pulse takes place.

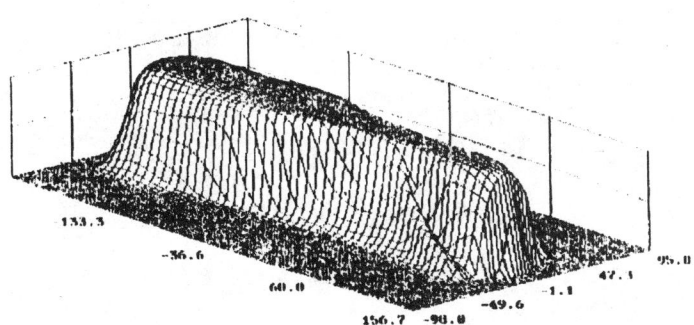

FIGURE 10. Plasma density distribution $N(x, z)$ at $t = 320$.

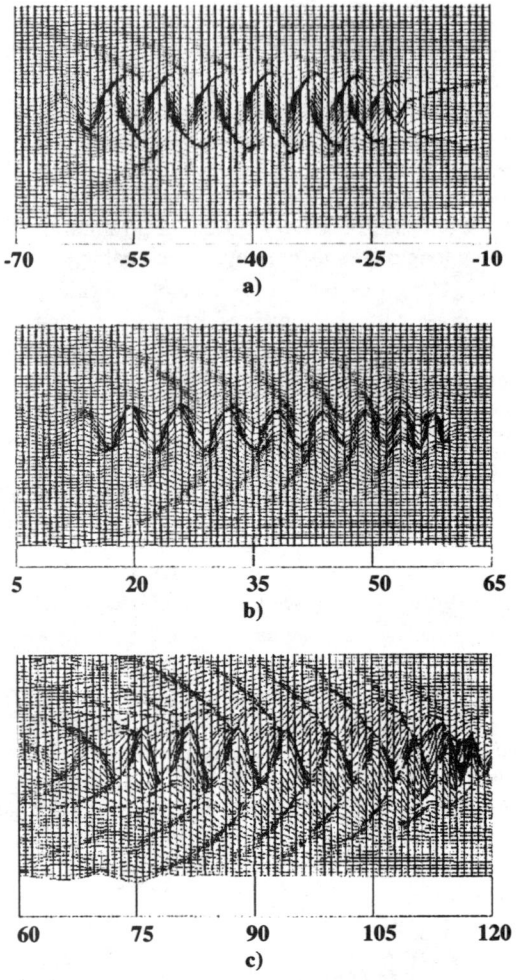

FIGURE 11. Spatio-temporal evolution of the electric field $E(x, z, t)$.
(a) – (c) correspond to $t = 160, 240, 300$, respectively.

ACKNOWLEDGMENTS

The work of A.M.S., V.B.G. and A.V.K. was supported in part by the Russian Basic Research Foundation.

REFERENCES

1. J. Zhou, G. Taft, C.-P. Huang, M.M. Murnane, and H.C. Kapteyn, *Opt.Lett.*, **19**, 1149 (1994).
2. A. Stingl, C. Spielman, F. Krausz, and R. Szipocs, *Opt.Lett.*, **19**, 204 (1994).
3. J. Zhou, J. Peatross, M.M. Murname et al., *Phys.Rev.Lett.*, **76**, 752 (1996).
4. A.M. Sergeev, A.V. Kim, and E.V. Vanin, *SPIE*, **2701**, 416 (1996).
5. I.P. Christov, M.N. Murnane, and H.C. Kapteyn, *Phys.Rev.Lett.*, **78**, 1251 (1997).
6. P.B. Corkum, N.H. Burnett, and M.V. Ivanov, *Opt.Lett.*, **19**, 1870 (1994).
7. P. Antoine, A. L'Huillier, and M. Lewenstein, *Phys.Rev.Lett.*, **77**, 1234 (1996).
8. K.J. Schafer and K.C. Kulander, *Phys.Rev.Lett.* **78**, 638 (1997).
9. I.P. Christov, J. Zhou, J. Peatross et al., *Phys.Rev.Lett.*, **77**, 1743 (1996).
10. E.V. Vanin, A.V. Kim, M.C. Downer, and A.M. Sergeev, *JETP Lett.*, **58**, 900 (1993).
11. S.C. Rae, K. Burnett, and J. Cooper, *Phys.Rev.A*, **50**, 3438 (1994).
12. M.I. Dyakonov and I.V. Gornyi, *Phys.Rev. Lett.*, **76**, 3542 (1996).
13. P. Antoine, B. Piraux, and A. Maquet, *Phys.Rev.A*, **51**, R1750 (1995).
14. J.M. Combes, A. Grossman, and Ph. Tchamitchian, in *Wavelets*, Springer-Verlag, Berlin (1989).
15. L.D. Landau and E.M. Lifshitz, *Quantum mechanics*, 3rd Ed. (Pergamon Press, 1997).
16. W.M. Wood, C.W. Siders, and M.C. Downer, *Phys.Rev.Lett.*, **67**, 3523 (1991).
17. B.M. Penetrante, J.N. Bardsley, W.M. Wood et al., *J.Opt.Soc.Am.*, **B9**, 2032 (1992).
18. S.P. LeBlanc, R. Sauerbrey, S.C. Rae, and K. Burnett, *J.Opt.Soc.Am.*, **B10**, 1801 (1993).
19. W.M. Wood, C.W. Siders, and M.C. Downer, *IEEE Trans. Plasma Sci.*, **21** 20 (1993).
20. V.B. Gildenburg, A.V. Kim, and A.M.Sergeev, *JETP Lett.*, **51**, 104 (1990).
21. V.B. Gildenburg et al., *IEEE Trans. Plasma Sci.*, **21**, 34 (1993).
22. W.M. Wood, G. Focht, and M.C. Downer, *Opt.Lett.*, **13**, 984 (1988).
23. V.B. Gildenburg, V.I. Pozdnyakova, and I.A. Shereshevskii, *Phys.Lett. A*, **203**, 214 (1995).
24. V.B. Gildenburg and A.V. Kim, *Sov.Phys. JETP*, **47**, 72 (1978).
25. A.V. Kim, L.A. Abramyan, and A.M. Sergeev, in *High Field Interaction and Short Wavelength Generation*, OSA Technical Digest Series, 1994, Vol. 16, p. 168.
26. M. Lontano, G. Lampis, A.V. Kim, and A.M.Sergeev, *Physica Scripta*, **T63**, 141 (1996).
27. A.M. Sergeev, M. Lontano, and A.V. Kim, in *Application of High Field and Short Wavelength Sources VII*, OSA Technical Digest Series, 1997, Vol.7, p.118.

Parametric Instabilities and Electron Heating in Ultra-Intense Laser-Plasma Interaction

P. Mora, J. C. Adam, A. Héron, S. Guérin,
G. Laval, and B. Quesnel,

*Centre de Physique Théorique (UPR 14 du CNRS),
Ecole Polytechnique, 91128 Palaiseau Cedex, France*

Abstract.
The general dispersion relation for electron parametric instabilities of an ultra-intense circularly polarized laser wave is established for arbitrary plasma density. It corresponds to a generalization of the stimulated Raman scattering instability, the relativistic modulational instability, the relativistic filamentation instability, and the two plasmons decay instability. In the relativistic regime the generalized instability is characterized by a wide extent of the unstable region in the wave vector space, with growth rates reaching a fraction of the laser frequency, and a strong harmonic generation. One-dimensional and two-dimensional particle-in-cell simulations confirm these results. In particular a systematic study of the propagation of very intense laser pulses through slabs of plasma of several tens of microns are presented. The instability leads to a rapid longitudinal and transverse electron heating, and to filamentary structures which progressively merge in a nonlinear stage. The heating results in highly energetic electrons with energy of several tens of MeV. Correlatively, a strong attenuation rate of the electromagnetic wave is observed.

INTRODUCTION

Due to the rapid developments of high intensity lasers [1], relativistic regimes for laser-plasma interaction are now available, with intensities ranging from 10^{17} to 10^{20} Wcm^{-2}. The interaction of such laser beams with plasmas is of particular interest in the "fast ignitor" context [2] or in the laser plasma advanced accelerator context [3], and is expected to display new physical phenomena.

In this review we stress the importance of electron parametric instabilities of such ultra-intense short laser pulses propagating in plasmas and discuss the resultant electron heating. We deal with ultra-short laser pulses for which ion motion can be neglected so that only electron instabilities have to be considered. At moderate intensities, these instabilities are clearly identified as the stimulated Raman scattering (SRS), the relativistic modulational instability (RMI), the relativistic filamentation instability (RFI), and the two-plasmons decay (TPD). In the weakly relativistic regime [4–6] forward SRS and RMI have been shown to merge in a rarefied plasma [4]. In the fully relativistic regime, the one-dimensional (1D) dispersion relation has been established recently [7–9]. Its analytical and numerical solution shows a wide variety of regimes depending on the parameters a_0 and n/n_c where $a_0 = eA_0/mc^2$ is the normalized amplitude of the laser wave, n is the electron plasma density, and $n_c = m\omega_0^2/4\pi e^2$ is the critical density corresponding to the laser frequency ω_0 [8].

A study of the fully relativistic 2D case was made by Sakharov and Kirsanov in the underdense plasma case [10]. A general 2D dispersion relation for circularly polarized waves in a cold plasma, valid for any laser intensity and plasma density was presented in Ref. [11]. The dispersion relation includes the SRS, RMI, RFI and TPD instabilities as limiting cases at low intensities, and the previous relativistic 1D results.

The results of a systematic study of the propagation of very high-intensity laser pulses through slabs of plasma of several tens of microns were presented in Ref. [12]. The one dimensional and two-dimensional electron parametric instabilities were evidenced. They lead to a rapid longitudinal and transverse heating, and to filamentary structures which progressively merge. The heating results in highly energetic electrons with energy of several tens of MeV. Correlatively, a strong attenuation rate of the electromagnetic wave was observed.

ONE-DIMENSIONAL RELATIVISTIC REGIME

Maxwell equations written for the vector potential $\mathbf{A}$ and the scalar potential Φ and fluid equations written for the electron density n and electron momentum $\mathbf{p}$ admit a zero order equilibrium solution [13] in the form of a circularly polarized wave propagating along the z direction in a uniform plasma ($n = n_0$),

$$\mathbf{A}_0 = \mathbf{e}_p A_0 \exp[i(k_0 z - \omega_0 t)] + c.c., \tag{1}$$

$$\phi_0 = 0, \tag{2}$$

$$\mathbf{p}_0 = e\mathbf{A}_0/c, \tag{3}$$

$$\gamma_0 = (1 + 2e^2 A_0^2/m^2 c^4)^{\frac{1}{2}}, \tag{4}$$

with $\mathbf{e}_p = (\mathbf{e}_x \pm i\mathbf{e}_y)/\sqrt{2}$ and

$$\omega_0^2 = \omega_{p0}^2/\gamma_0 + k_0^2 c^2, \tag{5}$$

where $\omega_{p0} = (4\pi n_0 e^2/m)^{\frac{1}{2}}$. With these definitions, $a_0 = eA_0/mc^2 = 1$ corresponds to an intensity of 5×10^{18} Wcm^{-2} for a 1.06 μm wavelength. The inequalities $\omega_{p0}/\sqrt{\gamma_0} < \omega_0 < \omega_{p0}$ corresponds to the so-called induced transparency regime [14,15]. If one perturbs the equilibrium state with first order perturbations n_1, $\mathbf{p}_1$ and $\mathbf{A}_1$, with in particular (other perturbed quantities are easily deduced from this one)

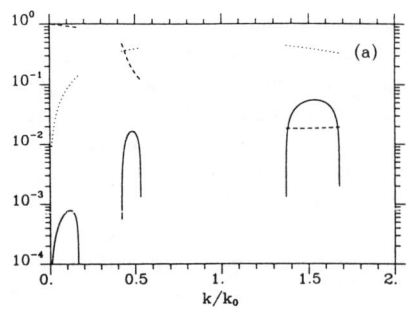 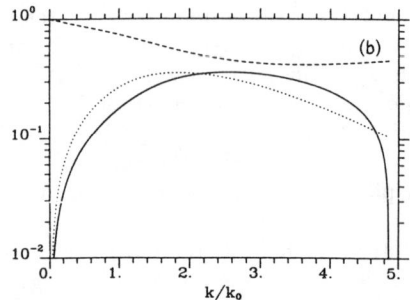

FIGURE 1. The solid lines are the growth rate (normalized to ω_0) as a function of the normalied wave number (normalized to k_0). The dotted lines correspond to the real part of the frequency of the unstable solutions of the relation dispersion [Eq. 7]. The dashed lines are the ratio of the amplitude of the anti-Stokes electromagnetic wave to the amplitude of the Stokes electromagnetic wave. (a) $a_0 = 0.1$ and $n_0/n_c = 0.15$; (b) $a_0 = \sqrt{3/2}$ and $n_0/n_c = 1.5$.

$$n_1 = \eta \exp\left[i(k_z z - \omega t)\right] + \text{c.c.}, \tag{6}$$

the solution of the set of perturbed Maxwell equations and electron fluid equations lead to the following dispersion relation: [7–9]

$$D_+ D_- = \frac{\omega_{p0}^2 a_0^2}{\gamma_0^3}\left(\frac{k_z^2 c^2}{D_p} - 1\right)(D_+ + D_-), \tag{7}$$

where

$$D_\pm = (\omega \pm \omega_0)^2 - (k_z \pm k_0)^2 c^2 - \omega_{p0}^2/\gamma_0 \tag{8}$$

and

$$D_p = \omega^2 - \omega_{p0}^2/\gamma_0. \tag{9}$$

For k real, the dispersion relation is sixth order in ω. It has four roots which are real and two roots which are either real or complex conjugates. This last case corresponds to the instability. If one plots the growth rate $\Gamma = \text{Im}(\omega)$ as a function of the wave number k_z, one can distinguish different regimes depending on the number of unstable branches [we consider that two branches are different when they are separated by a stable part in the $\Gamma(k)$ representation]. Figure 1 shows the growth rate and the corresponding value of $\text{Re}(\omega)$ (in dotted line) as functions of (k_z/k_0) for $a_0 = 0.1$ and $n_0/n_c = 0.15$ [Fig. 1(a)], and for $a_0 = \sqrt{3/2}$ and $n_0/n_c = 1.5$ [Fig. 1(b)].

Figure 1(a) corresponds to the moderate intensity regime, and one can distinguish, for increasing values of k_z, successively the RMI, the forward SRS, and the backward SRS. Figure 1(b) corresponds to the induced transparency, fully relativistic regime. The various branches of instability have merged into a single one, covering a large domain of wave number, and corresponding to growth rates reaching a significant fraction of the zero-order wave frequency ω_0.

THREE-DIMENSIONAL RELATIVISTIC REGIME

Three-dimensional theory [11] further includes the generalization of Raman side scattering, TPD, and RFI. The zero-order solution considered here is the same as in previous section, but one now includes a dependence of the perturbations along the perpendicular direction. The first order quantities are now expanded as

$$f = \sum_{l=-\infty}^{+\infty} f_l \, e^{i(\mathbf{k}+l\mathbf{k}_0)\cdot\mathbf{r}-i(\omega+l\omega_0)t} + \text{c.c.} \tag{10}$$

where $\mathbf{k} = k_z \mathbf{e}_z + \mathbf{k}_\perp$. Using these expansion in the Maxwell and fluid equations, one ends up with an infinite set of coupled equations, which can be truncated if one neglects all terms with $l > L$ or $l < -L$, so that the problem can be transformed into an eigenvalue problem,

$$Q \begin{pmatrix} X \\ \omega X \end{pmatrix} = \omega \begin{pmatrix} X \\ \omega X \end{pmatrix}, \tag{11}$$

where X is a $7 \times (2L+1)$ vector and Q a $(14 \times (2L+1))^2$ matrix. For given physical parameters a_0 and n_0, the system is parametrically unstable to perturbations at

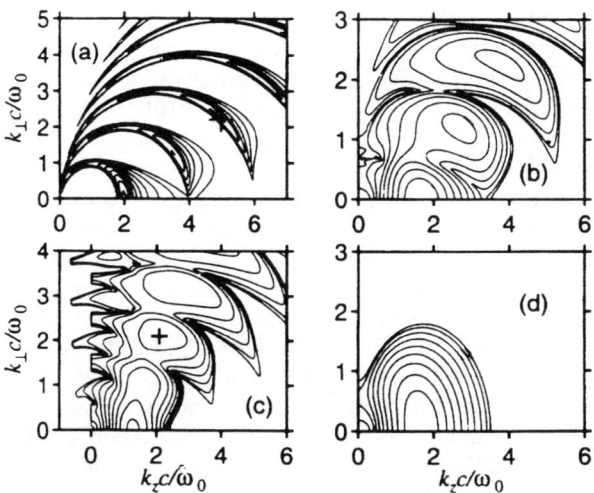

FIGURE 2. Contour plot of the growth rate as a function of the wave number $(k_z, k_\perp)$. The parameters are: (a) $a_0 = 1.46$ and $n_0/n_c = 5 \times 10^{-3}$, (b) $a_0 = \sqrt{3/2}$ and $n_0/n_c = 0.5$, (c) $a_0 = \sqrt{3/2}$ and $n_0/n_c = 1.5$. (d) shows the growth rate for the same parameters as (b) but calculated form the naïve 2D extension of Refs. [7–9]. There are 10 contours on each plot, the minimum and maximum being respectively: (a) 0.01, 0.1; (b) and (d) 0.04, 0.4; (c) 0.035, 0.35. The grey areas are the domains of wave vector used to draw Fig. 3.

wave vector **k** if the matrix Q has a complex eigenvalue. Figure 2 shows contour plots of the corresponding growth rate as a function of the wave number $(k_z, k_\perp)$ for various values of the parameters a_0 and n_0.

At low density and high intensities, one finds new zones of instability in the $(k_z, k_\perp)$ plane. Their shape is semi-circular and well described by the equation $(k_z - Nk_0)^2 + k_\perp^2 = N^2(\omega_0/c)^2$, $N \geq 2$, as can be seen in Fig. 2(a), for which $a_0 = 1.46$ and $n_0/n_c = 5 \times 10^{-3}$. Each lobe corresponds to a resonance with a different harmonic of the vector potential through a three-wave process, as pointed out in Ref. [10].

Our analytical calculation also describes the TPD in the non relativistic limit: for $\omega_0 = 2\omega_{p0}$, we only need to consider the resonant $l = 0$ and $l = -1$ terms and a straightforward calculation gives the well-known formula for the growth rate [11,16]. In the relativistic regime ($a_0 \gtrsim 0.5$), the TPD unstable region in the $(k_z, k_\perp)$ plane breaks into different zones reminiscent of the half circles zones described earlier, while the domain of densities where TPD occurs extends widely around $n_0 = \gamma_0 n_c/4$. A typical result is shown in Fig. 2(b), which corresponds to $a_0 = \sqrt{3/2}$ and $n_0/n_c = 0.5$ (i.e. $n_0/\gamma_0 n_c = 0.25$). We emphasize here that the naïve 2D extension of the 1D fully relativistic dispersion relation [7–9] (in which one simply replaces $k_z \mathbf{e}_z$ by **k**) doesn't predict any of the successive TPD lobes, as can be clearly seen on Fig. 2(d) which shows the growth rate as predicted by this naïve extension for the same parameters as Fig. 2(b). Correlatively, the electrostatic (ES) character of the instability evolves towards a mixed electromagnetic (EM)/ES character. Although they decrease with $|\mathbf{k}|$, the growth rates on the successive lobes remain a significant fraction of the pump wave frequency, even for large $|\mathbf{k}|$ [$\gamma = 0.3\omega_0$ for $(k_z c/\omega_0, k_\perp c/\omega_0) = (2.7, 1.2)$ in Fig. 2(b)].

In the induced transparency regime [14,15] ($n_c \leq n_0 < \gamma_0 n_c$), the different zones of instability already described tend to merge and new lobes appear that extend in the $k_z < 0$ region [See Fig. 2(c) for which $a_0 = \sqrt{3/2}$ and $n_0/n_c = 1.5$.] The successive lobes in the wave-vector plane are no more associated with a single harmonic as in the very underdense case [10], but a wide range of harmonics are simultaneously excited. As in Fig. 2(b), we have verified that the products of the instability have a mixed feature (ES/EM). In all regimes, the growth rate is maximum on axis, but takes significant values for high $k_\perp$. It must be stressed here that the instability produces backscattered light even for $0.25 < n_0/\gamma_0 n_c < 1$, an effect which has no equivalent in terms of classical theory of Raman backscattering.

We were able to confirm these results using a 2D$\frac{1}{2}$ particle-in-cell (PIC) code with periodic boundary conditions. The system was chosen to be $128c/\omega_0 \times 128c/\omega_0$, with $dx = dz \sim 0.17c/\omega_0$, allowing a sufficient number of modes in **k**-space and the correct handling of modes with large wave numbers up to $k \sim 4\omega_0/c$. The simulation used 10 particles per cell and about 6×10^6 particles in the system. The initial temperature was 2 keV. We set at $t = 0$ in the whole simulation box a circularly polarized plane wave propagating in the z direction and corresponding to the zero order solution [13] given above. The physical parameters are $a_0 = \sqrt{3/2}$ and $n_0/n_c = 0.5$, which corresponds to Fig. 2(b).

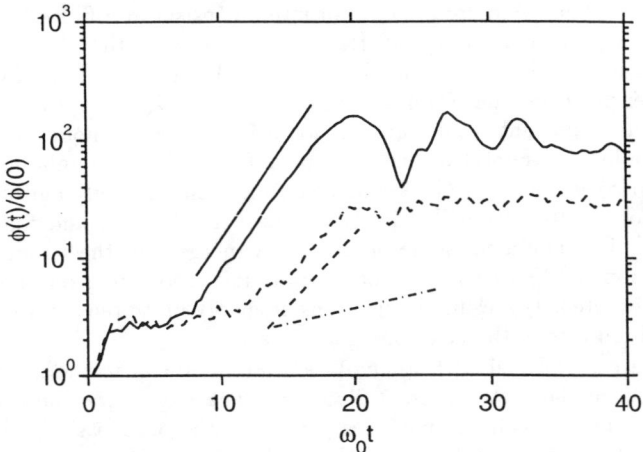

FIGURE 3. Time evolution of the scalar potential in two different zones of the wave-vector plane in the PIC simulation for $a_0 = \sqrt{3/2}$ and $n_0/n_c = 0.5$. Solid line: $1.5 \leq k_z c/\omega_0 \leq 2$ and $0 \leq k_\perp c/\omega_0 \leq 0.25$. Doted line: $2.5 \leq k_z c/\omega_0 \leq 3$ and $1 \leq k_\perp c/\omega_0 \leq 1.3$. The corresponding domains are shown as grey areas on Fig. 2. The slopes of the straight lines correspond to the theoretical value of the growth rate, respectively $0.38\omega_0$ and $0.28\omega_0$. The dashed line corresponds to the simple 2D extension of Refs. [7–9] for the second domain. Its slope is $0.06\omega_0$.

The results of the simulation are illustrated in Fig. 3, which shows the time evolution of the scalar potential averaged over two different zones of the wave-vector space corresponding to the first two lobes in Fig. 2(b). The unstable modes grow out of the numerical noise. The development of electrostatic perturbations with growth rates in close agreement with our calculation is a clear signature of the instability. Note also that the naïve 2D extension of Ref. [7–9] predicts a much smaller growth rate for the zone corresponding to our second lobe ($\gamma = 0.06\omega_0$ instead of $\gamma = 0.28\omega_0$).

For $\omega_0 t > 20$, the non-linear effects of the saturation of the instability can't be ignored. They lead to an important heating of the electrons, which temperature reaches 100 keV at time $\omega_0 t = 30$. Such a high value of the temperature is not surprising, as the initial jitter energy of an electron in the field of the pump wave is 500 keV.

PROPAGATION THROUGH PLASMA SLABS

The studies of previous sections are somewhat academic in the sense that a zero-order solution was assumed to exist before the instability develops. However the growth rates of the instability are so high, attaining a fraction of the wave frequency, that the zero-order solution cannot in practice be established before the instability develops. A more realistic situation consists in a finite laser beam impinging onto a plasma slab. In Ref. [12], we have studied the case of an electromagnetic plane wave incident from the left on a plasma slab of thickness $60\lambda_0$ and electron density n ranging from $0.5n_c$ to $1.5n_c$. Here $\lambda_0 = 2\pi/k_0$ is the laser wavelength. Our important results are the following: (i) even though transverse variations are allowed, the 1D strongly coupled Relativistic Stimulated Raman Scattering (RSRS) initially develops and heats electrons in the longitudinal direction; (ii) shortly after 2D RSRS leads to transverse heating and filamentary structures which in a nonlinear stage progressively merge; (iii) the strong absorption (ranging from 50 % to 90 % in our simulations depending on the plasma density and laser intensity) results in electrons of energy up to tens of MeV and in a large attenuation rate of the electromagnetic wave.

Simulations are performed with a simulation box of $100\lambda_0$ in the x (laser) direction and $37\lambda_0$ in the y (transverse) direction. The size of the mesh is about $0.05\lambda_0$ and there are typically 10 particles per cell in the plasma slab, which corresponds to a total of about 9×10^6 particles. The length of the vacuum region in front of the plasma slab is $24\lambda_0$. The initial temperature is $T_e = 1$ keV. We present the results of a simulation with $n/n_c = 0.5$ and a normalized laser pulse amplitude rising linearly from $a_0 = 0$ to $a_{0,max} = \sqrt{3/2}$ with a (rather long) rise time $t_{rise} = 310\omega_0^{-1}$, followed by a plateau (ω_0^{-1} corresponds approximately to 0.5 fs for a 1μm laser wavelength). Time $t = 0$ corresponds to the laser pulse reaching the edge of the plasma slab, defined by $k_0 x_{edge} \approx 150$.

The development of a large perturbation at the edge of the slab of plasma with a value of k_x close to wave number corresponding to the theoretical maximum growth rate of the 1D instability for the current value of the amplitude a_0 slightly inside the

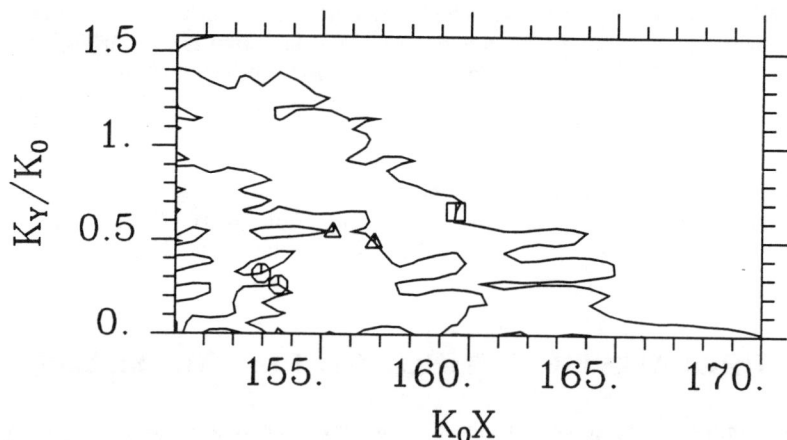

FIGURE 4. Contour plots of $|E_x(x,k_y)|$ for the slab simulation. The contours correspond to isovalues of $|E_x(x,k_y)|$ in units of a_0 [0.003 (□), 0.01 (△), 0.03 (O)].

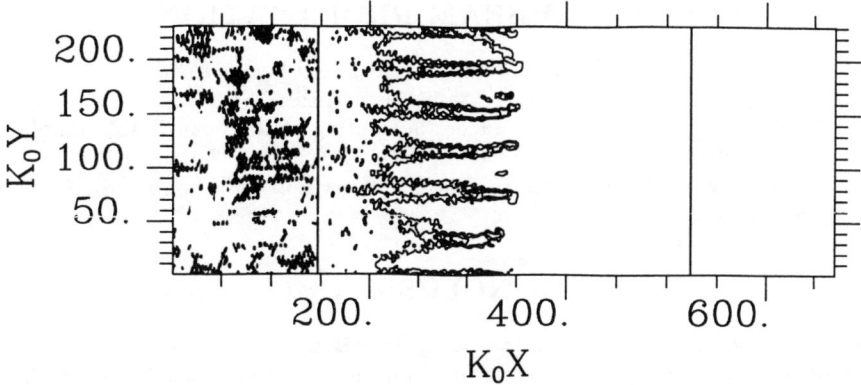

FIGURE 5. Contour plots of the flux of the Poynting vector in the x direction as a function of space for the plasma slab simulation, at time $t = 450\omega_0^{-1}$, for which $a_{edge} = a_{max} = \sqrt{3/2}$.

plasma is seen in the simulation, with an averaged growth rate comparable with the theoretical growth rates. Sidescattering instabilities ($k_y \neq 0$) have a slightly smaller growth rate than the longitudinal ones, but develops almost simultaneously leading to a strong heating in the y direction. This is the fundamental new feature of 2D simulations compared with the 1D case. In the simulation transverse perturbations appear as soon as $t = 125\omega_0^{-1}$. Figure 4 shows contour plots of $|E_x(x, k_y)|$ in the (x, k_y) space (the Fourier transform is taken only in the y direction). One can see that a broad spectrum has already developed with wave numbers extending up to $k_y = 1.5k_0$. The mode $k_y = 0$ is still dominant at this early time (its amplitude is 0.1 in the same unit as a_0), however modes with amplitude above 0.01 can be observed up to $k_y = 0.8k_0$.

After this initial phase the nonlinear structure of the 2D perturbations becomes filamentary. At $t = 150\omega_0^{-1}$ about 20 filaments can be counted corresponding to an average wavelength of $1.9\lambda_0$. The number of filaments decreases rapidly and at $t = 450\omega_0^{-1}$, for which $a_{edge} = a_{max} = \sqrt{3/2}$, there are only 6 of them, as shown in Fig. 5. Their coalescence continues then more slowly and at $t = 1050\omega_0^{-1}$ only 2 remain. The maximum value of the Poynting vector within a filament is approximately twice a_0^2. The heating which takes place in the filaments generates highly energetic particles with relativistic factor up to 40. Some electrons are penetrating ahead of the front of propagation of the laser light. Though they are preferentially accelerated in the x direction, a significant transverse heating also occurs. The accelerated electrons generate strong electric currents in the axial direction, which, together with return currents, give rise to quasi-static magnetic field in the 150 MG range.

FOCALIZED BEAM PROPAGATION

Finally we have done PIC simulations with focalized beams incident on plasma slab, showing the development of the same instabilities. Specific effects such as the self-focussing of the all beam and/or the spliting of the beam into a number of diverging filaments are also observed in some situations. The analysis of these simulations show a variety of effects which however appear beyond the scope of the present review.

CONCLUSION

The analytic calculations and the numerical results presented in this review show the crucial importance of electron instabilities in the context of ultra-intense laser-plasma interaction. We have shown that the growth rate of the instabilities can reach a significant fraction of the wave frequency. The nonlinear stage of the instability lead to a very strong heating of the plasma electrons and to a significant absorption of the electromagnetic energy.

REFERENCES

1. G. Mourou and D. Umstader, *Phys. Fluids B* **4**, 2315 (1992).
2. M. Tabak, J. Hammer, M. E. Glinsky, W. L. Kruer, S. C. Wilks, J. Woodworth, E. M. Campbell, M. D. Perry, and R. J. Mason, *Phys. Plasmas* **1**, 1626 (1994).
3. IEEE Trans. Plasma Sci. **PS-15**, No. 2 (1987), special issue on *Plasma-Based High-Energy Accelerators*, edited by T. Katsouleas, and references therein.
4. C. J. McKinstrie and R. Bingham, *Phys. Fluids B* **4**, 2626 (1992).
5. T. M. Antonsen, Jr. and P. Mora, *Phys. Fluids B* **5**, 1440 (1993).
6. N. E. Andreev *et al.*, *Physica Scripta* **49**, 101 (1994).
7. A. S. Sakharov and V. I. Kirsanov, *Phys. Rev. E* **49**, 3274 (1994).
8. S. Guérin, G. Laval, P. Mora, J. C. Adam, and A. Héron, *Phys. Plasmas* **2**, 2807 (1995).
9. C. D. Decker, W. B. Mori, K. C. Tzeng and T. Katsouleas, *Phys. Plasmas* **3**, 2047 (1996).
10. A. S. Sakharov and V. I. Kirsanov, *Plasma Phys. Rep.* **21**, 632 (1995).
11. B. Quesnel, P. Mora, J. C. Adam, S. Guérin, A. Héron, and G. Laval, *Phys. Rev. Lett.* **78**, 2132 (1997); B. Quesnel, P. Mora, J. C. Adam, A. Héron, and G. Laval, *Phys. Plasmas* **4**, 3358 (1997).
12. J. C. Adam, A. Héron, S. Guérin, G. Laval, P. Mora, and B. Quesnel, *Phys. Rev. Lett.* **78**, 4765 (1997).
13. A. I. Akhiezer and R. V. Polovin, *Sov. Phys. JETP* **3**, 696 (1956).
14. E. Lefebvre and G. Bonnaud, *Phys. Rev. Lett.* **74**, 2002 (1995)
15. S. Guérin, P. Mora, J. C. Adam and A. Héron, and G. Laval, *Phys. Plasmas* **3**, 2693 (1996).
16. J.F. Drake *et al*, *Phys. Fluids* **17**, 778 (1974); W. L. Kruer, *The Physics of Laser Plasma Interactions*, (Addison-Wesley, New-York, 1988).

Radiation at $2\omega_p$ from inverse two-plasmon decay in overdense plasma driven by ultra-short laser pulses

R. Lichters, J. Meyer-ter-Vehn, A. Pukhov

Max-Planck-Institut für Quantenoptik, Hans-Kopfermann-Strasse 1, D-85748 Garching

Abstract. We report on strong line emission from overdense plasma layers at the second (and also third) harmonic of the plasma frequency ω_p, when irradiated by intense ultra-short laser pulses. The results are obtained from one-dimensional particle-in-cell simulations. The emission is interpreted as conversion of two (or more) plasmons into a photon. Electron jets, generated by laser interaction at the surface, produce the plasmons. Phase matching requires counterpropagating plasmons which arise either from two-sided laser irradiation or electrons returning from the rear side.

INTRODUCTION

The decay of a photon into two plasmons (two plasmon decay) is known to contribute to laser light absorption in plasma [1]. This three-wave process has to satify the conditions

$$\Omega = \omega_1 + \omega_2, \quad \boldsymbol{K} = \boldsymbol{k}_1 + \boldsymbol{k}_2, \tag{1}$$

where $(\Omega, \boldsymbol{K})$, $(\omega_1, \boldsymbol{k}_1)$, $(\omega_2, \boldsymbol{k}_2)$ are frequency and wavevector of the photon and the 2 plasmons, respectively. In the present paper, we describe how the *inverse* of this process can be stimulated in thin plasma layers by irradiation with short-pulse high-intensity laser pulses. We consider overcritical plasmas close to solid densities with sharp edges and plasma frequencies $\omega_p = \sqrt{e^2 n_e/\varepsilon_0 m}$ (n_e electron density, e and m charge and mass of the electron) much larger than the frequency ω_0 of the driving laser. The plasmon frequencies ω_1 and ω_2 are close to ω_p, and the plasma layer therefore emits electromagnetic radiation at $2\omega_p$. In what follows, we show that this process is potentially an efficient source of coherent femtosecond high-intensity XUV light.

The role of the external laser pulses is only to produce jets of fast electrons at the surface [2–4] which penetrate the layer and excite the plasmons. The generation of these jets is particularly efficient for p-polarized laser pulses with oblique incidence. As described by Brunel [5], bunches of electrons are then extracted from the surface by the E_x-component of the laser field (normal to the surface) and are reinjected into the layer in the second half of each laser cycle. They excite Langmuir waves with phase velocities close to the maximum electron velocity. In order to satify the k-condition in Eq.(1), one needs plasmons propagating in opposite directions. Such plasmons naturally arise in thin layers , as we shall explain below. The process described here becomes efficient for laser intensities approaching the relativistic threshold (10^{18}W/cm^2). Short pulses (10 - 100 fs) of such intensity are now available at a number of places, using chirped-pulse-amplification (CPA) lasers [6].

Mechanisms to generate coherent electromagnetic radiation through electrostatic plasma waves have been discussed in detail in the context of solar radio-wave emission [7–10]. Such emission occurs in strongly driven plasmas due to non-linear wave coupling. The conversion of two plasmons into a photon has been identified as a process of particular importance leading to $2\omega_p$ emission. In the context of solar corona physics, the plasma waves are excited by electron beams emerging as bursts from the solar surface. Here, we discuss the same process, but for XUV rather than radio frequencies. The theory of non-linear coupling, given in [7,10], is not reproduced here; rather we present numerical results.

PIC SIMULATION

The numerical analysis is based on fully relativistic, electromagnetic particle-in-cell (PIC) simulations [12,13]. It is restricted to one space-dimension x (layer plane in y, z), but allows particles to move in three velocity-dimensions (1D3V). Oblique incidence of laser light is treated using Bourdier's method [11], i.e. a Lorentz transformation is performed from the laboratory frame L to a frame M such that the pulse is normally incident in frame M. For a plane wave incident in L under angle α, $(\omega^L, \mathbf{k}^L) = (\omega_0, k_0 \cos\alpha, k_0 \sin\alpha, 0)$ with $k_0 = \omega_0/c$, this is achieved with a frame velocity $\mathbf{v}_f = c \sin\alpha \, \hat{\mathbf{y}}$. The Doppler-shifted laser light in M is then described by $(\omega^M, \mathbf{k}^M) = (\omega_0 \cos\alpha, k_0 \cos\alpha, 0, 0)$, and its period is $\tau = \tau_0/\cos\alpha$ with $\tau_0 = 2\pi/\omega_0$. In M all physical quantities are supposed to depend only on x, neglecting possible plasma perturbations in (y, z) direction, and light can be emitted only in specular direction. Maxwell's equations in Coulomb gauge in frame M read

$$\left(\partial_x^2 - \frac{1}{c^2}\partial_t^2\right) \mathbf{A}(x,t) = -\frac{1}{\varepsilon_0 c^2} \mathbf{j}_\perp(x,t), \tag{2a}$$

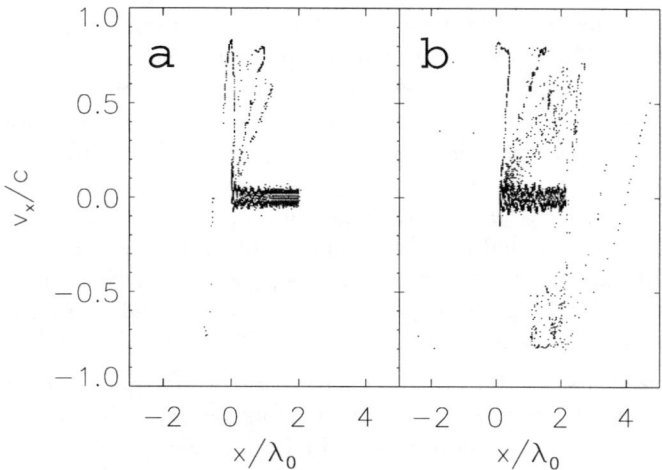

FIGURE 1. Electron phasespace (x, v_x) in the laboratory frame L at time (a) $t = 3.25\tau_0$ and (b) $t = 6.5\tau_0$ after the leading edge of the laser pulse has hit the surface. The p-polarized pulse with $a_0 = 1$ and pulse duration $t = 10\tau_0$ is obliquely incident from the left under $\alpha = 50.13°$ on a uniform plasma layer with $n_e/n_c = 27.5625$ and plasma frequency $\omega_p/\omega_0 = 5.25$.

$$\partial_x^2 \Phi(x,t) = -\frac{1}{\varepsilon_0}\varrho(x,t), \quad (2b)$$

where ρ and $\boldsymbol{j}_\perp$ are charge and transverse current densities, respectively. Electromagnetic fields are given by $\boldsymbol{E} = -\hat{\boldsymbol{x}}\partial_x\Phi(x,t) - \partial_t\boldsymbol{A}(x,t)$ and $\boldsymbol{B}(x,t) = \nabla \times \boldsymbol{A}(x,t)$. The equation of motion for electrons is $d_t(m\gamma\boldsymbol{v}) = -e(\boldsymbol{E} + \boldsymbol{v} \times \boldsymbol{B})$ with $\gamma = \sqrt{1 - \boldsymbol{v}^2/c^2}$. Ions are kept fixed. Results of PIC simulations based on these equations are presented in Figs. 1-4. They are obtained with typically 10^5 particles and 500 timesteps per laser period. Laser intensity $I\lambda_0^2 = a_0^2 \times 1.37 \cdot 10^{18}$ Wcm$^{-2}\mu$ m^2 is expressed in terms of the dimensionless amplitude $a_0 = eE_0/(m\omega_0 c)$. We use a p-polarized laser pulse with $a_0 = 1$, obliquely incident under $\alpha = 50.13°$ on a uniform plasma layer. The pulse duration is 10τ, including 3τ for linear rise and 3τ for linear fall. The plasma density is set to $n_e/n_c = 27.5625 = (\omega_p/\omega_0)^2 = (5.25)^2$ such that resonance between plasma frequency and laser harmonics is avoided. The ion density is uniform with sharp surfaces at $x = 0$ and $2\lambda_0$.

RESULTS

Figure 1 shows two snapshots of the electron phase-space (x, v_x) in the laboratory frame. It is seen that groups of fast electrons are generated at the irradiated (left) surface and are injected into the bulk of the plasma. In

Fig.1a, this is shown for the time $t = 3.25\tau_0$ after the pulse front has hit the plasma surface. There are three such groups corresponding to the three laser cycles that have passed, and one observes maximum velocities up to $v_0 = 0.8c$. When entering the plasma, they are sharply bunched in space, though with a broad velocity spectrum. These bunches disperse while running more deeply into the layer. In Fig.1b, we show the phase-space for $t = 6.5\tau_0$. At this time, the first jets have left the layer through the rear surface, have been reflected by the electrostatic potential, which they build up in the rear space, and are reinjected into the layer, now running in opposite direction. It is also seen that the jets originating from different laser cycles tend to mix and fill the phase space more homogeneously.

Due to beam instability, the jets excite electron plasma waves in the layer which can be identified in Fig.1 in the distribution of the low-velocity plasma electrons. They are seen more clearly in Fig.2a, where the electron density $n_e(x,t)$ is plotted in the x,t plane for two time windows. One observes high-amplitude waves ($\Delta n_e/n_e \approx 0.2 - 0.5$) in the bulk plasma. For times $3.5 < t/\tau < 5.5$ in Fig.2-a1, they propagate from left to right. Note that the grey scale for n_e has been chosen in a way to highlight the density contrast close to the bulk density, and that black colour is used also to display the low-density electron populations emerging from both surfaces of the layer. Near the laser driven (left) surface, one sees extraction and reinjection of electrons within

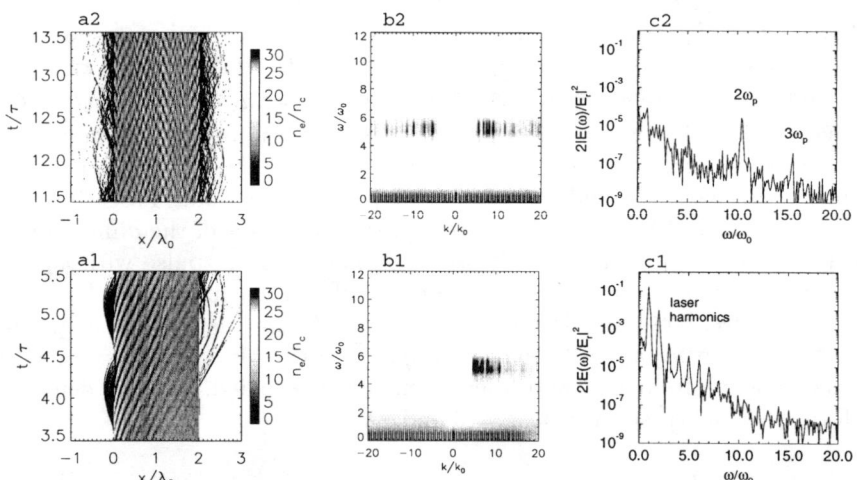

FIGURE 2. Same case as in Fig.1 showing (a) the electron density $n_e(x,t)$ (M-frame) in two time windows, (b) its Fourier transform $\tilde{n}_e(k,\omega)$, and (c) the power spectra emitted from the left surface in the time intervals (c1) $0 < t/\tau < 10$ and (c2) $10 < t/\tau < 20$. The units $k_0 = 2\pi/\lambda_0$, $\omega_0 = 2\pi/\tau_0$, $E_r = m\omega_0 c/e$ refer to the incident light, but note that its period is $\tau = \tau_0/\cos\alpha$ in the M-frame.

each laser cycle, demonstrating the Brunel effect. The first of these jets is seen to break out from the rear (right) surface after 4.1τ. Apparently, it has broken up into smaller bunches by interaction with the plasma wave, and each wave crest releases one of these bunches. Eventually they are reflected and start to drive waves from right to left inside the layer. The electrostatic field building up at both surfaces is of the same order as the E-field of the incident light, $E_r = m\omega_0 c/e \approx 30$ GV/cm. The second time window in Fig.2-a2 lies in the interval $11.5 < t/\tau < 13.5$ after the laser pulse interaction. A high level of wave excitation is observed with waves now propagating in both directions. Clouds of hot electrons extend in front of the surfaces on both sides.

Let us now look at the electromagnetic radiation generated by these charge dynamics. The power spectra emitted from the left surface are given in Figs. 2-c1 and 2-c2. They correspond to Fourier transforms over the time intervals $0 < t/\tau < 10$ and $10 < t/\tau < 20$, respectively. The key result of this paper is seen in Fig. 2-c2. A prominent emission line sticks out from the background at the frequency $2\omega_p = 10.5\omega_0$ and also another one at $3\omega_p$. These remarkable lines are missing during the earlier time interval in Fig. 2-c1, when the laser pulse is still on. Instead one identifies laser harmonics in the spectrum up to order 7 and higher at this earlier time. Actually, search for the laser harmonics had stimulated the present work, and corresponding results are discussed in [13], showing that they are generated promptly at the irradiated surface and can be understood as reflection from an oscillating mirror. Here we concentrate on the radiation at *harmonics* $n\omega_p$ $(n = 2, 3, \ldots)$ *of the plasma frequency* which is clearly of different nature and appears only at times, when the pattern of counterpropagating waves seen in Fig. 2-a2 has formed.

DISCUSSION

Let us further analyze these results in analytic terms. We refer to the cold plasma fluid equations, derived in [13] for the one-dimensional M-frame. Solving them within linear approximation, one obtains the dispersion relation for light waves in the standard form $\omega^2 = \omega_p^2 + c^2 k^2$; for longitudinal waves, however, one finds $\omega = \omega_p \cos \alpha$, $k_x = \omega_p \cos \alpha / v$, where v is the phase velocity in the M-frame. Supposing that two types of plasmons have been excited with phase velocities v_1 and v_2 in the M-frame, one gets photons with

$$\Omega = 2\omega_p \cos \alpha, \quad K_x = \omega_0 \cos \alpha (1/v_1 + 1/v_2) \qquad (3)$$

according to the matching conditions (1). Inside the plasma these photons have to satisfy the dispersion relation $\Omega^2 = \omega_p^2 + c^2 K_x^2$, which leads to the condition

$$0 \leq (4 - 1/\cos^2 \alpha) = (c/v_1 + c/v_2)^2 \leq 3. \qquad (4)$$

It implies $0 \leq \alpha \leq 60°$ for the angle, under which the photons can emerge from the layer. The cutoff at $\alpha = 60°$ corresponds to photons propagating under 90° in the plasma layer (index of refraction $n = \sqrt{3}/2$ at $2\omega_p$). Actually photons cannot escape from the layer in this limit. Because of $|c/v_1| > 1$ and $|c/v_2| > 1$, condition (4) also implies that v_1 and v_2 have to have opposite signs, i.e. counterpropagating plasmons.

The spectrum of Langmuir waves actually excited in the PIC-simulation can be determined from the density plots in column (a) of Fig. 2. Performing the Fourier transform of $n_e(x,t)$ in both space and time, we obtain the spectral density $\tilde{n}_e(k,\omega)$ which is plotted in column (b) of Fig.2. A strong feature is seen at $\omega = (5.25 \pm 0.50)\omega_0$ and $k > 4.2k_0$. It correponds to right-bounded Langmuir waves with phase velocities $v < 0.8c$. Apparently, an extended spectrum of k-values is excited, but since they all have the same sign, the matching condition for two-plasmon conversion into a photon cannot be satisfied. This possibility opens up only later (see Fig. 2-b2), when also negative k-values appear representing left-bounded waves with $v < -0.65c$. Due to the finite width (2τ) of the time windows transformed, the $\tilde{n}_e(k,\omega)$ patterns have also a width in ω. Let us further remark that the spectral band at $\omega = 0$ represents the static density profile of the layer and the weak feature at $\omega = \omega_0$ the laser driven oscillation of the surface.

In view of the distribution of right- and left-bounded waves, there is now a variety of possibilities to fulfill the matching condition (4) for different angles α. Also different signs of the photon wave-vector $K_x = k_{x1} + k_{x2}$ are possible such that one expects to see the $2\omega_p$ emission on both sides of the layer. Corresponding results are shown in Fig. 3. They are obtained for the same parameters as before, but varying α and plotting them for the directions of both the "reflected" and the "transmitted" light. Indeed we find $2\omega_p$ emission in all these cases with fluctuations as a function of α, a maximum somewhat below the cutoff at 60°, and slightly higher amplitudes in the forward (trans-

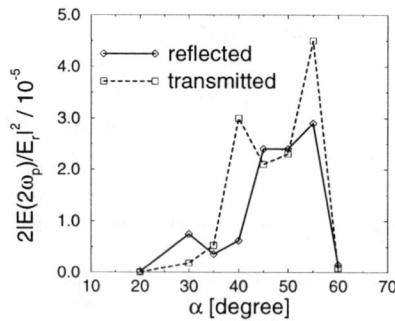

FIGURE 3. Peak power of $2\omega_p$ emission in the reflected and transmitted light for various angles of incidence α, time interval $10 < t/\tau < 20$, and otherwise same parameters as in Fig. 1.

mitted) direction. For the chosen case, the optimum corresponds to a flash of coherent light emitted at $\Omega = 10.5\omega_0$ under $\alpha = 55°$ with an intensity of about 50 TW/cm^2. We have checked the temporal coherence of the $2\omega_p$ emission and find a coherence time of about 10τ.

One might wonder wether the $2\omega_p$ emission can be enhanced by controlling the process of wave excitation more tightly, such that $\tilde{n}_e(k,\omega)$ is more localized in k-space while satisfying k-matching at the same time. In Fig. 4 we present a case in which the layer is driven by two laser pulses incident symmetrically from both sides. From the density plot it is evident that the counterpropagating waves are now produced directly during the interaction phase. Accordingly, the spectrum (reflected direction) shows the plasma radiation simultaneously with the laser harmonics. However, the spectral density $\tilde{n}_e(k,\omega)$ (not shown) is still quite extended in k, and we did not succeed to construct a sharply resonant case. The $2\omega_p$ peak, shown in Fig.4b for $\alpha = 55°$, has almost the same strength as the third laser harmonic, but still is not significantly stronger than for the single-sided irradiation discussed before. We also have varied the layer thickness within a factor 2 and have ramped the density profile of the irradiated surface over a distance of $0.5\lambda_0$ without any significant change of the $2\omega_p$ results. No systematic optimization has been performed, so far.

In conclusion, we predict, on the basis of PIC simulations, strong coherent plasma radiation at twice the plasma frequency to be generated in thin overdense plasma layers when irradiated by ultrashort high-intensity laser pulses. The idea is to produce such layers by converting solid foils into fully ionized plasma at near-solid density either with the help of a prepulse or by the interaction pulse itself. At such densities, the $2\omega_p$ radiation lies in the XUV spectrum. Temperatures in such layers are high enough to justify an approxi-

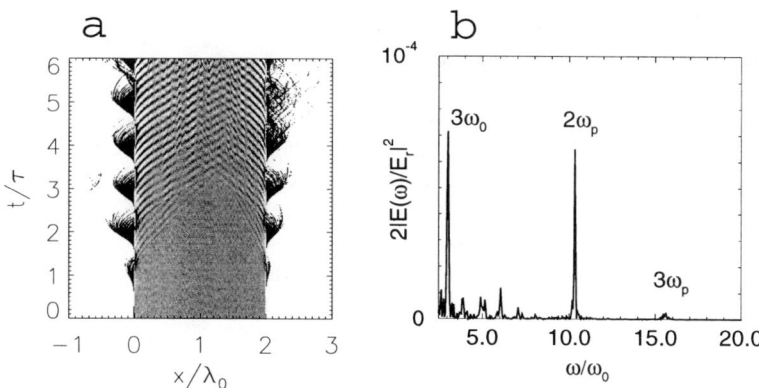

FIGURE 4. Two-sided irradiation with $a_0 = 1$ from left, $a_0 = 0.9$ from right, $\alpha = 55°$, and $n_e/n_c = 27.5625$; (a) electron density and (b) power spectrum (linear scale!) emitted from left surface during $0 < t/\tau < 20$.

mate description in terms of a classical plasma, as presented above. Of course, the ionization process as well as strong coupling effects need to be investigated, but this was beyond the possibilities of our present study. It is felt that experimental results are needed to further guide the theoretical investigations. If detected in experiments, the $2\omega_p$ radiation may become an important diagnostic tool for these short-pulse laser plasmas. Also, optimizing k-matching by fully exploiting three space dimensions (rather than one in the present study) may lead to substantial enhancement of the conversion rate and make the $2\omega_p$ light an attractive ultra-short pulse XUV source.

ACKNOWLEDGEMENTS

The authors would like to thank M. Zepf and U. Teubner for stimulating discussions. The work was supported in part by the Bundesministerium für Forschung und Technologie (Fördervorhaben 06 MM 364) and by EURATOM.

REFERENCES

1. W.L. Kruer, *The Physics of Laser Plasma Interactions*, Frontiers in Physics, vol. 73, Addison-Wesley 1988 (Reading/Mass.).
2. P. Gibbon and A.R. Bell, Phys. Rev. Lett. 68, 1535 (1992).
3. S. Bulanov et al., Phys. Plasmas **1** (3), 745 (1994).
4. H. Ruhl and P. Mulser, Phys. Lett. **A205**, 388 (1995); H. Ruhl, Phys. Plasmas **3**, 3129 (1996).
5. F. Brunel, Phys. Rev. Lett. **59** (1), 52 (1987).
6. D. Strickland and G. Mourou, Opt. Comm. **56**, 219 (1985); M.D. Perry and G. Mourou, Science **264**, 917 (1994).
7. V.L. Ginzburg and V.V. Zheleznjakov, Sov Astron. J. **2**, 653 (1958).
8. D.B. Melrose, Space Sci. Rev. **26**, 3 (1980).
9. P.H. Yoon, Phys. Plasmas **2**, 537 (1995).
10. A. Willes et al., Phys. Plasmas **3** (1), 149 (1996).
11. A. Bourdier, Phys. Fluids 26, 1804 (1983).
12. C.K. Birsall and A.B. Langdon, *Plasma Physics via Computer Simulation*, Series on Plasma Physics (Adam Hilger, New York, 1991).
13. R. Lichters et al., Phys. Plasmas **3** (9), 3425 (1996); R. Lichters and J. Meyer-ter-Vehn, in *Multiphoton Processes 1996* (Inst. Phys. Publ., Bristol, 1997), pp. 221-230; R. Lichters, PhD thesis, Techn. Univ. München (1997) and Report MPQ219, Max-Planck-Institut für Quantenoptik, 1997.

Emission of electromagnetic waves by accelerated short laser pulses in a plasma

N.L. Tsintsadze* and J.T.Mendonça

GoLP/Grupo de Lasers e Plasmas, Centro de Física de Plasmas
1096 Lisboa Codex, Portugal
* Tbilisi State University, Chavchavadze 3, Tbilisi, Georgia

INTRODUCTION

In the present paper we shall consider the interaction of a relativistically intense short laser pulse with a cold electron-ion inhomogeneous plasma, assuming that the ions are at rest. In 3 dimensions, the pulse shapes can significantly change along propagation and are very difficult to describe. We will restrict our analysis to pancake type solitons, because it was shown[2] that for such a case the shape changes in the direction perpendicular to the pulse propagation can be neglected. Taking this into account we could derive a one-dimensional Shroedinger equation for arbitrary laser pulse amplitude. Using this equation and assuming a slightly relativistic nonlinearity we obtained accelerated soliton solutions. Moreover, such accelerated soliton laser pulses can excite electron plasma waves, as well as low frequency electromagnetic waves.

BASIC EQUATIONS

Let us study the interaction of circularly polarized electromagnetic waves with the plasma electrons (the ion motion is neglected) by using Maxwell's equations complemented with the electron fluid equations. Using the multiple time-scale method [1], we obtain for the time averaged variables the following evolution equations:

$$\nabla^2 <\vec{p}_t> - \frac{\partial^2}{\partial t^2} <\vec{p}> - \frac{\partial}{\partial t}\nabla <\gamma> = \frac{<n>}{<\gamma>} <\vec{p}> \qquad (1)$$

$$\frac{\partial <n>}{\partial t} + \nabla \cdot \frac{<n>}{<\gamma>} <\vec{p}> = 0 \qquad (2)$$

$$\frac{\partial}{\partial t} <\vec{p}> = - <\vec{E}> - \nabla <\gamma> \qquad (3)$$

$$\nabla^2 \phi = <n> - 1 \qquad (4)$$

$$\nabla \times <\vec{p}> = <\vec{B}> \qquad (5)$$

On the other hand, the equation for the rapidly varying quantities are:

$$\nabla^2 \vec{p}_l - \frac{\partial^2 \vec{p}_l}{\partial t^2} = \frac{<n>}{<\gamma>} \vec{p}_l \qquad (6)$$

where we have used the dimensionless variables defined by $(p/mc) \to p$, $(e\phi/mc^2) \to \phi$, $(n/n_0) \to n$, $(\vec{r}\omega_p/c) \to \vec{r}$, $\omega_p t \to t$, $(e\vec{B}/mc\omega_p) \to \vec{B}$ and $(e\vec{E}/mc\omega_p) \to \vec{E}$.

From equation (3) we can easily get[2]:

$$\nabla_z <p_\perp^l> = \nabla_\perp <p_z^l> \qquad (7)$$

This relation allows us to compare $<p_\perp^l>$ and $<p_z^l>$. If the averaged physical quantities change faster along the direction of propagation of the laser pulse than in the perpendicular direction (which is the case for pancake-like shaped pulses) it follows from equation (7) that

$$<p_\perp^l> \ll <p_z^l> . \qquad (8)$$

Such a condition is satisfied in a variety of different physical situations related to very short laser pulses, such as self-focusing, self-channelling or the generation of wakefields. Condition (8) also corresponds to assuming that the laser pulses change mainly along the direction of its propagation and not in the perpendicular directions. Assuming that (8) stays valid, we can then write:

$$p_l = \frac{1}{\sqrt{2}} p_\perp(r_\perp) p_\parallel(z,t)(\hat{e}_x + i\hat{e}_y) e^{i(k_0 z - \omega_0 t)} \qquad (9)$$

where the perpendicular part of the vortex momentum perturbation can be written as $p_\perp(r_\perp = \sqrt{I_0} e^{-r_\perp^2/2r_0^2})$, and $p_\parallel(z,t)$ is an arbitrary function describing the longitudinal pulse profile, to be determined later. The electron plasma density can be written as:

$$<n> = n_0 + \Delta n(z) + \delta n(r_\perp, z, t) \qquad (10)$$

where n_0 is a constant, $\Delta n(z)$ characterizes the inhomogeneity profile and $\delta n(r_\perp, z, t) = n_\perp(r_\perp) n_\parallel(z,t)$ is the perturbation associated with the ponderomotive force effects.

Substituting p_t from equation (9) in (6), multiplying the resulting equation by $p_\perp 2\pi r_\perp dr_\perp$ and integrating over $r_\perp$, we obtain:

$$2i\omega_0 \left(\frac{\partial}{\partial t} + \frac{k_0}{\omega_0}\frac{\partial}{\partial z}\right)p_\| + \left(\frac{\partial^2}{\partial z^2} - \frac{\partial^2}{\partial t^2}\right)p_\| = [\beta(p_\|)n_\| - 1]p_\| \tag{11}$$

where we have used:

$$\beta(p_\|)n_\| = [1 + \Delta n(z)]\frac{1}{B}\int_0^\infty \frac{p_\perp^2}{<\gamma>}2\pi r_\perp dr_\perp + \frac{u_\|(z,t)}{B}\int_0^\infty \frac{p_\perp^2 u_\perp(r_\perp)}{<\gamma>}2\pi r_\perp dr_\perp \tag{12}$$

$$B = \int p_\perp^2 2\pi r_\perp dr_\perp \tag{13}$$

and $(k_0/\omega_0) = v_g$ is the "group" velocity along the axis Oz.

Introducing new variables ($\xi = z - v_g t, t$) instead of (z,t), and using $p_\| = a_\| \exp(i\phi)$ and taking the following conditions into account: $\omega_0 \gg |v_g(\partial/\partial\xi)|$, $|\partial\phi/\partial\xi| \gg |\partial\phi/\partial t|$ and $|\partial a_\|/\partial\xi| \gg |\partial a_\|/\partial t|$, we obtain from equation (11):

$$\omega_0 \frac{\partial a_\|^2}{\partial t} + \frac{\partial}{\partial x}\left(\frac{\partial \phi}{\partial x}a_\|^2\right) = 0 \tag{14}$$

and:

$$-2\omega_0 \frac{\partial \phi}{\partial t}a_\| + \frac{\partial^2 a_\|}{\partial x^2} - \left(\frac{\partial \phi}{\partial x}\right)^2 a_\| = [\beta(a_\|)n_\| - 1]a_\| \tag{15}$$

where $x = \gamma_g \xi$ and $\gamma_g = 1/\sqrt{1 - v_g^2}$. The solutions for these two equations can be taken in the form $a_\|(x - \bar{x}(t))$ and $\phi(x,t)$, where $\bar{x}(t)$ is the coordinate of the center of the laser pulse. The time evolution of this coordinate will determined in the following way. Since the amplitude is assumed to be a function only dependent on the self-similar argument $\eta = x - \bar{x}(t)$, and retaining only solutions which vanish at infinity ($\eta \to \infty, a_\| = 0$), we conclude from equation (14) that:

$$\phi(x,t) = \omega_0 \dot{\bar{x}}(t)x + F(t) \tag{16}$$

where $F(t)$ is a function of time which can be considered arbitrary, for the time being, but that will be specified later. Substituting equation (16) in (15) we obtain a closed equation for the amplitude $a_\|$:

$$\frac{\partial^2 a_\|}{\partial x^2} - \left\{2\omega_0 \dot{F}(t) + \omega_0^2 \dot{\bar{x}}^2 + 2\omega_0^2 x \frac{d^2\bar{x}(t)}{dt^2}\right\}a_\| = [\beta n_\| - 1]a_\| \tag{17}$$

Thus, starting from the fully relativistic three dimensional equations we could derive, for pancake-like shaped pulses, a one-dimensional nonlinear Schroedinger equation for arbitrary pulse amplitudes.

ACCELERATED SOLITON SOLUTIONS

We now consider the above equations in the weakly relativistic limit, when the analytical solutions for the problem can be carried out until the end. In this case we can write for δn:

$$(\frac{\partial^2}{\partial t^2} + 1)\delta n = \nabla^2 \frac{a^2}{2} \tag{18}$$

In this equation we can isolate that part of the density perturbation which is concentrated in the pulse region, δn_p, from the density perturbation associated with electron plasma waves left behind the pulse (usually called the wakefield), N. Equation (18) then leads to:

$$(\frac{\partial^2}{\partial t^2} + 1)N = -\frac{1}{2}\frac{d^2\bar{x}}{dt^2}\nabla^2\frac{\partial}{\partial x}a^2 \tag{19}$$

A solution of equation (19), in the case of a pancake-like pulse which satisfies the condition of absence of wakefield at $t \to -\infty$ takes the form:

$$N(\vec{r},t) = -\frac{1}{2}a_\perp^2(r_\perp)\frac{\partial^3}{\partial x^3}\int_{-\infty}^{t} dt' \frac{d^2\bar{x}}{dt^2} \sin(t-t')a_\parallel^2(x - \bar{x}(t')) \tag{20}$$

Now, we can come back to equation (17) and consider the case of a weakly relativistic correction of the electron motion in the electromagnetic field. Then, equation (17) becomes:

$$\frac{\partial^2 a_\parallel}{\partial x^2} - \{2\omega_0 \dot{F}(t) + \omega_0^2 \dot{x}^2 +$$

$$2\omega_0^2 x \frac{d^2 x(t)}{dt^2}\}a_\parallel - \{\Delta n(x) + \frac{a_{0\perp}^2}{2}(n_\parallel - \frac{a_\parallel^2}{2})\}a_\parallel = 0 \tag{21}$$

where we have used $n_\parallel = (\gamma_g^2/2)(\partial^2 a_\parallel^2/\partial x^2)$.

Taking the first integral of equation (21) and assuming that the maximum value of the amplitude $a_\parallel = a_m$ corresponds to the point $x = \bar{x}(t)$, we can obtain an equation for the function $\dot{F}(t)$ and then substitute it in equation (21) we obtain for the amplitude of $a_\parallel$:

$$\frac{\partial^2 a_\parallel}{\partial x^2} - \frac{a_{0\perp}^2 a_m^2}{8}a_\parallel + \frac{a_{0\perp}^2}{4}a_\parallel^3 - \frac{\gamma_g^2 a_{0\perp}^2}{4}\frac{\partial^2 a_\parallel^2}{\partial x^2}a_\parallel = 0 \tag{22}$$

If we neglect the last term in this equation, which described the variation of electron density due to ponderomotive force terms, the relativistic corrections alone lead to the following soliton solution:

$$a_\parallel = \frac{a_m}{ch(\frac{x-\bar{x}(t)}{d})} \tag{23}$$

where the soliton width is equal to $d = 2\sqrt{2}/a_{0\perp}a_m$ and the coordinate of the soliton center is given by:

$$\frac{d^2\bar{x}}{dt^2} = -\frac{1}{2\omega_0^2}\frac{\partial \Delta n}{\partial x}\Big|_{x=\bar{x}} \tag{24}$$

This equation shows that a plasma inhomogeneity implies an acceleration of the soliton. Replacing equations (23) and (24) in (20), we can find the explicit form of the density distribution associated with the emitted electron plasma waves.

We now analise the soliton motion for specific forms of the density profile of an inhomogeneous plasma. For a linear profile defined by $\Delta n(x) = 2\omega_0^2 \alpha_c x$, the motion of the soliton center is described, according to equation (24), by the formula:

$$\bar{x}(t) = \bar{x}_0 + v_g t - \frac{1}{2}\alpha_c t^2 \tag{25}$$

where $\bar{x}_0$ is the initial value of the $\bar{x}$ coordinate and the soliton acceleration is constant and equal to $-\alpha_c$.

We now write down the solution of equation (22) for the following asymptotic conditions: $a_{\|}(+\infty) = \partial a_{\|}/\partial x|_{x\to+\infty} = 0$. The first integral of equation (22) takes the form:

$$(1 - \alpha u^2)(\frac{\partial u}{\partial y})^2 - u^2(1 - u^2) = 0 \tag{26}$$

where we have used: $\alpha = \gamma_g^2 a_{0\perp}^2 a_m^2/2$, $y = a_{0\perp}a_m x/2\sqrt{2}$ and $u = a_{\|}/a_m$. Equation (26) is interesting because it leads to solutions distinct from equation (23) when $u^2 \simeq 1/\alpha$.

RADIATION MECHANISMS

We shall now give a detailed descriptions of the several different mechanisms of emission of low-frequency electromagnetic waves by accelerated short laser pulses in a plasma. We shall show that, even for a homogeneous plasma, relativistic effects due to a short laser pulse can generate low-frequency rotational electromagnetic fields.

In order to write down the equations for the low-frequency magnetic field, we take the rotational of equation (1) and use equation (5). The result is:

$$\{\nabla^2 - \frac{\partial^2}{\partial t^2} - \frac{<n>}{<\gamma>}\} <\vec{B}> = \nabla(\frac{<n>}{<\gamma>}) \times <\vec{p}> \tag{27}$$

where:

$$\frac{<n>}{<\gamma>} = \frac{1 + \Delta n(x) + \delta n(r_\perp, x, t)}{\sqrt{1 + a_\perp^2(r_\perp)a_\parallel^2(z,t)}} \qquad (28)$$

We shall consider equation (27) in two different cases. The first one corresponds to non-relativistic laser pulses for which we can take $<\gamma> \simeq 1$. We also assume that $|\Delta n(x)| \gg |\delta n|$. then, the emission of low-frequency electromagnetic waves is the inhomogeneity of the electron plasma density, or the associated acceleration of laser pulses:

$$\nabla(\Delta n(x)) \times \vec{v} \neq 0 \qquad (29)$$

The velocity here is determined by equation (3). If $\Delta n(x) = 0$ (homogeneous plasma) and if the variation of the density δn due to the ponderomotive force effects is negligible, the mechanism of electromagnetic wave emission is a pure relativistic effect:

$$\nabla \frac{1}{<\gamma>} \times <\vec{p}> \neq 0 \qquad (30)$$

We also want to mention that the result (30) is valid at a distance from the focus of the laser pulse. In the focal region we have $\delta <\gamma> \leq \delta n$ and we have to use the general expression (28).

CONCLUSION

We have studied the problem of the interaction of an intense short laser pulse with a plasma. Starting from the fully relativistic equations we derived, for pancake-like pulses, a one dimensional nonlinear Shroedinger equation. We have shown that the relativistic effects change the pulse shape along its propagation. In the weakly relativistic case we could describe these pulses by adequate soliton solutions. We also studied the rediation processes associated with the acceleration in an inhomogeneous plasma. Different kinds of mechanisms leading to radiation were identified.

REFERENCES

1. V.I. Berezhiani, N.L. Tsintsadze and D.D. Tskhakaya, *J. Plasma Phys.*, **24**, 15 (1980).
2. L. Tsintsadze, *Proceedings of the International Topical Conference on Research Trends in Coherent Radiation Generation and Particle Accelerators*, La Jolla, USA. edited by J.M. Buzzi, P. Sprangle and K. Wille (AIP, New York, 1992), p. 474.

Ionization of atoms in intense fields: semiclassical approach, fluid equations, and effects on laser propagation

F.Cornolti[1], A.Macchi[2] and E.C.Jarque[3]

[1]*Dipartimento di Fisica, Universitá di Pisa, Piazza Torricelli 2, 56123 Pisa, Italy*
[2]*Scuola Normale Superiore and INFM, Piazza dei Cavalieri 7, 56123 Pisa, Italy*
[3]*Departamento de Física Aplicada, Universidad de Salamanca, 37008 Salamanca, Spain.*

Abstract. We discuss a kinetic model for a system of free and bound electrons undergoing field ionization. Fluid equations are derived and used to study transient ionization effects on short laser pulse propagation. We finally discuss a suitable semiclassical fluid approach to heavy atoms ionization.

INTRODUCTION

The problem of an effective, self-consistent inclusion of transient ionization processes in kinetic or fluid models of intense laser-solid interaction is of interest, especially in the case of high-Z atoms and ultrashort pulses [1]. A correct ionization modelling seems also important in the interpretation of novel phenomena such as solid target transparency to femtosecond pulses [2]. For intense laser fields exceeding the atomic binding field, a semiclassical approach to the bound electrons dynamics may be appropriate, as shown from numerical simulations [3].

Here we discuss a kinetic model describing a system of free and bound electrons in the presence of an external electric field. The model leads to fluid equations where the inclusion of the ionization process introduces an additive force in the current equation, opposed to the external field. As an example we study the propagation of a short laser pulse in a thin slab solid target. Finally we propose a semiclassical fluid model for heavy atoms ionization.

KINETIC MODEL AND FLUID EQUATIONS

We describe bound electrons by a classical distribution function $\mathcal{F}_b = \mathcal{F}_b(\mathbf{r}, \mathbf{k}, t)$ where $(\mathbf{r}, \mathbf{k})$ are the canonical "internal" variables. The nuclei are taken to be immobile. The Hamiltonian for the bound electrons in an external electric field $\mathbf{E}_L$ is

$$H = \frac{k^2}{2m} - eV(r) + e\mathbf{E}_L \cdot \mathbf{r} = H_0 + e\mathbf{E}_L \cdot \mathbf{r} \qquad (1)$$

The classical Hamilton equations are $\dot{\mathbf{r}} = \partial_\mathbf{k} H = \mathbf{k}/m$, $\dot{\mathbf{k}} = -\partial_\mathbf{r} H$. Electrons are assumed to become free if the atomic field at their position becomes negligible with respect to the external field $\mathbf{E}_L$. We thus describe ionization as an electron "evaporation" through a sphere Σ of radius r_b with $|V(r_b)| \ll E_L r_b$ and $e|V(r_b)| \ll I_Z$, where I_Z is the ionization potential. We let $\mathcal{F}_b = 0$ outside Σ. The flux Φ of particles through Σ with velocity $\dot{\mathbf{r}}$ is given by the sum of incoming and outcoming fluxes

$$\Phi = \Phi^+ + \Phi^- \ ; \ \Phi^\pm \equiv \frac{1}{2}(\dot{\mathbf{r}} \cdot \mathbf{n} \pm |\dot{\mathbf{r}} \cdot \mathbf{n}|)\mathcal{F}_b(\mathbf{r}, \dot{\mathbf{r}}, t)\delta(r - r_b) \qquad (2)$$

The kinetic Boltzmann-like equation for $\mathcal{F}_b$ is

$$\partial_t \mathcal{F}_b + \dot{\mathbf{r}} \cdot \nabla \mathcal{F}_b + \dot{\mathbf{k}} \cdot \partial_\mathbf{k} \mathcal{F}_b = \Phi^- \qquad (3)$$

where we have subtracted a "condensation" source term Φ^- on the r.h.s.: recombination is neglected. Introducing the definitions

$$\langle g \rangle \equiv \int d\mathbf{r} \int d\mathbf{k}\, g \mathcal{F}_b \ ; \ \lambda_g^\pm \equiv \int d\mathbf{r} \int d\mathbf{k}\, g \Phi^\pm \qquad (4)$$

we obtain the following couple of fluid equations for the "macroscopical" bound electron density $n_b \equiv \langle 1 \rangle$ and the "polarization" current $\mathbf{j}_b \equiv \langle -e\dot{\mathbf{r}} \rangle$[1]:

$$\begin{aligned} \partial_t n_b &= -\lambda_1^+ \\ \partial_t \mathbf{j}_b &= \tfrac{e^2}{m}(\mathbf{E}_L - \langle \nabla V \rangle)n_b + \tfrac{e}{m}\lambda_\mathbf{k}^+ \end{aligned} \qquad (5)$$

The particle loss is due to the outgoing electrons.

The ejected electrons enter the free electron distribution function $f_e = f_e(\mathbf{x}, \mathbf{p}, t)$ with the momentum owned at $r = r_b$

$$\partial_t f_e + \dot{\mathbf{x}} \cdot \nabla f_e - e\mathbf{E}_\mathbf{L} \cdot \partial_\mathbf{p} f_e = \int_\Sigma d\sigma \Phi^+_{|\mathbf{k}=\mathbf{p}} \qquad (6)$$

The equations for the density n_e and the free electron current $\mathbf{j}_e$ are obtained in the standard way:

---

[1] Note that for $g = g(\mathbf{k})$, $\int d\mathbf{r} \int d\mathbf{k}\, g\dot{\mathbf{r}} \cdot \nabla \mathcal{F} = \int d\mathbf{k} \int_\Sigma d\sigma\, g\dot{\mathbf{r}} \cdot \mathbf{n}\mathcal{F}_b = \int d\mathbf{k} \int d\mathbf{r}\, g\Phi = \lambda_g$

$$\partial_t n_e + \nabla \cdot (n_e \mathbf{u}_e) = \lambda_1^+$$
$$\partial_t \mathbf{j}_e - \nabla \cdot \left(\frac{\mathbf{j}_e \mathbf{j}_e}{e n_e}\right) = \frac{e}{m} \nabla \cdot \mathrm{P} + \frac{e^2}{m} \mathbf{E_L} n_e - \frac{e}{m} \lambda_\mathbf{p}^+ \qquad (7)$$

The equation for the *total* current $\mathbf{j}_{tot} \equiv \mathbf{j}_e + \mathbf{j}_b$ is

$$\partial_t \mathbf{j}_{tot} = \frac{e^2}{m} \mathbf{E_L} n_e + \frac{e^2}{m}(\mathbf{E}_L - \langle \nabla V \rangle) n_b + \frac{e}{m} \nabla \cdot \mathrm{P} + \nabla \cdot \left(\frac{\mathbf{j}_e \mathbf{j}_e}{e n_e}\right) \qquad (8)$$

We notice that $\lambda_\mathbf{p}^+ = \lambda_\mathbf{k}^+$ because $\mathbf{k}$, $\mathbf{p}$ appear as dummy variables in their definition. Correctly, there are no "evaporation" terms in the *total* current equation. The term containing $\langle \nabla V \rangle$ is in the direction opposite to the laser field. The term $(\mathbf{E}_L - \langle \nabla V \rangle) n_b$ takes the complete dynamics of the bound electrons into account, including polarization dynamics (eventually with memory effects) and describes all effects on the current due to the ionization process. Energy conservation equations can be obtained in a similar way to eq.(8).

STUDY OF LASER PULSE PROPAGATION

Our aim is to use the above fluid equations to study ionization effects on laser pulse propagation. The term $(\langle \nabla V \rangle)$ is not known in terms of fluid quantities and thus we need a suitable approximation for it. Since the bound electrons motion is confined in the sphere $r < r_b$ where $r_b \ll r_q$, being r_q the quiver elongation of the free electrons, the bound polarization current is small compared to the free electron current, except for the very early stages of the interaction, and does not affect deeply the pulse propagation. So we may assume $\mathbf{j}_{tot} \approx \mathbf{j}_e$ and use eq.(7) instead of eq.(8). The only relevant effect on propagation is the depletion of the beam energy due to the loss for ionizing "evaporating" electrons, which can be modelled by adding to $\mathbf{j}_e$ an effective term $\mathbf{j}_b^{(eff)}$, such that[2]

$$\mathbf{j}_b^{(eff)} \cdot \mathbf{E}_L = \lambda_{(k^2/2-I)}^+ \approx (I_Z + \epsilon_I) \lambda_1^+ \qquad (9)$$

Here we added to the ionization potential the non-zero ejection energy $\epsilon_I = \epsilon_I(E)$, which has a lower bound of $1/2$ for the hydrogen atom [3]. In a quasi-static model $\mathbf{j}_b$ is parallel to $\mathbf{E}_L$ and thus we assume

$$\mathbf{j}_b^{(eff)} = (I_Z + \epsilon_I) \frac{\lambda_1^+ \mathbf{E}_L}{E_L^2} \qquad (10)$$

In eq.(7) we evaluate the evaporation term as

$$\lambda_\mathbf{p}^+ \approx \mathbf{p}_I \lambda^+ ; \quad \mathbf{p}_I = -\sqrt{2\epsilon_I} \frac{\mathbf{E}_L}{E_L} \qquad (11)$$

---

[2] Atomic units ($e = \hbar = m_e = 1$) are used in this and the following section.

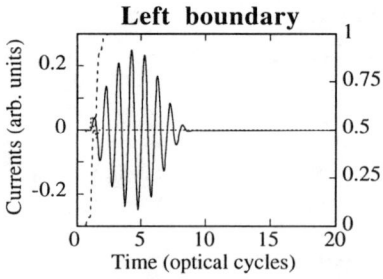 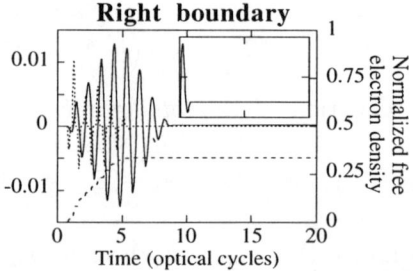

FIGURE 1. Temporal profiles of j_e (solid line), $10 \times j_b^{(eff)}$ (dotted line) and n_e (dashed line) at the left and right boundaries of the slab.

Assuming a one-dimensional geometry and completely neglecting pressure and convection effects, the continuity and current equations reduce to

$$\partial_t n_e = -\partial_t n_b = \lambda_1^+ \equiv \nu_I n_b = -\nu_I (n_e - n_o)$$
$$\partial_t j_e = n_e E + \sqrt{2\epsilon_I} \nu_I n_b = n_e E - \sqrt{2\epsilon_I} \nu_I (n_e - n_o) \frac{E}{|E|} \quad (12)$$

where we used $n_e + n_b = n_o$ and we made the ansatz $\lambda_1^+ = \nu_I n_b$. For a propagating EM wave, the electric field obeys the wave equation

$$\left(\partial_x^2 - \frac{1}{c^2} \partial_t^2 \right) E = \frac{4\pi}{c} \partial_t (j_e + j_b^{(eff)}) \quad (13)$$

where $j_b^{(eff)} = (I_Z + \epsilon_I) \nu_I n_b / E_L$. For solving equations (12,13) we employied a previously developed numerical code [4]. As a simplified case we simulated the interaction of a 10 optical cycles long pulse with a solid hydrogen thin slab target having a thickness of 0.1 times the laser wavelength. The plasma frequency ω_p corresponding to the maximum electron density was chosen to be higher than the laser frequency ω, so that the laser-produced plasma became eventually overdense during the interaction. For $\nu_I(E)$ we used the Landau-Lifschitz formula for $E < 0.084$, while for $E \geq 0.084$ we took $\nu_I = 2.4 E^2$, a formula proposed by Bauer [5] as a fit from numerical simulations. For ϵ_I we took a linear dependence upon the field intensity.

In fig.1 we have depicted the temporal evolution of the free and effective currents and the ionized electron density at the two boundaries of the slab for the case $E_{max} = 1.0$, $\omega = 0.05$, $\omega_p/\omega = 5$. The laser pulse is impinging on the left side. In spite of the very short thickness, the slab is not fully ionized: at the right boundary only a 30 percent of the electrons are free. The effective evaporation current is always lower than the free one, but its relative importance is greater at the right boundary. The main effect of this new term is the existence of a non-zero steady current which persists even when the laser field is disappeared, as is visible in the insert of the right graph.

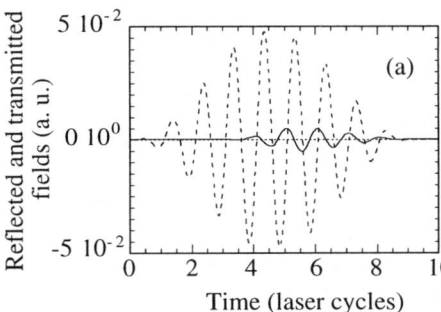 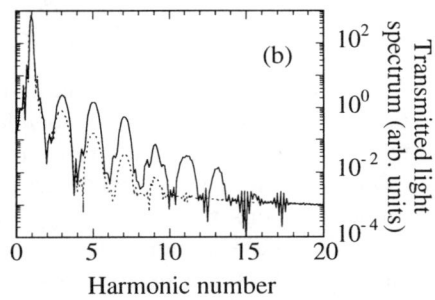

FIGURE 2. (a) Temporal profiles of transmitted (dashed line) and reflected (solid line) fields and (b) high harmonics spectrum of the transmitted field for the case $E_{max} = 0.05$, $\omega_p/\omega = 10$. Dotted line corresponds to the case of no "polarization" contributions.

Such steady current originates by the dephasing between the bound and free electron currents and may generate a quasistatic magnetic field that affects the motion of the plasma electrons and hence the propagation of the laser.

In the case of weak fields the target remains eventually optically thin to the laser. Fig.2(a) shows the temporal profiles of the transmitted and reflected pulses for $E_{max} = 0.05$, $\omega_p/\omega = 10$. The reflected pulse is much weaker and has a shorter duration than the incident one. As an interesting difference generated by the introduction of the polarization terms, high spectral harmonics result stronger. Fig.2(b) shows the transmitted pulse spectrum for the same parameters as (a) with and without these contributions. Due to increase of the refractive index during ionization there is also a blueshift of harmonics wavelength, which is, however, weak due to the little target thickness.

A SEMICLASSICAL FLUID MODEL FOR HEAVY ATOMS IONIZATION

For heavy atoms with $Z \gg 1$, the polarization contributions should be more important. In this case, a suitable model of ionization dynamics yielding empirical formulas for ν_I, ϵ_I may be given by the Bloch hydrodynamic equations:

$$\partial_t \rho + \nabla \cdot (\rho \mathbf{u}) = 0$$
$$\rho(\partial_t \mathbf{u} + \mathbf{u} \cdot \nabla \mathbf{u}) = -\nabla P + \rho(\nabla V(\mathbf{r}) - \mathbf{E}(t)) \qquad (14)$$

Here ρ and $\mathbf{u}$ are the number density and the fluid velocity of the electron fluid, V includes the nucleus potential and the electron-electron repulsion, and P is the "localization pressure" originating from the uncertainty and exclusion principle. The simplest choice for P is to evaluate it for a Fermi ideal gas:

$$P_F = -\left(\frac{\partial U}{\partial V}\right)_S = \frac{(3\pi^2)^{2/3}}{5}\rho^{5/3} \tag{15}$$

We have used the adiabatic approximation to claim that the entropy S can be taken as constant; the time-dependent potential only change the Fermi level but not the electron occupation numbers.

The semiclassical approximation underlying eq.(14) is appropriate for atomic electrons at intermediate radii such that $1/Z \ll r \ll 1$. These restrictions make the model useful for heavy atoms with $Z \gg 1$. Correction for quantum effects can be performed by including high-order local-density approximation (LDA) terms from the formal expansion of exact quantum-fluid equations [6] or by appropriate modifications to both the atomic potential and pressure terms [7]. As an example, a LDA approximation to the complete, nonlocal electron-electron repulsion terms which improves calculations of atomic binding energies has been proposed:

$$V_{ee}^{LDA} \approx 2^{-4/3}(N-1)^{2/3}\rho^{4/3}(r) \tag{16}$$

This term can be added to the Fermi pressure, eq.(15), to account for electron-electron repulsion and exchange terms.

The Bloch hydrodynamics model has been employed recently, in the linear approximation, for dynamic polarizability studies [8]. In the present case a fully nonlinear calculation, using standard numerical methods of fluid dynamics, seems necessary and is presently in progress.

ACKNOWLEDGEMENTS

We are grateful to D. Bauer for giving us the results about the ionization formula. We acknowledge fruitful discussions with P. Mulser. This work was supported by the European Commission through the TMR Networks SILASI and GAUS-XRP.

REFERENCES

1. Cornolti, F., Mulser, P. and Hahn, M. *Las. Part. Beams* **9**, 465 (1991).
2. Giulietti, D. *et al*, *Phys. Rev. Lett.* **79** (1997), in press.
3. Bauer, D., *Phys. Rev. A* **55**, 2180 (1997).
4. Malyshev, V., Conejero Jarque, E., and Roso, L., *J. Opt. Soc. Am. B.* **14**, 163 (1997).
5. Bauer, D., *Phd Thesis*, TH-Darmstadt (1997)
6. Deb, B.M. and Ghosh, S.K. *J. Chem. Phys.* **77**, 342 (1982)
7. Parr, R.G. and Yang, W., *Density Functional Theory of Atoms and Molecules*, Oxford: University Press, 1989, Chap.6, and references therein.
8. Felderhof, B.U., Bleński, T., and Cichocki, B., *Physica A* **217**, 175 (1995)

NONLINEAR REFLECTION OF HIGH INTENSITY PICOSECOND LASER PULSE FROM OVERDENSE PLASMA

A.A.Andreev, V.I.Bayanov, A.B.Vankov, A.A.Kozlov,
I.V.Kurnin, K.Y.Platonov, N.A.Solovyev, S.A.Chizhov, V.E.Yashin

Research Institute for Laser Physics, SC "Vavilov State Optical Institute"
12, Birzhevaya line, St. Petersburg, 199034, Russia

Abstract. The interaction of 1.5 ps FWHM laser pulses with solid targets at intensity 10^{15}- 10^{17} W/cm^2 and contrast ratio 10^6 is studied. Red shift of a "mirror" reflected fundamental wave and its second harmonic depending on the incident laser pulse energy and angle of incidence are observed. They are associated with Doppler shift corresponding to inward movement of the critical density surface from laser pondermotive pressure. Back scattered light has nonlinear dependence from laser intensity connected with SBS and changing of plasma surface.

High intensity picosecond laser pulse interaction in vacuum with a solid targets attracts significant interest during last years. Such an interest is caused by the possibility to realize the source of fast particles or hard X-ray emission and even the scheme of "fast ignition" for laser inertial confinement fusion [1]. The investigation of short laser pulse reflection from plasma can give the information about plasma parameters and it's processes[2,3].
In this work we have studied the picosecond joule laser pulse interaction with the different plane targets in vacuum. Laser setup, used in these experiment, uses the scheme of the chirped pulse amplification with the consequent its compression in diffraction gratings compressor (see Fig. 1). The scheme, based on the electrooptical deflector, mounted at the regenerative amplifier output, improved the amplified
pulse contrast.

Fig. 1. CPA Nd:GLASS LASER SYSTEM

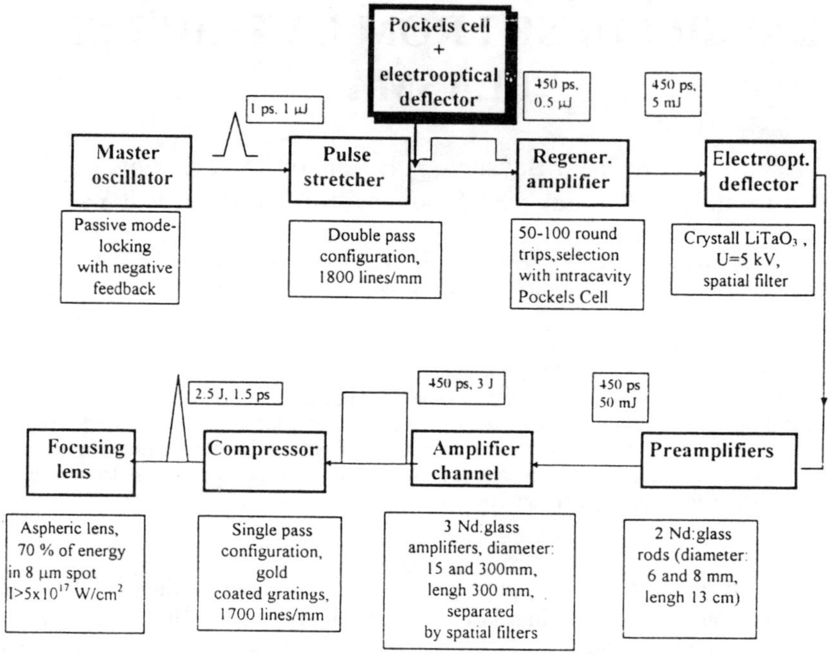

Master oscillator	→ 1 ps, 1 µJ →	Pulse stretcher	→ Pockels cell + electrooptical deflector (450 ps, 0.5 µJ) →	Regener. amplifier	→ 450 ps, 5 mJ →	Electroopt. deflector
Passive mode-locking with negative feedback		Double pass configuration, 1800 lines/mm		50-100 round trips, selection with intracavity Pockels Cell		Crystall LiTaO$_3$, U=5 kV, spatial filter

Focusing lens	← 2.5 J, 1.5 ps ←	Compressor	← 450 ps, 3 J ←	Amplifier channel	← 450 ps 50 mJ ←	Preamplifiers
Aspheric lens, 70 % of energy in 8 µm spot I>5x10^{17} W/cm^2		Single pass configuration, gold coated gratings, 1700 lines/mm		3 Nd.glass amplifiers, diameter: 15 and 300mm, lengh 300 mm, separated by spatial filters		2 Nd:glass rods (diameter: 6 and 8 mm, lengh 13 cm)

The polished flat target mounting in the center of vacuum chamber made it possible to vary the target angle with respect to the laser beam and to its polarization plane. Laser radiation was focused onto the target by the aspherical lens. Reflected light energy and spectrum for first and second harmonics were measured in the "mirror" and "back" directions. (see Fig. 2).

Fig. 2. EXPERIMENTAL SETUP

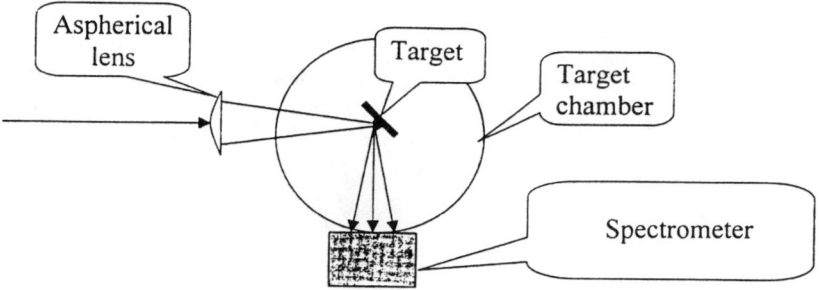

Calculations of the hydrodynamic parameters of the plasma formed by a laser pulse incident on a solid target in vacuum were made employing the "SKIN" Lagrangian code[4]. This code relies on a two-temperature hydrodynamic model and it takes account of the nonlocal electronic thermal conductivity and of the ionic viscosity. The absorption coeffitient of the S- and P-polarized laser radiation were found by solving stedy state wave equations for instantaneous hydrodynamic profiles. Fast electron generation from Landau dampint takes into account. The interaction of laser pulses with aluminium target was simulated. The plasma charge composition was determined by solving the equation of the balance of the ion composition in which the ionization of the ions by electron impact, as well as photorecombination and three-body recombination were taken into accont.

For second harmonic emissivity from plasma we used the model [5] in our code:

It was experimentaly found out that for p-polarized radiation significant part of the energy is converted into second harmonic of the light, emitted in the direction mirror-like to the laser radiation beam incidence direction with respect to the flat target. (see Figs. 3 and 4).

$$\frac{d^2 H_2}{dz^2} - \frac{d}{dz}\ln\varepsilon_2 \frac{d}{dz}H_2 + \left(\frac{2\omega}{c}\right)^2 (\varepsilon_2 - \sin^2\theta)\cdot H_2 =$$

$$= -\frac{ek\sin\theta}{2mc\omega}\cdot\frac{d}{dz}\cdot\left(F^2 - \frac{1}{k}H\frac{d}{dz}F\right)$$

$$\eta_2 = \left|\frac{H_2}{H}\right|^2_{vac} \cdot \cong \cdot 3\left(\frac{v_{osc}}{c}\right)^2 \sin^2\theta \cdot (kL)^2 \cdot \eta_r$$

$$\eta_r \cong \pi kL \cdot \frac{\sin^2\cos\theta}{\cos^2\theta + 2(kL)^{1/3}}$$

Fig 3. SH Conversion efficiency vs Laser Intensity

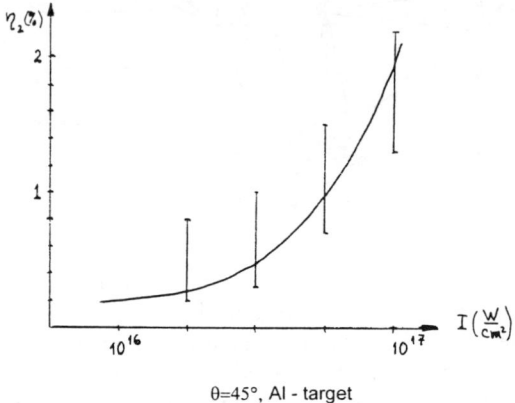

θ=45°, Al - target

Fig. 4. 2w emission from Al target, angle of includence 45

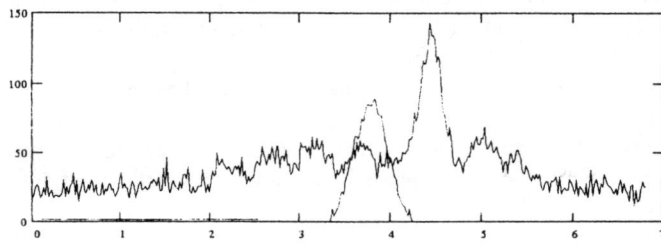

On all figures solid lines denote the results of simulations and the points are the experimental results.

Back reflected component which we observed in all experiments can be explained by SBS at low intensity of laser radiation (see Figs. 5 and 6). Our calculations with help of code "SCAT" [7] have a good coexistence with experimental results up to 10^{16} W/cm² when we used flat plasma. At more high intensities the ponderomotive pressure could create a nonuniform· distribution of the plasma density over the laser spot, which would give decreasing of the back reflected light component.

Fig. 5. Backscattering coefficient vs Laser Intensity

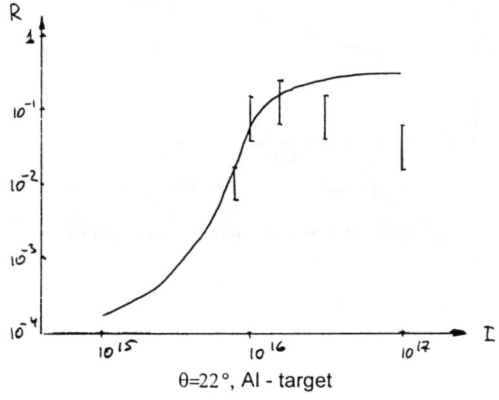

θ=22°, Al - target

Fig. 6. Spectrum

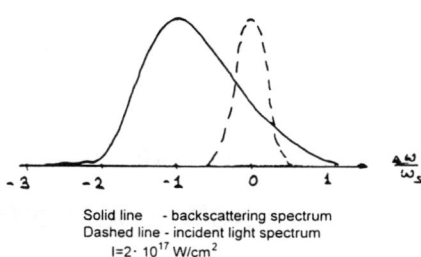

Solid line - backscattering spectrum
Dashed line - incident light spectrum
I=2· 10^{17} W/cm^2

Reflection coefficients (Fig. 7), shifting (Fig. 8) and broadening (Fig. 6) of the mirror-like reflected radiation spectral components can be explained by the movement and reflecting plasma surface distortion due to the intense laser radiation ponderomotive pressure [6].

Fig. 7. **Absorption Coefficient vs Laser Intensity**

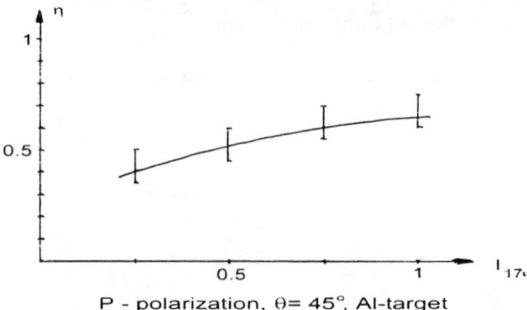

P - polarization, θ= 45°, Al-target

Fig. 8. **Wavelength Shift vs Laser Intensity**

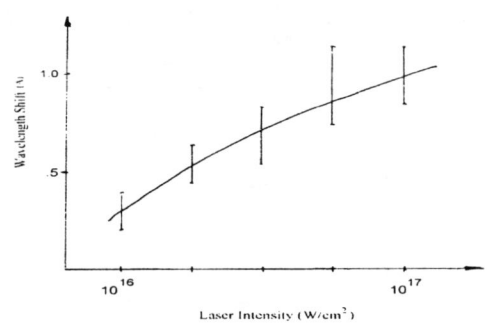

Laser Intensity (W/cm^2)

Experimental data coexist with simulation results of code SKIN.

ACKNOWLEDGMENTS

This' work was partly supported by INTAS grant 94-934 and by Centre National de la Recherche Scientifique.

REFERENCES

1. M.Tabak et al., Phys. Plasmas, **1**, 1626, (1994)
2. M.P.Kalashnikov et al., Phys. Rev. Lett., **73**, 260, (1994.)
3. S.D.Baton et al., Phys. Rev. E, **49**, R3602, (1994)
4. A.A.Andreev et al., Quantum Electronics, **26**, 884, (1996)
5. N.S.Erohin et al., Nuclear Fusion, **14**, 333, (1974)
6. A.Andreev et al., Optika I Spektroskopiya, **59**, 847, (1985)
7. A.Andreev et al., Quantum Electronics **27**, 132, (1997)

Study of a Plasma Diffraction Grating Induced by Super Strong Crossed Laser Beams.

Luis Plaja and Luis Roso

Dept. de Física Aplicada. Universidad de Salamanca.
37008 Salamanca. Spain.

Abstract. In this contribution we analyse the possibility of generating a grating structure in a plasma surface through the interaction of two crossed laser beams. We present one-dimensional particle-in-cell calculations that show that surface electron-density fringes can be induced by the beams if they are s-polarised. Finally, we give approximated analytical expressions for the time-averaged electron density which enclose information about the fringe width and maximum and minimum electron densities (i.e. fringe "visibility").

INTRODUCTION

In the past decade, the availability of very intense sources of coherent electromagnetic radiation gave rise to a new field of study in optics: the interaction of matter with light well beyond the perturbative limit. Sigle-atom theoretical approaches gave good insight on the physics underlying the new phenomena.

Although interesting as a tools for understanding fundamental physics, single atoms or rare gases are unactractive due to the overall weak response to the interaction. More interesting for practical purposes is to increase the number of light scatterers to obtain a more intense collective response. Hence, rencently the attention has been directed to multiparticle systems, like solids. The increase of the intensity of the laser sources leads to situations in which complete ionisation at the material surface is achieved after a few cycles. Surface plasma dynamics becomes, therefore, a major source of non-linearities specially when the charge density exceeds the critical [1]. For this case, experiments [2] and theory [3] show the presence of high-order harmonics in the reflected light.

Being ionised all the solid surface, the local electronic density exeedes the optical critial density in many cases. The plasma at the interface, thus, acts as a highly reflecting mirror, preventing most of the incident laser pulse to enter into the bulk material. It is, in fact, even worse: usually the critical density is formed in the pre-pulse stage, thus reflecting almost completely the main pulse. To our point of view, this is a situation in which plasma-induced diffraction gratings may be useful. At the pre-pulse stage, when the plasma density is close to the critical, the two crossed laser beams may reorganize the surface charge forming underdense grating fringes, alternated spatially with overdense ones. When the main pulse arrives, the underdense regions will allow for deeper field penetration.

In this contribution we would like to demonstrate that the ponderomotive force associated with the surface electromagnetic field associated to two crossed s-polarized laser beams is able to reorganize the surface electron density in the form of fringes, inducing a grating-like structure [4].

FRINGE FORMATION

Consider the interaction geometry depicted in Fig.1, where two s-polarised laser beams, I and II, are aimed at a preformed plasma at incidence angles of α and $-\alpha$. The electric field at the plasma surface ($y = 0$) is described by

$$\vec{E}(x,t) = \vec{E}_I(x,t) + \vec{E}_{II}(x,t)$$
$$= 2E_0 cos(\omega t) cos(kx \sin \alpha) \; \vec{e}_z \qquad (1)$$

where E_0 is the field amplitude, the same in both beams. The total magnetic field at the plasma surface is given by

$$B_x(x,t) = -2E_0 \cos \alpha \; \cos(\omega t) \cos(kx \sin \alpha)$$
$$B_y(x,t) = 2E_0 \sin \alpha \; \sin(\omega t) \sin(kx \sin \alpha) \qquad (2)$$

For now on we will consider a heavy ion plasma and, therefore, concentrate on the electron dynamics. The electromagnetic force on the electrons is described by the Lorentz equation

$$\vec{F} = q(\vec{E} + \frac{\vec{p}}{\gamma m_0 c} \times \vec{B}) \qquad (3)$$

where p is the particle's momentum, m_0 the rest mass and

$$\gamma = \sqrt{1 + \frac{p^2}{m_0^2 c^2}} \qquad (4)$$

A first order approximation for the electron momentum at position x may be calculated neglecting the magnetic field contribution

$$p_z(x,t) \simeq q \int_0^t E_z(x,t') = \frac{2qE_0}{\omega} \sin \omega t \; \cos(kx \sin \alpha) \qquad (5)$$

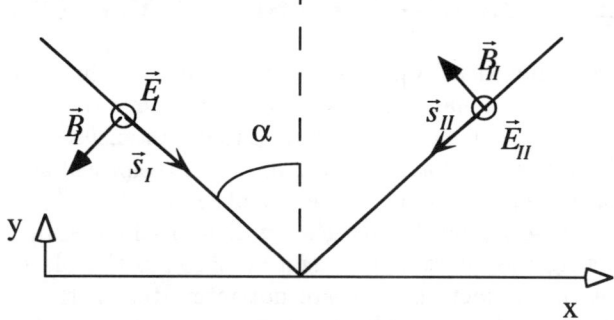

FIGURE 1. Interaction geometry for the two crossed beams.

and $p_x(x,t) \simeq p_y(x,t) \simeq 0$. If we insert Eq. (5) into Eq. (3), we have a second order approximation for the electromagnetic force, valid provided we are not in the ultra-relativistic regime ($v \simeq c$).

The electron dynamics described by equations (5) and (3) leads to a complex electron motion. The electron will quiver mostly in the z direction, with a small drift along the x axis due to the ponderomotive force. If we compute the cycle-averaged force, the quiver term averages to zero, and the motion is governed by the ponderomotive term

$$\langle \vec{F} \rangle = \left\langle \frac{p_z}{\gamma m_0 c} B_x \right\rangle \vec{e}_y - \left\langle \frac{p_z}{\gamma m_0 c} B_y \right\rangle \vec{e}_x \qquad (6)$$

Regarding Eqs. (2) and (5), the y component of the ponderomotive force averages to zero, therefore

$$\langle \vec{F} \rangle = \frac{2q^2}{m_0 c \omega} E_0^2 \sin \alpha \; \sin(2kx \sin \alpha) \left\langle \frac{\sin^2 \omega t}{\gamma(x,t)} \right\rangle \vec{e}_x \qquad (7)$$

A picture of the phenomena enclosed in eq.(7) can be drawn in the non-relativistic limit, i.e. $\gamma \simeq 1$. The simple spatial dependence of the ponderomotive force shows that the electrons will be drifted to the equilibrium points at

$$x_m = \frac{2m+1}{4 \sin \alpha} \lambda \qquad (8)$$

where stability conditions exist. This negative charge accumulation forms fringes along the y direction, giving rise to a diffraction grating structure at the plasma surface.

PARTICLE-IN-CELL SIMULATIONS.

To test the preceding ideas, we have performed particle-in-cell calculations (1D3V PIC), one-dimensional in the real space and three dimensional in the velocity space, for different incident angles and field intensities.

Our PIC code solves the Poisson equation in the reciprocal space, and the Lorentz equation for each particle. Since our plasma slab is assumed to have thickness 0, the field acting on the charges is well approximated by the incident field, i.e., neglecting the effect of propagation through the plasma medium. This, together with the fact that we are not interested in the study of the reflected and transmitted field in this paper, makes unnecessary the integration of the other three Maxwell equations.

The electromagnetic fields are described by plane wave expressions, with time-dependent envelope consisting of 4 turn-on cycles followed by 12 of constant amplitude E_0. The turn-on term, $\sin^2(\pi t/T_{on})$, induces our system adiabatically into a quasi-stationary regime.

Figure 2 shows the negative charge density along a region of width $2\pi/k_x$ in the x direction, as a function of time. The s-polarised laser pulses have a turn-on of 4 cycles followed by 12 cycles of constant amplitude and are aimed at incident angles (α) of $\pi/3$ and $-\pi/3$ respectively. The plasma is assumed to be preformed before the pulse interaction, and has a density equal to the critical ($1.76 \times 10^{21} cm^{-3}$). Electric field amplitude is $5.28 a.u.$ (intensity of $10^{18} W/cm^2$), and frequency $0.057 a.u.$ ($\lambda = 0.8 \mu m$).

FRINGE DESCRIPTION.

It is possible to obtain an analytical approximation for the averaged electron density. To do this, first we have to compute the time average term of the ponderomotive force given in Eq. (7). Using equations (4) and (5), the cycle time-average can be integrated to give the ponderomotive force

$$\langle \vec{F} \rangle = \frac{q^2 \sin\alpha}{m_0 c \omega} E_0^2 \sin(2kx \sin\alpha) F\left(\frac{1}{2}, \frac{3}{2}; 2; -K(x)\right) \vec{e}_x \qquad (9)$$

where $K(x) = \left[\frac{2qE_0}{\omega m_0 c} \cos(kx \sin\alpha)\right]^2$ and $F(a, b; c; z) =_2 F_1(a, b; c; z)$ is the hypergeometric function

Let us now study a neighbour region of the plasma surface around the stability point x_m of width $2x$. Let $Q(x)$ be the time-averaged amount of charge enclosed in such region. The electrostatic field at the edge of the region $x_m + x$ can be calculated through Gauss theorem to be $E_{st} = 4\pi Q$. For a stable situation to occur, the electrostatic force must be compensated by the ponderomotive, therefore $qE_{st}(x_m + x) + \langle F \rangle(x_m + x) = 0$. Therefore the enclosed charge will be

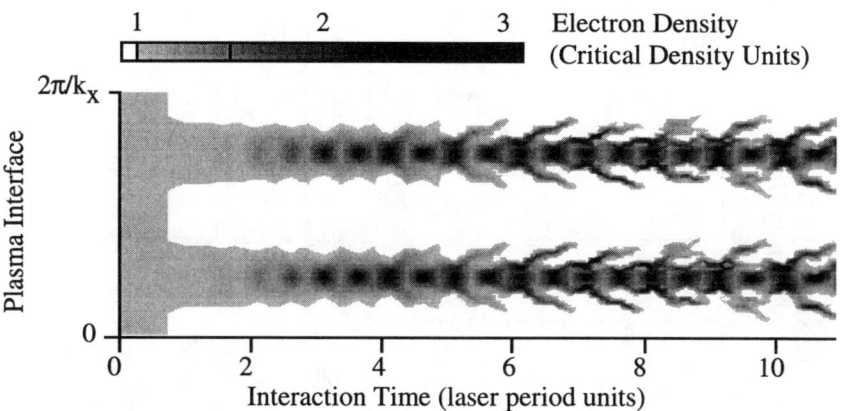

FIGURE 2. Time evolution of the electron density as calculated from the PIC code. The interaction time is plotted in the horizontal axis and a space region of length equal to the wavelength of the projected wavenumber on the x coordinate is plotted in the vertical axis. The beam intensity is $10^{18} W/cm^2$, wavelength $0.8 \mu m$, and the angle of incidence $\alpha = \pi/3$ rad. The plasma density is initially equal to the critical $n_c = 1.76 \times 10^{21} cm^{-3}$. To enhance fringe visibility, electron density above the critical is plotted in a scale of grey tones, while densities below the critical are plotted in white colour.

$$Q(x) = \frac{q \sin \alpha}{4\pi m_0 c \omega} E_0^2 \sin(kx \sin \alpha) \, F\left(\frac{1}{2}, \frac{3}{2}; 2; -K(x)\right) \qquad (10)$$

The electron charge density may be found as the spatial derivative of the enclosed charge $Q(x)$ plus the unperturbed charge density, ρ_0. The final expression reads

$$\rho(x) = \frac{1}{2\pi} \frac{q}{m_0 c^2} E_0^2 \sin^2 \alpha \cos(2kx \sin \alpha) F\left(\frac{1}{2}, \frac{3}{2}; 2; -K(x)\right)$$
$$+ \frac{3}{8\pi} \frac{q}{m_0 c^2} \left(\frac{qE_0}{m_0 c \omega}\right)^2 E_0^2 \sin^2 \alpha \sin^2(2kx \sin \alpha) F\left(\frac{3}{2}, \frac{5}{2}; 3; -K(x)\right) + \rho_0 \quad (11)$$

A more compact visualisation of these results is shown in Fig. 3, where the negative charge density, obtained from the PIC calculation and averaged over the total calculation time, is plotted for different field amplitudes (solid line) compared to the result of the analytical expresion, Eq. [11] (circles). The analytical expression for the average charge density shows a good agreement with the PIC simulation results for intensities below $2 \times 10^{18} W/cm^2$. As the electric field amplitude increases, the fringes become more pronounced and their width shrinks (Fig. 3a-c) . For very intense fields, however, the amount of charge stored in the fringes diminishes (Fig. 3d).

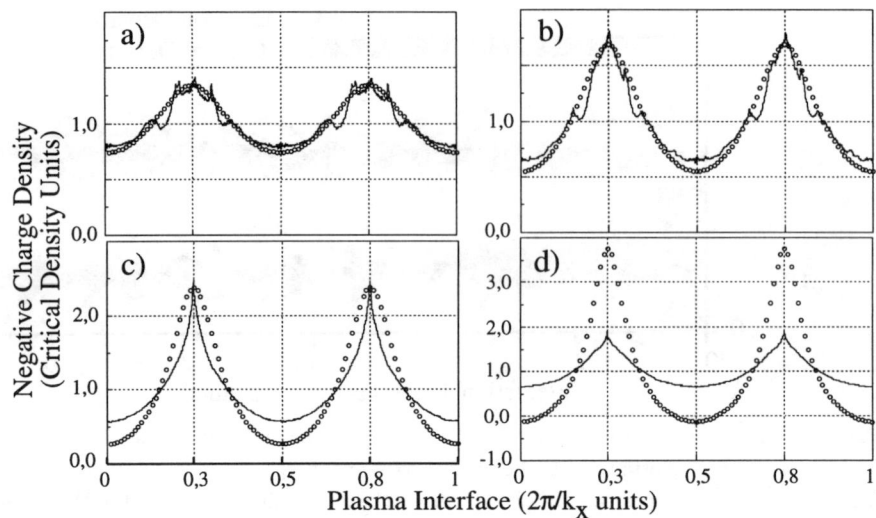

FIGURE 3. Time-averaged electron density in function of the spatial coordinate. The solid line shows the result of the PIC simulations while circles show the analytical prediction. All parameters are the same as in Fig. 2, except for the field intensity: (a) 5×10^{17}, (b) 10^{18}, (c) 2×10^{18} and (d) $4 \times 10^{18} W/cm^2$.

ACKNOWLEDGEMENTS

Partial support from the Spanish Dirección General de Investigación Científica y Técnica (grant PB95-0955), from the Junta de Castilla y León (grant SA81/96) and from the European Comission Training and Mobility of Researchers Program (under contract ERBFMRXCT96-0080) is acknowledged.

REFERENCES

1. P. Gibbon, E. Förster, Plasma Phys. Control. Fusion **38**, 769 (1996).
2. G. Farkas, C. Toth, S. D. Moustaizis, N. A. Papadogiannis and C. Fotakis, Phys. Rev. A **46**, R3605 (1992). D. von der Linde. T. Engers, G. Jenke, P. Agostini, G. Grillon, E. Nibbering, A. Mysyrowicz, and A. Antonetti, Phys. Rev. A**52**, R25 (1995).
3. R. Lichters, J. Meyer-ter-Vehn, and A. Pukhov, Phys. Plasmas, **3**, 3425 (1996). P. Gibbon, Phys. Rev. Lett. **76**, 50 (1996).
4. L. Plaja, L. Roso, Accepted for publication in PRE (1997)

Surface Electromagnetic Waves at the Interface between Vacuum and Plasma produced by Ultrashort Laser Pulses

S.A.Magnitskii, V.T.Platonenko, A.V.Tarasishin.

International Laser Center of Moscow State University, Moscow, Russia.

Abstract. The mathematical scheme that allows to calculate the dispersion characteristics of surface waves propagating along the boundary of a plasma with arbitrary fixed density profiles is proposed. Collisional and resonance damping are taken into consideration. Properties of surface wave propagation for different target materials have been investigated. Appropriate conditions of direct excitation of surface waves have been suggested.

INTRODUCTION

High-density near-surface plasma produced by intense femtosecond pulses is a novel unique object for fundamental investigations in the field of laser-matter interaction. One important feature of such femtosecond-pulsed plasma is the existence of the thin plasma-vacuum interface at the front surface of the target. As a result, one can supposes that surface electromagnetic waves (SEW) can propagate along the boundary of a plasma. Before proceeding to the investigation of surface plasmons it is necessary to define parameters of the plasma itself. The key parameters of a such plasma are its temperature and electron density. These parameters determinate the characteristics and behavior of the plasma. As to electron density is concerned it can be put equal to the product of solid state density by ionization degree of the target which is taken to be known. To estimate the temperature let us use the self-consistent decisions obtained by Rozmus and Tichonchuck (1):

$$T_e[\text{keV}] = 0.4 Z^{-4/25} I^{12/25} [10^{18} \text{W/cm}^2] \lambda^{-8/25} [\mu\text{m}] t^{6/25} [\text{fs}], \qquad (1)$$

here, Z is the degree of ionization, I is the laser intensity, λ is the laser wavelength, and t is the pulse duration.

Calculations show that the electron temperature of a such plasma is proved to be in the range 100eV - 10 keV (Fig.1).

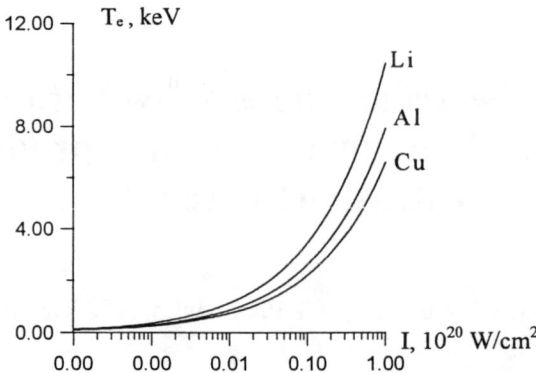

Fig.1. Calculated electron temperature as a function of laser intensity, $\lambda = 0.78$ μm, t=100 fs.

This is near-surface solid-state high-temperature femtosecond-pulse plasma which is an object of our investigation. It is easily to make sure that such femtosecond-pulse laser induced plasma can be treated in nondegenerate ideal plasma approximation. Indeed, typical Fermi energy $E_F \sim 0.1$ eV that is significantly less than plasma temperature and plasma parameter $\eta \sim 10^{-4}\text{-}10^{-6} \ll 1$. Confines of the range where anomalous skin effect takes place are given by following expression (2):

$$\left[\frac{E_c}{1\,\text{keV}}\right]^2 \cdot \left[\frac{N_e}{10^{24}\,\text{cm}^{-3}}\right]^{1/2} > 16 \qquad (2)$$

It's not difficult to see that in such a plasma where electron concentration is of $\sim 10^{23} - 10^{24}\,\text{см}^{-3}$ both normal ($T_e <$ 4-6 keV) and anomalous skin effects take place. Thus, in electron temperature range from a few tens of electron-volts to ~ 1 keV femtosecond-pulse plasma is an ideal, non degenerate one. Moreover, it should be possible to neglect the spatial dispersion of dielectric function. Dielectric function of a such plasma can be rather good described by Drude model. Hereafter, we will operate within the framework of this approximation.

THEORETICAL MODEL

We use the approach where a continuous density profile of the plasma-vacuum boundary is approximated by multilayer interface (3). The density profile is assumed to be fixed and dielectric function ε is assumed to be constant within each layer. H-waves are considered. In such approximation the following dispersion relation for n-layer interface has been obtained (4):

$$D_n(\beta_T) = 0 \qquad (3)$$

The complex function $D_n(\beta_T)$ is presented by the recurrence formula:

$$D_n(\beta_T) = (\beta_n+\beta_T)D_{n-1}(\beta_n) e^{-ik_{nz}d_n} +(\beta_n-\beta_T)D_{n-1}(-\beta_n) e^{ik_{nz}d_n} \qquad (4)$$

here, $\beta_i = \varepsilon_i / k_{iz}$, ε_T and ε_0 are plasma and vacuum dielectric constants, k_{iz} is a normal component of the wave vector in i-th layer, $D_0 = \beta_0+\beta_1$. The interface lies in the xy plane. Note, that for given SEW frequency and given dielectric function profile $D_n(\beta_T)$ is the function of one variable k_x because $k_{iz} = \sqrt{\varepsilon_i \frac{\omega^2}{c^2} - k_x^2}$. In the case of thin plasma-vacuum interface the equation (1) is simplified (4) and if to neglect the collisions it can be modified into analytical form which is identical to expression for resonance damping of SEW which propagates in the vicinity of the thin plasma-vacuum interface (5).

The dispersion relation (3) is valid for any fixed density profile and for arbitrary imaginary part of dielectric function in local approximation.

RESULTS OF NUMERICAL SIMULATION

The aim of our investigation is to find the dispersion characteristics of SEW which propagate along the boundary of femtosecond laser -induced plasma. The basic computational procedure is to approximate the continuous density profile by large number of homogeneous layers and then to calculate the complex root of equation (3). This complex root $k_x(\omega)$ is the approximate value of the surface wave vector. The greater the number of layers the more accuracy of approximation. However, increasing the number of layers leads to the serious calculation difficulties. This is due to the fact that the function $D_n(\beta_T)$ crosses from large negative values to large positive values. To overcome this difficulties we normalized formula (4). On each i-th step expression (4) for $D_i(\beta_{i+1})$ is divided by β_i. This normalization procedure has made possible to calculate the surface wave vector with any predefined accuracy. The Newton method is used for finding the roots.

Here we present the results of numerical calculations of real and imaginary parts of complex wave vector of SEW for various target materials. All calculations were carried out for the same electron temperatures T_e = 1keV and laser pulse wavelength λ_0 = 0.6 μm. Carbon and three different metals: lithium, natrium, aluminium were chosen for analysis. The choice is caused by different atomic numbers and, therefore, different degree of ionization of materials. After being ionized by superstrong field of femtosecond laser pulse the density of electron and ion components of created plasma is strongly varied with target material. We believed that ionization occurred if the ionization potential was three times less than electron temperature of a plasma.

Three types of electron density profiles n_e have been investigated: linear ($n_e(z)$= n_{solid} for z<0, $n_e(z)$= $n_{solid}(1 - z / L)$ for 0 <z<L, $n_e(z)$= 0 for z>L), exponential ($n_e(z)$= n_{solid} for z<0, $n_e(z)$= $n_{solid} \exp(-z / L)$ for z>0) and gaussian ($n_e(z)$= n_{solid} for z<0, $n_e(z)$= $n_{solid} \exp(-z^2 / a^2)$ for z>0).For all profiles maximum electron density n_{solid}

corresponding to the solid-state region was supposed to be equal to the product of ionization degree and solid state-density of the target.

The dielectric function of continuous plasma-vacuum interface is given by Drude formula

$$\varepsilon = 1 - \frac{\omega_{pe}^2}{\omega^2 + \nu^2} + \frac{i\nu\omega_{pe}^2}{\omega(\omega^2 + \nu^2)} \quad (5)$$

where $\omega_{pe} = \frac{4\pi e^2 N_e}{m_e}$ is the electron plasma frequency, ω is the frequency of electromagnetic field and ν is the collision rate. Ion-electron collisions play major role. Their rate is defined by following expression

$$\nu = 4/3\sqrt{2\pi/m_e} \, Ze^4 N_e \ln\Lambda / T_e^{3/2} \quad (6)$$

where $\Lambda = 3 T_e^{3/2}/(2\sqrt{\pi} Z N_e^{1/2} e^3)$ is the Coulomb logarithm. It is supposed that electron temperature T_e and ionization degree Z are constant everywhere over the region of the plasma-vacuum interface.

The results are shown in Fig.2-4. Both resonance and collisional damping have been taken into account.

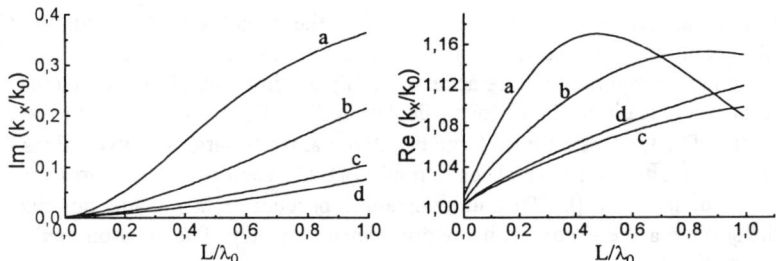

FIGURE 2. Imaginary and real parts of the surface wave vector as a function of the scale length for a linear density profile. Curves a, b, c and d represent the results for Li, Na, Al and C, respectively. T_e = 1 keV, λ_0 = 0.6 µm.

Fig.2 demonstrates significant difference in values of surface wave vector for different materials for the case of linear profile. Real parts of all wave vectors are proved to be grater than k_0. It should be noted the magnification of SEW decay with increasing the material atomic number. This phenomenon is in contrast with the case of sharp boundary where the lighter the material the less is imaginary part of the wave vector. In the case of sharp boundary collisional effects play the major role in SEW damping and the rate of collisions increases with atomic number. In the opposite case of fuzzy plasma-vacuum interface the main damping mechanism is the resonance absorption which is defined by size of near-critical density region. If the linear profiles with fixed scale length are considered one can see that the density gradient increases with the atomic number.

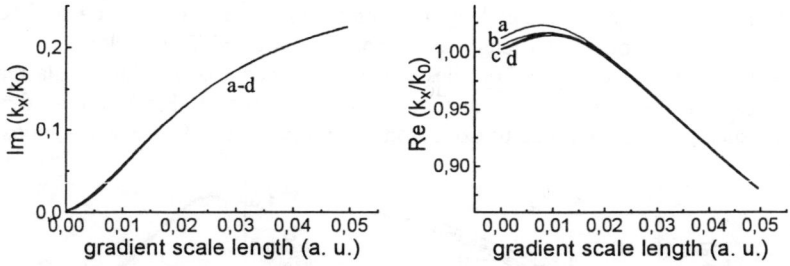

FIGURE 3. Imaginary and real parts of the surface wave vector as a function of the normalized (L/λ_0) gradient scale length for an exponential density profile. Curves a, b, c and d represent the results for Li, Na, Al and C, respectively. $T_e = 1$ keV, $\lambda_0 = 0.6$ μm.

The results obtained for exponential and gaussian profiles (Fig. 3-4) indicate very close values of surface wave vector for different materials for equal gradient scale lengths in the case of exponential and gaussian profiles. This is due to the fact that magnitude of resonance damping of SEW is completely defined by corresponding value of gradient scale length in the case of exponential profile. In the case of gaussian profile resonance damping additionally logarithmically depends on electron density.

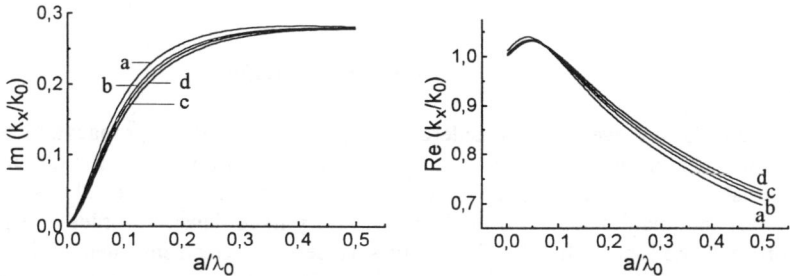

FIGURE 4. Imaginary and real parts of the surface wave vector as a function of the parameter a for gaussian density profile ($n_e(z) = n_{solid} \exp(-z^2/a^2)$). Curves a, b, c and d represent the results for Li, Na, Al and C, respectively. $T_e = 1$ keV, $\lambda_0 = 0.6$ μm.

It is known from model calculations (7) that a plasma density profile created by femtosecond laser pulse can be presented as a superposition of exponential and gaussian profiles. Density gradient scale length at incident laser intensities less than 10^{18} W/cm^2 is generally determined by hydrodynamic ion expansion. The hydrodynamic scale length could be roughly put to the product of the speed of the longitudinal sound wave v_s by the time of expansion, where $v_s = \sqrt{Z\frac{T_e}{M}}$. It follows that heavier targets should have smaller density gradient scale lengths. Therefore, they must exhibit smaller attenuation of surface waves.

In closing, we'll consider the conditions of direct excitation of SEW. In the cases of exponential or gaussian profiles real part of the surface wave vector can be smaller than in the bulk wave. This means a possibility for direct excitation of SEW. But in fact this occurs on relatively large scale lengths when resonance absorption of a surface wave is so strong that propagation length is comparable with SEW wavelength. We suggest here to use such laser pulse profiles which are capable of producing plasma density profiles with the 'pedestal' in the subcritical density region. Fig.5 shows advantage in using such profiles regarding to excitation and propagation of surface plasmons.

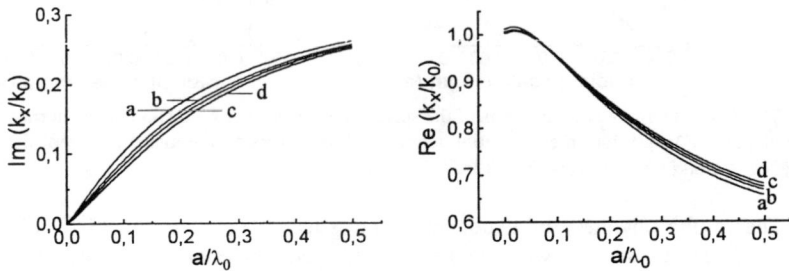

FIGURE 5. Imaginary and real parts of the surface wave vector as a function of the parameter a for the gaussian density profile with the 'pedestal' ($n_e(z) = n_{solid} \exp(-z^2/a^2)$) for $0<z<z(0.4n_{cr})$, $n_e(z)=0.4n_{cr}$ for $z(0.4n_{cr})<z<z(0.4n_{cr})+a$, $n_e(z)=0$ for $z> z(0.4n_{cr})+a$). Curves a, b, c and d represent the results for Li, Na, Al and C, respectively. $T_e = 1$ keV, $\lambda_0 = 0.6$ μm.

CONCLUSION

In conclusion, we have considered the problem of SEW propagation along the plasma-vacuum interface. Properties of surface wave propagation for different target materials have been investigated. It was shown that multilayer model allows to calculate the real and imaginary parts of SEW for arbitrary fixed density profiles of a plasma. It worth to note that spatial distribution of surface wave electromagnetic field inside the interface could be easily obtained by solving D'Alamber's equation with the use of previously computed dispersion characteristics.

REFERENCES

1. W. Rozmus, V.T. Tikhonchuk, Phys.Rev. **A 42**, 740 (1990).
2. E. G. Gamaly, Phys. Rev. **A 44**, 6828 (1991); Phys. Rev. **E 48**, 516 (1993).
3. Surface Polaritons: Electromagnetic Waves at Surfaces and Interfaces, edited by V. M. Agranovich and D. L. Mills, Amsterdam: North-Holland, 1982.
4. V.M.Gordienko, S.A. Magnitskii, T.Yu. Moscalev, V.T. Platonenko, Izv. Ross. Akad. Nauk. ser. fiz., **60**, 10 (1996). Proc. SPIE, v.2770, p.126 (1996).
5. K.N.Stepanov, Zh. Tech. Fiz. **35**, 1002 (1965) [Sov. Phys. Tech. Phys. **10**, 773 (1965)].
6. E. M. Lifshitz and L.P. Pitaevskii, Physical Kinetics, Oxford: Pergamon, 1981.
7. E. G. Gamaly, Phys. Fluids **B 5**, 944 (1993).

Self-Confinement Plasma Effect in Intense Laser Interaction with a Cluster Gas

A.V.Kim[†], D.Anderson[‡], M.D.Chernobrovtseva[†], and M.Lisak[‡]

[†]*Institute of Applied Physics, Russian Academy of Sciences,
Nizhny Novgorod 603600, Russia*
[‡]*Institute for Electromagnetic Field Theory,
Chalmers University of Technology, S-412 96 Göteborg, Sweden*

Abstract. When an intense laser pulse propagates through a cluster gas it is shown that a plasma channel with a refractive index higher than its surrounding region is produced. In the high intensity regime, that leads to a novel self-confinement effect which keeps the laser beam from diverging. This effect is of importance for ultrashort pulse durations, less than 100 - 300 fs, depending on ion mass, and is limited by the expansion process caused by strong heating in the near-solid density cluster microplasma. A computational study of the self-confinement effect has been performed where long plasma channels were demonstrated.

INTRODUCTION

Over the past decade, much effort has been dedicated to the understanding of the interaction of intense ultrashort laser pulses with matter (1). Applications include ultrahigh-gradient electron accelerators, generation of high harmonics, soft x-ray lasers and incoherent x-ray production extending to the megavolt range. Recent studies (2-4) of intense laser interactions with individual clusters reveal new properties of laser-matter interaction. They have shown that clusters of atoms provide a unique nonlinear medium with potentially useful applications. The purpose of this work is to present a novel nonlinear optical effect associated with high intensity ultrashort laser pulse propagation in a cluster gas medium. This effect results from the collective nonlinear interaction of high intensity ultrashort laser pulses with a medium containing a number of large clusters (> 5 nm in diameter) and consists of self-confinement of the laser beam in a self-generated waveguide. This waveguide is a dielectric channel with a plasma core having a significant contribution from cluster microplasmas and having a refractive index higher than its surrounding region.

We note that in the high intensity range, plasma waveguide production is one of the major goals in laser-matter interaction physics by providing a large interaction length (5). The predicted cluster-plasma nonlinearity can be considered as a new focusing type of nonlinearity which is additional to the two well-known nonlinearities: the relativistic and the cavity plasma density production due to the ponderomotive force which both play exceptionally important roles in intense laser-plasma interaction.

CLUSTER GAS NONLINEARITIES

The main idea of this work may be represented as follows. Let us consider the propagation of intense laser beam in a gas containing clusters. According to classical nonlinear optics in order to obtain a self-guiding effect, a positive variation of the refractive index must be generated in the high field region that will remain the laser beam from diverging. The nonlinear response of the cluster gas to intense electromagnetic radiation includes contributions from free electrons (plasma response) and ionized clusters which have dimensions much less than the laser wavelength. In the following we will consider cluster microplasmas with dimensions exceeding the Debye length and we will refer such plasma particles as *plasmoids*. Assuming also plasmoids to be spherical particles, their dipole moment may be taken in the form as (6)

$$P_\omega = \frac{\varepsilon - 1}{\varepsilon + 2} a^3 E_\omega, \qquad (1)$$

where $\varepsilon = 1 - \omega_p^2/\omega(\omega - i\nu)$ is the plasma dielectric permittivity, ω is the laser angular frequency, ν is the electron-ion collision frequency, $a = a(t)$ is the radius of the plasmoid which depends on time (at initial time $a(0) = a_{cl}$ - cluster radius) due to heating and microplasma expansion, $\omega_p = (4\pi e^2 n_e/m)^{1/2}$ is the plasma frequency, and n_e is the electron density. For simplicity, considering a collisionless plasma, we note that the electrodynamic response of a medium containing plasmoids, $P_\omega \equiv \chi_{pl} E_\omega = P_\omega N_{pl}$ (where N_{pl} is the ionized cluster density), depends dramatically on the electron density in the plasmoid. The plasmoid gas susceptibility is given by

$$\chi_{pl} = -\frac{n_e}{(3N_{cr} - n_e)} a^3 N_{pl}. \qquad (2)$$

Obviously, this expression for the susceptibility is not valid close to the resonance field ($n_e \sim 3N_{cr}$) where absorption effects must be take into account. At $n_e/N_{cr} < 3$ (where the critical density is $N_{cr} = m\omega^2/4\pi e^2$), the polarization of the plasmoid gas has the same property as an ordinary plasma (the plasma susceptibility has negative sign) and hence such an ionized nonuniform plasma structure acts as a negative lens increasing the divergence of the beam. *In the opposite case, when $n_e/N_{cr} > 3$, the response of the plasmoid gas, given by Eq. (2), has positive sign and, therefore, the refractive index will be increased.* This can lead to the self-confinement of the laser beam. Thus, at intense laser-cluster gas interaction it is possible to create an unusual plasma channel which is capable of guiding laser radiation.

This mechanism for nonlinearity production is appropriate for ultrashort pulses since the electron density decreases in time due to the expansion process. We can estimate the time of the expansion process if we assume ion-sonic expansion and require the electron density to drop from the solid density, n_{es}, to $3N_{cr}$. The resulting expansion time is approximately $\tau^* \approx a_{cl}(M/ZkT_e)^{1/2}(n_{es}/3N_{cr})^{1/3}$, where T_e is the cluster electron temperature, and M and Z are the mass and charge number representing the ion. For an Argon cluster with initial radius $a_{cl} \sim 10$ nm, an initial temperature of 1 keV, $Z \sim 8$, and $N_{cr} = 10^{21}$ cm^{-3} (laser wavelength $\lambda = 1$ μm), the expansion time will be approximately $\tau^* \sim 100$ fs. We also note that an important feature of the plasmoid gas response is the strong dependence of the susceptibility on its radius, as $\chi_{cl} \sim a^3$, which strengthens this effect for larger clusters. Though the average density of a gas containing clusters is low, due to the large susceptibility of the plasmoids, cluster contribution to the refractive index is very large. Obviously, in order to describe the behavior of the laser beam all contributions to the refractive index must be taken into account.

BASIC EQUATIONS

In order to make qualitative predictions about the nature of the self-confinement effect we develop a model for intense laser pulse interaction with a cluster gas that includes the various processes which occur on the time scale of the laser pulse. In most practical situations, the contributions to the refractive index from clusters, plasma, and plasmoids are very small as compared with unit. This allows us to use the slowly varying approach for the amplitude of the laser pulse, $E = 1/2 x(Ee^{(i\omega t-ikz)} + \text{c.c.})$. The following equation for the envelope of the pulse propagating in the z direction is obtained in the paraxial approximation

$$2ik\partial_z E - \Delta_\perp E - 4\pi k^2 (\chi_g + \chi_i)E = 0 \qquad (3)$$

where $\Delta_\perp = \partial_x^2 + \partial_y^2$ is the transverse Laplace operator, $k = \omega/c$ is the laser wavenumber in vacuum, χ_g, χ_i are the dielectric susceptibilities of the neutral and ionized components of the medium, respectively. Next, we calculate the difference of the refractive index between the intense laser beam region where the cluster gas is almost ionized, and the non-ionized region outside the beam.

In the non-ionized region ($\chi_i = 0$), the dielectric susceptibility of the gas is the sum of the atomic (χ_a) and the cluster (χ_{cl}) susceptibilities:

$$\chi_g = \chi_a + \chi_{cl} = \alpha N_a + \frac{n_o^2 - 1}{n_o^2 + 2} a_{cl}^3 N_{cl} \qquad (4)$$

where α is the atomic polarizability, N_a is the density of free atoms not involved in the cluster formation process, N_{cl} is the neutral cluster density, n_o is the linear refractive index of the clusters, for example, for Ar atoms $\alpha = 1.6 \times 10^{-24}$ esu, $n_o = 1.23$ (for a cluster in the liquid phase).

In the ionized region ($\chi_g = 0$), χ_i includes the plasma response (susceptibility of free electrons χ_p) and the plasmoid gas susceptibility (χ_{pl}):

$$\chi_i = \chi_p + \chi_{pl} = -\frac{N_e}{4\pi N_{cr}} + \frac{\varepsilon-1}{\varepsilon+2} a^3 N_{pl} \qquad (5)$$

where N_e is the electron density of the background plasma (for a fully ionized gas $N_e = N_a$). In Eq. (5) we neglect electron-ion collisions in the background plasma which has low density and very low absorption efficiency (<1%).

The nonlinear index difference in the following Eqs. (4) and (5) for intense laser radiation may be estimated as $\Delta n \approx [\eta a_o^3 - (1-\eta)/4\pi N_{cr}] N_g$, where η is the fraction of atoms involved in the cluster formation (7), N_g is the total gas density ($N_a = (1-\eta)N_g$), a_o is the initial lattice spacing in clusters. We note that numerical estimations of Eqs. (4) and (5) for atoms like Ar, Kr, Xe which are favorable for the production of large clusters show that the main effect counteracting the self-confinement of the laser beam comes from background plasma production due to ionization of free atoms. We may expect that when $\Delta n > 0$, i.e.

$$\eta > \frac{1}{1 + 4\pi N_{cr} a_o^3}, \qquad (6)$$

the laser beam will be kept from diverging by the self-confinement effect and therefore it will be possible to deposit a significant part of the laser energy into cluster microplasmas. Condition (6) corresponds to $\eta > 0.6$ for a Ti:Sa laser ($\lambda = 800$ nm) in a Ar gas ($a_o = 3.8$ A). Thus, one of the necessary condition for realizing the self-confinement effect considered here, is that more than half of the atoms in a gas jet must be in the form of clusters.

For description of the nonlinear propagation dynamics of the laser beam, Eq. (3) must be considered self-consistently together with the ionization rate equation describing the processes of plasma as well as plasmoid generation. The ionization rate of electron generation in a cluster will be a combination of direct optical ionization and electron impact ionization. Here we take into account that for the problem under of interest, due to the high density of atoms in clusters latter ionization rate is very fast compare to optical ionization which delivers, at least, first electrons in clusters for plasmoid production. Therefore, the electron density N_e and the plasmoid density N_{pl} may be governed by the same dynamical equations

$$\partial_\tau N_{e,pl} = (N_{a,cl} - N_{e,pl}) \cdot w(|E|), \qquad (7)$$

that describe the optical-field-induced ionization process with the field-dependent rate $w(|E|)$. The time $\tau = t - z/v_{gr}$ has been counted from the pulse arrival at a given point along the propagation path z. In Eq. (5) the electron density in the plasmoids n_e which may be produced during the ionization processes is considered as constant. At initial stage this value is close to the near-solid density n_{es}. For example, for Ar atoms, values $n_{es} \approx 2\times 10^{23}$ cm^{-3} are typically produced.

SIMULATION RESULTS

We present some simulation results in the case of tunneling ionization with the well known dependence of the ionization rate on the field amplitude: $w(|E|) = \gamma(E_a/|E|)^{1/2} \exp(-E_a/|E|)$. The values of the atomic field E_a, and the frequency γ may range in the wide intervals depending on the concrete kind of the ionized species. For numerical study we used the following dimensionless variables: $z/z_f \to z$, $r_\perp/a_f \to r_\perp$, $\tau/\tau_p \to \tau$, $N_{e,a,cl,pl}/(ka_f)^{-2}N_{cr} \to N_{e,a,cl,pl}$, where $z_f = ka_f^2$ is the Rayleigh length, and τ_p is the laser pulse duration. Here we present results obtained for case: $\gamma/\tau_p = 10^4$, $N_{cl} = 0.034$, and the collimated incident laser pulse with the Gaussian temporal and transverse distributions of laser intensity $|E|^2(z=0, x, y, \tau) = E_o^2 \, E_a^2 \cdot \exp[-(x^2+y^2)-\tau^2]$, $E_o = 0.5$.

Figure 1 presents the maximum intensity of the pulse at the axis along the propagation distance z at various densities of free atoms not involved in the cluster formation.

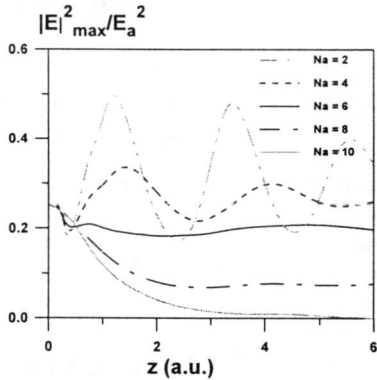

Figure 1. Maximum intensity of the laser pulse at the axis of the plasma channel as a function of propagation distance z measured in Rayleigh lengths for different gas density.

In this picture the self-confinement effect of the ionizing laser pulse in the long plasma channel exceeding many Rayleigh length is distinctly seen. At the higher N_a (more than 8), when the cluster contribution to the refractive index is small, the strong defocusing effect of the laser beam due to plasma production takes place. In this case, effective laser-gas interaction length is usually less than the Rayleigh length. The self-confinement effect leading to the long plasma channel formation and, therefore, to the high absorption efficiency of the laser energy occurs when a significant part of atoms will be in clusters. Next figures represent spatial distributions of the laser beam and the plasma density in the self-confinement regime. In figure 2 the transverse distribution of the intensity in the middle of the temporal profile is shown for different longitudinal distances from the gas boundary at $N_a = 6$. Contours of the background electron density in the plasma channel are presented in figure 3 at the gas density $N_a = 6.5$. The contours of the ionized cluster distribution have the same form. In the central region of the plasma channel atomic gas is completely ionized.

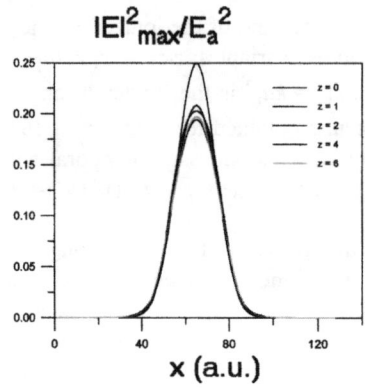

Figure 2. Transverse distributions of the intensity in the plasma channel for different distances z at the gas density $N_a = 6$.

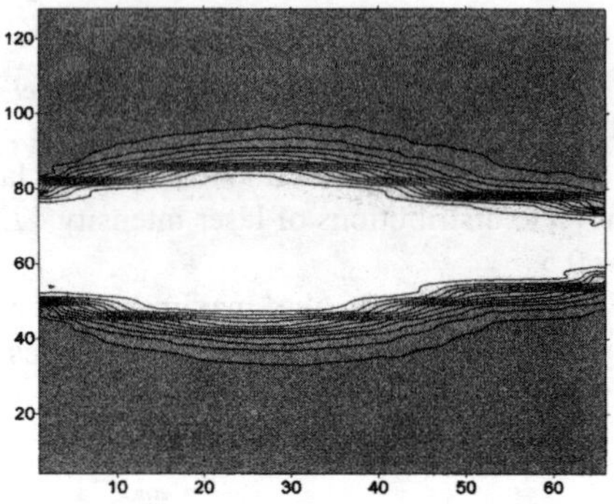

Figure 3. Contours of electron density in the plasma channel at the gas density $N_a = 6.5$. Here the Rayleigh length equals 10.

In summary, we have proposed a new updated effect of the self-confinement of high intensity laser pulse propagating in a cluster gas medium and have shown that this effect may be very useful for applications of laser-cluster interaction where high absorption efficiency of laser energy is needed.

ACKNOWLEDGMENTS

The work of A.V.K. and M.D.C. was supported in part by the Russian Basic Research Foundation Grant No. 95-02-05936.

REFERENCES

1. Perry, M.D., and Mourou G., Science **264**, 917 (1994).
2. McPherson, A., Luk, T.S., Thompson, B.D. et al., Phys. Rev. Lett. **72**, 1810 (1994).
3. Ditmire, T., Donnelly, T., Falcone, R.W., and Perry, M.D., Phys. Rev. Lett. **75**, 3122 (1995).
4. Ditmire, T., Tisch, J.W.G., Springate, E. et al., Phys. Rev. Lett. **78**, 2732 (1997).
5. Milchberg, H.M., Clark, T.M., Durfee, C.G. et al., Phys. Plasma 3, 2149 (1996).
6. Landau, L.D., and Lifshitz, E.M., *Electrodynamics of Continuous Media*, Oxford, Pergamon, 1984.
7. Hagena, O.F., Rev. Sci. Instrum. **63**, 2374, (1992).

Ionization Induced Self-Guiding of Femtosecond Laser Pulses in Gases

M. Lontano, M. Chernobrovtseva*, A.V. Kim*, A.M. Sergeev*

Istituto di Fisica del Plasma "Piero Caldirola"
Consiglio Nazionale delle Ricerche, EURATOM-ENEA-CNR Association
Via R. Cozzi 53, 20125 Milano, Italy
(lontano@ifp.mi.cnr.it)

*Institute of Applied Physics, Russian Academy of Sciences, Nizhny Novgorod, Russia

Abstract. It is shown that regimes of laser-gas interaction exist, characterized by the ionization saturation, where the laser pulse is partially channeled over several Rayleigh lengths. The process of self-channeling driven by the gas ionization has been found under tunnelling ionization conditions and is accompanied by the leakage of part of the laser pulse energy.

INTRODUCTION

In the last decade the development of compact sources of ultrashort $\left(\tau_p \approx 10-100\,\text{fs}\right)$ and highly intense $\left(I \approx 10^{15}-10^{20}\,\text{W}/\text{cm}^2\right)$ laser pulses has opened completely new fields of investigation of the laser-matter interaction [1]. Among the different applications of the relevant new interaction regimes which can now be afforded, the acceleration of relativistic electrons to GeV energies over a distance of few meters, the development of intense coherent X-ray sources, the implementation of high gain inertial confinement fusion schemes require that the focused laser be channeled over distances much in excess to the Rayleigh length ($Z_R = \pi w_0^2/\lambda$, where w_0 is the FWHM of the laser intensity distribution at the focus and λ the laser wavelength in vacuum). At relativistic intensities ($I > 10^{18}\,\text{W}/\text{cm}^2$, for $\lambda \approx 1\,\mu\text{m}$) it is known that an efficient self-channeling of the laser is accomplished due to the relativistic ponderomotive force resulting from the velocity dependence of the electron mass. On the contrary, at lower intensities (in the range $10^{15}-10^{17}\,\text{W}/\text{cm}^2$) it is expected that the vacuum diffraction and the ionization of the target gas result in the formation of a diverging phase front soon after the initial strong laser-gas interaction.

Here we demonstrate the existence of a particular laser-gas interaction regime in which a saturable ionization nonlinearity leads to the partial self-channeling of the laser pulse and to the formation of a long (longer than Z_R) and uniform electron density filament.

SATURABLE IONIZATION INDUCED LASER CHANNELING

Let us consider the laser electric field in the slowly varying envelope approximation

$$E(r,t) = \mathbf{E}(r,t)\exp[-i(\omega t - \mathbf{k}\cdot\mathbf{r})] + c.c. \tag{1}$$

where $\mathbf{E}(r,t)$ is the complex envelope of the field, and "c.c." means "complex conjugate". If T and L are the typical temporal and spatial scales over which $\mathbf{E}(r,t)$ varies appreciably, then the inequalities $2\pi/\omega T \ll 1$ and $2\pi/|k|L \ll 1$ hold. Under the above assumptions the field equation reads [2]:

$$2ik_0 \frac{\partial \mathbf{E}}{\partial z} + \frac{\partial^2 \mathbf{E}}{\partial x^2} + \frac{\partial^2 \mathbf{E}}{\partial y^2} - \frac{4\pi e^2}{mc^2} N\mathbf{E} + ik_0 \Gamma \mathbf{E} \approx 0 \tag{2}$$

where $k_0 = \omega/c$ and a linearly polarized transverse electric field propagating along the z-axis has been considered. Γ is the spatial damping coefficient of the field along the z-direction. The laser induced ionization of the gas introduces the nonlinear term proportional to $N\cdot\mathbf{E}$. Then, the dynamic of the laser pulse envelope is fully determined once the consistent evolution of the plasma density is considered. It is assumed that the electron density is produced by the pulse itself due to the ionization of the background gas with the following rate

$$\frac{\partial N}{\partial \tau} = [N_{sat} - N(r,\tau)]\cdot F(E), \tag{3}$$

where for tunneling ionization,

$$F(E) = \frac{4\gamma}{(E/E_{at})^{2n-3/2}} \exp\left(-\frac{E_{at}}{E}\right). \tag{4}$$

Eq.(3) describes the electron density production, with a rate which strongly depends on the field intensity, up to a maximum saturation density value N_{sat}, where the efficiency of the process goes to zero. The saturation mechanism may coincide with the detachement of all the electrons of a given energy level, say n. For the sake of comparison, the critical density for a Ti:Sa laser ($\lambda = 0.8\,\mu m$) is $N_{cr} = m\omega^2/4\pi e^2 \cong 1.7 \times 10^{21}\,cm^{-3}$. Since we are interested in the propagation of femtosecond laser pulses, no density transport process is included in the model. In the above equations, $\gamma = v_{at} k_0^2 a_p^2$, and $v_{at} = me^4/\hbar^3 \cong 4 \times 10^{16}\,s^{-1}$ and $E_{at} = m^2 e^5/\hbar^4 \cong 5 \times 10^9\,V/cm$ are typical intra-atomic frequency and electric field, respectively. n is the principal quantum number of the optical electron, and the "local time" $\tau = t - z/v_{gr}$ has been defined.

We treat the system of Eqs.(2) and (3) as an intial value problem and give the electric field distribution at $z = 0$ (where the laser-gas interaction begins) as a well localized distribution in x, y, τ, that is

$$E(x,y,z=0,\tau) = E_0 \exp\left(-\frac{x^2+y^2}{a_p^2} - \frac{\tau^2}{\tau_p^2}\right), \tag{5}$$

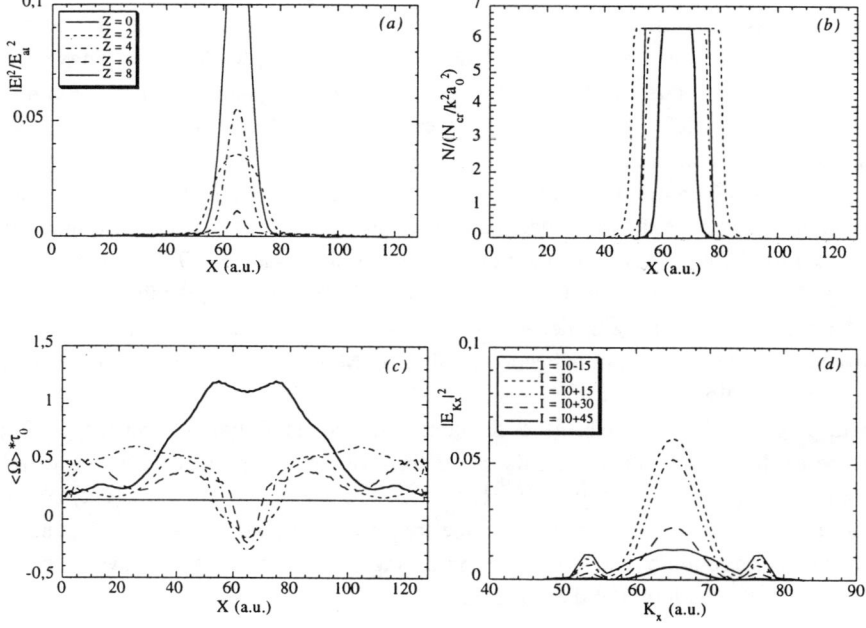

Fig.1

where, a_p and τ_p are the typical transverse spatial and temporal scales of the initial laser pulse. Moreover, at $z = 0$ the laser pulse is assumed to be at its waist, that is with a plane phase front. Fig.1 refers to the following choices of the physical parameters. The initial Gaussian laser pulse (Ti:Sa* with $\lambda = 0.8\,\mu m$) has a width $a_p = 20\,\mu m$, a time duration $\tau_p = 60\,fs$, and a peak electric field amplitude $E_0 = 0.4\,E_{at}$. It interacts with an initially neutral gas with $N_{sat} \approx 2 \times 10^{18}\,cm^{-3}$ and $n = 3$. Figs.1 a, b, and c refer to the normalized electric field amplitude $|E|^2/E_{at}^2$, electron density $N/(N_{cr}/k_0^2 a_0^2)$, and the frequency shift $\Delta\omega \cdot \tau_0$, respectively, at several z-positions during the pulse propagation. Here, $a_0 = 10\,\mu m$ and $\tau_0 = 10\,fs$ are reference values. The field amplitude (a) is taken at τ corresponding to its maximum value when $z = 0$. The density value (b) is taken after that the bulk of the pulse has already passed. Moreover, z-values are normalized over $\sqrt{2} \cdot Z_R = \sqrt{2\pi w_p^2/\lambda} \cong 1\,mm$. It is seen that after the focal plane, placed at $z = 0$, the pulse starts to spread (see $z = 2$) and then part of its energy is trapped inside a narrow channel (a), while a slight energy leakage occurs in the radial direction. A density channel extending from $z = 0$ to $z = 6$ is created with a typical radial scale a_n larger than the laser pulse dimension $a(t)$ (b). The transverse distribution of the laser frequency (c) shows that the electromagnetic (e.m.) radiation is channeled untill an axial minimum of the frequency persists [3]. This could be interpreted as the formation of a converging phase front. In fact, close to the axis the ionization saturates and no frequency blue-shift is produced, while at the wings of the laser pulse ionization is effective and the laser frequency is upshifted (see also Ref.4).

In Fig.2 the peak field intensity is plotted versus z, along the pulse axis. In Fig.2a the laser intensity is shown for vacuum propagation (dotted line) and for an initial gas pressure of $p = 0.1\,atm$ (full line), for $a_p = 20\,\mu m$ and $\tau_p = 30\,fs$. The horizontal dot-dashed line shows the saturated electron density level. In Fig.2b the peak laser intensity is shown for different values of the initial electric field value: $E = 0.4$, 0.5, 0.6, 0.7, and $0.8\,E_{at}$. Here, $a_p = 10\,\mu m$. It is observed that, after an initial spreading of the pulse with a consequent appreciable decrease of the laser intensity over a Rayleigh length, the partial trapping of the radiation becomes effective. Moreover, it is seen that the more is the initial radiation intensity, the longer is the channeling length.

By inspection of Fig.1c a red shift of the laser frequency is observed at small radii. This can be explained by substituting the polar form of the complex scalar electric field amplitude $E(x, y, z, \tau) = A(x, y, z, \tau) \exp[i\varphi(x, y, z, \tau)]$, where $A(\mathbf{r}, \tau)$ and $\varphi(\mathbf{r}, \tau)$ are two real functions, into Eq.(2). By defining the frequency shift $\Omega(\mathbf{r}, \tau) \equiv -\partial\varphi/\partial\tau$, and the wavevector variation $\chi(\mathbf{r}, \tau) \equiv \nabla\varphi$, arising during the laser propagation, we obtain an equation for the evolution of the real amplitude

$$\frac{\partial A^2}{\partial z} = -\frac{1}{k_0}(\nabla \cdot \chi_\perp) A^2 - \frac{1}{k_0} \chi_\perp \cdot \nabla A^2, \qquad (6)$$

and an equation for the frequency shift

$$\frac{\partial \Omega}{\partial z} = \frac{k_0}{2N_{cr}} \frac{\partial N}{\partial \tau} - \frac{1}{2k_0} \frac{\partial}{\partial \tau}\left(\frac{1}{A}\nabla_\perp^2 A\right) + \frac{\chi_\perp}{k_0} \cdot \frac{\partial \chi_\perp}{\partial \tau}. \tag{7}$$

Eq.(6) describes the spatial distribution of the e.m. energy during the propagation of the laser pulse. Eq.(7) gives the spatial rate of variation of the laser frequency as a consequence of the temporal changes in the system parameters, namely, the electron density increase due to gas ionization, the variation of the transverse pulse shape, and the time variation of the transverse wavevector, respectively. By assuming that the last contribution is negligible, we can see that where the ionization is effective the first term in the r.h.s. of Eq.(7) dominates giving an increasing frequency shift, i.e. a blue-shift. On the other side, if the ionization is inhibited, as in the presence of saturation, the evolution of the frequency is driven by the second term. It represents the time variations of the radial e.m. energy distribution inside the pulse. For instance, if the real amplitude is of the form given in Eq.(5), with $A(x,y,z,\tau) = E_0(z,\tau)\exp[-(x^2+y^2)/a^2(\tau)]$, then the second term in the r.h.s. of Eq.(7) writes $-(2k_0)^{-1}\partial/\partial\tau[A^{-1}\nabla_\perp^2 A] \cong -(4/k_0 a^3)\partial a/\partial t$, which gives a laser red-shift for a spreading pulse (i.e., increasing a(t)).

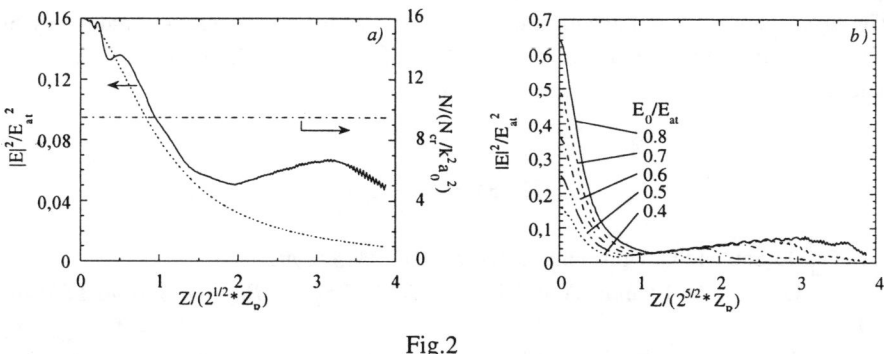

Fig.2

However this sort of quasi stationary state is accompanied by a small radial leakage of radiation from the pulse [3], as can be seen by the k_x-spectra (see Fig.1d) shown for $z = 4$, at different (local) times τ: at the front of the pulse (for $I < I_0$, where I and I_0 are the mesh indexes corresponding to the generic τ-value and to the center of the τ-interval, respectively), at the initial peak of the pulse ($I = I_0$), and at the rear part of the pulse (for $I > I_0$). The two minor lateral peaks in the spectra indicate that part of the energy is lost during the pulse propagation. The situation resembles that of light rays propagating into a waveguide made of a low refractive index medium. The reflection of the light beam at grazing incidence each time it hits the separation surface between the two media is accompanied by the transmisssion of part of the e.m. energy into the outer less dense medium. However it should be noticed that in our case the laser channeling is an intrinsically dynamical process.

Finally, it has been argued that the channeling of the radiation is effective if the following condition is satisfied:

$$\frac{Z_L}{Z_R} \approx \frac{k_0 a_n^3}{3a^2} \sqrt{\frac{N_{sat}}{N_{cr}}} \gg 1, \qquad (8)$$

where $Z_L \approx k_0^2 a^3 (N_{sat}/N_{cr})^{1/2}/6$ is the *leakage distance*, that is the distance over which the laser energy inside the channel decreases appreciably. Eq.(8) holds for $a_n \gg a$, and $(k_0 a)^{-2} \ll N_{sat}/N_{cr} \ll 1$ [5].

We have investigated the possibility of the partial self-channeling of a short laser pulse, induced by the tunneling ionization of the backgroung gas in which it propagates. The trapping of the radiation is possible whenever the ionization reaches a saturation regime; it is accompanied by the leakage of part of the laser energy, and lasts untill this quasi-steady state is no longer sustained.

The investigation of the ionization induced laser channeling is presently under way by means of a full Maxwell equation code which allows to treat laser pulses of arbitrary lengths.

REFERENCES

[1] - G. Mourou and D. Umstadter, *Phys. Fl. B* **4**, 2315 (1992).
[2] - M. Lontano, G. Lampis, A.V. Kim, and A.M. Sergeev, *Physica Scripta* **T63**, 141 (1996).
[3] - A.M. Sergeev, A.V. Kim, and M. Lontano, *Application of High Field and Short Wavelength Sources VII*, Vol.7, 1997 OSA Technical Digest Series (Optical Society of America, Washington DC, 1997).
[4] - A.M. Sergeev, V.B. Gildenburg, A.V. Kim, M. Lontano, and M. Quiroga-Teixeiro, "*Few-Optical-Cycle Pulse Interactions with Plasmas: Models and Nonlinear Effects*", this Conference.
[5] - A.A. Babine, *et al., Izv. VUSov Radiofizika* **39**, 713 (1996).

2. RELATIVISTIC NONLINEAR OPTICS OF ULTRASHORT LASER PULSES IN PLASMAS

Relativistic Nonlinear Optics in Plasmas by 3D PIC Simulations

A. Pukhov* and J. Meyer-ter-Vehn

Max-Planck Institut fur Quantenoptik, D-85748 Garching, Germany

Abstract. Using the multi-dimensional Particle-in-Cell (PIC) code **VLPL** [1], we investigate interaction of relativisticly strong laser pulses with under- and overdense plasmas. We study acceleration of background electrons to multi-MeV energies, generation of 100 MG magnetic fields, and dynamics of ion channel boring. We show that the magnetic fields strongly influence transport of the accelerated relativistic electrons, which, in turn, modify the plasma refraction index and contribute to guiding of the laser pulse. We present spectra of the accelerated electrons and ions and the laser energy conversion efficiency. This physics is crucial for the **Fast Ignitor** concept in Inertial Confinement Fusion.

INTRODUCTION

We are witnessing an explosion-like progress in short-pulse laser technology, which has resulted in a dramatic increase of achieved laser powers and intensities [2]. When such multi-terawatt laser pulses at intensities $I > I_{18} = 10^{18}$ W/cm^2 interact with plasmas, a wide spectrum of interesting nonlinear effects occurs. Background plasma electrons are accelerated to multi-MeV energies [3] forming 100 kA currents [4,5] and generating 100 MG magnetic fields [4,6]. Change in the plasma refraction index due to these relativistic electrons concomitant with the laser pulse contributes to the pulse self-guiding and leads to filament coalescence [4,7]. The underlying physics is essentially kinetic and strongly nonlinear, thus precluding the use of hydrodynamic plasma models and/or paraxial/quasistatic description of the laser pulse [8]. In this situation one is forced to use a fully kinetic description of the plasma.

The most powerful tools of plasma computer simulations applicable here are relativistic fully electromagnetic Particle-In-Cell (PIC) codes [9]. The PIC codes are based on fundamental equations for the particles and fields dynamics and provide the most detailed kinetic description of plasmas. Being very much detailed, PIC simulations are expensive both in computer storage and CPU time and push existing computers to their ultimate limits [4,11], demanding use of Massively Parallel Processing (MPP). Fortunately enough, the super-intense laser pulses are extremely short, 100 fs – 1 ps, i.e., only tens to hundreds of laser periods for 1 μm radiation wavelength. It is this short pulse duration that makes direct PIC simulations of relativistic laser – plasma interactions feasible, and the first three-dimensional PIC simulations of actual experiments have been reporterd recently [7,10].

Considering a PIC simulation one can speak about a numerical or "virtual" plasma experiment. Such virtual plasma experiments are of a great value for understanding of what is going on in the real experiments. In the present paper we report results obtained with electromagnetic relativistic PIC code **VLPL** (Virtual Laser Plasma Laboratory) developed at MPQ, Garching [10]. The code **VLPL** exists in 2D and 3D geometry and runs on the 512 processor CRAY-T3E at Rechenzentrum Garching.

SELF-FOCUSING IN UNDERDENSE PLASMAS

A laser beam propagating in underdense plasma with a plasma frequency ω_p smaller than the laser frequency ω undergoes relativistic self-focusing as soon as its total power P exceeds the critical value [12]

$$P_c \approx 17 \left(\omega/\omega_p\right)^2 \; GW; \tag{1}$$

The self-focusing is due to the relativistic mass increase of plasma electrons and the ponderomotive expulsion of electrons from the pulse region. Both effects lead to a local decrease of plasma frequency and an increase in refractive index. The medium then acts as a positive lens.

Laser power close to critical: multi-MeV electrons

Self-focussing of a laser with $P > P_c$ in three-dimensional geometry leads to strong intensity enhancement at the axis. Simultaneously the three-dimensional self-modulation of the laser pulse sets in [13] generating high amplitude plasma waves. These waves break due to 1D [14,15] or 2D [16] mechanism producing fast electrons that can be trapped and accelerated to high energies as it has been observed recently in experiments [3,17]. Detailed studies show that fast electrons appear only above some threshold in the laser power [17] and plasma density [18], and this threshold is about $P \approx P_c$.

We have carried out a series of 3D PIC simulations of a laser pulse with radius $r = 8.5$ μm, and duration 460 fs propagating through plasma with density $n = 0.036 n_c$ over 0.6 mm distance. The laser power was varied from 0.16 to 6 P_c, where $P_c = 470$ GW under these conditions.

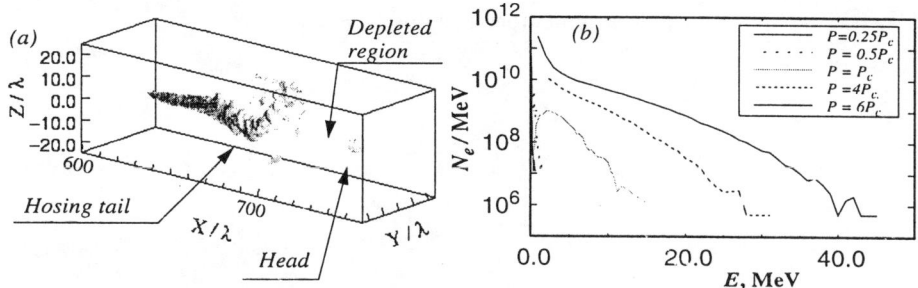

FIGURE 1. (a) Energy spectra of "real" electrons accelerated in plasma with density $n = 0.036 n_c$ for different laser powers; each "macro" electron in simulation represented 2.5×10^5 real electrons. (b) Laser pulse with $P = 4P_c$ initially, when having passed 0.6 mm plasma.

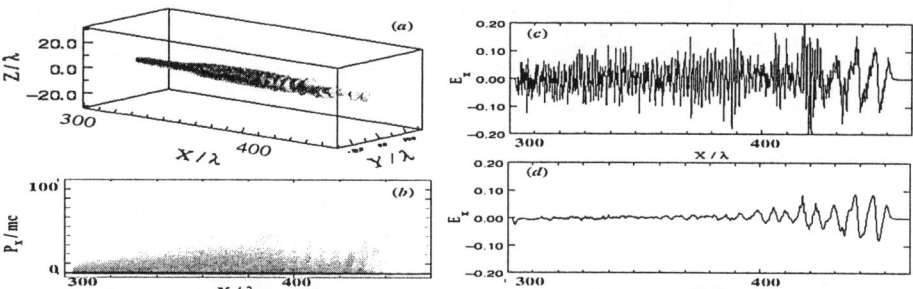

FIGURE 2. More details on electron acceleration mechanism (a) (b) Laser pulse with $P = 6P_c$, after 0.3 mm plasma. (b) Electron phase space. (c) Electric field $E_x = e\mathbf{E}_x/mc\omega$. (d) E_x, all modes with wavelengths shorter than λ are filtered out.

The characteristic evolution of laser pulse with $P = 4P_c$ is shown in Fig. 1a. The pulse has split into (1) its low-power head, diffracting in the plasma; (2) depleted region behind it, where the laser energy has been spent producing the palsma wake; and (3) the focussed and modulated tail of the pulse which experiences hosing [19].

The energy spectra of accelerated electrons are shown in Fig. 1b. The energetic electrons appear first at $P > P_c$. Surprisingly, *no sharp cutoff* associated with the dephasing limit is observed in the energy spectrum. Instead, the energy distibution rather resembles a Boltzmann one with a long smooth roll-off at high energies. The "effective temperature" of the distribution increases with laser power.

The Fig. 2 gives more insight in the process of acceleration for the high-power case $P = 6P_c$. Only the head of the laser pulse is modulated with the plasma wavelength, and a regular plasma wave exists only here, Fig. 2d. This reflects the electron phase space, Fig. 2b. It is also strongly modulated with the plasma wavelength in this region. However, the phase space shows

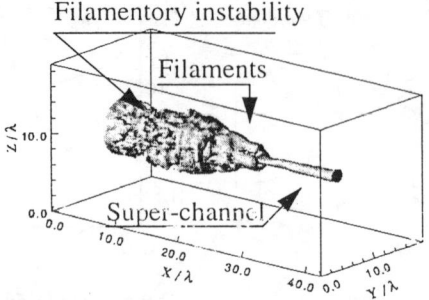

FIGURE 3. 3D perspective view of the self-focussing laser pulse, $I_0 = 1.2 \times 10^{19}$ W/cm^2, radius $r = 6$ μm, in plasma $n = 0.36 n_c$.

also another, unmodulated region containing fast electrons. It corresponds to the tightly focussed tail of the laser pulse, where no regular wake exists. Electron acceleration here looks more like a direct anisotropic heating by the laser pulse.

High laser power: magnetic fields and filamentation

When a laser pulse with a power far above threshold (1) propagates through slightly underdense plasmas, it is subject to the relativistic filamentation instability, that can prevent energy accumulation in a single channel. However, at intensities far above I_{18}, i.e., when $a > 1$, a new effect appears. The laser pulse drives strong currents of relativistic electrons in the direction of light propagation, magnetizing the plasma [4,15]. The generated quasistatic magnetic field is $B_\perp^s = (e n_e) 2\pi r$ at distance r from the beam axis. The field $B_\perp^s$ may become as strong as the magnetic field of the light wave itself, which is $B = a B_0$ in units of $B_0 = mc\omega/e$. For light of wavelength $\lambda = 2\pi c/\omega = 1\mu$m, one obtains $B_0 = 107.1$ MG. In units of B_0, the quasistatic magnetic field is

$$B_\perp^s / B_0 = (n_e/n_c) \pi r / \lambda. \qquad (2)$$

The magnetic field *pinches the concomitant relativistic electrons*, changing the plasma refraction. This leads to nonlocal interaction between light filaments, their mutual attraction, and final coalescence into a single "Super-Channel".

3D PIC simulations, see Fig. 3, reveal this effect [4]. The laser pulse sequentially passes an unstable filamentory stage, formation of a few filaments and their coalescence into the single super-channel due to the self-generated magnetic fields. In the plotted case, intensity in the super-channel rises from initial $I_0 = 1.2 \times 10^{19}$ W/cm^2 to 2×10^{20} W/cm^2, i.e., about 18 times.

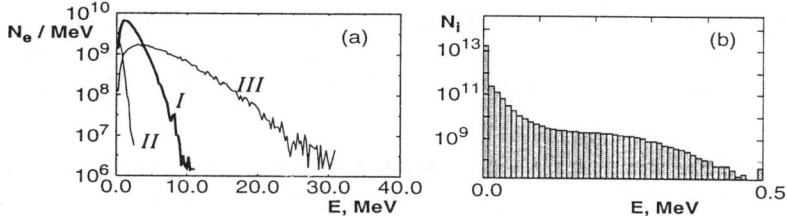

FIGURE 4. (a) Fast electron spectra for three runs (see text). I: $P = 1$ TW, $d_p = 30$ μm; II: $P = 1$ TW, $d_p = 3$ μm; III: $P = 10$ TW, $d_p = 30$ μm (b) Deuteron energy spectrum in run I. Laser is incident on a nonuniform plasma with density $n_e = n_c \exp(-x/d_p)$.

PREFORMED PLASMAS IN FRONT OF SOLID STATE SURFACE

Fast electrons accelerated in underdense plasmas and interacting with solid-state material can be used as a source of ultra-short X-ray pulses [20], γ−rays, and nuclear radiation [21]. The simplest experimental configuration consists of a solid state target with an ablatively preformed plasma in front of it.

Scale length of the preformed plasma is one of the most crucial parameters which determines the energy spectrum of the fast electrons and the laser energy conversion efficiency. To show the importance of the underdense region we performed 3D PIC simulations of plasma with an exponential density profile $n_e = n_c \exp(-x/d_p)$, where $n_c = 10^{21}$ cm^{-3} is the critical density for 1 μm wavelength laser pulse. The laser pulse had a Gaussian shape in time with a FWHM duration $T_{laser} = 200$ fs, and a FWHM diameter of the focal spot $D_{laser} = 9$ μm The spectra of accelerated electrons obtained in three runs for different density ranges d_p and peak laser intensities are presented in Fig. 4. In run I the maximum intensity was $I_0 = 1.5 \times 10^{18}$ W/cm^2, and the peak laser power about 1 TW.

The effective temperature of the suprathermal electrons $T_I \approx 0.83$ MeV should be compared with the "pessimistic" estimation $T_e = 0.06$ MeV for the case of sharp plasma density profile [22].

In the run II we have used a 10× steeper plasma density profile with $d_p = 3$ μm. The obtained spectrum of fast electrons, line II in Fig. 4, has an effective temperature $T_{II} \approx 0.2$ MeV, roughly 4 times lower than T_I. The total number of accelerated electrons was also significantly lower: $N_{II} = 3.0 \times 10^9$ electrons versus $N_I = 17.3 \times 10^9$ in the first run. In the third run, III, the plasma parameters were the same as in run I, but the laser intensity was 10 times higher, $I_0 = 1.5 \times 10^{19}$ W/cm^2. Thus, the peak laser pulse power was 10 TW. The resulting spectrum is shown as line III in Fig. 4. We find the effective temperature $T_{III} \approx 3.3$ MeV, much larger, than T_I.

As the laser pulse channels in the underdense region, it leads to electron cavitation, and then to the *ion acceleration* transverse to the channel axis.

The ion (deuteron mass) energies obtained in the run I, Fig. 4b, reach 0.5 MeV. These hot deuterons *make fusion reactions* that manifest themselves in experiment as neutron emission at the characteristic 2.45 MeV energy [23].

OVERDENSE PLASMAS: FAST IGNITOR PHYSICS

The concept of fast ignition of ICF targets with lasers [24] involves hole boring into overdense plasma [22] and generation of relativistic electrons that are to heat the ignition spot in the precompressed fuel core. Such external ignition may strongly relax driver requirements for ICF. Still, the underlying physics is not well understood.

One of the central issues for the Fast Ignitor concept is the magnetized electron transport through the overdense region. The relativistic electrons have to transport powers of $P = 10^{15}$ W. Taking electrons with 5 MeV mean energies, $\gamma = 10$, the corresponding current J amounts to 0.2 GA, or $1.1 \times 10^3 \, J_A$, where $J_A = \gamma 1.7 \times 10^4$ A is the Alfvén limit. Certainly, such a current cannot be transported as a simple beam, but it involves collective plasma transport in self-generated magnetic fields.

This problem is studied here by 2D (x,y) PIC simulations. A laser pulse propagating in x- and polarized in y-direction is normally incident on a plasma layer 20λ wide and 30λ thick with uniform density $n_e/n_c = (\omega_p/\omega)^2 = 10$, where n_c is the critical density, $\omega = 2\pi c/\lambda$ the laser frequency and λ the wavelength. Pulses have a Gaussian transverse profile of 6λ width and peak intensity I_0 in vacuum. They are semi-infinite in time and rise linearly over 20τ with $\tau = 2\pi/\omega$. The interaction physics scales with $I_0 \lambda^2$. We set $\lambda = 1\mu m$ corresponding to a cycle time of $\tau = 3.3$ fs. Intensities $I_0/(10^{19} \text{W/cm}^2) = 0.5, 2.2, 10, 15$ are considered such that electrons are driven to relativistic energies $E_{kin} = (\gamma - 1)m_e c^2$ with $\gamma = 1/(1-\beta^2)^{1/2} \gg 1$ and $\beta = v/c$. This enhances electron mass, $m_e \gamma$, and decreases plasma frequency, $\omega_p \langle \gamma^{-1} \rangle^{1/2}$, where $\langle ... \rangle$ denotes a cell average. A 2D version of the PIC code **VLPL** was run on 32 processors of CRAY-T3D at Rechenzentrum Garching. We use a spatial mesh of 1500×1200 cells with $7 \cdot 10^6$ electrons and $1.8 \cdot 10^6$ ions. The ion mass is $m_i = 1836 m_e$.

The results are presented in Fig. 5-9. We measure the conversion efficiency as the ratio of power carried by electrons in (x-direction per unit length along z) over that of incident light, $\eta = \int n v_x \varepsilon dy / \int I_L dy$. This quantity, η, and corresponding average energy $\langle \varepsilon \rangle = \int n v_x \varepsilon dy / \int n v_x dy$ of electrons carrying this power are shown in Fig. 6c-d. Both η and $\langle \varepsilon \rangle$ rise to peak values at $I_0 = 10^{20}$W/cm^2 and then fall again. All following results relate to this optimum intensity case The Fig. 5 shows two columns of snapshots, taken after 330 fs and 660 fs. At 330 fs, the light has punched a Δ-shaped crater into the overdense plasma about 4λ in depth (Fig. 5a-1). At 660 fs, it has changed its shape into a straight hole 12λ deep and 3λ in diameter (Fig. 5a-

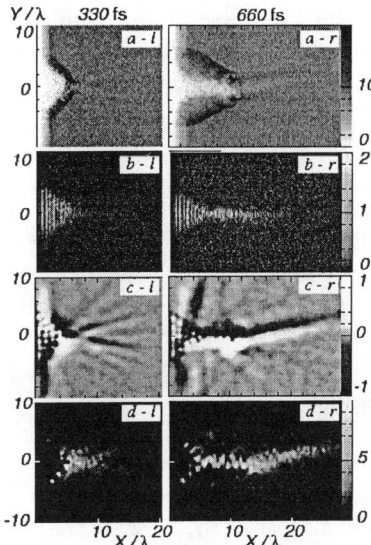

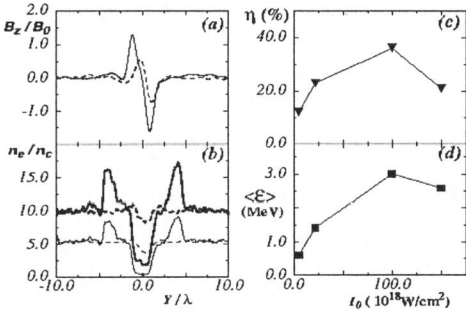

FIGURE 6. Transverse profiles of (a) cycle-averaged magnetic field and (b) electron density at $x = 7 \lambda$ (solid lines) and at $x = 16 \, l$ (dashed lines) at time 660 fs. The thinner lines in (b) give $n_e \langle \gamma^{-1} \rangle / n_c$. (c) Power of forward electron flow η in percent of incident laser power and (d) average energy ε defined in text, vs. intensity I_0, taken at 660 fs.

FIGURE 5. Channel boring in x,y plane by a $\lambda = 1$ mm laser pulse incident in x- and polarized in y- direction with $I_0 = 10^{20}$ W/cm² at time 330 fs (left column) and 660 fs (right column): (a) ion density, (b) instantaneous light intensity I (10^{19} W/cm²); (c) cycle-averaged magnetic field B/B_0; (d) X-component of the electron energy flux in units of 10^{19} W/cm².

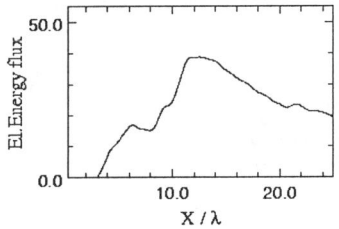

FIGURE 7. Power transported in X-direction by electrons in percents of incident laser power at t=660fs

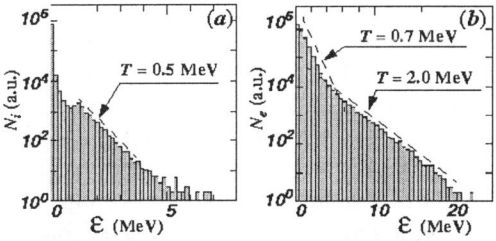

FIGURE 8. (a) Ion and (b) electron distribution vs. energy at 660 fs.

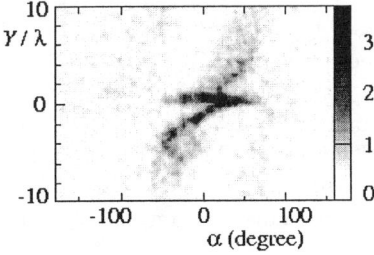

FIGURE 9. Electron energy flux distribution $F(x,y,\alpha,t)$ in y, α plane for t = 660 fs and $x = 16 \, \lambda$, angle α gives the electron direction in respect to x-axis.

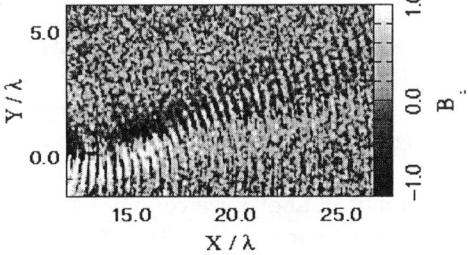

FIGURE 10. Non-averaged magnetic field in the jet region showing collimated propagation of the third harmonic.

r). Hole boring is driven by light pressure. It expells electrons from the interaction volume. This builds up an electric field that drags the ions. The hole boring velocity and its scaling with I_0 is found to be in fair agreement with Wilks et al. [22].

We have plotted the quantity η in Fig. 7 vs. x at time 660 fs. Conversion η reaches its maximum of 40% at the channel head and then decreases in the overdense plasma region to 20% at the right boundary. This strong absorption of elecron power is remarkable as it is obtained within a collisionless plasma model. Apparently, the absorption is due to collective effects. The incident laser power is $P_L = \pi r^2 I_0 \approx 28$ TW. Electrons transport 40% of it and this amounts to the current $J_e \approx 3.7$ MA, or $37 J_A$.

In Fig. 5a-r, one sees a collisionless shock of conical shape corresponding to ions running outwards from the channel region. They appear in Fig. 8b as a hot component in the ion spectrum with $T_i = 0.5$ MeV. At the shock front the density peaks up to 1.8 times the background density. It decreases toward the channel boundary and then sharply drops at the channel core, see Fig. 6b. Inside the channel, the electron density is $n_e/n_c \approx 2$. Yet, the laser light can propagate due to relativisticly induced transparency. We have plotted $n_e \langle \gamma^{-1} \rangle / n_c$ which indeed falls well below 1. In the rest of the plasma, $n_e \langle \gamma^{-1} \rangle / n_c \approx 5$ such that the third and higher harmonics can freely propagate all over the volume.

Light intensity is plotted in Figs. 5b. At 330 fs, the incident light just fills the crater; after 660 fs, it is funneled into the density channel, seen in Fig. 5b-r as a single light filament 10 λ length. It oscillates in width around 2 λ, similar to what has been observed in underdense plasma [7]. In addition, electromagnetic energy is scattered all over the simulation plane, more so at 660 fs. It may correspond to harmonic light and also to electromagnetic fluctuations of thermal nature, indicating plasma heating. At 660 fs, the electron energy distribution in Fig. 8a shows two populations with temperatures 0.7 and 2.0 MeV which are attributed to heated background plasma and laser-driven fast electrons, respectively.

The cycle-averaged magnetic fields plotted in Fig. 5c give evidence for the electron currents driven by the incident light. We distinguish three regions. The largest field occurs in the channel (region I). It is an azimuthal field produced by a strong current of relativistic electrons comoving with the laser light [4]. Its transverse profile at $x = 7\lambda$ is given in Fig. 6a. Within the chosen x, y geometry, it corresponds to a channel current $J/J_0 = (B/B_0)L/\lambda$ per length L in z-direction, where $J_0 = 17$kA. Taking $L = 2\lambda$ and $B/B_0 = 1.5$, we find $J = 3J_0 \approx 50$ kA, close to the Alfvén current J_A. Comparing this with the estimation above, we conclude about 97% of the current compensation by a cold return current.

Region II of the magnetic field is seen in Figs. 5c along the plasma surface ($x/\lambda \approx 2 - 3$). It corresponds to surface currents which feed the channel current. It reaches values of $0.5 B_0$ at 330 fs and falls to $0.1 - 0.2 B_0$ at 660 fs.

The most interesting is region III of the magnetic field located to the right of the channel in the overdense part of the plasma. In Fig. 5c-l at 330 fs, one observes a bunch of current filaments emerging like rays from the crater region, each marked by its magnetic field. The pattern correponds to electrons accelerated by the laser field into a cone that opens to the right. The electron current penetrates the overdense plasma, where it induces a return current and breaks up into filaments due to the Weibel instability [25]. The filaments are also seen as an imprint in the ion density (see shocked region in Fig. 5a-l).

The pattern changes when the channel gets deeper and elongated. At 660 fs, only one strong filament is left. The azimuthal B-field then traps all electrons below a certain transverse momentum in the channel. This is illustrated in Fig. 9. Here, we plot $F(x,y,\alpha,t) = \sum_i \varepsilon_i v_i/(\Delta V \Delta \alpha)$ in the y, α plane at $t = 660$ fs and $x = 16\lambda$, summing over all electrons moving within a given angle interval $\alpha \pm \Delta\alpha$ relative to the x-axis. For electrons leaving the channel head with a certain angular divergence, one expects F at an upstream position x to be distributed along a sloping straight line, as partially seen in Fig. 9. Evidence for trapping in the magnetized jet is given by the horizontal bar in the distribution, representing electrons that are confined within a narrow y-region (the jet), about 1λ in size, and have a large angular spread ranging between ± 40 degrees. Nonetheless, they can propagate in the vicinity of the axis, where the magnetic field vanishes. Laser acceleration of these trapped electrons leads eventually to a well collimated electron beam, which is injected into the overdense plasma at the channel head. This may explain the formation of the single jet observed in Fig. 5c-r. The jet is somewhat tilted relative to the x-axis. We find that this tilt changes with time and decreases, as the channel becomes longer. Obviously, the directional stability is an important issue for fast ignitor applications. From the present results, we cannot exclude that the direction may vary a bit statistically from shot to shot. The formation of filaments and subsequent coalescence of these filaments is also discussed in [26].

An additional feature of this jet is that it works as a waveguide for the higher harmonics of the insident laser pulse. Particularly, in our simulation we observe the third laser harmonic propagating in the density deep associated with the jet, Fig. 10. Moreover, the harmonics of a sufficiently powerful laser pulse, themself can experience relativistic self-focussing in the overdense region.

As to the overall energy balance at 660 fs, about 42.6% of the incident laser energy is found in the simulation volume, 33.0% in form of electron kinetic energy, 1.5% in ions, 4.3% in electric fields, and 3.8% in magnetic fields. The residual 53.4% are in reflected light, in harmonics and fast electrons that have left the simulation volume.

This work was supported by EURATOM and BMBF/Bonn.

REFERENCES

* Permanent address: Moscow Institute for Physics and Technology, Dolgoprudnyi, Moscow Region, Russia.
1. WWW-homepage: http://mpqibmr1.mpq.mpg.de:5000/ jxm1/vlpl.html
2. M. D. Perry and D. Mourou, Science **264**, 917-924 (1994); D. Mourou and D. Umstadter, Phys. Fluids B **4**, 2315 (1992); C. P. J. Barty et al., Opt. Lett., **21**, 668 (1996).
3. A. Modena et al., Nature (London) **377**, 606 (1995); D. Umstadter (et al., Science **273**, 472 (1996).
4. A. Pukhov and J. Meyer-ter-Vehn, Phys. Rev. Lett. **76**, 3975 (1996).
5. A. Pukhov and J. Meyer-ter-Vehn, Phys. Rev. Lett., 1997 (Accepted).
6. G.A. Askar'yan et al., JETP Lett. **60**, 251 (1994).
7. M. Borghesi et al., Phys. Rev. Lett. **78**, 879 (1997).
8. E. Esarey, P. Sprangle, J. Krall, and A. Ting, IEEE Trans. Plasma Science **24**, 252 (1996)
9. C. K. Birdsall and A. B. Langdon, *Plasma physics via computer simulations* (Adam Hilger, New York, 1991); J. Dawson, Phys. Plasmas **2**, 2189 (1995).
10. A. Pukhov and J. Meyer-ter-Vehn, APS Bulletin **41**, 1502 (1996).
11. W. B. Mori et al., Phys. Rev. Lett. **60**, 1298 (1988).
12. C. E. Max, J. Arons, and A. B. Langdon, Phys.Rev. Lett. **33**, 209 (1974); G.Sun et al., Phys. Fluids **30**, 526 (1987); P. Sprangle et al., IEEE Trans. Plasma Sci. **PS-15**, 145 (1983); G.Schmidt and W. Horton, Comments Plasma Phys. & Contr. Fusion **9**, 85 (1985); A. B. Borisov et al., Phys. Rev. **A45**, 5830 (1992).
13. E. Esarey, J. Krall, and Ph. Sprangle, Phys. Rev. Lett. **72**, 2887 (1994).
14. A. I. Akhieser, and R. V. Polovin, JETP **3**, 696 (1956).
15. S. V. Bulanov, F. Pegoraro, and A. Pukhov, Phys. Rev. Lett. **74**, 710 (1995).
16. S. V. Bulanov, F. Pegoraro, A. Pukhov, and A. S. Sakharov, Phys. Rev. Lett. **78**, (1997).
17. R. Wagner et al., Phys. Rev. Lett. **78**, 3125 (1997).
18. R. Fedoseevs, X. F. Wang, and G. D. Tsakiris, Phys. Rev. E (1997).
19. G. Shvets, and J. Wurtele, Phys. Rev. Lett **73**, 3540 (1994).
20. J.D. Kmetec et al., Phys. Rev. Lett. **68**, 1527 (1992); M. Schnürer et al., Phys. Plasmas **2**, 3106 (1995).
21. P. L. Shkolnikov et al., *submitted to Appl. Phys. Lett.*
22. S. C. Wilks et al., Phys. Rev. Lett. **69**, 1383 (1992).
23. G. Pretzler et al., *to be published.*
24. M. Tabak et al., Phys. Plasmas **1**, 1626 (1994).
25. E.W. Weibel, Phys. Rev. Lett. **2**, 83 (1959); F. Pegoraro et al., Physica Scripta **T63**, 262 (1996).
26. R. Lee and M. Lampe, Phys. Rev. Lett. **31**, 1390 (1973).

Nonlinear Optics in Relativistic Plasmas

D. Umstadter, S.-Y. Chen, G.S. Sarkisov, A. Maksimchuk, and R. Wagner

Center for Ultrafast Optical Science[1]
University of Michigan
Ann Arbor, MI 48109-2099

Abstract. We discuss various nonlinear optical processes that occur as an intense laser propagates through a relativistic plasma. These include the experimental observations of electron acceleration driven by laser-wakefield generation, relativistic self-focusing, waveguide formation and laser self-channeling.

INTRODUCTION

Due to recent advances in laser technology [1], it is now possible to generate the highest electromagnetic and electrostatic fields ever produced in the laboratory [2]. The interactions of such high-intensity and ultrashort-duration laser pulses with plasmas permits for the first time the study of optics in relativistic plasmas. Technological applications include advanced fusion energy, x-ray lasers, and table-top ultrahigh-gradient electron accelerators.

We have recently demonstrated that by simply focusing a table-top size, high-power (terawatt) laser into a gas, a beam of relativistic electrons is produced. It is accelerated up to an energy of 40 MeV by a laser-wakefield plasma wave in a distance less than one millimeter. The accelerated electron beam appeared to be naturally-collimated with a low-divergence angle and had over a nanocoulomb of charge in a 1-picosecond duration pulse. These characteristics are comparable to those of state-of-the-art radio-frequency linacs, but the acceleration length is ten-thousand times shorter. This enormous field gradient (2 GeV/cm)—which is the highest terrestrial electrostatic field ever recorded—would be of limited use if its length were limited by laser diffraction. Fortunately, we also recently showed that—above a certain laser power—the

---

[1] This work was supported by NSF PHY 972 661, NSF STC PHY 8920108 and DOE/LLNL subcontract B307953.

plasma acts like a lens, which guides the laser beam by a process called relativistic self-focusing. The relativistically self-guided channel was found to extend the laser propagation distance to many Rayleigh lengths, decrease the electron beam divergence and increase the electron energy. Where the laser was guided, an on-axis density depression was also observed to form, created by electron cavitation due to laser pressure followed by Coulomb explosion of the ions. Another intense laser pulse that was delayed in time with respect to the first was shown to propagate down the channel. Finally, we discuss optical injection of electrons into wakefield plasma waves.

When an intense laser enters a region of gaseous-density atoms, the atomic electrons feel the enormous laser electromagnetic field, and begin to oscillate at the laser frequency ($\omega = 2\pi c/\lambda = ck$). The oscillations can become so large that the electrons become stripped from the atoms, or ionized. At high laser intensity (I), the free electrons begin to move at close to the speed of light (c), and thus their mass m_e changes significantly compared to their rest mass. This large electron oscillation energy corresponds to gigabar laser pressure, displacing the electrons from regions of high laser intensity. Due to their much greater inertia, the ions remain stationary, providing an electrostatic restoring force. These effects cause the plasma electrons to oscillate at the plasma frequency (ω_p) after the laser pulse passes by them, creating alternating regions of net positive and negative charge, where $\omega_p = \sqrt{4\pi e^2 n_e/\gamma m_e}$, n_e is the electron density, e is the electron charge and γ is the relativistic factor associated with the electron motion transverse to the laser propagation. γ depends on the normalized vector potential, a_o, by $\gamma = \sqrt{1 + a_o^2}$, where $a_o = \gamma v_{os}/c = eE/m_o \omega c = 8.5 \times 10^{-10} \lambda[\mu m] I^{1/2}[W/cm^2]$. The resulting electrostatic wakefield plasma wave propagates at a phase velocity nearly equal to the speed of light and thus can continuously accelerate hot electrons [3]. Up to now, most experiments have been done in the self-modulated laser wakefield regime [4–6], where the laser pulse duration is much longer than the plasma period, $\tau \gg \tau_p = 2\pi/\omega_p$. In this regime, the forward Raman scattering instability can grow; where an electromagnetic wave ($\omega_o, \boldsymbol{k_o}$) decays into a plasma wave ($\omega_p, \boldsymbol{k_p}$) and electromagnetic side-bands ($\omega_o \pm \omega_p, \boldsymbol{k_o} \pm \boldsymbol{k_p}$).

Wakefield acceleration, as well as most other applications, depends critically on long-distance propagation of laser pulses at relativistic intensities ($\sim 10^{18}$ W/cm² for 1-μm wavelength light). In order to reach such high intensities, laser pulses are usually focused tightly, which, due to diffraction, results in a short interaction length ($\sim$ 1–2 Rayleigh ranges, $Z_R = \pi r_0^2/\lambda$). Several methods have been proposed to extend the propagation distance of pulses beyond this diffraction limit, as reviewed in [9].

For optical guiding of laser pulses in plasmas, the radial profile of the index of refraction, $n(r)$, must have a maximum on axis, causing the wavefront to curve inward and the laser beam to converge. When this focusing force is strong enough to counteract the diffraction of the beam, the laser pulse can propagate over a long distance and maintain a small cross section. The index of refraction for a plasma is given by $n(r) = 1 - (\omega_p^2/\omega_0^2) \cdot (n_e(r)/n_{e0}\gamma(r))$, where ω_p is the plasma frequency for electron density n_{e0}, ω_0 is the laser frequency, $n_e(r)$ is the radial distribution of electron density, and $\gamma(r)$ is the relativistic factor associated with the electron motion transverse to the laser propagation. The factor γ depends on the normalized vector potential, a_o, by $\gamma = \sqrt{1 + a_o^2}$, where $a_o = \gamma v_{os}/c = eE/m_o\omega c = 8.5 \times 10^{-10} \lambda[\mu m] I^{1/2}[W/cm^2]$. It can readily be seen from this that an on-axis maximum of $n(r)$ can be created through modification of the radial profile of γ and/or n_e.

In the former case, $\gamma(0) > \gamma(r)$ can be created by a laser beam with an intensity profile peaked on axis. When a pulse guides itself by this mechanism, it is referred to as relativistic self-guiding, and should occur provided the laser power exceeds a critical power given by $P_c = 17(\omega_0/\omega_p)^2$. A somewhat higher threshold is expected if the ionization defocusing effect is taken into account. In the latter case, $n_e(0) < n_e(r)$ is predicted to occur when the laser's ponderomotive force expels electrons radially from the region of the axis, so called "electron cavitation." Ponderomotive self-channeling [10,11] is expected to enhance the effects of relativistic self-guiding. Self-guiding has previously been observed experimentally [12,30], but electron cavitation has not.

Eventually, after a period of sustained electron expulsion, the ions are predicted to also blow out of the axial region [14] due to Coulomb repulsion, forming a channel [15] that could be used to guide a second laser pulse. A plasma-channel waveguide preformed in this way can guide a relatively high laser intensity ($\geq 10^{17}$ W/cm^2) in a relatively high plasma density ($\geq 10^{19}$ cm^{-3}), as compared with waveguides formed by long-duration laser pulses via thermal hydrodynamic expansion followed shock front formation ($\leq 10^{16}$ W/cm^2, $\leq 10^{18}$ cm^{-3}).

EXPERIMENTS

Experimental Arrangement

In this experiment, we used a Ti:sapphire-Nd:glass laser system based on chirped-pulse-amplification that produces 3 J, 400 fs pulses at 1.053 μm. The 43 mm diameter beam was focused with an f/4 off-axis parabolic mirror to $r_o = 8.5$ μm (1/e^2), corresponding to vacuum intensities exceeding 4×10^{18} W/cm^2. This pulse was focused onto a supersonic helium gas jet with a sharp gradient (250 μm) and a long flat-topped interaction region (750 μm). The maximum density varies linearly with backing pressure up to the maximum backing pressure of 1000 PSI, and an underdense plasma at 3.6×10^{19} cm^{-3} is formed by the foot of the laser pulse tunnel-ionizing the gas. This plasma density corresponds to a critical power of $P_c = 470$ GW. A sharp gradient and long interaction region are found to be essential.

Wakefield Acceleration

Recently, we have shown that an accelerated electron beam appeared to be naturally-collimated with a low-divergence angle (less than ten degrees), and had over 1-nC of charge per bunch [16]. Moreover, as shown in Fig. 1, acceleration occured in this experiment [16] only when the laser power exceeded a certain critical value, P_c, the threshold for relativistic self-focusing.

The total number of accelerated electrons (at all energies) was measured using either a Faraday cup or a plastic scintillator coupled to a photomultiplier tube, and the results were found to be consistent with each other. There is a sharp threshold for electron production at $\sim 1.5 P_c$, and the total number of electrons increases exponentially and finally saturates beyond $4 P_c$ [16]. At $6P_c$, 6×10^9 accelerated electrons were measured coming out of the plasma in a beam. By using aluminum absorbers, we determined that 50% of the electrons detected have energy greater than 1 MeV (corresponding to 0.5 mJ of energy in the electron beam).

The electron energy spectrum (see Fig. 1) was measured using a 60° sector dipole magnet by imaging a LANEX scintillating screen with a CCD camera. The normalized distribution is found to have a functional form of $\exp(-\alpha\gamma)$ where α is a fitting parameter. In the low power case ($< 6P_c$, no channeling), the normalized distribution follows $\exp(-\gamma)$, and when the laser power

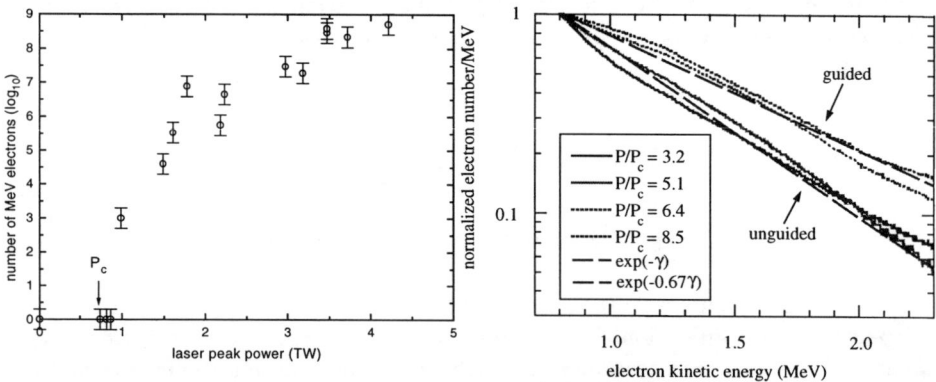

FIGURE 1. The number of relativistic electrons accelerated as a function of incident laser power focused in a gas of helium at atmospheric density. (right) Normalized electron kinetic energy spectrum as a function of laser power at fixed electron density. The upper curves represent the spectra obtained when self-guiding was observed; the lower curves represent unguided spectra.

increases ($> 6P_c$, with channeling), the electron energy distribution discretely jumps to follow $\exp(-0.67\gamma)$. The abrupt change in the electron distribution also occurs if the laser power is held fixed and the density is increased, as it should given the critical power threshold dependence on density. Below 850 PSI (3.1×10^{19} cm^{-3}, no channeling), the electron distribution follows the same trend as the lower power distribution, and above 850 PSI (with channeling) it follows the higher power distribution. For the electron energy distribution greater than 3 MeV, a significantly less steep slope that extends to 20 MeV was measured using aluminum absorbers. Even though the plasma wave amplitude increases as the laser power increases, the distribution only dramatically changes when self-guiding occurs. This indicates that extension in the accelerating length is the primary factor in determining the fitting parameter α.

Measurements of the satellites in the spectrum of the forward scattered light indicated that a self-modulated plasma wave occurred when the laser power exceeded $P_c/2$. Since then, two independent research groups have simultaneously reported direct measurements of the plasma wave amplitude with a Thomson-scattering probe pulse [28,29]. The field gradient was reported [29] to exceed that of a radio-frequency (RF) linac by four orders of magnitude ($E \geq 200$ GV/m). This acceleration gradient corresponds to an energy gain of 1 MeV in a distance of only 10 microns. The plasma wave was observed to exist for a duration of 1.5 ps or 100 plasma oscillations [29]. It was calculated that it damps only because all of the wave energy was converted to the accelerated electrons. Except for the large energy spread and low average power, these parameters compare favorably with medical linacs. In fact, the much smaller source size of a laser wakefield accelerator compared with that of a conventional linac, 10 microns compared with greater than 100 microns, may permit much greater spatial resolution for medical imaging.

Relativistic Self-Guiding

This enormous field gradient would be of limited use if the length over which it could be used to accelerate electrons were just the natural diffraction length of the highly focused laser beam, which is much less than a millimeter. Fortunately, we recently demonstrated that electrons can be accelerated beyond this distance [30]. At high laser power, the index of refraction in a plasma varies with the radius. This is both because the laser intensity varies with radius and the plasma frequency depends on the relativistic mass factor γ. Above the above-mentioned critical laser power P_c, the plasma should act like a positive lens and focus the laser beam, a process called relativistic self-focusing. This is similar to propagating a low power beam over an optical fiber optic cable, except in this case the intense laser makes its own fiber optic.

In order to diagnose the spatial extent of the plasma, a sidescattering imaging system with a spatial resolution of 15 μm was utilized. We were able to resolve the growth of the plasma channel as a function of both laser power and plasma density. Fig. 2 shows the sidescattered intensity distribution as a function of laser power, and the plasma channel clearly extends as the laser power increases. In the lower power cases ($< 2.6 P_c$), the channel length is only ~ 125 μm, which is smaller than the confocal parameter ($2Z_R$) of 430 μm. As the laser power increases for a fixed gas density, the channel length first jumps to 250 μm at $3.9 P_c$ and then reaches 750 μm at $7.2 P_c$. The maximum channel length was observed to be 850 μm at $9.1 P_c$. Note this is limited by the interaction length of the gas jet. At $5.5 P_c$, the sidescattered image formed has two distinct foci, and when the power exceeds $7.2 P_c$, either multiple foci or a channel are observed, depending on shot-to-shot fluctuations and the gas jet position. A similar channel extension occurs if the gas density is varied at fixed laser power. For a 3.9 TW laser pulse, the channel extends to 250 μm at 400 PSI backing pressure (1.4×10^{19} cm^{-3}, $3.2 P_c$) and 750 μm at 800 PSI (2.9×10^{19} cm^{-3}, $7.0 P_c$). The consistent behavior at specific values of P_c for varying laser power or plasma density indicates that the channeling mechanism is relativistic self-focusing.

The sidescattered light was spectrally analyzed by an imaging spectrometer, and the bulk of the emission comes from incoherent Thomson scattering of the blue-shifted laser pulse. We were unable to obtain any information about the plasma density or temperature from this measurement. The divergence of the laser beam transmitted through the plasma was measured using a diffusing screen and a CCD camera with a 1.053 μm narrow bandpass filter. At all laser powers, the laser expands to twice the vacuum divergence, and we attribute this expansion to ionization defocusing. This is consistent with the strong

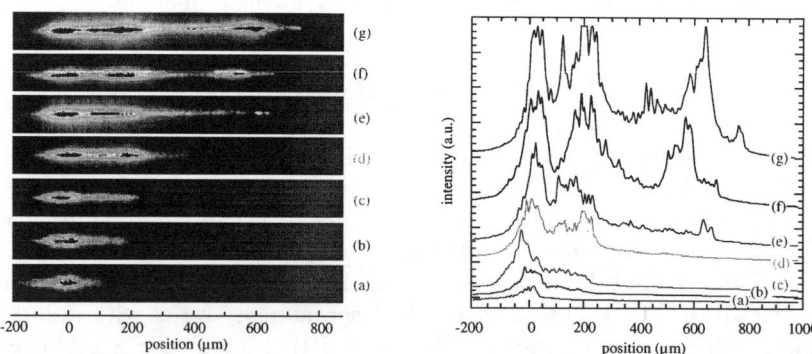

FIGURE 2. On-axis images (left) and corresponding lineouts (right) of sidescattered light at various laser powers and a fixed initial electron density of 3.6×10^{19} cm^{-3}. The various images and lineouts represent laser powers of $P/P_c =$ (a) 1.6, (b) 2.6, (c) 3.9, (d) 5.5, (e) 7.2, (f) 8.4, and (g) 9.1. Note: the curves have been displaced vertically for ease of viewing.

blue-shifting we observe in the scattered spectra. Even though simulations indicate that the laser focuses to $\sim 2~\mu$m [19], the complex dynamics that occur as the laser continually focuses and defocuses in the plasma make it impossible to determine the minimum self-focused beam width from the far field divergence angle.

A Maxwellian-like energy distribution has been observed in many previous experiments [20] and simulations [21], however no theoretical justification for it has been found to date. Because the energy distribution is exponential, a temperature in the longitudinal direction can be defined. The temperature of the low energy distribution changes from 500 keV (without guiding) to 750 keV (with guiding). In these plasmas, many different plasma waves can grow from various instabilities and local conditions. The interactions between these waves can lead to stochastic heating of the electron beam, so by extending the plasma length, the various waves will interact longer and heat the beam more. However, the dephasing length, $L_d = \lambda(\omega_o/\omega_p)^3$, which gives the maximum distance over which acceleration can occur (170 μm for our conditions), is significantly shorter than our accelerating length. From this expression, we would think that there would be no noticeable change in the electron spectrum when we extend the plasma length from 250 μm to 750 μm. Recent PIC simulations [21] indicate that this expression is too conservative for these highly nonlinear plasma interactions, and, in fact, the actual dephasing length may be many times longer. Consistent with our experimental results, these simulations indicate that the electron temperature, as well as the maximum energy, increase as the electrons propagate beyond the conventional dephasing length.

The relativistically self-guided channel was found to increase the laser propagation distance by a factor of four (limited thus far only by the length of gas), decrease the electron beam divergence by a factor of two (as shown in Fig. 3), and increase the electron energy.

The electron beam profile was measured using a LANEX scintillating screen imaged by a CCD camera [16]. The LANEX is placed behind an aluminum sheet which blocks the laser light, so only electrons greater than 100 keV can be imaged. Analysis of the electron spectrum indicates that the bulk of the electrons that create an image on the screen are in the 100 keV to 3 MeV range. We have found, using aluminum absorbers, that the electron divergence does not depend on electron energy in this range. At low power ($< 5 P_c$), the electron beam has a Gaussian-like profile with a 10° radius at half-maximum (see Fig. 3). As the laser power increases and the plasma channel length increases to $\sim 250~\mu$m, a second peak seems to grow out of the low-power profile. Ultimately at the highest laser powers and longest channel lengths, the divergence decreases to 5°, and the profile becomes more Lorentzian-like. The electron beam divergence should decrease as the longitudinal energy of the electrons increases since space charge will be less and the relative transverse momentum decreases due to the longer accelerating length. However, there should be a minimum divergence due to the space charge effect after the electrons leave the plasma. This effect is significant since the electrons are in the few MeV range (small γ) and the peak current is high (large number of electrons in a short bunch). We have roughly estimated the space charge divergence to be 6° by assuming 10^9 electrons at 1 MeV in a 1 ps bunch (note: $\theta_{hwhm} \propto \sqrt{N/\tau_e(\beta\gamma)^3}$, where N is the number of electrons, τ_e is the electron bunch duration, and $\beta\gamma$ is the normalized momentum of the electrons) [22]. The electron beam emittance can be found from the measured divergence angle and the radius of the plasma channel, and in the best case (5° half-angle and 5 μm half-max radius), the calculated emittance ($\epsilon = r_o \theta_{hwhm}$) is 0.4 π-mm-mrad. To verify that the reduction in the beam emittance is due to the extension of the plasma channel, another gas jet with a narrower width was used and the same measurements repeated. In this case, the sidescattered images show that the channel length is limited to 360 μm and the electron beam divergence is fixed at 12° for all laser powers.

Waveguide Formation

In order to observe the formation and evolution of the plasma-waveguide structure, probing interferometry and shadowgraphy were used. 3-D images of the plasma-density distribution were obtained in this way at different times. A probe pulse (400 fs, 1.053 μm) is obtained by splitting 5% of the pump pulse, sending it into a delay line, and crossing it perpendicularly with the pump pulse in the interaction region. A lens was used to image the probe pulse in the plasma region to a CCD camera, forming shadowgrams. Inter-

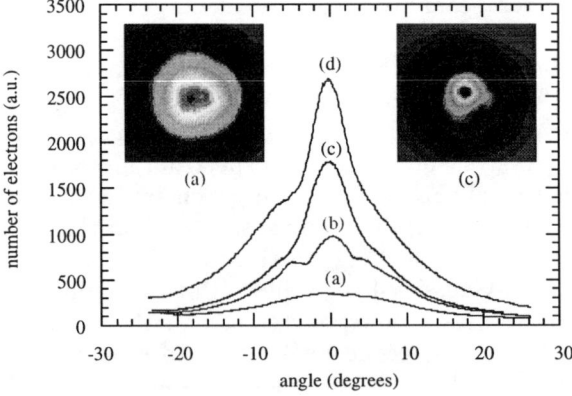

FIGURE 3. Electron beam divergence as a function of laser power. The various curves represent laser powers of $P/P_c =$ (a) 3.4, (b) 5.0, (c) 6.0, and (d) 7.5. The two insert figures show the complete beam images for curves (a) and (c).

ferograms were obtained by use of two glass wedges forming a vacuum wedge gap [23] after the lens. To get quantitative measurements of the evolution of the plasma waveguide, a 3-D plasma density distribution was obtained by means of a fringe tracking program (to get a 2-D phase-shift distribution) in conjunction with an abelization algorithm. Fig. 3 shows the 3D plasma density distribution (cylindrically symmetric) evolving in time. As can be seen, the density depression on axis become deeper and the width of the channel becomes larger as time goes by. At about 40 ps delay, a plasma waveguide of 800 μm in length is formed, which has an on-axis plasma density less than 10^{18} cm^{-3} and a channel width of 30 μm. In Fig. 3, the waveguide length is equal to the self-channeling length of the pump pulse.

REFERENCES

1. P. Maine, D. Strickland, P. Bado, M. Pessot and G. Mourou, *IEEE J. Quantum Electron.* **QE-24**, 398 (1988).
2. G. Mourou and D. Umstadter, *Phys. Fluids B* **4**, 2315 (1992).
3. T. Tajima and J. M. Dawson, *Phys. Rev. Lett.* **43**, 267 (1979).
4. N. E. Andreev, L. M. Gorbunov, V. I. Kirsanov, A. Pogosova and R. R. Ramazashvili, *Pis'ma Zh. Eksp. Teor. Fiz.*, **55** 551 (1992) [*JETP Lett.*, **55**, 571 (1992)].
5. T. M. Antonsen, Jr., and P. Mora, *Phys. Rev. Lett.*, **69**, 2204 (1992).
6. P. Sprangle, E. Esarey, J. Krall, and G. Joyce, *Phys. Rev. Lett.*, **69**, 2200 (1992).

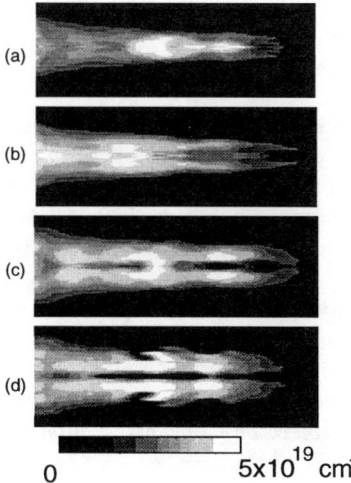

FIGURE 4. 2-D plasma density distribution for 2.5 TW laser power and 2×10^{19} cm^{-3} gas density at different times: (a) 5 ps, (b) 15 ps, (c) 30 ps, and (d) 40 ps.

7. D. Umstadter *et al.*, Science **273**, 472 (1996).
8. M. Tabak *et al.*, Phys. Plasmas **1**, 1626 (1994).
9. E. Esarey *et al.*, IEEE Trans. Plasma Sci. **PS-24**, 252 (1996).
10. G. Z. Sun *et al.*, Phys. Fluids **45**, 526 (1987).
11. A. B. Borisov *et al.*, Phys. Rev. A **45**, 5830 (1992).
12. P. Monot *et al.*, Phys. Rev. Lett. **74**, 2953 (1995).
13. R. Wagner *et al.*, Phys. Rev. Lett. **78**, 3125 (1997).
14. A. Pukhov and J. Meyer-ter-Vehn, Phys. Rev. Lett. **76**, 3975 (1996).
15. D. C. Barnes, T. Kurki-Suonio, T. Tajima, IEEE Trans. Plasma Sci. **PS-15**, 154 (1987).
16. D. Umstadter, S.-Y. Chen, A. Maksimchuk, G. Mourou, and R. Wagner, *Science* **273**, 472 (1996).
17. K. Krushelnick *et al.*, Phys. Rev. Lett. **78**, 2373 (1997).
18. S. P. Le Blanc *et al.*, Phys. Rev. Lett. **77**, 5381 (1996).
19. A. Chiron *et al.*, Phys. Plasmas **3**, 1373 (1996).
20. C. Rousseaux, *et al.*, Phys. Fluids B, **4**, 2589 (1992).
21. K.-C. Tzeng, *et al., Advanced Accelerator Concepts 1996*, to be published.
22. S. Humphries Jr., *Principles of Charged Particle Accelerators* (Wiley, New York, 1986).
23. G. S. Sarkisov, Instruments and Experimental Techniques **39**, 727 (1996).
24. H. Tawara *et al.*, Atomic Data and Nuclear Data Tables **36**, 167 (1987).
25. C. A. Coverdale, C. B. Darrow, C. D. Decker, W. B. Mori, K. -C. Tzeng, K. A. Marsh, C. E. Clayton, and C. Joshi, *Phys. Rev. Lett.* **74**, 4659, (1995).
26. K. Nakajima, D. Fisher, T. Kawakubo, H. Nakanishi, A. Ogata, Y. Kato, Y. Kitagawa, R. Kodama, K. Mima, H. Shiraga, K. Suzuki, K. Yamakawa, T. Zhang, Y. Sakawa, T. Shoji, N. Yugami, M. Downer and T. Tajima, *Phys. Rev. Lett.* **74**, 4428, (1995).
27. A. Modena, Z. Najmudin, A. E. Dangor, C. E. Clayton, K. A. Marsh, C. Joshi, V. Malka, C. B. Darrow, C. Danson, D. Neely and F. N. Walsh, *Letts. Nature* **377**, 606, (1995).
28. A. Ting, K. Krushelnick, C. I. Moore, H. R. Burris, E. Esarey, J. Krall, and P. Sprangle, *Phys. Rev. Lett.* **77**, 5377 (1996).
29. S. P. Le Blanc, M. C. Downer, R. Wagner, S.-Y. Chen, A. Maksimchuk, G. Mourou and D. Umstadter, *Phys. Rev. Lett.* **77**, 5381 (1996).
30. R. Wagner, S.-Y. Chen, A. Maksimchuk and D. Umstadter, *Phys. Rev. Lett.* **78**, 3122 (1997).
31. D. Umstadter, E. Esarey, and J. Kim, Phys. Rev. Lett. **72**, 1224 (1994); D. Umstadter *et al.*, Phys. Rev. E **51**, 3484 (1995).
32. D. Umstadter, J. K. Kim, and E. Dodd, *Phys. Rev. Lett.* **76**, 2073 (1996).
33. E. Esarey and M. Pilloff, Phys. Plasmas **2** 1432 (1995).
34. See *e.g., Advanced Accelerator Concepts, Fontana, WI, 1994*, Amer. Inst. of Conf. Proc. No. 335, P. Schoessow, ed., (AIP Press, New York, 1995) and references cited therein.

Ultra intense magnetic fields in laser plasma interaction: their generation and influence on light propagation

F. Pegoraro[1], S.V. Bulanov[2], F. Califano[3],
T.Zh. Esirkepov[4], M. Lontano[5], N.M. Naumova[2],
A.M. Pukhov[4,6], and V.A. Vshivkov[7]

[1] Physics Department, Pisa University and INFM, Pisa, Italy.
[2] General Physics Institute, RAS, Moscow, Russia
[3] Scuola Normale Superiore and INFM, Pisa, Italy
[4] Moscow Institute for Physics and Technology, Dolgoprudny, Russia.
[5] Institute for Plasma Physics - CNR, Milan, Italy.
[6] Max-Planck-Institute for Quantum Optics, Garching, Germany.
[7] Institute for Computational Technologies, RAS, Novosibirsk, Russia

Abstract

Extremely large, quasi-stationary magnetic fields can be generated in plasmas by high intensity laser pulses. These fields can change the plasma dynamics and the pulse propagation. Several aspects of their generation and of their effect on the plasma and on the laser pulse are discussed in the relativistic pulse amplitude regime: (a) the formation of magnetic wakes, (b) the development of longitudinal and transverse Langmuir wake wavebreaks and (c) the magnetic field generation on a thin foil, viewed as a model for overdense plasmas with sharp boundaries.

1 Introduction

Intense electromagnetic radiation in plasmas can generate quasi-stationary magnetic fields spontaneously due to nonlinear effects [1,2,3,4]. In the interaction of a super intense laser pulse with a plasma, the electrons become relativistic in the field of the laser pulse. In addition, wakes of Langmuir waves can be excited and can accelerate electrons to very large energies. In regimes where these waves break, they can produce highly energetic electron beams in a colder background. Under these relativistic conditions the oscillatory magnetic field in the pulse and the quasi-stationary self-consistent magnetic field generated in the plasma change the dynamics of the electrons.

Relativistic regimes can be characterized in terms of the dimensionless amplitude of the laser pulse by the condition $a \equiv (eA/m_e c^2) > 1$ where A is the amplitude of the pulse vector potential. An order of magnitude estimate of the largest quasi-stationary magnetic field B_0 that can be generated by a laser pulse with $a \gg 1$ is obtained[5] from the relativistic upper bound on the current density $J_{max} \approx nec$ and from the characteristic spatial scale in a relativistic plasma $d_p = c/\omega_p$. Here n is taken of the order of the unperturbed plasma density, and $\omega_p \equiv (4\pi n e^2/m)^{1/2}$ and m are the relativistic plasma frequency and electron mass. Taking $m \approx m_e a^2$ we obtain

$$B_0 \approx a m_e \omega_{pe} c/e \qquad (1)$$

This corresponds to $\Omega_c \approx \omega_p$ where $\Omega_c \equiv eB_0/mc$ is the relativistic electron gyration frequency and $\omega_p \equiv \omega_{pe}/a$. Even in an underdense plasma, if its density is not too low, the resulting characteristic electron-gyroradius can be as short as a few laser wavelengths. These estimates indicate that the self-generated magnetic field not only modifies the particle energy transport in the plasma, as emphasized by Bell [6] for lower amplitudes, but can also change the motion of the electrons and thus the propagation properties, e.g., the focalisation, of the laser pulse and the formation of Langmuir wake fields in the plasma.

Indeed the actual magnitude of these fields is huge: for a laser pulse with $a > 1$, $\lambda = 1 \mu m$ in a plasma with density, e.g., half its critical value, B_0 is of order 100 MG. Magnetic fields of such intensity might make it possible to explore completely new regimes in matter. Magnetic forces at least as important as electric forces inside atoms will lead to different atomic structures. Their study will provide information on extreme astrophysical conditions such as those that are expected to occur, e.g., on the surface of a neutron star.

2 Dynamical role of the quasi-steady magnetic field

In Ref.[5] we highlighted the dynamical guiding effect on the pulse propagation of the quasi-steady magnetic field. Investigating the propagation of high power pulses in an underdense plasma well above the relativistic self-focusing threshold $P_{cr} \equiv m_e^2 c^5 \omega^2/4\pi e^2 \omega_{pe}^2 = 17(\omega^2/\omega_{pe}^2) GW$, we saw in 2-D PIC simulations[5,7,8] that the pulses split into self-focused filaments that later coalesce. We proposed that the coalescence of the filaments can be explained on the basis of a magnetic interaction between them. Inside these filaments electrons are accelerated and produce a current that is rapidly cancelled by a cold electron current of opposite sign which arises in order to maintain plasma quasi-neutrality. These opposite currents repel each other and generate a quasi-static magnetic field. Due to the current repulsion, this field extends over a region wider than the channel radius. This "long range" interaction allows filaments to coalesce as the magnetic field of one filament contributes to deflect the accelerated electrons in the adjacent filaments, moving their "heads" closer and thus interrupting the return current flowing between them. This effect was later confirmed in 3-D simulations in Ref.[9].

3 Magnetic vortex wake

In relativistic regimes where the Langmuir waves produced behind an ultrashort pulse in an underdense plasma break, we observed[10] the formation of a magnetic wake. PIC simulations indicated two different mechanisms of magnetic field generation, one inside the pulse and the other behind it. Both fields vanish at the pulse axis. The field behind the pulse was shown by phase space plots[8] to be well correlated to the onset of the break of the Langmuir waves. The resulting magnetic wake does not propagate with the pulse, it propagates at a speed much smaller than c, and takes the form of a row of magnetic vortices of opposite polarity. The vortex row takes an approximately skew symmetric configuration far from the pulse, similar to the von Karman row in hydrodynamics[11].

The physical role of the break of the Langmuir wake waves in the generation of the magnetic field is to extend the current repulsion mechanism mentioned in the previous section well behind the region occupied by the pulse. The current of the electrons accelerated in the forward direction outside the pulse by the large electric field at the break of the Langmuir waves and the return current repel each other and generate a quasistatic magnetic field. This mechanism has been analyzed in detail[8] in the frame-work of the nonlinear development[12] of the Weibel instability[13,14] driven by the "anisotropy" of the two counterstreaming electron populations. For relativistic plasmas, the characteristic inverse time of current separation, which is related to the growth rate of the Weibel instability, is of the order of the plasma frequency. This makes this mechanism of magnetic field generation occur quite fast on the time scale of the plasma and laser pulse dynamics.

Electron fluid vortices are also formed, due to the direct connection between plasma current and the electron velocity in a plasma for time scales short on the characteristic ion time. The dynamics of these vortices in the 2-D limit and the origin of their skew symmetric configuration is described[10,8,15] by a nonlinear equation of the Hasegawa-Mima[16] type which accounts for the conservation of the generalized vorticity $B - d_e^2 \nabla^2 B$ into the electron fluid. In particular the symmetry of the vortex row is interpreted in terms of its stability against tilting motions[15].

This magnetic wake may provide a mechanism for producing extremely strong magnetic fields in spatial domains with a transverse size of the order of several tens of microns. In these relativistic regimes, the time scale of the magnetic wake dissipation is expected to be determined by anomalous resistivity due to the interaction between the fast electrons and the ions in the plasma.

4 Transverse break of the Langmuir wake waves

Depending on the plasma regime and laser pulse parameters, the Langmuir wake waves behind a short relativistic pulse can either break longitudinally (1-D break) or transversely[17] (2-D, 3-D breaks). In both cases the Weibel mechanism leads to the generation of a quasi-steady magnetic field. However, the spatial structure of the field is different in the two cases. In particular 2-D and 3-D breaks produce spatial structures in the plasma density and magnetic field distribution that are strongly inhomogeneous in the direction transverse to the laser pulse propagation[18]. If transverse break occurs in the case of a long laser pulse, these inhomogeneities can affect the symmetry of the pulse propagation and create spots of high electromagnetic energy density.

The longitudinal break appears when the electron displacement ξ in the nonlinear Langmuir wake wave is of the order of the wake wave length. However, in the case of finite width pulses, transverse wave break can occur for much lower pulse amplitudes. Relativistic nonlinearities in the dielectric constant (the increase of the effective electron mass and the decrease of the electron density where the laser pulse amplitude is greatest) curve the front of the the Langmuir wake waves behind a finite width, short laser pulse. The constant phase surfaces assume a specific "horse-shoe"[19] (or "D-shape") structure and their curvature increases with the distance from the pulse. Additional causes of the transverse inhomogeneities in the dielectric constant that lead to the curving of the wave front are the plasma nonuniformity, if the pulse propagates in a channel, or the presence of a self generated quasistatic magnetic field vanishing at the pulse axis. Transverse wave break occurs when the curvature radius R decreases until it is comparable to the electron displacement ξ. This leads to self-intersection of the electron trajectories mostly in the direction transverse to the propagation of the laser pulse. The spatial structure of the plasma density and magnetic field in the case of transverse break can be understood in terms of the configurations obtained by shifting the D-shaped (locally parabolic) phase fronts with the nonlinear displacement ξ. The configurations of interest are those that are structurally stable with respect to small changes in the form of the phase fronts and of the nonlinear displacement, as suggested by catastrophe theory[20]. In Fig. 1 an explicit model evolution[17] of the transverse wave break structure in 2-D is given, showing the formation of a "swallow tail" configuration[20].

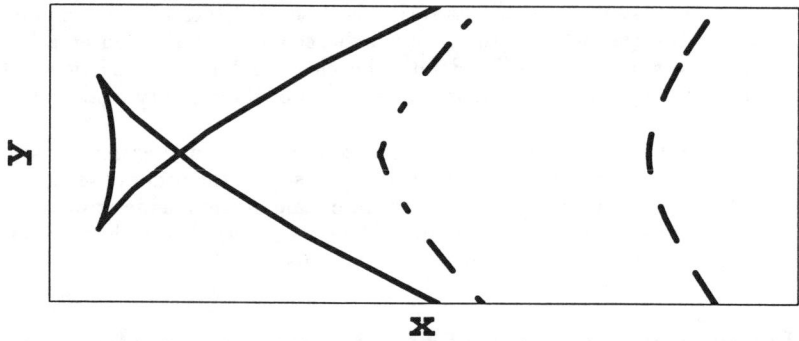

Fig.1 Constant phase fronts $y(x)$ for different values of ξ/R. The curve on the right, $\xi/R = 0.84$, resembles a parabola, the central curve is close to the transverse wavebreak condition and shows a sharp joint at $x = 0$, the curve on the left, $\xi/R = 2.1$, is known as the "swallow tail"

A swallow tail configuration in the electron density and magnetic field is apparent in the PIC simulation shown in Fig 2. A linearly polarized pulse with $a = 2.5$ propagates in an underdense plasma with $\omega/\omega_{pe} = 6$ along x. The pulse width (along y) and length (along x) are 8λ and 5λ. For such parameters the values of ξ, R and of the wavelength $2\pi/k_l$ of the Langmuir wake wave are of the same order. A phase space plot shows that electrons are accelerated both in the longitudinal and in the transverse directions. This provides an effective mechanism for injecting a relatively small fraction of electrons in the acceleration region of the wake Langmuir wave. The pattern of electron distribution (top frame) is related to the structure of the self-generated quasistatic magnetic field along the z-axis (middle frame).

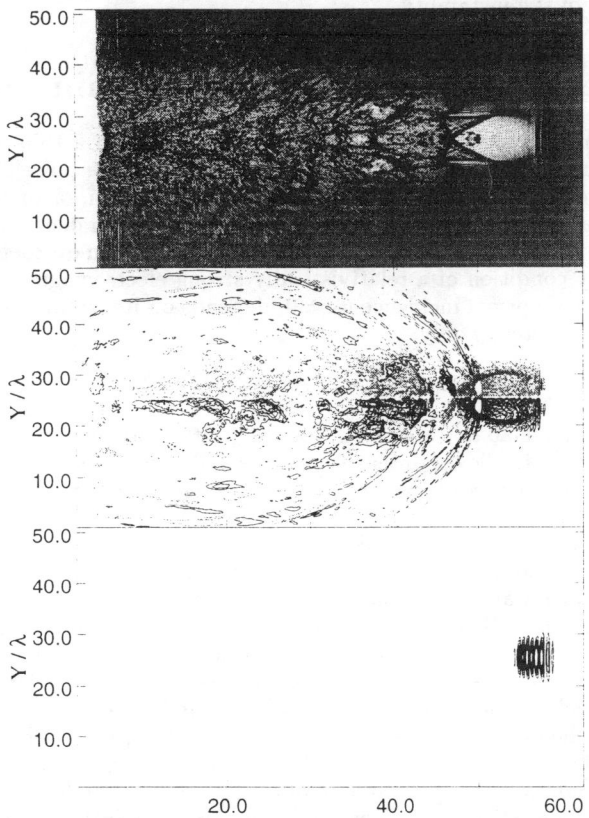

Fig.2 2-D structure of the break: distribution in the (x, y) plane of the electron density (top), of the quasistatic magnetic field (middle) and of the pulse electromagnetic energy density (bottom).

The field dimensionless amplitude is ≈ 0.1. We can identify several spatial regions with different mechanisms of magnetic field generation. Behind the laser pulse, between its rear edge and the Langmuir wave front, $50 < x < 58$, electrons are almost completely expelled and a regular dipole magnetic field, with size of the order of the Langmuir wave wavelength, is formed. It can be explained simply as arising from the moving positive charge due to the electrons expelled from this region. Further back, $x < 45$, $18 < y < 32$, a quasistatic magnetic field is excited near the axis as a result of the Weibel instability due to the fast electron beam accelerated in the longitudinal direction. The swallowtail in the magnetic field distribution, $x \approx 45$, $20 < y < 30$, is due to the freezing of the generalized vorticity in each electron beam. Outside both the laser pulse and the wake Langmuir wave, $x < 50$, $y < 20$, $y > 30$, a magnetic field with small scale structure is formed as a result of the development of the Weibel instability due to the multi-stream motion of the electrons thrown out of the wake in the transverse direction. The leading edge of this structure has the form of a parabola, with its vertex in the region where the break starts. The alternating polarity of the magnetic field in this region demonstrates that the main part of this field is not due to advection, but to the development of the instability.

5 Magnetic field generation on a thin foil

The nonlinear interaction of an ultraintense laser pulse with a thin foil modifies the shape, the frequency content and the polarization of the pulse[21,22,23]. A thin slab of overdense plasma, with thickness l smaller than, or of the order of, both the laser wavelength λ and the plasma collisionless skin depth d_e, exhibits features[24] that are not encountered either in underdense or in overdense plasmas. The transparency condition of a relativistically strong electromagnetic wave, e.g., in the simple case of normal incidence $\theta_0 = 0$, depends on its amplitude and on the dimensionless parameter ϵ_0 only:

$$a \gg \epsilon_0 = \omega_{pe}^2 l/(2\omega c) \equiv \lambda l/(4\pi d_e^2) \qquad (2)$$

An overdense plasma slab is transparent for the part of the pulse with sufficiently high intensity, while the lower intensity part is reflected This relativistic transparency leads to steepening of the front and rear edges of the pulse. Thus a thin foil provides a convenient method of producing steep laser pulses[21,25], which are useful for exciting strong regular wake waves in the Laser Wake Field Acceleration scheme. In addition, a thin foil may be an interesting model for studying the generation of a quasistatic magnetic field[3,26,9] on the surface of an overdense plasma with sharp boundaries. A dense foil, $\epsilon_0 \gg 1$, reflects the laser light almost completely. In a (semi-infinite) plasma with density much larger than the critical density, the light penetrates a distance small compared to its wavelength and the electromagnetic field is localized in the vicinity of the boundary. This narrow region can be modelled as a foil with thickness equal to the penetration depth. The advantage is that the analysis of the laser foil interaction can be simplified and reduced to the formulation of appropriate nonlinear boundary conditions for the field outside the foil. This 1-D model allows us to find simple solutions of problems related to relativistic transparency, laser pulse shaping, generation of high harmonics and of quasi-stationary magnetic fields. It is valid for an arbitrary incidence angle θ_0, since a Lorentz boost to a reference frame moving along the foil can be used to reduce the problem of oblique to that of normal incidence. The foil is assumed to be infinitely thin and the wave equation is written in the form

$$\partial_{t't'}\mathbf{a}' - \partial_{xx}\mathbf{a}' = \delta(x)\,\mathbf{j}'(\mathbf{a}'), \qquad (3)$$

where ' denotes, in the case of oblique incidence, quantities measured in the boosted frame, $\mathbf{j}'(\mathbf{a}')$ is the dimensionless electric current in the foil and is a nonlinear functional of the vector potential $\mathbf{a}'(0,t')$ at the foil. The delta function, models the current localization. Charge separation is neglected. The dimensionless magnetic field on the two sides of the foil is[25]

$$\mathbf{B}'(x,t') = \mathbf{B}'_0(x,t') + \mathbf{e}_x \times \mathbf{j}'(\mathbf{a}'(0,t'-|x|))\,\text{sign}(x)/2. \tag{4}$$

An analogous expression holds for the electric field. The subscript $_0$ denotes quantities of the incident pulse. The vector potential at the foil satisfies the ordinary differential equation

$$d\mathbf{a}'(0,t')/dt' - \mathbf{j}'(\mathbf{a}'(0,t'))/2 = d\mathbf{a}'_0(0,t')/dt'. \tag{5}$$

Equation (5) acts as a nonlinear boundary condition. Using the conservation of the y-z components of the canonical electron momentum in the boosted frame, we find [25] that the nonlinear electric current $\mathbf{j}'$ takes the form

$$\mathbf{j}'(\mathbf{a}') = -2\epsilon_0(1+\tan^2\theta_0)^{1/2}\delta(x)\left[\frac{-\tan\theta_0\mathbf{e}_y + \mathbf{a}'}{(1+(-\tan\theta_0\mathbf{e}_y + \mathbf{a}')^2)^{1/2}} + \sin\theta_0\mathbf{e}_y\right]. \tag{6}$$

From Eq.(4) the magnetic field in the wave can be written as

$$\mathbf{B}' = \mathbf{B}'_0 - \epsilon_0 \mathbf{e}_x \times \left[\frac{(-\tan\theta_0\mathbf{e}_y + \mathbf{a}'(0,t'-|x|))(1+\tan^2\theta_0)^{1/2}}{(1+(-\tan\theta_0\mathbf{e}_y + \mathbf{a}(0,t'-|x|))^2)^{1/2}} + \tan\theta_0\mathbf{e}_y\right]\text{sign}(x) \tag{7}$$

A quasistatic magnetic field arises at oblique incidence for both s and p-polarized pulses. It originates from the change of the electron velocity along y, i.e., from the first term inside the square brackets in Eq.(7), caused by the time variation of the amplitude of the incident pulse. The characteristic time scale of formation and decay of the quasistatic magnetic field is proportional to ϵ_0^{-1} and depends on the incidence angle θ_0. Its decay after the pulse has left the foil corresponds to the radiation of low frequency waves by the plasma surface.

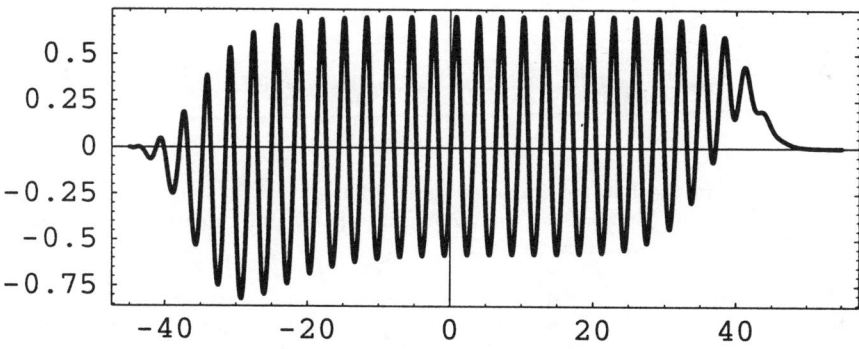

Fig.3 p-polarized component of the magnetic field at the foil surface versus time.

The dependence versus time of the magnetic field in the z direction (p-polarization) at the foil surface, as given by the above model, is shown in Fig.3 for an s-polarized incident pulse with $a = 6$, $\epsilon_0 = 3$, $\theta_0 = \pi/3$ The field is dominated by the 2nd harmonic of the incident pulse frequency. The quasi-steady magnetic field is negative at the beginning of the pulse and positive at the end. Its amplitude can be shown to increase with the incidence angle.

The generation of a quasistatic magnetic field on the foil surface is also apparent in 1-D and 2-D PIC simulations for both p and s polarized incident pulses. In all these simulations, the pulse amplitude is $a = 5$, its length is $l_{\|} = 25\lambda$ and the incidence angle $\theta_0 = \pi/2$. In the 2-D simulations the pulse width is $l_\perp = 10\lambda$. The foil thickness is $l = 0.375\lambda$ and its density corresponds to $\omega_{pe}/\omega = 1.8$. In terms of the above analytical model these parameters correspond to $\epsilon_0 = 3.82$. These simulations show relativistic transparency, steepening of the transmitted pulse, a change in the polarization of the transmitted and reflected pulses and the generation of high harmonics. The magnetic field along z (corresponding to the p-polarized component) is shown in Fig.4 (1-D simulations) for a p- and for an s-polarized incident pulse, and in Fig.5 (2-D simulations) for an s-polarized pulse.

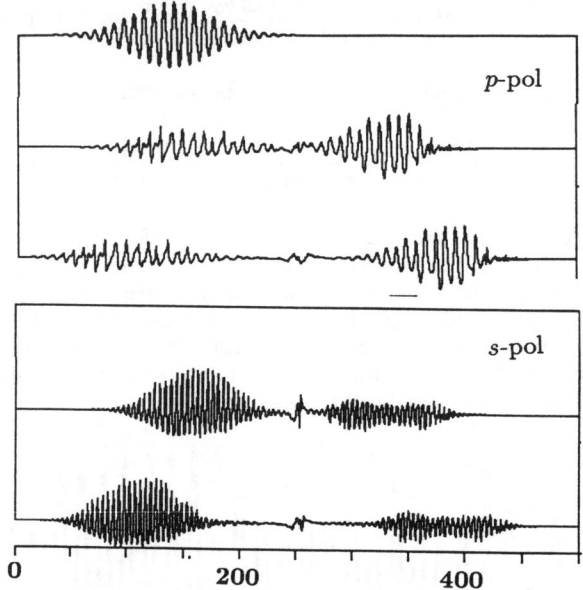

Fig.4 $B'_z(x)$ in the boosted frame at three different times, before and after the interaction with the foil (at $x = 250$), for a p-polarized (top) and for an s-polarized pulse (bottom). Lengths are normalized on $\lambda/(2\pi)$.

In the case of the s-polarized incident pulse, a B_z component is present in the reflected and transmitted pulses, indicating a polarization change. In addition a spatially structured, long lived, magnetic field at the foil is apparent for both polarizations. In the 2-D simulations (Fig 5, right frame) this field exhibits regions of different polarity with an approximate upside-down, left-right symmetry. The electromagnetic energy energy distribution in Fig.5, left frame, displays a "peeling" of the transmitted pulse due to relativistic transparency.

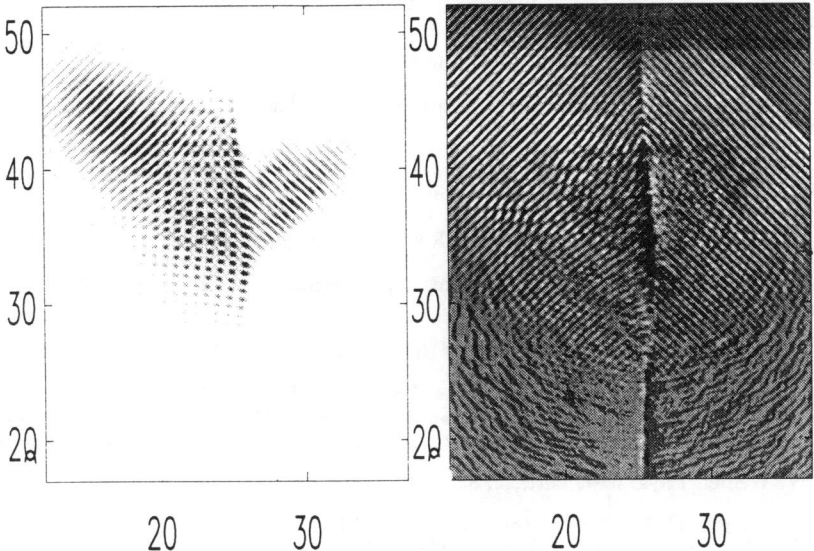

Fig.5 2-D spatial distribution in the laboratory frame of the electromagnetic energy density (left) and of the z-component of the magnetic field in the (x,y) plane. The foil is at $x = 25$. Lengths are normalized on λ.

6 Conclusions

Different mechanisms, active in different regimes and in different spatial regions, can generate quasistatic magnetic fields during the interaction of an ultra intense (relativistic) laser pulse with a plasma. The magnitude of these fields, their geometry and their formation and decay time constants have been discussed in the case of the magnetic vortices produced by longitudinal and/or transversal Langmuir wake wave breaks. The dynamical role of these fields on the laser pulse propagation has been stressed.

References

[1] D.W. Forslund, J.M. Kindel, W.B. Mori, C. Joshi, and J.M. Dawson, Phys. Rev. Lett., **54**, 558, (1985).

[2] J. Stamper, Laser and Particle Beams, **9**, 841, (1990).

[3] S.C. Wilks, W.C. Kruer, M. Tabak, and A.B. Langdon, Phys. Rev. Lett., **69**, 1383, (1992).

[4] R.N. Sudan, Phys. Rev. Lett., **70**, 3075, (1993).

[5] G.A. Askar'yan, S.V. Bulanov, F. Pegoraro, A.M. Pukhov, JETP Letters, **60**, 240, (1994); G.A. Askar'yan, S.V. Bulanov, F. Pegoraro, A.M. Pukhov, Comm. Plasma Physics Contr. Fusion, **17**, 35, (1995).

[6] A. R. Bell, Phys. Plasmas, **1**, 1643, (1994).

[7] G.A. Askar'yan, S.V. Bulanov, F. Pegoraro, A.M. Pukhov, Plasma Physics Reports, **21**, 835, (1995).

[8] S.V. Bulanov, T.Zh. Esirkepov, M. Lontano, F. Pegoraro, A.M. Pukhov, Physica Scripta, **T 63**, 280, (1996).

[9] A. Pukhov, J. Meyer-ter-Vehn, Phys. Rev. Lett. **76**, 3975, (1996).

[10] S.V. Bulanov, T.Zh. Esirkepov, M Lontano, F. Pegoraro, A.M. Pukhov, Phys. Rev. Lett., **76**, 3562, (1996).

[11] H. Lamb, *Hydrodynamics* (Cambridge University Press, 1975).

[12] F. Califano, F. Pegoraro, S.V. Bulanov, Phys. Rev.,**E 56**, 963, (1997); F. Califano, R. Prandi, F. Pegoraro and S.V. Bulanov, Non-linear Weibel instability in the interaction of an underdense plasma with a "relativistic" laser pulse, this conference.

[13] E.W. Weibel, Phys. Rev. Lett., **2**, 83, (1959).

[14] A. D. Steiger, C. H. Woods, Phys. Rev. A, **5**, 1467 (1971);
V. Yu. Bychenkov, V. P. Silin, V. T. Tikhonchuk, JETP, **98**, 1269, (1990).

[15] S.V. Bulanov, T. Zh. Esirkepov, M. Lontano, F. Pegoraro, Plasma Physics Reports **23**, 284, (1997).

[16] A. Hasegawa, K. Mima, Phys. Fluids,**21**, 87, (1978).

[17] S.V. Bulanov, F. Pegoraro, A.M. Pukhov, A.S. Sakharov, Phys. Rev. Lett., **78**, 4205, (1997).

[18] S.V. Bulanov, F.Pegoraro, and J.-I.Sakai, "Variety of Nonlinear Wave-Breaking", Joint ICFA/JAERI-Kansai International Workshop, Kyoto, July, (1997).

[19] S.V. Bulanov, F. Pegoraro, A.M. Pukhov, Phys. Rev. Lett., **74**, 710, (1995).

[20] V.I. Arnold, *The Theory of Singularities and its Applications* (Scuola Normale Superiore, Pisa 1991).

[21] S.V. Bulanov T.Zh. Esirkepov, N. Naumova, F. Pegoraro, I. Pogorelsky, A.M. Pukhov, IEEE Transaction on Plasma Science, **24**, 453, (1996).

[22] S.V.Bulanov, N.M.Naumova, F.Pegoraro, Phys. Plasmas, **1**, 745 ,(1994).

[23] R.Lichters, J. Meyer-ter-Vehn , A.M.Pukhov, Phys. Plasmas, **3**, 3425, (1996).

[24] J. Denavit, Phys. Rev. Lett., **69**, 3052, (1992).

[25] S.V. Bulanov, V.A. Vshivkov, G.I. Dudnikova, N.M. Naumova, F. Pegoraro, I.V. Pogorelsky, Plasma Physics Reports, **23**, 300, (1997).

[26] M. Tabak, Y. Hammer, M. E. Glinsky, W.L. Kruer, S.C. Wilks, J. Woodworth, E.M. Campbell, M.D. Perry, R.J. Mason, Phys. Plasmas, **1**, 1626, (1994).

Non-linear Weibel instability in the interaction of an underdense plasma with a "relativistic" laser pulse

F.Califano[1], R.Prandi[2], F.Pegoraro[2,3] and S.V.Bulanov[4]

[1] Dip. Astronomia, Università di Firenze, Italy and
Scuola Normale Superiore and INFM, Pisa, Italy
[2] Dip. Fisica Teorica, Università di Torino, Italy
[3] Dip. Fisica, Università di Pisa and INFM, Italy
[4] General Physics Institute, RAS, Moscow, Russia

Abstract.
Particle in cell numerical simulations of high-intensity laser pulses propagating in underdense plasmas show, in the wake of the laser pulse, a current layer structure characterized by one "fast" central current and two "slow" return currents, coupled to a quasi-static magnetic field. Using a relativistic two-fluid description, we show that the observed dynamics and the magnetic fields can be explained by the development of the "inhomogeneous" Weibel instability coupled to the two-stream instability.

Introduction

The study of the dynamics of a plasma interacting with ultra-short and ultra-intense laser pulses is a very recent and rich topic in which non-linear processes under relativistic conditions are not well understood or even explored. These conditions are realized, for example, in laboratory plasmas interacting with subpicosecond, multi-terawatt laser pulses up to 10^{19-21} W/cm^2 [1], and can also be of interest for high energy astrophysics [2,3].

One of the most interesting processes occurring in such conditions is the generation of quasi-static magnetic fields with important consequences for the plasma dynamics, for energy transport and for the propagation and focalization of the laser pulse itself (see [4] and references therein, these proceedings). These magnetic fields have been observed, for example, in particle in cell numerical simulations (PIC) in the wake of the laser pulse after the breaking of Langmuir plasma waves generated by the laser-pulse itself. Due to the wave-breaking, beams of accelerated electrons arise naturally in the plasma; then, magnetic fields are fed by the free energy stored in the electron anisotropy by means of the Weibel instability [5].

The Weibel instability is an electro-magnetic instability driven by the occurrence of temperature anisotropies or by the presence of counter-streaming electron beams with velocities comparable to the speed of light. In the laser-plasma case, the two electron beams are a "fast" one, directly generated by the breaking of plasma waves, and the opposite "slow" one carried by the cold component of the plasma electrons generated in order to maintain the plasma quasi-neutrality. These two currents are compenetrating and the densities of the beams are such that the total net current is zero. In these conditions, any small transversal disturbance displacing the two currents is reinforced since two opposite directed currents repel each other. As a consequence, a magnetic field grows exponentially in time with zero frequency. The characteristic "fast" time scale of this mechanism is the inverse of the electron plasma frequency, so that the ions can be assumed to be at rest. In the more general case of a 2D disturbance, the Weibel instability is coupled to the two-stream instability (see [6] for the dispersion relation) which remains the only one at play in the limit of an initial perturbation varying only in the direction of the electron beams.

In this paper, we study the dynamics of two inhomogeneous counter-streaming electron beams in the presence of a 1D/2D initial perturbation in the framework of a fluid-type description. This is motivated by the understanding of the physical mechanism responsible for the plasma dynamics observed in PIC simulations in the wake of the laser pulse where, at the end of the simulations, a central "fast" current together with two "slow" return currents are coupled to a quasi-static magnetic field.

The two-fluid electron equations

We assume the ions to be at rest providing a uniform neutralizing background and we normalize all quantities with a characteristic density $\bar{n}$, the speed of light c and the electron plasma frequency $\bar{\omega} = (4\pi \bar{n} e^2/m)^{1/2}$. Then, the dimensionless equations read:

$$\frac{\partial \mathbf{p}_a}{\partial t} = -\mathbf{v}_a \cdot \nabla \mathbf{p}_a - (\mathbf{E} + \mathbf{v}_a \times \mathbf{B}), \quad \frac{\partial n_a}{\partial t} = \nabla \cdot \mathbf{j}_a, \qquad (1)$$

$$\frac{\partial \mathbf{B}}{\partial t} = -\nabla \times \mathbf{E}, \quad \frac{\partial \mathbf{E}}{\partial t} = \nabla \times \mathbf{B} - \sum_a \mathbf{j}_a, \qquad (2)$$

where

$$\mathbf{v}_a = \frac{\mathbf{p}_a}{(1+p_a^2)^{1/2}}, \quad \mathbf{j}_a = -n_a \mathbf{v}_a, \quad a = 1, 2.$$

Notice that the normalized electron skin depth is equal to one.

At the initial time the two electron beams are directed in opposite directions along the x-axis and are concentrated in the central region around the $y = 0$ point,

$$\mathbf{v}_{0,1} = v_0 \cosh^{-2}(y/l) \mathbf{e}_x, \quad \mathbf{v}_{0,2} = -\mathbf{v}_{0,1} n_{0,1}/n_{0,2}, \tag{3}$$

where the subscript zero refers to zero order (equilibrium) quantities and l is the typical width of the beams. This equation models the two compenetrating currents, the current of fast electrons generated by the breaking of plasma waves and the return "cold" current produced by the plasma. For the sake of mathematical simplicity, we assume in the following that the initial densities are homogeneous and that $n_{0,1} + n_{0,2} = 1$. All results discussed in this paper are obtained for two initially non-symmetric beams with $n_{0,1} = 0.2 n_{0,2}$. We limit our analysis to the unidirectional magnetic field parallel to the z-axis $\mathbf{B} = (B_z)$ generated by the Weibel instability and to the fields $\mathbf{p}_a = (p_{0,a} + p_{a,x}, p_{a,y})$ and $\mathbf{E} = (E_x, E_y)$. At the initial time we introduce a "small" magnetic perturbation,

$$B_z = 10^{-3} R(x) e^{-y^2/\sigma} \sin(k_y y + \phi), \tag{4}$$

where $R(x)$ is a numerical random function in the range $(-\pi, \pi)$.

The linear dispersion relation of the Weibel instability for two homogeneous electron beams (i.e. $l \to \infty$) was presented in [6,7]. Here we recall that the Weibel mode (i.e. $R(x) = cost.$) has zero frequency and that its growth rate increases linearly for wavenumbers smaller than one and then saturates at $k_y \simeq 1$. In the relativistic limit, saturation occurs at slightly lower values of k_y, due to the relativistic increase of the effective electron skin depth. The growth rate is an increasing function of the beam velocity v_0.

1D evolution

To study the separation of the two initial currents produced by the Weibel instability, we have taken, as a first approach, $R(x) = 1$ and varied the most relevant physical parameters v_0, k_y and ϕ in Eqs. (3)-(4). The results can be summarized as follows.

After a rapid transient, longer for smaller values of k_y, the "resonant" Weibel mode (see [6] for a detailed description) is excited with a growth rate independent of the wavenumber and of the initial phase. In the non relativistic limit, at the end of the linear phase (see the first three frame at the left hand side of Fig. 1), the structure of the total current (which was zero at $t = 0$) depends on the phase of the initial perturbation. However, as soon as the non-linear phase starts producing larger and larger spikes due to the presence of singularities (see [8]), we observe that the current system becomes practically independent of the phase and is characterized by a central "fast" current with two "slow" return currents on both sides (see the first three frame at the right hand side of figure 1).

In the relativistic case, $0.95 \leq v_0 \leq 0.995$, the current structure is independent of the phase of the initial perturbation. In the last two frames of Fig. 1, we show as an example the total current at the end of the linear phase (left hand side) and during the beginning of the non-linear phase (right hand side) for $v_0 = 0.995$. From this figure it is evident that in the relativistic limit the currents system is different from that observed in the non-relativistic one, showing a double current layer structure each layer very similar to that observed in the non-relativistic case.

2D evolution

When the initial perturbation is 2D, the Weibel instability is coupled to the two-stream instability which amplifies the electric field parallel to the electron beams. In the homogeneous symmetric case (i.e. $n_1 = n_2$), the dispersion relation of the two-stream instability shows that only the modes for which $(k_x v_0)^2 < 1/\Gamma^3$ are unstable (here Γ is the Lorentz factor). This is approxi-

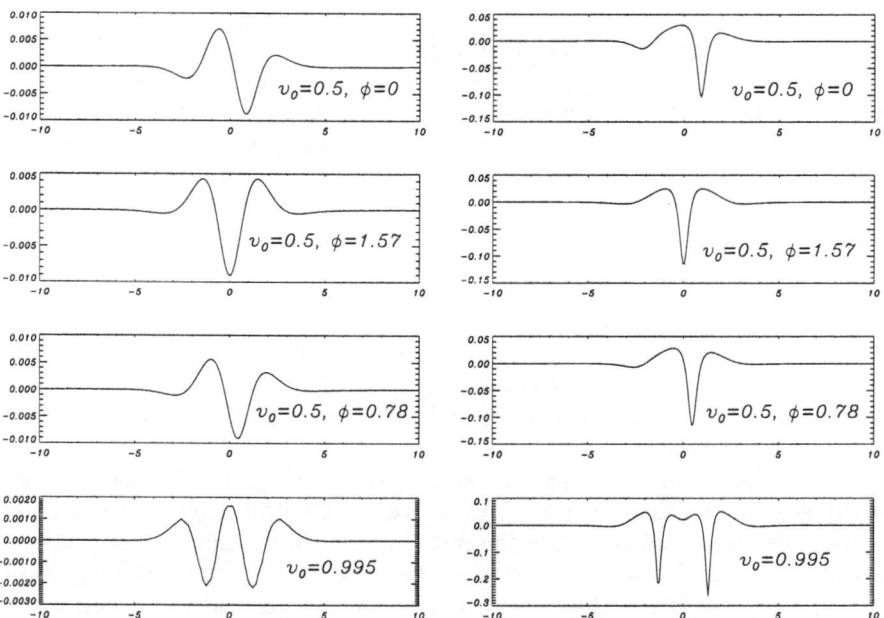

FIGURE 1. The total current in the x-direction at the end of the linear phase (left hand side) and during the beginning of the non-linear phase (right hand side).

mately valid in the presence of two strongly inhomogeneous electron beams. Therefore, in the relativistic case only the largest wavelengths are unstable. Here we give a very brief summary of the 2D results.

In the non-relativistic regime, $v_0 = 0.5$, there is a competition between the Weibel instability, which tends to separate the currents in the transversal y-direction, and the two-stream instability which tends to stretch the currents in the longitudinal x-direction. Since the electron streams are inhomogeneous, the total current as well as the other physical quantities are characterized in the (x, y) plane by an arrow-like structure as shown in figure 2. This is no longer the case in the relativistic regime, $0.95 \leq v_0 \leq 0.995$, in which the Weibel instability dominates more and more; as a consequence the fields are practically homogeneous in the x-direction. Finally, as observed in the 1D case, the current system is characterized by a double current layer structure.

Conclusions

Using a relativistic two-fluid approach, we have studied the development of the Weibel and two-stream instabilities for two initial non-symmetric inhomogeneous counter-streaming electron beams. In the non relativistic case, the

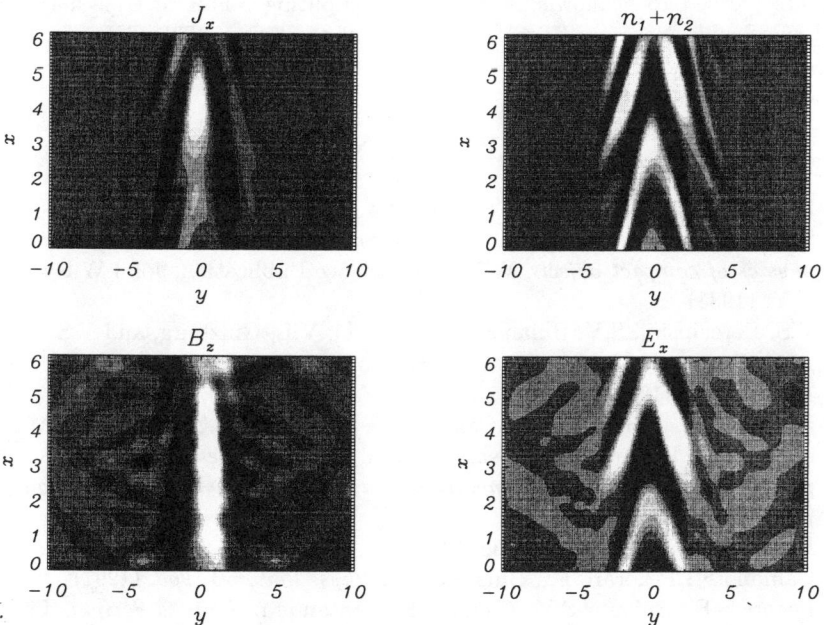

FIGURE 2. The shaded isocontours of the total current, the total density, the magnetic field and the electric field parallel to the stream direction.

resulting current structure is characterized by a central "fast" current and two "slow" return currents with an arrow-like shape in the stream direction. On the other hand, in the relativistic case the currents are practically homogeneous in the stream direction and a double current layer is formed even during the linear stage. A quasi-static magnetic field is observed both in the non-relativistic and relativistic regimes. These results support the concludion that the current structure and magnetic field observed in the wake of a "strong" laser-pulse impinging on a underdense plasma is explained by the development of a Weibel-type instability.

It is worth noticing that our simulations cannot follow the full non-linear evolution of the Weibel instability due to the formation of singularities where all quantities are characterized by the presence of larger and larger spikes, as shown in figure 1 (see [8] for a detailed discussion). In fact, as soon as the singularities come into play, smaller and smaller scales are generated on a characteristic (dimensionless) time scale $t_{nl} \sim 1$ and the fluid approximation becomes meaningless. An extension of the present work in the non-linear stage after the formation of the singularities is already in progress by numerical integration of the Vlasov-Maxwell equations.

Acknowledgement

We are pleased to acknowledge the supercomputing center of Cineca (Bologna) and the Scuola Normale (Pisa) for the use of their Cray T3D-T3E and CM200, respectively.

REFERENCES

1. Perry M.D. and Mourou G., *Science* **264**, 917, (1994)
2. S.L. Shapiro, S.A. Teukolsky, *Black Holes, white dwarfs and neutron stars. The physics of compact objects*, Wiley-Interscience Publication, John Wiley & Sons, N.Y. (1983)
3. V.S. Berezinskij, S.V. Bulanov, V.A. Dogiel, V.L. Ginzburg, and V.S. Ptuskin, *Astrophysics of Cosmic Rays*, North Holland Publ. Co., Elsevier Sci. Publ., Amsterdam, (1990).
4. Pegoraro F., Bulanov S.V., Califano F., Esirkepov T.Zh., Lontano M., Naumova N.M., Pukhov A.M., Vshivkov V.A, *Ultra intense magnetic fields in laser plasma interaction: their generation and influence on light propagation*, these proceedings
5. E. Weibel *Phis.Rev.Lett* **2**, 83, (1959).
6. Califano F., Pegoraro F.,, Bulanov S.V. *Phys. Rev.*, **56**, *963*, (1997)
7. Pegoraro F., Bulanov S.V., Califano F., Lontano M., *Physica Scripta*, **T63**, 262, (1996)
8. Califano F., Pegoraro F.,, Bulanov S.V., Mangeney A, *Phys. Rev., submitted*, (1997)

Relativistic Effects in Laser-Atom Interactions at Ultra-High Intensities

N.J. Kylstra, A.E. Ermolaev and C.J. Joachain

Physique Théorique, Faculté des Sciences CP 227,
Université Libre de Bruxelles, B-1050 Bruxelles, Belgium

Abstract. We discuss some of the results of our investigations of relativistic effects in laser-atom interactions at ultra-high intensities. We have solved numerically the time-dependent Dirac equation for a model, one-dimensional atom which is subjected to an ultra-intense, high-frequency laser field. Our method of computation consists of utilizing a B-spline expansion for the wavefunction in momentum space. We demonstrate that for a peak electric field strength of 175 atomic units (a.u.) and for an angular frequency of 1 a.u., relativistic effects are apparent. Even under these extreme conditions the wavefunction remains localized in a superposition of field-free bound states and very-low energy continuum states. Comparing our results with the numerical solution of the time-dependent Schrödinger equation, we find that the Dirac wavefunction is slightly more stable against ionization. It is shown that the relativistic quiver motion of the electron wavepacket differs substantially from the nonrelativistic quiver motion and that the energy distribution of the ionized electrons is strongly concentrated near threshold.

INTRODUCTION

Various aspects of the stabilization of atoms (i.e. the increasing lifetime of atoms as a function of increasing laser intensity) in high frequency, high intensity laser fields have been studied theoretically within the Floquet framework (see e.g. Gavrila 1992) and with time-dependent calculations (see e.g. Kulander *et al* 1991, Latinne *et al* 1994). These studies have been performed in the dipole approximation using the non-relativistic quantum theory. At sufficiently high intensities, a number of interrelated issues arise concerning the validity of these approaches, and hence the degree of stabilization of atoms. These include the modification of the electron's quiver motion by the magnetic field component and retardation effects, both of which are not present in the dipole approximation, relativistic effects which involve the dressing of the mass of the electron due to its relativistic motion in the laser field, the coupling of the field-free positive and negative energy states by the field and spin effects.

Relativistic effects become important when the ratio of the ponderomotive energy U_p (i.e. the nonrelativistic cycle-averaged quiver energy) of the electron in the field to the electron's rest mass energy becomes comparable to unity. In atomic units (a.u.), which we use, this ratio is $q = U_p/c^2 = \mathcal{E}_0^2/4\omega^2 c^2$, where $\mathcal{E}_0$ is the peak strength of the electric field component, ω is the angular frequency of the laser light and c is the velocity of light.

We discuss recent results of our calculations of the time evolution of a one-dimensional model Dirac atom in an intense, high frequency laser field (Kylstra et al 1997). Magnetic field and retardation effects are not included in a one-dimensional model since their description requires an additional spatial dimension. Therefore, we are concerned with relativistic effects due to the dressing of the electron mass and the influence of the negative energy states.

METHOD AND RESULTS

In momentum space, the model, one-dimensional Dirac equation we are considering is given by

$$i\frac{\partial}{\partial t}\begin{bmatrix}\hat{\psi}(p,t)\\ \hat{\eta}(p,t)\end{bmatrix} = \begin{bmatrix} 0 & cp+A(t) \\ cp+A(t) & -2c^2 \end{bmatrix}\begin{bmatrix}\hat{\psi}(p,t)\\ \hat{\eta}(p,t)\end{bmatrix} + \int dp'\begin{bmatrix}\hat{V}(p-p')\hat{\psi}(p',t)\\ \hat{V}(p-p')\hat{\eta}(p',t)\end{bmatrix}. \tag{1}$$

We have chosen the potential in configuration space $V(x)$ to be

$$V(x) = -\frac{\exp(-\epsilon|x|) - \exp(-|x|)}{|x|}, \tag{2}$$

with $\epsilon = 10^{-3}$. The laser field is described classically.

We obtain a numerical solution of the time-dependent Dirac equation by expanding the wavefunction in terms of a B-spline basis. The time-dependent Dirac equation is thereby reduced to a set of ordinary differential equations for the time-dependent expansion coefficients.

In figure 1 we show the probability density in momentum space after each cycle of the four cycle turn-on. The plots on the left correspond to the solution of the Dirac equation while the plots on the right are for the Schrödinger equation. The maximum electric field strength is $\mathcal{E}_0 = 175$ a.u. and the angular frequency is $\omega = 1$ a.u., giving $q = 0.4$, and we have used a four cycle $\sin^2$ turn-on. Striking peaks are seen in the probability density in each of the plots. These peaks are associated with the essentially classical motion of the

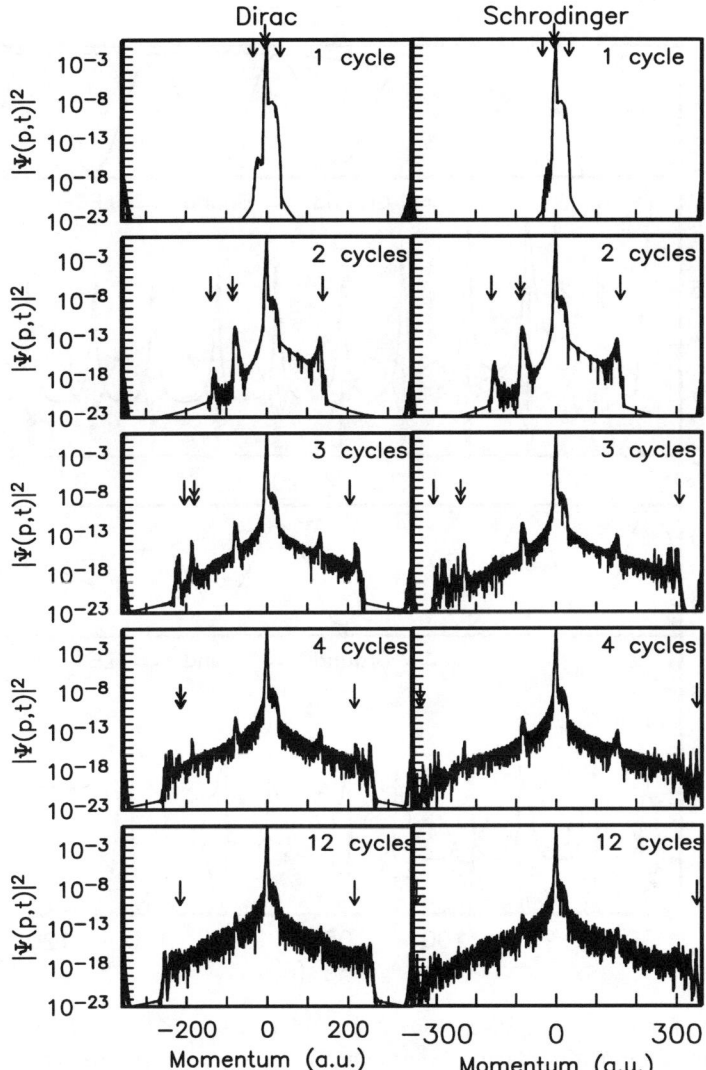

FIGURE 1. The probability density in momentum space is shown after each cycle of the four cycle turn-on and after 12 cycles. The angular frequency is 1 a.u. and the maximum electric field strength is 175 a.u. The left column shows the Dirac density while the right column shows the Schrödinger density. After each cycle, three new peaks are seen in the probability densities. These are indicated by arrows.

electron in the laser field and correspond to the maximum classical momentum which the electron can attain during each half-cycle of the turn-on of the pulse.

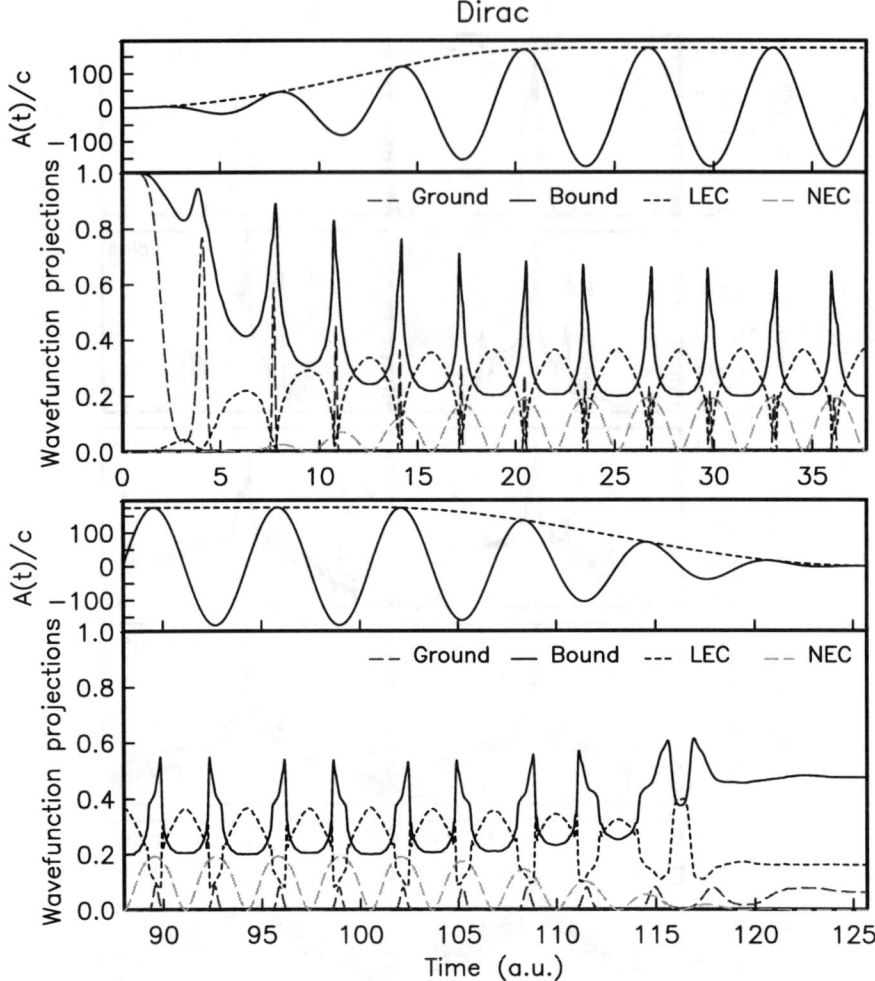

FIGURE 2. The time dependent vector potential divided by c of the laser pulse and the corresponding time evolution of the Dirac wavefunction. Shown are: the magnitude squared of the projection of the total wavefunction onto the ground state (the dashed lines), onto all the bound states (solid lines), onto the low energy continuum (LEC) states (dotted lines), i.e. states having energy less that 0.06 a.u., and onto the negative energy continuum (NEC) states (grey dashed lines).

After each cycle, a double arrow indicates the maximum momentum after the first half of each cycle, while the two single arrows indicate the maximum momentum after the second half of each cycle. After each half cycle, there are two peaks: one associated with the electron ejection in the direction of

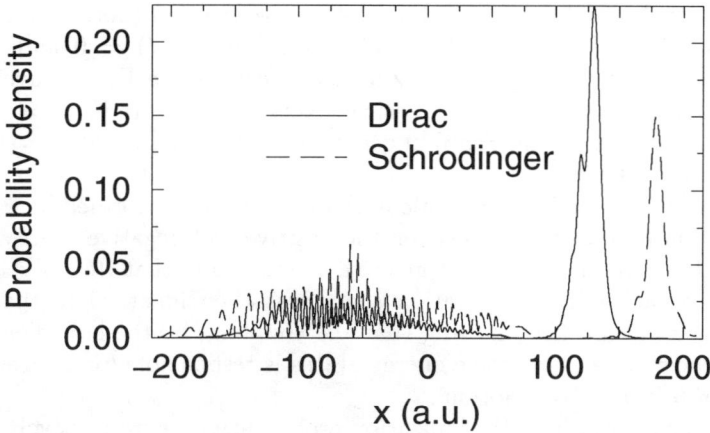

FIGURE 3. The Dirac (solid line) and Schrödinger (dotted line) probability densities $|\Psi(x,t)|^2$ in configuration space. Shown are the densities at the end of the 9th cycle. The peaks correspond to, respectively, the relativistic and nonrelativistic classical excursion amplitudes.

the gradient of the electric field potential and the other with electron ejection in the opposite direction (which is the preferred direction). The second peak corresponding to the first half of each cycle is not visible since it is 'washed out' in the second half of the cycle. Additionally, we see that: i) the difference in the position of the peaks in the Dirac and Schrödinger probability density is a clear manifestation of the relativistic mass shift which limits the momentum of the electron in the field, ii) the magnitude of the peaks decreases with the increasing electric field strength during the pulse turn-on, iii) the peaks, which are associated with the high energy components of the the wavefunction, are many orders of magnitude below the maximum wavefunction density at $p = 0$ and iv) slightly past the maximum allowed classical momentum, the magnitude squared of the wavefunction drops abruptly.

In figure 2 the time evolution of the Dirac wavefunction for $\omega = 1$ a.u. is monitored by showing the modulus squared of the projection of the total wavefunction onto the ground state (the dashed lines), onto all the bound states (solid lines), onto the continuum states having energy less that 0.06 a.u. (dotted lines) and onto the field-free negative energy states (grey dashed line). It is seen that the wavefunctions are composed, for the most part, of a superposition of the field-free bound states and very low energy continuum states, which implies the existence of a localized wavepacket. Indeed, during e.g. the 12th cycle, 99 percent of the momentum probability distribution of the wavefunction is found within the interval of -1.7 a.u. $< p < 1.7$ a.u. We also find a sizable negative energy component in the Dirac wavefunction.

In figure 3 we show the magnitude squared of the configuration space Dirac wavefunction (solid line) and Schrödinger wavefunction (dotted line) at the end of the 9th laser cycle. At this point in time $A(t) = 0$ and the corresponding electric field is maximum. The peak in the Dirac wavefunction corresponds to the relativistic classical excursion amplitude, $x = 124$ a.u. Likewise, the peak in the Schrödinger wavefunction occurs at $x = 175$ a.u., the nonrelativistic excursion amplitude. We find that there is no clear dichotomy of the wavefunction. Calculations were also performed in which the laser field-induced coupling between the positive and negative energy components of the wavefunction was neglected. It was found that the quiver motion of the wavepacket in configuration space was then identical to that of the Schrödinger wavepacket, showing that the coupling, by the laser field, of the field-free positive and negative energy states is responsible for the relativistic correction to the quiver motion.

At the end of the pulse, the ionization probabilities are, respectively, 0.52 for the Dirac wavefunction and 0.58 for the Schrödinger wavefunction, indicating that the Dirac wavefunction appears to be more stable against ionization. This is in agreement with the results of Protopapas *et al* (1996) who found that the relativistic Schrödinger wavefunction is more stable against ionization than the nonrelativistic Schrödinger wavefunction. In accordance with the discussion above, we find that the overwhelming majority of the electrons are emitted with very low energies. If the population of the negative energy states, after the laser pulse has been turned off, is to be interpreted as the pair-production probability, our calculated value for this probability is 6×10^{-15}. For the present laser parameters pair-production is highly improbable.

This work has been supported by the European Commission HCM Programme under contracts ERB CHRXCT940470 and ERB CHBGCT940552. Support from the Belgian Institut Interuniversitaire des Sciences Nucléaires is also gratefully acknowledged.

REFERENCES

1. N J Kylstra, A M Ermolaev and C J Joachain 1997. *J. Phys.* B **30** L449.
2. M Gavrila 1992, *Atoms in Intense Laser Fields, Adv. At. Mol. Opt. Phys., Suppl. 1* (Academic Press: San Diego) p 435.
3. K Kulander, K J Shafer and J L Krause 1991. *Phys. Rev. Lett.* **66** 2601.
4. O Latinne, C J Joachain and M Dörr 1994. *Europhys. Lett.* **26** 333.
5. M Protopapas, C H Keitel and P L Knight 1996. *J. Phys.* B **29** L591.

Formation of ionization fronts by intense short laser pulses propagating in a neutral gas

C.C.Rosa, L.Oliveira e Silva, N.Lopes, and J.T.Mendonça

GoLP/Grupo de Lasers e Plasmas, Centro de Física de Plasmas
1096 Lisboa Codex, Portugal

Abstract. Propagation of short laser pulses in gases resulting in relativistic ionization fronts is studied numerically using a kinetic formulation based in the photon number phase-space distribution function. With this approach we are able to follow the dynamics of the laser pulse both in time and spectral content. The advance of the photon number is obtained by solving a Klimontovich type equation. The properties of the emergent laser pulse, responsible for the ionization front, such as duration, chirp and spectrum are continuously monitored by adequate diagnostics of the photon number phase-space distribution. In particular, a detailed analysis of the evolution of the laser pulse velocity is presented.

INTRODUCTION

The propagation of short intense laser pulses is a central problem in Plasma Physics. The superstrong fields, delivered by lasers based on the chirped pulse amplification concept [1], generate a new class of nonlinear phenomena in gases and plasmas. Among these, laser wakefield generation for new electron and photon accelerators is a subject of paramount importance [2]. Photon acceleration, or the frequency upshift of electromagnetic radiation by plasma waves or ionization fronts [3], assumes an important role in this area not only as tunable source of radiation but also as a diagnostic tool for plasma accelerators.

Recent experimental results demonstrated the photon acceleration concept by ionization fronts [4]. A strong dependence on the physical parameters of the ionization front (velocity, maximum electron density, transverse profile) was found. A clear description of the ionization front dynamics (formation and propagation) is, therefore, a crucial point.

In this paper, the propagation of a short intense laser pulse in a backgroung gas is studied. For such pulses, the background gas is ionized and the pulse propagates in a gas-plasma interface, the so called ionization front. Due to the distinct dielectric features of the gas and the plasma, the ionizing laser pulse will be distorted and subsequently, the profile and the velocity of the ionization front will also denote a time dependence. Our analysis is based on a kinetic description of the laser pulse. The photon number, describing the phase-space distribution of the pulse, evolves in time according to a K-limontovich type equation, which is coupled to the time evolution equations of the ionic species of the background gas/plasma through the laser electric field strength and the dispersion relation of the electromagnetic waves in a plasma. We first introduce the most important features of the photon kinetics. The ionization model is briefly discussed as well as a non self-consistent calculation of the ionization front profile. We then present the numerical results of our model. A special emphasis is given to the time evolution of the spectral features of the laser pulse and the ionization front propagation, namely the time dependent electron density profile and the front velocity. A brief discussion of the implication of these results is given in the conclusions.

PHOTON KINETICS

The concept of the number of photons has been used in Plasma Physics since the 60's [5]. This definition is accurate for plane waves but it is not valid for laser pulses. Recent work [6] shows, however, that this concept can be extended in order to describe more general electromagnetic fields such as short laser pulses. In this case the number of photons is given by:

$$\mathcal{N}(\boldsymbol{k},\boldsymbol{r},t) = \frac{\epsilon_0}{8\hbar}\left(\frac{\partial D}{\partial \omega}\right)_{\omega_0} \mathcal{F}(\boldsymbol{k},\boldsymbol{r},t) \tag{1}$$

where $\mathcal{F}(\boldsymbol{k},\boldsymbol{r},t)$ obeys

$$\mathcal{F}(\boldsymbol{k},\boldsymbol{r},t) = \int d\boldsymbol{s}\, \boldsymbol{E}(\boldsymbol{r}-\boldsymbol{s}/2,t)\cdot \boldsymbol{E}^*(\boldsymbol{r}+\boldsymbol{s}/2,t)\exp(i\boldsymbol{k}\cdot\boldsymbol{s}) \tag{2}$$

The number of photons can be regarded as a distribution function of quasi-particles, the photons, in phase space $(\boldsymbol{k},\boldsymbol{r})$. The $\boldsymbol{k}$-dependence describes the fast time scale (field fast phase) and the $\boldsymbol{r}$-dependence the evolution of the slow time scale (electromagnetic field envelope). The number of photons $\mathcal{N}$ is a conserved quantity in phase space, up to corrections of the order of $\alpha^2 = \omega_p^2/\omega_0^2$ [6], where $\omega_p \propto \sqrt{n_e}$ is the electron plasma frequency, ω_0 is the laser central frequency, and n_e is the electron density. Therefore, in phase space, the number of photons evolves according to:

$$\frac{d\mathcal{N}}{dt} = \frac{\partial \mathcal{N}}{\partial t} + \frac{d\boldsymbol{k}}{dt}\cdot\frac{\partial \mathcal{N}}{\partial \boldsymbol{k}} + \frac{d\boldsymbol{r}}{dt}\cdot\frac{\partial \mathcal{N}}{\partial \boldsymbol{r}} = \frac{\partial \mathcal{N}}{\partial t} - \frac{\partial \omega}{\partial \boldsymbol{r}}\cdot\frac{\partial \mathcal{N}}{\partial \boldsymbol{k}} + \frac{\partial \omega}{\partial \boldsymbol{k}}\cdot\frac{\partial \mathcal{N}}{\partial \boldsymbol{r}} = 0 \tag{3}$$

where we have discarded the contributions of higher order in α^2, and we replaced the equations of motion for the phase space coordinates $\boldsymbol{r}$ and $\boldsymbol{k}$ by the corresponding eikonal equations obtained from the linear dispersion relation $\omega = \omega(\boldsymbol{k}, \boldsymbol{r}, t) = \sqrt{k^2 c^2 + \omega_p^2(\boldsymbol{k}, \boldsymbol{r}, t)}$. Coupling with the gas/plasma occurs through changes of ω_p^2 due to, for instance, ionization of the background gas or collective plasma oscillations. By calculating the several moments of this evolution equation, a set of the fluid equations describing the photon gas can also be derived [7]. Furthermore, the average of any physical observable, f, can be written as:

$$< f > = \int \frac{d\boldsymbol{k}}{(2\pi)^3} d\boldsymbol{r} \, f \mathcal{N} \qquad (4)$$

where the integral is taken over all phase space. Another important feature of $\mathcal{N}$, as given by eq.(1), which is closely related to its formal equivalence to the Wigner function, is the physical meaning of the following marginals:

$$\int \frac{d\boldsymbol{r}}{d\boldsymbol{k}} \, \mathcal{N}(\boldsymbol{k}, \boldsymbol{r}, t) = \frac{\epsilon_0}{8\pi} \left(\frac{\partial D}{\partial \omega} \right)_{\omega_0} |E(\begin{smallmatrix} \boldsymbol{k} \\ \boldsymbol{r} \end{smallmatrix}, t)|^2 \qquad (5)$$

giving the spectral intensity and the spatial intensity of the electromagnetic field described by $\mathcal{N}$.

IONIZATION MODEL

As it propagates in the gas, an intense laser pulse ionizes the gas, leading to a change in the refractive index of the laser pulse. In fact, the refractive index can be written as $n = \sqrt{1 - \omega_p^2/\omega_0^2}$ thus meaning that an increase in the electron density, due to photoionization, will lead to a decrease of n. This evolution will, in turn, change the characteristics of the ionizing laser propagation.

The photoionization process for very strong fields is commonly described by the tunneling ionization theory of Ammosov et al [8]. If $n_j(\boldsymbol{r}, t)$ is the number density of the charge state j at time t and position $\boldsymbol{r}$, the evolution of $n_j(\boldsymbol{r}, t)$ is given by a set of $Z_{max} + 1$ first-order ODE's:

$$\dot{n}_0(\boldsymbol{r}, t) = -W_1(\boldsymbol{r}, t) n_0(\boldsymbol{r}, t) \qquad (6)$$
$$\dot{n}_j(\boldsymbol{r}, t) = W_j(\boldsymbol{r}, t) n_{j-1}(\boldsymbol{r}, t) - W_{j+1}(\boldsymbol{r}, t) n_j(\boldsymbol{r}, t) \qquad (7)$$
$$\dot{n}_{Z_{max}}(\boldsymbol{r}, t) = W_{Z_{max}}(\boldsymbol{r}, t) n_{Z_{max}-1}(\boldsymbol{r}, t) \qquad (8)$$

where $W_j(\boldsymbol{r}, t)$ is the ionization rate for the production of charge state j, as given by the theory of Ammosov et al and Z_{max} is the maximum charge state of the background gas. On one hand, we have the evolution equation (3) for

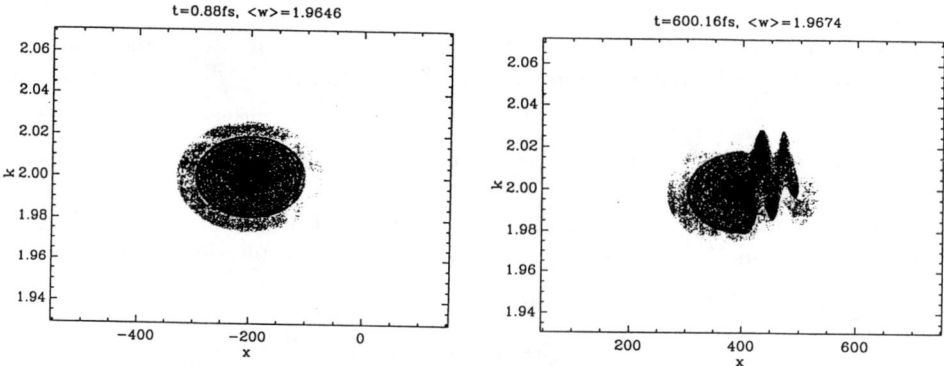

FIGURE 1. Photon number phase space distribution.

$\mathcal{N}$ (electromagnetic field) and, on the other hand, the set of equations describing the optical-field ionization (background gas/plasma). The ionization rate depends on the electric field [8]; this dependence establishes the coupling with the kinetic equation for $\mathcal{N}$. The linear dispersion relation, implicitly present in eq.(3), depends on the electron density and, therefore, it is coupled with the set of eqns.(6-8), since n_e verifies $n_e = \sum_{j=0}^{j=Z_{max}} j n_j$. With eqns.(3,6-8) we achieve a self-consistent picture of the laser pulse propagation when ionization of the background media is included.

NUMERICAL RESULTS AND DISCUSION

We have numerically solved the set of eqns.(3,6-8) for a short laser pulse propagating in a background gas region. The number of photons evolution equation is solved by the Lax-Wendroff algorithm for initial value problems in PDF's. Eqns.(6-8) were solved using the fourth order Runge-Kutta method. For simplicity, 1D laser pulse propagation was considered. The neutral gas density grows from zero (vacuum) to a maximum value $n_{0_{max}}$ according to $n_0(x) = 0.5 n_{0_{max}}(1 + \tanh(x/L_{gas}))$, where L_{gas} is the typical length scale of the neutral density gradient.

In our simulations, we have considered a laser pulse of 100 fs, $\lambda_0 = 800 nm$, propagating in Argon. By considering Argon, we are able to look for signatures of the several ionization steps of this gas. The gas has its half maximum density at $t = 200 fs$.

This can be seen in fig.1 (a and b) where we plot the photon number distribution for inital time and $t = 600 fs$. The initial photon number presents a gaussian 2-D phase-space distribution (fig 1-a). When the laser pulse penetrates the gas producing its ionization the photon number distribution evolves to a structure with blue-shifted peaks equal to the number of present ioniza-

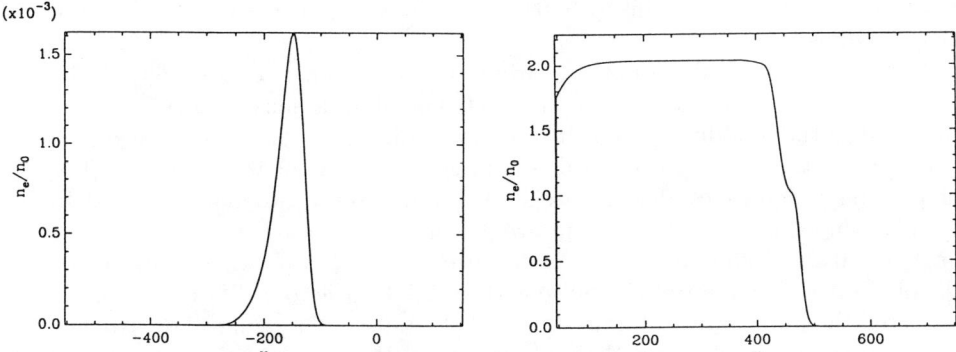

FIGURE 2. 1-D Ionization front shape.

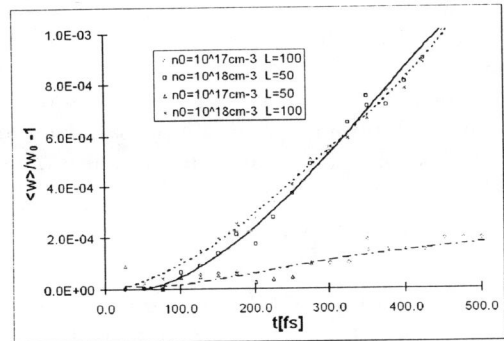

FIGURE 3. Up-shift.

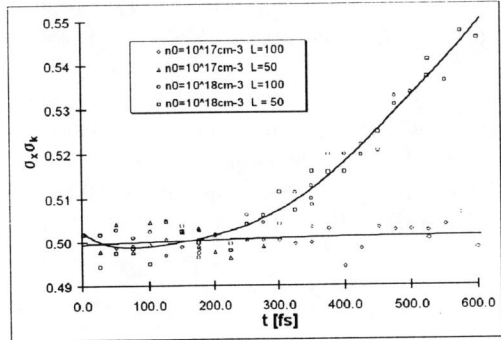

FIGURE 4. Pulse enlargement.

tion states. As we can see in fig. 1-b, the Argon was ionized twice as expected at this intensity.

The 1-D shape of the ionization front (moving from left to right) can be seen in fig.2 (a, b) where we plot the electronic plasma density normalized to the local initial gas density. We can observe (fig. 2-b) the appearing of the two steps in the moving front which is in good agreement with the fact that photon ionization takes place in two steps leading to the two peaks of fig. 1-b.

Pulse blue-shift and enlargement are presented respectively in figs. 3 and 4 for four runs at different initial gas densities and L_{gas} values. As we can see the qualitative behavior of the pulse is the same for the four diferent runs. As expected the pulse centroid blue-shift and the pulse enlargement are stronger for a bigger plasma density. We had also found that ionization front velocity goes assymptoticaly to the laser pulse group velocity in the plasma.

CONCLUSIONS

In conclusion we have presented a new laser pulse propagation-interaction simulation technique suitable for studies of laser pulse propagation in underdense regime. Future inclusion of ionization energy losses and plasma movements will make this technique a powerful and competitive tool for the simulation of a very important class of laser-plasma interaction problems.

REFERENCES

1. D.Strickland and G.Mourou, Opt.Comm. **56**, 219 (1985).
2. T.Tajima and J.M.Dawson, Phys.Rev.Lett. **43**, 267 (1979); L.M.Gorbunov and V.I.Kirsanov, Sov.Phys. JETP **66**, 290 (1987).
3. S.C.Wilks, J.M.Dawson, W.B.Mori *et al.* Phys.Rev.Lett. **62**, 2600 (1989); M.Lampe, E.Ott, J.H.Walker, Phys.Fluids **21**, 42 (1978); J.T.Mendonça, J.Plasma Phys. **22**, 15 (1978).
4. J.M.Dias, C.Stenz, N.Lopes *et al.* Phys.Rev.Lett. **78**, 4773 (1997).
5. R.Z.Sagdeev and A.A.Galeev, *Nonlinear Plasma Theory*, (W.A.Benjamin Inc., New York, 1969).
6. L.Oliveira e Silva and J.T.Mendonça, submitted for publication (1997); L.Oliveira e Silva, PhD. Thesis, IST/Lisbon, 1997 (unpublished).
7. N.L.Tsintsadze and J.T.Mendonça, to be submitted (1997).
8. M.V.Ammosov, N.B.Delone, and V.P.Krainov, Sov.Phys. JETP **64**, 1191 (1986); B.M.Penetrante and J.N.Bardsley, Phys.Rev. A **43**, 3100 (1991).

LPIC++ a Parallel One-dimensional Relativistic Electromagnetic Particle-In-Cell Code for Simulating Laser-Plasma-Interaction

R.E.W. Pfund, R. Lichters, and J. Meyer-ter-Vehn

Max-Planck-Institut für Quantenoptik, Hans-Kopfermann-Strasse 1, D-85748 Garching

Abstract. We report on a recently developed electromagnetic relativistic 1D3V (one spatial, three velocity dimensions) Particle-In-Cell code for simulating laser–plasma interaction at normal and oblique incidence. The code is written in C++ and easy to extend. The data structure is characterized by the use of chained lists for the grid cells as well as particles belonging to one cell. The parallel version of the code is based on PVM. It splits the grid into several spatial domains each belonging to one processor. Since particles can cross boundaries of cells as well as domains, the processor loads will generally change in time. This is counteracted by adjusting the domain sizes dynamically, for which the use of chained lists has proven to be very convenient. Moreover, an option for restarting the simulation from intermediate stages of the time evolution has been implemented even in the parallel version. The code will be published and distributed freely.

INTRODUCTION

Particle-In-Cell (PIC) codes are well established tools for kinetic simulations in plasma physics and astrophysics [1]. Recently, the progress in producing intense ($I > 10^{18}$Wcm^{-2}) ultra-short (< 100fs) laser pulses [2] calls for a kinetic description of the interaction of such laser pulses with plasmas. It involves high intensities, short time scales and large density gradients, and conventional hydrodynamic approaches assuming nonrelativistic dynamics, local thermodynamic equilibrium, etc. become insufficient. Meanwhile, PIC simulations have provided new insight into absorption of short laser pulses [3–5], the propagation of short pulses in underdense plasma, wake field generation, fast electron production [6–8], magnetic field generation [7], harmonic generation at overdense plasma surfaces [9–14], and also with respect to inertial confinement fusion (ICF), the fast ignitor concept [15,16,8].

The code `LPIC++` presented here, is a one-dimensional, electromagnetic, relativistic PIC code that has originally been developed for kinetic simulations of high harmonic generation from overdense plasma surfaces [12–14]. The code uses essentially the algorithm of Birdsall and Langdon [1], and Villasenor and Bunemann [17]. It is written in C++ in order to be easily extendable and has been parallelized to be able to grow in power linearly with the size of accessable hardware, e.g. massively parallel machines like Cray T3E. The parallel `LPIC++` version uses `PVM` for communication between processors. `PVM` is public domain software, can be downloaded from the world wide web [18].

Advantages of `LPIC++` are its clear program and data structure, which uses chained lists for the organization of grid cells and enables dynamic adjustment of spatial domain sizes in a very convenient way, and therefore easy balancing of processor loads. Also particles belonging to one cell are linked in a chained list and are immediately accessable from this cell. In addition to this convenient type of data organization in a PIC code, the code shows excellent performance in both its single processor and parallel version.

The code will be published and distributed freely [19].

ALGORITHM

In PIC codes ionized plasmas are simulated by macro particles with positive or negative charge. Macro particles represent groups of electrons or ions containing an extensive number of real particles. `LPIC++` solves the Maxwell equations for the fields and the equations of motion for macro particles simultaneously. The relativistic equations of motion for a collisionless plasma

$$\dot{\boldsymbol{p}} = q_s(\boldsymbol{E} + \boldsymbol{v} \times \boldsymbol{B}), \qquad \boldsymbol{p} = m_s \gamma \boldsymbol{v},$$
$$\dot{\boldsymbol{r}} = \boldsymbol{v}, \qquad \gamma = \sqrt{1 + (\boldsymbol{p}/m_s c)^2} \qquad (1)$$

are solved for each macro particle once per time step Δt. The particles contribute to charge and current densities ρ and $\boldsymbol{j}$ on a spatial grid with spacing Δx. The Maxwell equations

$$\nabla \times \boldsymbol{E} = -\partial_t \boldsymbol{B}, \quad \nabla \times \boldsymbol{B} = \frac{1}{\varepsilon_0 c^2} \boldsymbol{j} + \frac{1}{c^2} \partial_t \boldsymbol{E},$$
$$\nabla \cdot \boldsymbol{B} = 0, \qquad \nabla \cdot \boldsymbol{E} = \frac{1}{\varepsilon_0} \rho \qquad (2)$$

are then solved on this grid. This procedure is iterated leading to the selfconsistent evolution of plasma and fields. The PIC-cycle and our grid structure are shown in Fig. 1 and Fig. 2, respectively.

DATA STRUCTURE

A macro particle is represented by a C data structure. In addition to physical parameters like specific charge z/m, particle position x etc., there are three pointers in this structure.

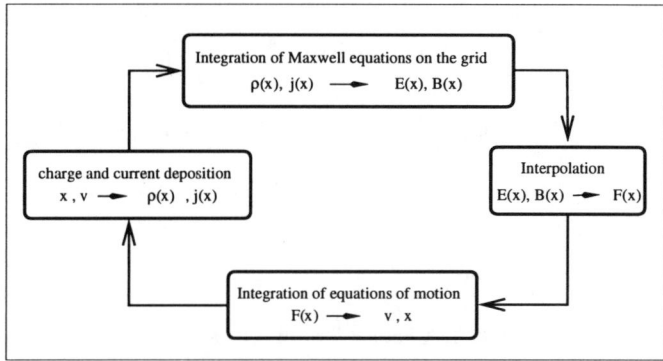

FIGURE 1. PIC code cycle.

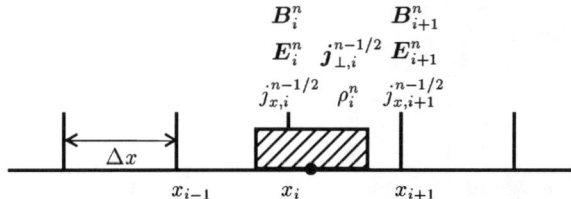

FIGURE 2. Grid structure: The macro particle belongs to cell i if its center x lies in $x_i \leq x < x_{i+1}$. Charge and transverse currents are located in cell centers, electromagnetic fields and longitudinal currents are located at cell boundaries.

Particles belonging to one cell, are linked in a chained list of particle structures, where each particle points to the preceeding and following particle in the list, see Fig. 3. Moreover, each particle points to the cell it belongs to.

The cell itself is represented by a C data structure. It contains physical parameters like the position of the left cell boundary, total charge, currents, fields and particle densities. For book-keeping, a unique cell number and the number of macro particles within this cell have been added. From each cell the list of these particles can be accessed using pointers to the first and last particle. The cells in turn are also linked in a chained list, see Fig. 3, so that they contain pointers to adjacent cell structures. Here, adjacent cell structures correspond to adjacent cells in the one-dimensional coordinate space.

In the parallel version of LPIC++, the whole grid (simulation box) is split into several domains each containing a fraction of the grid represented by chained lists of cells. At domain boundaries, two buffer cells are added, respectively, which are necessary for exchanging particles, fields and currents between adjacent domains.

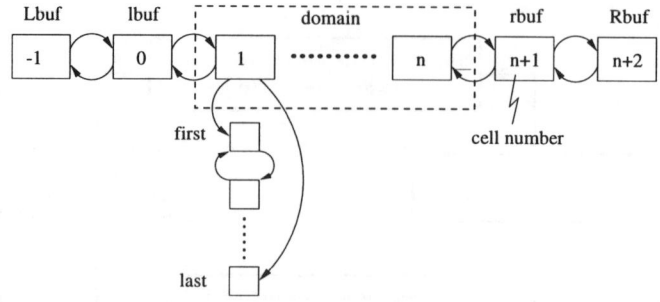

FIGURE 3. Chained lists of cells and particles.

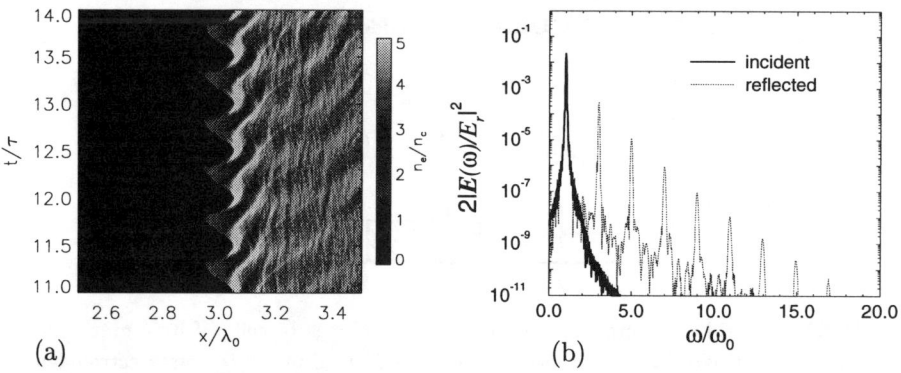

FIGURE 4. Generation of laser harmonics. Parameters: dimensinal amplitude $a_0 = 0.5$, electron density in units of the critical density $n_e/n_c = 4$. The plasma slab is located between $x/\lambda_0 = 3$ and $x/\lambda_0 = 4$. (a) n_e/n_c versus coordinate and time. (b) Power spectrum of incident and reflected light.

EXAMPLES

A Harmonic Generation at overdense surfaces

This code was used to simulate the generation of laser harmonics by interaction of an ultrashort laser pulse with a step boundary of a plane overdense plasma layer at intensities $I\lambda^2 = 10^{17} - 10^{19}$W cm^{-2} μm^2 [12,13]. See Fig. 4 for an electron density plot and the power spectrum of the reflected light made with an example input file distributed with LPIC++.

For another example see the contribution in the proceedings of this conference on generation of radiation at $2\omega_p$ from inverse two-plasmon decay in overcritical plasma [14].

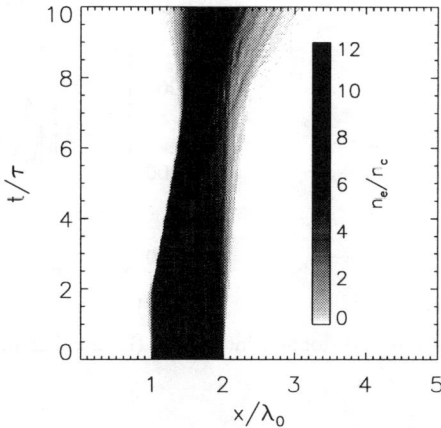

FIGURE 5. A laser pulse with amplitude $a_0 = 0.5$ is normally incident on a plasma slab with $n_e/n_c = 6$ and equal electron and 'ion' masses and charges. Electron density versus coordinate and time. Electron and ion density are identical.

B Parallel version and Restart

This example shows the effect of reorganizing the simulation box. We use a strong laser pulse from the left and 'ions' with *electron* mass, so that the whole plasma is pushed by the laser to the right within a short time period (20 cycles). The simulation box is split into three domains, and the box is reorganized once per period, starting at $t = 0$. Fig. 5 shows the electron density versus coordinate and time. This result is also obtained with killing and restarting LPIC++ at some intermediate stage of the simulation, and also when using the plain LPIC++ version. Fig. 6 (a) shows the box decomposition versus time, where the lines are the cell numbers at domain boundaries. Fig. 6 (b) shows the total number of particles in each domain. It is clearly seen that reorganizing the simulation box periodically helps to balance the processor load.

USING LPIC++

LPIC++ is written in C++, it is supposed to run on any Unix platform like Linux, Solaris, SunOS, AIX, etc.. It has been tested and used extensively under AIX.

Program and postprocessor will be freely available together with a detailed manual describing how to install and prepare input for LPIC++. It includes a description of output and postprocessing, discusses several examples and presents the main structure of the code.

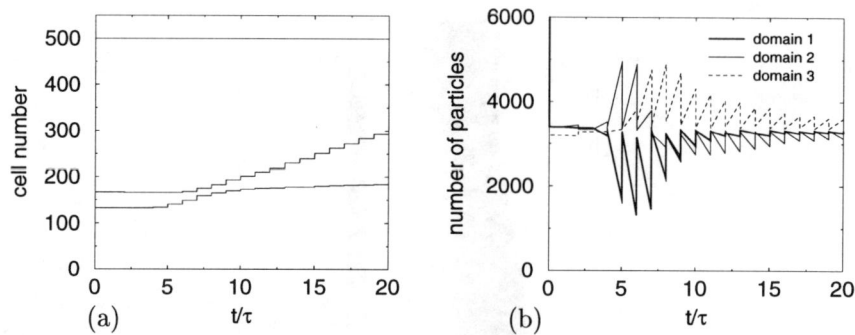

FIGURE 6. Reorganization: (a) domain interfaces, (b) number of particles in each domain.

REFERENCES

1. C.K. Birdsall and A.B. Langdon, *Plasma physics via computer simulation* (Adam Hilger, New York, 1991).
2. D. Strickland and G. Mourou, *Opt. Comm.* **56**, 219 (1985); M.D. Perry and G. Mourou, *Science* **264**, 917 (1994).
3. F. Brunel, *Phys. Rev. Lett.* **59** (1), 52 (1987).
4. P. Gibbon and A. Bell, *Phys. Rev. Lett.* **68** (10), 1535 (1992).
5. H. Ruhl and P. Mulser, *Phys. Lett. A* **205**, 388 (1995); H. Ruhl, *J. Opt. Soc. Am. B* **13**, 388 (1996).
6. P. Gibbon, *Phys. Rev. Lett.* **73** (5), 664 (1994).
7. A. Pukhov and J. Meyer-ter-Vehn, *Phys. Rev. Lett.* **76** (21), 3975 (1996).
8. A. Pukhov and J. Meyer-ter-Vehn, submitted to *Phys. Rev. Lett.* (1997).
9. S. Wilks, W. Kruer, and W. Mori, *IEEE Trans. Plasma Sci.* **21** (1), 120 (1993).
10. S. Bulanov, N. Naumova, and F. Pegoraro, *Phys. Plasmas* **1** (3), 745 (1994).
11. P. Gibbon, *Phys. Rev. Lett.* **76** (1), 50 (1996).
12. R. Lichters, J. Meyer-ter-Vehn, and A. Pukhov, *Phys. Plasmas* **3** (9), 3425 (1996).
13. R. Lichters and J. Meyer-ter-Vehn, in *Multiphoton Processes 1996* (Institute of Physics Publishing, Bristol and Philadelphia, 1997), pp. 221-230.
14. R. Lichters et al., submitted to *Phys. Rev. Lett.* (1997).
15. M. Tabak, et al., *Phys. Plasmas* **1**, 1626 (1994).
16. A. Pukhov and J. Meyer-ter-Vehn, Gesellschaft für Schwerionenforschung, Report GSI-95-06, ISSN 0171-4546 (1995).
17. J. Villasenor and O. Buneman, *Computer Physics Communications* **69**, 306 (1992).
18. A. Geist et al., *PVM: Parallel Virtual Machine System* (MIT Press, Cambridge, USA, 1994), http://www.netlib.org/pvm3.
19. R. Lichters, R.E.W. Pfund, and J. Meyer-ter-Vehn, submitted to *Computer Physics Communication* (1997)

Enhancement of ponderomotively generated wakefields in 1D

R. J. Kingham and A. R. Bell

Plasma Physics Group, Imperial College, London SW7 2BZ

Abstract. Nonlinear, relativistic 1D envelope equations predict that the amplitudes of wakefields ponderomotively excited by smooth laser pulses much longer than $\lambda_p/2$ far exceed the linear prediction at relativistic pulse intensities. Wakefield enhancement, for long pulses, is relevant to the seeding of pulse modulation in the 'self-modulated' LWFA when using pre-ionized plasmas hot enough to suppress Raman back scattering. Estimates show that significant modulation can be attained in a fraction of a Rayleigh length when seeding with enhanced wakes while several Rayleigh lengths may be needed with a linear amplitude wake.

I INTRODUCTION

Linear theory [1] of ponderomotively excited wakefields, in 1D, relates the wake amplitude to the Fourier transform of the pulse intensity profile at wavenumber $k = k_p$. The 'standard' LWFA (Laser Wake-Field Accelerator) uses a short 'matched' pulse of length $L \approx \lambda_p/2$. The amplitude of the intensity spectrum for a matched pulse is large at $k = k_p$ because such a pulse has a wide bandwidth. Consequently, matched pulses can excite wakefields of wave breaking amplitude at sufficient intensities; $I > 10^{18}$ Wcm^{-2}. Long smooth pulses (e.g. Gaussian) such as those used in the 'self-modulated' LWFA [2] have narrow bandwidths (compared to $k_p = \omega_p/c$) and therefore excite vanishingly small amplitude wakefields, according to linear theory.

In recent work [3] we showed that nonlinear, relativistic 1D envelope equations predict that smooth laser pulses much longer than $\lambda_p/2$ ponderomotively excite wakefields of much larger amplitude than predicted by linear theory at relativistic pulse intensities. We attributed this wakefield enhancement to the following mechanism: the plasma electrons (interacting with the laser) effectively experience an increased pulse bandwidth due to the nonlinearites. At realisable pulse intensities ($I \sim 10^{19}$ W cm^{-3}) and for pulse lengths of $L > 5\lambda_p$ we found that the enhanced wake remains below the wave breaking amplitude but is boosted (over the linear amplitude) enough to make it a viable seeding source for pulse modulation (via RFS [4] or coupled RFS & self-focusing [2]).

Here, we present new numerical results on wakefield enhancement. We also discuss more fully why wake enhancement is important to seeding pulse modulation.

OVERVIEW OF WAKE ENHANCEMENT

In [3] we used the following nonlinear, relativistic, electron fluid, envelope equation,

$$\frac{\partial^2 \phi}{\partial \zeta^2} = -\gamma_g^2 \left[\frac{(1-\phi)\beta_g}{\{(1-\phi)^2 - (1+2\alpha|a|^2)/\gamma_g^2\}^{\frac{1}{2}}} - 1 \right] \quad (1)$$

to determine the amplitudes of wakefields excited by long laser pulses with Gaussian intensity profiles. In this model ions are assumed to be immobile over the time scales of interest, the electrons are taken to be cold and collisionless and the unperturbed plasma density is uniform. Equation 1 uses the quasistatic approximation [5] and is valid when the change in the laser pulse is negligible during the time it takes a background electron to pass through (and interact with) the pulse: $\tau_L \omega_p \ll 2\gamma |n_o/n| (\omega_o/\omega_p)$, where $\gamma = (1+2\alpha|a|^2)^{1/2}/(1-\beta_z^2)^{1/2}$ is the electron relativistic factor. In equation 1, $a(\zeta, \tau)$ is the laser-pulse vector-potential envelope, $\phi(\zeta, \tau)$ is the scalar potential of the plasma, ζ, τ are the laser pulse frame coordinates (related to the laboratory coordinates z, t by $\zeta = z - \beta_g t$, $\tau = t$), $\gamma_g^2 = 1/(1-\beta_g^2)$ is the relativistic gamma factor associated with the laser pulse group velocity β_g, and $\alpha = 1$ or 2, is a parameter that specifies linear or circular polarization. All the quantities are normalised: $a(m_e c/e) \to a$, $\phi/\phi_{cwb} \equiv \phi(m_e c^2/e) \to \phi$, $\omega/\omega_p \to \omega$, $\zeta(\omega_p/c) \to \zeta$ and $\omega_p \tau \to \tau$. $a(\zeta, \tau)$ is related to the intensity by $I = 5.5 \times 10^{18} \alpha |a(\zeta, \tau)|^2/(\lambda_o/1\mu\text{m})^2$ W cm^{-2}. Equation 1 has been used by others in the context of high intensity matched pulse wakefield excitation [6]. An equation equivalent to 1 has been used in the context of wave breaking [7].

In [3] we showed how an improved prediction of the wakefield amplitude can be obtained by expanding the square root factor in equation 1. The argument of the square root can be rearranged to read $\beta_g^2 [1+\epsilon]$ where $\epsilon = [\phi^2 - 2\phi - 2|a|^2/\gamma_g^2]/\beta_g^2$ is a small expansion parameter (with $\alpha = 1$). After a binomial expansion of the square root term plus expanding the potential as $\phi = -|a|^2 + \phi_w$ we obtained the following equation,

$$\frac{\partial^2 \phi}{\partial \zeta^2} + k_p^2 \phi = -k_p^2 |a|^2 + \left\{ k_p^2 \frac{|a|^4}{2} - k_p^2 |a|^6 + k_p^4 \frac{3}{8} |a|^8 (5\beta_g^2 - 1) - \ldots \right\}$$
$$+ \phi_w \{\ldots\} + \phi_w^2 \{\ldots\} + \ldots \quad . \quad (2)$$

The $-|a|^2$ term in the 'ϕ expansion' represents the slow linear response of the plasma for long smooth pulses where $\partial^2\phi/\partial\zeta^2$ is small compared to $k_p^2\phi$. ϕ_w contains the plasma oscillation at ω_p and for long smooth pulses we expect $|a|^2 \gg |\phi_w|$ in the region of the pulse. Ignoring terms in ϕ_w^n, $(n \geq 1)$ equation 2 looks like the linear equation, $\left(\partial^2/\partial\zeta^2 + k_p^2\right)\phi = -k_p^2|a|^2$, but with extra driving terms involving higher orders of intensity ($I \propto |a|^2$). The linear equation predicts a wakefield amplitude of [1]

$$\phi_o = k_p \left|\mathcal{F}\{|a(\xi)|^2\}_{k=k_p}\right|, \qquad (3)$$

where $\mathcal{F}\{|a(\xi)|^2\}_{k=k_p}$ is the Fourier transform of the intensity profile, evaluated at wavenumber $k = k_p$. Including the extra driving terms results in an improved prediction of the wake amplitude: $\phi_o = \left|\sum_n C_n \mathcal{F}\{|a(\xi)|^{2n}\}_{k=k_p}\right|$ (where C_n are constants). The effective increase in the laser pulse bandwidth experienced by the electrons is due to the extra driving terms. For long smooth intensity profiles, the bandwidth of $I(\zeta)^m$ increases with m ($\Delta k = \sqrt{2m}/\sigma_g$ for a Gaussian profile: $a = a_{max}\exp(-\zeta^2/2\sigma_g^2)$) while the peak value of $\mathcal{F}\{I(\zeta)^m\}$ scales as a_{max}^{2m}.

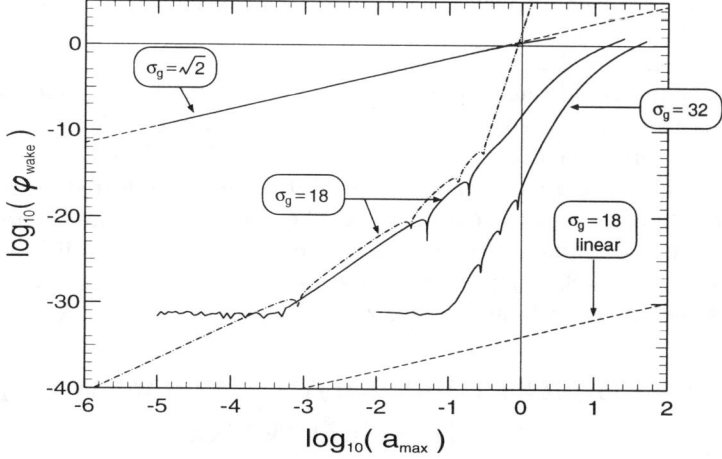

FIGURE 1. Peak-to-peak wakefield amplitude against pulse height for Gaussian pulses of widths $\sigma_g = \sqrt{2}$, 18 and 32. Solid curves correspond to numerically determined amplitudes. Dashed curves represent the linear prediction while the dashed-dotted curve is an improved analytical prediction based on expansion of the nonlinear equation.

RESULTS

We now present new results. Equation 1 was integrated numerically over $\zeta_{max} \geq \zeta \geq -\zeta_{max}$ using the boundary conditions $\phi(\zeta_{max}) = 0$ and $[\partial\phi/\partial\zeta]_{\zeta_{max}} = 0$ (plasma ahead of pulse is unperturbed) for a Gaussian laser profile: $a(\zeta) = a_{max} \exp\left(-\zeta^2/2\sigma_g^2\right)$ centred at $\zeta = 0$. The peak-to-peak wakefield amplitude was measured near to $-\zeta_{max}$. A leapfrog integration scheme with 100 points per plasma wavelength was used. ζ_{max} was chosen so that $a(\zeta_{max})$ dropped to 10^{-64} of the peak value.

Figure 1 shows curves of ϕ_{wake}, the peak-to-peak wake amplitude against pulse height, for Gaussian pulses of widths $\sigma_g = \sqrt{2}, 18$ and 32 ($L \approx 1/2, 6$ and $11\,\lambda_p$). A value of $\omega_o/\omega_p = 8.2$ was used, corresponding to a $\lambda = 1.0\,\mu$m laser and an electron density of $n_e = 1.5 \times 10^{19}$ cm^{-3}. As noted in [3], for long pulses the numerically determined wake amplitude (solid curves) is many orders of magnitude larger than the linear prediction (dashed curves), especially at relativistic intensities. The amount of enhancement over the linear amplitude ($\phi_{wake} = 2\,a_{max}^2\sigma_g\sqrt{\pi}\,\exp[-k_p^2\sigma_g^2/4]$) increases with intensity. The short 'matched' $\sigma_g = \sqrt{2}$ pulse shows no enhancement. Figure 2 shows curves of ϕ_{wake} against pulse half-width σ_g at 3 intensities: $a_{max} = 0.1, 1.0$ and 10.0 ($I = 6 \times 10^{16}, 6 \times 10^{18}$ and 6×10^{20} W cm^{-3}) for $\omega_o/\omega_p = 8.2$. The drop-off in ϕ_{wake} with σ_g becomes less pronounced with increasing intensity. Notice that the curves obtained numerically flatten out and oscillate randomly at $\phi_{wake} \sim 10^{-30}$ in figure 1 and at $\phi_{wake} \sim 10^{-12}$ in figure 2. This effect is not physical but due to the precision limit of Fortran quad (or double) precision variables being reached.

We have increased the resolution of the numerically determined curves since [3]. This has brought to light 'cusps' in the numerically determined curves, not seen before. For the $\sigma_g = 18$ pulse, the position of the 2 cusps are close to the corresponding cusps in the improved prediction (dotted dashed curve) from equation 2 (which occurs because the driving terms alternate in sign). We note that the quasistatic approximation is not strictly valid for the long pulses at $\omega_o/\omega_p = 8.2$. This can be remedied by increasing ω_o/ω_p and hardly changes the curves in figures 1 and 2.

SEEDING PULSE MODULATION WITH ENHANCED WAKES

Previously, plasma waves that are ponderomotively excited by smooth long pulses have been dismissed as a viable seeding source for pulse modulation in favour of Raman back scattering or ionization induced pulse front steepening [4]. According to linear theory, too many e-foldings of RFS growth are required for ponderomotively excited plasma waves to grow into large amplitude wakes. Figures 1 and 2 show that wake enhancement can boost the

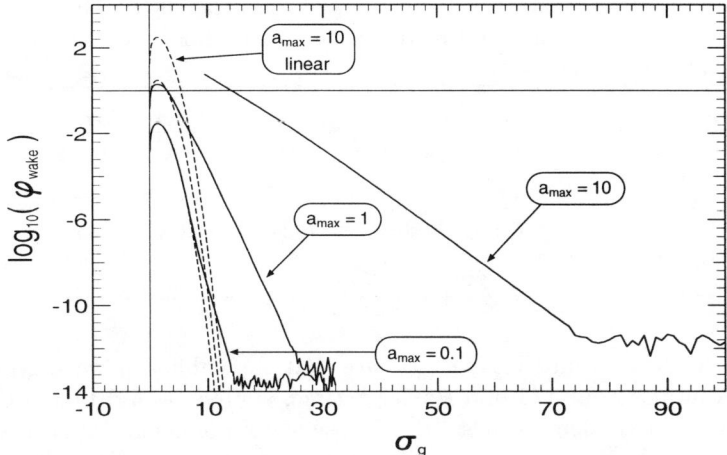

FIGURE 2. Peak-to-peak wake amplitude against pulse half-width for Gaussian pulses of peak intensity $a_{max} = 0.1, 1.0$ and 10.0 ($I = 6\times 10^{16}, 6\times 10^{18}$ and 6×10^{20} W cm^{-3}). Solid curves correspond to numerically determined amplitudes, while dashed curves represent the amplitudes predicted by linear theory.

wakefield amplitude from almost nothing (eg. 10^{-30} or less!) to a few orders of magnitude below wave breaking, in the case of long intense pulses. This greatly reduces the number of e-foldings of RFS growth required to reach wave breaking and under suitable experimental conditions, enhanced wakes should be large enough to act as the dominant seeding mechanism. Ionization induced seeded cannot occur if the plasma is pre-ionized. RBS can be suppressed by Landau damping while still permitting RFS at a suitable electron temperature; $100 \gg T_e > 1$ KeV when $n_e = 1.6\times 10^{19}$ cm^{-3} and $\lambda_o = 1$ μm. Enhanced wakes should be a 'cleaner' seeding source than either RBS or ionization. The large accelerating wakefield that grows from an enhanced wake seed should be of superior quality than the accelerating wakefield that grows from a noisy seed (such a RBS, etc.). Growth into an accelerating wakefield should also be more reliable when using an enhanced wake.

We now illustrate how seeding with enhanced wakes might work. Consider two co-propagating pulses, each having the parameters listed in table 1, passing through a pre-ionized plasma hot enough to suppress RBS. The first pulse excites an enhanced wake of amplitude e^{-18} times the cold wave breaking amplitude, ϕ_{cwb}: linear theory predicts an amplitude of e^{-79} ϕ_{cwb}! This enhanced wake will seed RFS in the trailing pulse. Spatial-temporal gain for 4-wave RFS is given by $G = [8(P/P_{crit})(\tau/\tau_R)(\omega_p/\omega_o)k_p\psi]^{1/2}$ [4], where $k_p\psi$ is the

TABLE 1. Parameters of a pulse that can fully modulate by RFS in under a Rayleigh length when seeded by the enhanced wake from a similar pulse co-propagating ahead.

Parameter	SI Value	Normalised value
Peak intensity	6×10^{18} W cm^{-2}	$a_{max} = 1$
Laser wavelength	1 μm	
e$^-$ density	1.7×10^{19} cm^{-3}	$\lambda_p/\lambda_o = 8.2$
Pulse length	$6.2\,\lambda_p \approx 50\,\mu\text{m} \equiv 170$ fs	$2\,k_p\sigma_g = 36$
Pulse radius	14μm	$Z_R \approx 600\,\mu$m
Peak power	40 TW	$40\,P_{crit}$

laser pulse length. From this, we estimate that the enhanced wake grows to wave breaking in about 1/4 of a Rayleigh time, while it would take the non-enhanced wake 4 Rayleigh times. In the absence of nonlinear 2D effects only the enhanced wake would be a suitable seed. One pulse may be able to seed RFS in the rear portion of itself because the ponderomotively excited plasma wave should attain part of its final amplitude by the middle of the pulse.

Because the the pulse power exceeds P_{crit} in this case relativistic self-focusing will be important. Self-focusing will increase (decrease) the pulse peak intensity (radius) while differential self-focusing (along the length of the pulse) may significantly sharpen up the front and rear of the pulse. The effects of radial plasma motion and self-focusing on wake enhancement are areas for further work. Also the quasistatic approximation is not fully satisfied for the above parameters. The effect of retaining the $\partial/\partial\tau$ terms in the electron fluid equations (which are dropped with the quasistatic approximation) is another area for further work.

ACKNOWLEDGEMENTS

We are greatful to the EPSRC for funding this work.

REFERENCES

[1] L. M. Gorbunov and V. I. Kirsanov, Sov. Phys. JETP **66**, 290 (1987)
[2] J. Krall, A. Ting, E. Esarey, and P. Sprangle, Phys. Rev. E **48**(3), 2157 (1993)
[3] R. J. Kingham and A. R. Bell, "Enhanced Wake-Fields for the 1D Laser Wake-Field Accelerator", submitted to PRL.
[4] C. D. Decker, et al, Phys. Plasmas **3**(4), 1360 (1996)
[5] P. Sprangle, E. Esarey, and A. Ting, Phys. Rev. A **41**, 4463 (1990)
[6] S. Dalla, and M. Lontano, Phys. Lett. A **173**(6), 456 (1993)
[7] D. Teychenné, and G. Bonnaud, Phys. Rev. E **48**(5), 3248 (1993)

Relativistic Electron Dynamics in Intersecting Laser Pulses

Zheng-Ming Sheng and Jürgen Meyer-ter-Vehn

*Max-Planck-Institut für Quantenoptik, Hans-Kopfermann-Str.1,
D-85748 Garching, Germany (e-mail: meyer-ter-vehn@mpq.mpg.de)*

Abstract. We study analytically the dynamics of a single electron in the field of two intersecting planar light waves. A simple relation is found between the scattering angle and the energy of the escaping electron. The relation is valid for both P and S-polarized laser pulses and holds even for relativistically intense laser pulses when stochastic motion of electrons sets in during the interactions. The momentum and energy of escaping electrons are calculated numerically as a function of the phase difference between the laser pulses. A transition from regular to stochastic electron motion occurs when the pulse intensity increases.

INTRODUCTION

The interaction of single electrons with lasers received renewed interest owing to the recent advent of high intensity, ultrashort laser pulses. One purpose of this study is concerned with particle acceleration by use of the intense laser pulses [1–3]. The other aspect is that the knowledge of single electron dynamics provides an important basis for the investigation of relativistic laser plasma interaction in the more complicated situations [4]. In some case, the electron dynamics in plasma is similar to that in vacuum [5,6]. Electron interaction with a single plane electromagnetic wave is well studied by Sarachik and Schappert [7]. Recent studies include the interaction of electrons with a focused intense laser beam, with special beam profiles or pulse shapes was pursued [1–3]. It was found that if electrons are injected with an initial velocity large enough; An acceleration mechanism so called the ponderomotive scattering occurs was suggested [1] The interactions of electrons with a standing wave or two counter-propagating electromagntic plane waves were also studied by a number of authors [5,8]. In general, stochastic motion of electrons develops when certain thresholds of the wave amplitudes are reached.

The present study is devoted to the interaction of electrons with two intersecting planar laser pulses. Different from the case of a single pulse interactions

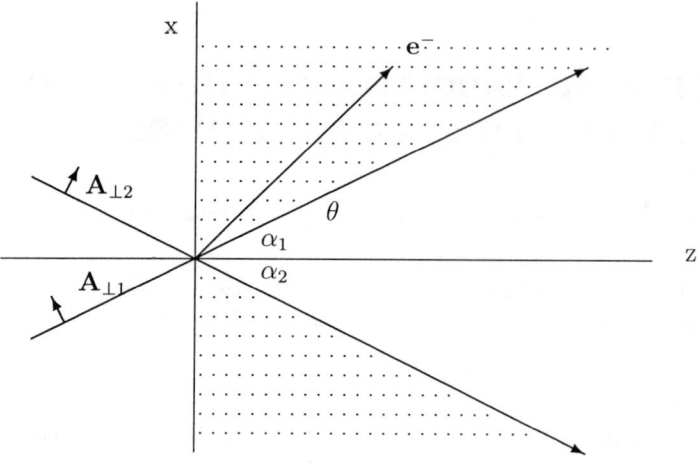

FIGURE 1. Schematic plot of electron scattering in two laser pulses. Electrons initially at the origin are scattered in the dotted region.

with electrons, electrons can absorb energy from the laser pulses in this situation, and the angular direction of the scattered electron is simply to its initial and final energy. This relation is valid even for very intense laser pulses when stochastic motion sets in.

ELECTRON SCATTERING IN TWO PULSES

Without loss of generality, we consider a geometry shown in Fig.1, where two laser pulses propagating at relative intersecting angle 2α interact with a electrons initially at the origin of the given coordinate system. Assuming that the vector potential of the P-polarized planar pulses in vacuum is in the form of

$$\mathbf{A}_{\perp i} = A_{0i}(\xi_i)\cos(\xi_i + \psi_i)(\cos\alpha_i\hat{\mathbf{x}} - \sin\alpha_i\hat{\mathbf{z}}), \quad i = 1, 2,$$

where $\alpha_1 = -\alpha_2 = \alpha$ and $0 \leq \alpha \leq \pi/2$, $\xi_i = k_{0i}(z\cos\alpha_i + x\sin\alpha_i) - \omega_{0i}t$ is the propagation coordinate, x (z), t, k_{0i}, and $\omega_{0i}(= k_{0i})$ are normalized to some k_0^{-1}, ω_0^{-1}, k_0 and $\omega_0(= k_0 c)$, respectively, and ψ_i is a constant phase. In terms of the vector potential, the equation of motion for electrons is

$$\frac{d\mathbf{p}}{dt} = \frac{d\mathbf{A}}{dt} - \nabla(\mathbf{v}\cdot\mathbf{A}) + \nabla\phi, \quad (1)$$

where the momentum $\mathbf{p}$ is normalized to mc, the velocity $\mathbf{v}$ normalized to c, $\mathbf{A} = \mathbf{A}_{\perp 1} + \mathbf{A}_{\perp 2}$ normalized to mc^2/e, and the first ∇ acts on $\mathbf{A}$ only, ϕ is the scalar potential normalized to mc^2/e. It is convinent to transform all variables into a moving frame with $\mathbf{V} = c\cos\alpha\hat{\mathbf{z}}$. Assuming $\phi = 0$ in the laboratory

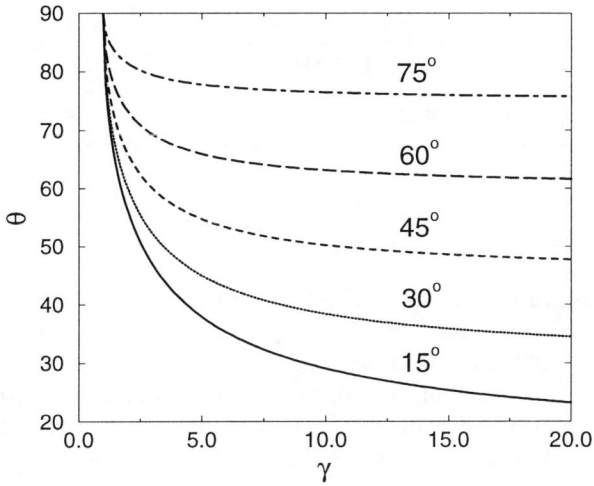

FIGURE 2. Scattering angles as a function of the escaping energy for $\alpha = 15°$, $30°$, $45°$, $60°$, and $75°$.

frame, where $A_i = A_{0i}(\xi'_i)\cos(\xi'_i + \psi_i)$, $\xi'_i = k_{0i}x'\sin\alpha_i - \omega_{0i}t'\sin\alpha$. In the moving frame, the x and z-components of the equation of motion reduce to

$$\frac{dp'_x}{dt'} = -v'_z \frac{\partial}{\partial x'}(A_2 - A_1), \tag{2}$$

$$\frac{dp'_z}{dt'} = \frac{d}{dt'}(A_2 - A_1). \tag{3}$$

Eq.(3) gives

$$p'_z = p'_{z0} + A_2 - A_1, \tag{4}$$

where p'_{z0} is the initial velocity of electron in the moving frame before it interacts with the laser pulses. Obviously, in the moving frame, only the x-component of the motion equation has to be solved. Meanwhile, Eq.(4) also gives the scattering angle as a function of the escaping energy. To show this, by transforming back to laboratory frame $p'_z = (p_z - \gamma\cos\alpha)/\sin\alpha$ and $p'_{z0} = (p_{z0} - \gamma_0\cos\alpha)/\sin\alpha$, we obtain

$$p_z - \gamma\cos\alpha + (A_1 - A_2)\sin\alpha = p_{z0} - \gamma_0\cos\alpha \equiv C. \tag{5}$$

Assume $p_y = 0$ and $p_x/p_z = \tan\theta$, the above equation leads to

$$\tan\theta = \pm\left(\frac{\gamma^2-1}{(\gamma\cos\alpha+C)^2}-1\right)^{1/2} \qquad (6)$$

when the laser pulses propagate away. For the simple case when the particle is initially at rest, i.e. $p_{x0}=0$, $p_{z0}=0$ and $\gamma_0=1$, Eq.(6) is simplified to

$$\tan\theta = \pm\left(\frac{2}{\gamma-1}+\frac{\gamma+1}{\gamma-1}\tan^2\alpha\right)^{1/2}, \qquad (7)$$

which is shown in Fig.2 for some given α value. Generally, the scattering angle θ ranges from α to $90°$ (or $-\alpha$ to $-90°$) with corresponding γ value from ∞ to 1. When $\alpha=90°$, i.e. two pulses counter-propagate along x-axis, one gets $\theta=\pm90°$, i.e. $p_z=0$. When $\alpha=0$, i.e. two pulses co-propagate along z-axis, a well-known relation is recovered [1,9]. It can be shown that for the case of S-polarization, the same relation as Eq.(7) can be derived [10].

NUMERICAL RESULTS

In the moving frame, one needs only to solve the x-component equation of motion to obtain the momentum and energy, then transforms them back to the laboratory frame. In the following, we study only for the simple case when $\omega_{01}=\omega_{02}=1$. For the P-polarization, the x-component equation of motion in the moving frame can be rewritten as

$$\frac{dv'_x}{dt'} = -\gamma'^{-2}(p'_{z0}+A_2-A_1)\left(\frac{\partial}{\partial x'}(A_2-A_1)+v'_x\frac{\partial}{\partial t'}(A_2-A_1)\right), \qquad (8)$$

where $\gamma'^{-1} = (1-v'^2_x)^{1/2}/(1+p'^2_y+p'^2_z)^{1/2}$. This equations is solved by the Rung-Kutta-Fehlberg method with variable time-step. We assume the pulse profile in the form of $A_{0i}=a_i\sin^2(\xi'_i/\tau_i)$ for $0\le\xi'_i\le\pi\tau_i$, $i=1,2$.

Fig.3(a) shows v_x as a function of time for $a_1=a_2=0.4$, $\alpha=90°$, i.e. the two pulses counter-propagate in the x-direction. A guiding center equation can be derived to describe this velocity component in this case [10]. However, when a_1 or a_2 are large enough or the two pulse do not counter-propagate, the guiding center equation is not applicable. Fig.3(b) shows a typical result.

Fig.4 shows the x-component momentum of the escaping electron as a function of the phase difference of the two pulses, where the electron is initially at rest. The momentum shows a smooth function of the phase difference for $a_1=a_2=0.1$. But it is not smooth again near $\psi_1-\psi_2=180°$ for $a_1=a_2=0.5$, showing that stochastic motion sets in. With the increase of the pulse intensity, the regime of the phase difference where stochastic motion develops is enlarged, until in the whole regime of $360°$. More calculation examples show that the threshold for stochastic motion depends both on the intersecting angle and polarization. The criterion of stochastic motion has been estimated [10] with the well-known resonance overlap theory [11], and showing quite in agreement with the numerical calculations.

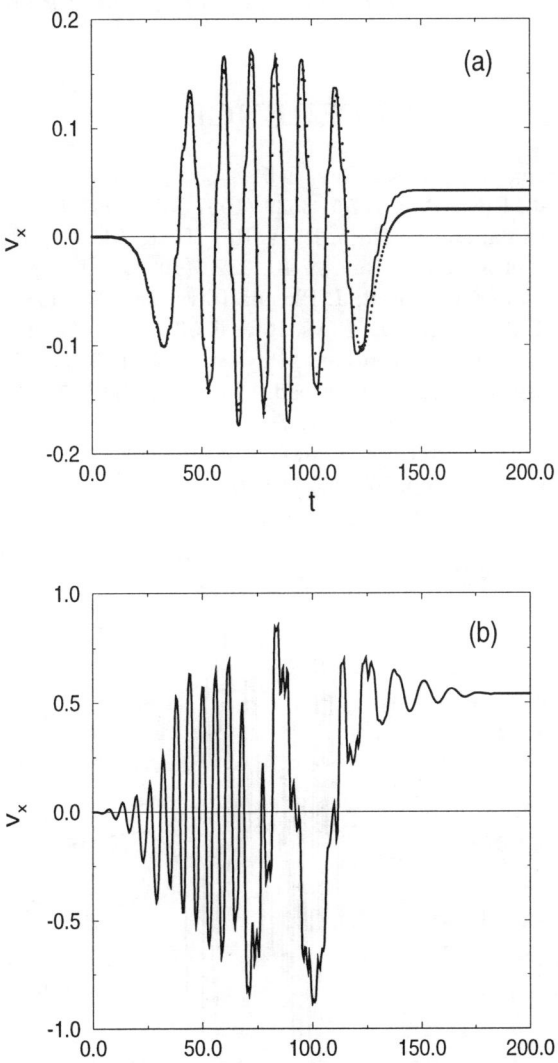

FIGURE 3. Velocity as a function of time. (a) $a_1 = a_2 = 0.4$, $\alpha = 90°$. The solid and dotted lines are obtaind from the equation of motion and from the corresponding guiding center equation, respectively. (b) $a_1 = a_2 = 1.0$, $\alpha = 60°$. In both cases, $\tau_1 = \tau_2 = 50.0$, $\psi_1 = 90°$, $\psi_2 = 0$.

ACKNOWLEDGEMENTS

This work was supported in part by BMBF (Bonn), Euratom, and by the EC network "High Energy Density Matter" under contract Number CT 930327. Z.M.S. would like to thank the hospitality of the Max-Planck-Institut für Quantenoptik.

REFERENCES

1. F. V. Hartmann, et al., Phys. Rev. E **51**, 4833 (1995).
2. B. Hafizi, et al., Phys. Rev. E **55**, 3539 (1997); ibid, **55**, 5924 (1997).
3. B. Rau, T. Tajima, and H. Hojo, Phys. Rev. Lett. **78**, 3310 (1997).
4. J. N. Bardsley, et al., Phys. Rev. A **40**, 3823 (1989).
5. J. T. Mendonca and F. Doveil, J. Plasma Phys. **28**, 485 (1982).
6. P. Mora, T. M. Antonsen, Jr., Phys. Rev. E **53**, R2068 (1996).
7. E. S. Sarachik and G. T. Schappert, Phys. Rev. D **1**, 2738 (1970).
8. D. Bauer, P. Mulser, and W. H. Steeb, Phys. Rev. Lett. **75**, 4622 (1995).
9. C. I. Moore, et al., Phys. Rev. Lett. **74**, 2439 (1995).
10. Z.-M. Sheng and J. Meyer-ter-Vehn, submitted.
11. B. V. Chirikov, Phys. Rep. **52**, 263 (1979).

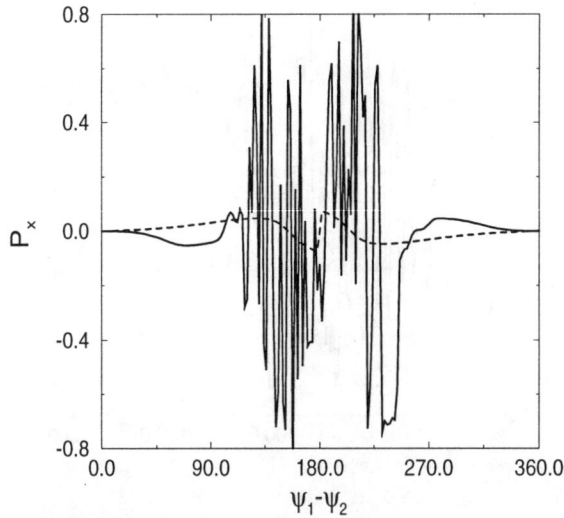

FIGURE 4. Dependence of the escaping momentum as a function of the phase difference for $a_1 = a_2 = 0.1$ (- - -), $a_1 = a_2 = 0.5$ (—), and $\alpha = 60°$. The electron is initially at rest in laboratory frame, $\tau_1 = \tau_2 = 50.0$ for the pulses and $\psi_2 = 0.0$.

Magnetic Field Generation in a Low Density Plasma Wake of a Short Laser Pulse

Z.-M. Sheng, J. Meyer-ter-Vehn, and A. Pukhov

*Max-Planck-Institut für Quantenoptik, Hans-Kopfermann-Str.1,
D-85748 Garching, Germany (e-mail: meyer-ter-vehn@mpq.mpg.de)*

Abstract. Magnetic field generation in the plasma wake of a relativisticly intense short pulse is studied, both analytically and by PIC-code simulation. We show that the magnetic field scales like $(m\omega_p c/e)dI_L/dr_\perp$ in this case, different from $(m\omega_p c/e)dI_L^2/dr_\perp$ for the low intensity case. The magnetic field profiles are calculated both for $\beta = 1$ and $\beta < 1$, which are found to be different in these two cases; here β is the normalized phase velocity of the wake plasma wave.

INTRODUCTION

Recently, the generation of quasistatic magnetic field in the wake of a laser pulse was predicted. One of the mechanisms is due to wave-breaking when an intense pulse propagates in plasmas at nearly critical density, which produces a net current of fast electrons moving with the pulse [1]. Meanwhile, it was showed by PIC simulation that the magnetic field can be generated without wave-breaking as well [2], which has been attributed to the nonvanishing $\nabla n \times \mathbf{v}$ recently [3,4], where n and $\mathbf{v}$ are the electron density and velocity, respectively. More recently, the calculation of the magnetic field generated in the wakefield has been extended to the case when the driving laser pulse propagates in a plasma channel [5]. A magnetic field is supposed to be important for the formation of fast ignition related plasma channels and plasma based particle accelerators.

We extend early works on magnetic field generation by a laser wake-field to the case when the driving pulse is at relativistic intensity while wave-breaking still does not occur [6]. In general, we show that the source of the magnetic field appears in the form of $\nabla(n/\gamma) \times \mathbf{p}$, which is a modification to the nonrelativistic case. We derive a set of equations to describe the wakefield with coupled electric and magnetic fields for arbitrary phase velocity of the wake plasma wave. Our calculation is found in good agreement with PIC-code simulations [7].

GENERAL FORMULATION

In the quasi-static approximation, we has the following coupled equations for the wakefield

$$\frac{\partial^2 \psi}{\partial \xi^2} = n - n_0 - \nabla_\perp^2 \gamma + \beta \frac{\partial}{\partial \xi} \nabla_\perp \mathbf{p}_\perp, \tag{1}$$

$$n = \frac{\gamma}{1 + \psi - \gamma \gamma_\beta^{-2}} \left(\beta^2 n_0 + \nabla_\perp^2 (\psi - \gamma \gamma_\beta^{-2}) + \beta \gamma_\beta^{-2} \frac{\partial}{\partial \xi} \nabla_\perp \mathbf{p}_\perp \right), \tag{2}$$

$$\left(1 - \gamma_\beta^{-2} \frac{\gamma}{n} \frac{\partial^2}{\partial \xi^2}\right) \mathbf{p}_\perp = \frac{\gamma}{\beta n} \frac{\partial}{\partial \xi} \nabla_\perp (\psi - \gamma \gamma_\beta^{-2}). \tag{3}$$

where $\psi = \phi - \beta a_x$, $\phi = e\Phi/mc^2$ is the scalar potential, $\mathbf{a} = e\mathbf{A}/mc^2$ is the slowly varying vector potential related to the magnetic field, a_x is the x-component of $\mathbf{a}$, $n_0 = 1$ for homogeneous plasmas, $\xi = k_p(x - \beta ct)$, p_x and $\mathbf{p}_\perp$ are respectively the longitudinal and transverse momenta of electrons normalized to mc. The normalized phase velocity β is equal to the group velocity of the laser pulse, which can be expressed as $\beta = 1 - q\omega_p^2/\omega_0^2$ for $\omega_p \ll \omega_0$, and $q > 0$ which is related both to pulse profiles and intensities [8]. In the following calculations, we assume β to be a constant parameter and $\gamma_\beta^2 = 1/(1 - \beta^2)$. We point out here that, when taking the limit that $\beta = 1$, these equations had been derived earlierly by Sprangle et al., however, without discussing the magnetic field generation in the wakefield [9]. In addition to above equations, we also have $p_x = (\gamma - 1 - \psi)/\beta$ and

$$\gamma = \gamma_\beta^2 \left(1 + \psi - \beta\sqrt{(1+\psi)^2 - \gamma_\beta^{-2}(1 + p_\perp^2 + a_L^2/2)}\right)$$

for linearly polarized laser pulse, where $a_L = e|\mathbf{A_L}|/mc^2$ is the amplitude of the vector potential for the laser pulse. Once the equations for the plasma wave are solved, the quasistatic magnetic field is calculated either by

$$\mathbf{B} = \nabla \times \mathbf{p}, \tag{4}$$

or by

$$\left(\nabla_\perp^2 + \gamma_\beta^{-2} \frac{\partial^2}{\partial \xi^2} - \frac{n}{\gamma}\right) \mathbf{B} = \mathbf{S}, \tag{5}$$

where $\mathbf{B}$ and $\mathbf{S}$ are normalized to $m\omega_p c/e$, the source $\mathbf{S} = \nabla(n/\gamma) \times \mathbf{p}$. Eq.(4) is derived from the equation of motion and Eq.(5) is derived from the Maxwell's equation. As given below, Eqs.(4) and (5) produce the same results.

MAGNETIC FIELD IN THE WAKE

One may note that all of the quantities can be expressed in terms of ψ, which is governed by Eq.(1). This equation is, however, cannot be solved because we do not know its boundary conditions at $\xi \to -\infty$. In the following, we consider the special case in which the intensity of the laser pulse changes slowly in the transverse direction. We assume that $|\nabla_\perp a_L| \sim \epsilon a_L$ where $\epsilon \ll 1$, and

$$g = g_0 + \epsilon^2 g_2 + \cdots, \quad \mathbf{h} = \epsilon \mathbf{h}_1 + \epsilon^3 \mathbf{h}_3 + \cdots,$$

where g stands for ψ, n, γ, or p_x and $\mathbf{h}$ for $\mathbf{p}_\perp$, $\mathbf{S}$, or $\mathbf{B}$. In the following, we just show some results to the first order $O(\epsilon)$. Substituting these into Eqs.(1)-(3), one obtains

$$\frac{\partial^2 \psi_0}{\partial \xi^2} = n_0 - 1, \quad (6)$$

$$n_0 = \frac{\beta^2 \gamma_0}{1 + \psi_0 - \gamma_0 \gamma_\beta^{-2}}, \quad (7)$$

$$\left(1 - \gamma_\beta^{-2} \frac{\gamma_0}{n_0} \frac{\partial^2}{\partial \xi^2}\right) \mathbf{p}_{\perp 1} = \frac{\gamma_0}{\beta n_0} \frac{\partial}{\partial \xi} \nabla_\perp (\psi_0 - \gamma_0 \gamma_\beta^{-2}), \quad (8)$$

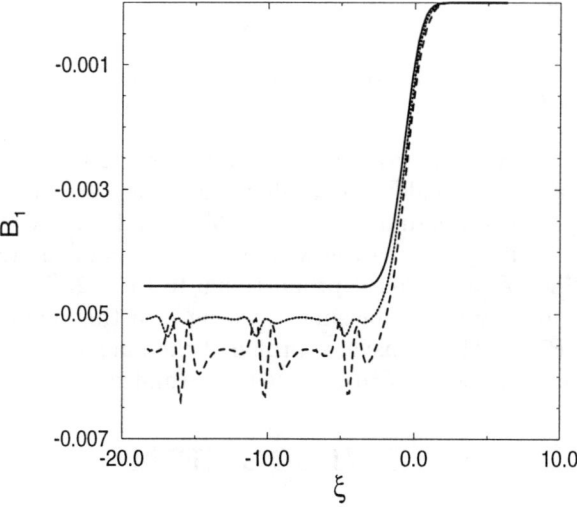

FIGURE 1. Longitudinal profile of the first order of the magnetic field (in unit of $m\omega_p c/e$) for a Gaussian pulse $a_L^2 = a_0^2 \exp[-(\xi^2/L^2 + y^2/W^2)]$ with $a_0 = 1.0$, $L = 0.25 k_p \lambda_p$, and $W = 2.0 k_p \lambda_p$ at $y = k_p \lambda_p$ for $\beta = 1$ (—), 0.95 ($\cdots$), and 0.9 (- - -).

up to order $O(\epsilon^3)$, where $\gamma_0 = \gamma_\beta^2[1+\psi_0 - \beta\sqrt{(1+\psi_0)^2 - \gamma_\beta^{-2}(1+a_L^2/2)}]$, and $p_{x0} = [\gamma_0 - (1+\psi_0)]/\beta$. In this approximation, the two-dimensional partial differential equation of ψ reduces to a serious of ordinary differential equations, which can be solved with the boundary conditions at $\xi \to +\infty$. Figure 1 shows the longitudinal profile of the order magnetic field for various β values. For $\beta = 1$, the magnetic field in the wake is a constant along the propagation; While for $\beta < 1$, it is no longer a constant, but changes periodically at the plasma wavelength. The variation amplitude increases with the decrease of β value.

To see the effect of finite phase velocities, We do further expansion for the quantities related to the wakefield in the form of

$$g_0 = g_{00} + \gamma_\beta^{-2} g_{02} + \cdots, \quad \mathbf{h}_1 = \mathbf{h}_{10} + \gamma_\beta^{-2} \mathbf{h}_{12} + \cdots,$$

assuming that $\gamma_\beta^{-1} \ll 1$ for underdense plasmas with $\omega_p \ll \omega_0$, where g_0 stands for n_0, γ_0, ψ_0, or p_{x0}, and $\mathbf{h}_1$ stands for $\mathbf{p}_{\perp 1}$, $\mathbf{S}_1$, or $\mathbf{B}_1$. After some tedious calculations, both Eqs.(4) and (5) lead to

$$\mathbf{B}_1 = \frac{d}{dr_\perp}\left[C_0 + \gamma_\beta^{-2}\left(\frac{1}{2w_0^2}C_0 + C_0'\right)\right]\hat{e}_x \times \hat{e}_\perp, \quad (9)$$

in the wake, where both C_0 and C_0' are integration constant independent of ξ, coming from the first integration of Eq.(6), and $w_0 = 1+\psi_{00}$. The value of C_0 depends on the longitudinal profile of the pulse and the intensity. For a rectangular pulse, Berezhiani et al. showed that [10], for given light intensity, the maximum value of $C_0 = \gamma_{0\perp}^2 + \gamma_{0\perp}^{-2}$, where $\gamma_{0\perp}^2 = 1 + a_L^2/2$. Therefore, one obtains the intensity scaling of the magnetic field

$$B_1 \sim \begin{cases} da_L^2/dr_\perp, & \text{for } a_L \gg 1 \\ da_L^4/dr_\perp, & \text{for } a_L \ll 1 \end{cases} \quad (10)$$

according to Eq.(9). Although the above scalings are derived for a rectangular pulse shape, numerical calculations show that they are also valid to the Gaussian pulses for corresponding intensities. Meanwhile, according to Eq.(9), when $\beta = 1$, the magnetic field in the wake is a function of the transverse coordinate only. When $\beta < 1$, a term proportional to $(\gamma_\beta^{-2}/2w_0^2)\nabla_\perp C_0$ appears, which varies periodically at the plasma wavelength in the longitudinal direction as shown in Fig.1. The transverse profile of the magnetic field is given in Fig.2, which is just consistent with the scaling formulas.

ACKNOWLEDGMENTS

This work was supported in part by BMBF (Bonn), Euratom, and by the EC network "High Energy Density Matter" under contract Number CT 930327. Z.M.S. acknowledges the support and hospitality of the Max-Planck-Institut für Quantenoptik.

REFERENCES

1. A. Pukhov and J. Meyer-ter-Vehn, Phys. Rev. Lett. **76**, 3975 (1996).
2. D. W. Forslund, J. M. Kindel, W. B. Mori, C. Joshi, and J. M. Dawson, Phys. Rev. Lett. **54**, 558 (1985).
3. L. Gorbunov, P. Mora, and T. M. Antonsen, Jr., Phys. Rev. Lett. **76**, 2495 (1996).
4. A. R. Bell and P. Gibbon, Plasma Phys. Controlled Fusion **30**, 1319 (1988).
5. N. E. Andreev, L. M. Gorbunov, V. I. Kirsanov, K. Nakajima, and A. Ogata, Phys. Plasmas **4**, 1145 (1997)
6. Z.-M. Sheng, J. Meyer-ter-Vehn, and A. Pukhov, GSI Annual Report 1996, Darmstardt/Germany.
7. Z.-M. Sheng, J. Meyer-ter-Vehn, and A. Pukhov, to be submitted.
8. C. D. Decker and W. B. Mori, Phys. Rev. Lett. **72**, 490 (1994).
9. P. Sprangle, E. Esarey, J. Krall, and G. Joyce, Phys. Rev. Lett. **69**, 2200 (1992).
10. V. I. Berezhiani and I. G. Murusidze, Phys. Lett. A **148**, 338 (1990).

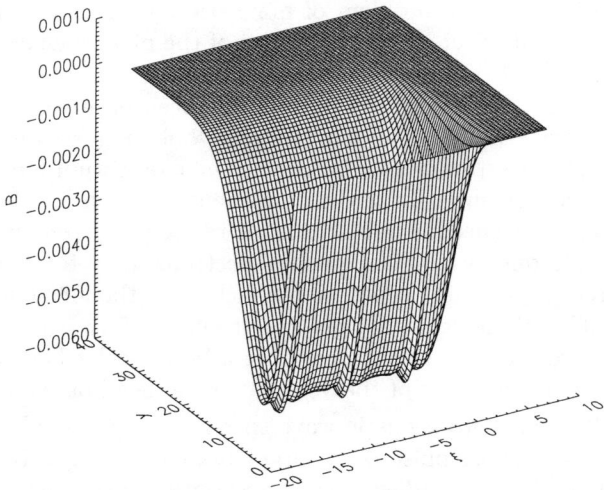

FIGURE 2. Longitudinal and transverse profile of the first order of the magnetic field (in unit of $m\omega_p c/e$) for a Gaussian pulse with $a_0 = 1.0$, $L = 0.25 k_p \lambda_p$, and $W = 2.0 k_p \lambda_p$ for $\beta = 0.95$.

Magnetic Fields in highly compressed Plasma formed at the Front of a Laser-driven Channel

A.V. Tatarinov and T.J.M. Boyd

Department of Physics, University of Essex, Colchester CO4 3SQ, UK

Abstract. Dynamics of two-dimensional steady-state supersonic magnetized plasma flows around the head of the hole bored in short-pulse laser-target interactions has been studied. An ideal MHD simulation is used to examine the general structure of the flow and the dependence of the strong magnetic fields on its characteristics is considered.

INTRODUCTION

The development of compact terawatt lasers with subpicosecond pulse durations has opened a new research area of ultrahigh intensity phenomena. Recently in the context of the fast ignitor for inertial confinement fusion attention has been drawn to the problem of magnetic field generation, since this may have a significant effect on the dynamics of the plasma channel bored by an intense laser pulse.

Boring of the light beam into the plasma has been observed recently in 2-3D PIC simulations [1] - [2]. The length of the hole increases with constant velocity, which can be estimated by balancing the momentum flux of the mass flow with the light pressure. In slightly overdense plasma for the intensities above 10^{19} W/cm^2 the speed is about $c/30$. A radial ponderomotive force due to the transverse intensity gradient pushes electrons away from the centre of the channel, creating a charge separation which pulls the ions out.

As the channel with depleted plasma density moves forward, a highly compressed region is formed ahead of it, creating a bow shock. The ion velocities in this region are of the order of the velocity of the head of the channel, and thus far exceeding magneto-acoustic wave speed: $U > \sqrt{v_A^2 + c_s^2}$.

Magnetic fields in short pulse laser-target plasmas are mostly initiated in the region of steep density gradient. There are several mechanisms, which can generate B-fields: $\nabla N \times \nabla T$ [3], [4], DC currents caused by the ponderomotive force [5] and the hot electron surface currents [6].[1] These mechanisms generate the high magnetic fields at the moving head of the channel. The laser beam propagates through the low density channel until it is reflected from the steep density gradient at the head of the channel. Within this thin layer magnetic fields are being constantly generated.

PIC simulations [1] revealed the fields of the order of 100 MG a few microns ahead of the hole surface. In this work we present an idealized phenomenological model which describes the plasma dynamics ahead of the front of the hole, i.e. inside a steady-state bow shock region. Magnetic field lines are everywhere transverse to the plasma motion. Numerical simulations have been performed in the framework of an ideal MHD model.

PROBLEM FORMULATION

One of the difficulties in the problem of shock formation at the head of the channel, driven forward by light pressure, is the lack of a clearly defined boundary between the channel and the surrounding plasma. One needs a kinetic treatment to balance the fluxes of energetic ions and electrons, which can pass through this boundary. We simplify the problem by neglecting the passage of these fast particles through the boundary. The evolution process, which leads to a steady-state bow shock ahead of the channel, is not of concern here. We take a prescribed hemispherical shape of the head of the channel with a certain magnetic field on its surface and investigate the structure of two-dimensional flow within the bow shock region. We assume that the channel moves with a constant speed in the overdense plasma and that the shape of the channel remains unchanged. The model is phenomenological, designed to emphasize the importance of ion dynamics in the process of penetration of magnetic fields inside the overdense plasma.

We assume that the channel behaves in some sense as a blunt body with a certain distribution of magnetic field at the front surface. The body moves along the direction of the laser beam through the undisturbed plasma of density $n_e > n_{cr}$ at a constant speed [1]

$$\frac{U}{c} = \left[\frac{n_{cr}}{2n_e}\frac{Zm}{M_i}\frac{I[W/cm^2]\lambda^2[\mu m]}{1.37 \times 10^{18}}\right]^{1/2} \quad (1)$$

The diameter of the nose of the channel is taken equal to the half-width of the focal spot of the laser beam.

When the blunt body moves in the fluid at a supersonic speed the ions and electrons also move with supersonic velocities inside the disturbed compressed

---

[1] Magnetic fields are also produced by the current of fast electrons which is partly balanced by the return current of the bulk plasma. This phenomena is not considered in this paper.

region between the nose of the blunt body and the bow shock surface. The latter represents the boundary between the undisturbed and highly compressed region, where the ions move around the nose, then leave the compressed region, decelerate and eventually stop.

The detached bow shock always represents a surface of strong discontinuity. The values of the flow parameters across this surface are related by the following conditions (see for example [7]) derived from the laws conservation of mass, momentum and energy flux in the frame moving with a body

$$\rho_1 U_{1n} = \rho_2 U_{2n} \tag{2}$$

$$\rho_1 U_{1n}^2 + p_1 + B_1^2/8\pi = \rho_2 U_{2n}^2 + p_2 + B_2^2/8\pi \tag{3}$$

$$\frac{\gamma p_1}{(\gamma-1)\rho_1} + \frac{U_{1n}^2}{2} + \frac{B_1^2}{4\pi \rho_1} = \frac{\gamma p_2}{(\gamma-1)\rho_2} + \frac{U_{2n}^2}{2} + \frac{B_2^2}{4\pi \rho_2} \tag{4}$$

where the subscripts 1 and 2 correspond respectively to the conditions in front of and behind the discontinuity surface; $U_{1n} = -U_0 \cos \vartheta$, where ϑ is the angle between the normal unit vector to the discontinuity surface $\mathbf{n}$ and the flow velocity of undisturbed plasma $\mathbf{U_0}$; γ is the ratio of the specific heat at constant pressure to the specific heat at constant volume. The tangential velocity remains unaltered on crossing the surface of discontinuity, i.e. $U_{1\tau} = U_{2\tau} = U_0 \sin \vartheta$, $\boldsymbol{\tau}$ is the tangent unit vector to the discontinuity surface.

It is shown for example in [7] that for large Mach numbers, that is

$$U_{1n}^2 \gg v_a^2 + c_s^2 = \frac{B_1}{4\pi \rho_1} + \frac{\gamma p_1}{\rho_1} \tag{5}$$

the solutions are close to hydrodynamic case, i.e. for a given B_1 which satisfies (5) one has

$$\frac{U_{1n}}{U_{2n}} = \frac{\rho_2}{\rho_1} = \frac{B_2}{B_1} \approx \frac{(\gamma+1)}{(\gamma-1)}$$

$$\frac{p_2}{p_1} \approx \frac{2\gamma M_1^2}{\gamma+1} \qquad \frac{T_2}{T_1} \approx \frac{2\gamma(\gamma-1)}{(\gamma+1)^2} M_1^2$$

We use an ideal MHD description of a steady-state plasma flow around the blunt body. In terms of $\mathbf{U}$, $\mathbf{B}$, ρ and P we have

$$\nabla \cdot (\rho \mathbf{U}) = 0 \tag{6}$$

$$\rho(\mathbf{U} \cdot \nabla)\mathbf{U} = -\nabla P + \frac{1}{4\pi}((\nabla \times \mathbf{B}) \times \mathbf{B}) \tag{7}$$

$$\nabla \times (\mathbf{U} \times \mathbf{B}) = 0 \tag{8}$$

$$P\rho^{-\gamma} = \text{const} \tag{9}$$

The system (6) - (9) with the boundary conditions (2) - (4) can be integrated along the streamlines to obtain the two-dimensional picture for plasma characteristics in the compressed region.

NUMERICAL PROCEDURE

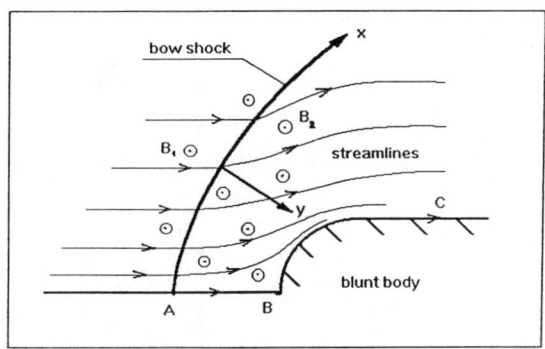

FIGURE 1. Sketch of supersonic plasma motion around the blunt body.

Numerical integration proceeds from the given bow shock surface, marching ahead along the lines perpendicular to this surface towards the nose of the body. Since the bow shock is not known in advance, several integrations were performed varying the geometry of the bow shock to achieve (after the marching procedure) the desired shape and size of the body.

Equations (6) - (9) were rewritten in the shock-oriented orthogonal curvilinear system of coordinates moving with the body [8]. We assume azimuthal symmetry of the problem, i.e. $\partial/\partial\phi = 0$ and that $\boldsymbol{U} = (u, v, 0)$, $\boldsymbol{B} = (0, 0, B(x,y))$ (see Fig. 1).

The x-derivatives were defined using the values on the $y =$ const plane. The expressions for y-derivatives of all the variables were obtained from equations (6) - (9). The step in y-direction is made, using a semi-implicit predictor-corrector scheme [9]. The surface of the body was traced from the stagnation point along the line for which the entropy had the same value as that of the stagnation line.

A similar approach to that outlined in this Section was adopted by Morozov and Savel'ev [10] in examining supersonic plasma flows in magnetic fields.

RESULTS

The results from several simulations are shown in Fig. 2-3. In Figs. 2-3(a) the contour plots for density, magnetic field and velocity distribution are shown for the case of low intensity $I = 10^{17}$ W/cm^2 and low Mach number $M = 6$. The magnetic field lines tend to follow the streamlines. In Figs. 2-3(a) the contours for the density and the perpendicular component of velocity v_y for the case when the magnetic field is switched off are shown as dashed lines. The presence of the field enlarges the stagnation region, slowing down the ions

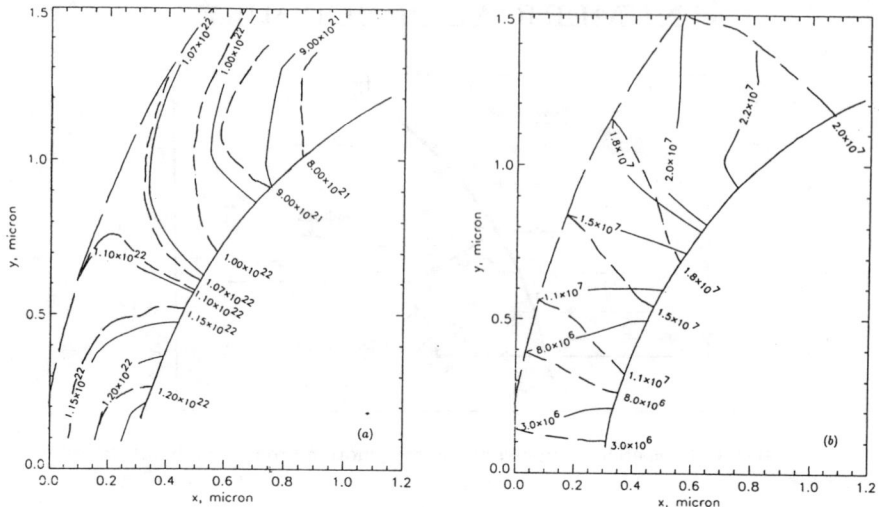

FIGURE 2. Contours of plasma density (a) and transverse velocity v_y (b) inside a bow shock. Dashed lines: magnetic field is switched off. $I = 10^{17}$ W/cm^2, $Z = 6$, $n_e^0 = 3 \times 10^{21}$ cm^{-3}, $T_e^0 = 100$ eV, $U_0 \sim c/500$, $M \simeq 6$.

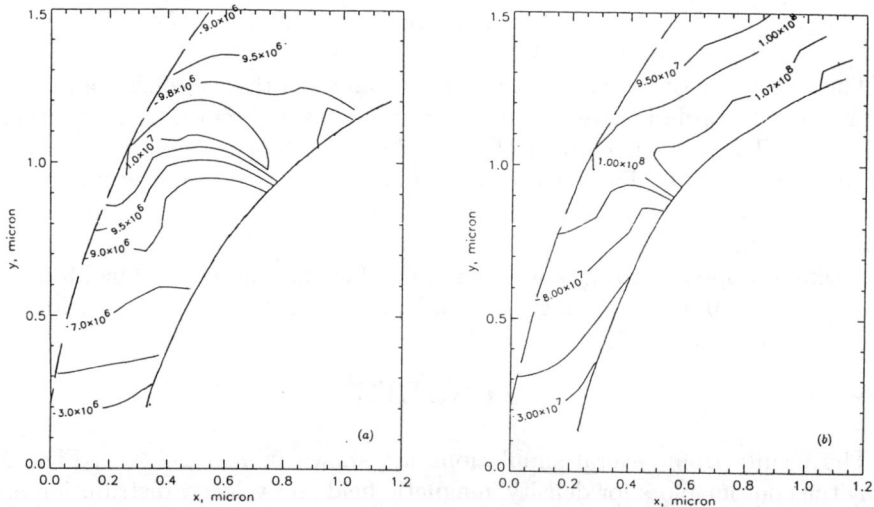

FIGURE 3. Contours of magnetic field within a bow shock region. (a): initial parameters correspond to those in Fig. 2; (b): $I = 10^{20}$ W/cm^2, $Z = 6$, $n_e^0 = 2 \times 10^{21}$ cm^{-3}, $T_e^0 = 100$ eV, $U_0 \sim c/12$, $M \simeq 250$.

in the direction transverse to the laser beam, and accumulates more ions on the top of the blunt body.

For large Mach numbers $M > 8$ magnetic fields that satisfy the condition $U_0 > \sqrt{v_A^2 + c_s^2}$ do not contribute considerably to the ion dynamics. We found a small changes in ion velocities only in a very thin layer close to the body surface.

Strong magnetic fields can be frozen in plasma at high intensities. For example for the case: $I = 10^{20}$ W/cm^2, $n_e^0 = 2 \times 10^{21}$ cm^{-3}, $T_e = 100$ eV one finds fields of the order of 100 MG (see Fig. 3(b)).

REFERENCES

1. Wilks, S. C. et al., *Phys. Rev. Lett.* **69**, 1383-1386 (1993).
2. Pukhov, A. M. and Meyer-ter-Vehn, J., *Phys. Rev. Lett.* **6**, 3975-3980, (1996)
3. Stamper, J., *Laser and Particle Beams*, **9**, 841-862, (1991)
4. Boyd, T. J. M., *ICPP94, AIP Conf. Proc.*, **345**, 238-248 (1995)
5. Sudan, R., *Phys. Rev. Lett.* **70**, 3075-3078, (1993)
6. Brunel, F., *Phys. Rev. Lett.* **59**, 52-55, (1987)
7. Boyd, T. J. M. and Sanderson, J. J., *Plasma Dynamics*, Nelson, London, 1969, ch. 6, pp. 120-124.
8. Hayes, W. D. and Probstein, R. F., *Hypersonic Flow Theory*, II ed., Academic Press, New York & London, 1966, ch. 5, pp. 266-267.
9. Press, W. H. et al. *Numerical Recipes*, II ed., Cambridge Univ. Press., 1992, ch. 16, pp. 730-745.
10. Morozov, A. I. and Savel'ev, V. V., *Plasma Phys. Rep.*, **22**, 4, 288-95, (1996)

A New Type Longitudinal Plasma Waves Induced in an Isotropic Plasma With a Superstrong Short Laser Pulse

Levan N. Tsintsadze and Nodar L. Tsintsadze*

Institute of Laser Engineering, Osaka University, Osaka, Japan
*Institute of Physics, Georgian Academy of Sciences, Tbilisi, Georgia

Abstract. A description of propagation of relativistically intense short laser pulse into an isotropic plasma is presented. A kinetic equation for a spectral function of electromagnetic waves is derived for an arbitrary amplitude of pump wave, fully relativistic case is considered. The resulting kinetic equation of spectral function is used along with the set of equations of the plasma to study the several type of instabilities and the importance of relativistic effects is pointed out. In the case of a superstrong short laser pulse a novel Langmuir waves, with phase velocities larger than the speed of light, are revealed in an underdense plasma, and the waves of ion-sound type, which damp only on ions, are found in an overdense plasma.

The development of ultra-intense short pulse lasers allows exploration of fundamentally new parameter regimes for nonlinear laser-plasma interaction. In fact, a number of experiments have been carried out in which plasmas are irradiated by laser beams with intensities up to $10^{19} W/cm^2$. At such intensities the electron quiver velocity rapidly approaches the speed of light, and a host of new phenomena have been predicted (1). Recently a number of works (2) have been devoted to the investigation of the relativistically intense electromagnetic (em) wave propagation into an isotropic plasma, when the radiation pressure becomes larger than the plasma pressure. The above treatments where mostly restricted to the case of monochromatic em waves. However, the bandwith of an initially "narrow" (short pulse) spectrum may eventually broaden, either as a result of the several kind of instability process itself, or as the result of other nonlinear wave-wave interaction processes. In the present paper, we consider certain problems involving the interaction of relativistically intense nonmonochromatic radiation bunches with an isotropic plasma.

We start from Maxwell equations (3,4) for a circularly polarized wave,

$$\Delta p - \frac{\partial^2 p}{\partial t^2} = \frac{n}{\gamma} p , \qquad (1)$$

where the following dimensionless quantities have been introduced:

$$p \to \frac{p}{m_o c}, \quad t \to \omega_p t, \quad \mathbf{r} \to k_p \mathbf{r}, \quad k_p = \frac{\omega_p}{c},$$

$$n \to \frac{n}{n_o}, \quad \gamma = (1+p^2)^{1/2}.$$

We shall consider Eq.(1) in two different points. There is some correlation between the values of $p(\mathbf{r},t)$ at different instants. This means that the value of $p(\mathbf{r},t)$ at a given instant t_1 affects the probabilities of its various values at a later instant t_2. Thus we can characterise the space and time correlation by the mean value of the product $<p(\mathbf{r}_1,t_1)p(\mathbf{r}_2,t_2)>=\Pi(\mathbf{r}_1,t_1,\mathbf{r}_2,t_2)$. Introducing new variables,

$$\mathbf{R} = \frac{1}{2}(\mathbf{r}_1+\mathbf{r}_2), \quad \mathbf{r} = \mathbf{r}_1 - \mathbf{r}_2, \quad t = \frac{1}{2}(t_1+t_2), \quad \tau = t_1 - t_2, \tag{2}$$

we obtain then from Eq.(1) the following relation for the correlation function, $\Pi(\mathbf{r}_1,t_1,\mathbf{r}_2,t_2) = \Pi(\mathbf{R}+\mathbf{r}/2, t+\tau/2, \mathbf{R}-\mathbf{r}/2, t-\tau/2)$,

$$\left(\nabla_\mathbf{R}\nabla_\mathbf{r} - \frac{\partial^2}{\partial t \partial \tau}\right)\Pi = \frac{1}{2}\left\{\rho\left(\mathbf{R}+\frac{\mathbf{r}}{2}, t+\frac{\tau}{2}\right) - \rho\left(\mathbf{R}-\frac{\mathbf{r}}{2}, t-\frac{\tau}{2}\right)\right\}\Pi, \tag{3}$$

where

$$\rho = \frac{n(\mathbf{r},t)}{\gamma(\mathbf{r},t)}.$$

The correlation function Π in Eq.(3) can be expanded as Fourier integral with respect to the time τ and the coordinate $\mathbf{r}$, or for the spectral function

$$\mathbf{P}(\mathbf{R},t,\mathbf{k},\omega) = \int d\mathbf{r} \int d\tau \Pi(\mathbf{R},t,\mathbf{r},\tau) expi(\mathbf{k}\mathbf{r} - \omega t). \tag{4}$$

From here it follows

$$\Pi(\mathbf{R},t) = p^2(\mathbf{R},t) = \int \frac{d\mathbf{k}}{(2\pi)^3} \int \frac{d\omega}{2\pi} \mathbf{P}(\mathbf{R},t,\mathbf{k},\omega). \tag{5}$$

Taking the Fourier transformation of Eq.(3) and expanding the $(\rho_1 - \rho_2)$ by Taylor series, we obtain after integration the equation for the spectral function

$$\left(\omega\frac{\partial}{\partial t} + \mathbf{k}\cdot\nabla_\mathbf{R}\right)\mathbf{P} = \sum_{l=0}^\infty \frac{(-1)^l}{(2l+1)!} \frac{1}{2^{2l+1}}\left\{\nabla_\mathbf{R}\cdot\nabla_\mathbf{k} - \frac{\partial}{\partial t}\cdot\frac{\partial}{\partial \omega}\right\}^{2l+1} \rho(\mathbf{R},t)\cdot\mathbf{P}. \tag{6}$$

We now linearize Eq.(6) with respect to the perturbation, which are represented as

$$\rho = \rho_o + \delta\rho expi(\mathbf{q}\mathbf{r} - \Omega t), \quad \mathbf{P}(\mathbf{R},t,\mathbf{k},\omega) = \mathbf{P}_o(\mathbf{k},\omega) + \delta\mathbf{P} expi(\mathbf{q}\mathbf{r} - \Omega t). \tag{7}$$

Use of (7) in Eq.(6) yields the following relation

$$(\mathbf{qk} - \Omega\omega)\delta\mathbf{P} = \delta\rho \sum_{l=0}^{\infty} \frac{1}{(2l+1)!} \frac{1}{2^{2l+1}} \left(\mathbf{q}\nabla_k + \Omega\frac{\partial}{\partial\omega}\right)^{2l+1} \cdot \mathbf{P}_\circ(\mathbf{k},\omega), \qquad (8)$$

or after summation we obtain

$$(\mathbf{qk} - \Omega\omega)\delta\mathbf{P} = \delta\rho\left\{\mathbf{P}_\circ^+\left(\mathbf{k} + \frac{\mathbf{q}}{2}, \omega + \frac{\Omega}{2}\right) - \mathbf{P}_\circ^-\left(\mathbf{k} - \frac{\mathbf{q}}{2}, \omega - \frac{\Omega}{2}\right)\right\}. \qquad (9)$$

Then from Eq.(5) we have for the perturbation of Π,

$$\delta\Pi = \int \frac{d\mathbf{k}}{(2\pi)^3} \int \frac{d\omega}{2\pi} \cdot \frac{\mathbf{P}_\circ^+ - \mathbf{P}_\circ^-}{\mathbf{qk} - \Omega\omega} \cdot \delta\rho \qquad (10)$$

and $\delta\rho$ can be derived as

$$\delta\rho = \frac{\delta n}{\gamma_\circ} - \frac{1}{2\gamma_\circ^3}\delta\Pi. \qquad (11)$$

From Eqs.(10) and (11) follows the relation between $\delta\Pi$ and δn

$$\left\{1 + \frac{1}{2\gamma_\circ^3} \int \frac{d\mathbf{k}}{(2\pi)^3} \int \frac{d\omega}{2\pi} \frac{\mathbf{P}_\circ^+ - \mathbf{P}_\circ^-}{\mathbf{qk} - \Omega\omega}\right\}\delta\Pi = \frac{\delta n}{\gamma_\circ} \int \frac{d\mathbf{k}}{(2\pi)^3} \int \frac{d\omega}{2\pi} \frac{\mathbf{P}_\circ^+ - \mathbf{P}_\circ^-}{\mathbf{qk} - \Omega\omega}. \qquad (12)$$

In the absence of the density perturbation δn we get from Eq.(12) the dispersion equation due to the relativistic selfmodulation

$$1 + \frac{1}{2\gamma_\circ^3} \int \frac{d\mathbf{k}}{(2\pi)^3} \int \frac{d\omega}{2\pi} \frac{\mathbf{P}_\circ^+ - \mathbf{P}_\circ^-}{\mathbf{qk} - \Omega\omega} = 0. \qquad (13)$$

Eq.(13) has been studied in Ref.(5) for the monochromatic waves, also for the case with $\delta n \neq 0$.

We now write the relativistic expression for the ponderomotive force

$$\mathbf{F} = -\nabla\gamma = -\nabla(1 + \Pi(\mathbf{R},t))^{1/2}. \qquad (14)$$

After linearization with respect to the perturbation we have

$$\delta\mathbf{F} = -\frac{1}{2\gamma_\circ}\nabla\delta\Pi exp i(\mathbf{qR} - \Omega t), \qquad (15)$$

or using Eq.(12) we obtain

$$\delta\mathbf{F} = -\frac{1}{2\gamma_\circ^2} \frac{\int d\mathbf{k}/(2\pi)^3 \int (d\omega/2\pi)(\mathbf{P}_\circ^+ - \mathbf{P}_\circ^-)/(\mathbf{qk} - \Omega\omega)}{1 + (1/2\gamma_\circ^3) \int d\mathbf{k}/(2\pi)^3 \int (d\omega/2\pi)(\mathbf{P}_\circ^+ - \mathbf{P}_\circ^-)/(\mathbf{qk} - \Omega\omega)} \times$$
$$\nabla\delta n exp i(\mathbf{qr} - \Omega t). \qquad (16)$$

Some interesting relativistic feature follows from the expression of the ponderomotive force (16). First, in the case when the dominator goes to zero, δF increase, or $\delta n \to 0$. The second when the integral in dominator becomes much greater than unity, in this case we have

$$\delta \mathbf{F} = -\gamma_o \nabla \delta n expi(\mathbf{qR} - \Omega t) . \tag{17}$$

This expression (17) of the ponderomotive force coincides formally with the gas-dynamic force, only instead of the temperature we have $m_o\gamma_o c^2$ in Eq.(17), and exist only in the relativistic motion of electrons in a superstrong short laser pulse.

If we write kinetic equations for electrons and ions with the ponderomotive force (14) and linearize them, taking into account the relation (12), we obtain the general dispersion relation

$$\varepsilon\left(1 + \frac{1}{2\gamma_o^3}\int \frac{d\mathbf{k}}{(2\pi)^3}\int \frac{d\omega}{2\pi}\frac{\mathbf{P}_o^+ - \mathbf{P}_o^-}{\mathbf{qk} - \Omega\omega}\right) + (1 + \delta\varepsilon_i) \times$$
$$\delta\varepsilon_e \frac{q^2c^2}{\omega_{pe}^2}\frac{1}{2\gamma_o^3}\int \frac{d\mathbf{k}}{(2\pi)^3}\int \frac{d\omega}{2\pi}\frac{\mathbf{P}_o^+ - \mathbf{P}_o^-}{\mathbf{qk} - \Omega\omega} = 0 , \tag{18}$$

where

$$\varepsilon = 1 + \delta\varepsilon_e + \delta\varepsilon_i, \quad \delta\varepsilon_\alpha = \frac{4\pi e^2}{q^2}\int \frac{q(\partial f_{o\alpha}/\partial \mathbf{p})}{\Omega - \mathbf{qv}}d\mathbf{p} .$$

Eq.(18) has several kind of complex solutions for Ω, resulting in a different type of instability. But here we focus our attention on the case of the stationary longitudinal wave propagation in a plasma due to a strong laser pulse. Such a possibility exist, if

$$\frac{1}{2\gamma_o^3}\int \frac{d\mathbf{k}}{(2\pi)^3}\int \frac{d\omega}{2\pi}\frac{\mathbf{P}_o^+ - \mathbf{P}_o^-}{\mathbf{qk} - \Omega\omega} \gg 1 \tag{19}$$

and from Eq.(18), then we have

$$(1 + \delta\varepsilon_i)\left(1 + \delta\varepsilon_e \frac{q^2c^2}{\omega_{pe}^2}\right) + \delta\varepsilon_e = 0 . \tag{20}$$

The dispersion relation (20) describes the stationary longitudinal wave propagation in the presence of the relativistically intense em waves.

Let us now consider some special cases. First, the case, when only electrons participate in the oscillation, i.e. $\delta\varepsilon_i = 0$, for $\Omega \gg qv_{tre}$, where $v_{tre} = (T_e/m_o)^{1/2}$, we obtain from Eq.(20)

$$\Omega^2 = \omega_{pe}^2 + q^2c^2 . \tag{21}$$

Note that this is a novel Langmuir wave due to the strong relativistic effect.

Next in the case, when $\delta n_i \neq 0$, the two frequency ranges can be considered for Ω. One is $kv_{tri} \ll \Omega \ll kv_{tre}$, and the other is $kv_{tre} \ll \Omega \ll \omega_{pe}$. For both cases we obtain from Eq.(20) a new type of ion-sound solution

$$\Omega = \left(\frac{m_{oe}\gamma_o}{m_{oi}}\right)^{1/2} \frac{qc}{(1+q^2c^2/\omega_{pe}^2)^{1/2}} = \frac{qv_s}{(1+q^2c^2/\omega_{pe}^2)^{1/2}} . \tag{22}$$

As follows from Eq.(22) the Maximum value of the frequency is ω_{pi}. The difference of these two cases is that, in the first case this wave damp slowly only on ions. We specifically note here that, now $v_s = c(m_{oe}\gamma_o/m_{oi})^{1/2}$ depends, not only on mass of the particles, but also on the intensity of the laser pulse ($\gamma_o = (1+\Pi_o)^{1/2}$). Therefore, it is possible to observe these waves in experiment by changing the intensity of the laser pulse and the sort of gas.

We now try to understand physically existence of the solutions (21) and (22). First note that we can get the result (17) for the ponderomotive force without linearization of equation (6), in stationary case, when the laser pulse propagates with a constant velocity as $\mathbf{v} = \mathbf{k}c^2/\omega$, or $\mathbf{P}(\mathbf{R},t,\mathbf{k},\omega) = \mathbf{P}(\mathbf{k},\omega, R-vt)$. In this case, the left-hand side of Eq.(6) becomes zero and one of the solutions of equation (6) is

$$\rho = \frac{n(\mathbf{R},t)}{\gamma(\mathbf{R},t)} = constant . \tag{23}$$

Equation (23) can be expressed as $n/(m_e(\gamma)) = constant,$ which shows that, the plasma density and the mass of electron satisfy the "frosen-in" condition. This means that there is localization of energy of the laser pulse in the region of high density. In other words, the behaviour of the system plasma-photon gas is similar to the one component fluid. The solution identical with Eq.(23) was found in Ref.(6), considering the strong em wave propagation in an electron-positron plasma. In the case when expression (23) is valid, we obtain the simple expression for the ponderomotive force from Eq.(14) for the arbitrary variation of the density

$$\mathbf{F} = -m_o\gamma_o c^2 \nabla \frac{n}{n_o} . \tag{24}$$

As an example, for the electron Langmuir wave Eq.(21). Taking into account Eqs.(23) and (24), from the hydrodynamic equations ($m_o\gamma_o c^2 \gg T_e$, where T_e is the temperature of electrons) for the arbitrary variation of the density, we obtain the linear equation

$$\left(\frac{\partial^2}{\partial t^2} + \omega_{pe}^2 - c^2\Delta\right)\frac{n}{n_o} = 0 \tag{25}$$

In conclusion, we note, that Eq.(1) at (23) becomes the linear equation and the em wave momentum with arbitrary power will always be spread out in a plasma.

REFERENCES

1. Tsintsadze,N.L., Zh. Eksp. Teor. Fiz. **59**, 1250 (1970); Tsintsadze,N.L., Phys. Lett. **50A**(1), 33 (1974); Shukla,P.K., Rao,N.N, Yu,M.Y., and Tsintsadze,N.L., Phys. Rep. **138**, 1-49 (1986); Berezhiani,V.I., Tsintsadze,N.L., Tskhakaya,D.D., Proc. *Inter. Conference on Plasma Physics* (Kiev, 1987), ed. by A.G.Sitenko (Kiev, 1987), p.575.

2. Tsintsadze,N.L., Sov. Phys. JETP **34**, 1809 (1964); Rosenbluth,M.N., and Liu,C.S., Phys. Rev. Lett. **29**, 702 (1972); Max,C.E., Arons,J., and Langdon,A.B., Phys. Rev. Lett. **33**, 209 (1974); Tsintsadze,N.L., Proc. *Phenomena in Ionized Gases* (Minsk, 1981), p.395; Tsintsadze,L.N., Proc. *Inter. Topical Conf. on Research Trends in Coherent Radiation Generation and Particle Accelerators* (La Jolla, 1991), ed. by J.M.Buzzi, P.Sprangle, K.Wille (AIP, New York, 1992), p.474; Tsintsadze,L.N., and Nishikawa,K., Phys. Plasmas **3**, 511 (1996).

3. Garuchava,D.P., Rostomashvili,Z.I., and Tsintsadze,N.L., Fiz. Plazmi (Soviet) **12**, 1341 (1986).

4. Garuchava,D.P., Murusidze,I.G., Suramlishvili,G.I., Tsintsadze,N.L., Tskhakaya,D.D., Sov. J.Plasma Phys. Rep. **22**(10), 841 (1996).

5. Tsintsadze,L.N., Sov. J. Plasma Phys. **17**, 872 (1991).

6. Berezhiani,V.I., Tsintsadze,L.N., and Shukla,P.K., J.Plasma Phys. **48**, 139 (1992).

On the Theory of Relativistically Intense EM Wave Propagation in Overdense Plasma and the Boundary-Value Problem

Levan N. Tsintsadze, Kunioki Mima and Kyoji Nishikawa*

Institute of Laser Engineering, Osaka University, Osaka, Japan
*Research Institute of Industrial Technology, Kinki University, Japan

Abstract. A generalized set of equations of the plasma hydrodynamics in superstrong radiation fields is presented. A dispersion relation for an arbitrary amplitude of pump field is found and new branches for the modulational instability are disclosed. The stationary solution describing a solitary wave propagation, which has the character of a compression wave, with a constant velocity is studied and the amplitude dependence of the propagation speed of the wave is found. The irradiance threshold for the transition is obtained as a function of plasma density. The plasma compression effects are also shown by the analysis of more realistic three dimensional electromagnetic (em) wave case. The boundary conditions for the fundamental equations, which define the field in a plasma are derived and the boundary-value problem is discussed.

Propagation of an intense laser beam through a high-density plasma is an important issue in connection with the fast ignition of the compressed core in the laser fusion research. In order for this FI scheme to be feasible, the laser pulse has to penetrate into a region of plasma density higher than the critical density. Two important mechanism are considered for the penetration, one is the self-focusing of the laser beam by digging a hole inside the high-density plasma (1), and the other is the relativistic effect due to the high electron quivering velocity acquired by the intense laser electric field (2). The present paper deals with the latter mechanism, namely the possibility for the propagation of an intense laser pulse to an overdense plasma region by relativistic effects.

Considering a circularly polarized intense em wave propagation in the z-direction, taking into account the relativistic nature of the electron motion, we begin with Maxwell equation and the plasma hydrodynamics equations neglecting the hydrodynamic pressure force as compared with the ponderomotive one,

$$c^2 \frac{\partial^2 \mathbf{E}}{\partial z^2} - \frac{\partial^2 \mathbf{E}}{\partial t^2} - \frac{\omega_p^2}{\gamma} \frac{n}{n_o} \mathbf{E} = 0 , \qquad (1)$$

$$\frac{\partial n_i}{\partial t} + \frac{\partial}{\partial z}(n_i v_i) = 0 , \quad \frac{\partial v_i}{\partial t} + v_i \cdot \frac{\partial v_i}{\partial z} = -\frac{m_o}{m_i} c^2 \frac{\partial \gamma}{\partial z} , \qquad (2)$$

using the traditional approach, $E_- = E_x - iE_y = E exp(-i\omega t) = a(z,t) exp(-i[\omega t - \phi(z,t)])$, we obtain from Eq.(1)

$$2i\omega \frac{\partial E}{\partial t} + c^2 \frac{\partial^2 E}{\partial z^2} + \omega^2 E - \frac{\omega_p^2}{\gamma} \frac{n}{n_o} E = 0 , \qquad (3)$$

which yields the following two equations:

$$\frac{\partial a^2}{\partial t} + \frac{c^2}{\omega} \frac{\partial}{\partial z}\left(a^2 \frac{\partial \phi}{\partial z}\right) = 0 , \qquad (4)$$

$$2\omega a \frac{\partial \phi}{\partial t} - c^2 \frac{\partial^2 a}{\partial z^2} + c^2 a \left(\frac{\partial \phi}{\partial z}\right)^2 - \omega^2 a + \frac{\omega_p^2}{\gamma} \frac{n}{n_o} a = 0 . \qquad (5)$$

These equations together with Eqs.(2) form the closed set of equations describing the behavior of a plasma in the field of an intense em wave.

We first discuss the modulational instability (3) in the case of strong relativism. To this end, we linearize Eqs.(2),(4),(5) with respect to the perturbation and assume $\delta n, \delta \phi, \delta a$ vary like $exp[i(Kz - \Omega t)]$, then we obtain the dispersion relation

$$\left\{4(X-1)^2 - \left(\frac{K}{k}\right)^2 + \frac{\omega_p^2}{k^2 c^2} \frac{\gamma_o^2 - 1}{\gamma_o^3}\right\} X^2 = \frac{\omega_p^2}{k^2 c^2} \frac{\gamma_o^2 - 1}{\gamma_o^2} \left(\frac{K}{k}\right)^2 \frac{m_o}{m_i} , \qquad (6)$$

where $X = \omega\Omega/(kKc^2)$. Eq.(6) has a couple of complex solutions in the wavenumber region, $K^2 c^2 < \omega_p^2 (\gamma_o^2 - 1)/\gamma_o^3$, which gives the maximum growth rate

$$\Gamma_{MAX} = \frac{\omega_p^2}{4\omega} \frac{\gamma_o^2 - 1}{\gamma_o^3} , \quad at \quad (Kc)^2 = \frac{\omega_p^2}{2} \frac{\gamma_o^2 - 1}{\gamma_o^3} . \qquad (7)$$

In the case in which ion dynamics plays an important role we have for the growth rate

$$\Gamma = \frac{kc\omega_p}{2\omega} \left(\frac{\gamma_o^2 - 1}{\gamma_o^3}\right)^{1/2} \left(\frac{2\omega_p^2}{k^2 c^2} \frac{\gamma_o^2 - 1}{\gamma_o^{5/2}} \left(\frac{m_o}{m_i}\right)^{1/2} - 1\right)^{1/2} . \qquad (8)$$

We next consider a solitary wave solution. A supersonic solitary wave solution of compressional character was obtained in Ref.(2). In the following we generalize the analysis to the case of arbitrary relativism and the arbitrary density perturbation and investigate the properties of the solitary wave. To this end, we first examine a general stationary solution of the basic equations by assuming a propagation with constant speed u in the z-direction. Transformation to the moving frame, $\xi = z - ut$ and $\tau = t$, yield the relation

$$\frac{n_i(\xi)}{n_o} = \frac{1}{\left\{1 - \beta\left([1 + a^2(\xi)]^{1/2} - 1\right)\right\}^{1/2}} , \qquad (9)$$

where $\beta = 2m_o c^2/(m_i u^2)$. Eq.(9) clearly shows a compressional character, i.e. $n_i(\xi) > n_o$. It also gives a lower limit for the propagation velocity u,

$$\frac{u^2}{c^2} > \frac{2m_o}{m_i}\{(1+a^2)^{1/2} - 1\} . \tag{10}$$

From Eq.(4) we obtain

$$\{\frac{c^2}{\omega}\frac{\partial \phi}{\partial \xi} - u\}a^2 = const. \tag{11}$$

For solitary waves, $a(\xi \to \pm\infty) = 0$, so that we have

$$\frac{\partial \phi}{\partial \xi} = \frac{\omega u}{c^2} . \tag{12}$$

If we further impose the local charge neutrality, i.e. $n(\xi) = n_i(\xi)$, and substitute Eq.(12) into Eq.(5), for the case $u^2/c^2 \ll 1$, we have

$$\frac{d^2 a(\xi)}{d\xi^2} + \{\frac{\omega^2}{c^2} - \frac{\omega_p^2}{c^2}\frac{n_i(\xi)}{n_o \gamma(\xi)}\}a(\xi) = 0 . \tag{13}$$

In the weak relativistic limit, $\gamma \approx 1$, $n_i(\xi)/n_o = 1 + m_o/(2m_i)(c^2/u^2)a^2$, we recover the nonrelativistic result, which admits no solitary wave solutions. Situation changes if the proper expression for the relativistic nonlinear term is used (2). Indeed, substituting Eq.(9) into Eq.(13), and introducing the notation, $(\omega_p/c)\xi = \eta$ and $\omega_p^2/\omega^2 = \alpha$, we obtain

$$\frac{d^2 a}{d\eta^2} + \frac{a}{\alpha} - \frac{1}{(1+a^2)^{1/2}} \frac{a}{\{1 - \beta([1+a^2]^{1/2} - 1)\}^{1/2}} = 0 , \tag{14}$$

from which we have

$$\left(\frac{da}{d\eta}\right)^2 + \frac{a^2}{\alpha} + \frac{4}{\beta}\{1 - \beta([1+a^2]^{1/2} - 1)\}^{1/2} = \frac{4}{\beta} , \tag{15}$$

where the right-hand side was determined by the condition that the solitary wave solution must satisfy the boundary conditions that $a = 0$ and $da/d\eta = 0$ at $\eta \to \pm\infty$. The height a_M of the solitary wave is determined by the condition $da/d\eta = 0$ at $a = a_M$,

$$\frac{a_M^2}{\alpha} + \frac{4}{\beta}\{1 - \beta([1+a^2]^{1/2} - 1)\}^{1/2} = \frac{4}{\beta} , \tag{16}$$

which gives the amplitude dependence of the propagation speed u,

$$\frac{u^2}{c^2} = \frac{m_o}{4m_i} \cdot \frac{a_M^4/\alpha^2}{(a_M^2/\alpha) - 2\{(1+a_M^2)^{1/2} - 1\}} , \tag{17}$$

valid for arbitrary amplitude of the electromagnetic wave. This relation, however, does not uniquely determine the amplitude a_M as a function of the propagation speed. Indeed, we have to select the solution for which $d^2a/d\xi^2 < 0$ at $a = a_M$. In order to check this condition, we first note two other conditions to be satisfied. One comes from Eq.(16) $a_M^2/\alpha \leq 4/\beta$, or using Eq.(17),

$$a_M^2 \leq 8\alpha(2\alpha - 1) . \qquad (18)$$

This condition corresponds to the condition for the absence of wave breaking, i.e. $n(\xi) < \infty$. The other comes from Eq.(17),

$$a_M^2 > 4\alpha(\alpha - 1) , \qquad (19)$$

The two conditions (18) and (19) can alternatively be written as $(2\alpha - 1)^2 < a_M^2 + 1 \leq (4\alpha - 1)^2$. We therefore write $a_M^2 + 1 = (s\alpha - 1)^2$, s being slowly dependent on α with $2 < s \leq 4$. After some calculations, we then find the following relation for the amplitude limit

$$(2\alpha - 1) < (1 + a_M^2)^{1/2} < \frac{(3\alpha - 1) + \{(9\alpha - 1)(\alpha - 1)\}^{1/2}}{2} . \qquad (20)$$

Let us now discuss some special cases. First, in the case of weak relativism where $a_M^2 \ll 1$, we can immediately recover the result of Ref.(2),

$$\frac{u^2}{c^2} = \frac{m_o \, \omega^2}{m_i \, \omega_p^2} \frac{a_M^2}{(\omega_p^2/\omega^2)a_M^2 - 4 \, |\varepsilon|} . \qquad (21)$$

where $\varepsilon = 1 - \omega_p^2/\omega^2$. In the other limit of strong relativism where $2a_M \gg \alpha$,

$$\frac{u^2}{c^2} = \frac{m_o \, \omega^2}{m_i \, \omega_p^2} \frac{a_M^2}{4} . \qquad (22)$$

Finally, for the case $\beta K_o \ll 1$ where $K_o = (1 + a^2)^{1/2} - 1$, Eq.(14) is reduced to

$$\left(\frac{da}{d\eta}\right)^2 - \frac{\omega^2}{\omega_p^2} |\varepsilon| a^2 + \left\{1 - \frac{m_o \, c^2}{m_i \, u^2}\right\} K_o^2 = 0 , \qquad (23)$$

which recovers the result of Ref.(2) for $u^2/c^2 \ll 1$. For the case $a^2 \ll 1$, we have $K_o^2 \approx a^4/4$, so that we obtain a solitary wave solution in the form

$$a(\xi) = A \mathrm{sech}\left(\frac{\xi}{\sigma}\right) , \qquad (24)$$

where $A = 4(\omega_p^2 - \omega^2)/\omega_p^2\left(1 - (m_o/m_i)(c^2/u^2)\right)^{-1}$ and $\sigma = c/(\omega_p^2 - \omega^2)^{1/2}$.

We now briefly discuss 3D effects. A radially symmetric propagation of a solitary wave can be described by replacing $\partial/\partial z$ in the basic equations by $\partial/\partial z + \nabla_\perp$,

$$\frac{\partial^2}{\partial \tau^2}\frac{\delta n}{n_o} = \beta\Big(\frac{\partial^2}{\partial \tau^2} + u^2 \Delta_\perp\Big)\Big((1 + a^2(r_\perp, \tau))^{1/2} - 1\Big), \qquad (25)$$

where $\tau = t - z/u$, $a^2 = a_o^2(r_\perp)\big(\Theta(\tau - \tau_1) - \Theta(\tau - \tau_2)\big)$, $a_o^2(r_\perp) = a_o^2(0)exp(-r_\perp^2/2r_o^2)$, $\Theta(x)$ is the unit step function, r_o is the radial width of the beam and $\tau_o = \tau_2 - \tau_1$ is the pulse length. Integrating Eq.(25) we obtain

$$\Big(\frac{\delta n}{n_o}\Big)_{r_\perp = 0} = \beta\Big((1 + a_o^2)^{1/2} - 1 - \frac{a_o^2}{1 + a_o^2}\frac{u^2(\tau - \tau_1)^2}{2r_o^2}\Big), \qquad (26)$$

which shows a compressional character, i.e. $\delta n > 0$, with the following conditions, $r_o > u\tau_o$ in the case of weak relativism, and $r_o \sim u\tau_o$ for the strong relativism.

We next consider a boundary value problem by considering an overdense plasma filling the half space $z \geq 0$, the other half space being in vacuum. The em wave in this half space consist of an incident and reflected waves

$$E_{ix} = E_i(t)cos\omega(t - z/c), \quad E_{iy} = E_i(t)sin\omega(t - z/c),$$
$$E_{rx} = E_r(t)cos\big(\omega(t + z/c) - \psi(t)\big), \quad E_{ry} = E_r(t)sin\big(\omega(t + z/c) - \psi(t)\big), \qquad (27)$$

and for the transmitted wave

$$E_x = E(z,t)cos\big(\omega t - \phi(z,t)\big), \quad E_y = E(z,t)sin\big(\omega t - \phi(z,t)\big), \qquad (28)$$

From the continuity conditions for the electric and magnetic fields, taking into account $\partial \mathbf{B}/\partial t = -c\,curl\,\mathbf{E}$, we obtain

$$E_i + E_r cos\psi = E_o cos\phi_o, \quad E_r sin\psi = E_o sin\phi_o$$
$$\frac{\omega}{c}E_i\Big(1 - \frac{E_r}{E_i}cos\psi\Big) = E_{oz}sin\phi_o + E_o\phi_{oz}cos\phi_o,$$
$$\frac{\omega}{c}E_r sin\psi = E_{oz}cos\phi_o - E_o\phi_{oz}sin\phi_o, \qquad (29)$$

where $E_o = E(0,t)$, $E_{oz} = (\partial E/\partial z)_{z=0}$, $\phi_{oz} = (\partial \phi/\partial z)_{z=0}$, $E_i, E_r, E \sim (E_x^2 + E_y^2)^{1/2}$. Introducing the reflection coefficient $R(t) = E_r^2(t)/E_i^2(t)$ and using (29)

$$R = 1 - \frac{c}{\omega}\frac{E_o^2}{E_i^2}\phi_{oz}, \qquad (30)$$

$$\frac{E_o^2}{E_i^2} = 4\Big(1 - \frac{c^2}{4\omega^2}\frac{E_{oz}^2}{E_i^2}\Big)\Big(1 + \frac{c}{\omega}\phi_{oz}\Big)^2, \qquad (31)$$

$$cos\phi_o = \frac{E_o}{2E_i}\Big(1 + \frac{c}{\omega}\phi_{oz}\Big), \quad sin\phi_o = \frac{c}{2\omega}\frac{E_{oz}}{E_i}. \qquad (32)$$

These relations, together with the conditions $E(z,t) \to 0$ $(z \to \infty)$, $\int_0^\infty E^2(z,t)dz < \infty$ form a complete system of boundary conditions for our basic equations. To calculate the reflection coefficient we must find ϕ_{oz}. Using Eq.(3) we obtain

$$R(t) = 1 - \frac{1}{cE_i^2(t)} \frac{d}{dt} \int_0^\infty E^2(z,t)dz . \tag{33}$$

We again assume that $E_i(t) \neq 0$ $(t_1 < t < t_2)$, and $E_i(t) = 0$ $(t < t_1, t > t_2)$. We now calculate the total energy penetrating the plasma, using Eq.(33)

$$W(t) = \frac{c}{8\pi} \int_{t_1}^t dt\prime E_i^2(t\prime)\left(1 - R(t\prime)\right) = \frac{1}{8\pi} \int_0^\infty E^2(z,t)dz . \tag{34}$$

Thus Eqs.(31),(33),(34) are generalized, valid for arbitrary amplitudes, including relativism. In the special case in which the solitary wave is given by Eq.(24),

$$W = \frac{c}{(\omega_p^2 - \omega^2)^{1/2}} \frac{E_{MAX}^2}{8\pi} . \tag{35}$$

In the case of weak relativism, using Eq.(12) we can estimate the quantities appearing in Eq.(31), which yields the relation $E_o^2 \approx 4E_i^2$. From Eq.(30)

$$(1-R)\frac{c}{8\pi}E_i^2 = \frac{u}{8\pi}E_o^2(E_o^2 \approx 4E_i^2) = \frac{u}{2\pi}E_i^2 \text{ and } W(t) = \frac{u}{2\pi}\int_{t_1}^t E_i^2(t\prime)dt\prime \tag{36}$$

where u is determined by Eq.(15) at a_{MAX}.

REFERENCES

1. Garuchava,D.P., Rostomashvili,Z.I., and Tsintsadze,N.L., Fiz. Plazmi (Soviet) **12**, 1341 (1986); Lee,Y.C., Liu,C.S., Chen,U.H., and Nishikawa,K., Plasma Phys. and Controlled Nuclear Fusion Research **3**, 207 (1974); Garuchava,D.P., Murusidze,I.G., Suramlishvili,G.I., Tsintsadze,N.L., Tskhakaya,D.D., Sov. J.Plasma Phys. Rep. **22**(10), 841 (1996).

2. Tsintsadze,N.L., and Tskhakaya,D.D., Sov. Phys. JETP **45**, 252 (1977).

3. Berezhiani,V.I., Tsintsadze,N.L., Tskhakaya,D.D., J.Plasma Phys. **24**, 15 (1980); Tsintsadze,L.N., Sov. J. Plasma Phys. **17**, 872 (1991); Tsintsadze,N.L., Proc. Inter. Conference on Plasma Physics (Foz Do Iguacu, 1994), ed. by P.H.Sakanaka (AIP, 1995). p.290; Tsintsadze,L.N., and Nishikawa,K., Phys. Plasmas **3**, 511 (1996).

Breaking of the Laser Excited Wakefield Due to Electron Beam

Levan N. Tsintsadze, Kunioki Mima and Kyoji Nishikawa*

Institute of Laser Engineering, Osaka University, Osaka, Japan
*Research Institute of Industrial Technology, Kinki University, Japan

Abstract. Effects of electron beam, accelerated by a superstrong short laser pulse, on the electron plasma waves are investigated. Shock formation and subsequent wave breaking are discussed and the time and the place of the shock formation are determined for two different cases, namely the first, electron beam is modulated by a wakefield and the second, electron beam is considered with Gaussian distribution.

The relativistic mass oscillation in the presence of a large amplitude electromagnetic wave can strongly enhance the anomalous absorption of the wave in plasma. Tsintsadze [1] has shown that this mass variation can parametrically amplify plasma waves. Drake et al. [2], studing the nonlinear evolution of large amplitude waves near cutoff including the relativistic mass changes of the electrons, shown that, due to the relativistic motion of the electrons, electrostatic perturbations of electromagnetic waves "slow down" in regions of high intensity, steepen, and then break. Electrons are heavier in the region of the highest wave intensity so the wave "slows down" in this region, allowing the lower intensity portion of the wave to "catch up".

In the present paper, the system of electron beam and electrostatic wakefield is described. Namely, the physical situation we have in mind is the following. An intense laser pulse incident on a high density plasma ejects electrons by the ponderomotive force in the transverse direction and at the same time accelerates the remaining electrons in the propagation direction and produces an electron beam. This electron beam is left behind the laser pulse and interects with the laser excited wakefield. If the time under consideration is short compared with the time needed for the radial redistribution of the ejected electrons, then we can consider a one-dimensional problem of beam and the electrostatic wakefield.

The beam electrons are represented by the momentum p and density n , while the electrostatic wake field by the electric field E. We use ω_p^{-1}, c/ω_p, $m_o c$, $m_o c^2$ and n_i as units of the time, length, momentum, energy and density where n_i is the ion density, m_o is the electron rest mass, c is the speed of light in vacuum and ω_p is the electron plasma frequency at the ion density, i.e. $\omega_p^2 = 4\pi e^2 n_i/m_o$. Then the basic equations can be written as follows:

$$\frac{dp}{dt} = -E(x,t), \tag{1}$$

$$\frac{\partial E(x,t)}{\partial t} = nu, \quad u = \frac{p}{(1+p^2)^{1/2}}, \tag{2}$$

$$\frac{\partial E(x,t)}{\partial x} = 1 - n, \tag{3}$$

$$\frac{\partial n}{\partial t} + \frac{\partial(nu)}{\partial x} = 0, \tag{4}$$

from which we obtain

$$\frac{d^2 p}{dt^2} + \frac{p}{(1+p^2)^{1/2}} = 0. \tag{5}$$

We transform to the Lagrangean frame

$$x_o = x - \int_0^t dt' u(x_o, t'), \tag{6}$$

which reduces Eqs. (4) and (5) to

$$n(x,t)dx = n_o(x_o, t=0)dx_o, \tag{7}$$

$$\frac{\partial^2 p(x_o, t)}{\partial t^2} + \frac{p}{(1+p^2)^{1/2}} = 0. \tag{8}$$

We now discuss the implifications of these equations. First, Eq. (7) implies that at $(\partial x/\partial x_o)_t = 0$, $n(x,t) \to \infty$, corresponding to the breaking of the electrostatic wave. Using Eqs. (6) and (2), we can write

$$\left(\frac{\partial x}{\partial x_o}\right)_t = 1 + \frac{\partial}{\partial x_o}\int_0^t dt' u(x_o, t') = 1 - \frac{\partial^2 p(x_o, t)}{\partial x_o \partial t}. \tag{9}$$

The time t_s and place x_s of the shock front is determined by the following simultaneous equations

$$\left(\frac{\partial x}{\partial x_o}\right)_{t=t_s} = 0, \quad \left(\frac{\partial^2 x}{\partial x_o^2}\right)_{t=t_s} = 0. \tag{10}$$

On the other hand, Eq. (8) can be written as

$$\left(\frac{\partial p}{\partial t}\right)^2 + 2(1+p^2)^{1/2} = H(x_o) \equiv 2\left(1+p(x_o)^2\right)^{1/2}, \quad (11)$$

which implies a periodic oscillation of $p(x_o, t)$ as a function of time with the period given by

$$T \equiv \frac{2\pi}{\omega} = \frac{2}{2^{1/2}} \int_{p_{xo}}^{-p_{xo}} \frac{dp}{\left(H(x_o) - (1+p^2)^{1/2}\right)^{1/2}}. \quad (12)$$

We therefore express the momentum as $p(x_o, t) = h(H, \phi)$ where $H = H(x_o)$ and $\phi = \omega(H)t$. Then the first equation of Eq. (10) can be written as

$$1 = \left(\frac{\partial^2 p(x_o, t)}{\partial x_o \partial t}\right)_{t=t_s} = \{t_s h_{\phi\phi}\omega \frac{\partial \omega}{\partial H} + h_{\phi H}\omega + h_\phi \frac{\partial \omega}{\partial H}\}\frac{\partial H}{\partial x_o}, \quad (13)$$

where from Eq. (8), $h_{\phi\phi}\omega^2 = -p/(1+p^2)^{1/2}$. Using these relations we get

$$1 = \{-t_s \frac{p_M}{(1+p_M^2)^{1/2}} \frac{\partial \ln \omega}{\partial H} + h_{\phi H}\omega + h_\phi \frac{\partial \omega}{\partial H}\}\frac{\partial H}{\partial x_o}, \quad (14)$$

where $p_M = p(x_o, t_s)$. Since the last two terms inside the bracket of the right-hand side of Eq. (14) are periodic function of time, the first term is dominant provided that $t_s \gg T$. In this case, we have

$$t_s \simeq \frac{(1+p_M^2)^{1/2}}{p_M}\left(\frac{\partial \ln \omega}{\partial x_s}\right)^{-1} \quad (15)$$

We now consider the ultra-relativistic case, i.e. $p^2 \gg 1$. In this case the oscillation period can be calculated by neglecting unity as compared with p^2 as

$$T = 2\left(p(x_o)\right)^{1/2}, \quad \text{or} \quad \omega = \frac{\pi}{\left(p(x_o)\right)^{1/2}}, \quad (16)$$

so that we have

$$t_s \simeq \frac{2p_M}{|\partial p_M/\partial x_s|}. \quad (17)$$

Let us assume that the wakefield modulates the beam as $p_M = p_o + a\sin x_s$. Then

$$t_s \simeq \frac{2(p_o + a\sin x_s)}{a\cos x_s}, \quad (18)$$

which is much larger than T if $p_o \gg a$. The second equation of Eq. (10) can then be calculated as

$$\left(\frac{\partial^2 x}{\partial x_o^2}\right)_{t=t_s} = -\left(\frac{\partial^3 p(x_o, t)}{\partial x_o^2 \partial t}\right)_{t=t_s} = -\frac{\partial t_s}{\partial x_o} \frac{p_M}{(1+p_M^2)^{1/2}} \frac{\partial \ln \omega}{\partial x_o} = 0, \quad (19)$$

which gives for $p_o \gg a$

$$a - p_o \sin x_s \simeq 0 \quad or \quad \sin x_s \simeq -\frac{a}{p_o}, \tag{20}$$

from which we finally obtain using Eq. (18)

$$t_s \simeq \frac{2p_o}{a}\left(1 - \frac{a^2}{p_o^2}\right)^{1/2}, \quad x_s \simeq \pi - \frac{a}{p_o}. \tag{21}$$

We next consider the case of small beam velocity, i.e. $p_o^2 \ll 1$. Then Eq. (8) describes a linear harmonic oscillation of frequency unity. We therefore write $p(x_o, t) = a(x_o)\sin(kx_o - t)$. Equation (13) then becomes

$$-a\prime(x_s)\cos(kx_o - t) + a(x_s)k\sin(kx_o - t) = 1. \tag{22}$$

The solution of this equation can be written as $a(x_s) = A\cos(kx_s)$, with

$$Ak\sin(2kx_s - t_s) = 1. \tag{23}$$

The second equation of Eq. (10) then becomes

$$2Ak^2\cos(2kx_s - t_s) = 0, \quad or \quad 2kx_s - t_s = \frac{\pi}{2}, \tag{24}$$

from which we have from Eq. (23) $Ak = 1$. Now the condition $p_o^2 \ll 1$ requires $A \ll 1$, or $k \gg 1$. Thus we conclude that in this case only a short wavelenght wakefield suffers from breaking.

Finally, we consider electron beam with Gaussian distribution

$$p = p_o exp\left(-\frac{x_o^2}{L_o^2} - \frac{t^2}{t_o^2}\right). \tag{25}$$

In this case, we obtain from Eq. (9) the following relation for the time of the shock front

$$t_s = \frac{L_o^2 \ t_o^2}{4x_s \ p_M}. \tag{26}$$

Using now Eq. (25), the second equation of Eq. (10) gives

$$x_s = 2^{1/2} L_o. \tag{27}$$

In conclusion, we note that the results of our investigations could be useful for the fast ignition program in the laser inertial fusion, for the problem of new plasma-based high-energy particle accelerators, as well as for explaining certain processes in astrophysical and cosmological plasmas.

REFERENCES

1. Tsintsadze,N.L., Zh. Eksp. Teor. Fiz. **59**, 1250 (1970).

2. Drake,J.F., Lee,Y.C., Nishikawa,K., and Tsintsadze,N.L., Phys. Rev. Lett. **36**, 196 (1976).

On Self-Consistent Stationary Propagation of Relativistically Coupled Electromagnetic and Electrostatic Waves in Three Component Plasma

Levan N. Tsintsadze and Kyoji Nishikawa[*]

Institute of Laser Engineering, Osaka University, Osaka, Japan
[*]Research Institute of Industrial Technology, Kinki University, Japan

Abstract. A general self-consistent theory is presented for one-dimensional stationary propagation of relativistically coupled electromagnetic and electrostatic waves in cold electron-ion plasma and the effects of electron-positron pair creation on this propagation are studied. A novel solution describing an envelope shock which represents a wakefield excitation by a solitary elecromagnetic pulse is obtained by taking into account the trapping of the pair-created particles in the upstream region.

Recent progress in high power laser technology has opened a new possibility for experimental investigation over the parameter range of the laser field energy comparable to or exceeding the electron rest mass energy. Under such conditions, strong laser-plasma interactions can take place (1,2), accompanied by the electron-positron pair production (3). Since the lifetime ($\tau_+ > \omega_p^{-1}$) of the positron in such a plasma is sufficiently long, we shall have a plasma which is an admixture of electrons, positrons, and positive ions. In the present paper, we report a general approach to the self-consistent treatment of the problem by restricting ourselves to one-dimensional stationary wave propagation. Considering a circularly polarized electromagnetic wave propagation in the z-direction, the relevant equations, for the vector potential $\mathbf{A}_\perp$ of the electromagnetic wave and for the electrostatic potential ϕ of the plasma wave are as follow:

$$\frac{\partial^2 \mathbf{A}_\perp}{\partial z^2} - \frac{1}{c^2}\cdot\frac{\partial^2 \mathbf{A}_\perp}{\partial t^2} = \frac{4\pi}{c}\mathbf{J}_\perp, \text{ and } \frac{\partial^2 \phi}{\partial z^2} = 4\pi e(n_- - n_+ - n_\circ), \tag{1}$$

and the perpendicular current density is given by

$$\mathbf{J}_\perp = -\frac{en_-\mathbf{P}_{\perp -}}{m_\circ \gamma_-} + \frac{en_+\mathbf{P}_{\perp +}}{m_\circ \gamma_+}, \text{ with } \gamma = \left\{1 + \frac{|\mathbf{P}_\perp|^2}{m_\circ^2 c^2} + \frac{P_\parallel^2}{m_\circ^2 c^2}\right\}^{1/2}, \tag{2}$$

where $n_-(n_+)$, $\mathbf{P}_-(\mathbf{P}_+)$ are the density and momentum of the electron (positron) fluid, n_o is the density of the ion which is for simplicity assumed to be singly ionized and the other notation is standard. Using the usual normalization and introducing new variables, i.e. $\xi \equiv z - v_g t$, $\tau \equiv t$, we obtain the closed system of equations when supplemented by the equations of motion and continuity:

$$\left\{\frac{1}{\gamma_g^2}\frac{\partial^2}{\partial \xi^2} + 2v_g \frac{\partial^2}{\partial \tau \partial \xi} - \frac{\partial^2}{\partial \tau^2}\right\}\mathbf{A}_\perp(\xi,\tau) = \left(\frac{n_-(\xi,\tau)}{\gamma_-(\xi,\tau)} + \frac{n_+(\xi,\tau)}{\gamma_+(\xi,\tau)}\right)\mathbf{A}_\perp(\xi,\tau), \quad (3)$$

$$\frac{\partial^2 \phi(\xi,\tau)}{\partial \xi^2} = n_-(\xi,\tau) - n_+(\xi,\tau) - 1, \quad (4)$$

$$\frac{\partial P_{\|\pm}(\xi,\tau)}{\partial \tau} + \frac{\partial}{\partial \xi}[\gamma_\pm(\xi,\tau) \pm \phi(\xi,\tau) - v_g P_{\|\pm}(\xi,\tau)] = 0, \quad (5)$$

$$\frac{\partial n_\pm(\xi,\tau)}{\partial \tau} + \frac{\partial}{\partial \xi}\left\{n_\pm(\xi,\tau)\left(\frac{P_{\|\pm}(\xi,\tau)}{\gamma_\pm(\xi,\tau)} - v_g\right)\right\} = 0, \quad (6)$$

where v_g is the group velocity of the electromagnetic wave normalized by the speed of light c and $\gamma_g^2 = (1-v_g^2)^{-1}$. We now look for stationary solutions of the above set of equations. To this end, we set $\mathbf{A}_\perp(\xi,\tau) = \{A(\xi)cos[\theta(\xi,\tau)], A(\xi)sin[\theta(\xi,\tau)]\}$, and $\partial P_\|/\partial \tau = 0$, $\partial n/\partial \tau = 0$. Under the boundary condition $P_\| = \phi = 0$ at $A_\perp = 0$, we obtain the set of equations for the background electron-ion plasma:

$$\gamma(\xi) - \phi(\xi) - v_g P_\|(\xi) = 1, \quad (7)$$

$$P_\|(\xi) = v_g \gamma_g^2[1 + \phi(\xi)] - \gamma_g^2\{[1 + \phi(\xi)]^2 - a^2(\xi)\}^{1/2}, \quad (8)$$

$$\frac{n(\xi)}{\gamma(\xi)} = v_g\{[1 + \phi(\xi)]^2 - a^2(\xi)\}^{-1/2}, \quad (9)$$

$$\frac{n(\xi) - 1}{\gamma_g^2} = \frac{v_g[1 + \phi(\xi)]}{\{[1 + \phi(\xi)]^2 - a^2(\xi)\}^{1/2}} - 1, \quad (10)$$

where $a^2(\xi) = [1 + A^2(\xi)]/\gamma_g^2$. As for the pair-created electrons and positrons, we assume that they are created as a cold electron-positron plasma of density σ (in the normalized unit) when the electromagnetic potentials are given by $A = A_o$ and $\phi = \phi_o$. Then for the pair-created electrons and positrons we have:

$$\gamma_\pm(\xi) - v_g P_{\|\pm}(\xi) \pm \phi(\xi) = E_\pm, \quad (11)$$

$$\frac{n_\pm(\xi)}{\gamma_\pm(\xi)} = \sigma v_g \{[E_\pm \mp \phi(\xi)]^2 - a^2(\xi)\}^{-1/2}, \tag{12}$$

$$\frac{n_-(\xi) - n_+(\xi)}{\gamma_g^2} = \sigma v_g \Big\{ \frac{[E_- + \phi(\xi)]}{\{[E_- + \phi(\xi)]^2 - a^2(\xi)\}^{1/2}} - \frac{[E_+ - \phi(\xi)]}{\{[E_+ - \phi(\xi)]^2 - a^2(\xi)\}^{1/2}} \Big\}, \tag{13}$$

where $E_\pm$ is given by the boundary condition that $P_{\parallel\pm} = 0$ at $A = A_o$ and $\phi = \phi_o$ as $E_\pm = [1 + A_o^2]^{1/2} \pm \phi_o$. Using the representation $\mathbf{A}_\perp$ in Eq.(3) we obtain

$$\theta(\xi,\tau) = \omega\tau + \chi(\xi) \qquad (\omega = const.), \tag{14}$$

$$2\Big(\frac{d\chi(\xi)}{d\xi} + \gamma_g^2 v_g \omega\Big) \frac{dA(\xi)}{d\xi} + \frac{d^2\chi(\xi)}{d\xi^2} A(\xi) = 0, \tag{15}$$

$$\frac{d^2 A(\xi)}{d\xi^2} - \Big\{\Big(\frac{d\chi(\xi)}{d\xi}\Big)^2 + 2\gamma_g^2 v_g \omega \frac{d\chi(\xi)}{d\xi} - \gamma_g^2 \omega^2\Big\} A(\xi) = \gamma_g^2 \Big\{\frac{n(\xi)}{\gamma(\xi)} + \frac{n_\pm(\xi)}{\gamma_\pm(\xi)}\Big\} A(\xi). \tag{16}$$

From Eq.(15), we have $[d\chi(\xi)/d\xi + \gamma_g^2 v_g \omega] A^2(\xi) = const.$, or assuming $d\chi(\xi)/d\xi$ is finite at $A \to 0$, $d\chi(\xi)/d\xi = -\gamma_g^2 v_g \omega \equiv -k$. This gives the group velocity with the relativistic correction as $\omega/k = 1/(\gamma_g^2 v_g) = (1 - v_g^2)/v_g$. The self-consistent stationary propagation of electromagnetic and electrostatic waves can then be described by the following set of equations:

$$\frac{d^2 A(\xi)}{d\xi^2} + \gamma_g^2 \Big\{\frac{1}{\alpha} - \frac{n(\xi)}{\gamma(\xi)} - \frac{n_+(\xi)}{\gamma_+(\xi)} - \frac{n_-(\xi)}{\gamma_-(\xi)}\Big\} A(\xi) = 0, \tag{17}$$

$$\frac{d^2\phi(\xi)}{d\xi^2} = n(\xi) - 1 + n_-(\xi) - n_+(\xi), \tag{18}$$

where $\alpha = 1/(\gamma_g^2 \omega^2)$. These equations can alternatively be written in terms of a two-dimensional Sagdeev potential $U_\sigma(X, Y)$, if we introduce the following transformation of variables $iA(\xi)/\gamma_g \equiv X(\xi)$, $1 + \phi(\xi) \equiv Y(\xi)$, as

$$\frac{d^2 X}{d\xi^2} = -\frac{\partial U_\sigma(X,Y)}{\partial X}, \tag{19}$$

$$\frac{d^2 Y}{d\xi^2} = -\frac{\partial U_\sigma(X,Y)}{\partial Y}. \tag{20}$$

$$U_\sigma(X,Y) = U(X,Y) + \sigma \Delta U(X,Y), \qquad (21)$$

$$U(X,Y) = -\gamma_g^2\left\{v_g\left([X^2+Y^2]-\frac{1}{\gamma_g^2}\right)^{1/2} - Y - \frac{X^2}{2\alpha} + \frac{1}{\gamma_g^2}\right\}, \qquad (22)$$

$$\Delta U(X,Y) = -\gamma_g^2 v_g\left\{\left[(E_+ + 1 - Y)^2 + X^2 - \frac{1}{\gamma_g^2}\right]^{1/2} + \left[(E_- - 1 + Y)^2 + X^2 - \frac{1}{\gamma_g^2}\right]^{1/2}\right\}. \qquad (23)$$

Let us first investigate the finite amplitude plane wave solution given by $A = A_o$ and $\phi = \phi_o$. Since $(E_\pm \mp \phi_o)^2 - a_o^2 = v_g^2(1 + A_o^2)$, where $a_o^2 = (1 + A_o^2)/\gamma_g^2$, we find $n_+ = n_- = \sigma$, so that we obtain from $d^2\phi/d\xi^2 = 0$, $\phi_o = (1+A_o^2)^{1/2} - 1$. Then Eq.(17) with $d^2A/d\xi^2 = 0$ yields the relation

$$A_o^2 = \alpha^{*2} - 1, \qquad \phi_o = \alpha^* - 1, \qquad (24)$$

where $\alpha^* = (1+2\sigma)\alpha$, for electron-ion plasma $\alpha^* = \alpha$. We see that the amplitudes of both electromagnetic and electrostatic waves are enhanced due to the increase of the plasma density.

We now show the possibility for a modulated plane wave described by $\phi(\xi) = \phi_o + \delta\phi \exp[iq\xi]$, $A(\xi) = A_o + \delta A \exp[iq\xi]$. Linearizing Eqs.(17) and (18) with respect to $\delta\phi$ and δA and using Eq.(24) we find that the frequency or amplitude range for such a wave falls in the region

$$1 > \frac{\gamma_g^2 \omega^2}{1+2\sigma} \geq \frac{2v_g}{1+v_g} \quad \text{or} \quad 0 < A_o^2 \leq \frac{1 - v_g}{2v_g}, \qquad (25)$$

We next consider a solitary wave solution, first the case of $\sigma = 0$. To this end, we note that $d^2A/d\xi^2 = d^2\phi/d\xi^2 = 0$ at $A = \phi = 0$, and that for $\alpha > 1$ the Sagdeev potential assumes an extremum (corresponding to a saddle point) at $A = \phi = 0$ and becomes infinite as A and ϕ go to infinity. Therefore, for the case $\alpha > 1$ one can expect a solitary wave solution, starting from and ending at $A = \phi = 0$ at $\xi \to \pm\infty$. The maximum values are determined by the condition

$$U\left(X = \frac{iA_M}{\gamma_g}, Y = 1 + \phi_M\right) = U(X = 0, Y = 1) = 0. \qquad (26)$$

Substituting Eq.(22) into this equation yields the relation,

$$\frac{2\alpha\phi_M}{A_M} = A_M \pm v_g\left\{A_M^2 - 4\alpha(\alpha - 1)\right\}^{1/2}. \qquad (27)$$

Thus we conclude that the solitary wave solution must satisfy the condition $A_M^2 \geq 4\alpha(\alpha - 1)$. The critical condition $A_M^2 = 4\alpha(\alpha - 1)$, corresponds to $\phi_M =$

$A_M^2/(2\alpha) = (1 + A_M^2)^{1/2} - 1$. Substituting $\phi = \phi(A)$ into Eq.(9), we can write Eq.(17) in the form $d^2A/d\xi^2 = -\partial V(A)/\partial A$ where $V(A) = \gamma_g^2(1 + A^2/2\alpha - [1 + A^2]^{1/2})$. One can estimate the pulse width by $2\pi/Q$, where

$$Q = \left(\frac{d^2V(A)}{dA^2}\right)^{1/2} \quad (at\ A = A_\circ) = \gamma_g\left(\frac{\alpha^2 - 1}{\alpha^3}\right)^{1/2} \tag{28}$$

The pulse duration time is then estimated as

$$\frac{2\pi k}{Q\omega} = \frac{2\pi\gamma_g v_g}{[(\alpha^2 - 1)/\alpha^3]^{1/2}}. \tag{29}$$

We now examine the effect of pair-created particles on the solitary wave which satisfies the boundary condition $d^2Y/d\xi^2 = 0$ for $\xi \to \pm\infty$. This condition yields the boundary value of the electrostatic potential, ϕ_∞, as

$$\phi_\infty = \sigma\gamma_g^2 v_g^3 \left\{ \frac{E_-}{[E_-^2 - 1/\gamma_g^2]^{1/2}} - \frac{E_+}{[E_+^2 - 1/\gamma_g^2]^{1/2}} \right\}. \tag{30}$$

Since $U(X = 0, Y = 1) = [\partial U(0, Y)/\partial Y]_{Y=1} = 0$, then

$$U_\sigma(0, 1 + \phi_\infty) = \sigma\Delta U(0, 1) = -\sigma\gamma_g^2 v_g\left\{ \left(E_+^2 - \frac{1}{\gamma_g^2}\right)^{1/2} + \left(E_-^2 - \frac{1}{\gamma_g^2}\right)^{1/2} \right\}. \tag{31}$$

We assume that $A = A_M$ ($A_M = -i\gamma_g X_M(\sigma)$) and $\phi = \phi_M$ ($\phi_M = Y_M(\sigma) - 1$), at $\xi = 0$. We choose $A_\circ$ and $\phi_\circ$ to be $A_\circ = -i\gamma_g X_M(0)$ and $\phi_\circ = Y_M(0) - 1$, and denote $X_M(\sigma) = X_\circ + \sigma\Delta X_M$, $Y_M(\sigma) = Y_\circ + \sigma\Delta Y_M$, where $X_\circ = X_M(0)$ and $Y_\circ = Y_M(0)$. Then at maximum

$$U_\sigma(X_M, Y_M) = \sigma\left\{ \Delta U(X_\circ, Y_\circ) + \Delta X_M \frac{\partial U(X_\circ, Y_\circ)}{\partial X_\circ} \right\}. \tag{32}$$

From condition $U_\sigma(X_M, Y_M) = U_\sigma(0, 1 + \phi_\infty)$, we obtain noting Eq.(31)

$$\Delta X_M = \frac{\Delta U(0, 1) - \Delta U(X_\circ, Y_\circ)}{\partial U(X_\circ, Y_\circ)/\partial X_\circ}. \tag{33}$$

$$\Delta U(X_\circ, Y_\circ) = -2\gamma_g^2 v_g^2[1 + A_\circ^2]^{1/2}, \tag{34}$$

$$\frac{\partial U(X_\circ, Y_\circ)}{\partial X_\circ} = \gamma_g^2 X_\circ \left\{ \frac{1}{\alpha^*} - \frac{v_g}{((1 + \phi_\circ)^2 - (1 + A_\circ^2)/\gamma_g^2)^{1/2}} \right\}. \tag{35}$$

Noting that $d^2A/d\xi^2 = i\gamma_g(\partial U(X,Y)/\partial X) < 0$ at $\xi = 0$, and that $\Delta U(0, 1) < 0$ for $\phi_\circ > 0$, we find $\Delta A_M = -i\gamma_g\Delta X_M > 0$. As $\partial U(X_M, Y_M)/\partial Y_M = 0$, then we can calculate $\Delta Y_M = \Delta\phi_M$ from the condition $\Delta X_M(\partial^2 U(X_\circ, Y_\circ)/\partial X_\circ\partial Y_\circ) +$

$\Delta Y_M(\partial^2 U(X_\circ, Y_\circ)/\partial Y_\circ^2) = 0$, which yields the relation $\Delta \phi_M = A_\circ/[\gamma_g^2(1+\phi_\circ)]\Delta A_M$. Since $\Delta \phi_M > 0$ and $\phi_\infty < 0$, the potential height of the solitary wave, $Y_M - (1 + \phi_\infty)$, is increased by the presence of the pair-created particles. This is due to the enhanced charge separation caused by the ponderomotive force.

We next consider an envelope shock solution which represents a self-consistent wakefield excitation due to the trapping of the pair-created particles. Namely, we consider the situation in which there are no pair-created particles in the downstream region, i.e. $\sigma = 0$ at $\xi < 0$, as they are trapped in the upstream region, i.e. $\sigma > 0$ at $\xi > 0$. In the upstream region, the Sagdeev potential is given by Eq.(21), which at $\xi \to +0$ takes on the value

$$U_\sigma(X_\circ, Y_\circ) = \sigma \Delta U(X_\circ, Y_\circ) = -2\sigma \gamma_g^2 v_g^2 [1 + A_\circ^2]^{1/2}. \tag{36}$$

At sufficiently large values of ξ, $A \simeq 0$ and the Sagdeev potential becomes $U_\sigma(0, 1+\phi)$ which gives the equation

$$\frac{1}{2}\left(\frac{d\phi}{d\xi}\right)^2 + U_\sigma(0, 1+\phi) = U_\sigma(X_\circ, Y_\circ). \tag{37}$$

For $\sigma \ll 1$, we have

$$U_\sigma(0, 1+\phi) = -\sigma \gamma_g^2 v_g \left\{\left(E_+^2 - \frac{1}{\gamma_g^2}\right)^{1/2} + \left(E_-^2 - \frac{1}{\gamma_g^2}\right)^{1/2}\right\} + \frac{\phi^2}{2v_g^2}. \tag{38}$$

Substitution of Eq.(38) into Eq.(37) yields a harmonic oscillation of $\phi(\xi)$ with frequency or wavenumber Q given by $Q = 1/v_g$ and amplitude ϕ_∞ given by

$$\phi_\infty = \left\{2\sigma v_g^3 \gamma_g^2 \left[-2v_g(1+A_\circ^2)^{1/2} + \left(E_+^2 - \frac{1}{\gamma_g^2}\right)^{1/2} + \left(E_-^2 - \frac{1}{\gamma_g^2}\right)^{1/2}\right]\right\}^{1/2}. \tag{39}$$

We specifically note here that the oscillation amplitude is proportional to $\sigma^{1/2}$.

REFERENCES

1. Tsintsadze,L.N., Proc. *International Topical Conf. on Research Trends in Coherent Radiation Generation and Particle Accelerators* (La Jolla, 1991), ed. by J.M.Buzzi, P.Sprangle, K.Wille (AIP, New York, 1992), p.474.

2. Tsintsadze,L.N., and Nishikawa,K., Phys. Plasmas **3**, 511 (1996).

3. Tsintsadze,L.N., Kusano,K., and Nishikawa,K., Phys. Plasmas **4**, 911 (1997).

Propagation of Relativistic Ultrashort Laser Pulse in Near-critical Density Plasma

K. Nagashima, Y. Kishimoto, and H. Takuma

Advanced Photon Research Center, Japan Atomic Energy Research Institute, Tokai-mura, Naka-gun, Ibaraki-ken, 311-01, Japan

Abstract. Two-dimensional characteristics of propagation of the relativistic ultrashort laser pulse in a thin plasma layer have been studied. When the electron density is nearly equal to the critical density, some structures of the electron density are generated. As the laser pulse penetrates into the plasma layer, the structure changes from a wall-like to a bubble-like one. Moreover, it was found that the thin plasma layer is useful for reducing the pulse length of the incident laser as a non-linear optical material.

Recent progress in laser technology has made possible the generation of ultrashort pulse lasers, of which intensities have reached above 10^{19} W/cm^2. In such an intense laser field, electrons oscillate with relativistic quivering energy and are moved by strong ponderomotive and v×B forces. In high density plasma, the electrons can not move freely because of a strong electrostatic restoring force. Therefore, the interaction of an ultra-intense laser pulse with high density plasma is very complicated and many studies have been reported on propagation and absorption of the intense laser pulse in underdense [1-5] or near-critical density plasmas [6-10]. An electromagnetic wave can not propagate in plasma where the electron density is higher than the critical density, $n_{cr} = m_e \varepsilon_0 \omega_0^2/e^2$ where ω_0 is the angular frequency of the wave. However, when the laser pulse has a relativistic intensity, the pulse can propagate beyond the critical density because the effective plasma frequency decreases due to the relativistic effect. The propagation condition is obtained as $\omega_0^2 > \omega_p^2/\gamma$ for a circular polarized wave, where $\gamma = (1+a_0^2)^{1/2}$ is the relativistic factor of an electron quivering in the laser field and ω_p the plasma frequency [8]. This condition was obtained assuming the constant electron density. However, in a real situation, the electron density is modified by the interaction with the laser pulse. As a result, the electron density increases significantly in front of the laser pulse and this high density region interrupts the laser propagating forward. Therefore, propagation of the laser pulse is more complicated, in particular, when the pulse is focused at a small spot size.

Here, we study propagation of the relativistic ultrashort laser pulse in a thin plasma layer with near-critical density using two-dimensional PIC simulation code. The numerical procedure used in the code is similar to that in Ref. [11]. The laser pulse is normally incident on a thin plasma layer and has a wavelength $\lambda_0 = 800$ nm, a pulse length τ_{pulse} = 16 fs, a peak intensity $I_0 = 10^{19}$-10^{20} W/cm^2 and a linear polarization. The pulse shape is assumed to be Gaussian in both longitudinal and transverse directions. The envelope of the laser electric field is given by $E_0 \exp(-(t/\tau_{pulse})^2)$. Simulations are performed with a simulation box of $40\lambda_0$ in the x (longitudinal) direction ($-20\lambda_0 \leq x \leq 20\lambda_0$) and $20\lambda_0$ in the y (transverse) direction ($-10\lambda_0 \leq y \leq 10\lambda_0$). The size of the spatial mesh is $0.1\lambda_0$. The plasma consists of electrons and ions with the ion mass of m_i = 1836m_e. Initially, the particles are placed in a region of $-2\lambda_0 \leq x \leq 2\lambda_0$ uniformly and have zero energy. The laser pulse is placed at $x = -12\lambda_0$ and $y = 0$ initially and propagates toward the plasma layer.

First, typical two-dimensional characteristics of propagation was examined. The laser pulse was s-polarized, the electric field being in the direction perpendicular to the x-y plane, and was focused on a plasma layer which has an uniform initial electron density equal to the critical density. The focal spot diameter was $6\lambda_0$. When the laser pulse penetrated into the plasma, electrons were pushed forward because of the strong longitudinal ponderomotive force. As the result, the electron density increased in front of the laser pulse. Figure 1 shows contours of the laser field (the absolute value of E_z) (a) and electron density (b) at $t = 9\lambda_0/c$. The high density region seemed to be a wall of electrons. Since the wall of electrons had a density of several times the critical value, it affects the relativistic transparency. Then, the wall was broken in the peripheral region and the structure of the electron density was changed from a wall-like to a bubble-like one. The change propagated from the peripheral region to the central region of the wall. Figure 2 shows contours of the laser field (a) and electron density (b) at $t = 12.5\lambda_0/c$. The bubble-like structure of the electron density was observed and a typical scale of the bubble was 1-1.5 times the laser wavelength. This phenomenon is similar to the corrugated plasma surface found in Ref.[6], which is the structure of the ion density in the ramped density region in front of the overdense plasma. In our simulation, no similar structure was observed in the ion density. As can be seen in Fig.2(a), the structure of the electric field was also distorted significantly and the laser pulse was broken into small pieces.

It is thought that transverse distribution of the laser pulse influences the formation of the bubble-like structure. So, it was examined how the structure depends on the laser focal size. We compared two cases with an infinite focal size (a plane wave) and the finite focal size. In the case with an infinite focal size, clear filamentation with an average transverse length of $1.7\lambda_0$ was observed. This phenomenon is essentially equal to that found in Ref.[10]. The filamentation appeared at $t = 12\lambda_0/c$, while the bubble appeared at $t = 10\lambda_0/c$. In the case with a finite focal size, there is transverse distribution of the laser pulse. On the other hand, in the case with an infinite focal size, there is no initial non-uniformity except for the random variation of the initial particle density. It is thought that the transverse variation influences not only the structure of the electron density but also the time when the structure appears.

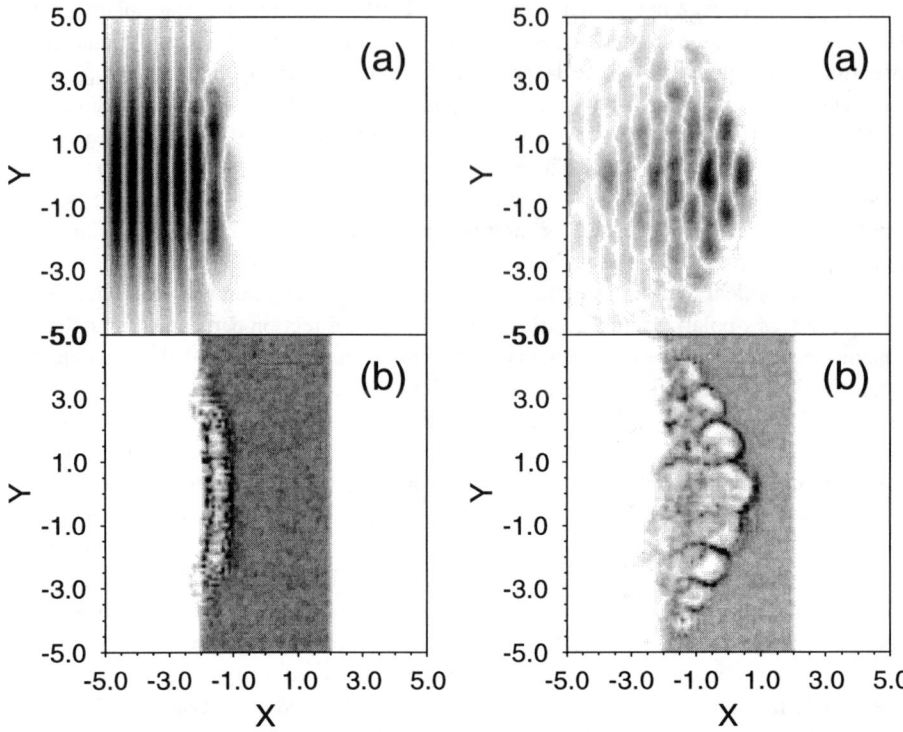

FIGURE 1. Contours of an absolute value of the laser field, E_z (a) and the electron density (b) at $t = 9\lambda_0/c$. A wall-like structure can be observed in the electron density.

FIGURE 2. Contours of an absolute value of the laser field, E_z (a) and the electron density (b) at $t = 12.5\lambda_0/c$. A bubble-like structure can be observed in the electron density.

We have examined absorption, reflection and transparency of the incident laser pulse with a focal spot diameter of $6\lambda_0$. Figure 3 shows reflectivity, transparency, kinetic energies of electrons and ions as a function of the electron density. The kinetic energy of electrons increased by a factor of 50 with increasing the density from $n_e/n_{cr} = 0.1$ to 1. It means that the absorption is enhanced near the critical density. In this case with the peak intensity of 10^{20} W/cm^2, the kinetic energy of electrons had a maximum value at $n_e/n_{cr} = 2$, while it had a maximum at $n_e/n_{cr} = 0.8$-1 in a case with a lower peak intensity of 10^{19} W/cm^2. On the other hand, the reflectivity increased with increasing the electron density and was equal to the transparency at $n_e/n_{cr} = 1.3$ in the case of 10^{20} W/cm^2, while it was at $n_e/n_{cr} = 0.7$ in the case of 10^{19} W/cm^2. Although the relativistic factor is considerably larger than one, the transition from reflection to transparency occurs at a low density near the critical value. This is due to the high density wall of electrons generated in front of the laser pulse. It was also examined how polarization of

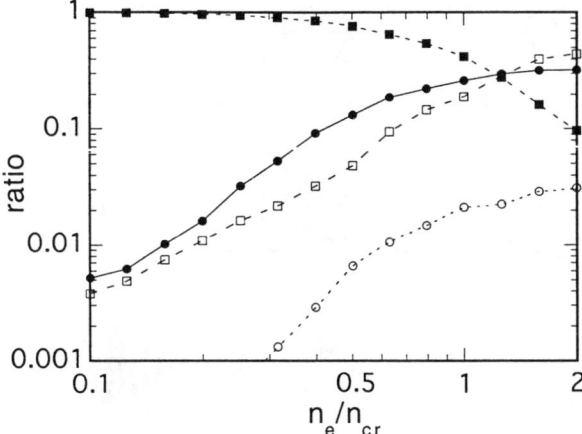

FIGURE 3. Reflectivity (the open square), transparency (the solid square), kinetic energies of electrons (the solid circle) and ions (the open circle) as a function of the electron density. The values were calculated at $t = 24\lambda_0/c$.

the incident pulse influences the absorption and reflection. The kinetic energy of electrons for the p-polarized wave was 30-50 % higher than that for the s-polarized wave in a density range of $n_e/n_{cr} = 1$-2. It is thought that the difference is due to resonance absorption and/or vacuum heating for oblique incidence [12]. On the other hand, the reflectivity of the p-polarized wave was about 40 % lower than that of the s-polarized wave in all density range of $n_e/n_{cr} = 0.1$-2.

Next, the thin plasma layer was examined as a non-linear optical material. In marginally overdense plasma, a high intensity portion of the laser pulse can propagate due to the relativistic effect, but a low intensity portion can not propagate. Therefore, it is supposed that the thin plasma layer works as a non-linear optical material which reduces the pulse length. On the other hand, in highly overdense plasma, transparency of the incident pulse decreases significantly as shown in Fig.3. Therefore, it is not suitable for working as the optical material. Figure 4 shows two shapes of the electric field, E_z before penetrating into the plasma layer (a) and after passing the plasma layer (b). The shaded region shows the initial plasma layer, which has a width of $4\lambda_0$ and an initial electron density equal to the critical density. It was found that the pulse length is reduced by a factor of about two. Figure 4(c) shows a contour of an absolute value of the electric field. It was observed that the focal size is significantly reduced in the range of $5\lambda_0 \leq x \leq 8\lambda_0$. This is caused by the relativistic focusing in the plasma layer. In this case, the plasma layer is thinner than the focal length and works like as an optical convex lens. It was found that with increasing the electron density, the focused intensity increases and the focal length is shortened.

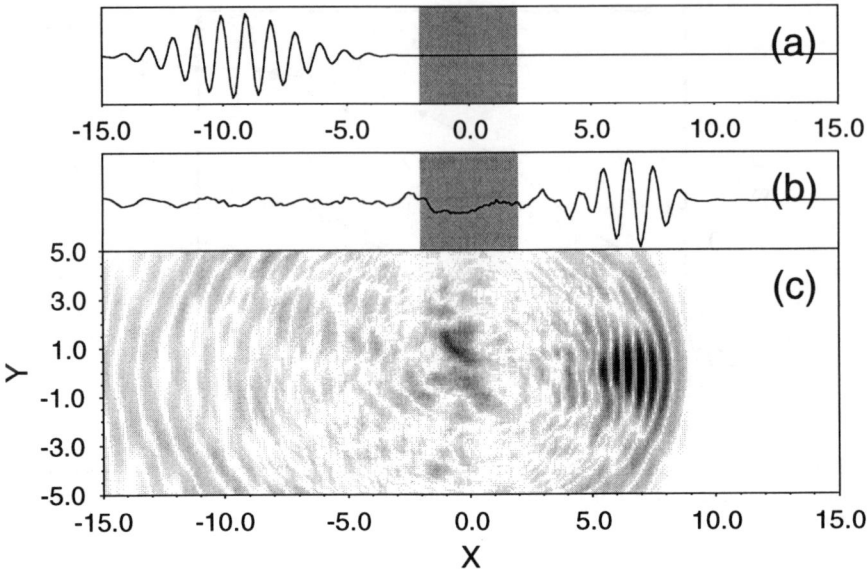

FIGURE 4. Two shapes of the electric field, E_z; (a) is the case at $t = 3\lambda_0/c$ before penetrating into the plasma layer and (b) is the case at $t = 21.5\lambda_0/c$ after passing the plasma layer. The shaded region shows the plasma layer, which has a width of $4\lambda_0$ and an initial electron density equal to the critical density. A contour of an absolute value of the electric field at $t = 21.5\lambda_0/c$ is also shown in (c).

In summary, we have studied two-dimensional characteristics of propagation of the relativistic ultrashort laser pulse in a thin plasma layer. When the electron density is nearly equal to the critical density, some structures of the electron density are generated by the strong ponderomotive force. As the laser pulse penetrates into the plasma layer, the structure changes from a wall-like to a bubble-like one. The energy of the electromagnetic wave is substantially converted to the kinetic energy of electrons near the critical density. Moreover, we have examined the thin plasma layer as a non-linear optical material. It was found that this plasma layer is useful for reducing the pulse length of the incident laser and for a focusing lens.

ACKNOWLEDGMENTS

It is a pleasure to acknowledge the support of the staff at the Advanced Photon Research Center. The authors would like to thank Dr. T. Tajima for his many helpful comments.

REFERENCES

[1] W. B. Mori, C. Joshi, J. M. Dawson, D. W. Forslund, and J. M. Kindel, Phys. Rev. Lett. **60**, 1298 (1988)
[2] P. Sprangle, E. Esarey, and A. Ting, Phys. Rev. Lett. **64**, 2011 (1990); P. Sprangle, E. Esarey, J. Krall, and G. Joyce, Phys. Rev. Lett. **69**, 2200(1992)
[3] T. M. Antonsen, Jr. and P. Mora, Phys. Rev. Lett. **69**, 2204 (1992)
[4] S. V. Bulanov, F. Pegoraro, and A. M. Pukhov, Phys. Rev. Lett. **74**, 710 (1995)
[5] K-C. Tzeng, W. B. Mori, and C. D. Decker, Phys. Rev. Lett. **76**, 3332 (1996)
[6] S. C. Wilks, W. L. Kruer, M. Tabak and A. B. Langdon, Phys. Rev. Lett. **69**, 1383 (1992); S. C. Wilks, Phys. Fluids B **5**, 2603 (1993)
[7] E. Lefebvre and G. Bonnaud, Phys. Rev. Lett. **74**, 2002 (1995)
[8] S. Guerin, P. Mora, J. C. Adam, A. Heron and G. Laval, Phys. Plasmas **3**, 2693 (1996)
[9] A. Pukhov and J. Meyer-ter-Vehn, Phys. Rev. Lett. **76**, 3975 (1996); M.Borghesi et al., Phys. Rev. Lett. **78**, 879 (1997)
[10] J. C. Adam et al., Phys. Rev. Lett. **78**, 4765 (1997)
[11] C. K. Birdsall and A. B. Langdon, in Plasma Physics via Computer Simulation (McGraw-Hill Book Company, 1985), p. 351.
[12] F. Brunel, Phys. Rev. Lett. **59**, 52 (1987); F. Brunel, Phys. Fluids **31**, 2714 (1988)

3. SOLID DENSITY PLASMAS AND OTHER ULTRAINTENSE LASER DRIVEN EXOTIC STATES OF MATTER

Ultraintense Laser-Solid Interaction Phenomena

P. Mulser, D. Bauer, and S. Hain
Theoretical Quantum Electronics (TQE), Technische Universität,
Hochschulstr. 4A, D-64289 Darmstadt, FRG
and
F. Cornolti
Dipartimento di Fisica, Università di Pisa, Piazza Torricelli 2, 56100 Pisa, Italy

Abstract

In the variety of processes occuring in matter under the action of superstrong laser fields rapid field ionization and ponderomotive effects play a dominant role. We investigate field ionization by solving the time-dependent Schrödinger equation numerically and compare the results with classical calculations. We obtain the ionization rates of hydrogen-like ions, the ejection energy spectrum of the electrons, and we show the transition to classical behaviour at high energies. In addition, we are able to give a physical eyplanation of non-sequential ionization. Our ionization rates differ by more than an order of magnitude from the standard expressions. A model is presented which allows to calculate the back action of field ionization on the laser beam. Finally, we present the most general relativistic derivation of ponderomotive forces on single particles and present their transformation properties. Ponderomotive forces are of non-Newtonian character.

1. Introduction

Experiments with table top lasers delivering up to $I\lambda^2 = 10^{21}$ Wcm$^{-2}\mu$m^2 have become feasible now. The accompanying electric field amplitude amounts to $\hat{E} = 10^{12}$ Vcm^{-1}, thus exceeding considerably the Coulomb fields encountered in standard atomic physics. When such a pulse impinges onto matter it is rapidly highly ionized, regardless of whether it is an insulator or a conductor. A simple estimate shows that ionization due to field emission or non-linear multiphoton absorption occurs in a fraction of a laser cycle. Unfortunately, above $I\lambda^2 = 10^{16}$ Wcm$^{-2}\mu$m^2 no standard formulas allowing the reliable determination of the ionization rates are available. For calculating the subsequent heating by inverse bremsstrahlung the knowledge of the energy spectrum of the ejected electrons is relevant. Regarding these quantities the uncertainty is even much higher. So, it is a widely spread opinion that the ejection energy is zero although the arguments in favour of such a hypothesis were never very convincing from a physical point of view. Given a certain laser intensity it would be interesting to determine the degree of ionization by field emission. At present, this is too complex a problem to be tackled. All which can be done with our massively parallel computers is to calculate field emission from hydrogen and hydrogen-like ions. Here we present such calculations for the first time and we show the ejection energy spectra. A comparison of our ionization rates with those obtained from formulas commonly used so far shows deviations up to a factor of 30.

Rapid ionization leads to electromagnetic field energy reduction and drastic nonadiabatic changes in the refractive index. Both together lead to sensitive pulse deformations. Here we present a model in terms of non-standard fluid dynamics which treats such changes in an energetically correct manner and we show in one case how field ionization acts back on the laser pulse.

The ponderomotive force or light pressure has been recognized since a long time ago as a quantity of central interest for understanding laser-plasma interaction and nonlinear plasma optics. Correspondingly, much effort has been concentrated in the past to obtain correct non-relativistic and relativistic expressions for it. Only now we are in the position to formulate the problem correctly in arbitrarily strong laser fields. As a byproduct we shall present the simplest and most basic derivation of the ponderomotive force on a single point system without and with internal degrees of freedom.

2. Field ionization at high intensities

2.1 Ejection energies and ionization rates of hydrogen-like atoms

In conducting and semiconducting matter the free electrons are immediately excited to high oscillation energies by the impinging laser pulse and absorb beam energy by individual electron-ion collisions and by irreversible collective motion. In insulators generally single photon excitation and subsequent ionization does not occur owing to the low photon energy. Above intensities $I = 10^{15}$ Wcm^{-2} nonlinear multiphoton ionization or field emission is rapid enough to become the dominating process of plasma formation. Only after a high number of electrons has been set free, collisional heating and further ionization by electron impact takes place. At flux densities above $I = 10^{16}$ Wcm^{-2} field ionization in the outer shells of the atoms occurs during a fraction of a light cycle (1).

Numerous studies have been undertaken and a whole variety of methods has been employed to study field ionization in the tunneling regime and slightly above, the so-called barrier suppression (BS) regime, at intermediate intensities, $I < 10^{16}$ Wcm^{-2} (2). As the laser pulses have become more intense and shorter, accompanied by a much steeper electric field increase, there is a real need for exact calculations of ionization rates and ejection energies of the electrons in the high intensity regime, $I > 10^{16}$ Wcm^{-2}. Generally, owing to the lack of a theory for this regime, existing theories of the intermediate intensity range are extended to high intensities to obtain ionization rates. To investigate their validity, here we have solved the fully time-dependent Schrödinger equation in the length gauge numerically (3). Such a procedure is feasible at present in two dimensions only if a typical run on massively parallel computers should not exceed the order of one hour. By limiting ourselves to hydrogen-like systems the problem assumes cylindrical symmetry which allows to split off the azimuthal coordinate,

$$\Psi(r,\varphi,x,t) = \psi(r,x,t)e^{im\varphi}$$

$$-i\hbar\frac{\partial}{\partial t}\Psi = \left\{\frac{\hbar^2}{2m_e}\left(\frac{1}{r}\frac{\partial}{\partial r}r\frac{\partial}{\partial r} - \frac{m^2}{r^2} + \frac{\partial^2}{\partial x^2}\right) + \frac{Ze^2}{4\pi\varepsilon_0(r^2+x^2)^{1/2}} - exE\right\}\Psi. \quad (1)$$

If one starts from the ground state of hydrogen, with a sin^2 laser pulse of the form

$$E(t) = \hat{E}\sin^2\left(\frac{\pi}{T}t\right)\sin\omega t, \quad (2)$$

the probability distribution $|\psi(x,t)|^2 = \int |\psi(r,x,t)|^2 2\pi r dr$ evolves in time as shown in Fig. 1(a) for $\hat{E} = 0.25$, $\omega = 0.2$, $T = 6 \times 2\pi/\omega$ in atomic units ($\hat{E} = 1$ a.u. corr. to $I = 3.5095 \times 10^{16}$ Wcm^{-2}, $\omega = 1$ a.u. corr. to 4.1341×10^{16} s^{-1}, $x = 1$ a.u. corr. to Bohr radius $a_0 = 0.5292$ Å). The original Gaussian broadens in time up to 280 a.u. during 6 laser cycles in the laser field direction and only up to 0.8 a.u. perpendicular to it. At $\tau = 1.71$ laser cycles the critical laser field E_c (i.e. ground state energy $E_0 = V_{\max}$) is reached for the first time (4th half cycle) in atomic hydrogen. Since the probability distribution is represented in the Kramers-Henneberger representation (reference system co-moving with the classical oscillating free electron) the maxima and minima become straight lines as they (1) move away from the nucleus and (2) the rescattering process becomes inefficient at higher energies. Both effects are particularly well observed in the Volkov representation of $|\psi(x,t)|^2$ (Fig. 1 (b); see perturbations at $t = 3, 4, 5$ laser cycles). The maxima and minima are spaced by the photon momentum $\hbar k$. There is a one-to-one correspondence between the maxima and minima in the pictures (a) and (b). It shows how the early quantum dynamics evolves into a classical behaviour with increasing oscillation energy of the electron.

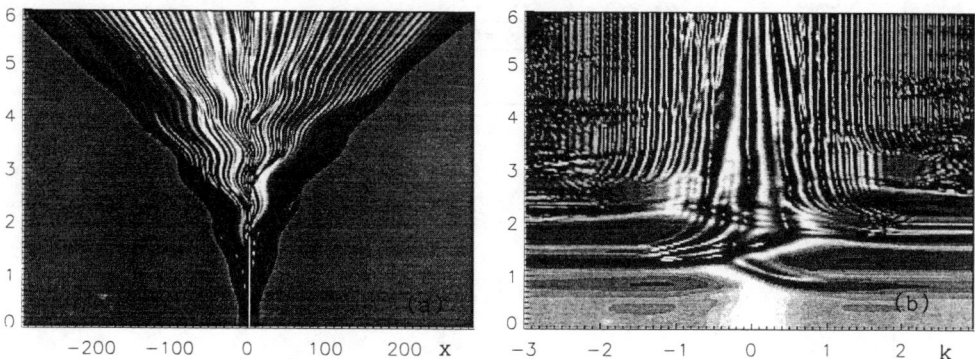

Figure 1. (a): Probability distribution $|\psi(x,t)|^2$ as a function of time (6 laser cycles) in the Kramers-Henneberger representation. (b): The same probability distribution in time in the Volkov momentum space. a.u. = atomic units. Maximum laser amplitude $\hat{E} = 0.25$ a.u., $\sin^2$ pulse, laser frequency $\omega = 0.2$ a.u.

The transition to classical dynamics allows a simple interpretation of the above threshold ionization (ATI) spectra in terms of the initial or ejection energy of the freed electron. In a simplified picture the electron is set free with the velocity v_0 at time t_0, hence its velocity is

$$v(t) = v_0 + \frac{e\hat{E}}{m_e\omega}\left(\cos\omega t - \cos\omega t_0\right), \qquad (3)$$

with $v_i = v_0 - e\hat{E}\cos\omega t_0/m_e\omega$ the ATI speed. From a quantum mechanical point of view one could object that determining v_i is impossible because ionization is a continous process and when one part of the wave function has already sufficiently moved away from its origin and starts being accelerated by the laser field the other part is still sitting in the neighbourhood of the nucleus and undergoes a very reduced oscillatory motion. As a consequence ionization rates are not defined either. In practice, these difficulties can be removed. So, it can be shown that ionization defined as $\eta_I(t) = 1 - |\langle\psi|0\rangle|^2$, $|0\rangle$ ground state, and as that portion of the probability distribution which is outside a fixed radius r_0, e.g. $r_0 = 4a_0$, both lead to almost the same shape of $\eta_I(t)$, the major difference lying in a time shift which equals r_0/v, with v the local group velocity of the wave packet (3). Furthermore, ionization in the barrier suppression regime can also be calculated classically by averaging over suitable ensembles of electrons (3). In Fig. 2 three quantities are reported as functions of time: the quantum mechanical expectation value of energy $\langle\mathcal{E}_{\mathrm{qm}}\rangle$, the classical ensemble average kinetic energy $\langle\mathcal{E}_c\rangle$, and $\langle\mathcal{E}_c\rangle - m_e v_0^2/2$. All quantities are well-defined entities. The figure clearly shows that $\langle\mathcal{E}_{\mathrm{qm}}\rangle$ agrees with $\langle\mathcal{E}_c\rangle$ very well as soon as the electron becomes nearly free, and that v_0 differs from zero, in contrast to a widely used hypothesis. In Fig. 3 the classical ejection energy spectra $\mathcal{E}_{j\perp}$ (perpendicular to the laser field), $\mathcal{E}_{j\|}$ (in laser field direction) and $\mathcal{E}_j = \mathcal{E}_{j\|} + \mathcal{E}_{j\perp}$ are

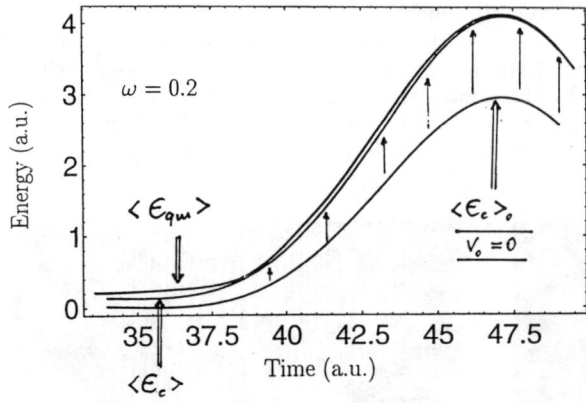

Figure 2: Quantum-mechanical expectation value of energy $\langle \mathcal{E}_{\rm qm}(t) \rangle$ agrees well with the classical quantity $\langle \mathcal{E}_c(t) \rangle$, if the classical initial energy $\langle m_e v_0^2/2 \rangle$ is added to $\langle \mathcal{E}_c \rangle_0$.

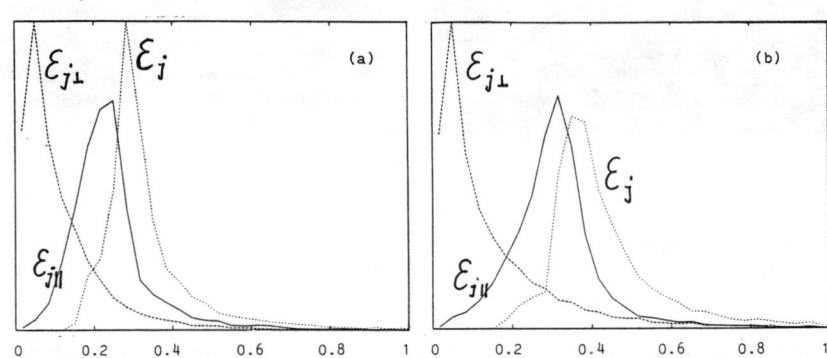

Figure 3. Classical ejection energy distributions $\mathcal{E}_{j\parallel}$, $\mathcal{E}_{j\perp}$ and $\mathcal{E}_j = \mathcal{E}_{j\parallel} + \mathcal{E}_{j\perp}$, in field direction and perpendicular to it (a.u.). (a): $\hat{E} = 0.25$ a.u., (b): $\hat{E} = 1.0$ a.u.; $\omega = 0.2$ a.u. .

shown for $\hat{E} = 0.25$ and $\hat{E} = 1.0$ at $\omega = 0.2$. The average values are $\langle \mathcal{E}_j \rangle = 0.6\, E_I$ for $\hat{E} = 0.25$ and $\langle \mathcal{E}_j \rangle = 0.8\, E_I$ for $\hat{E} = 1.0$. The ratio $\langle \mathcal{E}_{j\perp} \rangle / \langle \mathcal{E}_{j\parallel} \rangle$ is typically 0.1. With increasing laser intensity the width of $\mathcal{E}_j$ also increases. In Fig. 4 the classical distribution of positions of the ionizing electrons in phase space, (r, x) and (p_r, p_x), is presented. It can be shown analytically that there exists a maximum distance $r_{\max}$ beyond which a classical electron is no longer bound, and a minimum energy of ejection $\mathcal{E}_{j\min}$, whereas no upper bound for $\mathcal{E}_j$ can be found (4),

$$r_{\max} = 3a_0, \quad \mathcal{E}_{j\min} = \frac{E_I}{3}, \quad \mathcal{E}_{j\max} = \infty. \tag{4}$$

The second picture (RHS) shows particularly well the existing gap of momenta and energies in the direction of the laser field whereas the perpendicular quantity p_r is continous around $p_r = 0$. As a rule, $\langle \mathcal{E}_j \rangle = E_I/2$ can be assumed.

From Eq. (1) the ionization rate $\lambda_I(E)$ was calculated in the intensity interval $10^{14} - 10^{18}$ Wcm^{-2} for pulses of the form of Eq. (2) for $\omega = 0.1$ and $\omega = 0.2$. There is no recognizable frequency dependence of λ_I in this interval. In the intensity interval considered the numerical results of $\lambda_I(E)$ can be fitted well by

$$\lambda_I(E) = 2.4 \times E^2(t). \tag{5}$$

Several theories provide formulas for λ_I. Landau's tunneling formula (5), the Keldysh rate (6), the ADK formula (Ammosov, Delone, Krainov (7)) and its extension to the BS

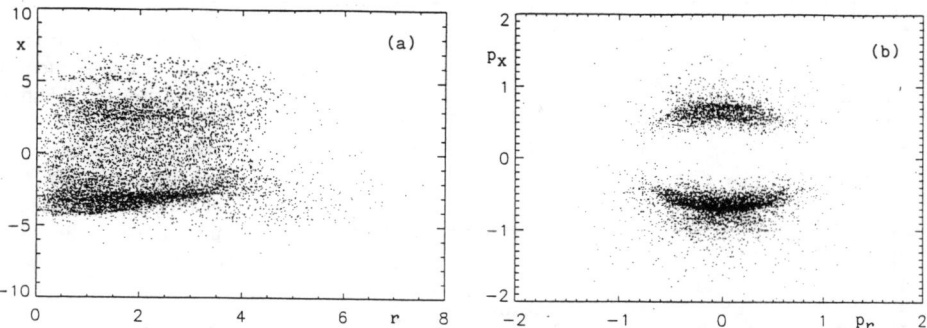

Figure 4. Distributions of positions (a) and momenta (b) in a.u. at the instant of ionization.

regime as well as an angle averaged tunneling formula (1), and the classical expression by Postumus et al.(8) were evaluated for atomic hydrogen and compared with Eq. (5). The agreement is very poor; deviations by an order of magnitude are typical. The analytical expressions deviate from each other up to a factor as large as 30 (9). If sometimes excellent agreement was found between one of the expressions above mentioned (e.g. ADK) and experiments it may have its natural explanation in the fact that measurements refer to intensities below 10^{14} Wcm^{-2} or a free parameter was chosen in a convenient way. It should be mentioned that ealier numerical results obtained by Kulander (10) in our intensity interval agree well with Eq. (5). From $\dot{\eta}_I = \lambda_I(E)(1 - \eta_I)$ the ionization degree

$$\eta_I(t) = 1 - \exp\left(-\int_0^t \lambda_I(E(t'))dt'\right) \qquad (6)$$

results. In Fig. 5 the quantity $1 - \eta_I(t)$ is presented for 5 test pulses of the form of Eq. (2) with $\hat{E} = 0.1, 0.375, 0.75, 1.0$ and 70.0 (solid curves) and compared with the numerical curves (dashed lines). The agreement is very good.

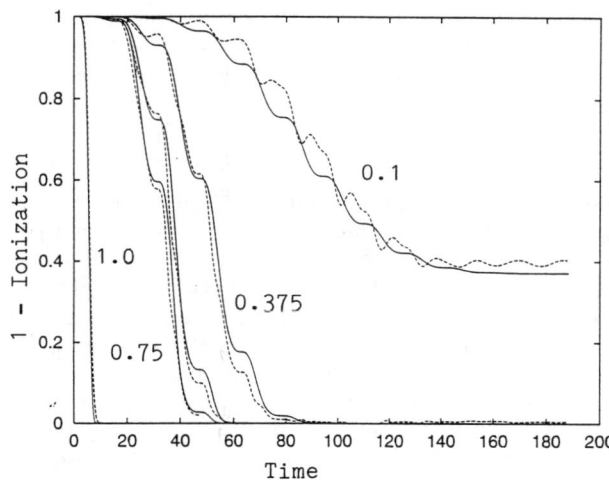

Figure 5. Ionization degree $\eta_I(t)$ from Eq. (6) (solid) and from Eq. (1) (dashed) for 5 sin^2 pulses, $\hat{E} = 0.1, 0.375, 0.75, 1.0$ and 70.0. Agreement is surprisingly good.

2.2 Physical interpretation of non-sequential ionization

Multiple ionization is generally treated on the basis of the single active electron model (SAE (11)) in which one electron is extracted after another and no apparent correlation exists among them (12). In 1992 Fittinghoff et al. observed a "knee" or "shoulder" in

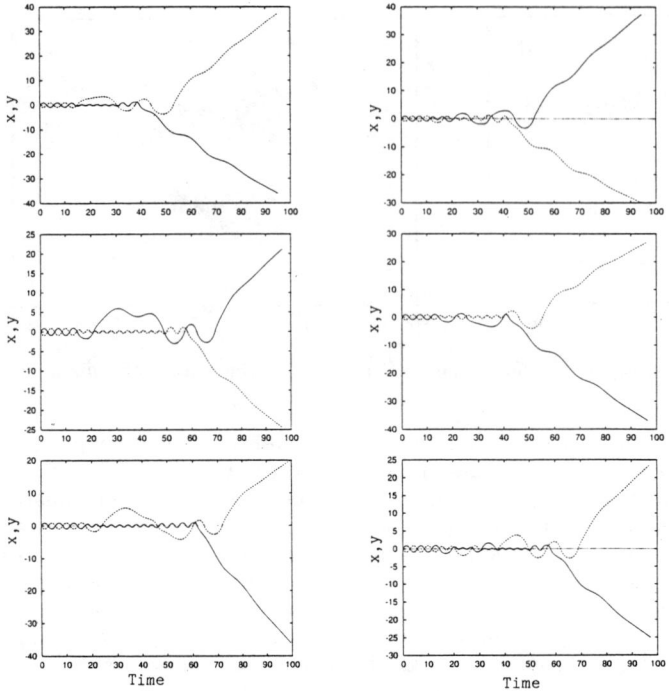

Figure 6. Classical scenario of non-sequential ionization for 6 different initial conditions, starting from the same ground state. The two electrons are ejected into opposite directions (a.u.).

their ionization measurements of He$^+$ → He^{++}. At 10^{15} Wcm^{-2} for example He^{++} is by a factor 10^6 times more present than one would calculate if ionization occured sequentially. In the subsequent years several hypotheses were formulated to explain this case of non-sequential ionization (NSI). In calculations in which the outer electron follows the SAE model, the second one, however, feels the Coulomb force of the first one, the knee could be reproduced (13) but the physical mechanism producing the effect remained unclear. It will be explained in the following. In a numerical study helium can be treated in one dimension only. The two electron Hamiltonian is used in the form

$$H(x,y,t) = \frac{1}{2}(p_x^2 + p_y^2) - \frac{2}{(x^2+\varepsilon)^{1/2}} - \frac{2}{(y^2+\varepsilon)^{1/2}} + \frac{1}{[(x-y)^2+\varepsilon]^{1/2}} + (x+y)E(t). \quad (7)$$

The parameter ε in the soft Coulomb potential is taken equal to 0.55 in order to produce the correct ground state energy of He. The Schrödinger equation is solved with this Hamiltonian. In order to have a guide how to interpret the results, classical calculations with the Hamiltonian (7) were also perfomed (14). In Fig. 6 the classical orbits of the two electrons are plotted as functions of time for $\hat{E} = 0.25$ for 6 different initial conditions of the two electrons starting from the helium ground state. Among 1000 case studies all NSI events belong to the scenario of the figure: First the two electrons move both into the same direction, hence giving origin to a mutual repulsion over a long time. In this way the second electron helps the first one to be ionized. By the repulsive force it is slowed down and is ready to take so much energy from the laser field to be ionized one or more half cycles later in the opposite direction. This scenario is in agreement with the corresponding time-dependent Schrödinger calculations. We may conclude that NSI is not a pure quantum effect: rather is it Coulomb repulsion assisted acceleration of the second electron in the laser field.

2.3 Feedback of field ionization on pulse propagation in dense matter

It is important to take fast field ionization and its back action on pulse shape and pulse propagation into account when problems like prepulse suppression, pulse front steepening, building up of surface plasma layers, self-induced transparency of thin foils or important applications, e.g. optical shutters are to be investigated. If the generation rates $g_I(\boldsymbol{E}, \boldsymbol{v})$ of free electrons of velocity $\boldsymbol{v}$ under the action of the laser field $E(t)$ are known, field ionization is easily incorporated in a Vlasov-Boltzmann description. Unfortunately, these rates are not yet known enough. On the other hand, in many cases of relevance all we want to calculate is the laser field distribution in space and time and no detailed microscopic picture is needed. The laser field distribution is uniquely determined by the wave equation,

$$\nabla \times \nabla \times \boldsymbol{E} + \frac{1}{c^2}\frac{\partial^2}{\partial t^2}\boldsymbol{E} = -\frac{1}{\varepsilon_0 c^2}\frac{\partial \boldsymbol{j}}{\partial t}, \tag{8}$$

once the macroscopic quantity of the total current density $\boldsymbol{j}(\boldsymbol{x},t)$ is known. The latter must be such as to fulfill Poynting's theorem which relates the $\boldsymbol{j} \cdot \boldsymbol{E}$ work spent for field ionization to a corresponding reduction of the Poynting vector, which, in turn, is a function of $\boldsymbol{E}(\boldsymbol{x},t)$. For this purpose we consider three kinds of currents. The first one originates from those electrons which are in their ground states or slightly excited. Consequently, their current density $\boldsymbol{j}_b$ is determined to a good approximation from the refractive index η_b of the undisturbed matter (the index b stands for "bound"). It merely contributes to change the phase velocity from its vacuum value c to c/η_b. The second contribution $\boldsymbol{j}_I$ is intimately connected with the field ionization process. In fact, during the ionization process the binding field $\boldsymbol{E}_b$ is gradually lowered as the electrons move away from their nuclei. Finally, there is the fluid of the free electrons contributing by the current density $\boldsymbol{j}_e$ to the total current density $\boldsymbol{j}$ in Eq. (8). The ions can be assumed immobile and hence, besides causing scattering of the partially bound and the free electrons, they do not directly contribute to $\boldsymbol{j}$. We introduce the quantities

$$m_e\langle\boldsymbol{v}\rangle = \frac{m_e}{\lambda_I}\int g_I(\boldsymbol{E},\boldsymbol{v})\boldsymbol{v}\,d\boldsymbol{v}, \quad \frac{m_e}{2}\langle\boldsymbol{v}^2\rangle = \frac{m_e}{2\lambda_I}\int g_I(\boldsymbol{E},\boldsymbol{v})\boldsymbol{v}^2\,d\boldsymbol{v}, \quad \lambda_I(\boldsymbol{E}) = \int g_I(\boldsymbol{E},\boldsymbol{v})\,d\boldsymbol{v}.$$

Then, the conservation equations of particles, momentum and energy read as follows,

$$\frac{\partial n_e}{\partial t} + \nabla \cdot (n_e \boldsymbol{u}_e) = n_b \lambda_I, \quad n_b = n_0 - n_e/Z,$$

$$\frac{\partial}{\partial t} m_e n_e \boldsymbol{u}_e + \nabla \cdot (m_e n_e \boldsymbol{u}_e \otimes \boldsymbol{u}_e + \boldsymbol{P}_e) = -e n_e \boldsymbol{E} - (\nu_{ei} + \nu_{en}) m_e n_e \boldsymbol{u}_e + \lambda_I n_b m_e \langle\boldsymbol{v}\rangle,$$

$$\frac{\partial}{\partial t} n_e(\varepsilon_e + \frac{m_e}{2}\boldsymbol{u}_e^2) + \nabla \cdot (n_e \boldsymbol{u}_e(\varepsilon_e + \frac{m_e}{2}\boldsymbol{u}_e^2) + \boldsymbol{u}_e \cdot \boldsymbol{P}_e) = \boldsymbol{j}_e \cdot \boldsymbol{E} + n_b \lambda_I \frac{m_e}{2}\langle\boldsymbol{v}^2\rangle - \nabla \cdot \boldsymbol{q}_e. \tag{9}$$

Besides the terms containing λ_I, $\langle\boldsymbol{v}\rangle$ and $\langle\boldsymbol{v}^2\rangle$, these are the standard conservation equations with $\boldsymbol{u}_e, \boldsymbol{P}_e$ and $\boldsymbol{q}_e$ the flow velocity, pressure tensor and heat flux density. ν_{ei} and ν_{en} are the collision frequencies with ions and neutrals. They can also be justified microscopically (15). As soon as $\langle\boldsymbol{v}\rangle$ is known, e.g. from a fully time-dependent solution of the Schrödinger equation or from a classical Monte Carlo simulation, the electric current density $\boldsymbol{j}_e$ is determined from the momentum equation of Eqs. (9). The ionization current density $\boldsymbol{j}_I$ is obtained in the most immediate way from energy conservation. The work spent by the laser field to ionize $n_b \lambda_I$ atoms (or ions) and to impart to them the ejection energy $\langle\mathcal{E}_j\rangle$ per unit volume and unit time is determined from Poynting's theorem by $\boldsymbol{j}_I \cdot \boldsymbol{E}$, hence

$$\boldsymbol{j}_I \cdot \boldsymbol{E} = n_b \lambda_I (E_I + \langle\mathcal{E}_j\rangle). \tag{10}$$

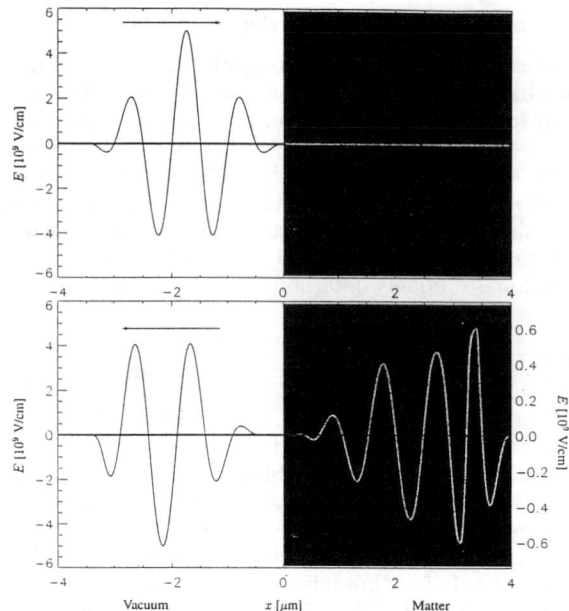

Figure 7. Nd laser pulse, 3 wavelengths long, with an intensity of $I = 3,3 \times 10^{16}$ Wcm^{-2} impinges normally on al liquid He layer of density $n_0 = 3 \times 10^{22}$ cm^{-3} from vacuum (upper picture; black region: He target). In the lower picture the reflected and transmitted pulse are shown. Due to the dominance of the field ionization current j_I on j_e in the pulse front and the onset of strong reflection afterwards the transmitted pulse shows more relevant crests than the incident signal (8 versus 5-7) and a wavelength varying locally.

It holds $\langle \mathcal{E}_j \rangle \ll m_e c^2$ up to very high laser fluencies ($I > 10^{18} - 10^{19}$ Wcm$^{-2}\mu$m^2). Therefore the $\boldsymbol{u}_e \times \boldsymbol{B}$ term is neglected in the momentum equation. Under this approximation, $\boldsymbol{j}_I$ is parallel to $\boldsymbol{E}$ and is given by

$$\boldsymbol{j}_I = n_b \lambda_I \frac{E_I + \langle \mathcal{E}_j \rangle}{\boldsymbol{E}^2} \boldsymbol{E}. \tag{11}$$

The neutral current $\boldsymbol{j}_b$ is obtained from $\boldsymbol{j}_b = i\varepsilon_0 \omega(1 - \eta_b^2)\boldsymbol{E}$, owing to $\eta^2 = 1 + i\sigma/\varepsilon_0 \omega$, $\boldsymbol{j} = \sigma \boldsymbol{E}$.

In order to give an impression of how $\boldsymbol{j}_I$ acts, we present a Nd laser pulse, 3 wavelengths long ($\lambda = 1\mu$m), impinging on a liquid He layer of density $n_0 = 3 \times 10^{22}$ cm^{-3}. The pulse amplitude is $\hat{E} = 5 \times 10^9$ Vcm^{-1} corresponding to an Intensity $I = 3.3 \times 10^{16}$ Wcm^{-2}. The ionization rates λ_I for the first ($E_I = 24.58$ eV) and the second electron ($E_I = 54.4$ eV) are determined from Landau's tunneling formula for simplicity. The collision frequencies are set zero. In Fig. 7, upper half, the pulse incident onto the target (black) is plotted. Due to rapid field ionization the He layer becomes overdense, i.e. $n_e > n_c$, $n_c = 10^{21}$ cm^{-3}, and highly reflecting (lower half). As a consequence of $\boldsymbol{j}_I$ from Eq. (11), the leading part of the transmitted pulse shows an interesting anomaly: it appears shortened in its wavelengths up to 40 % with respect to the vacuum value and the total pulse apparently consists of 9 relevant crests in contrast to the 6 ones of the incident pulse. Almost no change besides attenuation appears in the reflected pulse. At higher current densities $\boldsymbol{j}_I$ the wave crests of the pulse penetrating the medium assume a plateau-like structure since field ionization cuts the E-field maxima of the leading edge of the laser pulse. Sometimes the reflected pulse appears also distorted (15).

3. Relativistic ponderomotive forces, non-Newtonian character and transformation properties

The ponderomotive force on single point charges and on fluid elements has revealed as fundamental in high power laser-matter interactions and accelerator physics (16). Accordingly, much effort has been put, since decades now, in obtaining correct and general expressions for this force (17), and a whole variety of different methods was used. They are all based on perturbation procedures. For a rather complete review of the subject see (18). With the development of laser systems delivering sub-ps light pulses in the relativistic regime there was a real need to calculate the relativistic ponderomotive force f_p on a point charge beyond a perturbative analysis (19).

To perform this impressive variety of analysis, whether relativistic or nonrelativistic, the effort was immense and yet unsatisfactory. As far as the ponderomotive force on a single particle without or with internal degrees of freedom is concerned, the reason for such a failure came from a tendency to enter too early into the technical aspects of the problem instead of making exhaustive use first of the general properties of point-like forces, guarantied by relativity. As we shall see in the following, by generalizing Newton's concept of force on a point particle, the correct relativistic expression of f_p and its transformation properties follow in a very stringent way in full generality, all with minimum effort.

Let us assume that the change of momentum $\dot{p}$ of a point particle in its comoving inertial frame ("tangent frame") is given by the relation

$$m\ddot{x} = f^N(x, \tau), \tag{12}$$

with m the rest mass and τ the proper time. Then, f^N is named a Newtonian or Newton force. With this definition it is tacitly assumed that the rest mass does not depend on time. We found that throughout the relativistic literature this hypothesis is taken for granted by all text books (see for instance most excelling ones like (20) and (21)). When passing from the comoving system S' to another one, S, with respect to which the point particle moves at speed v the Newton force f^N becomes an Einstein force f^E.

In order to find the correct transformation law from f^N to f^E, there is the standard procedure which consists in constructing a four-force $\mathbf{F} = (F^\mu), \mu = 1, ..., 4$, the so-called Minkowski force. The requirement is that $\mathbf{F}$ transforms like the four-position vector $X' \rightarrow X$, $X = (x, ct)$, $X' = (x', ct')$,

$$x = x' + \frac{\gamma - 1}{v^2}(v \cdot x')v + \gamma v t', \quad ct = \gamma(ct' + v \cdot x'/c) \tag{13}$$

with $\gamma = (1 - v^2/c^2)^{-1/2}$. Differentiating the position vector X with respect to the proper time τ yields the four-velocity

$$\mathbf{U} = \frac{dX}{d\tau} = \gamma \frac{dX}{dt} = (\gamma v, \gamma c), \tag{14}$$

where $dt = \gamma d\tau$ holds. Multiplying $\mathbf{U}$ with the Lorentz scalar m generates the four-momentum $\mathbf{P}$,

$$\mathbf{P} = m\mathbf{U} = (\mathbf{p}, \gamma m c); \quad \mathbf{p} = \gamma m v. \tag{15}$$

Differentiating once more with respect to τ produces two more four-vectors, the four-acceleration $d\mathbf{U}/d\tau = d^2 X/d\tau^2 = \gamma d(\gamma v, \gamma c)/dt$ and the Minkowski force $\mathbf{F}$,

$$\mathbf{F} = (F^\mu) = \frac{d\mathbf{P}}{d\tau} = (\mathbf{f}, F^4) = \gamma \left(\frac{d\mathbf{p}}{dt}, mc\frac{d\gamma}{dt}\right). \tag{16}$$

For obvious historical reasons $d\mathbf{p}/dt = \mathbf{f}^E$ is given the name Einstein force. In the rest frame $\mathbf{F}$ becomes

$$\mathbf{F} = (\mathbf{f}^N, 0). \tag{17}$$

Hence, all we have to know is the Newton force $\mathbf{f}^N$ and that $\mathbf{F}$ transforms according to Eqs. (13) when passing from the comoving frame S' to a general system S,

$$\mathbf{F} = \left(\mathbf{f}^N + \frac{\gamma - 1}{v^2}(v \cdot \mathbf{f}^N)v, \frac{\gamma}{c}v \cdot \mathbf{f}^N\right) = \gamma\left(\mathbf{f}^E, \frac{1}{c}v \cdot \mathbf{f}^N\right), \tag{18}$$

$$f^{\mathrm{E}} = \frac{1}{\gamma}\left[f^{\mathrm{N}} + \frac{\gamma-1}{v^2}(v\cdot f^{\mathrm{N}})v\right], \quad v\cdot f^{\mathrm{N}} = v\cdot f^{\mathrm{E}}. \tag{19}$$

These are the desired transformation relations from a Newton force to its correct Einstein force f^{E} which holds in any inertial system. This is the concept to follow with f_p.

To this aim, the concept of point force has to be extended to a rest mass which changes in space-time, $m = m(X) = m(\boldsymbol{x}, t = \tau)$. In this case Eq. (16) becomes

$$\mathbf{F} = \frac{d\mathbf{P}}{d\tau} = (\boldsymbol{f}, F^4) = \gamma\left(\frac{d}{dt}(\gamma m\boldsymbol{v}), c\frac{d}{dt}(\gamma m)\right). \tag{20}$$

The Einstein force is again given by $\boldsymbol{f}^{\mathrm{E}} = d\boldsymbol{p}/dt = d(\gamma m\boldsymbol{v})/dt$. Spezializing to the rest frame,

$$\mathbf{F} = (\boldsymbol{f}, F^4) = \left(m\ddot{\boldsymbol{x}}, c\frac{dm}{d\tau}\right) \tag{21}$$

replaces Eq. (17). That is, F^4 does no longer vanish and the knowledge of the three-vector $\boldsymbol{f}$ is not any longer sufficient to built the four-force $\mathbf{F}$. Therefore, we speak of a non-Newtonian character of $\mathbf{F}$. It still holds $\mathbf{P}^2 = -m^2(X)c^2$ from which by differentiation with respect to the proper time τ follows

$$P^\mu \frac{dP_\mu}{d\tau} = P^\mu F_\mu = -c^2 m(X)\frac{dm(X)}{d\tau} = -c^2 m(X) U^\mu \partial_\mu m(X) = -c^2 P^\mu \partial_\mu m(X), \tag{22}$$

since $d/d\tau = U^\mu \partial_\mu$. From $P^\mu F_\mu = -c^2 P^\mu \partial_\mu m(X)$ we can conclude that the resulting Minkowski force F_μ consists of two parts

$$\mathbf{F} = {}^{\mathrm{L}}\mathbf{F} + {}^{\mathrm{N}}\mathbf{F}; \quad {}^{\mathrm{N}}\mathbf{F}\cdot \mathbf{U} = 0, \tag{23}$$

the non-Newtonian part ${}^{\mathrm{L}}\mathbf{F}$ and the Newtonian component ${}^{\mathrm{N}}\mathbf{F}$. The latter is orthogonal to U since the four-velocity and the four-acceleration are orthogonal to each other. Thus, without any loss of generality we may set

$${}^{\mathrm{L}}\mathbf{F} = ({}^{\mathrm{L}}F_\mu) = -\left(\partial_\mu mc^2\right) = -c^2\left(\nabla m, \frac{1}{c}\frac{\partial}{\partial t}m\right), \quad f^{\mathrm{E}} = -\frac{c^2}{\gamma}\nabla m. \tag{24}$$

The ponderomotive force is a secular, i.e. a dc force arising from spatial inhomogeneities of the electromagnetic field, and requires only the existence of an oscillation center. In an arbitrary reference system S an oscillation center $\boldsymbol{x}_o(t)$ moving at velocity $\boldsymbol{v}_o(t)$ exists if and only if a system S' can be found in which (i) the motion of a point charge $\boldsymbol{x}'(t') = \boldsymbol{x}'_o + \boldsymbol{\xi}'(t')$ is (nearly) closed, $\boldsymbol{\xi}'(t' + T') \simeq \boldsymbol{\xi}'(t')$, T' period, and (ii) $\boldsymbol{x}'_o$ is stationary, i.e. $\dot{\boldsymbol{x}}'_o = 0$. Only if $\dot{\boldsymbol{x}}'_o = 0$ holds the decomposition $\boldsymbol{x}(t) = \boldsymbol{x}_o(t) + \boldsymbol{\xi}(t)$ is done in an invariant way, that is, such that $\boldsymbol{x}_o(t)$ is the Lorentz transformed point $\boldsymbol{x}'_o(\tau)$ when changing from S' to S. This requirement is essential for $\boldsymbol{x}_o(t)$ to be a physical entity. It allows the oscillation center to assign an invariant rest mass m_{eff} in S',

$$m_{\mathrm{eff}}(\boldsymbol{x}'_o, \tau) = \frac{m}{T'}\int_\tau^{\tau+T'}\left[1 + \left(\frac{\boldsymbol{p}'(t')}{mc}\right)^2\right]^{1/2} dt', \quad \boldsymbol{p}' = m\gamma' \boldsymbol{v}', \tag{25}$$

a rest energy $E'_o = m_{\mathrm{eff}} c^2$, a four-velocity $\mathbf{U}_o = dX_o/dt'$ and a four-momentum $\mathbf{P}_o = m_{\mathrm{eff}} \mathbf{U}_o$. For a purely propagating monochromatic electromagnetic wave the effective mass is $m_{\mathrm{eff}} = m(1 + q^2 \mathbf{A}^2/m^2 c^2)^{1/2}$ if the four-potential $\mathbf{A}$ obeys the Lorentz gauge. In a pure hf field ${}^{\mathrm{N}}\mathbf{F}$ is zero and consequently, f^{E}_p reads

$$f^{\mathrm{E}}_p = -\frac{c^2}{\gamma_o}\nabla m_{\mathrm{eff}}; \quad \gamma_o = \gamma(\boldsymbol{v}_o). \tag{26}$$

It is not of Newtonian character. In terms of the components in the oscillation center system S', $\boldsymbol{f}_p^{\rm E}$ reads in S

$$\boldsymbol{f}_p^{\rm E} = \frac{1}{\gamma_o}\left[\boldsymbol{f}'_p + \frac{\gamma_o-1}{v_o^2}(\boldsymbol{v}_o^2\cdot\boldsymbol{f}'_p)\boldsymbol{v}_o + \gamma_o\boldsymbol{v}_o\frac{dm_{\rm eff}}{d\tau}\right]. \qquad (27)$$

This is the desired compact derivation of $\boldsymbol{f}_p$ on a point charge. It is rigorous, general and simple and at the same time probably the most fundamental derivation given so far. In some way the method we introduced here has an affinity to the method of the effective mass in solid state theory and the dressed particle concept in quantum optics and elementary particle physics.

One consequence of Eq. (27) is the following. If the amplitudes $\hat{A}$ or $\hat{E}$ are constant in space in one special reference system S', $\boldsymbol{f}_p^{\rm E}$ does not vanish in another frame S, in contrast to nearly all expressions for $\boldsymbol{f}_p$ given in the literature so far. The method presented here is susceptible to a variety of applications and extensions. So for instance it holds for particles with internal degrees of freedom if $m_{\rm eff}$ is replaced by

$$m_{\rm eff}(X) = m + (\overline{\mathcal{E}}_{\rm os} - \overline{\mathcal{E}}_{\rm in})/c^2, \qquad (28)$$

where the time-averaged internal energy $\overline{\mathcal{E}}_{\rm in}$ has to be subtracted from the total oscillatory energy $\overline{\mathcal{E}}_{\rm os}$ since internal forces do not influence the oscillation center.

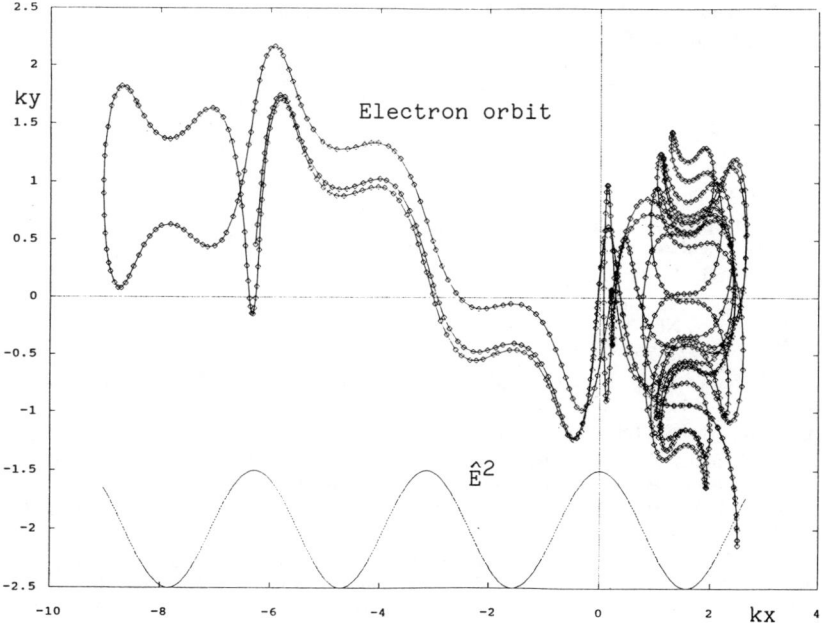

Figure 8. Chaotic motion of an electron in a standing wave of a Nd laser beam at $I = 6\times 10^{17}\,{\rm Wcm}^{-2}$. The electron starts at $kx = 0.04\,\pi$ with zero oscillation center speed. The corresponding secular motion at lower laser intensity is limited to the interval $(0, 2\pi)$.

From the superintense laser-matter interaction another new aspect emerged: Despite the monochromatic character of the laser beam the oscillatory motion of the electrons becomes chaotic above a certain threshold as soon as more than one wave vector $\boldsymbol{k}$ is present in the beam (19, 22, 23). The physical origin of this chaotic behaviour is the ponderomotively induced Doppler effect (19); it may lead to the elimination of any recognizable oscillation center in the single particle motion (Fig. 8). It will no longer be clear how a secular force can be determined in such a case. Perturbative methods would hardly be of any help since the motion looks similar to that in fully developed turbulence. The only knowledge that can be taken for granted is the overall (integrated) light pressure on the medium, $p_L = (1+R)I$; I intensity, R reflection coefficient. The concept of m_{eff} presented here may be successful again. We propose to take convenient clusters of particles in the medium and to calculate their effective mass $M_{\mathrm{eff}} = \rho_{\mathrm{eff}} dV$ by averaging over $T = 2\pi/\omega_{\mathrm{min}}$, where ω_{min} is the lowest relevant frequency of the chaotic motion. In addition to the oscillatory energy the thermal motion also contributes to ρ_{eff} now. The only condition for this procedure to work is that T is much shorter than any relevant time interval during which the secular force undergoes a substantial change.

This work has been supported by the European Commission through the TMR Network SILASI (Super Intense LAser pulse-Solid Interaction), No. ERBFMRX-CT96-0043.

References

1. Mulser, P., *Interaction of intense fs laser pulses with matter*, in More, R.M., ed., *Laser Interaction with Atoms, Solids and Plasmas*, Plenum Press, New York, 1994, p. 425.
2. Lambropoulos, P. and Walther, H., eds., 1997, *Multiphoton Processes 1996*, Proc. 7th Int. Conf. Multiphoton Processes, Garmisch, Sept. 30th - Oct. 4th 1996, Inst. Phys. Publ., Bristol 1997.
3. Bauer, D., *Dynamik der Feldionisation im intensiven Laserpuls (field ionization dynamics in the intense laser pulse)*, Thesis, Darmstadt Inst. Tech., 1997.
4. Bauer, D., Phys. Rev. A **55**, 2180 (1997).
5. Landau, L.D., and Lifshitz, E.M., *Quantum Mechanics*, 3rd ed., Pergamon, Oxford, 1977, p. 294.
6. Keldysh, L.V., Sov. Phys. JETP **20**, 1307 (1965).
7. Ammosov, M.V., Delone, N.B., and Krainov, V.P., Sov. Phys. JETP **64**, 1191 (1987); Krainov, V.P., in (2), p. 98.
8. Posthumus, J.H., Thompson, M.R., Frasinski, L.F., and Kodling, K., in (2), p. 298.
9. Bauer, D., and Mulser, P., *Exact field ionization rates in the barrier suppression regime from numerical TDSE calculations*, submitted for publication.
10. Kulander, Kenneth C., Phys. Rev. A **35**, 445 (1987).
11. Kulander, Kenneth C., Schafer, K.J., and Krause, J.L., in Garrila, M., ed. *Atoms in Intense Laser Fields*, Acad. Press, New York, 1992, p. 247 - 300.
12. Lambropoulos, P., and Tang, X., J. Opt. Soc. Am. B **4**, 82 (1987).
13. Watson, J.P., Sanpera, A., Lappas, D.G., Knight, P.L., and Burnett, K., Phys. Rev. Letters **78**, 1884 (1997).
14. D. Bauer, Phys. Rev. A **56**, (Sept. 1997).
15. Mulser, P., Cornolti, F., and Bauer, D., *Modelling Field Ionization, Ionization Dephasing and Nonstandard Fluid Dynamics*, submitted for publication to Phys. Plasmas.
16. Rubenchik, A., and Witkowski, S., *Physics of Laser Plasma*, Vol. 3 of *Handbook of Plasma Physics*, North Holland, Amsterdam, 1991, chaps. 8-11.
17. Boot, H.A.H., Self, S.A., and Shersby-Harvie, R.B.R., J. Electron. Control **4**, 434 (1958); Gapunov, A.V., and Miller, M.A., Sov. Phys. JETP **7**, 168 (1958); Kibble, T.W.B., Phys. Rev. **150**, 1060 (1966); Hopf, F.A., Meyestre, P., Scully, M.O., and Luisell, W.H., Phys. Rev. Lett. **37**, 1342 (1976).
18. Kentwell, G.W. and Jones, D.A., Phys. Rep. **145**, No. 6, 319 - 403 (1987).
19. Bauer, D., Mulser, P., and Steeb, W.-H., Phys. Rev. Letters **75**, 4622 (1995).
20. Weinberg, Steven, *Gravitation and Cosmology*, John Wiley, New York, 1972, chap. 2.
21. Misner, Charles W., Thorne, Kip S., and Wheeler, John A., *Gravitation*, Freeman, W.H., & Co., San Fransisco 1973, chaps. 2,3.
22. Bardsley, J.N. and Penetrante, B.M., Phys. Rev. A **40**, 3823 (1989).
23. Penetrante, B.M. and Bardsley, J.N., Phys. Rev. A **43**, 3100 (1991).

Measurement of Acceleration in Femtosecond Laser-Plasmas

R. Häßner, W. Theobald, S. Niedermeier, K. Michelmann, T. Feurer, H. Schillinger, and R. Sauerbrey

Institut für Optik und Quantenelektronik
Friedrich-Schiller-Universität Jena
Max-Wien-Platz 1
D-07743 Jena, Germany

Abstract. Accelerations up to 4×10^{19} m/s² are measured in femtosecond laser-produced plasmas at intensities of 10^{18} W/cm² using the Frequency Resolved Optical Gating (FROG) technique. A high density plasma is formed by focusing an ultrashort unchirped laser pulse on a plane carbon target and part of the reflected pulse is eventually detected by a FROG autocorrelator. Radiation pressure and thermal pressure accelerate the plasma which causes a chirp in the reflected laser pulse. The retrieved phase and amplitude information reveal that the plasma motion is dominated by the large light pressure which pushes the plasma into the target. This is supported by theoretical estimates and by the results of independently measured time integrated spectra of the reflected pulse.

INTRODUCTION

Femtosecond pulse generated laser-plasmas are an exciting topic of research since high power table-top laser systems provide intensities up to 10^{18} W/cm². At these intensities rapid ionization of a solid during the first few optical cycles leads to the formation of a high density plasma with electron densities $n_e \geq 10^{23}$ cm^{-3} exceeding that of solids [1–3]. The emission of hard x-ray radiation with energies of up to several MeV [4,5] and the production of fast particles have been observed [6,7]. The femtosecond laser pulse rapidly heats the electrons in the plasma and a pressure gradient is built-up. This drives a mass motion in the direction of the surface normal for laser intensities less than 10^{17} W/cm². The plasma actually experiences accelerations on the order of $\simeq 10^{18}$ m/s² during the interaction time [8].

It is well known that frequency resolved measurements of reflected laser light from a plasma surface reveal the mass motion [9,10]. A more complete picture of the plasma evolution can be obtained by measuring the phase and amplitude of the reflected pulse.

Frequency Resolved Optical Gating (FROG) detection is a method that provides a two dimensional temporally resolved spectrum from which phase and amplitude information can be retrieved [11]. It has been applied in measuring the dynamics of femtosecond laser pulses propagating through an underdense plasma [12]. We have used this technique to measure the reflected waveform in a high intensity laser-solid interaction experiment.

THEORETICAL MODEL

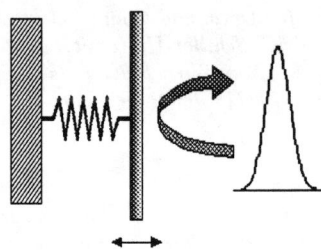

FIGURE 1. Principle of the moving-mirror model : a laser pulse is reflected by a movable mirror. The light pressure pushes the mirror until a restoring force - here graphically represented as a spring - leads the mirror to move back.

At intensities $I \geq 10^{18}$ W/cm^2 the light pressure reaches Gigabars and overwhelms the thermal plasma pressure so that the material is pushed back into the target. Extremely high accelerations are expected while the laser pulse impinges on a target. We assume for an order of magnitude estimate of the acceleration that the plasma motion is only affected by the light pressure $p_L = 2I/c$ which pushes the plasma like a piston. A force $F_L = p_L A$ is applied on the piston with surface A which results in an acceleration $b = F_L/m$ of the piston with mass m. The mass is related to the mass density ρ and the volume V by $m = \rho V$. The volume of the piston can be estimated by the product of the surface area and the skin depth δ. The penetration depth δ of the laser light intensity into a highly overdense plasma layer with a steep density gradient can be estimated by $\delta \simeq (1/2) c/\omega_p$. Here, $\omega_p = \sqrt{n_e e^2/(\epsilon_0 m_e)}$ is the plasma frequency, n_e is the electron density, and c is the speed of light in vacuum. This yields for the acceleration $b = 2I/(\rho \delta c)$ and finally b can be expressed by the formula $b = 4IZe/(M_i c^2 \sqrt{n_e \epsilon_0 m_e})$. The mass density was calculated by $\rho = M_i n_e/Z$ where M_i and Z are the ion mass and the degree of ionization, respectively. The laser light will be reflected in the density range between the critical density n_c and the maximum electron density of the fully ionized solid target. Values of 1.8×10^{20} m/^{2} and 9.8×10^{18} m/^{2} are obtained for $n_c = 1.8 \times 10^{21}$ cm^{-3} and for $n_e = 6 \times 10^{23}$ cm^{-3}, respectively, assuming an intensity of 10^{18} W/cm^2. An electron density of $n_e = 6 \times 10^{23}$ cm^{-3} is expected for a fully ionized solid carbon target which was used as the target material in our measurement.

On the other hand, the laser pulse itself is affected by the plasma, especially by the phase modulation induced by the motion of the reflecting surface. This is shown schematically in Figure 1 where a mirror symbolizes the reflecting plasma surface which is pushed forward by the pulse. A spring with a restoring force stands for the compressibility of the plasma which stops the mirror motion when the increasing internal plasma pressure eventually balances the light pressure.

Assuming plane incident waves the complex electrical field of the reflected pulse can be written in general as $E_r(t) = E_i(t)\ r(t)\ e^{i\varphi}$ where $E_i(t)$ is the field of the incidence pulse and $r(t)$ is the change in the amplitude due to the reflection. The phase modulation caused by a displacement $\Delta x_r(t)$ of the reflecting plasma surface is given by $\varphi(t)=2k_0\Delta x_r(t)\cos\theta$, where θ is the angle of incidence and k_0 is the wave number of the pulse. When the mirror is accelerated, i.e., by the light pressure or by the laser heating, a change in position $\Delta x_r \simeq \frac{1}{2}b\tau^2$ leads to a phase modulation $\Delta\varphi = 2k\,\Delta x_r \simeq k\,b\,\tau^2$ of the reflected pulse for light with normal incidence.

The 2-fluid-approximation [13] is used for the equation of motion which is numerically solved considering the influence of the light pressure $p_L(t) = 2\,I(t)/c$ with a Gaussian pulse profile. A constant acceleration affecting the plasma layer causes a τ^2-term in the phase, a so-called linear chirp. The chirp results in spectral broadening of the reflected light and can be measured by using a phase sensitive technique, e.g., Frequency Resolved Optical Gating. The phase of the reflected pulse is compared to that of the incident laser pulse and the acceleration b can be deduced from the difference in the term $\sim \tau^2$.

EXPERIMENTAL SET-UP

The experimental set-up is shown in Figure 2. The experiment was performed with a Chirped-Pulse-Amplification Ti:sapphire laser system delivering unchirped laser pulses of 110 fs pulse duration with a repetition rate of 10 Hz at a wavelength of 793 nm. The available pulse energy was 105 mJ on target with a contrast ratio of peak pulse power to that of the pedestal on the order of 10^5. The whole experiment took place inside a target chamber evacuated to a pressure of ≤ 1 mbar. The laser beam was focused with an angle of incidence of 12^0 by an off-axis parabolic mirror with an effective focal length of 17.9 cm onto a polished glassy carbon target (Sigradur). The target was mounted on a xyz-positioning table to provide a fresh surface for each shot. The focus diameter was measured with a microscope objective to be less than 10 μm and hence an intensity of about 10^{18} W/cm^2 was reached in the spot.

Due to the fast rise time of the impinging short laser pulse a dense plasma with an increasing electron density is generated by ionization of the carbon atoms. The trailing part of the pulse is transmitted through the plasma up to the turning point and the reflected fraction of the laser light is analyzed. The small angle of incidence of 12^0 provides a separation of the reflected laser pulse from the incoming pulse.

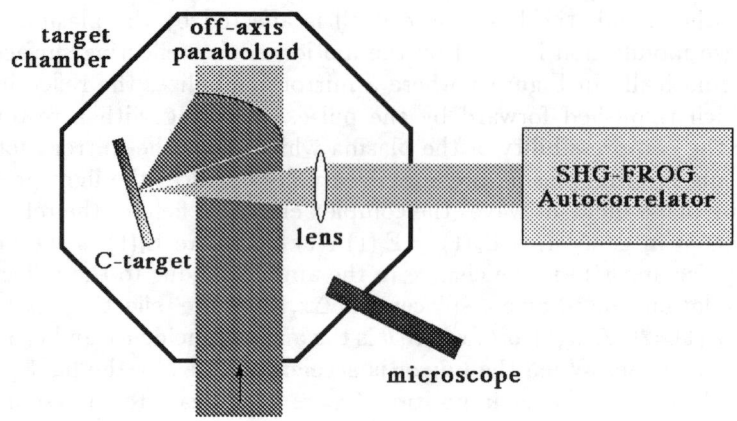

FIGURE 2. Experimental set-up

A collimating lens with a focal length of f = 250 mm is used to collect the reflected light into a parallel beam and send it to a Second-Harmonic-Generation (SHG) FROG-autocorrelator. The central part of the laser beam with a diameter of 5 mm is taken out and is split into two replicas which are focused by cylindrical lenses to two spatially overlapping lines on a BBO crystal.

Both beams are focused in the crystal in such a way that they are not exactly parallel but there is a small angle of 4^0 between them. This gives rise to a small temporal mismatch along the line focus. The temporal overlap is best in the middle of the beam while at the edges both pulses are almost separated. Hence, the intensity of the second harmonic signal at 397 nm produced by both beams together varies along the line focus. The second harmonic signal is now imaged onto the slit of a spectrograph. Along the slit the relative time delay between both pulses is parameterized. The one dimensional SHG signal is dispersed in its spectral components for each time delay (see examples in Figure 3) and a CCD camera detects the FROG trace. In one direction is the autocorrelation function of the laser pulse and perpendicular to it are the corresponding spectral components. Provided that the intensity distribution of the fundamental is constant along the line focus the generated SHG signal is symmetric and consequently the FROG trace is symmetric about the $\tau = 0$ axis.

By using a FROG retrieval algorithm [11] it is possible to reconstruct the temporal amplitude of the electric field and the absolute value of the phase of the laser pulse, respectively, out of the FROG traces in an iterative minimization process. The sign of the phase can not be determined with a SHG FROG but for calculating the acceleration only the absolute value of the phase is needed.

In addition, the spectrum of the reflected laser pulse has been measured separately by a spectrograph to investigate the expected broadening and the shift in the spectrum and to test the FROG trace quality.

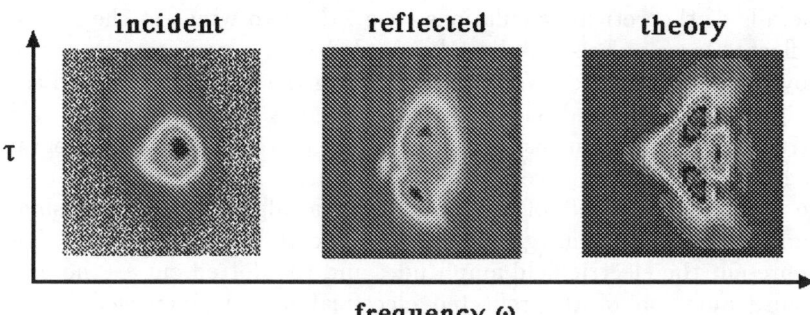

FIGURE 3. Measured FROG-trace of the incident laser pulse (left), the measured FROG-trace of the laser pulse after reflection on an accelerated plasma (middle) and the spectrogram which was calculated with the moving-mirror model (right).

RESULTS AND DISCUSSION

The FROG-trace of the incident laser pulse shown in Figure 3 has a compact structure with one nearly symmetric main peak and a small background which stems from a reflex of one imaging lens. In contrast, the measured FROG-trace of the pulse which is reflected from a plasma generated at an intensity of 10^{18} W/cm^2 is elongated along the time coordinate and shows two distinct peaks. Assuming the simple model of a moving mirror (see Figure 1) the measured reflected FROG spectrogram is theoretically reproduced and shows qualitative agreement with the measurement.

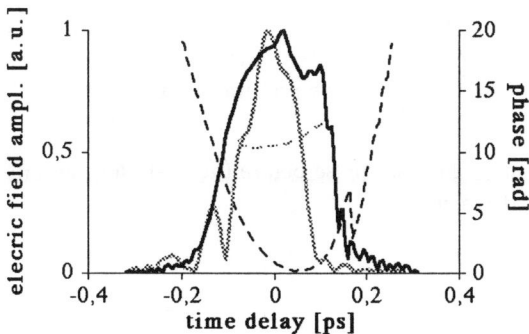

FIGURE 4. Results of the FROG retrieval algorithm: Electric field amplitudes (solid lines) and phases (dashed lines) of single shot incident laser pulse (gray) and laser pulse after reflection on a plasma (black).

The details of theoretical calculations - e.g., the two wings in the FROG trace of the reflected laser pulse - are directly coupled with a compression of the dense plasma by the light pressure. Without considering the light pressure and assuming a plasma that is accelerated outwards the two wings disappear in the calculated FROG trace. Hence, the measured and the calculated FROG spectrograms are only in agreement when an acceleration of the mirror into the target is assumed.

Figure 4 shows the result of a FROG retrieval algorithm for two single shot measurements of the incident (gray) and the reflected (black) waveform. The solid curves represent the electric field amplitudes and the dotted curves the phases.

The pulse duration of the reflected electrical field is increased by a factor of 1.7 compared to the incident laser beam and the phase exhibits a strongly parabolic curved shape. This is attributed to an accelerated plasma mirror moving with $\triangle x \simeq \frac{1}{2} b \tau^2$ which leads to the parabolic phase modulation $\triangle \varphi = 2 k \triangle x \simeq k b \tau^2$ of the reflected pulse. By fitting the phase to a second order polynomial function $\varphi = C \tau^2$ a value of $\varphi \simeq (3.2 \pm 0.1)*10^{26}$ rad/s² × τ^2 is obtained. The measurement yields for the 793 nm laser radiation an acceleration of b $\simeq$ C/k = $(4.0 \pm 0.1)*10^{19}$ m/s² for the plasma, which is in good agreement with theoretical estimates and with other experimental results.

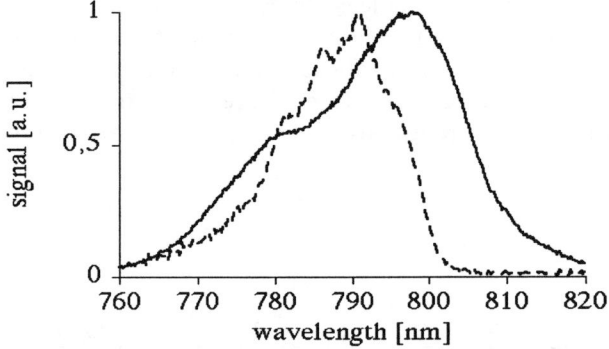

FIGURE 5. Spectrum of the laser pulse before (dashed line) and after (solid line) reflection on the accelerated plasma.

An acceleration on the same order is also obtained by analyzing the spectra of the reflected pulse with a time integrating spectrograph. The spectra shown in Figure 5 were measured independently from the FROG traces under the same experimental conditions.

A red shift of 6 nm is observed which is interpreted as the result of the reflection from a plasma that is pushed into the target by the light pressure. From both the spectral broadening [8] and the spectral shift of the reflected light an acceleration on the order of 1×10^{19} m/s^2 is estimated in agreement with the results of the FROG measurement.

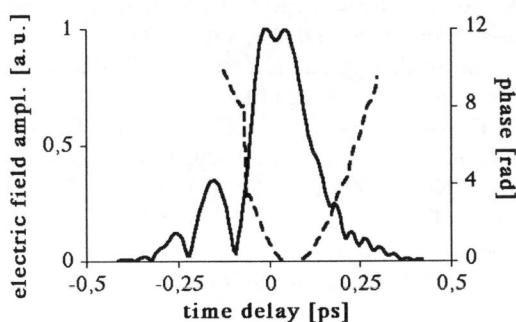

FIGURE 6. Electric field amplitude (solid line) and phase (dashed line) for a reflected laser pulse at intensities of 10^{15} W/cm^2

The same experiment was performed at a lower intensity of 10^{15} W/cm^2 and shows a clearly different FROG trace, which is also reproduced by the moving mirror model but now for a freely expanding plasma without taking the light pressure into account (see Figure 6). The phase still has a τ^2-term, which is slightly less than for 10^{18} W/cm^2. At 10^{15} W/cm^2 the acceleration calculated from the phase is still $(2.00 \pm 0.1)*10^{19}$ m/s^2 which is only a factor of 2 less than the acceleration at 10^{18} W/cm^2.

In summary we have presented the first direct measurement of accelerations on the order of $4 * 10^{19}$ m/s^2 in femtosecond laser-produced plasmas using the FROG technique. With the simple moving mirror model qualitative predictions for the expected FROG observations can be achieved for intensities of 10^{15} and 10^{18} W/cm^2.

REFERENCES

1. H.M. Milchberg, R.R. Freeman, S.C. Davey, Phys. Rev. Lett. **61**, 2364 (1988).
2. J. C. Kieffer, Z. Jiang, A. Ikhlef, C. Y. Cote, and O. Peyrusse, J. Opt. Soc. Am. B **13**, 132 (1996).
3. W. Theobald, R. Häßner, C. Wülker, and R. Sauerbrey, *Phys. Rev. Lett.* **77**, 298 (1996).
4. J. D. Kmetec, C. L. Gordon, III, J. J. Macklin, B. E. Lemoff, G. S. Brown, and S. E. Harris, *Phys. Rev. Lett.* **68**, 1527 (1992).
5. M. Schnürer, M. P. Kalashnikov, P. V. Nickles, Th. Schlegel, and W. Sandner, *Phys. Plasmas* **2**, 3106 (1995).
6. A. P. Fews, P. A. Norreys, F. N. Beg, A. R. Bell, A. E. Dangor, C. N. Danson, P. Lee, and S. J. Rose, *Phys. Rev. Lett.* **73**, 1801 (1994).
7. G. Malka and J. L. Miquel, *Phys. Rev. Lett.* **77**, 75 (1996).
8. R. Sauerbrey, *Phys. Plasmas* **3**, 4712 (1996).
9. O. L. Landen, W. E. Alley *Phys. Rev. A* **46**, 5089 (1992).
10. X. Liu and D. Umstadter *Phys. Rev. Lett.* **69**, 1935 (1992).
11. R. Trebino and D. J. Krane *JOSA A* **10**, 1101 (1993).
12. P. R. Bolton, A. B. Bullock, C. D. Decker, M. D. Feit, A. J. P. Megofina, and P. E. Young, *J. Opt. Soc. Am. B* **13**, 336 (1996).
13. W.L. Kruer, *The Physics of Laser Plasma Interactions* (Addision-Wesley, New York, 1988).

Generation of High Order Optical Harmonics in Steep Plasma Density Gradients

D. von der Linde

Institut für Laser- und Plasmaphysik, Universität Essen, D-45117 Essen, Germany

Abstract. During the interaction of an intense ultrashort laser pulse with solid targets a thin layer of surface plasma is generated in which the density drops to the vacuum level in a distance much shorter than the wavelength. This sharp plasma-vacuum boundary performs an oscillatory motion in response to the electromagnetic forces of the intense laser light. It is shown that the generation of reflected harmonics can be interpreted as a phase modulation experienced by the light upon reflection from the oscillating boundary. The modulation sidebands of the reflected frequency spectrum correspond to odd and even harmonics of the laser frequency. Retardation effects lead to a strong anharmonicity for high velocities of the plasma-vacuum boundary. As a result, harmonic generation is strongly enhanced in the relativistic regime of laser intensities.

INTRODUCTION

Generation of optical harmonics of very high order is a subject of great current interest. Harmonics almost up to order 300 have recently been observed during the interaction of an intense laser pulse of 25 fs duration with a jet of low pressure helium gas (1). Harmonic generation in *gases* has been studied extensively during the last years (2). The basic physical mechanism is the strong nonlinear response of the electrons in a laser field close to the field ionization limit of the atoms.

High order harmonic generation from *solid targets* was observed by Carman et al. (3, 4) many years ago. Using nanosecond pulses from a CO_2 laser to irradiate solid targets at intensities of the order of 10^{15} W/cm^2, they observed harmonics up to about order 35 with a distinct high frequency cutoff. A key point of the theoretical explanation (5, 6) of these observations was the assumption of a step-like plasma density gradient. Electrons driven across this steep gradient by the laser field experience strongly anharmonic forces. The anharmonicity of the electron motion leads to the generation of odd and even harmonics. The formation of a steep density gradient in a plasma produced by nanosecond laser-solid interaction can be attributed to the action of the ponderomotive force, which counteracts the hydrodynamic expansion. The high frequency cutoff was also explained. According to the theoretical models (5, 6) the cutoff frequency is given by the plasma frequency corresponding to the upper density value of the density step.

The interaction of ultrashort laser pulses with solid materials leads quite naturally to the formation of a steep plasma density gradient at the boundary to vacuum. During the formation of the plasma by properly shaped femtosecond laser pulses there is no time for significant plasma expansion. A thin layer of plasma is formed on the surface of the target. The plasma density in this layer drops from approximately solid density to vacuum in a very short distance. The interaction of intense ultrashort laser pulses with solid surfaces is thus expected to lead to generation of odd and even harmonics.

Progress in the generation of extremely intense ultrashort laser pulses has led to new interest in harmonic generation from solid targets. Specularly reflected coherent harmonics from solid surfaces up to the seventh order have been observed by Kohlweyer et al. (7), and up to the eighteenth by the author and coworkers (8). In both cases femtosecond laser pulses from a titanium sapphire CPA laser system have been used. Norreys et al. (9) reported generation of harmonics as high as 75 with picosecond laser pulses from a neodymium laser system. In the latter case the harmonic radiation had a much larger angular distribution, not being confined to the specular direction.

Recent particle-in-cell (PIC) simulations by Gibbons (10) and by Lichters et al. (11) provided new insight into high order harmonic generation from a plasma-vacuum boundary. In particular, these calculations demonstrated the importance of contributions to the electron anharmonicity from relativistic effects. It was shown that harmonic generation could be extended well beyond the high frequency limit of the earlier theories (5,6) by increasing the laser intensity to the relativistic regime.

Lichters et al. (11) also performed analytic calculations based on relativistic plasma fluid equations. They showed that their detailed numerical simulations of the complex collective electron dynamics are in excellent agreement with a simple model in which the harmonics are obtained simply upon reflection of light from an oscillating surface. This moving mirror model originally due to Bulanov et al. (12) turns out to be extremely useful for understanding high order harmonic generation at a plasma-vacuum boundary.

THE MOVING MIRROR MODEL

The threshold of plasma formation in typical dielectric solids is about 10^{13} W/cm^2 for a 100 fs laser pulse (13). When the laser intensity exceeds this threshold, condensed matter is very rapidly turned into a plasma. During the short time of interaction with the laser pulse the ions can be regarded as fixed positive background charges. Electrons in a skin depth layer experience strong electromagnetic forces from the laser pulse. These forces drive electrons back and force across the vacuum boundary. The basic approximation of the moving mirror model (11, 12, 14) is to neglect the details of the electron spatial distribution and to represent the collective electronic motion by the motion of some characteristic electronic boundary, e.g. the critical density surface. This boundary represents an *effective reflecting surface,* the moving mirror, performing an oscillatory motion.

Consider for a moment the spectrum of light reflected from an ordinary optical mirror oscillating at a frequency ω_m, much smaller than the optical frequency ω_0. The mirror motion produces a phase-modulation of the light, and the reflected spectrum exhibits

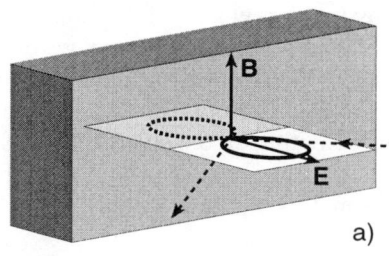 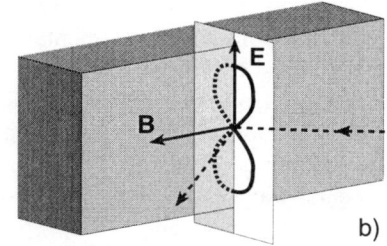

FIGURE 1. Directions of the electric and magnetic fields and "figure-of-eight" orbit of an electron: a) for p-polarized light, b) for s-polarized light. The dashed line indicates the incident and the reflected light.

sidebands at multiples of the modulation frequency. If the mirror could be made to vibrate at $\omega_m = \omega_0$, these modulation sidebands would represent optical harmonics of the fundamental optical frequency.

To gain some qualitative insight into what type of motion the reflecting electron surface might perform under the action of strong electromagnetic driving forces, let us consider the orbit of a single free electron. During one optical cycle the electron follows the well-known "figure-of-eight" motion in a plane spanned by the wavevector and the electric field vector (15). Figure 1 a and 1 b illustrate the situation for a slab of plasma with p-polarized and s-polarized incident light, respectively. For p-polarization the electron orbit lies in the plane of incidence, and the normal component of the electron motion oscillates at ω_0 (Fig. 1 a). For s-polarization the plane of the orbit is perpendicular to the plane of incidence. In this case we have a normal motion at $2\omega_0$, twice the optical frequency.

These considerations suggest the following conclusions. The electron surface oscillates at ω_0 for a p-polarized driving wave, and at $2\omega_0$ for a s-polarized wave. In the first case the sidebands of the reflected light represent even and odd optical harmonics. In the second case the frequencies of the sidebands are multiples of $2\omega_0$, and only odd harmonics are obtained. The $2\omega_0$ longitudinal component is driven by magnetic forces, which play a significant role in the relativistic regime, when the electron velocity approaches the speed of light.

Summarizing these conclusions the polarization rules shown in Table 1 can be formulated. It follows from the moving mirror model that the polarization of the fundamental and the harmonics are the same. P-polarized fundamental light generates p-polarized even and odd reflected harmonics. When the fundamental is s-polarized, the polarization of the harmonics is also s-type, but only odd harmonics are generated. These results agree with

Table 1. Polarization selection rules

	s-pol. harmonics	p-pol. harmonics
s-pol. fundamental	odd	even
p-pol. fundamental	forbidden	even & odd

the polarization selection rules derived rigorously by Lichters et al. (11). In addition, they showed that s-polarized fundamental light also produces p-polarized harmonics of even order (top of last column in Table 1). This case cannot be interpreted as phase-modulated reflected light. However, it is possible to extend the moving mirror model and include this situation. The motion of the electrons produces an oscillating dipole layer at the plasma-vacuum boundary. Production of p-polarized harmonics of even order by a s-polarized fundamental wave can be interpreted as radiation from this dipole layer (14).

CALCULATION OF THE REFLECTED LIGHT

It is quite helpful to distinguish the "mechanical" process causing the surface deformation from the "optical" process responsible for harmonic generation. Thus, the surface deformation could be due to a strong electromagnetic wave incident on the boundary. We can then consider the modulation experienced by a second weak wave upon reflection assuming that the intensity of this wave is too low to affect the deformation.

Referring to Fig. 2 let us assume a surface deformation with an excursion normal to the surface (z-direction) of the form

$$u(t,x) = u_o \sin(\omega_m t - qx + \phi) \qquad (1)$$

The phase angle ϕ takes into account the phase difference between the deformation wave $u(t,x)$ and the incident optical wave.

In (1) the original picture of a rigidly moving mirror surface has been modified to take into account the variation of the electromagnetic forces parallel to the surface (x-direction). For oscillation frequencies ω_m comparable with the optical frequency ω_0 the surface deformation has the form of a travelling wave. The rigidly moving mirror surface is recovered in the low frequency limit $\omega_m \ll \omega_0$, when the spatial dimensions of the mirror

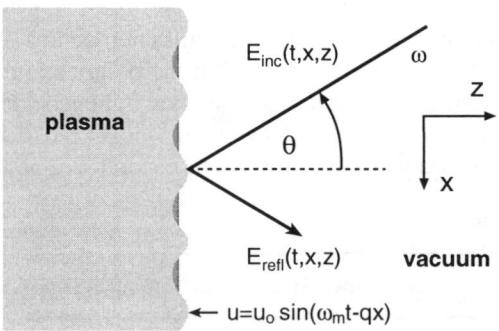

FIGURE 2. Schematic showing sinusoidal deformation of the spatial distribution of the electrons. The darker area indicates the fixed ion background charge.

are much smaller than the wavelength corresponding to ω_m. Note also that in (1) a purely sinusoidal deformation has been assumed for simplicity. Thus, the anharmonicity of the electron motion is neglected in our present discussion.

The reflected wave can be readily calculated for an arbitrary modulation frequency. In the following we restrict ourselves to $\omega_m = 2\omega_o$. Consider the reflection of a incident plane monochromatic wave of the form

$$E_{inc}(t,x,z) = E_o \exp[-i\omega_o(t - x/c\sin\theta - z/c\cos\theta)] \quad (2)$$

where the angle of incidence θ is chosen to satisfy the phase matching condition $\omega_o/c\sin\theta = q$.

The reflected wave is taken to be a plane wave of the form

$$E_{refl}(t,x,z) = G(t - x/c\sin\theta + z/c\cos\theta) \quad (3)$$

The functional form $G(\tau)$ of the reflected wave follows from the boundary condition that the total electric field at the surface vanishes:

$$E_{inc}(t,x,u(t,x)) + E_{refl}(t,x,u(t,x)) = 0 \quad (4)$$

The boundary condition (4) implies total reflection at the electron surface, again an assumption made just for convenience. With the substitutions

$$\tau = \xi + \chi/2 \sin(2\omega_o\xi) \quad (6a)$$

$$\chi = 2\omega_o u_o/c \cos\theta \quad (6b)$$

$$\xi = t - x/c\sin\theta \quad (6c)$$

the reflected wave is obtained in the following form

$$G(\tau) = -E_o \exp[-i(\omega_o\tau - \chi\sin(2\omega_o\xi))] \quad (5)$$

where $\xi = \xi(\tau)$ is the solution of (6a). The condition that the surface velocity must be less than the speed of light requires

$$\chi < \cos\theta < 1 \quad (6)$$

The spectrum of the reflected light is obtained by taking the Fourier transform of (5). However, before proceeding to the calculation of the harmonic spectra, it is quite instructive to discuss the reflected wave (5) in the time domain.

Attosecond Pulses

According to (2) the incident wave is purely monochromatic. However, for an observer in the laboratory frame the wave reflected from a rapidly oscillating mirror is strongly distorted. The stronger the deformation, the higher the harmonic content of the reflected spectrum, and the higher the number of harmonics. The actual shape of the reflected wave form is also critically dependent on the phase difference ϕ. For a suitable value of ϕ the maxima of the sinusoidal incident wave are compressed into a series of extremely short pulses with a duration very much smaller than the time of a fundamental optical cycle. The situation is illustrated by Fig. 3. The dotted line shows the intensity of the incident wave, which is proportional to $(\cos\omega_0 t)^2$. The dashed and the solid line represent the reflected wave form (5) for $\chi = 0.5$ and $\chi = 0.9$, respectively. In both examples the incident optical wave and the surface oscillation are exactly out of phase ($\phi = \pi$). It can be seen that for the highly relativistic case $\chi = 0.9$ the compression is quite substantial. The full width at half maximum of the pulses corresponds to approximately 3 percent of the fundamental optical cycle. For an optical wavelength of 800 nm the pulse duration would be 8×10^{-17} s or 80 attoseconds.

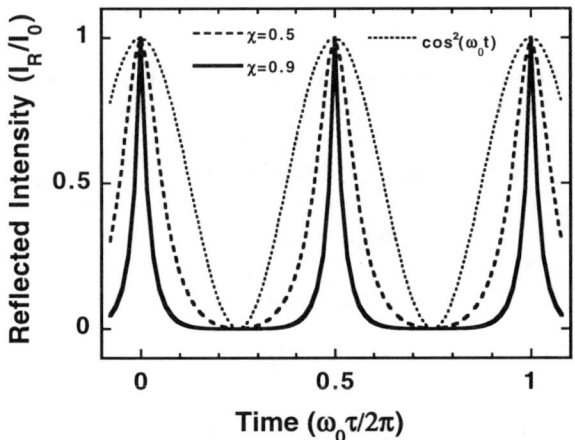

FIGURE 3. Solid curve and dashed curve: Train of attosecond light pulses obtained by reflecting a monochromatic wave (dotted curve) from an oscillating mirror.

Harmonic Spectra

It turns out that for a sinusoidal surface oscillation of arbitrary frequency analytic expressions for the reflected spectrum can be found (14). For example, the spectrum of the odd harmonics for $\omega_m = \omega_0$ is given by

$$S((2n+1)\omega_0) = (\pi E_0)^2 \left(\frac{J_n((n+1)\chi)}{(n+1)} - \frac{J_{n+1}(n\chi)}{n} \right)^2 \qquad (7)$$

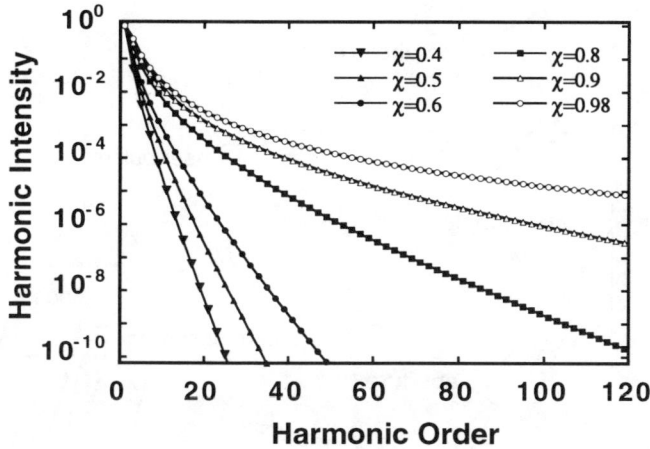

FIGURE 4. Example of spectra of odd harmonics for different surface excursions (parameter χ).

where $J_n(x)$ are the ordinary Bessel functions of order n.

Examples of the harmonic spectra are shown in Fig. 4 for different values of χ. It can be seen that the efficiency of harmonic generation increases strongly with χ, that is, with the amplitude u_0 of the surface excursion. Harmonics of the order of about 100 would be produced with an efficiency of $\approx 10^{-5}$ when χ approaches the relativistic limit $(\chi \approx 1)$.

For $\omega_m = \omega_0$ the maximum surface excursion corresponding to the relativistic limit is $u_0 = \lambda/4\pi$, where λ is the fundamental wavelength. It can be shown that in the non-relativistic intensity regime s-polarized light produces much smaller surface excursions than p-polarized light (14). Thus, at low intensities only p-polarized light is expected to generate high order harmonics.

MEASUREMENT OF HIGH ORDER HARMONICS

The experimental schematic for the observation of high order harmonic generation from solid targets is depicted in Fig. 5. Laser pulses from a titanium sapphire laser ($\lambda \approx$ 800 nm) produced peak intensities at the target surface between 10^{17} and 10^{18} W/cm². Pulse durations of 60 to 120 fs were used. The incident laser beam was linearly polarized in the plane of incidence (p-polarization) at an angle of incidence of approximately 65 degrees. The reflected harmonics were recorded using a toroidal grating with a phosphor screen at the output plane. The phosphorescent light excited by the harmonic photons was recorded by a photodiode array or a CCD camera. The targets were optically polished bare glass substrates and glass substrates coated with 200 nm of aluminium. The targets were raster-scanned to provide a fresh sample surface for each laser pulse.

Two examples of typical harmonic spectra from glass are depicted in Fig. 6 a and 6 b. The intensity on target was approximately 10^{17} W/cm². Even and odd harmonics were observed up to a maximum order of 18. The peak at 38 nm in Fig. 6 b is line emission from the plasma. Very similar spectra were observed with aluminium samples.

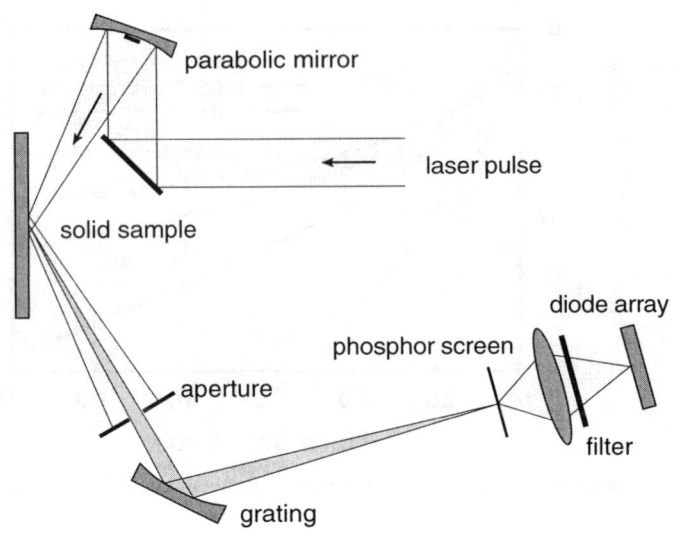

FIGURE 5. Schematic of the experimental setup.

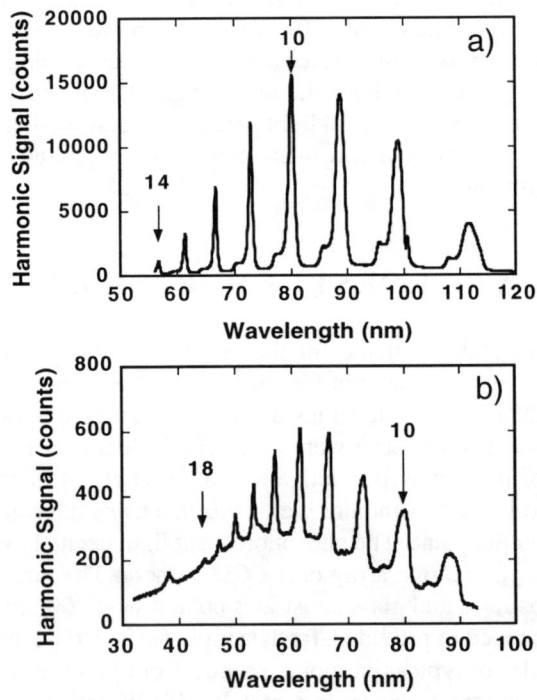

FIGURE 6. Examples of measured harmonic spectra from glass samples.

As shown in Fig. 6 b the harmonic signal was typically superimposed on some background. It was observed that the amount of background radiation and also the appearance of plasma line emission depends on the temporal profile of the laser pulses. For instance, clean harmonic spectra with little background were observed, when laser pulses of 120 fs duration and a temporal profile with an intensity drop of 10^{-5} in 1 ps were used. Figure 6 a is an example of very strong harmonic emission with a large harmonic-to-background ratio. On the other hand, the spectra were dominated by background radiation and plasma line emission with only very weak harmonic lines, when laser pulses of nominally 60 fs duration with an intensity drop of only about 3×10^{-3} in 1 ps were used. Optical emission of the plasma in the visible and ultraviolet undoubtedly contribute to the background, but these effects are not well understood as yet.

The relative strength of the harmonic peaks in the spectra of Fig. 6 a and 6 b is determined by the properties of the phosphor screen and optical aberrations of the toroidal grating. When these effects are taken into account a smooth roll-off corresponding to an exponential decrease of the harmonic signal in dependence of the harmonic order is obtained. A high frequency cutoff as seen in harmonic generation in rare gases and in the earlier harmonic work with solid targets (3, 4) was not observed.

The conversion efficiency of high order harmonic generation is relatively high. For the strongest harmonic spectra the number of photons per pulse is estimated to be about 10^{11} and 5×10^9 at the tenth and the fifteenth harmonic, respectively. The corresponding photon conversion efficiency is approximately 10^{-6} and 5×10^{-8}. These numbers are subject to some uncertainty, because of the error in estimating the properties of the phosphor screen.

DISCUSSION

For peak intensities of about 10^{17} W/cm^2 the shortest wavelength was 45 nm, which is the eighteenth order (photon energy of 27.5 eV). Harmonic generation could only be observed when p-polarized laser pulses were used. These results are in agreement with the theoretical expectations for harmonic generation in the non-relativistic interaction regime. On the other hand, the oscillating mirror model and the PIC simulations (10, 11) showed that harmonic generation by s-polarized light is possible and that harmonics of much higher order can be produced with relativistic laser intensities. This would require intensities greater than 10^{18} W/cm^2 at the wavelength of the titanium sapphire laser.

Experimentally, the following behavior was observed, when the intensity on target was increased to a few times 10^{18} W/cm^2. At low intensities a clean, undisturbed reflected beam from the target surface was observed. The harmonic spectra shown in Fig. 6 a and 6 b were obtained under these conditions. An increase of the fundamental intensity led to an increase of the plasma emission and of the background radiation. As a result, the ratio of harmonics to background decreased. At still higher intensities the reflected beam became unstable, and substantial beam distortions began to develop. For intensities around 10^{18} W/cm^2 the reflected light was no longer collimated but spread out over some large solid angle around the specular direction. Under these conditions the generation of coherent high order harmonics in the specular direction disappeared.

SUMMARY

High order harmonics from solid targets can be interpreted as a phase modulation imposed on the light upon reflection from an oscillating surface. It was shown that the reflected wave represents a train of attosecond pulses for certain values of the phase between the incident light and the surface deformation.

Experimentally, specularly reflected high order odd and even optical harmonics were observed for p-polarized laser pulses at oblique incidence, while s-polarized light was ineffective in the non-relativistic intensity regime. Evidence has been obtained that a sharp plasma-vacuum interface is crucial for harmonic generation from solid targets.

ACKNOWLEDGEMENTS

This work has been supported by the Deutsche Forschungsgemeinschaft and the European Union under "Training and Mobiltity by Research". The author gratefully acknowledges contributions of K. Razazewski, T. Engers, G. Jenke and the hospitality and support of the Laboratoire d'Optique Appliquée, Ecole Polytechnique/ENSTA, Palaiseau, France.

REFERENCES

1. Murnane, M., "Coherent Ultrashort Pulse X-Ray Generation below 3.6 nm", Intern. Conf. on Superstrong Fields in Plasmas, Varenna, Italy, Aug. 27 - Sept. 2, 1997.
2. Balcou, Ph., Salières, P., and L'Huillier, A., in *Super-Intense Laser-Atom Physics*, NATO ASI Series B 316: Physics, New York: Plenum Press, 1993.
3. Carman, R.L., Forslund, D.W., and Kindel, J.M., Phys. Rev. Lett. **46**, 29-32 (1981).
4. Carman, R.L., Rhodes, C.K., and Benjamin, R.F., Phys. Rev. A **24**, 2649-2663 (1981).
5. Bezzerides, B., Jones, R.D., and Forslund, D.W., Phys. Rev. Lett. **49**, 202-205 (1982).
6. Grebogi, C., Tripathi, V.K., and Chen H.H., Phys. Fluids **26**, 1904-1908 (1983).
7. Kohlweyer, S., Tsakiris, G.D., Wahlström, C.G., Tillman, C., and Mercer, I., Opt. Comm. **117**, 431-438 (1995).
8. von der Linde, D., Engers, T., Jenke, G., Agostini, P., Grillon, G., Nibbering, E., Mysyrowicz, A., and Antonetti, A., Phys. Rev. A **52**, R25-R27 (1995).
9. Norreys, P.A., Zepf, M., Moustaizis, S., Fews, A.P., Zhang, J., Lee, P., Bakarezos M., Danson, C.N., Dyson, A., Gibbon, P., Loukakos, P., Neely, D., Walsh, F.N., Wark, J.S., and Ongor, A.E., Phys. Rev. Lett. **76**, 1832-1835 (1996).
10. Gibbon, P., Phys. Rev. Lett. **76**, 50-53 (1996).
11. Lichters, R., Meyer-ter-Vehn, J., and Pukhov, A., Phys. Plasmas **3**, 3425-3437 (1996).
12. Bulanov, S.V., Naumova, N.M., and Pegoraro, F., Phys. Plasmas **1**, 745-757 (1994).
13. von der Linde, D., and Schüler, H., J. Opt. Soc. Am. B **13**, 216-222 (1996).
14. von der Linde, D., and Rzazewski, Appl. Phys. B **63**, 499-506 (1996).
15. Landau, L.D., and Lifshitz, E.M., *The Classical Theory of Fields*, Vol. 2, Oxford, Pergamon Press, 1983, p. 118.

Isochoric Heating of Solid Density Matter with Ultra-High Intensity Femtosecond Laser

Z.Jiang, P.Gallant, J.C.Kieffer, H.Pépin

INRS-énergie et matériaux, 1650 Boul.Lionel-Boulet, Varennes, Québec J3X 1S2, Canada

O.Peyrusse, J.L.Miquel

CEA Centre d'étude de Limeil-Valenton, France

Abstract. Ultra-high intensity femtosecond laser pulses have been used to irradiate foil targets to study the concept of isochoric heating of dense matter. The ionization dynamics and electron density evolution of Al targets (thickness between 500 Å and 2μm) irradiated at laser intensity between 10^{18} W/cm^2 and 10^{20} W/cm^2, were inferred from time-resolved keV x-ray spectroscopy with 750fs resolution.

INTRODUCTION

The recent introduction of a new amplification technique, namely the chirped pulse amplification technique, and new broadband amplifying materials has made possible the production of extremely compact high intensity lasers. This revolution in laser technology has extensively stimulated the research on the interaction of ultrafast, ultra-high intensity laser pulses with matter [1,2]. When a short laser pulse is focused on the targets, a bright burst of x-ray emission whose photon energy ranges from a few hundred eV to a few hundred keV is radiated with a duration of a few picoseconds [3] or even a fraction of picosecond if all conditions are optimized. The laser-based ultrafast x-ray sources could be a very unique tool for application like, for example, the observation of molecular dynamics during chemical reactions through time-resolved x-ray absorption or diffraction [4].

The production of very dense and very hot matter is necessary to achieve the shortest multi-keV x-ray pulses [5]. To produce such hot, solid density plasmas it was recently proposed to use judiciously the laser radiation pressure to prevent the hot plasma from expanding during the heating phase [6]. In such a regime where the thermal pressure is balanced by the radiation pressure, the plasma should be confined during the heating and the electron density should stay very close from the fully ionized solid density value during the interaction. In these conditions, foil targets seem to be the most appropriate targets to realize isochoric heating of solid matter, i.e. heating of matter at constant density, up to extreme temperatures [6]. Indeed thin foil targets will reach a much higher plasma temperature because of the limitation of the thermal diffusivity. With foil thickness much smaller than the thermal penetration depth, the entire foil could be heated up instantly by ultraclean, ultrashort laser pulse producing 1D, hot and uniform plasma slab [6,7] which will expand very rapidly and cool at low density. Such plasma conditions are of interest for astrophysics [8], strongly correlated plasma [9], non-stationary and non local plasma [10], line broadening and quasi-molecular states problems [11], relativistic effects and self induced transparency, LTE and non-LTE plasma problems [12].

Theoretical calculations of the time history of the various Al line emissions have been realized for thick target and 500 Å foil, in a regime where the light pressure is an important ingredient of the interaction [13], being high enough to exceed the thermal pressure but still too low to make a hole in the plasma. The emitted x-ray spectrum has been predicted with a multi-cell time dependent atomic model TRANSPEC coupled to a one fluid, two-temperature Eulerian code. This code includes a wave solver to model self-consistently the laser energy deposition, the ponderomotive force and a non-steady state atomic physics package to model ionization and recombination [13]. Simulations indicate that the duration of some lines can be very short, while some other lines display very long recombination tail. As an example, Fig.1 shows the Li-like pulses expected for an Al thick target and for a thin foil (500 Å) irradiated by a Gaussian 300 fs laser pulse of green light. The Li-like pulses are both sub-picosecond but the foil emission is shorter and rises faster than the thick target pulse.

Experimental conditions for efficient isochoric heating have to be optimized in terms of laser intensity, pulse duration, target material and target thickness. On one hand, laser intensity should be high enough to overcome plasma thermal expansion in order to keep plasma density constant during heating. On the other hand, laser intensity should not exceed the limit above which the 2D/3D density profile modification and hole boring may occur [14]. In this paper we present the experiments realized to test our concept of isochoric heating. Al foils have been irradiated with laser intensity between 10^{18} to 10^{20} W/cm^2. The Al foils were free standing with thickness ranging from 50nm to 300nm. This largest thickness is close to the penetration found in our previous experiments [15]. Results indicate that the x-ray emission from the foils is produced at near solid electron density with electron temperature much higher due to the limitation of thermal conduction. Time-resolved line broadening measurements indicate the electron density of thin foil targets (500 Å and 700 Å thick) remains around 4×10^{23} cm^{-3} during the interaction, even at the highest intensity, showing the effectiveness of the plasma confinement by the light pressure. At low intensity, the density decreases linearly with time after the heating showing a 1D dynamics. At higher intensity, around 10^{19}W/cm^2, the duration of Li-like emission from foils is shorter than the one from thick target and decreases when the thickness of foil decreases, demonstrating a faster cooling as the initial plasma temperature increases.

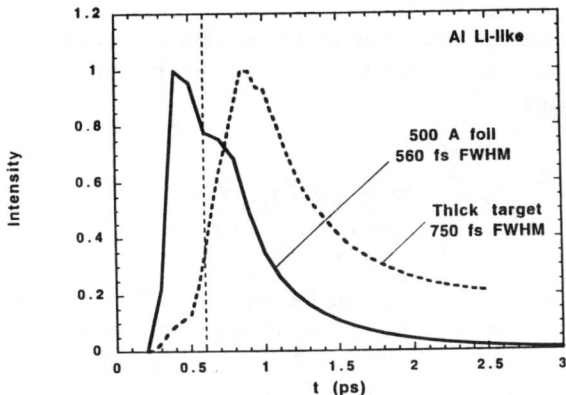

Figure 1. Calculated Al Li-like emissions. The Li-like emission from Al foil target (500 Å, solid line) is shorter and rises faster than the one from thick target (dot line) for the same conditions (300fs-530nm laser pulse at intensity of 10^{18} W/cm^2). The vertical line indicates the position of the peak of the laser pulse.

EXPERIMENTAL SETUP

Experiments were performed in two laser intensity regimes, 10^{18}W/cm^2 and 10^{20}W/cm^2, with the CPA T^3 laser systems at INRS (lower intensity regime) and at CEA-Limeil (higher intensity regime with P102 laser system). The contrast of laser pulse is crucial for these experiments since any prepulse will explode the foil target before the main laser pulses, producing a low density preplasma. High contrast ratio (prepulse/main pulse 10^{-10}:1) was realized in our experiments by using frequency doubled (0.53μm, 400fs pulse width) light of 1.05μm laser.

At INRS, the laser light was focused with an off-axis parabola (45° incidence angle) in a 12μm diameter focal spot, giving a laser intensity on target around 10^{18} W/cm^2. Time resolved keV spectra have been obtained with our modified Kentech x-ray streak camera, which has a temporal resolution of 1.5 ps, coupled to a Von-Hamos crystal spectrometer [16]. Spectral resolution was around 2mÅ.

At CEA-Limeil the 0.53μm laser light was focused on target along the target normal in a 4μm focal spot, giving a maximum laser intensity around 5×10^{19}W/cm^2. Time resolved keV spectra were recorded with our novel sub-picosecond x-ray streak camera (PX1 camera), which has a temporal resolution of 750fs in the keV range [17], coupled to a conical crystal spectrometer. The spectral resolution was around 4mÅ.

In both experiments, time-integrated spectrometer has also been used, as well as x-ray diode filtered with k-edge filters to monitor the integrated x-ray emission between 1-10keV range.

X-RAY YIELD

Figure 2 shows the time-integrated spectra in 6.5 to 8.5Å range detected for thick Al target and 700Å thin Al foil irradiated with laser pulse at intensity of 10^{18}W/cm^2. In both cases the Li-like satellites are very broadened indicating this emission is produced at near solid density [15,18]. The near solid electron density plasma found in 700Å foil target indicates also the prepulse level of the interaction laser pulse is extremely low. The K_α emissions in high ionization states and the cold $K\alpha$ emission have not been observed with thin foils, contrary to what is observed with thick targets. This is not surprising considering that the hot and dense plasma slab has a thickness which is extremely small compared to the range of the hot electrons. Furthermore, The Ly_α line is increasingly brighter as the foil thickness is decreased and the time integrated x-ray yield measured with foil is almost four times higher than the one measured with thick target as shown in Fig.3 for the yield in 1-10keV photon energy range. This indicates that the electron temperature reached with foils is much larger than the one reached in thick targets. The line intensity ratio of Ly_α/He_β gives us a rough estimation of 1400eV for the electron temperature in foil plasma , compared to about 600eV for the electron temperature in thick target [15].

TIME-RESOLVED SPECTROSCOPY

The x-ray duration from thin foils remains short, due to the fast cooling after the laser pulse. However the line duration changes significantly with the laser intensity and with the target thickness. Figure 4 shows the time-resolved spectra, He_α and Li-like line emissions, obtained with a temporal resolution of 1.5ps, when Al thin foil and thick targets are irradiated at 10^{18}W/cm^2. The duration of Li-like emission from foil is a few picosecond longer than the one from thick target, which is somehow different from expectations (Figure 1). However, the $Ly\alpha$ line from the foil rises in 1.5ps and has a 3ps FWHM (as shown in Figure 5), which is significantly shorter than the He_α FWHM (5 ps).

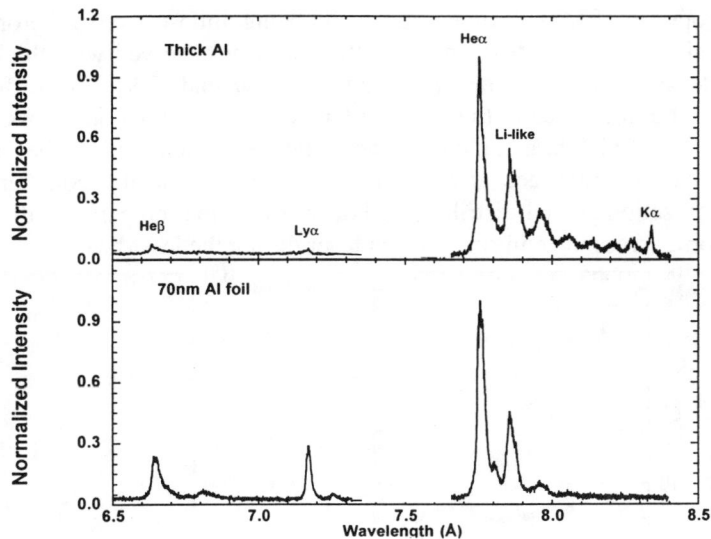

Figure 2. Time-integrated spectra for thick target (upper) and Al 700Å foil (bottom). The laser intensity is 7×10^{17} W/cm^2.

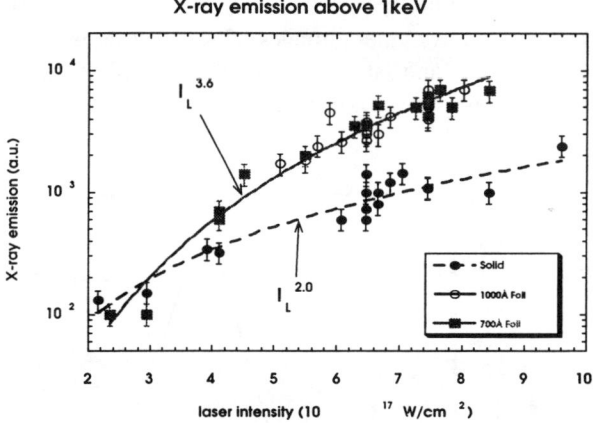

Figure 3. Spectrally integrated x-ray yield (1keV to 10keV) for solid Al target (full dot), 1000Å foil (open dot) and 700Å foil (full square) irradiated by 0.53μm laser pulses at various laser intensities.

The electron density has been deduced from line broadening calculations by using the standard method of Stark line broadening in plasma as described in [18]. The temporal evolution of electron density for the 700Å foil target and at 10^{18} W/cm^2, as deduced from time resolved Li-like line broadening, is shown in Figure 6. The density reaches a maximum value, close to the solid density value, during the laser heating and then drops very fast with time, the density being almost inversely proportional to the time. This behavior is different from the behavior observed with thick target for which the density has its maximum after a few picoseconds [19]. This difference may be related to the difference in the electron temperature distribution across the plasma emission layer. In thick target, there is a temperature gradient with higher temperature at the plasma-laser interface and lower temperature inside the cold target, which induces a ionization spatial distribution. For thin foil, the temperature distribution is nearly uniform across the emitting plasma layer during the laser heating.

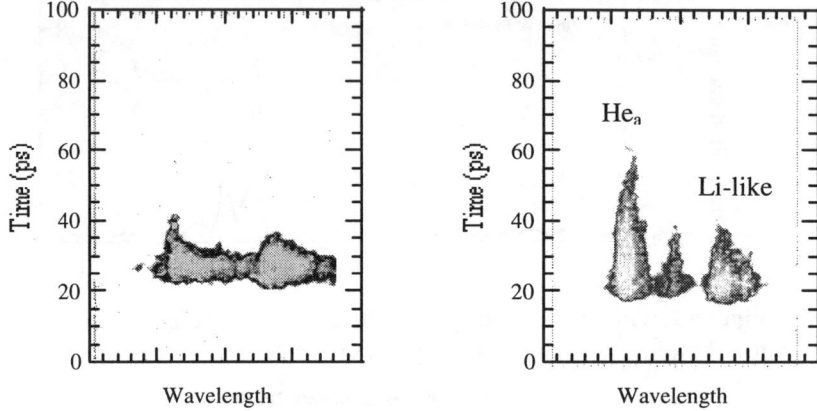

Figure 4. X-ray streak camera images of Al thick target (left) and Al thin foil (700 Å, right) and at laser intensity of 10^{18}W/cm^2. The time resolution of streak camera is 1.5ps.

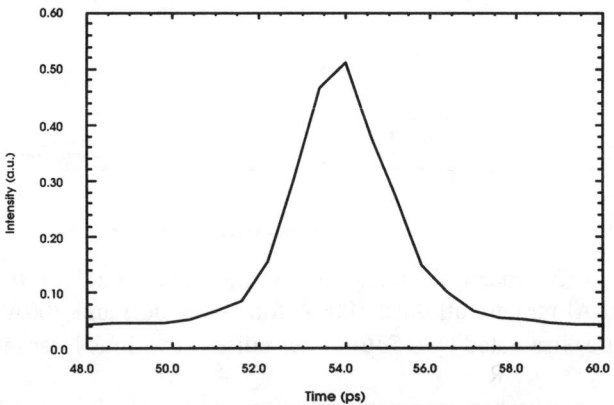

Figure 5. Time history of Ly$_\alpha$ line emitted by a 700Å Al foil irradiated at 10^{18}W/cm^2.

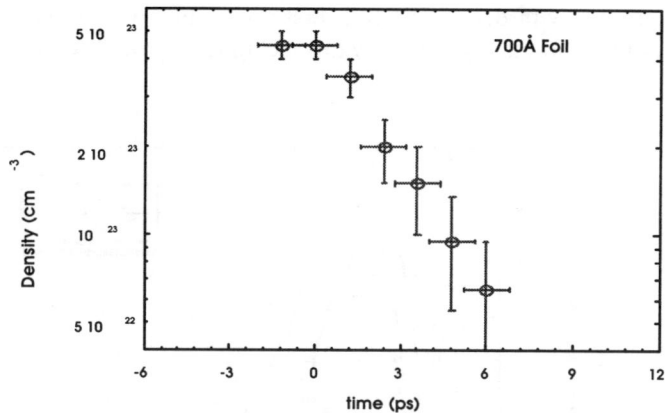

Figure 6. Time history of the electron density for 700Å Al foil irradiated by a 400 fs, 0.53μm laser pulse at 10^{18}W/cm^2.

At laser intensity of 10^{19}W/cm^2, the Li-like emission from foil is, contrary to what has been observed at lower intensity, shorter than the one from thick target. Figure 7 shows the time resolved spectra, recorded with the PX1 x-ray streak camera with a temporal resolution of 750fs, for 500Å Al foil and thick Al target at 10^{19}W/cm^2. The Li-like emission for the foil is extremely short and the He$_\alpha$ line displays a very long recombination tail.

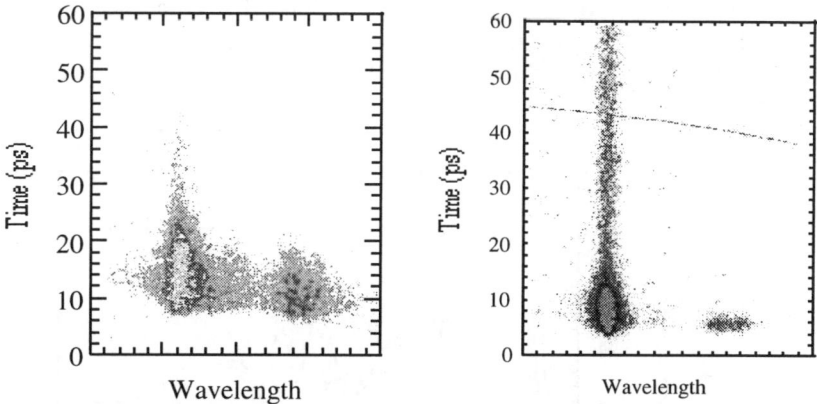

Figure 7. Time-resolved x-ray spectra from Al thick target (left) and 500Å Al foil (right) at laser intensity of 1×10^{19}W/cm^2, recorded with the novel INRS PX1 sub-picosecond x-ray streak camera (temporal resolution in the keV range is 750fs).

The time evolution of these Li-like satellites changes also strongly with target thickness. Figure 8 shows the time history of Li-like emission as a function of different foil thickness at laser intensity of 4×10^{19}W/cm^2. The duration of Li-like emission decreases with decreasing foil thickness, which is very encouraging for the production of ultrashort x-ray pulses by this technique of foil isochoric heating.

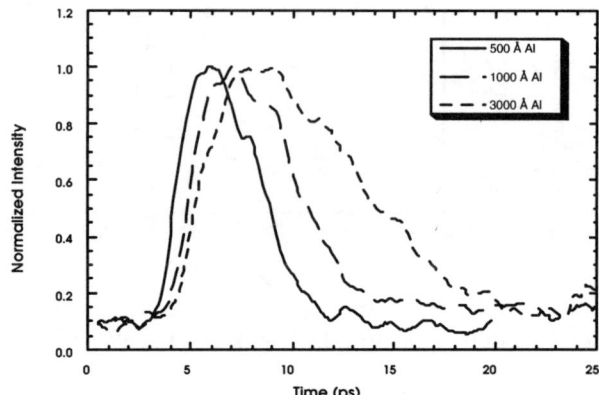

Figure 8. Time history of Li-like emission from Al foil target irradiated at laser intensity of 4×10^{19} W/cm^2. The duration decreases when the foil thickness decreases.

However, one may notice, from Fig.7, that emission lines from different ionization states undergo quite different evolution. He$_\alpha$ resonance line from foil targets irradiated at 4×10^{19} W/cm^2 shows a very long recombination tail which changes dramatically with the target thickness, as shown in Fig.9 (for the same irradiation conditions).

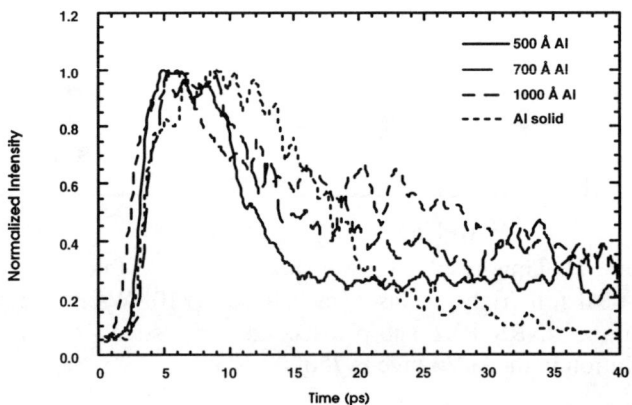

Figure 9. The time history of He$_\alpha$ resonance line for various foil targets irradiated at 5×10^{19} W/cm^2. This figure shows the complex recombination process as a function of foil thickness.

CONCLUSION AND DISCUSSION

These preliminary experiments indicate that isochoric heating can be achieved by using laser heated thin foil and a judicious choice of the laser and target parameters. The intensity of x-ray emission from foil and the foil electron temperature are much higher than those measured with thick target under the same laser irradiation conditions. The foil electron density is close to the solid density, due to the confinement of plasma by laser light pressure. Such hot dense plasma allows the production of bright and ultrashort burst of x-ray emission for certain ionization states.

Further studies are required to optimize the experimental conditions (target material and thickness, laser intensity and pulse duration) in order to get shorter x-ray pulse down to a few hundreds of femtoseconds and to clarify the behavior of hot electrons with thin foils. Indeed during the interaction of an ultrashort ultrahigh intensity laser with matter, a significant fraction of absorbed laser energy is carried out by hot electrons [20]. With a foil target the hot electron can not escape from the plasma due to the strong space charge effect and they would be bouncing several times within the plasma. This may affect the foil ionization dynamics.

ACKNOWLEDGEMENTS

The authors would like to thank F.Poitras, C. Sirois, N.Blanchot, A.Mens, D.Gontier and the P102 team for great technical support, and J.P.Matte, C.Y.Côte and J.Fuchs for many helpful discussions.

REFERENCES

1. M.Perry and G.Mourou, *Science* **264**, 917(1994)
2. C.Joshi, P.Corkun, *Phys.Today* **48**, 36(1995)
3. J.C.Kieffer, et al., *Phys. Fulid.* **B5**, 2676(1993)
4. F.Raksi, et al., *J.Chem.Phys.***104**, 6066(1996)
5. M.Murnane, et al., *Sciences* **251**, 531(1991); M.D.Rosen, *SPIE* **1229**, 160(1990)
6. J.C.Kieffer, et al., *Can.J.Phys.***72**, 802(1994)
7. Z.Jiang, *Ultrafast Phenomena IX, Springer Series in Chemical Physics*, Vol.**60**, 239(1994)
8. F.J,Rogers and C.A.Iglesias, *Science* **263**, 50(1994)
9. C.Darrow, et al., *Phys.Rev.Lett.***69**, 442(1992)
10 J.P.Matte, J.C.Kieffer, et al., *Phys.Rev.Lett.***72**, 1208(1994)

11. E.Leboucher-Dalimier, et al., *Phys.Rev.***E47**, 1467(1993)
12. C.Y.Cote, et al., *Phys.Rev.***E56**, 992(1997)
13. O.Peyrusse, et al., *Phys.Rev.Lett.***75**, 3862(1995)
14. M.Tabak, et al., *Phys.Plasmas* **1**, 1626(1994)
15. Z.Jiang, J.C.Kieffer, et al., *Phys.Plasma* **2**, 1702(1995)
16. J.C.Kieffer, et al., *Appl.Opt.***32**, 4247(1993)
17. P.Gallant, et al., *"Subpicosecond time resolved x-ray spectroscopy of plasma produced by high intensity ultrashort laser pulses"*, to be published in SPIE Vol.**3157**, 1997
18. O.Peyrusse, et al., *J.Phys.***B26**, L511(1995)
19. J.C.Kieffer, et al., *J.Opt.Soc.Am.***B13**, 132(1996)
20. B.Soom, H.Chen, et al., *J.Appl.Phys.***74**, 5372(1993)

X-Ray Production and SHG from Femtosecond Plasmas induced in Modified Solid Targets

V.M.Gordienko, M.S.Dzhidzhoev, M.A.Joukov,
A.B.Savel'ev, A.A.Shashkov, R.V.Volkov

International Laser Center, M.V.Lomonosov Moscow State University,
Moscow 119899, Russia

1. Introduction

During the past years a tendency of increasing a number of researcher groups involved in the problem of superintense laser fields SSLF and its applications was continuously growing. SSLF problem is based on the generation of femtosecond pulses by the use of so called table-top femtosecond laser systems with out put energy of the order 1-100mJ and intensity of 10^{15}-10^{18}W/cm^2. Under these intensities a thin layer (about hundreds angstroms) of a hot femtosecond laser plasmas FLP is produced on the surface of a solid target. These plasmas possesses unique parameters: extremely high temperature, electron density of the solid, life time of about 1ps. Solution a number of problems related to FLP opens a new horizon for nonlinear optics and nuclear physics, for new generation of ultashort UVU, X- and G-ray sources, to be used in a number of application including astrophysics [1-5].

A very efficient harmonic generation can be observed from plasmas generated on solid target [6]. Powerful incoherent X-ray bursts emitted by near solid density femtosecond laser plasmas FLP can be used for microscopic, spectral and other investigations of materials with subpicosecond temporal resolution, in particular as a source for the soft X-ray microscopy in the "water-window" region, for excitation of nuclear transition etc [7,8]. That is one main reasons why it is draw interest.

Among the principal directions in which femtosecond laser physics and laser-plasma femto-technology grow today are the control of FLP parameters. In the heated region the electron temperature can reach values of 100-1000eV depending on laser intensity, angle of incidence on the target and laser polarization.

One way to control plasma parameters lies in varying of intensity or pulse duration of the laser radiation. The other efficient way of affecting is to modify target surface, which is known to play an important role in the process of the interaction of laser radiation with matter. Obviously, the nature of the target material, roughness of the target surface, thickness of the target should also affect electrodynamics of the interaction of laser light with plasma, the amount of absorbed laser energy and thermal regime, and as a result affect the spectral range of X-ray emission, the conversion efficiency of harmonic generation.

In this paper we show experimentally that the use of modified targets is the promising way to optimize the process of interaction with a target of the subpicosecond laser pulse having moderate intensity 10^{15}-10^{16}W/cm^2 and to control plasma's parameters (electron temperature, nonlinear optics properties, etc).

In describe below experiments we consider two chief approaches:

1) resonant interaction of superintense laser radiation with corrugated plasma surface in the condition of the excitation of surface electromagnetic waves SEW which increase the effective nonlinearity of FLP and enhances its absorption coefficient;

2) "overheating" of FLP in the condition of reduced thermal conduction to control X-ray yield.

2. Resonant Second Harmonic Generation in High-Temperature Femtosecond Plasma Produced on a Target Surface Modified by Interfering Laser Beams.

Since femtosecond laser plasma has high density and relatively sharp interface with vacuum there can exist surface electromagnetic waves SEWs. This possibility is of interest for two main reasons. First, resonant SEW excitation is promising for increasing plasma absorption and X-ray yield [9]. Second, resonant SEW excitation on a corrugated target surface can significantly enhance the local field in the vicinity of the surface that leads to increasing effective plasma nonlinearity. Therefore the excitation of an SEW is a perspective to a considerable increase in the efficiency of harmonic generation. The use of two beam interactions of superintense femtosecond laser pulses makes it possible to modify the state of a surface and convert it from an initially flat to one which is spatially modulated. This creates a new quality which makes it possible to excite SEWs.

SEW is a solution of Maxwell equations on a boundary with the field amplitude falling down exponentially at both sides of it. SEW excitation is possible if $Re(\varepsilon)<-1$, where ε is permitivity (for homogeneous and isotropic media) [10]. A SEW cannot be excited directly by a vacuum electromagnetic wave because of the mismatch of their phase velocities. To compensate for this mismatch targets with suitable surface corrugation can be used [9]. In this case resonance can be obtained by tuning the angle of incidence. The resonance condition is: $k = k_{0t} \pm mg$ where k is SEW wave vector, k_{0t} is the component of the vacuum wave vector k_0 parallel to the target surface, $m = 0, \pm 1, \pm 2,...$, g is reciprocal grating vector. Performed calculation showed that for the nearsurface plasma relation $Re\varepsilon<-1$ holds for the wide temperature range. For example for plasma temperature of 140eV $\varepsilon`=-28$, $\varepsilon``=55$ [12]. In [13] was shown that plasma expansion into vacuum does not destroy existence conditions for surface plasmon-polariton up to interface thickness of 100A. This value corresponds to ~ 1ps time interval with typical velocity of plasma expansion ~10^7cm/s. At plasma temperature of 200eV free path length are proved to be ~10μm, that coincide with size of a heated spot by incident radiation. The life time of polaritons are in the range of tens of fs that comparable with pump pulse duration.

Here we present the results of investigation of high temperature near-surface plasmas HNFP generated by laser beams interfering on a target surface. Using two-beam plasma generation one can modify the target creating spatially modulated surface.

The experiments were carried out with the help of a femtosecond dye-laser system [14]. The system produced 200 fs pulses with the energy of 1 mJ (energy contrast 1000) at the wavelength of 600 nm. The laser pulse spectrum and autocorrelation function were measured in each shot. In our experiments we use pump-probe technique which provides femtosecond time resolution. The probing scheme is also based on the aberration-free optics that produces a magnified image of the plasma. This image is recorded with the help of a CCD-camera, and in our experiments we obtain both spatial and temporal resolution [14].

The experimental schematic of plasma pumping and probing is shown in **Figure 1**. The plasma was generated by two p-polarized beams focused with the help of lenses 4 and 5, and overlapping on the target surface. The beam focal spot diameters were 10 μm FWHM. The relative time position of the pumping pulses could be matched with the help of the optical delay system. The angles of incidence were 47^0 and 60^0 with the relative intensity 1:0.6 respectively. The total pump intensity on the target surface averaged over the beams cross-section was 10^{15} W/cm². The third beam (also p-polarized) with the variable time delay was used as a probing one. It was focused (with the objective 7) slightly off the target surface in such a way, that an area with the diameter of 60 μm round the plasma spot was illuminated. The magnified (×30) surface image was formed with the aberration free objectives 6 and 7 **(Figure 1)** and was recorded with the help of the CCD-camera 10. The image of the pumping beams interfering on the target surface or the pattern of diffraction on the modulated surface were recorded with the help of the aberration-free objective 8 and the second CCD-camera. The transparent target (fused silica) was used as a target. It was placed in a vacuum chamber with the background pressure of 10^{-3} Torr. The computer-controlled system of the target raster-scanning provided positioning with an accuracy of 5 μm, and target shift after each laser "shot".

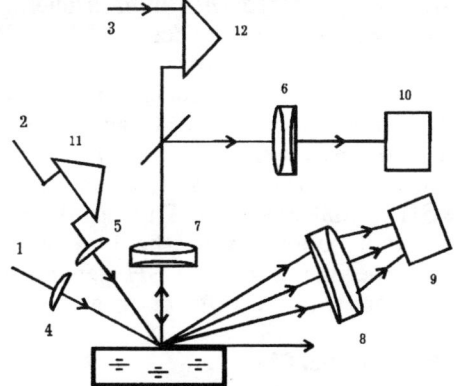
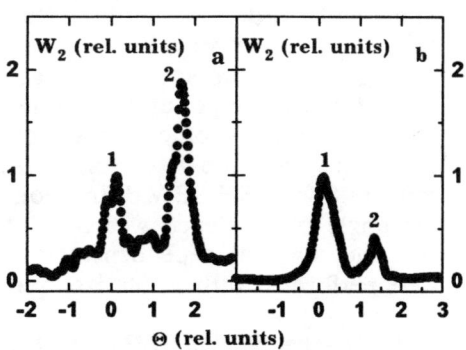

FIGURE 1. Schematic diagram of the apparatus: (1, 2) pump beams; (3) probe beam; (4,5) lenses with the focal length 10 cm; (6-8) aberration-free objective; (9,10) CCD linear arrays; (11,12) delay lines; (13) fused quartz target.

FIGURE 2. Angular distribution of the second-harmonic energy W_2 in the resonant (pump beam incident at 60^0 on the grating, a) and nonresonant (pump beam incident at 57^0 on the grating, b) cases, obtained for the zeroth (1) and first (2) diffraction orders.

The periodic structure, connected with the pumping beam interference was formed in the case of zero delay between the pulses. The target image obtained with the help of the CCD-camera. The structure period is 4.5 µm. The periodic structure is completely washed out if a time delay grater than the pulse duration is introduced.

From obtained experimental results we conclude that:
i) the periodical modulation of plasma parameters is strong enough for the noticeable diffraction of the probing beam starting from zero relative delay.
ii) when all relaxation processes are completed a "frozen in" grating is formed on the target surface.

Under zero relative delay between pumping pulses we have obtained noncollinear second harmonic (SH) generation. The conversion efficiency to SH was about 10^{-5}. Thus this scheme of noncollinear SH generation can be used in background-free measurements of superintense laser pulses duration (in wide range of wavelengths).

The "frozen in" grating, generated with the help of interfering beams makes it possible to resonantly excite SEW. As far as resonant SEW excitation leads to the possible enhancement of effective optical nonlinearity, we have carried out experiments on SH generation under the conditions of resonant SEW excitation.

In this experiments a single p-polarized pumping beam 1 (**Figure 1**) with the angle of incidence of 60^0 produced HNFP on a modified surface. The scheme was aligned in such a way that zero diffraction orders on the fundamental frequency and SH and the first diffraction order of SH entered the objective 8. The angle of incidence of the pumping beam could be matched according to the resonance condition with $m = 1$.

In the experiments we measured relative energy of zero and first orders of SH diffraction. Under resonant SEW excitation with the fundamental frequency a SEW with the doubled frequency should be also generated with high efficiency. The 0 and 1 orders of SH diffraction coincide with -1 and -2 orders of SH SEW diffraction respectively. **Figure 2** demonstrates SH angular distribution. It can be seen, that if pumping beam is detuned from the resonance, the energy distribution corresponds to a standard one for a small amplitude grating (the amplitude of the profile is small compared to the wavelength). Under resonance conditions the 0-order to 1-order energy ratio is opposite. This effect shows the contribution of SH SEW diffraction in the SH spatial spectrum. The variation of the angle of incidence was 3^0 (57^0 -60^0), small compared to the known SH efficiency angular dependence [10]. Similar experiments on the SH generation were performed with s-polarized pump using conventional diffraction grating as a target [15].

3. "Overheating" of FLP in freely suspended ultrathin films

Superintense fs pulses are used to heat target quickly before any significant expansion occur thus creating a hot plasma solid density. It means that during superintense fs pumping the energy is concentrated mainly inside the target and it is the nonlinear thermodiffusion that limits maximum plasma temperature. The rise in FLP temperature and consequently in the X-ray radiation power and shift of its spectra in the shorter wavelength region can be achieved by localizing the emitted heat in the thin layer near the surface by limiting heat conductivity.

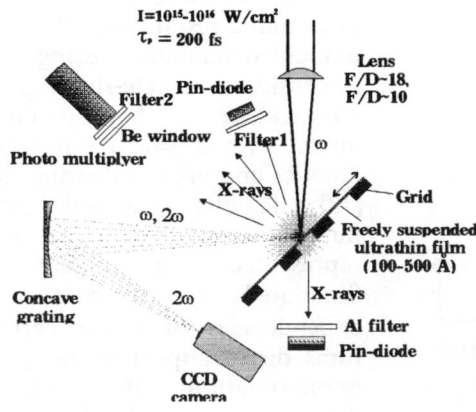

"Suppression" of thermal flux into the bulk of the target should lead to increase in specific energy per atom, and electron temperature could surpass 1keV, even when using "moderate" intensities of 10^{15}-10^{16} W/cm^2, with subsequent increase in ionization degree.

Freely suspended carbon ultrathin film UTF 10-60nm thick served as targets [16]. This materialis produced using dc glow-discharge technique. Amorphous diamond-like carbon-UTF is a result of free radical polymerization in plasma (initial mixture Ar-C_7H_8).

FIGURE 3. Experimental arrangement

Used technique characterizes by rapid growth of about of 2,5Å/s and possibility of large area coverage. The forbidden energy band of carbon-UTF is more than 2eV. As a "thick" target we used 1μm mylar film. In these experiments we used linear polarized radiation λ~600nm, W~100-500μJ incident on the target surface at the angle of Θ ~ 45° **(Figure 3)**. It was focused with a F/18 or F/10 lens, and intensity at the target reached 2x10^{16} W/cm^2. Target was placed in chamber with remaining pressure not worse than 10^{-5} Pa. Conversion efficiency η was measured with two x-ray p-i-n diodes. Main part of the emission from carbon plasma is provided by characteristic lines of C^{4+}(4.02 nm, 0.308 keV) and C^{5+}(3.37 nm, 0.368 keV) ions [17]. Filters used were 750nm Al or 1μm mylar films. On **Figure 4a** we show dependence of conversion efficiency η on films thickness D for p- and s- polarized light. Striking difference in dependencies for different polarization should be attributed for resonant absorption taking place for p-polarized light. Notable peculiarity of obtained data is the fall of η for the thinnest film of 10nm. It is connected with plasma fast ablation and "overheating". Judging from filter's thickness, mean x-ray quanta energy measured by the detector with and without Al filter can be estimated as 0.31±0.02keV, which fits the expected spectra of carbon plasma with ionization degree of 4-6. In the case of mylar filtered detector, signal appeared to be below detector noise level, what is related to low transmission of this filter in quanta energies region of 0.3-0.7keV. Thus x-ray spectra measured by reference detector without filters lies in the region of 0.3-0.4keV and is determined by the emission from H- He- like ions of carbon. For 10nm film signal was measurable even with mylar filter. It points at overheating of 10nm sample and effective generation of E>0,7keV X-ray quanta.

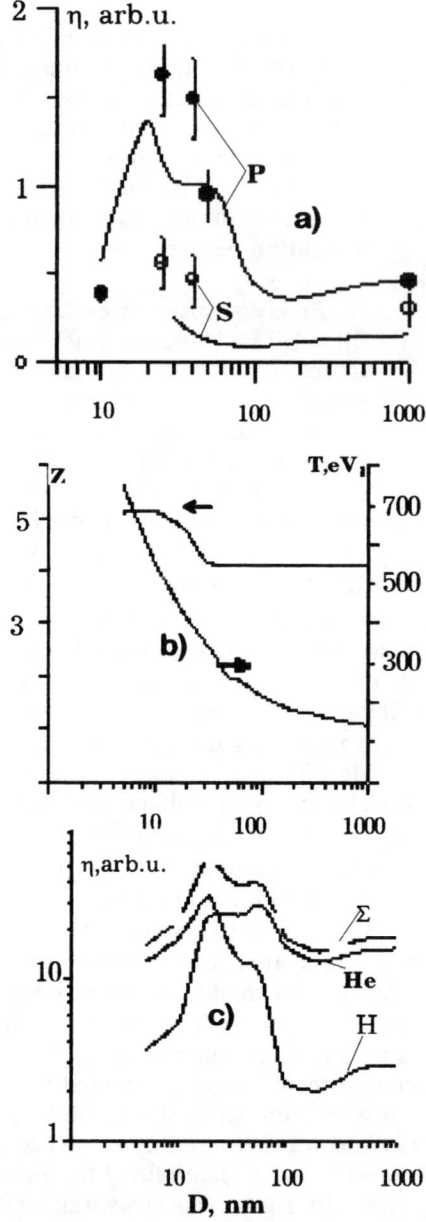

FIGURE 4. Dependence on target thickness D, nm of: a) to x-ray conversion efficiency η, b) electron temperature T and ionization degree Z, c) partial inputs of H- and He- like ions and sum of the both.

For complete interpretation of acquired results we conducted numerical simulation using one-dimensional, single-fluid, two-temperature hydrodynamic model with transient ionization kinetics and limited thermal flux [18]. Results acquired from numerical simulation are also represented on **Figure 4**. Simulated curves of dependencies of η on carbon films depth appear to be good approximation of measured data, once normalized for the "thick" film data and p-polarization.

According to the calculated data, when target's depth drops below 100nm, there starts a rapid increase in electron temperature and mean ionization degree (**Figure 4b**). This, in turn, leads to change in ratio of partial contribution of H- and He-like ions in x-ray yield (**Figure 4c**). For UTF maximum plasma density growth cuts off sharply and is replaced with its rapid fall, becoming under-critical for D~30nm at t~0.5ps, and t=0 for D~10nm (t=0 corresponds here to the maximum of laser pulse). Thus, decrease in conversion η is primarily connected with fast decrease in plasma density.

Hence, carbon-UTF are effective source of ultrashort pulses in the "water window" optimal for microscopy of living cells [7]. Numerical simulation for "heavy" targets as exemplified by Ni thin film (10nm thick) (see **Figure 5**) demonstrate considerable

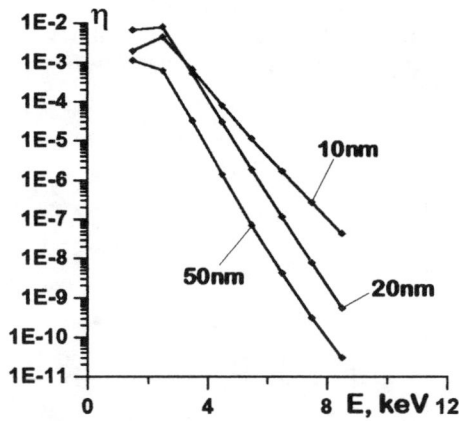

FIGURE 5. Simulated spectra of Ni film irradiated with $10^{16} W/cm^2$ for various film thicknesses.

increase in plasma ionization degree and hard x-ray (E>4keV) emission brightness.

4. FLP in porous silica

Porous materials found wide application in study of matter under extreme pressures and temperatures [18]. Porous matter consists of small solid particles separated by voids, morphology of porous material (particles size and shape, porosity, etc.) can be controlled.

In [19-20] so-called "black gold" and "black aluminum" (or colloidal targets) were used to achieve plasma "overheating". In this experiments there has been detected a considerable increase of conversion efficiency to soft x-ray region (E>1keV) for colloidal targets over solid targets, with serious duration increase of soft x-ray pulse [21]. It should also be noted that sort of analogous to metallic colloidal film is a jet of Xe clusters, where considerable increase in hot electron production as well as hard x-ray generation has been also observed [22,23].

Porous silica is one of the most interesting and easily accessible materials with good control over its morphology [18]. Simplicity and reliability of production of porous silica with desired parameters (density, structure, cluster size), ease of use to and high mechanical toughness are the main attractive points about using por-Si as target for conducting experiments on plasma overheating and hard x-ray generation instead of colloidal metallic films. Studying of electropolishing of silicon surface in hydrofluoric acid (HF) solution A.Uhlir obtained that under certain condition of substrate type and resistivity, electrolyte concentration and current density a localized attack of the subtrate occurred leading to the formation of porous structure within the bulk of the silicon substrate [24]. Since photoluminescence PL in the visible was found in porous silicon por-Si at room temperature the material became an object of grate attention for many research groups. An electrochemical treatment of crystalline Si in HF solution is a simple method to get Si-nanostructures. Porous silicon layers of thickness up to several micron can be obtained by this treatment. Por-Si consist of wires and/or clusters with characteristic size 2-10nm [26,27]. The porosity of the material depends on current density and solution. It can be varied between 20 and 90% [25, 26]. Porous silicon is a sample of a cross-linking systems close to fractal structure of matter with nanometer dimensions [26]. Porous silicon films have huge surface area of the order of 600-800m²/cm³. The thermal conductivity of the por-Si is sensitive to the structure of por-Si and falls with the porosity to 10^{-2}W/cm.K [27]. It should be noted that laser induced melting of ultrathin nanoporous silicon layers leads to a essential changes in the morphology at laser fluence of the order of 0,1J/cm² [27]. The formation of porous silica strongly depends on the type of the silica substrate and on the electrochemical parameters.

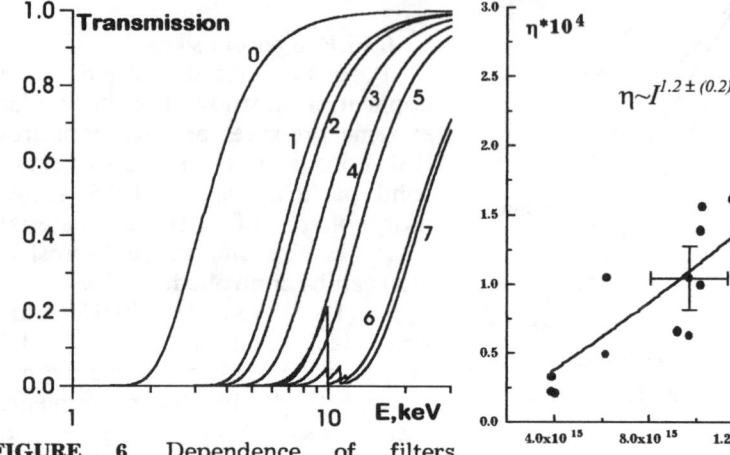

FIGURE 6. Dependence of filters transmission coefficient on quanta energy for different filters: 0 - 200μm Be; 1 - 180μm Teflon; 2 - 240μm Teflon; 3 - 100μm Al; 4 - 1mm Teflon; 5 - 300μm Al; 6 - 15μm Ta; 7 - 15μm Ta and 1mm Teflon.

FIGURE 7. Dependence of second harmonic conversion efficiency η on laser intensity I for por-Si (P=2.5).

Por-Si samples we used had the porosity (ratio of solid density to mean density of porous material) of P~1.5-7.0. Cluster size was d~2nm, and size of voids was determined by samples porosity. Porous layer depth was designed by preparation time and was ~20μm. To measure hard X-rays we used NaI scintillator detector combine with photomultiplier along with a set of different filters (**Figure 6**), simultaneously p-i-n diode measure soft X-ray yield E<1.5 keV. In addition with detecting plasma x-ray we were measuring the second harmonic generated in plasma (**Figure 7**). In the case of tight focusing, second harmonic signal was used as a reference for exact focusing. Efficiency of 2nd harmonic generation depends on intensity like as $I^{1.2 \pm 0.2}$ for both c-Si and por-Si samples with slightly larger value scattering in the latter case. Lack of dependence on the porosity and thus on temperature and ionization state agrees with harmonics generation by plasma mirror [28].

On **Figure 8** there is dependence of conversion into x-ray on porosity for the intensity $I=2\cdot10^{16}$W/cm². X-ray flux measured by NaI detector increases extremely rapidly. It should be noted that this behavior was observed only for highest porosities above P~5, for lower values no effect was detected. The higher conversion efficiency obtained is ~10^{-4} % for P=6.6 or above 10^3 quanta in the 5.0±0.5 keV bandwidth.

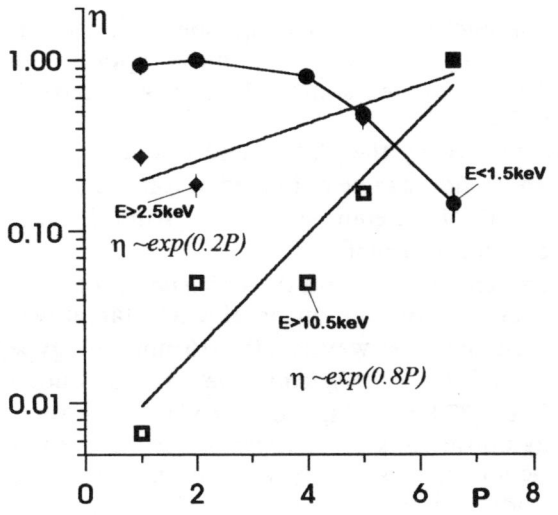

FIGURE 8. Dependence of x-ray conversion efficiency η on porosity P (for por-Si), $I \sim 2 \times 10^{16} W/cm^2$.

This hard x-ray emission from por-Si can be attributed for suprathermal electrons with the mean temperature of $T_e \sim 4$ keV for P=5 and $T_e \sim 7$ keV for P=6.6.

5. Excitation of nuclear transitioins by x-ray emission of high-temperature dense plasma

Nuclei excitation and stimulation of the nuclear reactions in the matter under extreme conditions is of great interest for the wide variety of applications in laser and nuclear physics, spectroscopy, and some technical applications [31]. One of the new promising directions here is connected with the plasma produced by superstrong femtosecond laser pulses in the thin surface layers of solids.

Excitation of nuclear transitions and stimulation of nuclear reactions in the laser plasma produced by the laser pulses with intensity in excess of 10^{19} W/cm^2 have been considered in [3]. However laser pulses of a such intensity should have the energy above 1J and can be produced only by the unique laser systems. This restricts seriously the area of the possible applications in scientific or applied researches.

Here we analyze the feasibility of the experimental observation of nuclei excitation of isotopes with appropriate energy level schemes, especially the ^{201}Hg isotope. Specifically we discuss the possibility of the nuclear transition excitation by x-ray emission of laser plasma produced by the laser pulses with "moderate" intensities of 10^{15} -10^{17} W/cm^2 , those are much easy to produce.

The plasma produced by the femtosecond laser pulse is more dense and the intensity of x-ray emission is higher than in the course of nanosecond laser pulses. As a result the rate of nuclear transition excitation should be significantly higher. It makes this situation more favorable for realization of nuclei excitation by those part of the continuos x-ray emission spectra which is resonant to the Doppler broadened nuclear transition.

With the help of Nuclear Data Sheets NDS [32] we searched for the nuclear isotopes possessing the low-lying excited states, E<5keV [8]. It is seen from NDS that the lifetimes, parity and momentum of the excited states are often not known. Therefore, the methods of plasma x-ray spectroscopy could be very valuable for the nuclear spectroscopy of the low lying isomer states. Those characteristics can hardly be acquired using conventional nuclear spectroscopy methods.

The analysis of the parameters of nuclei with low-lying isomer's state has shown that the stable isotope ^{201}Hg possesses the most appropriate characteristics [32]. The energy of the first excited state in ^{201}Hg is $\varepsilon \approx 1,56$ keV. The concentration spread of ^{201}Hg is 13,18%.

The number of the x-ray quanta emitted by the plasma and resonant to the nuclear transition with the Doppler width $\Delta\omega_D$ can be estimated as follows $N_p = (E \cdot \eta / \varepsilon) \cdot (\Delta\omega_D / \Delta\varepsilon)$, where $\Delta\varepsilon$ is the width of the emitted x-ray spectrum, E is the laser pulse energy, and η is the conversion coefficient.

We carried out preliminary experiments to estimate the efficiency η of laser pulse energy conversion into desired hard x-ray component. HgCdTe target was irradiated by dye-laser pulse having parameters: wavelength 616 nm, energy $\approx$ 300µJ, pulse duration $\approx$200fs, intensity of $10^{15} - 10^{16}$ W/cm^2. The x-ray pin-diode detector was filtered by Al-mylar filter (270nm - Al, 7µm - mylar). The filter transmits x-rays with quantum energy above 1,4 keV. It appears, that hard x-ray yield behaviour with focusing tightness coincides closely with second harmonic generation efficiency. In the above conditions the Doppler width of ^{201}Hg is $\Delta\omega_D \approx 3 \cdot 10^{-2}$ eV. We estimated the value of η as $\eta \approx 4 \cdot 10^{-4}$ for 1,4-2 keV energy band. Hence using laser pulses with energy of 10 mJ and duration $\approx$300fs from excimer XeCl system [33] it is possible to obtain the number of resonant x-ray quanta N_p of the order of 10^6 per pulse.

The absorption length of the pumping x-ray quantum is

$1/l_p$(cm^{-1})$=2\pi(c\hbar/\varepsilon) \cdot (R_0 / \Delta\omega_D \cdot T_{rad})$, where R_0 is the isotope density ($\approx 10^{22}$cm^{-3}), and T_{rad} is the radiative lifetime of the excited nuclei level, c is the speed of light. The radiational life time was estimated as $T_{rad} \approx 10^{-4}$s [8]. Hence one can get $l_p \approx 100$µm for $\varepsilon \approx 1,56$ keV. It means that the resonant x-ray quanta are effectively absorbed in the plasma volume. The energy of nuclei excitation is released by emitting γ-quanta or electrons of internal conversion. The input of the latter process grows up with plasma cooling due to filling of the atomic levels by electronic recombination. The crude estimations of the internal conversion coefficient show that it is very important to take into account the process of internal electron conversion speeding up the level's decay. But the excited level lifetime $\tau > 10^{-7}$ s is essentially greater than the time of plasma cooling $\tau \approx 10^{-11}$ s [8]. Thus it is possible to detect the excited nuclei decay both by registering the delayed γ-emission (1,56 keV) or x-ray emission arises in the process of internal conversion.

It should be noted that as it was shown above, by using modified targets one can effectively enhance the number of resonant x-ray quanta.

5. Conclusion

Thus, by modifying optical and thermal properties of near-surface target layer allows one to control interaction regime of powerful ultrashort, laser pulse with target as well as velocities of various processes, occurring in plasma and hence, parameters of emerging FLP.

For both s- and p-polarized pump an increase in second harmonic conversion efficiency under condition of SEW excitation in high temperature near-surface femtosecond plasmas, created on the corrugated modified surface of the solid target by laser pulses with duration of 200fs and intensity of about 10^{15}W/cm^2.

Usage of freely suspended films with thickness of the order of laser skin-layer in plasma as targets suppresses heat removal out of the region directly heated by laser radiation of moderate intensity I~10^{15}-10^{16}W/cm^2. We have shown that in case of carbon target 30nm thick one can get threefold increase in x-ray yield in the region of "water window" (2-4nm), and with 10nm target electron temperature ~ 700eV and ionization ~5.2 can be achieved.

Porous materials could serve as effective laser plasma hard x-ray source. For porous silica with porosities in excess of 5 small cluster size allow to produce hot electrons with temperatures of 5-7 keV and hence hard x-rays of the same quanta energy.

Acknowledgments

Authors greatly acknowledge support from Russian Fund for Basic Research (grants #96-02-19146a, #97-02-17013a) and State Scientific Program "Fundamental Metrology".

References

1. Akhmanov S.A., *Itogi nauki i teckhniki*, ser. "Modern problems of laser physics", Moscow, VINITI, **4**, 5-18 (In Russian), (1991).
2. Murnane M., Kapteyn H., Rosen M., Falcone R., *Science*, **251**, 531, (1991).
3. Luther Davies B., **Gamaly E.G.**, Wang Yanji et al. *Quantum Electronics*, **19(4)**, 317, (1992).
4. Platonenko V.T., *Laser Physics*, **2**, 852, (1992).
5. P.Gibbon P., R.Forster R., *Plasma Physics Control. Fusion*, **38**, 769, (1996).
6. Engers T., Fendel W., Shuler H., et al., *Phys. Rev. A*, **43**, 4564, (1991).
Von der Linde D., Jenke G., Engers T., et al. *Ultrafast Processes in Spectroscopy*, Ed. Svelto O, De Silvesti S., Denardo G., Plenum Press, NY and London, 1996, p.319.
7. Savel'ev A.B., Dzhidzhoev M.S., Gordienko V.M., Tarasevitch A.P. in: *Femtochemistry. Ultrafast chemical and physical processes in molecular systems.* Ed.:M.Cherqui, World Scientific, Singapore, 1996, pp. 675-679.
8. Andreev A.V., Gordienko V.M., Dykhne A.M., Savel'ev A.B., Tkalya E.V. *JETP Lett.*, **66(5)**, (1997).
9. Gauthier G., Bastiani S., Audebert P. et al., *Proc. SPIE*, **2523**, pp. 242-253, (1995).
10. Eds. Agranovich E.M.and Miles D.L., *Surface Polaritons: Electromagnetic Waves at Surface and Interfaces,*. Amsterdam: North-Holland, 1982.
11. Emel'yanov V.I., Seminogov V.N., Sokolov V.I., *Sov. J. Quant. Electr.*, **17**, p. 17, (1987).
12 Djidjoev M.S., Gordienko V.M., Joukov M.A. et. al. *Modern Problems of Laser Physics*, Eds. Bagaev S.N. and Denisov V.I., Novosibirsk, 1996, pp. 163-170.
13. Gordienko V.M., Magnitskii S.A., Moskalev T.Yu., Platonenko V.T., *Bulletin of RAN, Physics*, **60(3)**, 335-341, (1996).
14. Volkov R.V., Gordienko V.M., Dzhidzhoev M.S. et al., *Quant. Electr.*, **26(5)**, 524-528, (1996).
15. Volkov R.R., Gordienko V.M., Savel'ev A.B., Tarasevitch A.P. Timoshin A.O., *Laser Phys.*, **6(6)**, 1162-1168, (1996).
16. V.G.Babaev, Volkov R.V., Gordienko V.M. et al., *Quantum. Electr.*, **27(4)**, 283-284, (1997).
17. U.Teubner, W.Theobald, C.Wulker, *J. Phys. B*, **29**, 4333, (1996).
18. M.S.Dzhidzhoev M.S., Gordienko V.M., Kolchin V.V., et.al., *JOSA B*, **13**, 143-147, (1996).
18. Kormer S.B., Funtikov A.I., Urlin D.V., Kolesnikova A.N., *JETP*, **42**, 9, (1962).

19. Murnane M., Kapteyn H., Gordon S., et al., *Appl. Phys. Lett.*, **62 (10)**, 1068, (1993).
20. Wulker C., Theobald W., Gnass D., et al., *Appl. Phys. Lett.*, **68 (10)**, 1388, (1996).
21. Dzhidzhoev M.S., Gordienko V.M., Kolchin V.V., et al., *Proc. SPIE*, **2777**, 148-158, (1995).
22. McPherson A., Boyer K., Rhodes C., *J.Phys. B*, **27**, 637, (1994); **29**, 43; (1996), **29**, 113, (1996).
23. Shao Y., Ditmire T., Tisch J. et al., *Phys. Rev. Lett.*, **75(17)**, 3122; (1995), **77(16)**, 3343, (1996).
24. Uhlir A,. *Bell. System Tech. J.*, .**36**, 33, (1956).
25. Dittrich Th., Kashkarov P.K., Konstantinova E.A., Timoshenko V.Yu. *Thin Solid Films*, **255**, 74, (1995).
26. Bomchil G, Halimaoui A., Herino R. *Appl. Surf. Sci.*, **41/42**, 604, (1989).
27. Dittrich Th, Sieber I., Henrion W. et al, J.Rappich. *Appl. Phys.*, **A63**, 467, (1996).
28. Lichters R., Meyer-ter-Vehn J., Pukhov A., *Phys.Plasmas*, **3(9)**, 3425, (1996).
29. Chen H., Soom B., Yaakobi B., et al., *Phys.Rev.Lett.*,**70(22)**, 3431, (1993).
30. Rousse A., Audebert P., P.Geindre J., et al., *Phys. Rev. E*, **50(3)**, 2200, (1994).
31. Kirzchniz D.A., *Uspekhi Fizicheskikh Nauk*, **104**, 489, (1971).
32. Ed. Martin M., *Nuclear Data Sheets*, Academic Press, NY, (1966-1986).
33. Volkov R.V., Gordienko V.M., Dzhidzhoev M.S., et al., *Quantum Electronics*, **23(12)** in press, (1997).

HIGH INTENSITY 30 FEMTOSECOND LASER PULSE INTERACTION WITH THIN FOILS

A.Giulietti, A. Barbini, P.Chessa*,
D.Giulietti[#], L.A.Gizzi, D.Teychenné[#]

*Istituto di Fisica Atomica e Molecolare del CNR
Via del Giardino, 7, 56127 Pisa, Italy*

[#]*Dipartimento di Fisica, Università di Pisa, Italy*
* *Laboratoire d'Optique Appliquée, - ENSTA , Palaiseau , France*

Abstract. An experimental investigation on the interaction of 30 femtosecond laser pulses with 0.1 and 1.0 µm thick plastic foils has been performed at intensities from 5×10^{16} to 5×10^{18} W/cm^2. The interaction physics was found to be definitely different whether the nanosecond low intensity prepulses led to an early plasma formation or not. In the first case high reflectivity and very low transmittivity were observed, together with second and three-half harmonic generation. In absence of precursor plasma, with increasing intensity, reflectivity dropped to low values, while transmittivity increased up to an almost complete transparency. No harmonic generation was observed in this latter condition, while ultra-fast ionisation was inferred by the blue-shift of the transmitted pulse. Finally, intense hard X-ray emission was detected at the maximum laser intensity level. Current theories or numerical simulations cannot explain the observed transparency. A new model of magnetically induced optical transparency (MIOT) is briefly introduced.

INTRODUCTION

Since the last generation of powerful femtosecond lasers was available, a new, exciting class of experiments has begun. The virtual lack of hydrodynamic expansion during the pulse interaction with solids makes it possible to achieve the completely unexplored physical domain of extremely high fields in solid-density ionised matter. From an experimental point of view, a serious problem that can prevent interaction of short pulses with solid-density plasmas may arise from the laser prepulse originating from amplified spontaneous emission (ASE) in the laser amplifier chain. If the intensity on target due to the prepulse (typically of nanosecond duration) is higher than the threshold intensity for plasma formation on target, a precursor plasma is formed which prevents the main femtosecond pulse to interact directly with the solid. In a previous experiment ([1]), it was shown that the use of targets consisting of thin plastic foils may avoid formation of precursor plasma, enabling the interaction of the main femtosecond pulse with high density laminar plasmas characterised by ultra steep gradients.

In this paper we report novel experimental results on the interaction of 30fs laser pulses delivered by the Ti:Sapphire system of the Laboratoire d'Optique Appliquée, focused onto either 0.1 or 1.0 µm thick plastic foils at intensities ranging between 5×10^{16} and 5×10^{18} W/cm^2. The effect of the prepulse was accurately tested for each series of measurements. The interaction physics, and consequently the observed effects, resulted definitely different, whether the nanosecond low intensity prepulses led to an early plasma formation or not. In condition of precursor plasma (*preplasma*) formation by the prepulse, the high intensity femtosecond pulse interacted with the preplasma and was strongly reflected. In this condition the fraction of the laser light transmitted through the target, if any, was merged into the experimental background level, while generation of both 2ω and $3/2\omega$ harmonics of the laser light was observed. The space-resolved spectra of those harmonics showed interesting features.

A completely different scenario was observed in absence of preplasma, when the femtosecond pulse could interact with the unexploded foil. The most surprising result concerns the transmittivity ([2]). We found that when the laser intensity on target is greater than 10^{17} W/cm^2, the transmittivity goes above the experimental background level of $\approx 1\%$ and increases dramatically with laser intensity, approaching full transparency at intensities above 10^{18} W/cm^2. To our knowledge, this is the first time that such effect is observed in the laser interaction with a solid density plasma. As the transmittivity increases, the reflectivity drops and reflection is mostly restricted to the boundaries of the laser spot on target. This fact produces an interesting effect of spatial filtering of the transmitted pulse. At intermediate intensity the partially transmitted light shows a definite blue shift due to self phase modulation, from which the time scale of the ionisation was estimated. No harmonics were detected in absence of preplasma. Also the spectrum of the X-rays emitted during the interaction was considerably affected by the presence or absence of the preplasma. As the analysis of a large number of data related to the X-ray measurements is still in progress, this part of our measurements could not be included in the present paper.

It is very important to point out that the current theory of laser interaction cannot account for the almost full transparency of 0.1 and 1.0 µm plastic foils to laser radiation at 3×10^{18} W/cm^2. Several effects have been considered that predict enhanced propagation, including anomalous skin effect([3]), self induced transparency ([4]), hole boring([5,6]). Hole boring and self-induced transparency have been mostly studied for density higher than, but comparable with the critical density, while in our experiments the electron density is close to the solid density, in absence of preplasma. Self-induced transparency needs ultra-relativistic fields, while we observed quasi-transparency at an intensity corresponding to a weakly relativistic field. In what concerns the anomalous skin effect, a description of this process in plasmas has been given by Weibel ([7]). Recently the anomalous skin-effect in solid-density plasmas has been considered both analytically ([8]) and numerically ([9]). Transmittivity is predicted higher than the in the case of normal skin effect, but still very lower than what we have measured. In the last section of this paper we briefly introduce an original theoretical model ([10]) of magnetically induced optical transparency (MIOT). The model assumes the existence of a static magnetic field of high intensity parallel to the oscillating magnetic field, and accounts for the observed high transmittivity of solid density plasmas to ultra-short pulses of weakly relativistic intensity. The generation of such an intense magnetic field may be ascribed to the electron motion in the dense matter under the action of the ultra-short e.m. wave-packet.

EXPERIMENTAL TECHNIQUE

The experiment was performed using the advanced Ti:Sapphire laser system recently developed at LOA. The 815nm, 30fs laser pulse was focused f/7.5 onto either a 0.1 or a 1.0 µm thick plastic foil target, using an off-axis parabolic mirror, with an angle of incidence on target of 20 degrees. The laser pulse was linearly polarised with the electric field in the plane of incidence (*P*- polarised). The focal spot was 10μm in diameter; the intensity on target was varied from $5 \cdot 10^{16}$ to $5 \cdot 10^{18}$ W/cm^2, by varying the energy in the pulse. The transmitted pulse was studied by using a diffusing screen placed beyond the target, on the laser propagation direction, at a distance 1.8 times the focal length of the focusing optics. A de-magnified image of the screen was formed onto a CCD array and on the entrance slit of a spectrometer. A second CCD array was placed on the output focal plane of the spectrometer. An additional channel was set up on the specular reflection direction, with the object plane located at the target plane. Also this channel was equipped with both an imaging CCD and a spectrometer with CCD, giving a space resolved spectrum of the reflected light. In addition, we monitored the soft X-ray emission ($\approx$1-10keV) by means of a PIN diode and the hard X-ray yield (up to several MeV) by using six NaI(Tl) crystal scintillators, coupled to photomultiplier tubes.

The laser system was characterised by an ASE lasting approximately 10ns, forming a "pedestal" to the main pulse. The measured contrast ratio, i.e., the ratio between the power delivered in the fs pulse and that delivered in the ASE was $\geq 10^7$. A severe test on the effect of the ASE on target was performed by firing the laser system, without injecting the fs pulse in the amplifier chain. In this condition, for a distinct set of measurements, we observed no damage on target over the whole range of ASE intensities. In this case we concluded that in full shots the target does not explode before the arrival of the femtosecond pulse. This test is indeed, for two distinct reasons, a proof "a fortiori". Firstly, since no energy is spent in the amplification of the fs pulse, the level of ASE is greater than in the case of operation with fs pulse injection. Secondly, only the leading part of the ASE pulse prior to the arrival of the main fs pulse is relevant in determining the interaction conditions of the main pulse. Atomic excitation and partial ionisation in the target due to ASE cannot be excluded, but definitely the target does not explode prior to the high power pulse impact. So, we can state that the set of measurements described below as "interaction with no preplasma" are the actual result of interaction of the short pulse with the unexploded foil.

INTERACTION IN PRESENCE OF PRECURSOR PLASMA

In a few series of shots the prepulse was able to explode the thin target, allowing the formation and expansion of a preformed plasma, which then interacted with the 30 fs pulse. In this condition the short pulse was not transmitted though the plasma, while a considerable fraction of the pulse energy was specularly reflected. A typical monochromatic (at the laser wavelength) image of the target plane taken in the specular direction is shown in Fig. 1. The image evidences the distribution of the laser reflecting centres in the plasma. The pattern is similar to the far field pattern of the laser beam, including the diffraction rings. However in the reflected pattern there are two central maxima in place of one. This may be due to some misalignment in the focusing optics or to non-uniformity of the preformed plasma. Since the central part of the pattern was not reproducible shot by shot, the second explanation seems more likely.

In the same interaction condition also 2ω and 3/2ω harmonics generated specularly were observed. Space resolved spectra showed that blue and red shift of the second harmonic light originates in different regions, and may give useful information on the interaction of the ultra-short pulse with the precursor plasma. On the other hand, the observation of the 3/2ω harmonic is clear evidence that the plasma has expanded and developed an $n_c/4$ layer with a finite scalelength.

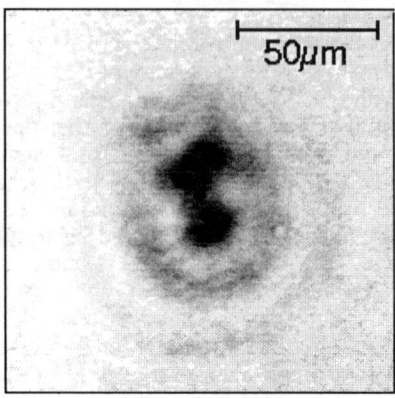

Figure 1. Monochromatic image (at the laser wavelength) of the target plane. The image was obtained in the specular direction in presence of precursor plasma. Laser intensity: 3×10^{18} W/cm^2; foil thickness: 1.0 μm.

INTERACTION WITH NO PRECURSOR PLASMA

A completely different scenario was found when the prepulse was not able to explode the foil before the arrival of the 30 fs pulse. No 2ω nor 3/2ω harmonics were observed. The most striking feature discovered in this condition of direct interaction of the short pulse with the solid foil was a high level of transmittivity above a given threshold of the laser intensity.

Transmittivity vs. laser intensity

The dependence of transmittivity as a function of the incident laser intensity is presented in the plot of Fig. 2. Open dots refer to 0.1 μm thick targets, the solid dot refers to the 1.0 μm case. Each data point was obtained by taking into account several interaction events for each laser intensity and by averaging the results. The error bar was estimated by the standard deviation of the set of data considered. The *background* line reported on the graph indicates the level at which the transmitted energy is comparable with the ASE energy (close to 1% of the main pulse energy) and consequently below this level the measurements cannot be entirely related to the main pulse. According to the plot of Fig.1, the transmitted fraction at incident intensities below 10^{17} W/cm^2 lies within the experimental background level. However, as the incident intensity increases, the transmitted fraction increases dramatically and the target becomes basically transparent at $3\ 10^{18}$ W/cm^2.

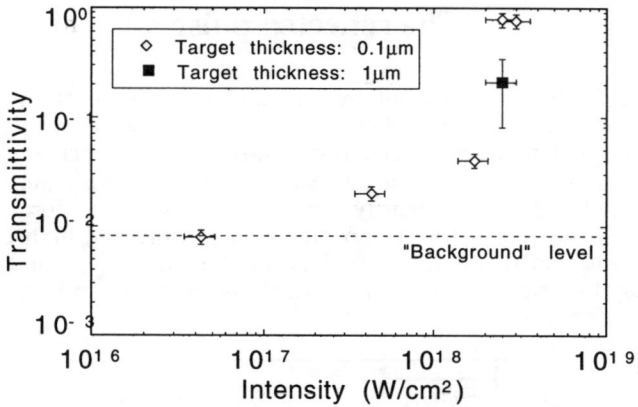

Figure 2. Transmittivity as a function of the intensity of the 30fs laser pulse. Open dots: 0.1 μm foil; solid dot: 1.0 μm foil. The background level indicates the level at which the energy in the pedestal (ASE) is comparable with the transmitted energy.

Transmittivity vs. target position

The transmittivity at the intensity of $2.5 \cdot 10^{18}$ W/cm^2 was measured for different target positions along the laser propagation axis. The results are plotted in Fig.3, where the position "0" corresponds to the nominal focus of the parabolic mirror. Open dots refer to 0.1 μm thick targets, solid dots refer to the 1.0 μm case. The maximum transmittivity is observed when the target is beyond the nominal focus. The range of positions allowing high transmittivity in the case of 1.0 μm thick foil is of the order of few 1-2 Rayleigh lengths.

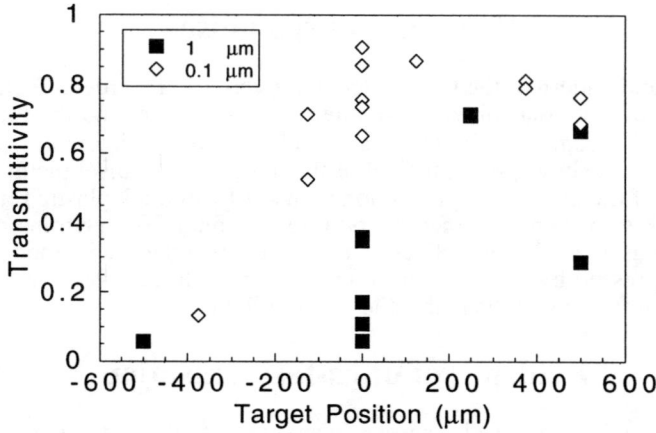

Figure 3. Transmittivity of the 30fs laser pulse, at an intensity of $2.5 \cdot 10^{18}$ W/cm^2, for different target positions. Open dots: 0.1 μm foil; solid dots: 1.0 μm foil.

The reflected pulse

The pulse reflected in absence of preformed plasma was spectrally analysed and no 2ω nor $3/2\omega$ harmonics were detected. Images of the target plane were also obtained. A typical monochromatic image taken in the specular direction at 3×10^{18} W/cm^2 is shown in Fig. 4. It is interesting to compare this pattern with the one of Fig.1, obtained at the same intensity, but in presence of the preformed plasma: the pattern of Fig.4, corresponding to no precursor plasma and high transmittivity, shows reflection mostly at the boundary of the laser spot, and is somehow complementary to the pattern of Fig.1, in which the dominant reflection comes from the centre of the spot.

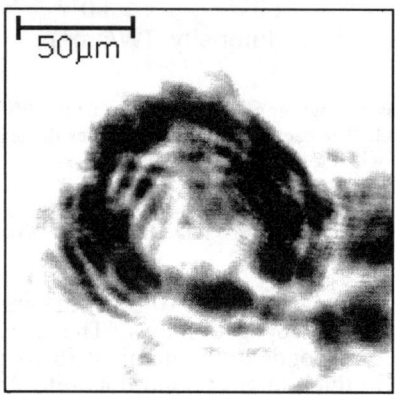

Figure 4. Monochromatic image (at the laser wavelength) of the target plane. The image was obtained in the specular direction in absence of precursor plasma. Laser intensity: 3×10^{18} W/cm^2; foil thickness: 0.1 μm.

The transmitted pulse

The near-field pattern of the transmitted pulse was also recorded. Pictures a) and b) of Fig. 5 show the transmitted pulse after propagation beyond the focus at high intensity (3×10^{18} W/cm^2); straight lines are spatial markers. Picture a) was obtained without target, and shows the near field of the unperturbed pulse; picture b) shows the near field of the pulse after interaction with a 0.1 μm thick plastic foil. Pictures c) and d) of Fig. 5 were obtained by two dimensional Fourier transform of the intensity patterns a) and b) respectively. It is evident that the spatial modes of higher order are suppressed by the 30 fs pulse interaction with the target, resulting in an effective spatial filtering of the high intensity laser light.

Evidence of ultra-fast ionisation

The transmitted pulse was also spectrally analysed at different intensities with and without target. The spectra of the pulse propagated without target have a bandwidth close to the Fourier limit for a pulse with a 30fs FWHM gaussian temporal profile. The spectra of the pulse after interaction with the foil target show that the interaction process produces a clear blue shift at moderate (5×10^{16} W/cm^2) and intermediate (4

10^{17} W/cm^2) intensities, while the spectrum of the pulse transmitted at the highest intensity (3 10^{18} W/cm^2) is basically unaffected. The spectral properties of the transmitted pulse were found to be stable shot to shot, except at the intermediate intensity, were shot-to-shot variations in shift and width were observed. The blue shift in the spectrum of the transmitted pulse is a clear signature of ultra-fast ionisation.

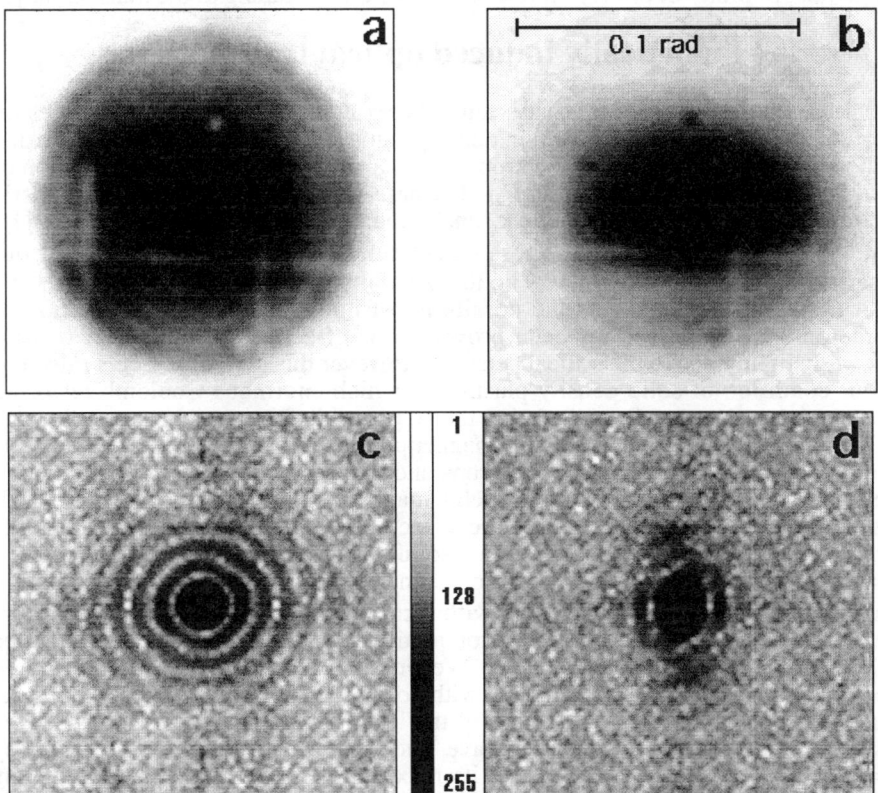

Figure 5. a) Near field pattern of the pulse propagated without target; **b)** with 0.1 μm foil. Laser intensity: 3 10^{18} W/cm^2. **c)** and **d)** are two-dimensional Fourier transforms of pictures a) and b) respectively. The log of the modulus of the FT is shown as a grey-scale.

This is well supported also by the amount of the blue shift, which is about 13 nm at 5×10^{16} W/cm^2, and about 20 nm at 4×10^{17} W/cm^2. Let us attribute the shift to self-phase modulation of the laser pulse, namely to the ultra-fast decrease in the refractive index due to the laser induced ionisation. Considering a change $\Delta\mu \approx -1$ in the refraction index, due to the transition from zero electron density to the critical density, we can evaluate the time scale Δt of such a transition from the approximate expression of the frequency shift produced by self-phase modulation ([11]):

$$\delta\omega \approx (L/c)\,(\Delta\mu/\Delta t)\,\omega_0.$$

Taking the interaction path L equal to the foil thickness 0.1 µm, we found $\Delta t \approx$ 20 fs and $\Delta t \approx 13$ fs for the low and intermediate intensity, respectively. The definite consistency of these values, confirms that an ultra-fast ionisation actually occurs. The absence of shift in condition close to the full transparency, namely at 3×10^{18} W/cm^2, suggests that in this case the ionisation involves a negligible portion of the pulse, while the spectral variability observed at intermediate intensity may be due to the proximity of a sort of threshold for the effect leading to the transparency.

Magnetically induced optical transparency

The observed high transmittivity cannot be explained by the current theories on propagation of high intensity light in dense plasmas. Self-induced transparency due to the relativistic change of the electron mass is a marginal effect in our condition of weakly relativistic intensity ($a_0 = 1.2$, where $a_0 = eE_0/m\omega c$ is the relativistic parameter). The transmission due the anomalous skin effect has been evaluated by Matte et al. for thin foils [7]. Those calculations give for the conditions of our experiment a transmittivity of 10^{-6} up to 10^{-5}, much lower than the values we have measured. In what concerns the possibility of hole-boring due to the action of ponderomotive forces, it has been proved to be effective for plasma of density below or slightly above the critical density. However this kind of effect is ruled out in our condition of solid density plasma, in which enormous Coulomb restoring forces prevent electrons from depleting the 10 µm wide focal region.

The creation of extremely intense magnetic fields in the interaction between the laser pulse and the foil may trap electrons and induce transparency. Even though the actual mechanism of creation of such a magnetic field has to be still clarified, it may be useful to shortly discuss the effect of a static magnetic field in the propagation of the short pulse through a solid density plasma. In particular, if we assume the presence of a quasi-static magnetic field parallel to the oscillating magnetic field of the e.m. wave [10], with some suitable physical assumptions that are not discussed here, the single electron motion may be studied, and the refractive index may be qualitatively evaluated. We find that the equation of motion of the electron will be a typical Hill equation, with no analytical solution, unless we restrict ourselves to a limited range of values of the leading parameters. We consider the relativistic parameter a_0 of the laser wave, and the normalised cyclotron frequency $\Omega = \omega_c/\omega = eB_s/m\omega c$. We fixed a_0 to its maximum value of our experiment, say 1.2, and we looked for the electron motion, in the direction of the oscillating electric field $\mathbf{E}_\omega$, for various values of Ω, during the 30 fs pulse. Two classes of solutions were found: the first one regular and with limited values of the momentum transferred by the e.m. field to the electron; the second one chaotic and with the momentum dramatically growing in time. We also considered the refractive index of the plasma

$$\mu = \sqrt{1 - \frac{n_e}{n_c} \frac{\boldsymbol{\beta}_\psi \cdot \mathbf{a}}{|\mathbf{a}|^2}}$$

expressed in terms of the normalised electron velocity along $\mathbf{E}_\omega$, $\beta_\psi = v_\psi/c$, and the normalised vector potential $\mathbf{a} = e\mathbf{A}/mc$.

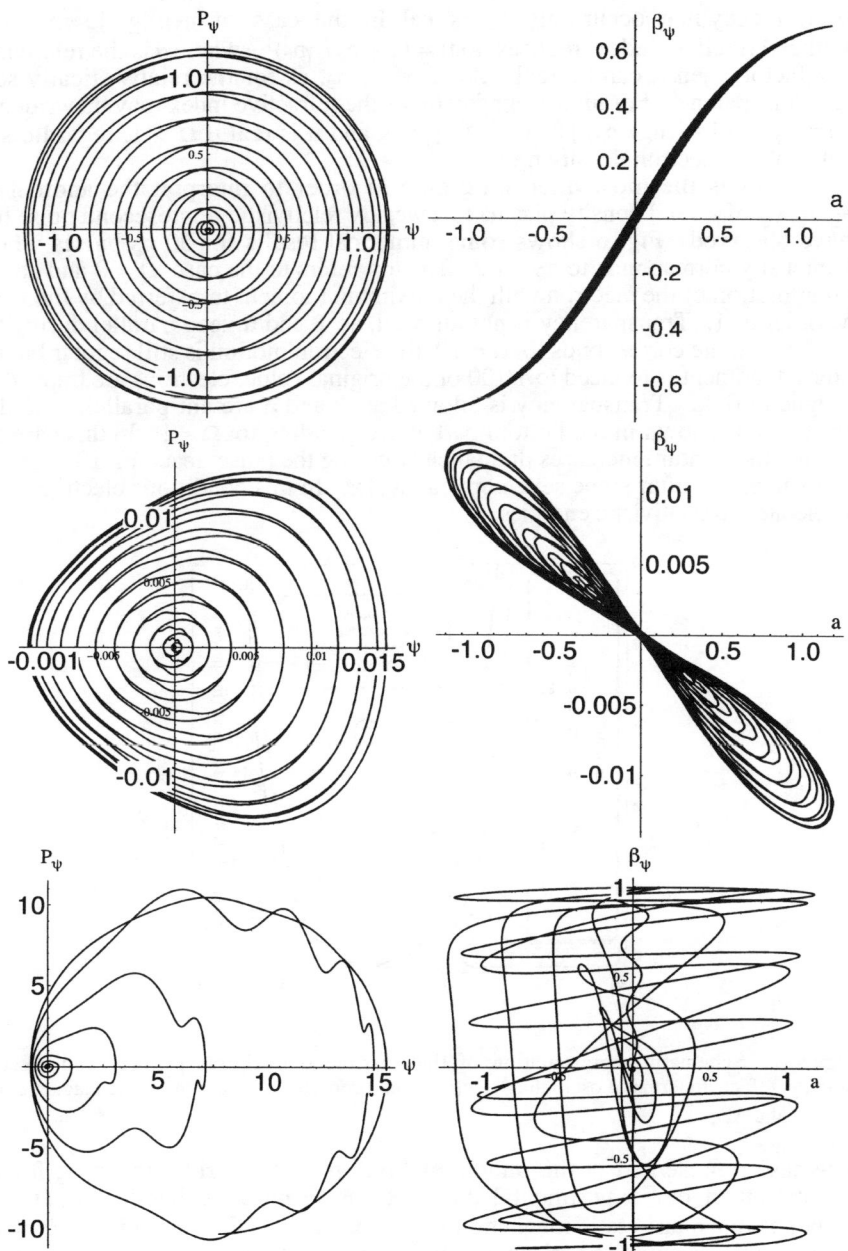

Figure 6. Left hand side: phase-space diagrams of the electron motion during the 30 ps pulse whose peak intensity corresponds to $a_0 = 1.2$. **Right hand side:** $\underline{\beta}$ versus $\underline{a}$ diagrams. **Top:** $\Omega = 0$. **Middle:** $\Omega = 10$. **Bottom:** $\Omega \approx 1$.

Transparency may occur only if μ is real. In the case of negligible magnetic field, the refractive index reduces to $\mu = (1 - n_e / \gamma n_c)^{1/2}$ where γ is the relativistic Lorenz factor; then μ can be real only if $\gamma n_c > n_e$, as for the relativistically self-induced transparency. For high magnetic fields the refractive index may be written in the form $\mu = \{1 - (n_e / n_c) [1 / (1 - \Omega^2)]\}^{1/2}$, and it is real if $\Omega > 1$, regardless of the value of the electron density n_e.

This latter is the most interesting case in order to interpret the anomalous transparency of a solid density plasma to a weakly relativistic femtosecond pulse that we have observed. Fig. 6 shows some numerical results for 30 ps pulses whose peak intensity corresponds to $a_0 = 1.2$. The top refers to the case $\Omega = 0$ and shows a regular motion of the electron with the maximum moment transferred by the wave of the order of 1. Transparency is not allowed, as β and **a** keep parallel during the pulse. The middle corresponds to $\Omega = 10$; the electron motion is still regular but the transferred moment is reduced to 1/100 of the original value: electrons are trapped by the magnetic field. Transparency is allowed as β and **a** are ant-parallel. Another relevant case is shown in the bottom part, corresponding to $\Omega \approx 1$. In this case the transferred momentum increases dramatically during the pulse, reaching 10 times the normal value, just after some seven optical cycles. In this conditions electrons may be accelerated to relativistic energies.

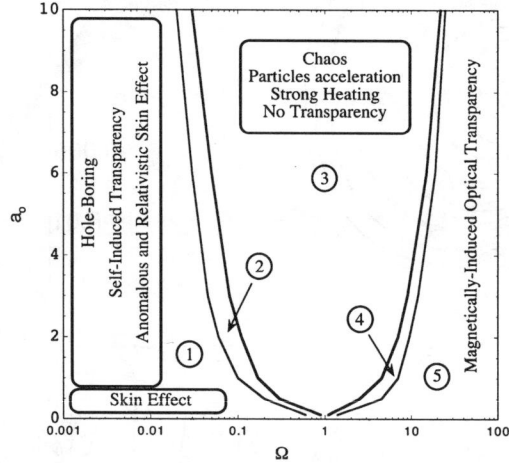

Figure 7. Schematic representation of the possible optical transparency and electron acceleration effects in terms of oscillating electric field (proportional to a_0) and static magnetic field (proportional to Ω).

The case represented in the middle of Fig.6 is particularly interesting for the interpretation of our experimental data. It has to be considered that $\Omega = 10$ corresponds to a quasi-static magnetic field exceeding 10^9 G, much higher than any magnetic field measured in a plasma so far. It may be the consequence of the ultra-fast volume ionisation in the solid, even though the real mechanism of production of such a field has still to be understood. In this case a new type of self-induced transparency will occur at lower laser intensity than expected by pure relativistic effects. This scenario is schematically represented in Fig. 7, were different physical phenomena occurring in the interaction of intense laser pulse with matter are labelled. Most of the theoretical work has been devoted so far to the effect of extremely high laser intensity ($a_0 \gg 1$). Our work suggest that also the region of high magnetic fields has to be investigated. A special consideration has to be also devoted to the resonant condition $\Omega \approx 1$, for which the cyclotron frequency associated to the magnetic field is close to the optical frequency of the laser pulse. The resonance condition may be suitable for particle acceleration.

CONCLUSION

The experimental technique using very thin plastic foils allows for the first time real interaction of an ultra-intense laser field with solid matter. In this unexplored condition ultra-fast volume ionisation was observed and above a given intensity threshold the highly ionised foil was found to be almost transparent to femtosecond pulses. The observed transparency is not explained by the current theories. The experimental results suggest that once the matter is highly ionised in times comparable with an optical cycle, its response cannot be described in terms of ordinary plasma behaviour. The tentative assumption of the creation of a very intense magnetic field in the interaction region seems to account for the anomalous transparency and the hard X-ray emission we have observed.

AKNOWLEDGMENTS

The experiment performed at LOA was partially supported by the EEC under the scheme of Access to Large Facilities. Authors are grateful to all the LOA scientific and technical team, and particularly to A. Mysyrowicz, who supervised and encouraged our activity at the *salle jaune* of LOA. Among the authors, P.C. is supported by LOA in the frame of the European Network GAUS-XRP. This work is part of the Pisa Group activity in the European network SILASI, which is also supporting one of the authors (D.T.)

REFERENCES

1. L.A.Gizzi, D.Giulietti, A.Giulietti, P.Audebert, S.Bastiani, J.P.Geindre, A.Mysyrovicz, Phys. Rev. Lett. **76**, 2278 (1996).
2. D.Giulietti, L.A.Gizzi, A.Giulietti, A.Macchi, D.Teychenné, P.Chessa, A.Rousse, G.Cheriaux, J.P.Chamberet, G.Darpentigny, Phys. Rev. Lett. (1997), in press.
3. W. Rozmus, V. Tikhonchuk, Phys. Rev. A. **42**, 7401 (1990).
4. E. Lefebvre, G.Bonnaud, Ohys. Rev.Lett **74**, 2002 (1995).
5. S.C. Wilks, W.L. Kruer, M. Tabak, A.B. Langdon, Phys. Rev. Lett. **69**, 1383 (1992); R.Kodama, K. Takanashi, K.A. Tanaka, M. Tsukamoto, H. Hashimoto,Y. Kato, K. Mima, Phys. Rev. Lett. **77**, 4906 (1996).
6. G.Malka, J.L.Miquel, Phys.Rev.Lett. **77**, 75 (1996).
7. E.S.Weibel, Phys. Fluids **10**, 741 (1967).
8. G. Ma, W. Tan, Phys. Plamas **3**, 349 (1996).
9. J. P. Matte, K. Aguenou, Phys. Rev. A **45**, 2558 (1992).
10. D.Teychenné, D.Giulietti A.Giulietti, L.A.Gizzi, submitted to Phys. Rev. Lett. (1997).
11. D.Giulietti, V.Biancalana, M.Borghesi, P.Chessa, A.Giulietti, E.Schifano, Optics Comm. **106**, 52 (1994).

Optimizing Harmonics from Solid Targets

M. Zepf [2], G. Pretzler [1], U. Andiel [1], D. M. Chambers [3],
A. E. Dangor [2], P. A. Norreys [4], J. S. Wark [3], I. Watts[2],
and G. D. Tsakiris[1]

[1] *Max-Planck-Institut für Quantenoptik, D-85748 Garching, Germany*

[2] *Blackett Laboratory, Imperial College of Science, Technology and Medicine, Prince Consort Road, London SW7 2AZ, United Kingdom*

[3] *Department of Physics, Clarendon Laboratory, University of Oxford, Parks Road, Oxford, OX1 3PU, United Kingdom*

[4] *Central Laser Facility, Rutherford Appleton Laboratory, Chilton, Didcot, Oxon, OX11 0QX, United Kingdom*

Abstract. One of the factors that currently limits harmonics from solids as a useful source of XUV radiation is the short temporal coherence time associated with spectral broadening in the underdense coronal plasma. In this paper we present data obtained at contrast ratios between $1:10^{10}$ and $1:10^3$ and demonstrate that the pulse contrast needs to be controlled carefully to optimize the bandwidth of harmonics from solid targets.

The search for a coherent, tunable source of X rays has been underway for a number of years in a variety of different fields. The first demonstration of x-ray lasing in a hot dense plasma from selenium was made at the United States of America's Lawrence Livermore National Laboratory in 1985 using the giant NOVA laser [1]. Since then, an enormous international effort has been attempted both to optimize these soft x-ray lasers by driving them to the saturation intensity limit and to reduce their wavelength to the biologically important water window of the spectrum which lies between $4.4nm$ and $2.6nm$ [2–4]. This research has been concentrated largely in national laboratories, due to the complex, expensive laser systems that are needed to generate the hot dense plasmas needed for x-ray lasers.

New possibilities for short wavelength generation have recently emerged through advances in short pulse technology. The most intensely pursued so far has been the harmonic generation from gases, which provides a reproducible source of short pulse XUV radiation [5,6] and have already found uses in several applications [7]. In addition to harmonic generation from gases, there has recently been a renewal in interest in generating high order harmonic radiation from high-power laser interactions with solids [8]. Here the radiation pressure associated with an intense laser pulse ensures that the plasma density profile remains extremely steep. Both odd and even harmonics are generated. This can be understood in terms of the fundamental light being reflected from the oscillating critical density surface [9]. The first experiment using $150fs$ pulses from Ti:sapphire lasers at $794nm$ generated harmonics up to the 8th order [10] which was later extended to the 15th harmonic [11] for focused intensities up to a few $10^{17}W/cm^2$. Moreover, in recent experiments using ultraviolet, KrF lasers the generation of up to 4th harmonic has been reported [12,13].

Simulations using Particle in Cell Codes (PIC) have suggested that very high harmonic orders (> 60) can be generated with conversion efficiencies $> 10^{-6}$ at values of $I\lambda^2 = 10^{19}W\mu m^2/cm^2$ [9,14,15]. This suggests that using very powerful UV drive lasers, very short wavelength radiation can be generated using this method. These predictions were recently confirmed in an experiment using the Chirped Pulse Amplification (CPA) beam line of the VULCAN laser at the Rutherford Appleton Laboratory [16]. This laser delivers pulses of up to $2.5ps$ duration at $1053nm$ with an intensity contrast of $1:10^6$. Up to the 68th harmonic (at $15.5nm$) was observed in the first order diffracted signal, with the 75th harmonic (at $14.0nm$) in the second order diffracted signal. The energy conversion efficiency was measured to be $> 10^{-6}$ for each harmonic up to the 68th.

After a comparatively short period of experimental investigation harmonics from laser solid interaction have already achieved a spectral brightness second only to collisional XUV lasers [17]. While the energy conversion efficiency of the harmonic is comparable to those of saturated collisional XUV lasers the pulse length is significantly shorter which is advantageous for many potential applications. A comparison of their respective source brightness shows that the current difference is largely due to the isotropic angular distribution and the large fractional bandwidth ($\Delta\lambda/\lambda \sim 10^{-2}$) of the harmonic source at high intensities. The parameters for XUV lasers are a divergence angle $\sim 10mrad$ and a fractional bandwidth of $\Delta\lambda/\lambda \sim 10^{-4}$. The experimental observations of the source parameters for the solid harmonics deviate quite significantly from theoretical expectations, which suggest that the harmonic radiation should be reflected in the specular direction and with near transform limited bandwidth. These deviations are probably due to the production of a preformed plasma, associated with the pedestal encountered in current ultra short pulse lasers. This pedestal is typically $1:10^{-6}$ of the peak intensity leading to plasma formation at times $\sim 100ps$ before the peak of the pulse for focused intensities $> 10^{17} W/cm^2$. Results by Bulanov et al. [19] suggest that high power laser pulses can be spectrally broadened by self-phase modulation in such an underdense plasma. Indeed, observations by Zhang et al. [18] showed bandwidth broadening in good agreement with Bulanov et al. [19] and also an isotropic angular distribution. The latter was attributed to Rayleigh-Taylor-like surface rippling [20] and distortions due to the ponderomotive pressure [21], as the preformed plasma density profile is locally steepened at the critical surface.

Optimizing the source brightness during high-intensity interactions remains a decisive challenge for this promising source of XUV radiation. Preliminary observations of second harmonic radiation have been performed by Marjoribanks et al. [22] showing the dependence of the second harmonic bandwidth on intensity contrast ratio. It is pertinent to perform experiments with XUV harmonics (which can clearly be identified as being generated by the surface electrons) to investigate the processes currently limiting the performance of this new XUV source and to devise methods of minimizing their impact. In this paper we report preliminary results on the optimization of harmonic generation from solids.

The experiment was conducted using the ATLAS laser at the Max-Planck-Institut für Quantenoptik. This laser is a Ti:sapphire operating at a wavelength of $790nm$ and produces $150fs, 200mJ$ pulses at a repetition rate of $10Hz$. The full width at half maximum (FWHM) of the laser spectrum was $10nm$. The laser intensity contrast ratio was measured on a routine basis by a second order autocorrelator in the $\pm 3ps$ time window and it was found to be $1:10^3$ at $1ps$. The prepulse from $30ns - 2ns$ before the main pulse was monitored using photodiodes with differential attenuation. The contrast in this time window was measured to be $> 1:10^7$ with a prepulse at 10^{-6} level $5ns$ before the main pulse. For parts of the experiment, the laser was frequency doubled in a $4mm$ thick KD*P crystal. This resulted in $60mJ$ at $395nm$ in approximately the same pulse duration. The infrared was rejected by passing the beam over four dielectric coated mirrors. This increased the contrast ratio to $> 1:10^{10}$ in the $30ns - 2ns$ time window. The laser was focused using an $f/2.5$ diamond turned, off-axis parabola. This resulted in a focused spot of $4\mu m$ (FWHM) diameter, which was measured by imaging the focal plane of the laser onto a charge coupled device (CCD). The peak focused intensity was $2.5 \times 10^{18} W/cm^2$ for the fundamental and $5 \times 10^{17} W/cm^2$ at $395nm$.

The laser was incident onto an optically polished glass flat target at 45^0, which was rotated between shots. The specularly reflected radiation was collected by a $5cm$ diameter, 45^0 spherical focusing mirror and imaged onto the entrance slit of an *Acton* vacuum spectrometer. The spectrometer was equipped with a $1200l/mm$ iridium coated grating. The dispersed radiation was then detected with double microchannel plate detector in chevron configuration in the $160nm$ to $40nm$ spectral range. The data was digitized using a CCD detector. The radiation between $600nm$ and $180nm$ was detected using an *Oriel Instaspec* 16-bit CCD detector. All the data presented here was taken on single shots, as the conversion efficiency was sufficiently high not to require integration of several shots.

Figure 1 shows a single shot, time integrated spectrum of the harmonics generated using the ATLAS laser. The harmonic radiation is sufficiently intense to be clearly visible above the time integrated plasma background. Although the detection system was not calibrated for this experiment, it shows that strong conversion into XUV harmonics can be obtained using table-top lasers. The highest harmonics observed on a single shot basis was the 16th harmonic of the fundamental at $46.5nm$ and the 7th harmonic of the frequency doubled beam at $56.4nm$. This is in good agreement with previous observations for similar values of $I\lambda^2$ [11,16].

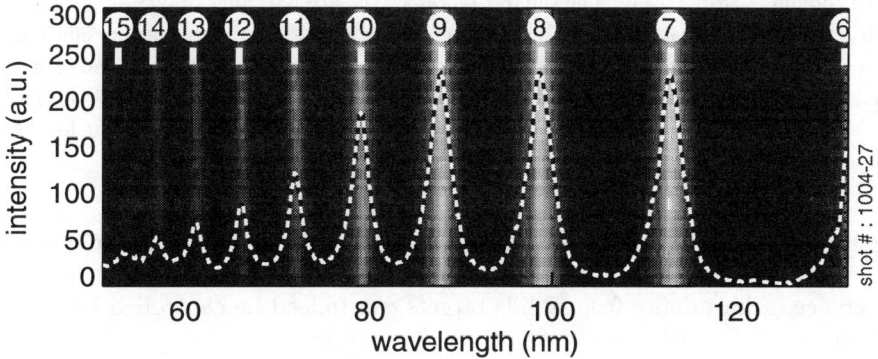

FIGURE 1. Single shot harmonic spectrum recorded with the MCP detector. The laser parameters for this shot are $2.5 \times 10^{18} W/cm^2$, and a pulse contrast of $1 : 10^6$. Note that the harmonics are significantly brighter than the time integrated background. The spectrum has not been corrected for the spectral response of the detection system.

Figure 2a shows the dependence of the bandwidth of the harmonic emission at $98.7nm$ plotted against incident $I\lambda^2$. This harmonic was chosen because it was visible above time integrated background for the large range of intensities of interest here. The bandwidth is shown for three different contrast ratios varying from $1 : 10^3$ over $1 : 10^6$ to $> 1 : 10^{10}$ whereby the large bandwidths are associated with the large prepulse levels. The $1 : 10^3$ prepulse was inserted using a $3mm$ quartz disk covering most of the fundamental beam resulting in a prepulse $5ps$ before the main pulse.

The bandwidth of the harmonic can be seen to depend strongly on the prepulse level of the laser. This is due to the reduction of preplasma production and thus minimizes the broadening due to self-phase modulation associated with the propagation of an intense laser through an underdense plasma. The bandwidth of the harmonics generated with the high-contrast frequency doubled pulse is close to the resolution limit of the spectrometer used in this

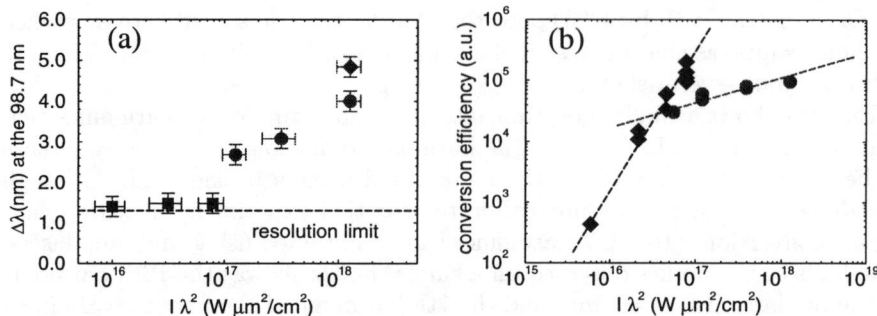

FIGURE 2. (a) Dependence of the bandwidth of the harmonic radiation at 98.7 nm. Diamond and circles represent the 8th harmonic of 790 nm produced with a $1:10^3$ and a $1:10^6$ prepulse, respectively. The squares represent the 4th harmonic of 395 nm generated with a prepulse contrast of $> 1:10^{10}$. (b) $I\lambda^2$ scaling of the relative conversion efficiency for a contrast of $1:10^{10}$, 395 nm fundamental (diamonds) and $1:10^6$, 790 nm fundamental (circles). The conversion efficiency for the high contrast case scales strongly with $I\lambda^2$ [dashed line best fit $\sim (I\lambda^2)^{2.1}$], while the lower contrast leads to a much weaker $I\lambda^2$ scaling [dashed line best fit $\sim (I\lambda^2)^{0.35}$].

experiment, and therefore is possibly significantly narrower than suggested by Figure 2a. This implies that the bandwidth and therefore the temporal coherence of harmonics from solids targets can indeed be controlled by using a high-contrast laser to avoid preplasma production.

It is also of interest to compare the $I\lambda^2$ scaling and the relative conversion efficiency of the harmonic emission for the 790 nm, $1:10^6$ contrast case and the high-contrast at 395 nm. Figure 2b shows the scaling of harmonic conversion efficiency with $I\lambda^2$ for the two cases. While the high-contrast case at 395 nm scales strongly with $I\lambda^2$, the low-contrast case shows only weak increase in conversion efficiency as the $I\lambda^2$ is increased. This is in contrast to theoretical predictions that the nonlinearity of harmonic generation process for the two cases should be similar (efficiency depends only on $I\lambda^2$) [9,14]. Again this can be understood in terms of the influence of the prepulse. At higher peak intensities the prepulse becomes more and more relevant and can lead to an increased distortion of the critical density surface. This in turn would increase the angular spread of the harmonics and reduce the intensity observed on the detector. Evidence for the onset of angular spreading when the prepulse reaches the plasma formation threshold has been found in a recent experiment conducted on the Titania KrF facility [13]. Additionally, the increasing prepulse levels will also lead to a different density profile, which in turn affects the conversion efficiency of the harmonic generation process [15]. To separate these effects it will be necessary to monitor the angular distribution of the harmonic emission.

In conclusion, we have investigated a regime in which the deterioration of the properties of solid harmonics can be observed. We have demonstrated that the properties of XUV harmonics from solid targets can be improved by carefully controlling the prepulse level of the driving laser. This leads to an increase in spectral brightness and temporal coherence and is a prerequisite for many applications, such as microscopy. We also demonstrated that the scaling of the conversion efficiency is adversely affected as soon as the prepulse becomes large enough to create a preplasma.

ACKNOWLEDGMENTS

The authors would like to acknowledge the support of the ATLAS laser staff at MPQ. The experiment was financed by the EC HCM programme Contract No. ERB CHGE CT92 006. This work was also partially supported by the Engineering and Physical Sciences Research Council of the United Kingdom. Two of us (D.M.C. and I.W.) would like to acknowledge financial support from EPSRC and in addition, one of us (D.M.C.) from AWE.

REFERENCES

1. D. L. Matthews, *et al.*, Phys. Rev. Lett. **54**, 110 (1985).
2. R. C. Elton, *X-ray lasers*; Academic Press: New York, 1990.
3. J. Zhang, *et al.*, Science, **276**, 1097 (1997).
4. L. B. Da Silva, *et al.*, Opt. Lett. **18**, 1174 (1993).
5. A. L'Huillier, *et al.*, Phys. Rev. Lett. **70**, 774 (1993).
6. K. C. Kulander, *et al.*, *Dynamics of short-pulse excitation, ionization and harmonic conversion*; Plenum Press: New York, 1993.
7. W. Theobald, *et al.*, Phys. Rev. Lett. **77**, 298 (1996).
8. G. D. Tsakiris, Physics World, p. 22, June 1996.
9. R. Lichters, *et al.*, Phys. of Plasmas **3**, 3425 (1996).
10. S. Kohlweyer, *et al.*, Optics Comm. **177**, 431 (1995).
11. D. von der Linde, *et al.*, Phys. Rev. A **52**, R52 (1995).
12. I. B. Földes, *et al.*, IEEE J. of selected topics in Quan. Elec. **2**, 776 (1996).
13. D. M. Chambers, *et al.*, submitted to Opt. Comm. (June 1997).
14. P. Gibbon, Phys. Rev. Lett. **76**, 50 (1996).
15. R. Lichters, *et al.*, In ICOMP, Inst. of Phys. Conference Series, 221-230 (1997).
16. P. A. Norreys *et al.*, Phys. Rev. Lett. **76**, 1832 (1996).
17. D. M. Chambers *et al.*, J. Appl. Phys. **81**, 2055 (1997).
18. J. Zhang *et al.*, Phys. Rev. A **54**, 1597 (1996).
19. S. V. Bulanov *et al.*, Physica Scripta **T63**, 258 (1996).
20. S. C. Wilks, *et al.*, Phys. Rev. Lett. **69**, 1383 (1992).
21. R. Fedosejevs, *et al.*, Physics of Fluids **24**, 537 (1981).
22. R. Marjoribanks, *Private Communication*; (1996).

Soft x-ray spectroscopy and heat transport in plasmas created by high-contrast fs-laser pulses

A. Saemann and K. Eidmann

Max-Planck-Institut für Quantenoptik
D-85748 Garching

Abstract. We measured the penetration depth of the heat wave in a laser plasma generated by 150fs pulses. The heat wave penetrates only layers of small thickness (≈ 300Å) at intensities of a few $10^{17} \frac{W}{cm^2}$. This is not explained by classical heat transport, even with a flux limited heat transport (f=0.08). Possible mechanisms are discussed.

INTRODUCTION

The new table top CPA laser systems allow the generation of extreme dense and hot plasmas [1] which so far have been reached in the laboratory only in inertial confinement fusion by large laser installations. It is thus possible to investigate matter under conditions which we find in stellar interiors. Because of the steep density and temperature gradients these plasmas are also short living sources of x-ray emission which opens an attractive new field to study fast material processes.

In spite of these challenging applications the basic laser matter interaction in this regime is not yet completely understood. We therefore investigated one aspect of the laser-matter interaction, namely the energy transport into the target which is crucial for the generation of dense hot plasmas.

EXPERIMENTAL SETUP

Our Ti-Sapphire laser delivers pulses of 150fs duration and 200mJ at $\lambda = 790$nm. To achieve a high contrast ratio, we frequency doubled the pulses and obtained 65mJ at $\lambda = 395$nm. Frequency doubling improves the prepulse contrast to 10^{-10} at 5ns-12ns before the main pulse. The intensity rises 6-8 orders of magnitude within the last picosecond before the pulse reaches its peak value. By focusing with an off axis parabola (f-number=2.7) we achieved an intensity of $5 \times 10^{17} \frac{W}{cm^2}$. All measurements were done with normal incidence. As diagnostic tools we utilized two grating spectrometers and filtered scintillators. The first spectrometer makes use of a Hitachi flat field reflection grating [2] with 2400 l/mm which has a wavelength range of 10Å to 70Å and a resolution of $\lambda/\Delta\lambda = 150$ at $\lambda = 19$Å. A thinned backside illuminated x-ray CCD camera is used to record the spectra. To get spatial resolution and to increase the solid angle of the spectrometer two collecting mirrors image the plasma onto the entrance slit and CCD camera respectively. The slit is 0.5mm wide and covered with a 1000Å aluminum filter to suppress stray light. In addition we have a transmission grating with 5000 l/mm and a 50μm wide slit covered by a 1000Å thick beryllium stray light filter. We operate the spectrometer at wavelengths $\leq$45Å with a resolution of $\lambda/\Delta\lambda = 40$ at 19Å. The spectra are also recorded by a x-ray CCD camera. Four filtered scintillators with cutoff energies of 1.7, 3.5, 8.0, 12.0keV combined with photo multipliers record the emission at higher energies. To study the energy transport into the target we utilize layered targets. They consist of two layers of different material evaporated onto a glass substrate. The laser energy is deposited in the topmost layer and the second layer below is heated by electron heat transport and shock heating. By measuring the emission of both layers as a function of the front layer thickness we are able to determine the penetration depth.

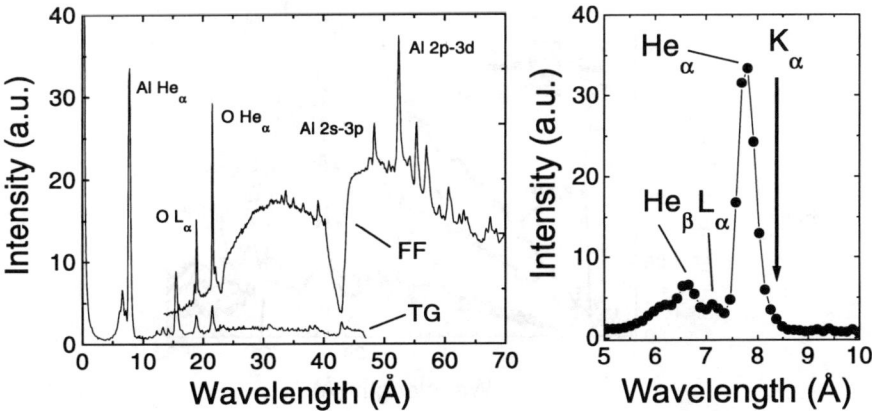

FIGURE 1. Typical aluminum spectra for high contrast pulses. The spectrum from 0Å to 45Å is recorded by the transmission grating (TG) and the other is recorded by the flat field spectrometer (FF). On the right hand side we plotted the aluminum He$_\alpha$, as measured by the transmission grating, on an extended wavelength scale. There is no significant contribution of cold K$_\alpha$ which is attributed to the high contrast ratio of the laser.

MEASUREMENTS

A typical aluminum spectrum recorded by the two spectrometers is shown in figure 1. The spectrum is characterized by the emission of He-like and Li-like lines of aluminum and of H-like and He-like lines of oxygen from the aluminum oxide. The spectrum recorded by the flat field spectrometer shows a dip near the carbon edge (at $\lambda=44$Å). It is caused by a thin carbon layer (pump oil) on top of the two collecting mirrors and the grating. This leads to a reduced reflectivity in this region.

The XUV lines are narrow. Their width is limited by the instrumental resolution. Due to Stark broadening we would expect broader lines when they are generated at high densities close to solid state density. For example, the prominent Al 2p-3d line should have a width of a few Å. The fact that we do not observe broader lines is attributed to the time integration of the measurement. We expect the XUV emission to last much longer than the laser pulse duration [5]. Therefore, we see mainly emission from the diluted expanding plasma.

Figure 1 also shows the aluminum He$_\alpha$ line (measured by the transmission grating) on an extended wavelength scale. In spite of the poor resolution of the transmission grating spectrometer it is evident that no significant contribution from the aluminum K$_\alpha$ line is present. K$_\alpha$ emission is produced by fast electrons entering the cold solid target. Fast electrons are generated efficiently

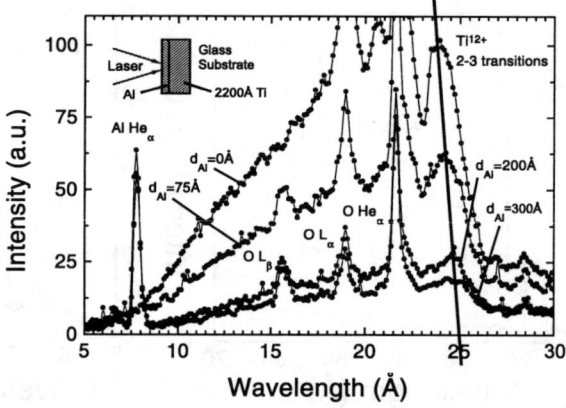

FIGURE 2. Measurement of the penetration depth recorded by the transmission grating. The targets consisted of aluminum on titanium both evaporated on a glass substrate (see inset). We used four targets with following Al thickness: $d_{Al} = 0$Å, 75Å, 200Å, 300Å. It is clearly seen that the heat wave penetrates only ≈ 300Å. In the spectrum we also see an additional shift of the peak Ti L-shell emission to longer wavelength. This is because the temperature decreases with increasing depth. The Al He$_\alpha$ emission saturates already after 75Å because this line is optically thick.

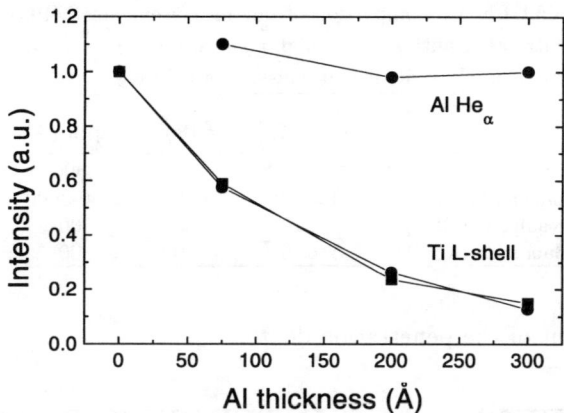

FIGURE 3. Summary of the penetration depth measurement. The circles are measured with the transmission grating and the squares with the flat field spectrometer.

when larger prepulses are present which flatten the density profile. The absence of K_α emission in our experiment is attributed to the high contrast of the laser pulse. This is of importance for the heat transport studies presented below, because prepulses strongly influence the ablated thickness.

Measurements of the space and time averaged electron temperature with the filtered scintillators show a two temperature plasma. We find a cold component of ≈ 700eV which comes from the cold expanding plasma or from regions inside the target which are heated by electron heat transport. The hot component of $\approx 5\text{-}10$keV originates from the plasma directly heated by the laser pulse.

The measurement of the energy penetration depth with a layered target is shown in figure 2. Here we used Al layers of varying thickness on top of a 2200Å thick Ti layer. Both layers were evaporated on a glass substrate. The target was moved between successive laser shots in order to irradiate a fresh surface in each shot. Typically, 300 shots were accumulated to generate a spectrum. In figure 3 we plotted the L-shell emission of Ti and the He$_\alpha$ emission of Al. We find that the Ti emission is reduced by a factor of 10 within 300Å. The spectra show a shift of the emission maximum (at ≈ 25Å) to longer wavelengths which indicates a colder plasma. The He$_\alpha$ of Al saturates already within the thinnest aluminum layer of 75Å thickness. This indicates that the line is optically thick which is confirmed by calculations of the absorption coefficient.

In table 1 we compare our results with results of other authors. It is important to notice that the highest penetration values are accompanied by a high level of K_α emission which shows the production of fast electrons. This could be an evidence for a prepulse which causes ablation and may influence

TABLE 1. Comparison of our measurements (MPQ) with other authors. The high penetration values are accompanied by significant emission of cold K_α.

Author	$I_L \left(\frac{W}{cm^2}\right)$	τ (fs)	Depth (Å)
Rousse (LULI) [3]	3×10^{16}	100	2500
Zigler (NRL) [4]	3×10^{16}	600	2500-3000
Kieffer (INRS) [5]	8×10^{17}	400	800
Saemann (MPQ)	5×10^{17}	150	300

the measurement of the penetration depth.

THEORY AND INTERPRETATION

The essence of our measurement is that the energy transport into the target is restricted to layers of small thickness ($l \approx 300$Å). In the following section we will show that this experimental result is not predicted by standard theories. We will therefore point out possible mechanisms which may be responsible for this result.

It is easily shown that the classical heat transport theory of Spitzer [6] is not applicable in our case. The reason is that the temperature gradient length L_T ($L_T^{-1} = \partial \ln T / \partial x$) is much smaller than the electron mean free path $\lambda_e = 3000$Å ($T_e \approx 5$keV, $n_e \approx 10^{23}$, $Z \approx 10$). It is then commonplace to make use of a flux limiter to reduce the heat flux to the observed value (see [7] and references therein). We therefore made simulations with the modified MULTI-fs code [8] and a flux limiter of $f = 0.08$ which is an accepted value in this regime [7]. The result is that even a flux limited heat transport predicts a penetration depth of ≈ 1000Å exceeding the measured value by at least a factor of 3.

In order to find a possible physical explanation, we will describe the energy transport by a diffusion equation. Although this picture might be too simple, it is justified afterwards by the small electron relaxation time τ_e and it points out the important mechanisms involved. We assume the rate at which the electron temperature changes to be proportional to the deviation from the equilibrium value T_0 and inverse proportional to the laser pulse duration $\tau_L = 150$fs. The diffusion equation then reads:

$$D\frac{\partial^2 T}{\partial x^2} \approx -\frac{(T-T_0)}{\tau_L}, \quad D = \frac{1}{3}v^2 \tau_e = \frac{1}{3}\lambda_e v \qquad (1)$$

whereby τ_e is the electron relaxation time and λ_e is the electron mean free path. Equation (1) leads to a diffusion length of

$$l = \sqrt{D\tau_L} \qquad (2)$$

which corresponds to our measured penetration depth of $l=300$Å. With our measurement of the electron temperature $T_e \sim v^2$ we can now calculate τ_e or λ_e from equation (2). We find:

$$\lambda_e = 4\text{Å} \text{ or } \tau_e = 1 \times 10^{-17}\text{s} \tag{3}$$

This immediately suggests two possible speculations. First, the measured mean free path equals the inter ionic spacing a ($a = n_i^{1/3}$). It is known from resistivity measurements at lower intensities ($\approx 10^{14} \frac{W}{cm^2}$), that there exists a minimum mean free path [9] [10] which equals a. This phenomenon is called "resistivity saturation" and is possible only because the electrons are heated far in excess of the Fermi temperature. The second possible explanation comes from the observation that the electron relaxation time τ_e is of the order of the inverse plasma frequency $\omega_p^{-1} \approx 1 \times 10^{-17}$. This is the characteristic time scale of the collective plasma modes and suggests electron collisions with waves and fields in the plasma.

We finally note, that in addition two-dimensional effects may be important and may cause lateral energy transport out of the focal region.

CONCLUSIONS

We demonstrated that we cannot efficiently heat target layers thicker than 300Å with high contrast 150fs-lasers and discussed possible mechanisms. This may be a problem for the attractive possibility to measure opacities of hot dense matter created by fs-lasers, as for example proposed in [11]. Therefore it would be helpful to find a way to couple the energy more efficiently into deep target layers without decreasing the density (isochoric heating).

REFERENCES

1. Mourou G. et al., *Phys. Fluids* **B4**, 2315 (1992)
2. Harada T. et al., *Appl. Opt.* **19**, 3987 (1980)
3. Rousse A. et al., *Phys. Rev. E* **50**, 2200 (1994).
4. Zigler A. et al.,*Appl. Phys. Lett.***59**, 534 (1991)
5. Kieffer J. C. et al.,*JOSA B*, **13**, 132 (1996).
6. Spitzer L. Jr., *Physics of Fully Ionized Gases*, Wiley, New York (1962)
7. Kruer W. L., *The Physics of Laser Plasma Interactions*, Addison-Wesley Publishing Company, Inc. (1988)
8. Ramis R. et al., *Comput. Phys. Commun.* **49**, 475 (1988)
9. Mott N. F., *Metal-Insulator Transitions*, Taylor and Francis, London (1974)
10. Milchberg H. M. et al., *Phys. Rev. Lett.* **61**, 2364 (1988)
11. Nazir K. et al., *Appl. Phys. Lett.* **69**, 3686 (1996)

Generation of a Train of Attosecond Pulses in the Reflected Field from a Laser-Plasma Interaction.

Luis Plaja[1], Luis Roso[1], Kazimierz Rzążewski[2] and Maciej Lewenstein[3]

(1) Departamento de Física Aplicada, Universidad de Salamanca, E-37008 Salamanca, Spain

(2) Center for Theoretical Physics and College of Science, Polish Academy of Sciences, Aleja Lotników 32/46, 02-668 Warsaw, Poland

(3) Commissariat à l'Energie Atomique, DSM/DRECAM/SPAM Centre d'Etudes de Saclay, 91191 Gif-sur-Yvette, France

Abstract. We present a mechanism for the generation of a chain of attosecond pulses through the interaction of a intense laser field with a solid surface. We include one-dimensional particle-in-cell calculations in support of this idea and we discuss the underlying physics in the light of a simple moving mirror model.

INTRODUCTION

During the past decade, laser pulse chirping techniques have been extensively used in the development of ultra-high intensity laser beams. The basic process is to stretch an incoming pulse, to amplify it and to recompress it in a final stage, this last step being crucial as the final intensity depends inversely on the pulse width. The short length of the final pulse may be, in fact, considered as a side-effect of the chirping amplification. However as we enter in the femtosecond width regime, laser pulses are not only interesting by their peak intensity but also by their short duration. Femto and sub-femtosecond physics might be a new source of interesting phenomena.

There have been recently several proposals for achieving light pulses in the femtosecond regime. One of them relies in the fact that harmonic generation under intense electromagnetic fields have a characteristic region in which several harmonics have similar intensities. Filtering some of the harmonics in

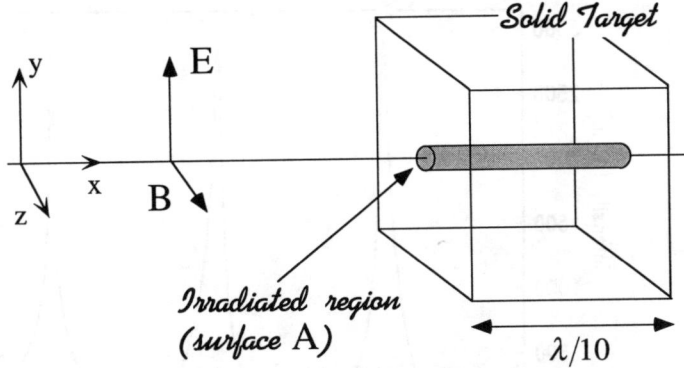

FIGURE 1. Interaction geometry for the laser-solid interaction.

this region we may obtain subfemtosecond pulses [1] provided they are phase locked by some propagation effect [2]. Other proposals involve harmonic generation from already ultra short pulses [3] or the combination of time-dependent ellipticity with chirped fundamental pulses [4].

We would like to propose a new mechanism for the generation of attosecond pulse trains by means of the non-linear reflectivity induced by the relativistic motion of the electric charges in the surface of a solid irradiated by a very intense laser beam.

MOVING MIRROR MODEL.

Consider the interaction geometry depicted in Fig. 1: an ultra-intense laser of wavelength λ, is aimed perpendicularly to a solid surface. The solid is a thin slab of matter of width $\lambda/10$, which we suppose completely ionised before the laser pulse arrives. For the short interaction times that we consider, the drift of the ions may be neglected, the dynamic of the negative charges being the main contribution.

A very simple model can teach us the basic phenomena we are interested. Let us describe the negative charges as particles driven by the electromagnetic field and bounded to the ionic background by an harmonic restoring fore $-\omega_p^2 x$, where ω_p is the plasma frequency and x is the longitudinal displacement from the equilibrium coordinate. Under these circumstances, the equations of motion should read as

$$\frac{dp_y}{dt} = qE_y(t - x/c) - qv_x B_z(t - x/c)/c \tag{1}$$

$$\frac{dp_x}{dt} = qv_y B_z(t - x/c)/c - \omega_p^2 x \tag{2}$$

where E_y and B_z are the electromagnetic field terms and q is the electron

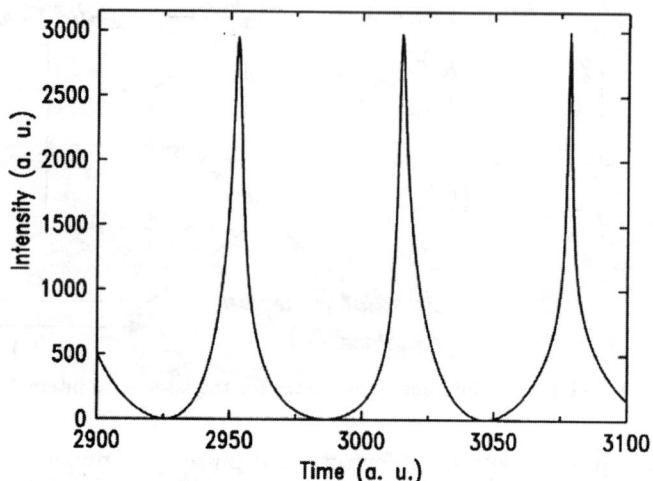

FIGURE 2. Reflected field calculated from the moving mirror model. $E_0 = 54.8$ a.u. and $\omega = 0.05$ a.u.

charge. If the electron density at the surface is sufficiently high, the plasma slab may be considered as a highly reflecting surface [8–10]. In the ideal mirror approximation, the reflected field may be computed from the requirement that the total field at the surface should vanish,

$$E_R(t + x(t)/c) + E_0 \cos\left[\omega\left(t - x(t)/c\right)\right] = 0 \qquad (3)$$

where we have assumed an incident field of constant amplitude and frequency ω, and $x(t)$ is the electron's coordinate, given by Eqs. (1)-(2).

In figure 2, we plot the resulting reflected intensity for $E_0 = 54.8 a.u.$ ($\simeq 10^{20} W/cm^2$) and $\omega = 0.05$ ($\lambda \simeq 0.8 \mu m$). As it is apparent, the reflected field consists in a succession of spikes of widths below the femtosecond. This can be understood simply by noting that a relativistic electron governed by Eqs. (1)-(2) describes a characteristic *number eight* shape in the x-y space. Every cycle, the electron is accelerated twice to a velocity close to c. When the velocity of the mirror becomes comparable to light velocity, we approach the shock wave instability (in which, as in the sonic boom for example, the velocity of the source becomes comparable to the phase velocity of the wave).

PARTICLE-IN-CELL CALCULATIONS

To test these ideas in a realistic system, we have performed particle-in-cell calculations, one dimensional in space but three dimensional in velocity (1D3V). Our code solves the four Maxwell equations coupled with the Lorentz

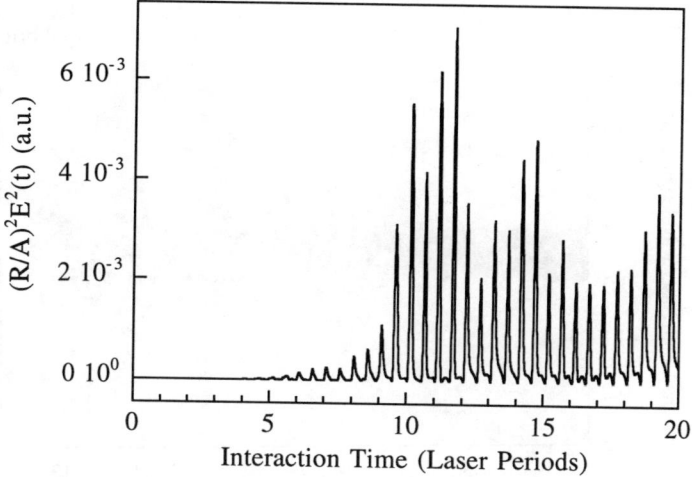

FIGURE 3. Reflected field in the far region, calculated from our PIC code for $E_0 = 25 a.u.$ ($I \simeq 2.2 \times 10^{19} W/cm^2$), $\omega = 0.057 a.u.$ ($\lambda \simeq 0.8 \mu m$).

equations for 30000 particles in a solid slab of $\lambda/10$ thickness. The laser pulse has a 5-cycle sinc turn-on followed by a constant amplitude. We assume that the PIC code describes reasonable well a 1D plasma region at the focal spot of our laser. The electric field calculated directly as a solution of the 1D Maxwell equations in the PIC code, corresponds to the near field, i.e. the field at a distance smaller that the laser spot size. Obviously, we should be interested in the reflected field computed at a distance R which is typically much greater than the laser spot size ($R > \sqrt{A}$, where A is the area of the focal spot). To do this we add up the far field radiated by every charge in the solid, calculated from the relativistic retarded expression:

$$\vec{E}(x,t) = \frac{Ae}{cR} \sum_{all\ charges} \left[\frac{\vec{e}_x \times \left\{ (\vec{e}_x - \vec{\beta}_j) \times d\vec{\beta}_j/dt \right\}}{(1 - \vec{\beta}_j \cdot \vec{e}_x)^3} \right]_{ret} \quad (4)$$

where j labels each charge.

Figure 3 shows the reflected far field for the same frequency as in Fig. 2 and a field amplitude of $25 a.u.$ (intensity $2.2 \times 10^{19} W/cm^2$). The plasma density is about $3 \times 10^{22} cm^{-3}$, which 16 times over critical. The presence of a sequence of ultrashort pulses is apparent.

Our numerical calculations show, however, that the amplitude of the chain of short pulses decreases in time. This happens when the plasma electric charges are injected into the vacuum, resulting in a diffuse plasma interface which

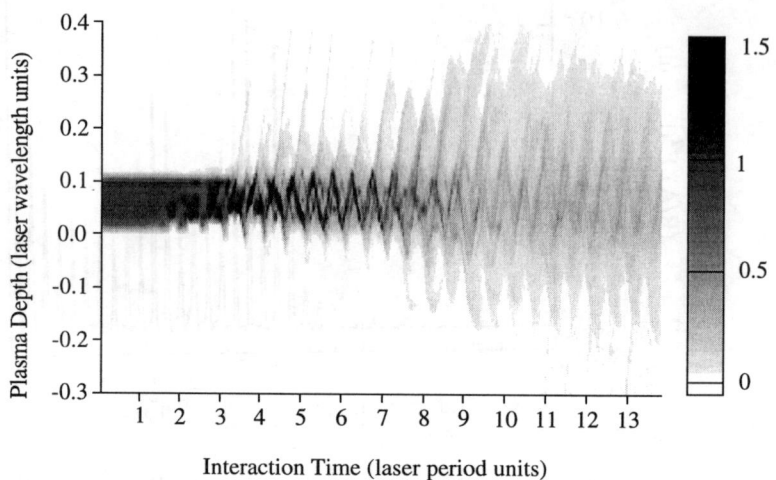

FIGURE 4. Negative charge density in function of time during the interaction with a laser aimed perpendicularly to the front surface. The plasma front interface is located at $x = 0$ and the plasma width is $\lambda/10$. Laser parameters are those of Fig. 3.

degrades dramatically its quality as a mirror. Figure 4 shows the negative charge density in function of time for the same case as in Fig. 3. The moving mirror behaviour of the charge density is apparent up to the 13th or 14th cycle. From this time on, the electrons injected into the vacuum become important enough to disturb the oscillatory motion. Consistently, Fig. 3 shows a decrease in the reflected electric field intensity after that time.

ACKNOWLEDGEMENTS

We acknowledge enlighting discussions with Th. Auguste, P. Monot, and A. Pukhov. K.R. acknowledges support of KBN Grant No.: 2P03 B04209. Partial support from the Spanish Dirección General de Investigación Científica y Técnica (grant PB95- 0955) and from the European Comission Training and Mobility of Researchers Program (under contract ERBFMRXCT96-0080) is acknowledged.

REFERENCES

1. G. Farkas and C. Toth, Phys. Lett. A **168**, 447 (1992); S. E. Harris *et al.*

2. Ph. Antoine, A. L'Huillier, and M. Lewenstein, Phys. Rev. Lett. **77**, 1234 (1996).
3. K. J. Schafer and K. C. Kulander, submitted to Phys. Rev. Lett. (1996).
4. C.-G. Wahlström *et al.*, in print in *Multophoton Processes VII*, Eds. P. Lambropoulos and H. Walther, (IOP Publishing, Bristol, 1997).
5. D. von der Linde. T. Engers, G. Jenke, P. Agostini, G. Grillon, E. Nibbering, A. Mysyrowicz, and A. Antonetti, Phys. Rev. **A52**, R25 (1995).
6. P. Gibbon, Phys. Rev. Lett. **76**, 50 (1996).
7. A. Pukhov and J. Meyer-ter-Vehn, Phys. Rev. Lett. **76**, 3975 (1996).
8. S. V. Bulanov, N. M. Naumova, and F. Pegoraro, Phys. Plasmas **1**, 745 (1994).
9. R. Lichters, J. Meyer-ter-Vehn, and A. Pukhov, Phys. Plasmas, **3**, 3425 (1996).
10. D. von der Linde and K. Rzążewski, Applied Phys. **B63**, 499 (1996).

Kinetic Approach to SuperIntense Laser-Solid Interaction

H.Ruhl[1], F.Cornolti[2], F.Califano[3] and A.Macchi[3]

[1] *Theoretical Quantum Electronics, TH-Darmstadt, Hochschulstrasse 4A, 64289 Darmstadt, Germany*
[2] *Dipartimento di Fisica, Universitá di Pisa, Piazza Torricelli 2, 56100 Pisa Italy*
[3] *Scuola Normale Superiore and INFM, Piazza dei Cavalieri 7, 56123 Pisa, Italy*

Abstract. Relativistic Vlasov simulation is employed for the study of ultrashort, superintense laser-solid interaction. A novel absorption mechanism is evidenced in 1D2V simulations. Analytical calculations elucidate the simulation results. Implementation of the Vlasov code on a parallel platform and progress towards a full Boltzmann approach are discussed.

INTRODUCTION

We present a study of superintense laser-solid interaction which is based on the numerical and analytical solution of relativistic kinetic equations. In the first section we describe a mechanism leading to enhanced laser absorption in the skin layer of ultrashort laser-produced plasmas, which is connected to the occurrence of a secular magnetic field. In the second section we discuss the implementation of Vlasov codes on a parallel architecture which makes possible to carry out very large numerical simulations. Finally, we briefly discuss some progress towards a Boltzmann description including collisional and ionization processes.

ENHANCED ABSORPTION IN SELFMAGNETIZED SKIN LAYERS

Numerical calculations

Recent high precision measurements by Sauerbrey *et al.* [1] for both s- and p-polarized light in ultrashort laser pulse produced plasmas showed enhanced absorption for small angles of incidence ($\theta \leq 25°$), for an irradiation of $I\lambda^2 =$

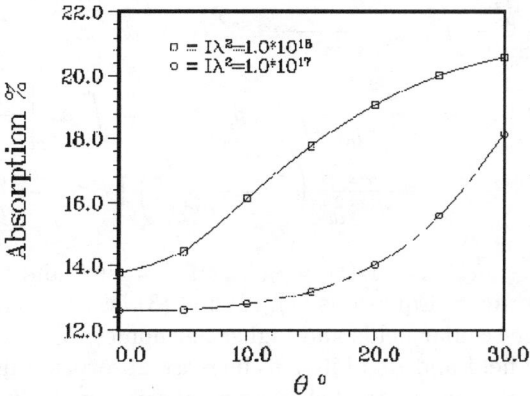

FIGURE 1. Numerical result for fractional absorption versus angle of incidence.

$2 \cdot 10^{19}$Wcm$^{-2}\mu$m^2. We simulated the experiment by solving numerically the relativistic Vlasov equation in a 1D2V phase space (x, p_x, p_y) using the well-known boosting technique to take oblique incidence of the laser pulse into account. The target was a very sharp boundary plasma of moving electrons with a Maxwellian initial velocity distribution and a fixed ion background.

Figure 1 gives fractional absorption obtained by the simulations versus angle of incidence for initial temperature $T_e = 10$keV, electron density $n/n_c = 25$, where n_c is the critical density corresponding to the laser wavelength, and density scalelength $L/\lambda = 0.023$. The irradiances are $I\lambda^2 = 10^{17}$ Wcm$^{-2}\mu$m^2 and $I\lambda^2 = 10^{18}$ Wcm$^{-2}\mu$m^2. We see that Vlasov simulations agree qualitatively with the experimental data, showing an increase of absorption with growing intensity.

Analytical model

We describe analytical calculations which suggest that the secular magnetic field in the skin layer of the irradiated material strongly influences the collective absorption mechanism [2]. Assuming a Maxwellian as the initial distribution function, we calculate the total charge and current densities at any times and keep only those terms which are at most quadratic in the field amplitudes. It can be shown that such an expansion is valid if $\omega_p v_{os}^2/\omega c v_{th} < 1$ [3] where ω_p is the plasma frequency, v_{os}^2 the quiver velocity in the laser field, ω the laser frequency and v_{th} the thermal velocity. Introducing the calculated charge and currents into Maxwell equations yields

$$\partial_x E_x \approx \frac{\omega_p^2 \bar{\gamma}}{\sqrt{2\pi} v_{th}} \int_{-\infty}^{\infty} dp_x \frac{p_x}{m^2 v_{th}^2} e^{-\frac{p_x^2}{2m^2 v_{th}^2}} \int_0^t d\tau \left[E_x^L + \frac{eA_y \partial_x A_y}{m\bar{\gamma}^3} \right] \quad (1)$$

$$\partial_t E_x \approx -\frac{\omega_p^2 \bar{\gamma}}{\sqrt{2\pi} v_{th}} \int_{-\infty}^{\infty} dp_x \frac{p_x^2}{m^3 v_{th}^2 \bar{\gamma}} e^{-\frac{p_x^2}{2m^2 v_{th}^2}} \int_0^t d\tau \left[E_x^L + \frac{eA_y \partial_x A_y}{m\bar{\gamma}^3} \right] \quad (2)$$

$$\left(\partial_x^2 - \frac{1}{c^2} \partial_t^2 \right) A_y \approx \frac{\omega_p^2 \bar{\gamma}}{\sqrt{2\pi} v_{th} c^2} \int_{-\infty}^{\infty} dp_x\, e^{-\frac{p_x^2}{2m^2 v_{th}^2}} \int_0^t d\tau \left[\frac{1}{m\bar{\gamma}^3} \partial_\tau A_y \right. \quad (3)$$

$$\left. + \frac{p_x}{m^2 v_{th}^2} \left(\bar{\beta} c - \frac{eA_y(t)}{m\bar{\gamma}^3} \right) \left(E_x^L + \frac{eA_y \partial_x A_y}{m\bar{\gamma}^3} \right) \right]$$

where $E_x^L = E_x - \bar{\beta} c \partial_x A_y$, $A_y(t) = A_y(x,t)$, $\bar{\beta} = \sin\theta$ and $\bar{\gamma} = (1-\bar{\beta}^2)^{-1/2}$. The quadratic terms in Equations (1), (2) and (3) lead to a cascade of coupled Equations for the secular fields and higher harmonics. We will concentrate on the dc magnetic field and its ability to increase absorption in the small angle range. The secular magnetic field enters absorption by a renormalization procedure of physical parameters. We insert the formal decompositions $E_x = E_{x0} + E_{x1}$ and $A_y = A_{y0} + A_{y1}$ into equations (1), (2) and (3). After a time scale separation we find for the renormalized dc fields

$$\partial_x E_{x0}^{(R)} \approx \frac{\omega_p^2 \bar{\gamma}}{\sqrt{2\pi} v_{th}} \int_{-\infty}^{\infty} dp_x \frac{p_x}{m^2 v_{th}^2} e^{-\frac{p_x^2}{2m^2 v_{th}^2}} \int_0^t d\tau \left[E_{x0}^{L(R)} + \frac{eA_{y1}^{*(R)} \partial_x A_{y1}^{(R)}}{2m\bar{\gamma}^3} \right] \quad (4)$$

$$\partial_x^2 A_{y0}^{(R)} \approx \frac{\omega_p^2 \bar{\gamma}}{\sqrt{2\pi} v_{th} c^2} \int_{-\infty}^{\infty} dp_x \frac{p_x}{m^2 v_{th}^2} e^{-\frac{p_x^2}{2m^2 v_{th}^2}} \int_0^t d\tau \left[\bar{\beta}^{(R)} c\, E_{x0}^{L(R)} \right. \quad (5)$$

$$\left. - \frac{e}{2m\bar{\gamma}^3} \left(A_{y1}^{*(R)}(t) E_{x1}^{L(R)} - \bar{\beta}^{(R)} c\, A_{y1}^{*(R)} \partial_x A_{y1}^{(R)} \right) \right]$$

For the renormalized first order fields we get

$$\partial_t E_{x1}^{(R)} \approx -\frac{\omega_p^2 \bar{\gamma}}{\sqrt{2\pi} v_{th}} \int_{-\infty}^{\infty} dp_x \frac{p_x^2}{m^3 v_{th}^2 \bar{\gamma}} e^{-\frac{p_x^2}{2m^2 v_{th}^2}} \int_0^t d\tau\, E_{x1}^{L(R)} \quad (6)$$

$$\left(\partial_x^2 - \frac{1}{c^2} \partial_t^2 \right) A_{y1}^{(R)} \approx \frac{\omega_p^2 \bar{\gamma}}{\sqrt{2\pi} v_{th} c^2} \int_{-\infty}^{\infty} dp_x\, e^{-\frac{p_x^2}{2m^2 v_{th}^2}} \times \quad (7)$$

$$\times \int_0^t d\tau \left[\frac{\partial_\tau A_{y1}^{(R)}}{m\bar{\gamma}^3} + \frac{p_x \bar{\beta}^{(R)} c}{m^2 v_{th}^2} E_{x1}^{L(R)} \right] \quad (8)$$

where $E_{x1}^{L(R)} = E_{x1}^{(R)} - \bar{\beta}^{(R)} c\, \partial_x A_{y1}^{(R)}$ is the renormalized first order electric field in the lab frame and $\bar{\beta}^{(R)} = \bar{\beta} - eA_{y0}^{(R)}/mc\bar{\gamma}^3$ the renormalized angle of incidence. The important feature is that the shape of the Equations (4), (5), (6) and (7) does not change while physical parameters like $\bar{\beta}$, the skin lengths and surface impedances are rescaled. In particular we observe that the effective angle of incidence $\bar{\beta}^{(R)}$ is different from the geometrical angle of incidence $\bar{\beta}$. The difference comes from the secular magnetic field A_{y0} which rescales $\bar{\beta}$. Writing [3]

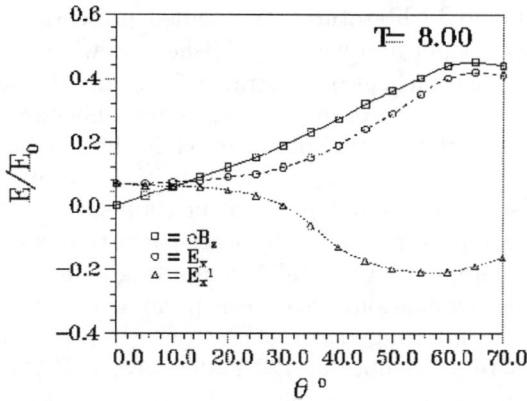

FIGURE 2. Secular electric and magnetic fields versus angle of incidence, at the position where $\bar{\gamma}E_{x0}^L$ reaches its maximum.

$$E_{x1}^L = \Re(\hat{E}_{x1}^L(\theta)e^{-i\omega t}[\Theta(x)e^{-x/l_e} - \Theta(-x)e^{x/l_e}] \quad (9)$$

$$A_{y1} = \Re(\hat{A}_{y1}(\theta)e^{-i\omega t}[\Theta(x)e^{-x/l_s} - \Theta(-x)e^{x/l_s}] \quad (10)$$

we find an approximate solution for A_{y0} [2]

$$A_{y0}(x) \approx -\left(\frac{e|\hat{A}_{y1}(\theta)|^2\bar{\beta}}{4mc\bar{\gamma}} + \frac{e\Re\left[\hat{A}_{y1}^*(\theta)\hat{E}_{x1}^L(\theta)\right]l_s}{4mc^2\bar{\gamma}}\right) \times \quad (11)$$

$$\times \frac{1}{1-\frac{4\lambda_D^2}{l_s^2}}\left[e^{-\frac{2x}{l_s}} - \frac{2\lambda_D}{l_s}e^{-\frac{x}{\lambda_D}}\right] + \quad (12)$$

$$+\frac{e\Re\left[\hat{A}_{y1}^*(\theta)\hat{E}_{x1}^L(\theta)\right]\lambda_D\,\bar{\gamma}}{2mc^2}\frac{1}{1-\frac{\lambda_D^2}{l_e^2}}\frac{l_e^3}{\lambda_D^3}\left[e^{-\frac{x}{l_e}} - \frac{\lambda_D^3}{l_e^3}e^{-\frac{x}{\lambda_D}}\right]$$

Since $A_{y0} < 0$ holds the magnitude of $\bar{\beta}^{(R)}$ is increased. As a consequence fractional absorption increases for small angles of incidence.

Figure 2 shows the simulation results for the secular field in the skin layer [3], evidencing the presence of a strong magnetic field which may affect absorption via the mechanism described above.

VLASOV SIMULATION ON PARALLEL SUPERCOMPUTERS

The numerical scheme we employed to integrate the Vlasov equations is based on a fully relativistic splitting scheme making use of only four splitting steps, which means a considerable gain in computational speed compared to similar schemes found in literature. A detailed presentation is beyond the scope of the present paper and will be published elsewhere [4].

The approach to parallelization is straightforward. We subdivide the 1D physical space into partitions of independent data which are distributed over the compute nodes of the parallel platform while the momentum space (2V) is kept entirely on the single nodes. The algorithm requires only local data to advance the distribution functions in momentum space and only data from the next nearest compute nodes in physical space; thus we are able to take the maximum advantage from a parallel computing architecture.

A number of test simulations have been perfomed on the T3E system at the CINECA supercomputing centre of Bologna (Italy). The CINECA T3E system employs a total number of 128 nodes with a RAM of 128 MB per node. To measure the scaling of the CPU speed with the total number of processors employed we simulated a same physical situation keeping the grid size constant and increasing the number of nodes used progressively up to 64. The computing speed was found to scale linearly with the number of nodes employed, giving evidence that the algorithm is fully parallel. To show explicitly that there is data transfer only between adjacent nodes we verified that the CPU time was invariant when varying the physical space grid size and the number of computing nodes by the same factor.

Since parallel computation can yield a computational speed more than two orders of magnitude faster with respect to ordinary computer, it looks now possible to carry-out multidimensional Vlasov simulations. This approach can be used as a suitable alternative to widely used Particle-In-Cell (PIC) codes, with the advantage of negligible numerical noise, high resolution and flexibility to include processes such as collisions and ionization. Development of a 2D3V code is in progress.

BOLTZMANN APPROACH

The non-collisional Vlasov approach can be extended to a Boltzmann approach by adding a "collision" operator to the r.h.s. of the Vlasov equation:

$$\frac{df_\alpha}{d\tau} = C(f_1, \ldots, f_n) \tag{13}$$

where $f_1, \ldots, f_n$ are the distribution functions for the various particle species. As an example, the covariant binary operator for Coulombian collisions between electrons, ions and neutrals can be written as

$$C_{kl} = \int \frac{d^3 p_l}{p_l^0} \int \frac{d^3 p_k'}{p_k'^0} \int \frac{d^3 p_l'}{p_l'^0} W_{k'l'kl} \left(f_k' f_l' - f_k f_l \right) \tag{14}$$

$$W_{k'l'kl} = s\sigma(s, \Psi)\, \delta^4(p_k + p_l - p_k' - p_l')\,;\ s = (p_k + p_l)^2 \tag{15}$$

where $k, l = (e, i, n)$ and $p_k = (p_k^0, \mathbf{p}_k)$ is the energy-momentum four-vector. In the case of $Z \gg 1$, $m_i \gg m_e$ and neglecting collisions between neutral particles the collison operator reduces to

$$C_{ei} = n_i v_e \int d\Omega' \sigma \left(f_e' - f_e\right) \tag{16}$$

The angle integration in eq.(16) can be numerically carried out [5].

It is also possible to derive appropriate collision operators for different, not particle-conserving processes. As a second example we introduce a collision term yielding an effective description of field ionization. We describe this process as a simple decay of one particle (the neutral atom) in two other particles (the ion and the electron). The influence of the field, considered as an external variable, is taken into account by modifying the conservation laws by the introduction of some phenomenological parameters. We write the collision operator as

$$C = \int \frac{d^3 p_n}{p_n^0} f_n(\mathbf{p}_n) \int \frac{d^3 p_i}{p_i^0} \delta^{(3)}(\mathbf{p}_e + \mathbf{p}_i - \mathbf{p}_n - \mathbf{\Pi}_I) \times$$
$$\times \delta(p_i^0 + p_e^0 - p_n^0 + p_I^0) \Gamma(\mathbf{p}_e, \mathbf{p}_i, \mathbf{p}_n; \mathbf{E}) \tag{17}$$

where $\mathbf{\Pi}_I(\mathbf{E})$ and $p_I^0(E)$ are the ejection momentum and ejection energy of the ionized electron, respectively, and can be given from atomic physics calculations. Integrating over the ion and neutral momenta in eq.(17) we again obtain a pure angular integration which can be carried out in a way similar to the case of eq.(16). Inclusion of collision and field ionization processes in the Boltzmann code is scheduled.

ACKNOWLEDGEMENT

We are grateful to CINECA supercomputing centre (Bologna, Italy) for the use of the Cray T3E system. We also acknowledge an INFM promotional supercomputing grant. This work was supported by the EC TMR Network SILASI, contract No. ERBFMRX-CT96-0043.

REFERENCES

1. Sauerbrey, R. *et al*, submitted for publication (1997)
2. Ruhl, H. and Cornolti, F., to be submitted to *Phys. Rev. E* (1997)
3. Ruhl, H. *Phys. Plasmas* **3**, 3129 (1996).
4. Ruhl, H., D'Avanzo, J., Califano, F., and Macchi, A. to be submitted to *J. Comp. Phys.* (1997)
5. Ruhl, H., TQE TH-Darmstadt internal report (1997), unpublished.

High Quality Shocks produced by Lasers: Application to Equations of State Measurements

S.Bossi, D.Batani, A.Bernardinello, L.Muller
INFM, via Celoria 16, 20133 Milan, Italy

A.Benuzzi, M.Koenig, B.Faral
Laboratoire pour l'Utilisation des Laser Intenses, Ecole Polytechnique, 91128 Palaiseau, France

T.A.Hall
University of Essex, Dept. of Physics, Wivenhoe Park, 504 3SQ Colchester, UK

Th.Löwer,
Max Plank Institut fur Quantenoptik, Garching, Munich, Germany

Abstract
High quality shock waves with direct and indirect laser drive were generated. We used Phase Zone Plate smoothing technique in the case of direct drive and thermal X-rays from laser heated cavities in the case of indirect drive. The possibility of producing homogenous, steady shock waves without significant preheating effects with both methods has been proved.
Such shocks have been used to test a new method for EOS experiments. Indeed the first simultaneous measurement of colour temperature and shock velocity in laser driven shocks is presented. The two parameters have been measured on each laser shot respectively from the target rear side emissivity in two spectral channels and by using stepped targets. A very good planarity of the shock has been ensured by using the Phase Zone Plate smoothing technique. A simple model describing the shock luminosity has been developed in order to estimate the shock temperature from the experimental rear side emissivities. Results have been compared to temperatures determined from shock velocity for materials of known equation of state.

INTRODUCTION

The study of shock wave dynamics is very important in the framework of laser-driven inertial confinement fusion. In order to obtain quantitative and reproducible results it is necessary to produce "quality" shock, i.e. very flat in space, with constant pressure in time and without preheating the material ahead the shock front. One of the main problems in realising such quality shocks is inherent to the use of coherent laser light which produces lack of uniformity in the irradiation due to beam modulations by interference effects. Such non uniform illumination produces strong pressure non-uniformities which, in the past, have prevented the use of lasers as a quantitative tool in high pressure physics, even if several experiments (1-3) have clearly demonstrated that very high pressures were indeed obtained. There are two possibilities of overcoming such a difficulty. The first one is based on the conversion of coherent laser radiation into an incoherent, very uniform, soft X-rays source inside a laser heated cavity (Hohlraum) which is then used as driver for the shock wave. The second solution, is to use direct-laser drive with optically smoothed laser beams.

In this paper, we first present a comparison of high pressure shocks generated either by indirect laser drive with optimised geometry cavities or by direct laser irradiation with the Phase Zone Plates (PZP) smoothing technique. Such comparison has been undertaken, in particular, by looking at the time history of the target rear side emissivity.

An important application of such high quality shocks concerns experiments on Equations of State for strongly compressed materials. Here we present some preliminary results on a new technique which allows to perform *absolute* EOS measurements. These are difficult to perform since they require the simultaneous determination of two shock parameters.

The shock velocity D is certainly the easier quantity to be measured with a good precision. Another quantity which can be measured (with optical techniques) is the shock temperature T_S. Direct measurements of T_S can be easily performed with materials that are transparent in their initial state (4). In the case of an opaque material, such measurements could only be made at the time the shock arrives at the solid-vacuum interface, before the plasma expands into the vacuum. The finite temporal resolution of the instruments makes such direct measurements of T_S impossible and leads to measuring the properties of the expanding plasma instead.

Here we present the first simultaneous measurements of target rear side colour temperature and shock velocity. Very planar shock fronts have been obtained using direct approach with PZPs. The colour temperature has been determined by recording the space-time resolved rear surface emissivity in two different spectral regions on the same detector. The imaging technique allowed us to check the shock quality shot-by-shot. From a simple model (presented in sec.II) which calculates the rear surface emission, we obtained an estimation of the shock temperature, noted T_S^m. On the other hand, from a precise measurement of the shock velocity and knowing the EOS it is possible to deduce the shock temperature T_S^{eos}.

It is for this reason that we used aluminium (Al) targets whose EOS is well known (e.g. from the SESAME library) in the range of pressures investigated in this paper ($P \leq 12$ Mbars). The value of the shock temperature T_S^{eos} can then be used to check the validity of T_S^m and than check the reliability of the method. The shock velocity has been measured using stepped targets.

I. Quality shocks: direct/indirect drive comparison

The experiment was performed using the ASTERIX iodine laser of the MPQ, which delivers a single beam, of diameter 27 cm, with an energy of 250 J per pulse at a wavelength of 0.44 μm. The temporal behaviour of the laser pulse is gaussian with a FWHM of 450 ps. Fig.1 shows the two different schematic experimental set-ups. An important aspect of the experiment was also the easiness of switching between direct and indirect drive configurations, achieved thanks to the particular cavity design and to the arrangement of experimental diagnostics.

In the direct laser drive configuration (Fig.1.a), the laser beam was focused directly onto the target with a $f_{3\omega} = 564$ mm lens. The characteristics of our optical system (PZP + focusing lens) were such that we produced a total focal spot of 400 μm FWHM, with a 250 μm wide flat region in the centre, corresponding to a laser intensity $I_L \leq 2\ 10^{14}$ W/cm^2.

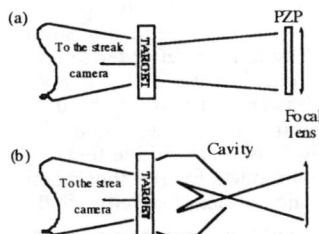

Fig.1: Schematic arrangement of the direct (a) and indirect (b) drive approaches used in the experiments.

In the indirect laser drive configuration (Fig.1.b) we focused the laser beam into a 1 mm size gold cavity through a small entrance hole. An isotropic radiation is then created (5) whose temperature depends upon the cavity size and the laser power. It can be determined by observing the velocity of a shock wave generated when radiation is absorbed in low-Z material (6). In our experiment it has been measured to be in the range of 100-150 eV. Our cavity (7) has been designed not only to reach such high temperatures, but also to optimise the irradiation uniformity when only one laser beam is used, and to minimise the preheating of the target, produced by direct primary X-rays. Here, a shield with a conical shape has been constructed so that the laser irradiated area and the shocked material were not in direct view of each other, as shown in Fig.1.b.

The diagnostic used to detect the shock emergence from the target rear face was the same in the two configurations. It consisted of an optical system imaging the rear face onto the slit of a streak camera, operating in the visible region. The temporal resolution was 8 ps and the imaging system magnification was M = 10, allowing a spatial resolution of 10 μm. A protection system was also used for the diagnostics light path, to shield the streak camera from scattered laser light.

Result analysis

Once we checked the spatial flatness of shock waves in the direct and indirect drive schemes, we focused our attention on time history of the target rear side emissivity in the two configurations. This point is important since it gives information about the preheating effects. In order to compare the emissivity in the two cases, we considered shocks in aluminium with the same pressure and targets approximately of the same thickness. We observe, as shown in Fig.2, that the emissivity is similar in the two cases. First, as mentioned in details in previous papers (8/9), we note that the shape of the two signals is typical of negligible preheating effects.

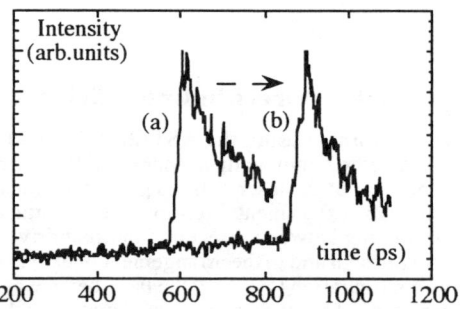

Fig.2: Emissivity profiles of targets rear face in the direct (a) and indirect (b) approach (the two profiles have been artificially shifted)

The rapid emissivity decay proves in fact that the peak corresponds indeed to the shock breakthrough at an unperturbed step density gradient of solid matter and that the plasma cools in the void without being heated by X-rays. In order to have a confirmation, we performed some tests on gold targets with indirect drive. Gold cannot be significantly preheated by our cavity blackbody radiation which is free of primary hard X-rays. We found again the same emissivity shape obtained in aluminium. Then, analysing in more detail the aluminium emissivity time behaviour of fig.2, one can notice the same growth time and a comparable relaxation. Up to now, theoretical models for the shock emissivity have not been developed because of the difficulties in the calculations of opacities in the visible region, for high densities ($\approx$ 1 - 4 times the solid density) and low temperatures ($\approx$ a few eV). However, preliminary calculations using a power-law for opacities have been performed (9) and suggest a $\approx t^{-0.55}$ dependence of the intensity as a function of time, which approximately corresponds to our experimental results both in the direct and the indirect drive scheme.

II. The EOS experiment

The experiment was performed at the LULI laboratory using the direct scheme. Aluminium targets were irradiated with a λ = 0.53 μm Gaussian laser pulse with a FWHM $\approx$ 600 ps. The laser pulse was focused by a f 2ω = 50 cm lens onto the target. The characteristics of our optical system (PZP + focal lens) were such that a focal spot of ≈ 350 μm FWHM with a flat region of ≈ 200 μm was produced. Spatially averaged intensities between $3 \cdot 10^{13}$ and 10^{14} W/cm^2 were obtained, depending on the number of laser beams we used to produce the shock.

An optical system made of an objective, two lenses and a biprism allowed the image of the target rear face to be split into two onto the slit of a visible streak camera. A different coloured filter was then used in front of each image. In this way, we measured on each shot the emissivities I_r and I_b of the red and blue regions of visible spectrum respectively. The streak camera temporal resolution was of ≈ 10 ps. The imaging system magnification was M = 10, resulting in a spatial resolution of 10 μm. The correct alignment of biprism was very crucial for the experiment. In order to check precisely this alignment, during the experiment we recorded systematically the double image (without filters) of a plastic target positioned in the centre of the chamber with the streak camera in static mode. We then verified that the intensities in the two images were well balanced. A schematic arrangement of the imaging system is showed in Fig.3.

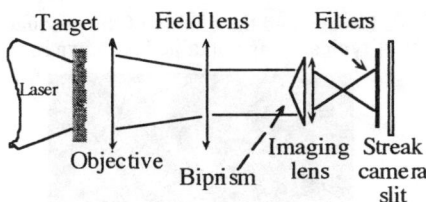

Fig.3: Schematic arrangement of the set-up. The field lens has been used in the imaging system to eliminate the vignetting effects caused by the aperture of the objective.

The convolution of the blue transmission filter with the spectral streak camera response Φ_b (λ) is a curve characterised by a maximum of 20% transmission around 400 nm. For the red "channel", the curve Φ_r (λ) has a maximum of 35% transmission around 600 nm. The filter transmission functions have been measured with a spectrophotometer. The streak camera has been spectrally calibrated *in situ* by using a spectrometer (11). In order to have a good precision on the time scale, we have also performed the temporal calibration of the streak camera by using the ultra-short laser pulse (≈ 250 fs) of the LULI (11). We have verified that for the optics used, the time delay between I_r and I_b due to the group velocity dispersion is still shorter than our time resolution.

From the ratio I_r/I_b we could then deduce the colour temperature T_c which is defined (12) as the temperature of a perfect blackbody, which would yield a ratio of brightness in two spectral regions equal to the experimentally measured ratio. The colour temperature is expected to be different from the real one because the emission is not a perfect blackbody. In, addition, the absorption between the point of emission and the observer will differentially affect the values of I_r and I_b.

By taking into account the transmission function of the filters we used, we deduced the curve $T_c = T_c (I_r/I_b)$. Such curve is very sharp for temperatures above ≈ 6 eV (this corresponds to I_r/I_b below ≈ 0.6). For such range of temperatures the maximum of the blackbody emission lies in the XUV region and hence the visible region is in the tail of the curve. This implies that a little variation of the ratio I_r/I_b gives a very large change in the temperature. For temperatures below ≈ 1 eV (I_r/I_b above ≈ 1.5) we have the opposite behaviour: the temperature is not very sensitive to a variation of the ratio I_r/I_b. Therefore, in our case, the colour temperature diagnostic can be reasonably applied in the 1 - 6 eV region.

Because of the finite streak camera resolution time, the colour temperature measurement at the shock breakout corresponds to a time when the plasma is already expanding into the vacuum and not to the instant of shock breakout. Hence, it gives a lower bound for the shock temperature. In fact, the expanding material forms a cool, optically thick layer obscuring the higher temperature material behind it.

In order to perform the shock velocity D measurement we focused the laser beam smoothed using the PZP technique on a grid aluminium target (Fig.1). The high quality shock produced and the use of grid targets allowed a direct and precise determination of the shock velocity shot by shot (10). Indeed D is obtained by measuring the shock breakthrough times from both the base and the step. The base thicknesses of the targets were in the range of 12-15 μm, while the step thicknesses were 5-6 μm. The choice of targets thicknesses was suggested by simulations, performed with the 1D hydrodynamic code FILM developed at Ecole Polytechnique, under the condition of ensuring constant shock pressure in the steps. The spacing between the steps was $\approx$ 140 μm, while the width of the steps was $\approx$ 40 μm. Therefore, such a geometry together with our focal spot dimension ensured there was always one step within the flat region of the focal spot.

Data Analysis

Once the use of PZPs has ensured the quality of the produced shocks, we used the grid targets for D measurement. A typical image obtained is shown in Fig.4.

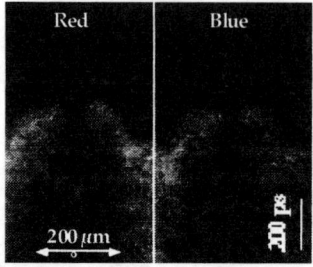

Fig.4: Typical streak camera record of the "red" and "blue" spectral light emitted by the rear side of a grid aluminium target. The base thickness is 13.55 μm and the step thickness is 5.7 μm.

As one can observe, the shock is perfectly planar on a dimension of $\approx$ 200 μm which corresponds to the focal spot flat region. The time resolved blue and red emissivities are shown in Fig.5a. First, as mentioned in previous papers (8,9), we note that the shapes of the two signals are typical of negligible preheating effects. Then, we observe that the two emissivities have a different time decay (indeed they cross). This happens because at early times, the temperature is of a few eV and hence the maximum of the emissivity is in the UV region, while at later times the plasma cools down and the maximum moves towards lower energies. The fact that we effectively see the crossing point is also due to the characteristics of the filters used. The colour temperature, corresponding to the emissivities of fig. 5a is shown in Fig.5b, as a function of time. It can be seen that the plasma cooling yields a decrease of T_c with time.

It is well known (12) that, when a shock wave reaches the interface between solid and vacuum the material begins to expands. In our case, the shock wave is so strong that the metal is completely vaporised on unloading and expands in a gas/plasma phase. The layer from which the radiation comes, enlarges during the time and simultaneously the material cools down and the emission relaxes.

In order to evaluate the shock temperature, we developed a simple model which reproduces the blue and red emission. This model is based on:
i) the usual assumption that the hydrodynamic expansion is an isentropic self-similar rarefaction wave (12). We preferred such a description instead of those given by usual lagrangian hydrodynamic codes because the sensivity to the cell size prevents a reliable description of rear side expansion. Recently (13,14) it has been pointed out that the expansion should be considered as isothermal in the first $\approx$ 5 ps after the shock breaktrough and becomes isentropic only after this characteristic time. We did not take into account this effect since it takes places on a time scale shorter than our temporal resolution.

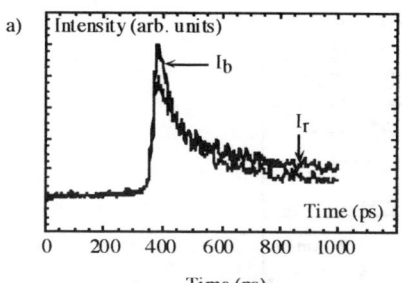 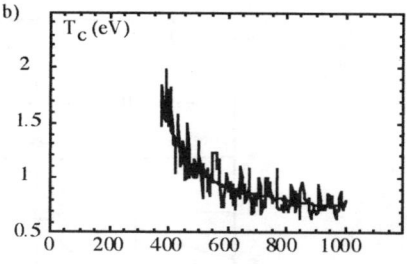

Fig.5: (a) Emissivities in the red and blue regions of the shocked target rear side as a function of time. (b) Colour temperature as a function of time deduced from the ratio of the two emissivities. The solid line represents the best fit.

ii) the assumption of the Kramers Unsöld's opacity, as suggested by ref. (12) for the absorption of visible light in metallic vapors. Such absorption coefficient accounts for free-free and bound-free transitions involving highly-excited states of hydrogen-like atoms. This formula describes accurately the absorption in the low density region, while it does not apply in high density regions. The problem of opacity calculations for the visible light in highly compressed ($\approx$ 1 - 3 times the solid density) and cold matter ($\approx$ a few eV) is a very difficult problem which has not been solved yet. As far as we know, all the shock visible luminosity calculations performed so far, have used a power law for the opacities (Kramers Unsöld's or free absorption formula) (9,15,16). In ref. (16) the authors used also Rosseland frequency-averaged opacities from the SESAME library which is rather questionable since frequency averaging leads to an important overestimation of the luminosity.

iii) the use of the More's (18) formula for the mean ionic charge based on Thomas Fermi model which gives a reliable value for degenerate plasmas.

The calculation of I_r and I_b has been undertaken using the hypothesis of local thermal equilibrium (in our case the thermalisation chararacteristic times are smaller than the hydrodynamical characteristic times). Hence at each time we used the formula:

$$I_{r/b} = \int_0^{+\infty} \Phi_{r/b}(\lambda) \int_0^{+\infty} k_\lambda U_{p\lambda} \exp\left(-\int_0^x k_\lambda dx'\right) dx \, d\lambda$$

i.e. the solution of the transfer radiation equation (12). $\Phi_{r/b}(\lambda)$ are the transmission functions of the red and blue "channels" (cf. sec. II), $U_{p\lambda}$ is the Planck's law, k_λ is the absorption coefficient, λ is the radiation wavelength and x the spatial variable.

We noticed that at early times ($\approx$ 10 - 50 ps after the shock breakthrough), the radiation can come from regions slightly overcritical (at most $\approx$ 0.1 μm beyond the critical layer position) since a screening layer has not completely developed yet. However, in our calculation we did not have to take into account any particular effects near the critical layer. Indeed the sharp cut off at $\omega = \omega_p$ occurs at low pressures only (when $v_{ei}/\omega \ll 1$; v_{ei} is the electron ion collision frequency, ω is the light frequency and ω_p the plasma frequency) and becomes less and less distinct as collisional damping increases (19). At early times, the plasma has not yet expanded and we are clearly in the opposite regime ($v_{ei}/\omega > 1$) (14).

We determined T_s^m by changing its value and the adiabatic coefficient γ in the self similar expanding profiles until we found the best interpolation of our experimental curves. We varied γ in the range $\approx$ 1.2 - 1.7 as expected in rarefaction after a strong shock (9). The results are shown in Fig.6 where the colour temperatures T_c and the shock temperatures T_s^m are compared with the shock temperatures T_s^{eos} given by SESAME equation of state.

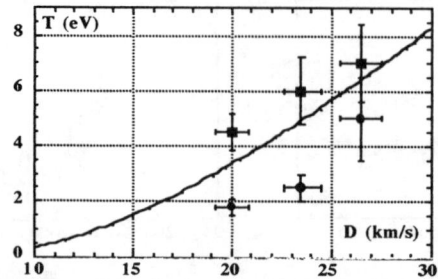

Fig.6: Aluminium shock temperature T_s vs shock velocity D. The solid line corresponds to T_s^{eos} given by SESAME EOS, the solid circles to colour temperature T_C measurements and the solid squares to the shock temperature T_s^m obtained by the model.

The calculated values of the shock temperature T_s^m are not too far from those obtained from the SESAME EOS. Part of difference may be due to a poor description of plasma opacities.
The error bar for D measurements has (± 4 %) been determined considering the uncertainties on: the step thicknesses, the shock breakthrough time and the streak camera sweep speed. While, the error bar for the colour temperature has been determined taking into account that the uncertainty comes principally from an imperfect alignment of the optical system. Obviously, the error on the ratio I_r/I_b implies an error on T_C which depends on the behaviour of the function T_C (I_r/I_b). According to the previous explanation, the error increases considerably with the shock strength. The error varies in the range ± 15% - 20% for the two points at lower temperatures, while the error for the third point is more important ($\approx$ ± 30%) since it lies near the limit of the diagnostic applicability.

CONCLUSIONS

In this paper, we have shown the possibility of producing shock with a comparable accuracy using the direct and indirect laser drive. In particular, in the two cases the target rear side emissivity showed that there was no preheating. In the direct drive case, a high quality shock has been reached by making use of the PZP smoothing technique. In the case of the indirect drive it has been obtained by taking advantage of the geometry of the cavity, where a shield protected the target from a primary X-ray irradiation. We have then presented some results on the applicability of theses high quality shocks to test a new method for EOS measurements. In this framework colour temperature and shock velocity have been measured on the same laser shot. By using a material of reference (Al), we had a direct comparison between the shock temperature deduced from SESAME EOS and the colour temperature. Using a simple model for the target rear side emissivity, we obtained an independent estimation of the shock temperature. If, in the future, opacity data are improved, the method presented here would provide a more precise EOS data in the 1-6 eV temperature regime.

ACKNOWLEDGEMENTS

This work was supported by EU Human Capital and Mobility program under contract CHRX-CT93-0338 and ERB-CHGE-CT93-0046. L.Muller took part to the experiments thanks to the ERASMUS programme coordinated by Prof.Lanz at the University of Milan.

REFERENCES

1. R.J.Trainor, J.W.Shaner, J.M.Auerbach, N.C.Holmes, Phys.Rev.Lett. **42**, 1154 (1979).
2. R.Fabbro, B.Faral, J.Virmont, H.Pepin, F.Cottet, J.P.Romain, Las.Part.Beams **4**, 413 (1986).
3. B.Faral, R.Fabbro, J.Virmont, F.Cottet, J.P.Romain, H.Pepin, Phys. Fluids B **2**, 371 (1990).
4. S. B. Kormer, Sov. Phys. Usp., **11**, 229 (1968).
5. Th. Löwer, R. Sigel, Contributions to Plasma Physics **33**, 355 (1993).

6. R. L. Kauffman, L. J. Suter et al., Phys. Rev. Lett. **7 3**, 2320 (1994)
7. Th. Löwer, R. Sigel, Proceedings APS Topical Conference on Shock Waves, Seattle (1995)
8. Th.Löwer et al., Phys.Rev.Lett. **7 2**, 3186 (1994).
9. S. Hüller et al. in Gesellschaft für Schwerionenforschung Project: *'High Energy Density in Matter.'* GSI-94 Annual Report (Darmstadt, 1994), ISSN 0171-4546.
10. M. Koenig et al., Phys. Rev. Lett. **7 4**, 2260 (1995).
11. A. Benuzzi et al. LULI Rapport Scientifique (1996) (unpublished).
12. Ya. B. Zeldovich and Yu. P. Raizer,*Physics of Shock Waves and High Temperature Hydrodynamic Phenomena* (Academic Press, New York, 1967).
13. D. Parfeniuk et al. Can. J. Phys. **6 6**, 662 (1988)
14. P. Celliers and A. Ng, Phys. Rev.E **4 7**, 3547 (1993).
15. L. Da Silva et al. J. Appl. Phys. **5 8**, 3634 (1985).
16. M. Mahdieh and T. Hall, LPB **1 4**, 149 (1996).
17. R.M. More, *Applied Atomic Collision Physics* (Academic Press, New York, 1982).
18 G. Bekefi, *Radiation Processes in Plasmas* (Jhon Wiley and Sons, New York, 1966).

Coherent X-Ray Generation at 2.7nm using 25fs Laser Pulses

Andy Rundquist, Zenghu Chang, Haiwen Wang, Ivan Christov, Henry C. Kapteyn, and Margaret M. Murnane

Center for Ultrafast Optical Science, University of Michigan, Ann Arbor, MI 48109-2099
*Permanent address: Department of Physics, Sofia University, 1126 Sofia, Bulgaria

Abstract. We demonstrate for the first time that coherent soft-x-ray pulses at wavelengths of 2.7nm can be generated using 25fs driving pulses. High-order harmonic generation in He is used to produce the femtosecond x-ray harmonics, which exhibit discrete individual orders up to 221, followed by a continuum of unresolved harmonics which extend up to at least the 299th order, corresponding to a wavelength of 2.7nm, or an energy of 450eV. The large ionization potential of He, together with the ultrashort nature of the driving field, results in this dramatic extension of the harmonic plateau, by approximately 200 orders more than has been observed previously. We also obtain excellent agreement with theoretical predictions.

INTRODUCTION

In recent years there has been much progress in developing reliable sources of coherent ultrashort light pulses in the visible and infrared regions of the spectrum.[1] At shorter wavelengths, the technique of high harmonic conversion (HHG) of an ultrashort pulse has proven most effective in generating coherent light.[2, 3] The great attraction of HHG as a source of coherent ultrafast x-ray pulses is that it is simple. A high peak-power femtosecond laser pulse is focused into an atomic gas, and the highly nonlinear interaction of the laser light with the atoms results in the emission of coherent high-order harmonics of the laser in the forward direction, as a low divergence x-ray beam. In past work, harmonic orders up to a cutoff of 135 were observed by L'Huillier et al. from the two lighter noble gases using 1ps, 1054nm laser pulses.[3] Macklin et al. observed harmonics up to a cutoff of the 109th order using 125fs, 806nm pulses in neon.[4] These previous results correspond to a minimum wavelength of $\approx$ 7nm, or a maximum energy of < 170eV.

In this work, we present an experimental and theoretical investigation of high harmonic generation using ultrashort, 25fs, driving pulses. We find that we can dramatically extend the cutoff using shorter excitation pulses over that observed previously using longer pulses. The minimum wavelength we observe is 2.7nm, which corresponds to a harmonic order of 299, or an energy of 450eV. This work dramatically extends the available wavelength range of coherent, ultrashort-pulse, x-rays well into the "water window" region, where water is less absorbing than carbon.

THEORETICAL

Quantum mechanical models are needed to fully describe the x-ray emission during HHG accurately.[5, 6] However, to a first approximation, quantum models can be used to describe the ionization process, while semiclassical theories can be used to describe the motion of the just-ionized electron during the first optical cycle after the ionization.[7, 8] In this picture, on the rising edge of a high intensity laser pulse, the ionization of atoms occurs via tunneling through the core potential. Once free, the electron moves in the laser field, and when the laser field reverses, the electron can return to the core with a maximum kinetic energy of 3.17 U_p. Some fraction of these electrons will undergo stimulated recombination with the core, and release their energy as high harmonics. Here, $U_p = E^2/4\omega^2$ is the ponderomotive or quiver energy (atomic units) of a free electron in an electric field E of frequency ω. The maximum kinetic energy of 3.17 U_p corresponds to an electron released at an optimum phase of the driving field - all other release phases result in lower electron energies (or even no subsequent re-encounters with the atom).

Therefore, from the above picture, the energy of the highest harmonic emitted from an atom of ionization potential I_p is predicted to be:

$$h\nu_c = I_p + 3.17\, U_p \qquad (1)$$

It is worth noting that U_p is evaluated at the maximum field that an electron may experience *before ionizion*, even though the laser field may subsequently increase. This is because once the atoms are ionized, the x-ray emission terminates (with the possible exception of harmonic emission from ions, which is not observed here). It is also worth noting that Eqn. 1 has also been confirmed by numerical and analytic calculations, and by experimental observations for 100fs excitation laser pulsewidths or longer.

For laser pulses < 100fs, the ionization process can be strongly affected by the ultrashort rising edge of the pulse, and in the presence of finite ionization rates, the atoms survive to higher laser intensities prior to ionization.[9, 10, 11, 12] The electron is then exposed to a stronger, rapidly increasing, laser field, which allows the electron to gain even more energy prior to re-encountering the parent ion. In order to derive an explicit dependence of the cutoff photon energy (ν_c) on the atomic and laser parameters, we developed a simple model to predict ν_c for given input parameters. Assuming that the atom is ionized on the leading edge of a linearly polarized laser pulse and following the method developed by Chang et al.,[13] an analytical expression of the saturation (ionization) intensity I_s can be obtained. Substituting $U_p = 9.33\,10^{14} I_s \lambda^2$ into Eqn. 1, and assuming ADK ionization rates,[14] we obtain:

$$h\nu_c = I_p + \frac{0.5 I_p^{(3+a)} \lambda^2}{\left(\ln\left(0.86\tau 3^{2n^*-1} G_{lm} C_{n^* l^*} I_p\right)/(-\ln(1-p))\right)^2} \qquad (2)$$

where $h\nu_c$ and I_p are in eV, a = 0.5 (to correct an approximation in the derivation of the analytical expression of I_s), λ is the laser wavelength in μm, and τ is the FWHM of the pulse in fs. Here p is the ionization probability for defining the saturation intensity (which is chosen to be 0.98 for our calculation), n* is the effective principle quantum number and $C_{n^*l^*}$ can be found in $G_{lm}=(2l+1)(l+|m|)!/6^{|m|}|m|!(l-|m|)!$,where l and m are the orbital and magnetic quantum numbers of the outermost electron. In Eqn. 2, $C_{n^*l^*} \approx 2$, $G_{lm}=3$ (except for He, where it is 1), and n* varies between 0.74 for He and 1 for Xe. Eqn. 2 clearly shows how the cutoff photon energy changes with the laser pulse duration and wavelength, the atomic species, and the electron quantum state. These predictions are summarized in Table 1, which demonstrate that shorter duration excitation pulses should result in the generation of higher order harmonics. For comparison, the experimentally observed cutoffs for 25fs excitation pulses are also listed in Table 1, as will be discussed below. Complete quantum mechanical calculations[10, 11] show further effects for very short pulses (<25fs). For rapid risetime pulses, the phase of the atomic dipole lags that of the field, which can result in further reduced ionization due to the non-sinusoidal shape of the electric field in time.

Table 1: Theoretical predictions from Eqn. 2 of harmonic cutoffs for all the noble gases, for 100fs and 25fs excitation pulses (assuming the same peak intensity).

Species	Ionization Pot. (eV)	100fs cutoff order (theory)	25fs cutoff order (theory)	25fs expt.
He	24.6	239	333	299
Ne	21.6	119	163	155
Ar	15.76	45	61	61
Kr	13.99	33	41	41
Xe	12.13	23	27	29

The pulse duration of the harmonics is predicted to be short,[11] because the emission occurs only on the rising edge of the pulse. For 25fs excitation pulses, harmonics are generated during ≈ 3 half-cycles of the laser pulse (i.e. 3.5fs), during which time the ionization probability varies from 10% to ≈ 100%. During this time also, the amplitude of the incoming pulse changes by ≈ 30%. For very short or excitation pulses (≈5fs), a *single-cycle* of the excitation pulse can drive harmonic emission over a range of adjacent harmonic orders. As a result, the temporal coherence of the harmonics is dramatically improved compared with longer excitation pulses, and more efficient x-ray pulses, with durations as short as 100 attoseconds, can be emitted under the proper focusing and gas jet pressure conditions.

EXPERIMENTAL

To observe the high harmonic cutoff energies for various gases, a setup shown in Fig. 1 was used. The Ti:sapphire laser system used for the experiments can generate TW-level, 26fs, pulses with a center wavelength of 800 nm.[15, 16] A 1cm diameter laser beam is focused onto the gas target using a 1m focal length curved mirror, which produces a ~100 μm diameter focal spot. The gas nozzle diameter is 1mm, while the gas pressure was approximately 8 torr (at the interaction region) for these experiments.

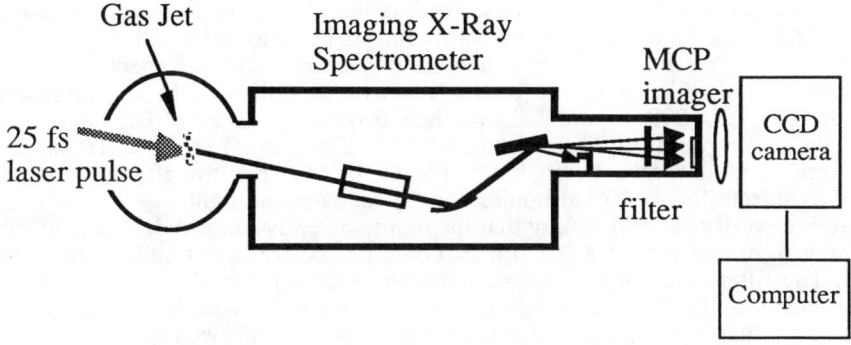

Figure 1: Experimental setup for high harmonic generation

Typically, 20mJ of laser energy is used to generate the harmonics, corresponding to an intensity of 6×10^{15} W/cm^2 at the focus. The x-rays are dispersed using a flat-field soft x-ray spectrometer, and then detected using an image intensifier with a pair of microchannel plates (MCPs). The spectrally dispersed image is recorded using a cooled CCD camera connected to a computer. It is essential to block the fundamental laser beam inside the spectrometer to prevent the generation of a significant ion background at the detector. X-ray filters must also be placed in front of the MCP to block the very bright scattered low-order harmonics.

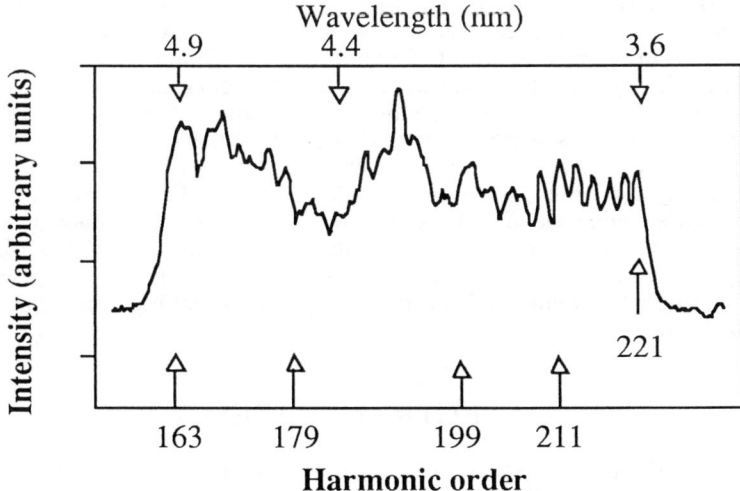

Figure 2: Discrete harmonic emission from helium for 25fs excitation pulses using a grating optimized for 5nm wavelengths. The cutoff at 3.6nm is instrumental.

For the experimental conditions described above, the harmonic spectra observed from Ne, Ar, Kr and Xe exhibit discrete harmonic peaks up to order 155, 61, 41 and 29 respectively, as listed in Table 1. For the case of He, the harmonic spectra observed had discrete, resolvable, peaks up to the 221st order, followed by an unresolved plateau up to order ≈ 299. Figure 2 shows the harmonic spectrum of He, taken using an efficient x-ray grating optimized around 50Å. The cut-off at the 221st order is instrumental, and due to a beam block placed in the spectrometer to eliminate scattered light from low-order harmonics and the fundamental light.

In order to verify experimentally that the harmonic emission is indeed due to short wavelength light, we placed a 0.4 µm carbon filter between our spectrometer and detector. The filtered harmonic emission spectrum, on a log scale, is shown in Fig. 3(a). The position of the carbon 4.37nm absorption edge can clearly be seen, as well as the discrete harmonics, and plateau region. This also allowed us to verify our spectrometer calibration, using the position of both the carbon and boron absorption edges.

To observe shorter wavelength radiation from He, we used a grating optimized for shorter wavelengths, which allowed us to block the fundamental beam without simultaneously obscuring the harmonic radiation. However, the lower efficiency of this grating results in much lower signal-to-noise and also lower resolution. Nevertheless, we can observe harmonic radiation transmitted through a 0.2µm Ti filter, terminating for wavelengths shorter than the Ti edge at 2.73nm, as shown in Fig. 3(b).

From Table 1, the observed cutoff harmonic orders is in excellent agreement with theory in all cases except for He, where there is still a slight 10% difference between our experimental observation (299th) and the theoretical prediction (333rd). The good agreement between theory and experiment is most likely because our gas densities are sufficiently low and our pulses sufficiently short that propagation effects do not play a major role in determining the output. In the case of He, there are several possible explanations for the small discrepancy between theory and experiment. First, we have not definitively observed the cutoff. Second, the laser intensity in the gas medium may be well below the intensity that we used in the calculation, due to the defocusing of the laser beam induced by the ionization. Third, Eqn. 2 does not take into account propagation effects - the phase mismatch induced by the free electrons may play a significant role for efficient production of harmonics below 2.7nm. Finally, the ADK approximation and/or other assumptions made in deriving Eqn. 2 may not be valid in the case of helium, since technically ADK is valid only for large values of n^*.

Very recently, radiation at wavelengths as short as 4.4nm was observed by another group using very short driving laser pulses.[17] In this case the non transform-limited bandwidth of the driving pulses precludes the observation of discrete harmonic orders.

SPECTRAL SHAPES

The observed harmonic spectra change dramatically with the chirp of excitation laser pulse, which can easily be varied by adjusting the separation of the stretcher gratings. As expected, the harmonic peaks shift to longer wavelengths for positive chirp, when the leading edge of the pulse is redder than the trailing edge. This result is qualitatively similar to our results with argon, but in this case the shift is larger (x2) and it can cover four harmonic orders (two peaks).

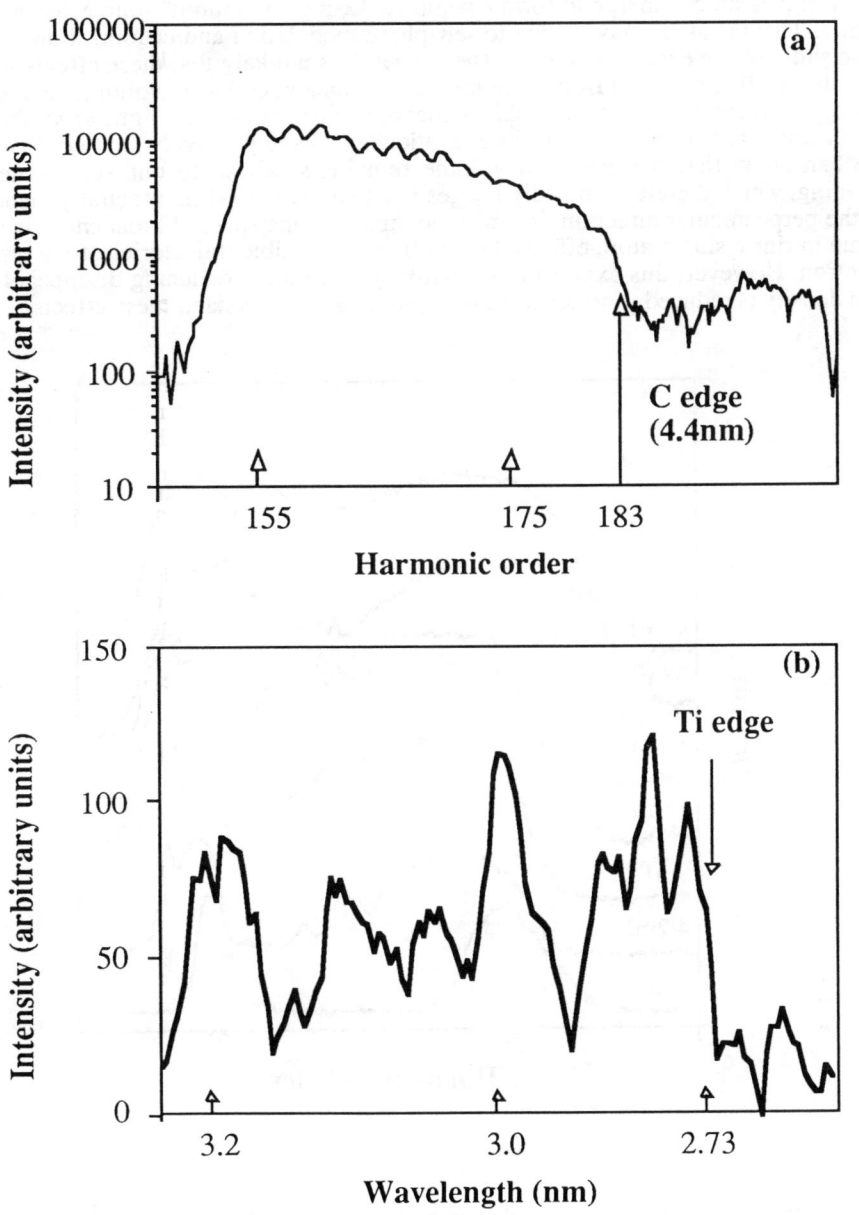

Figure 3: Harmonic emission from Helium, (a) filtered through a 0.4 μm carbon filter, and (b) filtered through a 0.2 μm titanium filter.

Finally, by increasing either the gas jet pressure or laser energy, the harmonics in the mid-plateau can merge to form a complete "x-ray continuum" source, as shown in Fig. 4. This behavior may be due to self-phase modulation and/or ionization-induced blue shifts of the excitation laser in the gas jet. It is unlikely that these effects are due to volume effects arising from different interaction and emission volumes as the laser energy is increased. The blue shift observed experimentally might arise from an increasing interaction volume if the excitation laser were positively chirped. However, we can show that the emission volume remains small since our spectrometer is imaging, which therefore spatially images in one direction, while spectrally dispersing in the perpendicular direction. It is also possible that the spectral broadening is due to some intrinsic single atom effects due to different possible trajectories of the emitting electron. However, this explanation is unlikely since the broadening disappears if the gad density is reduced. Further work is in progress to understand these effects.

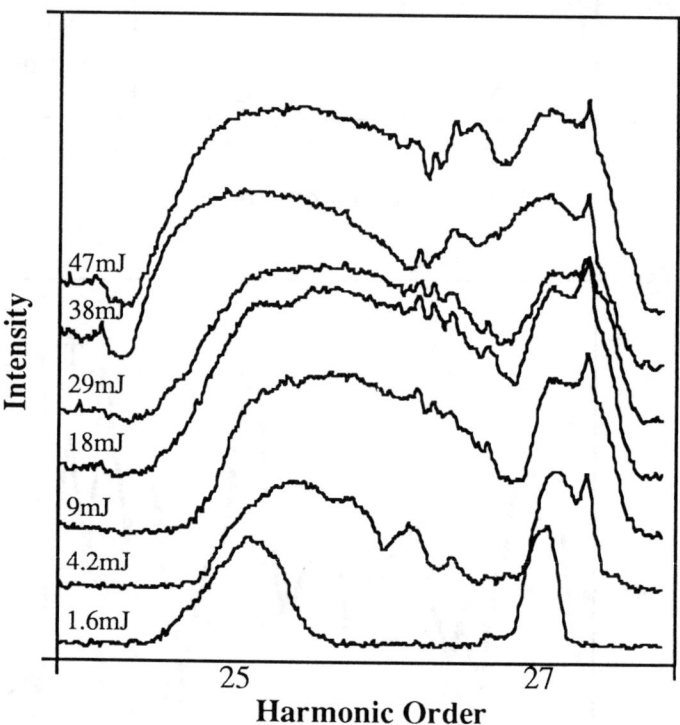

Figure 4: Spectral broadening and blue-shifting of the 25th harmonic in argon, as a function of incident laser energy. The cutoff on the right is due to a spectrometer aperture.

CONCLUSION

In conclusion, we have generated coherent x-ray pulses at wavelengths of 2.7nm, which is well within the "water window" region between 4.4nm and 2.3nm, where water is less absorbing than carbon. Our shortest observed wavelength to date of 2.7nm is the shortest wavelength coherent light generated to date. These x-ray pulses are possibly a few femtoseconds in duration. Therefore, using ultrashort excitation pulses, coherent, tunable, femtosecond, x-ray beams can be generated throughout the soft-x-ray region. In the future, this very compact femtosecond x-ray source, driven by kHz repetition rate lasers, may be very important for applications such as imaging through aqueous solutions, or time-resolved photoelectron spectroscopy of organic molecules and solids.

ACKNOWLEDGMENTS

The authors gratefully acknowledge support for this work from the National Science Foundation. We also thank Xi'an Institute of Optics and Precision Mechanics for collaboration on the microchannel plate detector used in this work. H. Kapteyn acknowledges support from an Alfred P. Sloan Foundation Fellowship.

REFERENCES

1. A. Rundquist, C. Durfee, Z. Chang, G. Taft, E. Zeek, S. Backus, M. M. Murnane, H. C. Kapteyn, I. Christov, V. Stoev, Applied Physics B **65**, 161 (1997).
2. A. McPherson, G. Gibson, H. Jara, U. Johann, T. S. Luk, I. A. McIntyre, K. Boyer, C. K. Rhodes, J. Opt. Soc. Am B **4**, 595 (1987).
3. A. L'Huillier, P. Balcou, Phys. Rev. Lett. **70**, 774 (1993).
4. J. J. Macklin, J. D. Kmetec, C. L. Gordon, III, Phys. Rev. Lett. **70**, 766 (1993).
5. J. L. Krause, K. J. Schafer, K. C. Kulander, Phys. Rev. Lett. **68**, 3535 (1992).
6. K. Schafer, K. Kulander, Physical Review Letters **78**, 638 (1997).
7. K. C. Kulander, K. J. Schafer, J. L. Krause, in *NATO 3rd Conference on Super Intense Laser-Atom Physics* Han-sur-Lesse, Belgium, 1993).
8. M. Lewenstein, P. Balcou, M. Y. Ivanov, P. B. Corkum, Physical Review A **49**, 2117 (1993).
9. J. Zhou, J. Peatross, M. M. Murnane, H. C. Kapteyn, I. P. Christov, Physical Review Letters **76**, 752 (1996).
10. I. P. Christov, J. P. Zhou, J. Peatross, A. Rundquist, M. M. Murnane, H. C. Kapteyn, Physical Review Letters **77**, 1743 (1996).
11. I. P. Christov, M. M. Murnane, H. C. Kapteyn, Physical Review Letters **78**, 1251 (1997).
12. Z. Chang, A. Rundquist, H. Wang, H. C. Kapteyn, M. M. Murnane, to be published in Physical Review Letters (1997).
13. B. Chang, P. Bolton, D. Fittinghoff, Phys. Rev. A **47**, 4193 (1993).
14. M. Ammosov, N. Delone, Sov. Phys. JETP **64**, 1191 (1986).
15. J. Zhou, C. P. Huang, C. Shi, M. M. Murnane, H. C. Kapteyn, Opt. Lett. **19**, 126 (1994).
16. J. Zhou, C. P. Huang, M. M. Murnane, H. C. Kapteyn, Optics Letters **20**, 64 (1994).
17. C. Spielmann et al., in *Quantum Electronics and Laser Science Conference*, Paper QPD 4 (OSA, Baltimore, MD, 1997)

Compression of High Energy Femtosecond Pulses by the Hollow Fiber Technique: Generation of sub-5-fs Multigigawatt Pulses

M. Nisoli[*], S. Stagira[*], S. De Silvestri[*], O. Svelto[*],
S. Sartania[†], Z. Cheng[†], M. Lenzner[†], Ch. Spielmann[†], and F. Krausz[†]

[*]Centro di Elettronica Quantistica e Strumentazione Elettronica - CNR, Dipartimento di Fisica, Politecnico, P.za L. da Vinci 32, 20133 Milano Italy
[†]Abteilung Quantenelektronik und Lasertechnik Technische Universität Wien, Gusshausstr. 27, A-1040 Wien, Austria

Abstract. Powerful techniques for spectral broadening and ultrabroad-band dispersion control, which allow compression of high energy femtosecond pulses to duration of a few optical cycles, are presented. Spectral broadening by propagation along hollow fiber filled with noble gases is studied under two excitation regimes with high energy input pulses of 140-fs and 20-fs duration respectively. With 20-fs input pulses and under optimum compression conditions we show a pulse shortening down to 4.5 fs with output energy up to 70 µJ using a high throughput prism-chirped mirror delay line.

Overview of Short Pulse Generation

Femtosecond laser pulses are essential to the investigation of many important processes in physics, chemistry, biology and electronics. A significant step forward in the generation of femtosecond pulses was made in 1981 with the development of the colliding pulse mode-locked (CPM) dye laser (1), with the first demonstration of sub-100-fs laser pulses. Pulses as short as 27 fs were generated in 1984 using a prism-controlled CPM laser (2). The evolution of the femtosecond dye laser technology during the 1980s has revolutionized molecular and condensed-matter spectroscopy. An extremely important breakthrough in the technology of femtosecond lasers was achieved in 1991 with the first demonstration of the self-mode-locked Ti:sapphire laser by Sibbett et al. (3). This development determined the renaissance of solid state lasers. Since then, a dramatic reduction in achievable pulse duration was obtained. Pulses as short as 7.5 fs have been directly generated by a Kerr-lens mode-locked Ti:sapphire oscillator by using chirped mirrors for intracavity dispersion control (4) and, more recently, self-starting 6.5-fs pulses were generated by using prism pairs in combination with a double-chirped mirror and a broadband semiconductor saturable absorber mirror (5). In parallel with this progress in femtosecond pulse generation, the introduction of the technique of chirped-pulse amplification (CPA) (6) has made possible the amplification of ultrashort pulses to unprecedented power levels. Ti:sapphire amplifiers seeded by femtosecond laser oscillators can now generate pulses of 20-30 fs with gigawatt (7,8) or terawatt (9,11) peak power at repetition rates in the kHz and 10 Hz regimes, respectively. Ultrashort pulses can also be generated by extracavity compression techniques. In 1981 Nakatsuka et al. (12) introduced a method for optical pulse compression based on the interplay between self-phase modulation (SPM) and group velocity dispersion (GVD) that arise during the propagation of short light pulses in single-mode optical fibers. The propagation of the pulse along the optical fiber spectrally broadens and chirps the laser pulse as a result of the combined action of SPM and GVD. The spectrally broadened pulse is subsequently compressed in a carefully designed optical dispersive delay line. The increased spectral bandwidth of the output pulse enables to generate a compressed pulse shorter in time than the input pulse. Using this technique pulses as short as 6 fs were obtained in 1987 from 50-fs pulses generated by a CPM dye laser (13). More recently 13-fs pulses from a cavity-dumped Ti:sapphire laser were compressed to 4.6 fs with the same technique using a prism chirped mirror Gires-Tournois interferometer compressor (14). However, the use of single-mode optical fibers limits the pulse energy to a few nanojoules. In 1996 a powerful pulse compression technique based on spectral broadening in a hollow fiber filled with noble gases has demonstrated the capability of handling high energy pulses (sub-mJ range) (15). This technique presents the advantages of a guiding element with a large diameter mode and of a fast nonlinear medium with high damage threshold (multiphoton ionization). The implementation of the hollow-fiber technique using 20-fs seed pulses from a Ti:sapphire system (7) and a high-throughput broadband prism chirped-mirror dispersive delay line has led to the generation of multigigawatt 4.5-fs pulses (16).

In this review paper we analyze the hollow-fiber technique in two different input pulse duration regimes: 140 and 20 fs. Compression experiments with high-energy femtosecond pulses are presented and the optimum conditions for pulse compression are outlined considering the role of SPM and gas dispersion.

Hollow-Fiber Mode Characteristics

Wave propagation along hollow fibers can be thought of as occurring by grazing incidence reflections at the dielectric inner surface. Since the losses caused by these reflections greatly discriminate against higher-order modes, only the fundamental mode, with large and scalable size, will be transmitted through a sufficiently long fiber. For fused silica gas-filled hollow fibers the fundamental mode is the EH_{11} hybrid mode, whose intensity profile as a function of the radial coordinate r is given by:

$$I_0(r) = I_0 \, J_0^2(2.405\, r/a) \tag{1}$$

where I_0 is the peak intensity, J_0 is the zero-order Bessel function and a is the capillary radius. The phase constant β and the field attenuation constant α for this mode are given by (17):

$$\beta = \frac{2\pi}{\lambda}\left[1 - \frac{1}{2}\left(\frac{2.405\lambda}{2\pi a}\right)^2\right] \quad (2)$$

$$\frac{\alpha}{2} = \left(\frac{2.405}{2\pi}\right)^2 \frac{\lambda^2}{2a^3} \frac{v^2+1}{\sqrt{v^2-1}} \quad (3)$$

where λ is the laser wavelength in the gas medium and v is the ratio between the refractive indices of the external (fused silica) and internal (gas) media. Since α is proportional to λ^2/a^3, the losses can be made arbitrarily small by choosing a sufficiently large capillary radius. By proper mode matching, the incident radiation can be dominantly coupled into the fundamental propagation mode. In this way the discrimination of the fundamental mode from higher order-modes is very high. Moreover, the hollow fiber preserves the polarization of the input radiation. The spatial coherence of the beam exiting the fiber was investigated by measuring the transverse profile at different distances from the end tip of the fiber using a CCD matrix detector. The measured beam profiles were compared with the calculated beam profiles assuming free-space propagation of a beam with an initial shape equal to that of the EH_{11} mode of the fiber. The full width at half maximum of the measured beam profiles is plotted in Fig. 1 as a function of the distance from the fiber end tip together with the calculated values. The good agreement between experimental and theoretical results demonstrates that the output beam is diffraction limited.

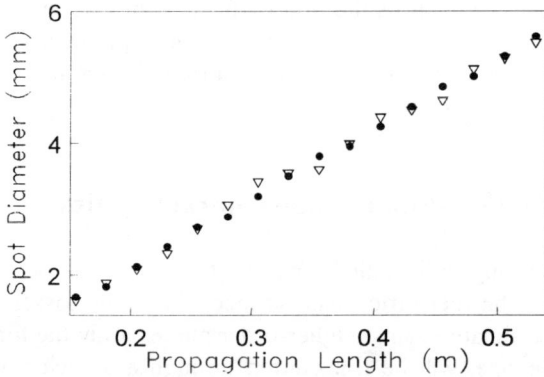

FIGURE 1. Measured full width at half maximum of the output beam intensity profile (triangle) as a function of the distance from the tip of fiber; calculated values (dots) from free space propagation of a beam with an initial profile equal to the EH_{11} mode of the fiber.

Pulse Propagation in a Gas-Filled Hollow Fiber

Pulse propagation in a gas-filled hollow fiber can be described by the same equations used for the case of optical fibers. Considering propagation along the z-direction, the electric field for the fiber modes can be written as follows (18):

$$E(\mathbf{r},\omega) = F(x,y) A(z,\omega) \exp[i\beta(\omega)z] \quad (4)$$

where $A(z,\omega)$ is the mode-amplitude, $F(x,y)$ the mode-transverse-distribution and $\beta(\omega)$ the mode-propagation constant. The propagation equation for the guided field splits in two equations for $A(z,\omega)$ and $F(x,y)$. Taking the inverse Fourier transform it is possible to write the propagation equation for the mode-amplitude in the time domain in a reference frame moving with the pulse at the group velocity v_g. Let us introduce a temporal scale normalized to the initial pulse half-width (at 1/e-intensity point) T_0 through $\tau = (t-z/v_g)/T_0$ and a normalized amplitude U using the definition:

$$A(z,\tau) = \sqrt{P_0} \exp(-\alpha z/2) U(z,\tau) \quad (5)$$

where P_0 is the peak power of the incident pulse and the exponential factor accounts for fiber losses. Assuming an instantaneous nonlinear response, the normalized amplitude U satisfies the following propagation equation (18):

$$i\frac{\partial U}{\partial z} = \frac{\mathrm{sgn}(\beta_2)}{2 L_d} \frac{\partial^2 U}{\partial \tau^2} + i\frac{\mathrm{sgn}(\beta_3)}{6 L'_d} \frac{\partial^3 U}{\partial \tau^3} - \frac{e^{-\alpha z}}{L_{nl}}\left[|U|^2 U + \frac{2i}{\omega_0 T_0}\frac{\partial}{\partial \tau}(|U|^2 U)\right] \quad (6)$$

where $\beta_2 = d^2\beta/d\omega^2$ is the GVD of the fiber, $\beta_3 = d^3\beta/d\omega^3$ is the third order dispersion, $\mathrm{sgn}(\beta_i) = \pm 1$ depending on the sign of β_i, $L_d = T_0^2/|\beta_2|$ and $L'_d = T_0^3/|\beta_3|$ are the second and third order dispersion lengths, $L_{nl} = 1/\gamma P_0$ is the nonlinear length. The nonlinear coefficient γ is given by $\gamma = n_2 \omega_0/c A_{\mathrm{eff}}$, where ω_0 is the center laser frequency, c is the speed of light in vacuum, n_2 is the nonlinear index coefficient ($n = n_0 + n_2 I$, where I is the field intensity) and A_{eff} is the effective mode area. The first term in the squared bracket of Eq. 6 describes the self-phase modulation process, the second one governs the self-steepening effect. The dispersion lengths L_d, L'_d and the nonlinear length L_{nl} give the length scales over which the dispersive or nonlinear effects become important for pulse evolution along the fiber. Optimum exploitation of the interplay between GVD and SPM for the generation of linearly chirped pulses calls for an optimum propagation length $L_{opt} \approx (6 L_{nl} L_d)^{1/2}$ (19), which shows the relative importance of both the GVD and SPM effects for good pulse compression. In the following section we report on the experimental results of spectral broadening in a hollow fiber filled with argon and krypton for 140-fs and 20-fs input pulses. Results of computer simulation of the pulse propagation along the fiber obtained by numerically solving Eq. 6 will be discussed.

Experimental Set-Up

A first set of experiments was carried out using a Ti:sapphire laser with chirped-pulse amplification, which produces 140-fs pulses at 780 nm, with energy up to 660 µJ at a 1-kHz repetition rate. A second set of experiments was performed using a 1-kHz multipass Ti:sapphire amplifier seeded by sub-10-fs pulses generated by a mirror-dispersion-controlled Ti:sapphire oscillator (7). The output pulses have a duration of 20 fs, energy up to 300 µJ and spectrum centered on 780 nm. Fused silica hollow fibers of 60-70 cm length and 70-80 µm capillary radius were used. The fiber was kept straight in a V-groove made in an aluminum bar placed in a pressurized chamber, with 1-mm thick fused silica windows coated for broadband antireflection. The chamber was first evacuated and then filled with different gases, such as argon and krypton at different pressures. The overall transmission of the pressurized chamber was 50-65 %, for the different fibers used in the experiments.

Spectral Broadening with noble gases

A first set of experiments was performed with argon. A typical shape of the output spectrum measured by using 140-fs input pulses at peak power P_0=3.5 GW and for a gas pressure p= 4 bar is shown in Fig. 2(a). The spectrum presents amplitude modulations which are characteristic of a pure SPM process. Assuming for argon n_2/p=9.8×10^{-24} m^2/W bar (20) one obtains $L_{nl} \approx$ 0.66 cm for a capillary radius of 70 µm. Upon considering the contributions to second order dispersion from both gas (21) and fiber (18) one gets a value of $\beta_2 \approx$ 48 fs^2/m which gives $L_d \approx$ 131 m. These values provide $L_{opt} \approx$ 2.3 m, which turns out to be much longer than the fiber length (70 cm). Therefore the effect of dispersion is almost negligible as observed experimentally. Using 20-fs input pulses the spectral broadening presents a different behavior. A typical shape of the spectrum measured at the output of the fiber for an argon pressure p=3.3 bar and an input peak power P_0 =4 GW is shown in Fig. 2(b). The shape of the spectrum is much more uniform than that obtained with 140-fs input pulses. This indicates that, at this shorter pulse duration, gas dispersion, besides SPM, is also playing an important role during pulse propagation. In fact, in this case one obtains $L_{nl} \approx$ 0.92 cm (for a capillary radius of 80 µm) and $L_d \approx$ 3.2 m ($\beta_2 \approx$ 40 fs^2/m), so that the optimum length turns out to be $L_{opt} \approx$ 42 cm, close to the fiber length used in the experiments.

A second set of experiments was performed with krypton. A typical broadened spectrum measured at p=2 bar and P_0=3.5 GW in the case of 140-fs input pulses is shown in Fig. 3(a).

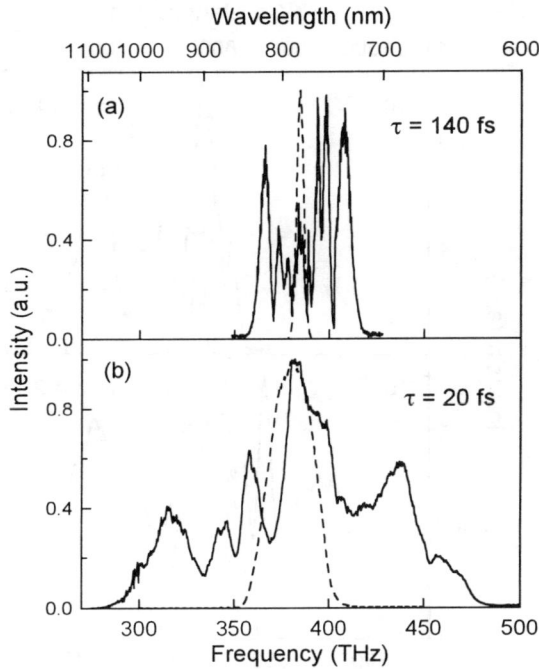

FIGURE 2. Spectral broadening in argon (a) at p=4 bar and P_0=3.5 GW obtained with 140-fs input pulses; (b) at p=3.3 bar and P_0=4 GW obtained with 20-fs input pulses. The spectra of the input pulses are shown as dashed lines.

The shape is somewhat asymmetric and more uniform compared to that observed with argon. The greater extension to the blue side of the input laser wavelength is due to self-steepening. Assuming for krypton n_2/p=2.78×10^{-23} m^2/W bar [20] one obtains $L_{nl} \approx$ 0.47 cm for a capillary radius of 70 µm. Upon considering the contributions to second order dispersion one gets a value of $\beta_2 \approx$ 60 fs^2/m which gives $L_d \approx$ 105 m. These values provide $L_{opt} \approx$ 1.7 m, which turns out to be longer than the fiber length used in the experiment. Therefore the effect of dispersion is almost negligible as in the case of argon. A typical broadened spectrum measured using 20-fs input pulse for a krypton pressure p=2.1 bar and an input peak power P_0=2 GW is shown in Fig. 3(b). In this case one obtains $L_{nl} \approx$ 1.1 cm (for a capillary radius of 80 µm) and $L_d \approx$ 2 m ($\beta_2 \approx$ 60 fs^2/m), so that the optimum length turns out to be $L_{opt} \approx$ 36 cm, which is closer to that used in the experiments (60 cm). Figure 4 shows the results of computer simulations of 20-fs pulse propagation along the hollow fiber, in the experimental conditions described above, obtained by numerically solving Eq. 6 using the split-step Fourier method. To obtain a qualitative agreement with the measured spectral broadening we have assumed a nonlinear index coefficient n_2 three times smaller than the quoted value.

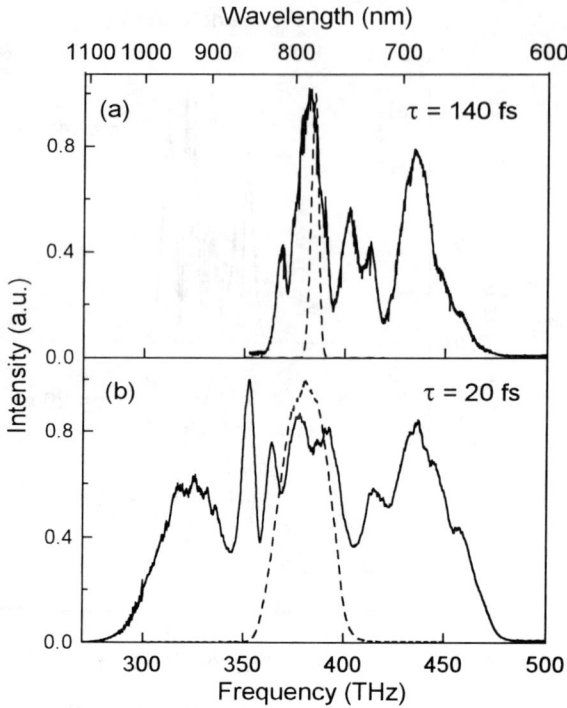

FIGURE 3. Spectral broadening in krypton (a) at p=2 bar and P_0=3.5 GW obtained with 140-fs input pulses; (b) at p=2.1 bar and P_0=2 GW obtained with 20-fs input pulses. The spectra of the input pulses are shown as dashed lines.

We can attribute the difference to the uncertainty of the quoted n_2 value, especially for femtosecond pulses.

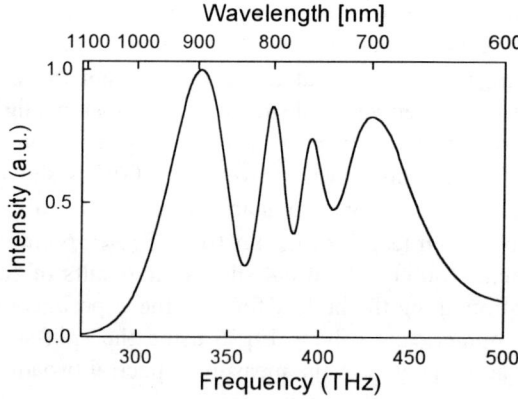

FIGURE 4. Spectral broadening obtained by computer simulations of the pulses at the output of a hollow fiber filled with krypton using 20-fs input pulses.

Pulse Compression

Pulse compression experiments with gas-filled hollow fibers were first performed with 140-fs input pulses. Using argon and krypton generation of transform-limited 18-fs and 10-fs pulses respectively, with energies up to 240 µJ was demonstrated (15). In the following we will report on pulse compression experiments performed using 20-fs input pulses.

The frequency broadened pulses emerging from the hollow fiber were collimated and propagated through a dispersive delay line which introduced an appropriate group delay $T_g = d\phi/d\omega$ ($\phi(\omega)$ is the phase retardation). To compress the spectrally broadened pulses to their transform limit, T_g must be precisely controlled over a bandwidth exceeding 130 THz. We have designed a compressor based on chirped mirrors and two pairs of fused silica prisms of small apex angle (20°). The calculated group-delay versus frequency curve fits the required delay variation well over a bandwidth of 120 THz, for a prism separation set at 1.8 m. The dispersive delay line has a high transmission (>85%) over the wavelength range of 630-1030 nm. The compressed pulses were measured by interferometric second harmonic autocorrelation. The second order interferometric autocorrelation signal is generated in a 15-µm-thick BBO frequency doubling crystal and recorded at a scan rate of 1 Hz. To evaluate the pulse duration, we took the inverse Fourier transform of the spectrum and assumed, as free parameter, some residual cubic phase distortion, ($d^3\phi/d\omega^3$).

By filling the hollow fiber with argon, pulses as short as 5.3 fs, with energy up to 40 µJ were obtained. The minimum pulse duration, as calculated upon assuming optimum phase distortion compensation, was 5.2 fs. Therefore the pulses can be considered to be almost transform limited. Figure 5 shows the measured second harmonic interferometric autocorrelation trace obtained by filling the hollow fiber with krypton.

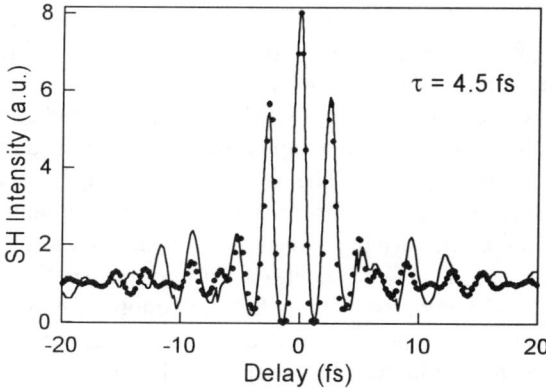

FIGURE 5. Measured (solid line) and calculated (dots) interferometric autocorrelation trace of the compressed pulses obtained respectively with krypton using 20-fs input pulses.

From this trace, a pulse duration of 4.5 fs (FWHM) was evaluated by the previously described method, upon assuming a residual cubic phase distortion $|d^3\phi/d\omega^3| = 10$ fs^3.
Pulse energy after compression was 20 µJ. These pulses represent the shortest generated to date at tens of microjoule energy level. The minimum pulse duration, estimated from the spectrum shown in Fig.3(b) was 4.3 fs; therefore the pulses are almost transform limited.

The potential scalability of this system to higher pulse energies is an important issue considering the current availability of 20-fs laser pulses with peak powers up to the terawatt level or more. An increase of the input energy and a reduction of the nonlinearity by lowering the argon pressure allowed in a recent experiment an increase of the compressed pulse energy by an order of magnitude. The input pulses with an energy of about 1 mJ were broadened in a 250 µm thick hollow waveguide filled with argon at a pressure of 0.5 bar. Ultrabroadband chirped dielectric mirrors compressed the pulses in five reflections. The setup produced 5-fs, 0.1-TW pulses at 1 kHz (22).

Applications

The generation of diffraction-limited multigigawatt light pulses in the single-cycle regime promises to be a powerful tool for precisely triggering and controlling the evolution of atomic systems in strong laser fields. Many applications will benefit from this capability in the future. Perhaps one of the most challenging one is the generation and control of high-order harmonic radiation in the soft X-ray spectral region (23). Coherent X-ray are significant for a number of fields in science and technology because of their potential for providing high spatial and temporal resolution. Intense, spatially coherent laser-generated X-rays in the "water-window" (2.33-4.36 nm) will make significant impacts on high-resolution X-ray microscopy and holography of living cell in biology (24). Since the duration of these X-ray pulses is expected to be shorter than that of the exciting light pulse, their use for time-resolved X-ray diffraction will allow studying chemical reactions and structural changes of condensed matter with never-before-achieved time resolution. Coherent extreme ultraviolet radiation extending into the water window has been generated at a repetition rate of 1 kHz by focusing the 5-fs, 0.1-TW pulses in a helium gas jet. The incident light field contains just a few oscillations, which results in the emission of a x-ray super continuum. Owing to the extremely short rise time of the driving pulses neutral atoms can be exposed to very high fields before ionization. As a result, the observed spatially coherent x-ray continuum extends to wavelength as short as 2.4 nm (photon energy: 0.5 keV) and is delivered in a well-collimated beam. This compact system holds promise as a laboratory-scale source for biological holography and nonlinear optics in the x-ray regime (25).

REFERENCES

1. Fork, R. L., Green, B. I., and Shank C. V., *Appl. Phys. Lett.* **38**, 671-672 (1981).
2. Valdmanis, J. A., Fork, R. L., and Gordon, J. P., *Opt. Lett.* **10**, 131-133 (1985).
3. Spence, D. E., Kean, P. N., and Sibbett, W., *Opt. Lett.* **16**, 42-44 (1991).
4. Xu, L., Spielmann, Ch., Krausz, F., and Szipöcs, R., *Opt. Lett.* **21**, 1259-1261 (1996).
5. Jung, I. D., Kärtner, F. X., Matuschek, N., Sutter, D. H., Morier-Genoud, F., Zhang, G., Keller, U., Scheuer, V., Tilsch, M., and Tschudi, T., *Opt. Lett.* **22**, 1009-1011 (1997).
6. Strickland, D., and Mourou, G., *Opt. Commun.* **56**, 219-221 (1985).
7. Lenzner, M., Spielmann, Ch., Wintner, E., Krausz, F., Schmidt, and A. J., *Opt. Lett.* **20**, 1397-1399 (1995).
8. Backus, S., Peatross, J., Huang, C. P., Murnane, M. M., and Kapteyn, H. C., *Opt. Lett.* **20**, 2000-2002 (1995).
9. Zhou, J., Huang, C. P., Murnane, M. M., and Kapteyn, H. C., *Opt. Lett.* **20**, 64-66 (1995).
10. Barty, C. P. J., Guo, T., Le Blanc, C., Raksi, F., Petruck, C. R. -P, Squier, J., Wilson, K. R., Yakovlev, V. V., Yamakawa, K., *Opt. Lett.* **21**, 668-670 (1996).
11. Chambaret, J. P., Le Blanc, C., Chériaux, G., Curley, P., Darpentigny, G., Rousseau, P., Hamoniaux, G., Antonetti, A., and Salin, F., *Opt. Lett.* **21**, 1921-1923 (1996).
12. Nakatsuka, H., Grischkowsky, D., and Balant, A. C., *Phys. Rev. Lett.* **47**, 910-913(1981).
13. Fork, R. L., Brito Cruz, C. H., Becker, and Shank, *Opt. Lett.* **12**, 483-485 (1987).
14. Baltuska, A., Wei, Z., Pshenichnikov, M. S., and Wiersma, D. A., *Opt. Lett.* **22**, 102-104 (1997); Baltuska, A., Wei, Szipöcs, R., Z., Pshenichnikov, M. S., and Wiersma, D. A., *Appl. Phys. B* **65**, 175-188 (1997).
15. Nisoli, M., De Silvestri, S., and Svelto, O., *Appl. Phys. Lett.* **68**, 2793-2795 (1996).
16. Nisoli, M., De Silvestri, S., Svelto, O., Szipöcs, R., Ferencz, K., Spielmann, Ch., Sartania, S., and Krausz, F., *Opt. Lett.* **22**, 522-524 (1997); Nisoli, M., Stagira, S., De Silvestri, S., Svelto, O., Sartania, S., Cheng, Z., Lenzner, M., Spielmann, Ch., and Krausz, *Appl. Phys. B* **65**, 189-196 (1997).
17. Marcatili, E. A. J., and Schmeltzer, R. A., *Bell Syst. Tech. J.* **43**, 1783-1809 (1964).
18. Agrawal, G. P., *Nonlinear Fiber Optics*, San Diego: Academic Press, 1995.
19. Tomlinson, W. J., Stolen, R. H., and Shank, C. V., *J. Opt. Soc. Am. B* **1**, 139-149 (1984).
20. Lehmeier, H. J., Leupacher, W., and Penzkofer, A., *Opt. Commun.* **56**, 67-72 (1985).
21. Dalgarno, A., and Kingston, A. E., *Proc. R. Soc. London* Ser. A **259**, 424-429 (1966).
22. Sartania, S., Cheng, Z., Lenzner, M., Tempea, G., Spielmann, Ch., Krausz, F., and Ferencz, K., *Opt. Lett.* **22**, October 15 (1997).
23. Corkum, P. B., *Opt. Photon. News* **6**, 18-24 (1995).
24. Stuhrmann, H. B., ed., *Uses of synchrotron radiation in biology*, New York: Academic Press, 1982.
25. Spielmann, Ch., Burnett, N. H., Sartania, S., Koppitsch, R., Schnurer, M., Kan, C., Lenzner, M., Wobrauschek, P., and Krausz, F., *Science*, October 24 (1997) to be published

The Interaction of Atomic Clusters with Intense Laser Fields

M. H. R. Hutchinson*, T. Ditmire, J. W. G. Tisch, E. Springate,
J. P. Marangos, M. B. Mason and R.A.Smith

*The Blackett Laboratory, Prince Consort Road,
Imperial College of Science, Technology, and Medicine
London, SW7 2BZ, UK*
**Central Laser Facility, Rutherford Appleton Laboratory,
Chilton, Didcot, OX11 0QX, UK*

Abstract. The dynamics of the heating and dissociation of noble gas clusters when irradiated by intense, femtosecond laser pulses is described. The clusters explode with the generation of highly energetic electrons and ions. A plasma model of the ionised cluster is discussed.

INTERACTION WITH ATOMIC CLUSTERS

Atomic clusters have been studied by chemists and physicists because of the unique position that clusters hold as an intermediate state between molecules and solids. Recently, there has been intense activity in extending these studies to very high intensity, ultrashort laser pulses with peak laser intensities $>10^{15}$ Wcm^{-2} and pulse widths of 0.1 to 10 ps (1-6). Since the short pulses used are comparable to or shorter than the disassembly times of a cluster in the laser field [3], the entire laser pulse interacts with an inertially confined body of atoms.

The dramatic difference between the nature of intense laser interactions with atoms and solid density plasmas raises the question. Do clusters of a few hundred to a few thousand atoms which are small compared to a laser wavelength behave like smaller molecules in strong fields or more like the energetic plasmas produced from solids? Initial studies of a variety of cluster sizes using lasers of wavelengths between 248nm and 1054 nm indicated that the intense laser interaction with clusters was much more energetic then interactions at similar intensities with atoms or small molecules. These initial studies found that intense irradiation of a medium of clusters resulted in very intense x-ray emission. In fact, X-rays with energies up to 5 keV were observed in a Xe cluster gas jet irradiated by 248 nm pulses at intensity $>10^{18}$ Wcm^{-2} (1). These studies indirectly indicated that the clusters were absorbing substantial fractions of laser energy and were producing hot electrons and high charged ions capable of producing the observed radiation.

Clusters of the heavy noble gases, Xenon, Krypton and Argon form clusters when the gas is cooled by adiabatic expansion through a nozzle and are thus convenient systems for experimental study. To study their dissociation, a beam of atomic clusters, produced in the expansion of a high-pressure gas into vacuum, is irradiated by a focused, high-intensity, femtosecond laser beam. The electrons and ions expelled from the clusters with velocities perpendicular to both the cluster beam and the laser beam propagate along a time-of-flight tube and are detected by a micro-channel plate detector (MCP). The ion energies may then determined by measurement of their flight time and the electron energies are found by measuring the decrease in the MCP signal when a retarding voltage is applied to grids placed between the laser focus and the MCP.

The laser used is a high-power Ti:sapphire CPA laser which delivers 40mJ pulses of 150 fs duration at a wavelength of 780 nm which is focused to yield a peak focused intensity of ~2 x 10^{16} Wcm^{-2} with 20 mJ of laser intensity. The noble gas clusters are produced in a jet by a pulsed solenoid gas valve.

The presence of clusters in the gas jet and a determination of their average size may be obtained from Rayleigh scattering measurements. These measurements indicate that clusters of a detectable size begin to form at pressures around 1000 mbar in xenon and 2000 mbar in krypton. The scattered signal from the Xe and Kr clusters displays a p_0^3 dependence with increasing backing pressure. This is consistent with simple scaling arguments from which it is possible to estimate the cluster size as a function of the gas pressure. It may be inferred that, at $p_0 = 4$ bar, an average xenon cluster will contain ~1600 atoms and have a radius $R_c \approx 46 \pm 10$ Å, while a krypton cluster will typically contain ~500 atoms and have a radius $R_c \approx 28 \pm 7$ Å.

Electron Energy Spectra

The measured energy spectrum of electrons emitted along the direction of the laser polarization during the irradiation of clusters of ~2100 xenon atoms with an intensity of 1.5 x 10^{16} Wcm^{-2} is shown in figure 1. There are two distinct features in the electron energy spectrum. The first, broad peak consists of so-called 'warm electrons' with energies ranging from 0.1 to 1 keV. A second, sharper peak (referred to here as the 'hot electrons) appears at 2.5 keV. The most remarkable aspect of this energy distribution are the high electron kinetic energies, with a large fraction of the electrons having energies between 2 and 3 keV. Previous measurements of ATI spectra from single atoms at this intensity and pulse duration have indicated that the vast majority of electrons produced have energies below 100 eV (7). Only a very small fraction of electrons (typically 10^{-3} to 10^{-4}) have higher energy, with no detectable electrons having an energy of above 1 keV. The spectrum observed from

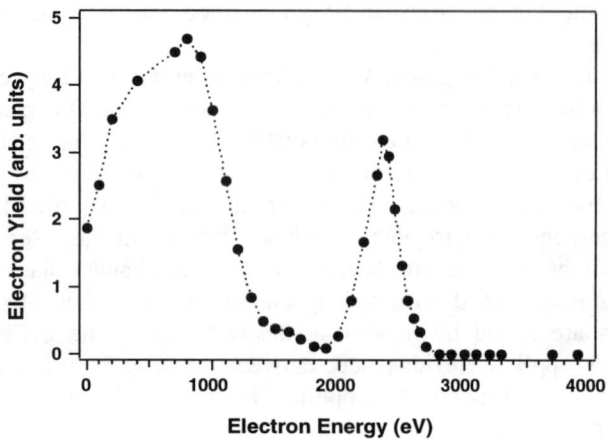

FIGURE 1. Electron energy distribution from xenon clusters for a peak intensity of 1.5×10^{16} Wcm^{-2}.

Xe clusters clearly indicates a much greater coupling of laser energy to electrons than is present during the irradiation of single atoms. Furthermore, this spectrum indicates that the laser - cluster interaction produces even hotter electron temperatures than a laser - solid interaction at this intensity, where electron temperatures of 100 - 500 eV are typical (8). The yield of hot electrons as a function of gas-jet backing pressure provides more evidence that this signal results from the interaction of the laser with clusters rather than atoms. There is close correlation in the pressure at the onset of hot electron production with the onset of clustering in xenon; both hot electron production and clustering have a sharp onset with a gas-jet backing pressure of 1 bar. The presence of two distinct peaks in the electron energy spectrum suggests that these two groups may be produced under different conditions at different times in the cluster expansion. This assumption is supported from the angular dependence of the electron emission with respect to the laser polarization. The hot electron emission is completely isotropic while warm electron emission is peaked along the laser polarisation with a width of about 60°. Both these distributions are significantly different from the angular distributions associated with single atoms; electrons from high-order ATI normally have a much narrower angular distribution. In high field tunnelling ionisation, the narrow angular distribution arises from the much higher tunnelling rate in the direction of the laser field. The electrons observed from clusters cannot therefore be interpreted as simply resulting from the tunnel ionisation of individual atoms. Some rescattering of the electrons by ions in the cluster is necessary to explain the broadening observed in the warm electron distribution. The isotropic distribution of the hot electrons indicates that these electrons have undergone many electron-ion collisions in the laser field, completely randomising their velocity distribution.

Ion Energy Spectra

The remarkably high energies of the electrons produced in the intense laser-cluster interaction suggests that highly charged ions with large kinetic energies may also be ejected from the cluster. Charge separation of these hot electrons will inevitably accelerate the cluster ions to high velocities.

The energy spectrum of ions resulting from the interaction of 2500-atom Xe clusters with a laser pulse of intensity ~2 x 10^{16} Wcm^{-2} is shown in figure 2.

The most remarkable aspect of this energy distribution is the presence of ions with energies up to 1 MeV. This energy is four orders of magnitude higher than has previously been observed in the Coulomb explosion of molecules (9) and about 1000 times higher than the energy of Ar ions ejected in the disintegration of small clusters of up to six argon atoms irradiated at 10^{15} Wcm^{-2} (2). The average ion energy of this distribution is 45 ± 5 keV. Thus the average laser energy deposited per ion is also substantial. In addition it is found that the ion energy distribution is isotropic with respect to the direction of laser polarisation, apparently a consequence of a spherically exploding cluster.

Charge State Distributions

A striking feature of the ions produced from the cluster explosion is ionisation to very high charge states, (1,10). The charge state distributions are shown in figure 3. For high energy Xe ions, the peak is at Z = 18^+ to 25^+, with some ions having charge states as high as 40^+. The peak for high energy krypton ions is at 12^+ to 17^+ with the highest charge state present being around Z = 25^+. These are much higher charge states than those expected from field ionisation of single atoms at these intensities.

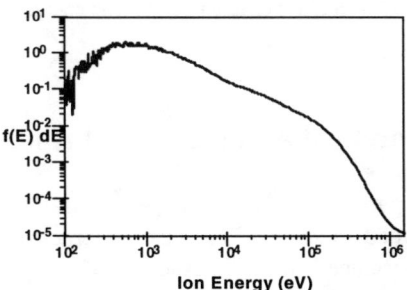

FIGURE 2. Ion energy spectrum from clusters of 2500 Xe atoms at 2 x 10^{16} Wcm^{-2}.

One would expect to see charge states up to 12^+ in xenon and 8^+ in krypton at an intensity of 2×10^{16} Wcm^{-2} (10). Ionisation to Xe^{40+} would require an intensity of nearly 10^{20} Wcm^{-2} if the ionisation were due to tunnel ionisation alone. In the cluster, high temperature electrons strip the ions to higher charge states by collisional ionisation. The ion charge state depends only weakly on ion kinetic energy, contrary to what would be expected from a simple Coulomb explosion.

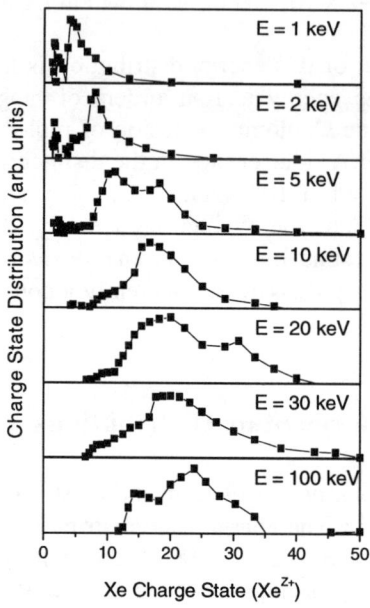

FIGURE 3. Measured charge state distribution of 2500-atom Xe irradiated with a peak intensity of 2×10^{16} Wcm^{-2} for different ion kinetic energies.

Theoretical And Numerical Analysis

The very high energy particles observed experimentally are dramatically different then those typically produced in strong field laser interactions with molecules and are more typical of particles produced in the interaction of a high intensity laser with solid density plasmas. This suggests that the appropriate way to explain the exploding cluster behaviour is to treat it as a small microplasma. Consequently one may treat the ionised cluster as a classical, spherical plasma ball of uniform density. This picture of the cluster implies a number of interesting consequences. Firstly, because of the high electron and ion densities within the cluster, electron collisional processes will be very important. In particular, the electrons will undergo rapid heating by the laser field due to electron-ion collisions (inverse bremsstrahlung). This process converts the coherent oscillation energy of the electron cloud to random thermal energy. Electron collisional ionisation will also be important, stripping the constituent atoms to very high charge states.

The second consequence is that the cluster, which becomes conducting once some electrons have been liberated by ionisation, will exhibit some of the optical properties of metallic clusters. In particular, a giant resonance in the optical absorption spectrum (11) occurs when the light frequency is near to the plasma frequency of the electrons in the cluster and is a result of a resonantly driven oscillation of the entire cluster electron cloud. (The actual plasma density at the resonance is dependent upon the shape of the cluster). The presence of this resonance is very important in the dynamics of the high intensity laser interaction with the cluster. The numerical model may be constructed (3) which treats the cluster as a spherical microplasma, subject to the standard processes of a laser heated plasma. Ionisation in the cluster is by tunnelling and both thermal and laser driven electron collisions. The model calculates the free streaming rate of electrons leaving the cluster, accounting for the mean free path of electrons in the cluster. Only electrons with energy sufficient to overcome the Coulomb attraction of the positively charged cluster are allowed to leave. The cluster expansion, assumed to be uniform and isotropic, is calculated accounting for hydrodynamic and Coulomb repulsion forces within the charged cluster and the electron energy distribution within the cluster is assumed to be Maxwellian.

A zero frequency approximation for the laser field may be used to account for the collective electron oscillation effects on the optical absorption of the cluster. This approximation is appropriate when the cluster is much smaller than the laser wavelength and assumes that the response of the cluster electron cloud is fast compared to the time scale of the cluster expansion dynamics. Hence, one can calculate the electric field inside the cluster using the expression for the electric field of a dielectric sphere in a uniform electric field. This is $E = E_0 \frac{3}{|\varepsilon + 2|}$, where E_0 is the laser electric field in vacuum. The cluster dielectric constant is given by the Drude model for a plasma, $\varepsilon = 1 - (n_e/n_{crit})(1 + i\nu/\omega)^{-1}$, where n_e is the electron density, n_{crit} is the electron critical density for a laser field of frequency w, and n is the electron-ion collision frequency. This formula predicts that when $n_e/n_{crit} > 6$ the electric field inside the cluster is shielded by the oscillating electron cloud and the field inside the cluster is smaller then the surrounding field in vacuum. On the other hand, E has a sharp maximum when $n_e/n_{crit} = 3$. At this point the oscillating laser field resonantly drives the cluster electron cloud and the field inside the cluster is enhanced. As a result, the free electrons in the cluster undergo rapid collisional heating because of the local increase in the field energy density.

An example of the dynamics of a xenon cluster are illustrated in figure 4 which shows the time history of a 50 Å Xe cluster (~2100 atoms) irradiated by a pulse with a peak intensity of 1×10^{16} W/cm^2. Here the electron temperature, laser pulse envelope and escape energy threshold are shown, along with the calculated rate at which electrons

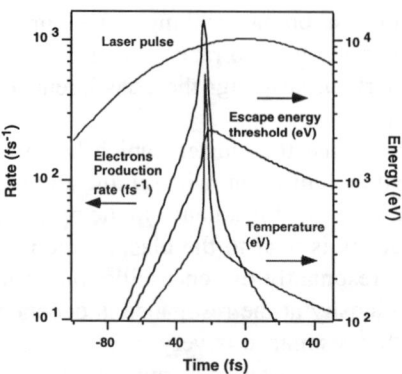

FIGURE 4. Time history of 50 Å Xe cluster (2100 atoms/cluster) at 1×10^{16} Wcm^{-2} showing the laser pulse envelope, the electron temperature, the rate at which electrons exit the cluster by free streaming and the escape energy threshold.

exit the cluster by free streaming. Some electrons escape from the cluster in the initial stages of the cluster interaction, as the electron temperature rises but, once the electron temperature peaks due to the heating by the giant resonance, the electron escape rate also peaks sharply since many of the electrons acquire enough thermal energy to overcome the space-charge forces of the cluster. This history implies that the electron energy spectrum might exhibit two features, one arising from the lower energy electrons which escape from the cluster early in the interaction and one from the hot electrons that escape during the resonance heating of the cluster, in qualitative agreement with the experimental observations.

The calculated electron distribution exhibits a two-lobed distribution and exhibits a close similarity to the measured electron distribution. The sharp is evidence for the giant resonance in the heating of the electrons in the cluster spherical microplasma. The warm electron peak is the result of collisional heating of electrons near the surface of the cluster on the rising edge of the laser pulse. These electrons have undergone a limited number of collisions, broadening the angular distribution from that of purely tunnel ionised electrons. The hot electrons result from rapid collisional heating of the remaining electrons in the bulk of the cluster later in the pulse when the electron density drops to a point to bring the heating into resonance. Consequently, their velocity distribution has been completely randomised, accounting for the isotropic distribution observed.

The production of hot electrons through inverse bremsstrahlung can drive a very energetic explosion of the cluster and charge separation of the hot electrons will inevitably drive a rapid expansion of the cluster. The explosion of the cluster can be driven by two forces. The first is the Coulomb repulsion between the highly charged ions in the cluster. If all the free electrons are retained in the cluster, the cluster is quasineutral and this force is negligible. However, the free streaming of electrons from the cluster will cause a charge build-up of the plasma sphere, and a Coulomb "pressure" will develop.

The second force important in driving the cluster explosion is the hydrodynamic pressure of the free electrons in the cluster. This force is present even if the cluster plasma remains neutral. The hot electrons in a plasma will set up a radial ambipolar potential which then accelerates the cluster ions. The pressure driving this expansion mechanism is simply given by $P_{hyd} = n_e k_B T_e$, where T_e is the electron temperature. The Coulomb explosion mechanism is similar to the mechanism that drives the explosion of small, optically ionised molecules. However, in bulk solid plasmas, the plasma remains quasineutral and the expansion is driven by the hydrodynamic force. This difference points to the question of which mechanism is responsible for the explosion of the clusters observed in these experiments. The scaling laws of these forces suggests that Coulomb explosion forces may dominate for small clusters whereas the hydodynamic force dominates for larger systems (>1000 atoms). The appearance of the high ion charge states observed (>20^+ for Xe ions) is predicted due to rapid collisional ionisation by the hot electrons within the cluster which can strip the ions to very high charge states (up to Xe^{40+}), in agreement with the experimental observations of figure 3.

REFERENCES

1. McPherson, A., Thompson, B.D., Borisov, A.B., Boyer, K., and Rhodes, C.K.,*Nature* **370**, 631, (1994).
2. Purnell, J., Snyder, E.M., Wei, S. and A.W.C. Jr., *Chem. Phys. Lett.* **229**, 333, (1994).
3. Ditmire, T., Donnelly, T., Rubenchik, A.M., Falcone, R.W. and Perry, M.D., *Phys. Rev. A* **53**, 3379, (1996).
4. Shao, Y.L., Ditmire, T., Tisch, J.W.G., Springate, E., Marangos, J.P., and Hutchinson, M.H.R., *Phys. Rev. Lett.* **77**, 3343, (1996).
5. Ditmire, T., Tisch, J.W.G. Springate, E., Mason, M.B., Hay, N., Smith, R.A., Marangos J., and Hutchinson, M.H.R., *Nature* **386**, 54, (1997).
6. Ditmire, T., Tisch, J.W.G., Springate, E., Mason, M.B., Hay, N., Marangos, J.P., and Hutchinson, M.H.R., *Phys. Rev. Lett.* **78**, 2732, (1997).
7. Walker, B., Sheehy, B., DiMauro, L.F., Agostini, P., Schafer, K.J., and Kulander K., *Phys. Rev. Lett.* **73**, 1227, (1994).
8. Shepherd R., Price, D., White W., Gordan, S., Osterheld, A. Walling R., Goldstein, W. and Stewart, R., *J. Quant. Spec. Rad. Trans.* **51**, 357, (1994).
9. Augst, S., Meyerhofer, D.D., Strickland, D., and Chin, S.L., *J. Opt. Soc. B* **8**, 858, (1991).
10. Ditmire, T., Donnelly, T., Falcone, R.W., and Perry, M.D., *Phys. Rev. Lett.* **75**, 3122, (1995).
11. Bregnacac, J. and Connerade, J.P., *J.Phys.B: At. Mol.Opt.Phys.* **27**, 3795, (1994).

Theoretical/Experimental Studies of Ultraviolet High-Power-Density Self-Trapped Channels

A. B. Borisov[1,2], B. D. Thompson[1], A. McPherson[1], F. Omenetto[1], T. Nelson[1], W. A. Schroeder[1], K. Boyer[1], and C. K. Rhodes[1,2]

[1]*Department of Physics (M/C 273), University of Illinois at Chicago, 845 W. Taylor Street, Chicago, IL 60607-7059, USA*
[2]*Center for Tsukuba Advanced Research Alliance (TARA), University of Tsukuba, 1-1-1 Tennodai, Tsukuba, Ibaraki 305, Japan*

Abstract. Experimental evidence indicates that the dynamics of channeled propagation arising from a relativistic/charge-displacement mechanism can produce exceptionally high intensities ($> 10^{20}$ W/cm^2) and power densities ($> 10^{19}$ W/cm^3) under spatially well-controlled conditions in the interior of the channel. This process opens up a new domain of high intensity interactions in which controllable field strengths in the 10 - 50 (e/a_0^2) range can be applied to materials. Furthermore, these field strengths can be developed on very fast rise times (~ 1 - 3 fs), since the dynamical response of the channel formation can be on the order of the plasma time ~$2\pi/\omega_p$. Under these conditions, matter is expected to respond in a highly complex and nonlinear way. Of particular significance are (1) the detailed dynamics of the channeling mechanism, including the wavelength scaling, and (2) the development of advanced analytical tools to represent the interactions theoretically.

I. INTRODUCTION

A mode of highly confined propagation arising from a relativistic/charge-displacement mechanism is enabling the production and study of ultrahigh power density states of matter under controlled conditions. These studies have revealed several new and unusual properties of the nonequilibrium plasmas associated with the channels. Among the most significant findings are (1) the development of stable channeled propagation for a distance exceeding 100 Rayleigh ranges [1], (2) the production of controlled power densities of ~ 3×10^{19} W/ cm^3 (~ 1 W/atom) in the channels [2], (3) evidence for an enhanced form of multiphoton coupling to clusters [3-6], and (4) strong x-ray emission with prominent nonthermal characteristics [4, 5, 7-9]. Previous studies [4, 5, 9, 10] conducted at 248 nm have examined the spatial characteristics of the channels produced by this process by two methods. They involved (a) the recording of images [9, 10] of the x-ray radiating zone produced from Xe clusters with an x-ray (~ 1 keV) pinhole camera with a resolution ~ 30 µm and (b) the measurement of spatially resolved Xe(L) spectra [4,

5]. An important conclusion derived from these data was that peak intensities of ~ 10^{20} W/cm^2 were produced in the channels. This finding was consistent with both analyses and experimental results on the characteristics of the propagation produced by the relativistic/charge-displacement mechanism [11-15].

II. STUDIES OF THE CHANNELING MECHANISM

A. Images of Channels

Images of the Thomson scattered light generated by the radiation (248 nm) propagating in the channels are now able to provide further data on the detailed morphology of the channels with substantially improved spatial resolution. Figure 1 schematically illustrates the experimental configuration used to record the 248 nm Thomson signal scattered transverse to the direction of propagation of the incident energy. The conditions of irradiation and the parameters of the pulsed-valve used to generate the Xe cluster target are fully described elsewhere [4, 5, 7, 9]. The Thomson scattered signal was imaged on a CCD array with a system that could provide a limiting spatial resolution of ~ 3 μm.

The measured Thomson profile corresponding to conditions for which strong channeling occurs is shown in Fig. 2. The image shows (1) a spurious reflection of the incident 248 nm pulse (~ 270 fs), which enters the zone of interaction through an aperture (2) positioned at the edge of the orifice of the pulsed valve, and the sudden collapse (3) into the narrow channel (4) with length ~ δ. The measured diameter of the narrow zone δ is limited by the spatial resolution of the imaging system. The high contrast of the signal characteristic of the region δ in which the isolated channel occurs indicates that the dynamics of the focusing efficiently conducts the energy into the channel and that the mechanism is stable [13]. These two important characteristics are in conformance with the corresponding theoretical analyses [11, 13, 14]. A thin dark central region (5) is seen to extend the channel into the region $Z > Z_0$. This feature is consistent with a reduction of the electron density in the central core of the high intensity zone of the propagating channel, the expected dynamical signature of the relativistic/charge-displacement mechanism [11-14].

Simultaneously recorded single-shot spatial images of the Thomson scattered 248 nm radiation and the Xe(M) x-ray emission (~ 1 keV) produced from Xe clusters [8-10], as illustrated in Fig. 3, provide further significant information. The morphology of the Thomson signal appearing in Fig. 3(a) is essentially identical to the image shown in Fig. (2). The length of the narrow feature representing the gap is designated as $\delta_{UV} \cong 275$ μm, a distance slightly less than ten Rayleigh ranges ($\delta_R \cong 28.5$ μm). The Xe(M) x-ray image depicted in Fig. 3(b) also exhibits a gap with a length $\delta_x \cong 430$ μm corresponding to the formation of a narrow channel [4, 5, 10] of high intensity. It is significant that $\delta_x - \delta_{UV} \cong 155$ μm, a distance exceeding $5\delta_R$. However, if we consider the length designated as δ_{Ch} in Fig. 3(a) associated with the extension represented by inclusion of the narrow dark

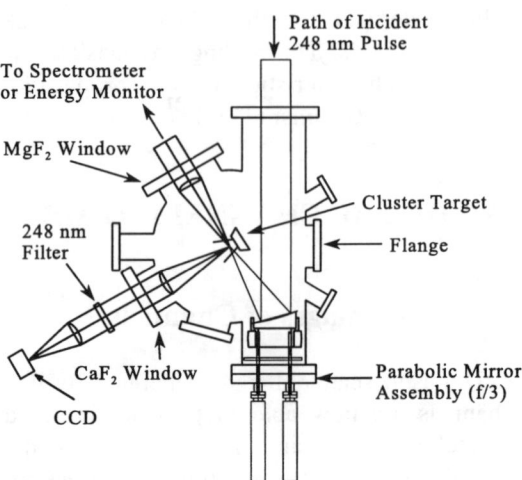

FIGURE 1. Schematic of the experimental configuration used to record the Thomson signal (248 nm) scattered transverse to the direction of propagation of the incident beam. The Thomson component is imaged onto a CCD and <u>single-shot</u> exposures are recorded. A narrow-band filter centered at 248 nm admits the Thomson signal and rejects interfering components produced by the plasma. The limiting spatial resolution of the imaging system was experimentally determined to be ~ 3 µm with the use of a calibrated test pattern, a value approximately ten-fold sharper than that characteristic of x-ray images of Xe(M) emission (~ 1 keV) recorded with a pinhole camera in experiments reported in Refs. [9] and [10].

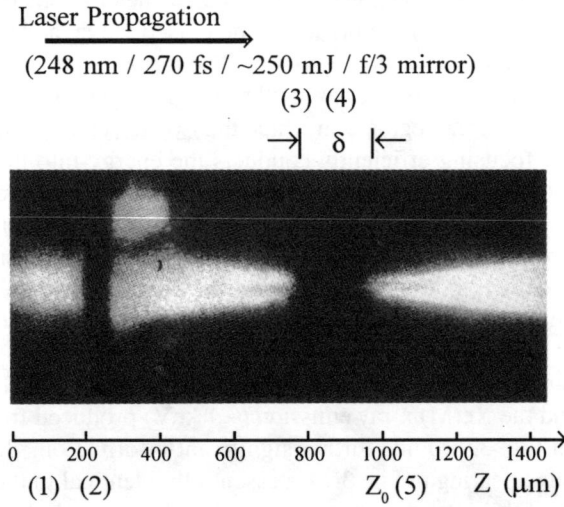

FIGURE 2. Image of Thomson scattered 248 nm radiation viewed transversely to the direction of propagation of the incident ~ 1 TW 248 nm pulse (~ 270 fs). The average xenon atom density was ~ 1.2 × 10^{19} cm^{-3}. A spurious scattering from the incident pulse (1) is shown and the position (2) of the aperture is visible at the left. The collapse (3) to the narrow channeled (4) region δ is apparent in addition to a slender dark zone (5) for Z > Z_0. This image was recorded with a magnification of 4.6, a value giving a spatial resolution of 5.5 µm. The Rayleigh range of the f/3 focusing system is δ_R = 28.5 µm.

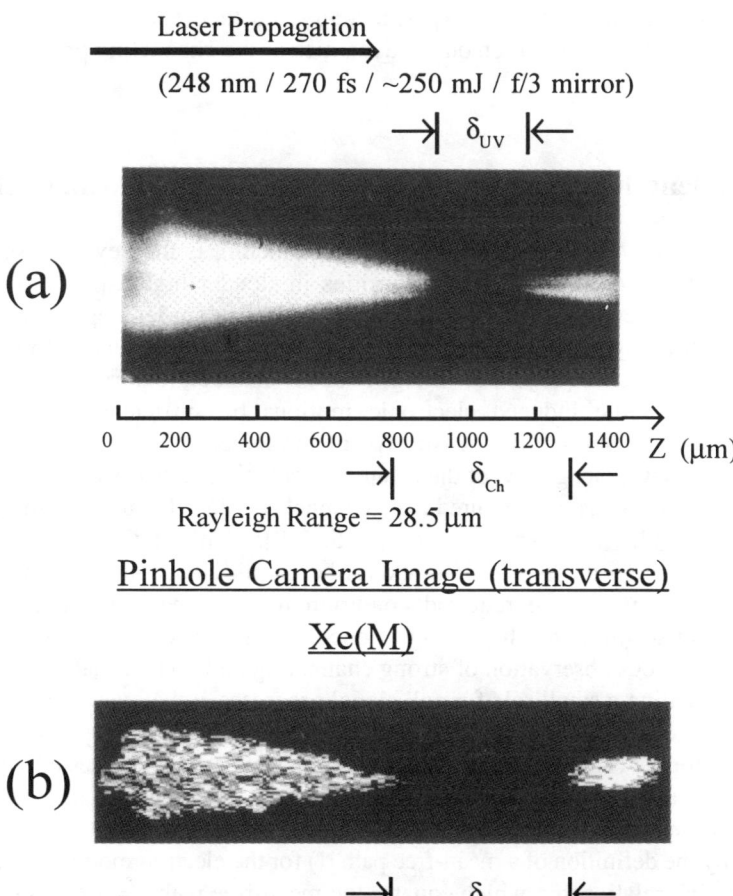

FIGURE 3. Simultaneously recorded single-shot images of Thomson scattered radiation (248 nm) and Xe(M) emission (~ 1 keV), both viewed transversely to the direction of propagation of the 248 nm pulse. The target conditions for the pulsed valve were: Xe gas at a pressure of 65 psia and plenum temperature of 20° C. The Rayleigh range of the f/3 focusing system is $\delta_R = 28.5$ µm.
(a). Thomson scattered 248 nm radiation obtained with the apparatus shown in Fig. 1. The image is similar to that illustrated in Fig. 2. The dimensions δ_{UV} and δ_{Ch} are shown. The conditions of imaging were the same as used in Fig. 2.
(b). X-ray pinhole camera image of the Xe(M) emission. The experimental arrangement for this camera is described in Refs. [3], [9], and [10]. The distance δ_x is indicated. The camera was equipped with a 25 µm diameter pinhole.

zones, we find $\delta_x \approx \delta_{Ch}$. This indicates that the true length of the narrow high-intensity channel exceeds the length of the gap δ_{UV} appearing in the Thomson image. Taken together, the two images shown in Fig. 3 give a direct visualization of the channel and provide detailed spatial information on (1) the dynamical distribution of the 248 nm radiation, (2) the electron density profile, and (3) the x-ray emitting ions. The further detailed study of such images, both experimentally and theoretically, should provide considerable quantitative information on the dynamics of the channeling process and the x-ray generation.

B. Wavelength Scaling Limits of Plasma Heating in Channels

The relativistic/charge-displacement self-channeling is the key process for the generation of highly compressed power densities in slender high-aspect-ratio spatial distributions. The operation of the ponderomotive mechanism leading to the charge-displacement, which dynamically develops as the channel propagates and provides a strong stabilizing influence [13] on the integrity of the channel, requires that collisional processes disturbing the induced electronic motions be sufficiently weak. This requirement depends upon (1) the density (ρ) of the material in the channel, (2) the intensity (I), and (3) wavelength (λ) of the channeled radiation. Also as mentioned above in Section I, the x-ray emission measured from channels produced with 248 nm radiation exhibits highly nonthermal characteristics [4, 5, 7-9]. This indicates that thermal processes are weak under these conditions and further implies that collisional interactions are correspondingly weak, the required condition for the operation of the charge-displacement mechanism in the channeling. Hence, experiments with 248 nm radiation lead to the simultaneous observation of strong channeling and nonthermal x-ray spectra, two features indicating a small role for collisional processes at that wavelength.

It is of considerable interest to understand the wavelength (λ) dependence of the collisional limit for the channeling mechanism. A simple experimentally tested model can be used to derive a scaling relationship and estimate the importance of the wavelength dependence on the achievable power density. In this picture, the effect of collisions is characterized by the definition of a mean-free path (ℓ) for the electron motions. This leads to a simple scaling relationship which equates the mean-free path $\ell = (\sigma\rho)^{-1}$, in which σ and ρ designate the effective collisional cross section and the particle density, respectively, to the magnitude of the driven excursion of a free electron $x_0 = K\lambda^2 \sqrt{I}$, evaluated nonrelativistically, in which λ, I, and K denote the wavelength, the intensity, and an appropriate constant, respectively. This procedure gives the relationship

$$\rho^2 I \leq \frac{1}{K^2 \lambda^4 \sigma^2} \tag{1}$$

which defines the domain of the collisionless regime in the ρ-I plane for the propagation. For the simultaneous attainment of high intensities (I) and high densities (ρ), the desired condition representing the high power density, the relationship expressed in Eq. (1)

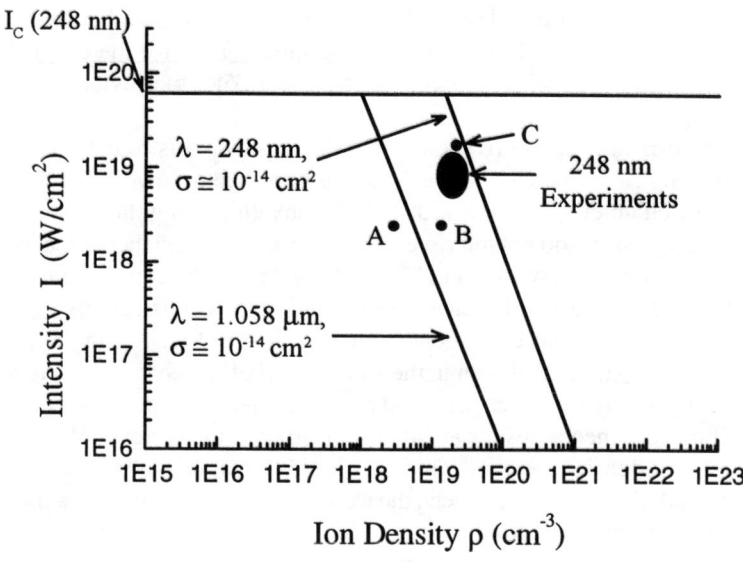

FIGURE 4. Wavelength scaling of collisionless regime. For the collisional cross section shown, the present data at 248 nm have not crossed the boundary although experiments conducted at λ = 1.058 μm appear to have probed the collisional regime. Data points A and B taken from Ref. [16]. Datum C taken from Ref. [17]. See text for discussion.

persuasively favors shorter wavelengths.

Figure 4 illustrates the scaling represented by Eq. (1) in connection with currently available experimental data [1, 2, 4, 5, 9, 10, 12, 16] at 248 nm and 1.058 μm. The conditions studied in experiments [1, 2, 4, 5, 9, 10, 12] performed at 248 nm are represented by the approximately elliptically shaped zone indicated. It is significant that this region lies essentially in the area which respects the collisional limit expressed by Eq. (1) for λ = 248 nm. It has also been found that <u>quantitative</u> agreement between the experimental results [1, 10, 12] and theory [13, 14] is possible within the framework of the collisionless picture for propagation under these conditions.

Data (points A and B) corresponding to experimental studies [16] of the channeling process at 1.058 μm are also placed on Fig. 4, along with the corresponding collisional limit from Eq. (1) for that wavelength. In this case, datum A falls in the collisionless region while datum B violates the restriction of Eq. (1) significantly.

The collisionless model [13, 14] has been used to calculate the expected behavior for both experimental conditions (data A and B). The comparison of the experimental and theoretical results corresponding to point A are shown in Fig. 5. Four data are shown in Fig. 5, namely, (a) the experimental axial Thomson profile [16], (b) the calculated on-axis Thomson profile [16], (c) the calculated intensity profile originally presented by Monot et al. [16], and (d) a second theoretical profile performed by us independently with the procedures described in Ref. [14]. The close agreement shown between Figs. 4(c) and 4(d) and the finding that all four data give the same length for the channeling region (~ 3 mm) indicate that the collisionless picture for $\lambda = 1.058$ µm provides a good description of the propagation for point A.

Datum B in Fig. 4 corresponds to a density which is four-fold greater than that used for datum A. An increase in the density is expected to lower the critical power (P_{cr}) governing the channeling process [13, 14], thereby enhancing the channeling process, if collisional mechanisms do not interfere. The experimental and theoretical correspondence for these conditions is represented in Fig. 6. In this case, the experimental channel length observed by Monot et al. [16] is clearly less than 2 mm, a considerably shorter value than observed for the conditions corresponding to datum A in Fig. 4. Significantly, this trend is exactly the opposite of that seen in the studies [10] of the channeling process conducted at 248 nm, apparently for which, as stated above, collisional processes were not important. The specific experimental results at 248 nm illustrating the increase of the channel length with increasing density are shown in Figs. 1 and 2, respectively, of Ref. [10]. There is also substantial disagreement between the experimentally-measured channel length shown in Fig. 6(a) and the corresponding calculated value (~ 3 mm) represented by the axial intensity profile shown in Fig. 6(b). Therefore, at the high density corresponding to datum B in Fig. 4, we find that the collisionless model is a poor representation of the experimental findings.

The reduction in the measured channel length and the difference in the observed and theoretical values shown in Fig. 6 suggest that significant collisional losses are present in the experimental situation represented by datum B in Fig. 4. Furthermore, it is found by us that the inclusion of simulated losses in the calculation readily brings the theoretically predicted channel length into significantly improved agreement with the experimental outcome. Overall, these results are fully consistent with the position of the collision limit expressed by Eq. (1), a boundary which separates the data points A and B in Fig. 4.

Additional results, recently obtained [17] at $\lambda = 800$ nm with a peak power of ~ 5 TW in Ar at a density of $\rho_{Ar} = 2 - 3 \times 10^{19}$ W/cm^3, can also be placed on Fig. 4. The peak focal intensity achieved in vacuum was estimated to be ~ 2×10^{19} W/cm^2 with an f/3 off-axis parabolic mirror. These parameters place the corresponding position in Fig. 4 at the top of the zone indicated for the extant studies conducted at 248 nm, a point specified as datum C. Therefore, these experimental conditions represent a point deep in the region expected for collisional processes to be important for an infrared wavelength (~ 800 nm). The reported observation of Ar(K) radiation and Bremsstrahlung emission for quantum energies well above 10 keV are fully consistent with the presence of considerable electron heating and the corresponding location of datum C in Fig. 4.

These preliminary results and analyses indicate that the limit established by collisional processes appears to favor strongly the use of short wavelength radiation for the production of high-power-density channels with the relativistic/charge-displacement mechanism. These findings are also consistent with other experimental work [18] showing that ultraviolet radiation can readily produce cold plasma environments.

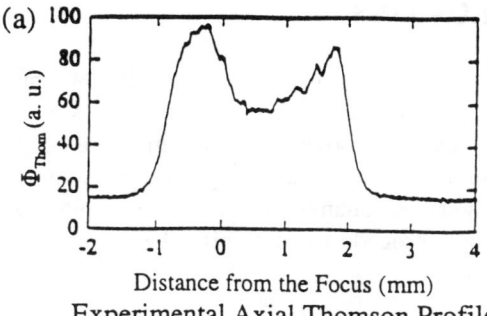
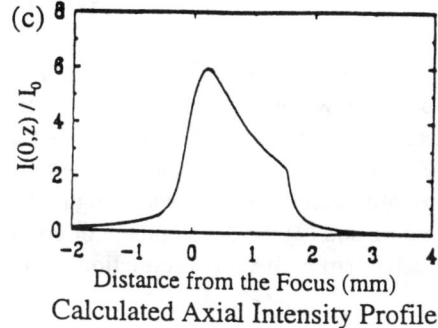
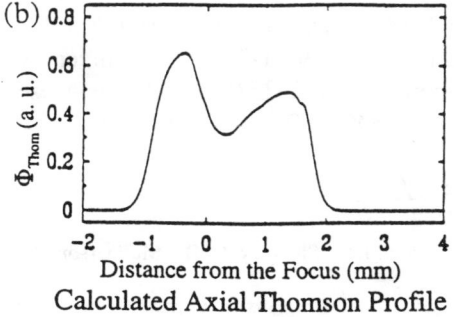
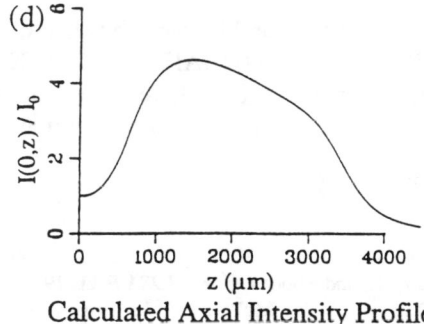

FIGURE 5. Comparison of experimental and theoretical values for relativistic self-channeling of 1.058 μm pulses in H_2 plasma reported in Ref. [16] for datum A shown in Fig. 4. (a) Experimental axial Thomson profile, (b) calculated on-axis Thomson profile, (c) calculated intensity profile, and (d) independently calculated axial intensity profile using the procedures described in Ref. [14]. Panels (a), (b), and (c) are adapted from Ref. [16].

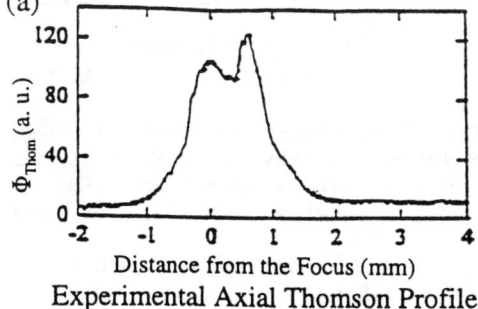
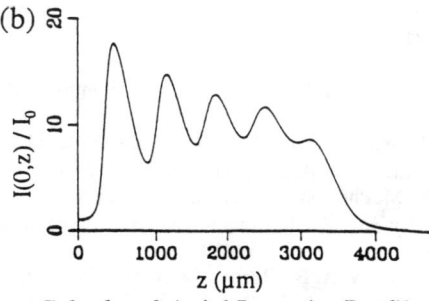

FIGURE 6. Comparison of experimental and theoretical values for relativistic self-channeling of 1.058 μm pulses in H_2 plasma reported in Ref. [16] for datum B shown in Fig. 4. (a) Experimental axial Thomson profile and (b) calculated axial intensity profile using the procedures described in Ref. [14]. Panel (a) is adapted from Ref. [16].

III. CONCLUSIONS

The production of self-trapped plasma channels carrying intensities of 10^{20} W/cm^2 opens a new domain of ultrahigh-power-density plasma phenomena to systematic study. Experimental data [4-6] indicate that power densities above ~ 10^{19} W/cm^3 have been achieved. Of particular relevance for understanding these high power-density environments are (1) the wavelength (λ) scaling of the dynamics and the limits of plasma heating and (2) the development of computational procedures combining the channeling mechanism with x-ray production from clusters.

IV. ACKNOWLEDGMENTS

Support for this research was provided under contracts with SDI/NRL (N00014-93-K-2004), ARO (DAAH04-94-G-0089), ARO (DAAG55-97-1-0310), the University of California/Lawrence Livermore National Laboratory (B328353), and the Japanese Ministry of Education, Science, Sport and Culture (#08405009 and #08750046).

REFERENCES

1. Borisov, A. B., Shi, X., Karpov, V. B., Korobkin, V. V., Solem, J. C., Shiryaev, O. B., McPherson, A., Boyer, K., and Rhodes, C. K., *JOSA B* **11**, 1941 (1994).
2. Borisov, A. B., McPherson, A., Boyer, K., and Rhodes, C. K., *Progress in Crystal Growth and Characterization of Materials* **33**, 217 (1996).
3. Boyer, K., Thompson, B. D., McPherson, A., and Rhodes, C. K., *J. Phys. B* **27**, 4373 (1994).
4. Borisov, A. B., McPherson, A., Boyer, K., and Rhodes, C. K., *J. Phys. B* **29**, L113 (1996).
5. Borisov, A. B., McPherson, A., Boyer, K., and Rhodes, C. K., *J. Phys. B* **29**, L43 (1996).
6. Boyer, K., and Rhodes, C. K., *J. Phys. B* **27**, L633 (1994).
7. McPherson, A., Thompson, B. D., Borisov, A. B., Boyer, K., and Rhodes, C. K., *Nature* **370**, 631 (1994).
8. Borisov, A. B., Longworth, J. W., McPherson, A., Boyer, K., and Rhodes, C. K., *J. Phys. B* **29**, 247 (1996).
9. McPherson, A., Borisov, A. B., Boyer, K., and Rhodes, C. K., *J. Phys. B* **29**, L291 (1996).
10. Borisov, A. B., McPherson, A., Thompson, B. D., Boyer, K., and Rhodes, C. K., *J. Phys. B* **28**, 2143 (1995).
11. Solem, J. C., Luk, T. S., Boyer, K., and Rhodes, C. K., *IEEE J. Quantum Electron.* **QE-25**, 2423 (1989).
12. Borisov, A. B., Borovskiy, A. V., Korobkin, V. V., Prokhorov, A. M., Shiryaev, O. B., Shi, X. M., Luk, T. S., McPherson, A., Solem, J. C., Boyer, K., and Rhodes, C. K., *Phys. Rev. Lett.* **68**, 2309 (1992).
13. Borisov, A. B., Shiryaev, O. B., McPherson, A., Boyer, K., and Rhodes, C. K., *Plasma Phys. and Control. Fusion* **37**, 569 (1995).
14. Borisov, A. B., Borovskiy, A. V., Shiryaev, O. B., Korobkin, V. V., Prokhorov, A. M., Solem, J. C., Luk, T. S., Boyer, K., and Rhodes, C. K., *Phys. Rev. A* **45**, 5830 (1992).
15. Pukhov, A., and Meyer-ter-Vehn, J., *Phys. Rev. Lett.* **76**, 3975 (1996).
16. Monot, P., Auguste, T., Gibbon, P., Jakober, F., Mainfray, G., Dulieu, A., Louis-Jacquet, M., Malka. G., and Miquel, J. L., *Phys. Rev. Lett.* **74**, 2953 (1995).
17. Rousse, A., Kien, D., Chériaux, G., Mysyrowicz, A., Antonetti, A., Stenz, C., Blasco, F., Stevefelt, J., Pellicer, J. C., Andebert, P., Geindre, J. P., and Gauthier, J. C., private communication, reported at the Conference Ultrafast Optics 1996.
18. Nagata, Y., Midorikawa, K., Kubodera, S., Obara, M., Tashiro, H., Toyoda, K., and Kato, Y., *Phys. Rev. A* **51**, 1415 (1995).

Phase-Matched Optical Parametric Conversion of Ultrashort Pulses in a Hollow Waveguide

Charles G. Durfee, Sterling Backus,
Margaret M. Murnane and Henry C. Kapteyn

Center for Ultrafast Optical Science, University of Michigan, Ann Arbor, MI 48109-2099

Abstract. We demonstrate for the first time nonresonant phase-matched frequency conversion of ultrashort pulses in gases. Broad-bandwidth ultrafast pulses, tunable around 270nm, were generated from a Ti:sapphire amplifier system using $2\omega + 2\omega - \omega$ parametric wave mixing in a capillary waveguide. Both the fundamental and the second-harmonic light were coupled into the lowest-order (EH_{11}) mode. The output pulses have an energy >4µJ at a 1kHz repetition rate, in the EH_{11} spatial mode. This method can be made to generate 10-20fs pulses, and is the first phase-matching technique which is applicable to frequency conversion into the deep- and vacuum-ultraviolet regions of the spectrum.

INTRODUCTION

In recent years there has been much progress in developing reliable sources of ultrashort-pulse light. Most of these new lasers are based on titanium-doped sapphire, and thus operate in the near infrared.[1] However, many applications of these light sources require photons in other wavelength ranges. For example, many studies of molecular dynamics require light in the deep-ultraviolet or even the vacuum ultraviolet. Ultrashort pulses in the infrared can be converted to the UV using nonlinear-frequency conversion in crystals such as BBO,[2] and parametric generation can be used to generate ultrashort pulses tunable from the near-UV into the mid-infrared. However, these techniques suffer from limitations inherent to the propagation of ultrashort light pulses in solid-density materials: opacity in the VUV, and very high group-velocity walk off which makes it difficult to generate pulses in the UV much shorter than a couple of hundred femtoseconds.

In this work, we present an entirely new method for the efficient, phase-matched conversion of ultrashort pulses into the UV, which is equally applicable in the VUV and possibly even in the XUV region of the spectrum.[3] In this technique, guided-wave optical parametric generation, the nonlinear-medium is a gas (in this case argon). Since many gasses are transparent from the VUV into the far-infrared, the potential tuning range of this technique is much larger than in the case of crystal conversion. Furthermore, dispersion in the gasses is very low, making it a technique suitable for the frequency conversion of extremely short (<10 fs) pulses.

THEORY

Frequency conversion in crystals typically relies on the use of $\chi^{(2)}$ materials to obtain high nonlinearity, and on birefringence to phase match propagation vectors of the fundamental and the harmonic. Since a gas is homogeneous and isotropic with respect to light propagation, neither of these concepts is applicable in gasses. $\chi^{(3)}$ or higher odd-order nonlinearities must be used, making it essential to increase the interaction length to make-up for the lower magnitude of the nonlinearity. Phase-matching in free-space propagation does not normally occur except in special cases near atomic resonances. [4,5]

In this work, we show theoretically and demonstrate experimentally that phase-matched conversion with increased interaction length can be accomplished using propagation of the light in a waveguide rather than in free space. Recent work has demonstrated the utility of hollow-core fibers (simple gas-filled capillary tubes) for self-phase modulation and compression of ultrashort pulses.[6] Propagation of light in a waveguide introduces a negative phase shift on the wave to counter the tendency of the wavefront to move forward due to diffraction. This phase shift is greater for longer wavelengths, resulting in a net negative dispersion in the hollow waveguide; i.e. a phase velocity that is inversely proportional to wavelength. This negative dispersion can be counterbalanced by the positive dispersion of the gas filling the center of the waveguide, resulting in a value of pressure at which dispersion at a particular wavelength vanishes.

Propagation in an optical waveguide offers several options for phase matching parametric processes, such as mixing of light propagating at different frequencies and/or in different spatial modes.[7,8] The modal properties of a dielectric capillary were given by Marcateli et al.[9] The propagation constant for a capillary filled with a medium of index n_g is given by -

$$k = \frac{2\pi n_g(\lambda)}{\lambda}\left[1 - \frac{1}{2}\left(\frac{u_{nm}\lambda}{2\pi a}\right)^2\left(1 + \text{Im}\left(\frac{v_{EH}\lambda}{\pi a}\right)\right)\right] \qquad (1)$$

where a is the core radius, u_{nm} is the modal constant, $v_{EH} = 1/2\,(v^2 + 1)/((v^2 - 1)^{0.5})$, and v is the ratio of the refractive index of the capillary material to that of its contents. Although $|v| > 1$, Fresnel reflections at the gas-glass interface allow lossy guiding of optical beams. The field loss rate for hybrid modes (EH_{nm}) is given by -

$$\alpha = \left(\frac{u_{nm}\lambda}{2\pi}\right)^2 \frac{\lambda^2}{a^3} \text{Re}(v_{EH}) \;. \qquad (2)$$

Consider the case of difference frequency mixing between two beams propagating in an optical fiber at central frequencies ω_1 and ω_2, with modal constants u1 and u2. The phase mismatch Δk in the case of difference frequency mixing, $\omega_3 = N\omega_2 - M\omega_1$, where $\omega_2 = R\omega_1$, $\Delta k = Nk_2 - Mk_1 - k_3$, may be written as -

$$\Delta k = \Delta k_{mode} - \Delta k_{material}$$

$$= \frac{\lambda_1}{4\pi a^2}\left[\frac{u_3^2}{RN-M} + Mu_1^2 - \frac{Nu_2^2}{R}\right] - \frac{2\pi p}{\lambda_1}((RN-M)\delta_3 + M\delta_1 - N\delta_2) \quad (3)$$

where the index is written as $n_g = 1 + p\delta_g$, with p the gas pressure. The phase mismatch is a balance of modal dispersion ($\Delta k_{mode} \propto 1/\pi a^2$) material dispersion ($\Delta k_{material} \propto p$). In gases (or plasmas) with normal dispersion, $\Delta k_{mat} > 0$. If the modes of the pump and signal are chosen such that $\Delta k_{mode} > 0$, there will exist a pressure for which the process is phase-matched. For the difference frequency-mixing case where two 400 nm photons are combined and one 800 nm photon is given-up, N=2, M=1, and $u_1 = u_2$,. Therefore $\Delta k_{mode} > 0$ for any signal mode. In the present experiment, R = 2, and any pair of modes ($u_1 = u_2$) can be phase matched to a particular mode, u_3. It is most desirable, and gives the largest model overlap and nonlinearity, when the TEM$_{00}$ free-space mode is coupled into the EH$_{11}$ mode of the waveguide.

EXPERIMENT

In the experiment, a laser system generating 20 femtosecond pulses with up to 4 mJ energy at 1 kHz repetition-rate was used.[10] A 100 µm LBO crystal was used to frequency double the 800 nm light, with 20% conversion efficiency. The two colors were then separated, and later recombined with a relative time delay, before being focused into the capillary (127 µm core diameter, 60 cm long). The divergence of the input telescope was adjusted to optimize the coupling of the 400 nm light (35% throughput); the fundamental beam was not as well optimized (<10% throughput). Figure 1 shows the 3ω output energy versus argon gas pressure. In the case of krypton gas, the output vs. pressure is has a similar structure, but is shifted to lower gas pressure due to the higher dispersion of krypton. The positions of peaks in conversion efficiency are in excellent agreement with the values calculated from Eqn. 3 for conversion into various modes. The highest conversion efficiency is observed into the EH$_{11}$ mode, at a measured and calculated optimum pressure of 85.1 and 89.7 Torr, respectively. The output mode is very nearly Gaussian. With 30 µJ at 2ω and 64 µJ at ω leaving the fiber, the output energy at 3ω was 4 µJ, giving 13% conversion from 2ω to 3ω.

Figure 2 shows the spectrum of the output under two conditions. Calculated phase-matching bandwidths (>20 nm) and values of group-velocity walk off (<10 fs) lead one to expect excellent characteristics for the frequency conversion of ultrashort pulses, as is observed. Figure 2(a) is the case of moderate intensity of the 800 nm idler field. The spectral bandwidth, which has a bandwidth corresponding to a 12 fs pulse, is consistent with what one would expect given the input bandwidths and a perturbative frequency-conversion process using $\chi^{(2)}$ in LBO and $\chi^{(3)}$ in argon. Figure 2(b) is the case where the intensity of the 800 nm idler light is increased to a level similar to that used in hollow-core fiber pulse compression experiments.[6] In this case, cross-phase modulation broadens the spectrum of the UV light to the point where the bandwidth could support a 4 fs pulse.

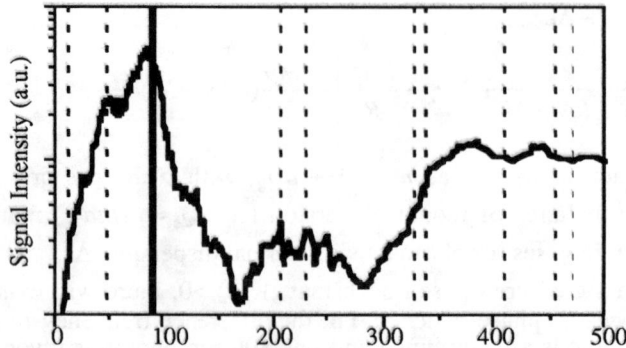

FIGURE 1: Signal intensity of third harmonic of 800 nm, as a function of argon gas pressure within a 127 μm diameter, 60 nm long argon-filled capillary waveguide. The solid line represents the calculated pressure for phase-matched conversion into the EH_{11} mode, while the dashed line represents phase-matching pressures corresponding to high-order modes.

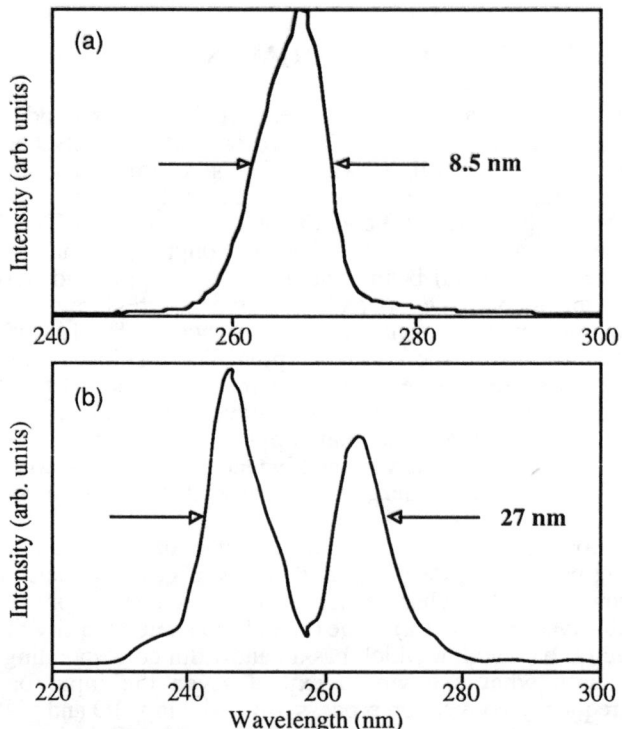

FIGURE 2: Output pulse spectrum (a) in the case of moderate intensity of the 800 nm "idler" pulse, showing a spectrum corresponding to a 12 fs pulse. (b) when the intensity of the 800 nm light is increased, cross-phase modulation of the UV by the nonlinearity caused by the 800 nm pulse broadens the spectrum of the output. The bandwidth could be compressed to ~4 fs duration.

In conclusion, we have demonstrated a new technique for efficient, phase-matched frequency upconversion of light. This is the first phase-matched conversion technique suitable for the generation of ultrafast pulses in the deep-UV, VUV, and possibly even the soft x-ray regions of the spectrum. This technique also lends itself well to the implementation of a tunable ultrafast UV source, with the use of tunable infrared from an optical parametric amplifier as the idler source.

REFERENCES

[1] A. Rundquist, C. Durfee, Z. Chang et al., "Ultrafast Laser and Amplifier Sources," Applied Physics B **65**, 161 (1997).

[2] J Ringling, O Kittelmann, F Noack et al., "Tunable femtosecond pulses in the near vacuum ultraviolet generated by frequency conversion of amplified Ti:sapphire laser pulses," Optics Letters **18** (23), 2035-7 (1993).

[3] C.G. Durfee, S. Backus, M.M. Murnane et al., "Ultrabroadband phase-matched optical parametric generation in the ultraviolet using guided waves," Optics Letters **To be published** (1997).

[4] T. Jong, G. Bjorkland, A. Kung et al., Physical Review Letters **27**, 1551 (1971).

[5] J.F. Reintjes, *Nonlinear Optical Parametric Processes in Liquids and Gases* (Academic Press, 1984).

[6] M. Nisoli, S. De Silvestri, and O. Svelto, "Generation of High-Energy 10 fs pulses by a new pulse compression technique," Applied Physics Letters **68** (20), 2793-2795 (1996).

[7] H. Milchberg, C. Durfee III, and T. McIlrath, Physical Review Letters **75**, 2494 (1995).

[8] G.P. Agrawal, *Nonlinear Fiber Optics* (Academic Press, 2nd Ed., San Diego, 1995).

[9] E.A.J. Marcateli and R.A. Schmeltzer, Bell System Techmical Journal **43**, 1783 (1964).

[10] S. Backus, C. Durfee, M.M. Murnane et al., "0.2 Terawatt laser system at 1 kHz: 4 mJ, 20 fs pulses.," Optics Letters **To be published** (1997).

Anomalous emission of soft X-rays from OFI-plasmas

Georg Pretzler and Ernst E. Fill

Max-Planck-Institut für Quantenoptik, D–85748 Garching, Germany

Abstract. Low-density plasmas produced by optical-field ionization (OFI) were investigated by X-ray spectroscopy under well-controlled conditions. While most features of the spectra agree well with a numerical model, several lines show anomalously high intensity. A systematic variation of parameters yields strong indications that excitation processes other than collisional excitation play an important role in OFI-plasmas.

INTRODUCTION

The process of optical field ionization (OFI) has been under intense investigation during recent years both theoretically and experimentally (e.g. (1,2)), and the observed parameters are, in general, well predicted by theory. We have focused our interest on the plasma generated by the OFI-process. A model of Corkum et al. (3) predicts that the residual energy of the released electrons is low if the laser pulses are linearly polarized and increases with increasing laser pulse ellipticity. Therefore, compared to other plasma creation methods, OFI-plasmas have the advantage that one has the electron temperature in hand as a free parameter for "shaping" the plasma.

X-ray spectroscopy is the method of choice for studying such features by investigating how much the intensities of X-rays in a certain wavelength range are effected. However, previous spectroscopic experiments were performed with gas jet targets (e.g. (4)), and the results reflect little of the OFI process because high-density effects (e.g. inverse bremsstrahlung heating, collisional ionization, clustering, laser self-defocusing) are dominating. Excluding these effects demands for low-density conditions. However, low density plasmas emit low numbers of X-ray photons, therefore a compromise has to be found between avoiding high-density effects in the plasma and obtaining enough intensity on the spectrometer.

EXPERIMENTS

The experiments were conducted using the ATLAS (Advanced Ti:sapphire Laser) system at MPQ Garching delivering pulses of 200 mJ in 200 fs at 10 Hz at a wavelength of 790 nm. The experimental setup is shown in Fig.1. The laser pulses were focused into a gas-filled cell (pressures in the range of 0.3 to 2 mbars) to an intensity of up to 3×10^{16} W/cm^2. Diagnostics perpendicular to the laser direction was provided by a transmission grating with a 50-μm slit and an X-ray CCD. Differential pumping in the diagnostics path (two slits and two pump sections) was necessary for minimizing absorption of the measured X-rays. The first slit (150 μm wide and 3 mm long), 2 mm from the laser focus, defined the observed plasma volume. A 0.1-μm Be-filter on the grating rejected visible stray light. Ellipticity was introduced by a low-order λ/4-platelet allowing ellipticity values (defined as the ratio of minimum to maximum electric-field amplitude) from 0 to 0.65. To get a sufficiently high number of counts, 6000 shots were accumulated on the CCD for each spectrum. The measured spectra are integrated over time and the observed volume.

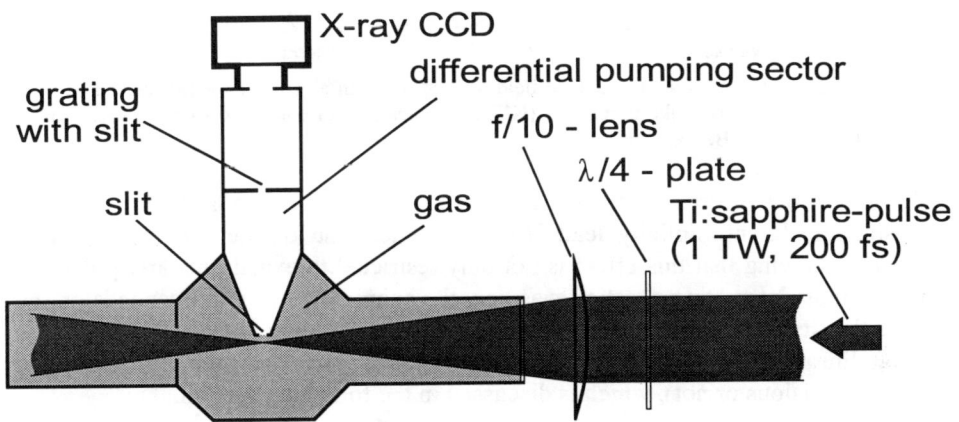

FIGURE 1. Experimental setup for X-ray spectroscopy of low-density OFI-plasmas

A first experimental result was that in general, elliptically polarized laser pulses produce more intense X-ray spectra than linearly polarized pulses do, as expected. These results are reported elsewhere in detail (5,6). As an example, spectra for nitrogen are shown in Fig.2. In these spectra, however, several transitions originating from high-energy levels of Li-like nitrogen (N^{4+}) show stronger emission with linear than with elliptic polarization, a result unexpected from theory. The present paper is focused on these anomalous emissions. Investigating different gases (He, Ar, CH$_4$, NH$_3$, N$_2$O, N$_2$, O$_2$, SF$_6$), we found anomaluosly high emissions from high livels with linear polarization in each of these gases.

A numerical model was developed which is based on ADK ionization rates (1) and subsequent hydrodynamic plasma expansion. It does all the bookkeeping for the population densities of the relevant excitation states by means of a collisional-radiative code and calculates spectra for our actual experimental situation (for details, see (6)). The overall agreement is quite good (see Fig.2), but the anomalous emissions are not reproduced. From the model, we learn that absorption plays no role within the plasma, three-body recombination is negligible (both due to the low plasma density), and the time-scale for the measured emission is in the range of 1 ns.

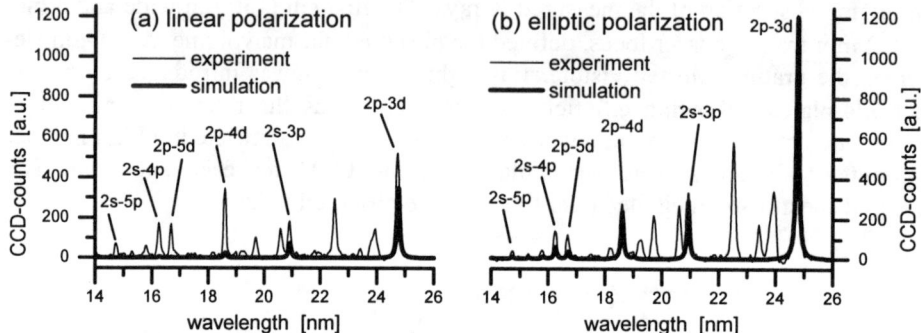

FIGURE 2. Soft X-ray spectra from optical-field ionized nitrogen at 1 mbar for (a) linear and (b) elliptic polarization. Main Li-like transitions (N^{4+}) are indicated and compared to simulations, other significant lines are from Be-like ions (N^{3+}).

The anomalous effect was investigated experimentally in more detail for nitrogen. Increasing the laser ellipticity leads to a slow decrease of the "anomalous" line intensities, showing that this effect is not only restricted to exactly linearly polarized light. The most interesting question was how the anomalous emission depends on the particle density. The results, shown in Fig. 3, show a line intensity dependence between linear and quadratic for all the transitions (whether their emission was declared anomalous or not), which is discussed in the following section.

DISCUSSION

The dependence of the transition line intensities I on the gas pressure p yields information on the envolved processes for generating excited states. The dependence is expected to be

- <u>linear</u> for single-atom excitation effects
- <u>quadratic</u> for two-particle effects (e.g. excitation by electron collisions).
- <u>cubic</u> for three-particle effects (e.g. generation of excited states by three-body recombination).

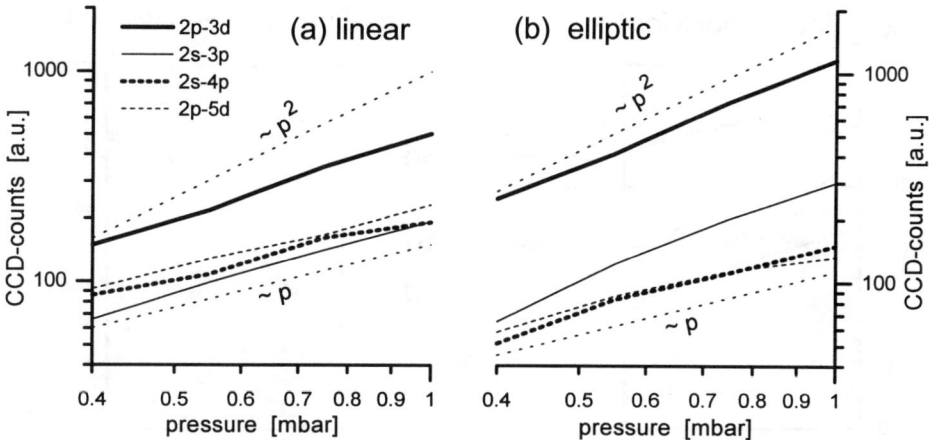

FIGURE 3. Li-like nitrogen soft X-ray emission as a function of pressure in the gas cell. The intensity of several transitions is shown for (a) linear and (b) elliptic polarization. The slopes for linear and quadratic pressure dependence are also indicated.

Since plasma self absorption is negligible (as we know from the simulations, see above), the experimental results (Fig.3) may be explained by assuming excitation due to competetive excitation mechanisms. Fitting the expression

$$I = A \cdot p + B \cdot p^2 + C \cdot p^3 \tag{1}$$

to the data gives good agreement and yields the linear and quadratic coefficients A and B (see Fig. 4). The cubic coefficient C was small in all cases demonstrating that the generation of excited states by three-body recombination plays no role, as expected from our simulations. The statistical errors preclude an exact quantitative analysis, but interesting conclusions can be drawn from the data displayed in Fig.4.

For all transitions, the quadratic coefficient B is considerably larger with elliptic polarization than with linear polarization (by factors around 4), reflecting the fact that collisional excitation is stronger due to the higher electron temperature.

The linear coefficient A is slightly larger for linear polarization, possibly according to the electric field amplitude which is larger by a factor of $\sqrt{2}$. Note that the linear excitation coefficient is always nonzero, even for lines which seem to behave "normally" in the spectra. The conclusion is that besides collisional excitation, a second excitation mechanism is present which is a single-particle effect because it depends linearly on the density. Evidently, this effect is driven by the optical field, and therefore may be called OFE (Optical Field Excitation).

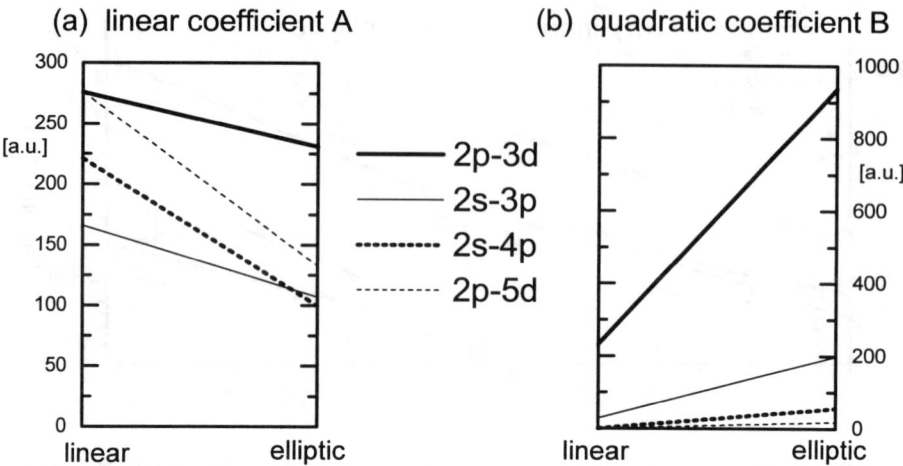

FIGURE 4. Dependence of the linear and quadratic coefficients A and B (describing the dependence of the line intensities from the gas pressure) from the laser pulse ellipticity, shown for several N^{4+} transitions. Errors for the end points of the lines are in the range of 30 a.u. for both diagrams.

It is not clear at the moment which mechanism is responsible for the observed linear (i.e. single-particle) excitation. A few possibilities have been discussed in the literature:

Excited ionic states may be populated by collisions of released electrons with their original nuclei as long as the field is on (7,8). This is a single-particle effect, but it will only be efficient for linear polarization and cannot explain the substantial values of coefficient A for the case of elliptic polarization.

In (9) a simple model is given for non-sequential ionization, postulating a *shake-off* process, i.e. fast ionization of one electron leading to reordering of the remaining electrons such that a second electron is left in an excited state from where it is assumed to be immediately ionized. If stabilization effects play a role, suppressing the last step, *shake-up* (without *shake-off*) may lead to the observed excited states.

Resonant multiphoton excitation is a possible mechanism for populating upper levels, since the ac Stark effect within the intense optical field will shift any level into resonance at a distinct intensity (see e.g. (10,11)). It has been demonstrated that these excited states play an important role for ionisation (e.g. (12)). Furthermore, it has been shown by means of electron spectroscopy that a fraction of atoms remain in excited states without being ionized (13). These experiments were done with Xe^+, and 6 photons were needed for resonant multiphoton excitation. Note that in our case, 40-50 photons are required.

However the process may work in detail, OFE (Optical Field Excitation) yields excited states within the duration of the laser pulse and is much faster than subsequent generation of excited states in the plasma (by collisions or recombination). Therefore,

OFE may be considered as a source for pulsed X-rays. Such an effect was reported in (14), where the time-resolved emission of the helium Ly-α transition from an OFI-plasma showed an initial spike. Furthermore, non-equilibrium population of excited levels after the laser pulse has been proposed as a mechanism for generating an X-ray lasing medium (15), inversion to the ground state being created by collisional relaxation into one distinct upper level.

In conclusion, we have shown by X-ray spectroscopy that single-atom excitation mechanisms play an important role in optical field ionization and may result in new ways to create intense X-rays. Our next investigations aim at clarifying the relevant excitation mechanism by time-resolved measurements and optimizing its conditions.

ACKNOWLEDGEMENTS

This work was supported in part by the Commission of the European Communities within the framework of the Association Euratom/Max-Planck-Institut für Plasmaphysik. G.P. was supported by the European Union "TMR"-program (contract ERB4001GT950525).

REFERENCES

1. Ammosov M.V., Delone N.B., and Krainov V.P., *Sov. Phys. JETP* **64**, 1191–1194 (1986).
2. Augst S., Strickland D., Meyerhofer D.D., Chin S.L., and Eberly J.H., *Phys. Rev. Lett.* **63**, 2212–2215 (1989).
3. Corkum P.B., Burnett N.H., and Brunel F., *Phys. Rev. Lett.* **62**, 1259–1262 (1989).
4. McPherson A., Luk T.S., Thompson B.D., Boyer K., and Rhodes C.K., *Appl. Phys. B* **57**, 337 (1993).
5. Pretzler G., and Fill E.E., *Opt. Lett.* **22**, 733–735 (1997).
6. Pretzler G., and Fill E.E., *Phys. Rev. E* **56**, 2112–2117 (1997).
7. Corkum P.B., *Phys. Rev. Lett.* **71**, 1994–1997 (1992).
8. Brabec T., Ivanov M.Yu., and Corkum P.B., *Phys. Rev. A* **54**, R2551–2554 (1992).
9. Fittinghoff D.N., Bolton P.R., Chang B., and Kulander K.C., *Phys Rev Lett.* **69**, 2642–2645 (1992).
10. Freeman R.R., Bucksbaum P.H., Milchberg H., Darack S., Schumacher D., and Geusic M.E., *Phys. Rev. Lett* **59**, 1092–1095 (1987).
11. Freeman R.R., and Bucksbaum P.H., *J. Phys. B* **24**, 325–347 (1991).
12. Rottke H., Wolff B., Brickwedde M., Feldmann D., and Welge K.H., *Phys Rev Lett.* **64**, 404–407 (1990).
13. De Boer M.P., Muller H.G., *Phys. Rev. Lett.* **68**, 2747–2750 (1992).
14. Borgström S., Fill E.E., Starczewski T., Steingruber J., Svanberg S., and Wahlström C.-G., *Laser Part. Beams* **13**, 459–468 (1995).
15. Chichkov B.N., Egbert A., Meyer S., Weichert A., and Wellegehausen B., "Ground State Lasers: Theory and Experiments", in *Proceedings of the X-ray Lasers 1996 Conference*, Lund, Sweden, IOP Conf. Ser. **151**, 227–233 (1996).

Laser-Plasma Harmonics with High-Contrast Pulses and Designed Prepulses

R.S. Marjoribanks,[1] L. Zhao,[1] F.W. Budnik,[1] G. Kulcsár,[1]
A. Vitcu,[1] H. Higaki,[1*] R. Wagner,[2] A. Maksimchuk,[2] D. Umstadter,[2]
S.P. Le Blanc,[3] and M.C. Downer[3]

(1) Department of Physics, University of Toronto
60 St. George St.; Toronto, Ontario; Canada M5S 1A7

(2) Center for Ultrafast Optical Science, University of Michigan
2200 Bonisteel Blvd.; Ann Arbor, MI 48109

(3) Department of Physics, University of Texas at Austin, Austin, Texas 78712

Abstract. One aspect of the complexity of mid- and high-harmonic generation in high-intensity laser-plasma interactions is that nonlinear hydrodynamics is virtually always folded together with the nonlinear optical conversion process. We have partly dissected this issue in picosecond and subpicosecond interactions with preformed plasma gradients, imaging and spectrally resolving low- and mid-order harmonics. We describe spatial breakup of the picosecond beam in preformed plasmas, concomitant broadening and breakup of the harmonic spectrum, presumably through self-phase modulation, together with data on the sensitivity of harmonics production efficiency to the gradient or extent of preformed plasma. Lastly, we show preliminary data of regular Stokes-like and anti-Stokes-like satellites to the harmonics, accompanied by modification of the forward-scattered beam.

INTRODUCTION

With the development of ultra-intense, high-contrast, sub-picosecond lasers, mid- and high-harmonic generation from thin, near-solid-density plasmas has become an productive topic in recent years, complementary to nanosecond laser-plasma harmonics research dating back nearly twenty years (1,2,3,4). Several experimental studies have been reported recently (5,6,7), different models have been developed (8,9,10,11), and new mechanisms suggested in this regime, largely based on particle-in-cell simulations.

One of the important issues in mid- and high-harmonic generation from solid target is the laser pulse contrast, because of its effect on the electron density profile. In earlier CO_2 experiments, using nanosecond pulses, the ponderomotive force of the laser pulse was seen to steepen the plasma density profile significantly (12). But in the picosecond laser/solid target interactions, the ponderomotive steepening is less effective because of the short interaction time. Therefore the plasma preformed by the pedestal or prepulse

---

* current address: Graduate School of Human and Environmental Studies, Kyoto University, Sakyo-ku, Kyoto, 606-01 Japan.

becomes very important. Although this preplasma issue in harmonic generation has been mentioned before, no systematic experiment on this effect has been reported.

It would be interesting and useful to characterize the optical participation of electrons without at the same time grossly modifying the ion density gradients during irradiation, principally by ponderomotive steepening. To this end, we have conducted a series of experiments in which a plasma gradient is prepared by a separate laser prepulse, and the effect of gradient on the harmonic production from an intense main pulse is examined.

EXPERIMENTS

Laser Systems Used

Two laser systems were used in this work, whose main parameters are summarized in Table 1. The FCM–CPA laser at the University of Toronto is a Nd:glass-based chirped-pulse amplification laser system, based on gratings-only expansion and compression of high-contrast 1-ps seed pulses which have energy greater than 1 µJ per pulse; these pulses are directly produced in a feedback-controlled Nd:glass oscillator (13). In the final output of the system, the intrinsic pulse-to-pedestal contrast is better than 5×10^7. Under some conditions, a Pockels-cell-leakage prepulse is observed in the output, at 10^{-4} of peak and 1.5 ns ahead of the main pulse. To eliminate these prepulses, we placed a saturable-absorber dye-cell in the final output line, producing a final 'clean' pulse contrast greater than 10^{10}. In experiments, this system delivered 1 ps pulses with energies to 200 mJ, focused at the target to intensities up to 10^{17} W cm^{-2}.

TABLE 1: Summary of laser parameters.

Laser system	λ (µm)	I_{max}(W cm^{-2})	τ (ps)	Contrast	Method
FCM-CPA	1.053	1×10^{17}	1.1	>10^{10}	dye-cell
T^3	0.527	1×10^{18}	0.4	>10^{10}	crystal

The T^3 laser at CUOS, University of Michigan, is a hybrid Ti:sapphire and Nd:glass CPA laser system operating at a wavelength of 1.053 µm, capable of producing 400-fs pulses with energies up to 3 J (14). Intrinsic pulse contrast is ~10^5; in order to increase it, the infrared pulse was converted to its second harmonic (λ=527 nm) using a 4-mm thick type-I KD*P crystal (15). Four harmonic beam-splitters and a BG39 glass filter were used after the crystal to filter out the infrared component in the laser beam. The contrast for the green pulse was estimated to be better than 10^{10}.

Picosecond, λ = 1 µm Experiments

First, a series of experiments on the second harmonic generation were made with the FCM-CPA laser system, concentrating on the effect of preplasma produced by small prepulses. Polished flat silicon targets were used. With the p-polarized laser incident at 35° from the target normal, the 2ω signal was collected in the specular direction. We compared the

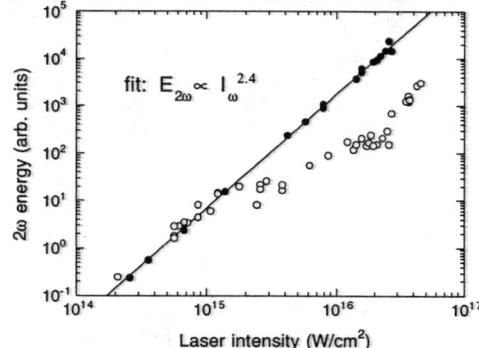

Fig. 1: 2ω yield from clean pulses (solid) and with fixed-fraction prepulse (open circles). fit: $E_{2\omega} \propto I_\omega^{2.4}$

2ω yield under two experimental conditions: 'clean pulse' (with dye-cell in) and 'intrinsic pulse' (with dye-cell out). Figure 1 shows the laser intensity scaling of the 2ω energies measured with the intrinsic pulse and with the dye-cell-cleaned pulses. No difference was observed while the laser intensity was below 2×10^{15} W cm^{-2}. As the laser intensity passed 2×10^{15} W cm^{-2}, the collected 2ω energy from the intrinsic-contrast pulses began to saturate, presumably as preplasma was formed.

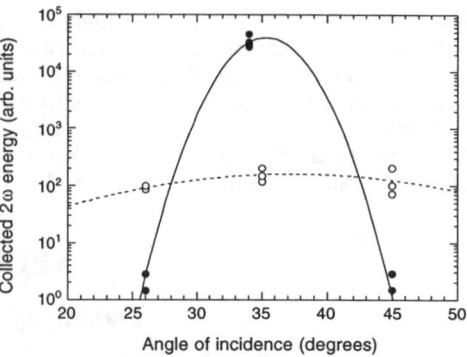

Fig. 2: 2ω yield: clean pulses (solid dots); fixed-fraction prepulse (circles). Deviation seen in Fig. 1 follows this spread to larger solid-angle, inferred from modest changes in the angle of incidence.

To examine this yield loss, the angular distribution of the second harmonic was measured at a laser irradiance of 2×10^{16} W cm^{-2} under these two pulse-contrast conditions. The 2ω angular distribution spread out and its intensity dropped, with prepulse present (Fig. 2). After correcting for the increased solid angle, the total yields are almost the same with and without preformed plasma.

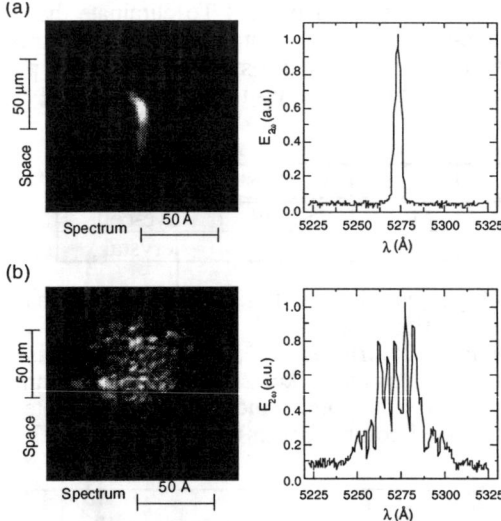

Fig. 3: Spatially resolved 2ω spectrum: (a) with clean pulse; (b) with 10^{-4} prepulse. At left is a spectrum (resolved horizontally) for a slit-image of the plasma (vertically), from center of the laser focus; at right is a lineout of the spectrum through centre.

Secondly, the details of the spatial and spectral structure of the harmonic source were examined, using a 30× image of the harmonic source relayed from the focal spot to an imaging spectrograph to give simultaneous spatial and spectral resolution. The results shown in Figure 3 illustrate the complex spatial and spectral changes which accompany beam breakup and self-phase modulation in the preformed plasma of Figs. 1 and 2.

350 fs, λ = 0.53 μm Experiments

For higher-harmonic experiments, the CUOS T^3 laser system was used. An f/3.0 23-cm focal length, off-axis (15°) parabolic mirror was used to focus the 43-mm diameter laser beam onto the target at an incidence angle of 60° from normal. The focal spot was measured to be 9 μm FWHM using a partially amplified beam. Harmonic spectra were recorded using a 1-m Seya-Namioka type VUV spectrometer (McPherson) equipped with a 1200 lines/mm grating. The laser focal spot on target was positioned on the Rowland circle and served as would an entrance slit for the spectrometer. At the output, a microchannel plate intensifier (Galileo) was used as the detector, lens-coupled to a 16-bit CCD camera (Photometrics) located outside the vacuum chamber.

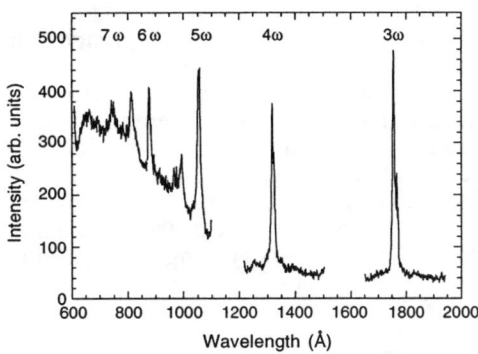

Fig. 4: Typical time-integrated harmonic spectrum from Si target, I ~ 3.2 × 10^{17} W cm^{-2}.

Several flat targets of various atomic-number materials (beryllium, CH, silicon, aluminum, nickel, gold) were used in this experiment. Harmonic spectra from all these targets exhibited similar features. Figure 4 shows a typical time-integrated harmonic composite spectrum recorded from a silicon target irradiated in p-polarization at a laser intensity of 3.2 × 10^{17} W cm^{-2}. Harmonics from 3rd to 7th, of both odd and even orders, can be easily identified, sitting on top of a broad Z-dependent plasma recombination background (Fig. 4).

By rotating a half-wave plate located after the doubling crystal, the 3ω yield from a beryllium target was measured both with p- and s-polarized incident laser pulses. With irradiances up to 1 × 10^{18} W cm^{-2} on target, we observed clear 3ω signal (S/N~30) for p-polarized irradiation but no 3ω signal for s-polarized irradiation (Fig. 5). Since there is no analyzing polarizer for the harmonic output in the current arrangement, we cannot draw clear conclusions about the polarization of the harmonics generated. It is likely that there is some polarization-selection inherent in diffraction from the grating of the spectrograph. If we assume that the p-polarized incident irradiation produced p-polarized harmonics (11) and that the grating efficiency of the spectrometer is 5 times greater for p-polarization than for s-polarization, we can conclude from our results that with s-polarized irradiation any 3ω yield is: a) if p-polarized, no more than 1/30th of that from the p-polarized laser pulse, and b) if s-polarized, 1/6th (5/30) of that from the p-polarized laser pulse.

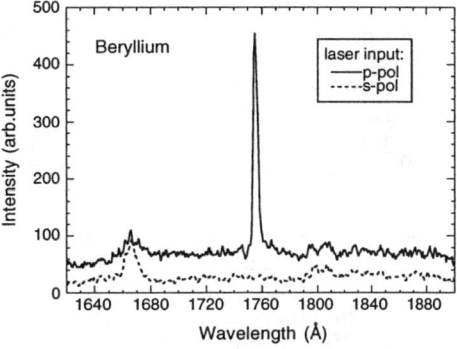

Fig 5: Spectra recorded using p-polarized laser pulses (solid), and s-polarized pulses (dashed).

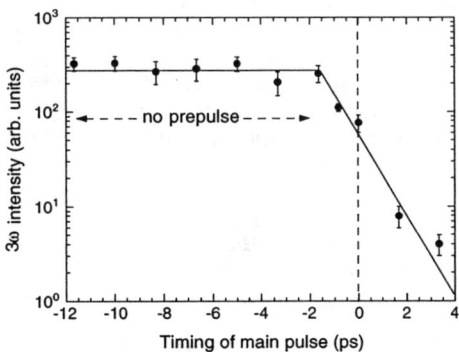

Fig. 6: Relative efficiency of 3ω from flat Si vs. prepulse/main pulse delay. Prepulse is 5% of the main pulse, which is 5 × 10^{17} W cm^{-2}.

Although a satisfactory model of the relation between harmonic generation and plasma-scalelength is still lacking for these laser-plasmas, it is generally believed that the efficiency of harmonic generation will vary with the gradient of the density profile (16). Experimentally, the gradient of the plasma density profile can be controlled by adding a deliberate prepulse at a controllable time before the main pulse arrives. In this way, the effect of density gradient on harmonic generation can be studied quantitatively.

After the compressor, we split out a small part of the infrared beam, frequency-doubled it in a crystal similar to that used in the main beam, and then co-propagated it with the infrared beam. The focussed intensity and spot-size of the prepulse were controlled by a an adjustable diaphragm. In this experiment, the main pulse intensity was 5×10^{17} W cm^{-2}, and the prepulse intensity was set to be 2.5×10^{16} W cm^{-2} — about 5% of the main pulse. The timing between peaks of these two orthogonally polarized green pulses was determined with a resolution of ±70 fs by frequency-domain interferometry. Figure 6 shows 3ω yield from separate shots onto a flat Si target, collected in the specular direction, as the delay between the main pulse and the small prepulse was increased. It can be seen that the harmonic efficiency starts to decrease once the prepulse moves ahead of the main pulse, and drops quickly by two orders of magnitude for 5% prepulses arriving 3 ps ahead of the main pulse.

Previously, it has been established that such intensities do produce motion of the critical density surface, near the peak of the pulse (17) and presumably some steepening occurs over 200 fs or so. A sense of the dependence of third-harmonic efficiency on the scalelength initially seen by a generating pulse can nonetheless be inferred from hydrodynamic modelling of the plasma expansion, following the prepulse. For each delay in Figure 6, the scalelength of the evolving plasma was calculated using the 1-D hydrocode MEDUSA (18), and the new correspondence plotted in Figure 7: as the scalength changes from about 0.1 λ to about 0.66 λ, the third-harmonic efficiency drops by two orders of magnitude.

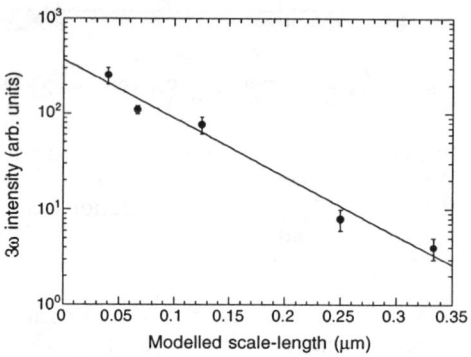

Fig. 7: Efficiency of 3ω vs. MEDUSA-modelled scalelength, from prepulse delays in Fig. 6.

In resolving the structure of harmonic lines at higher irradiances, we have observed the development of novel satellite lines, both red- and blue-shifted, which appear to have a regular Stokes- and anti-Stokes-like structure. These lines appeared around each of the 3rd–6th harmonics, apparently simultaneously but appreciably after the appearance of the harmonics themselves. The appearance intensity of these satellites, in our geometry, was mid-10^{17} W cm^{-2}. As the irradiance was increased to 1×10^{18} W cm^{-2}, we observed the sequential appearance of three such peaks: first a red-shifted peak, then on blue-shifted peak, then an additional red-shifted peak, each stepped in frequency by the same increment. The satellites were repeatable and spectrally narrow; in a few cases, the line on the longer-wavelength side was as intense as the harmonic line itself (Fig. 8).

These lines appeared for different atomic-number materials: Be, CH, Al, and Si, but were not seen under any of our conditions for the higher-Z elements, Ni and Au. For CH targets, the frequency step between satellites was ~7.4×10^{13} rad s^{-1}; initial analysis suggests that this shift may be weakly Z-dependent, with a possible 10% difference between Be and Si.

At a sensitivity of 10^{-4} of incident intensity, we detected no backscattered light. However, the spectrum of forward-

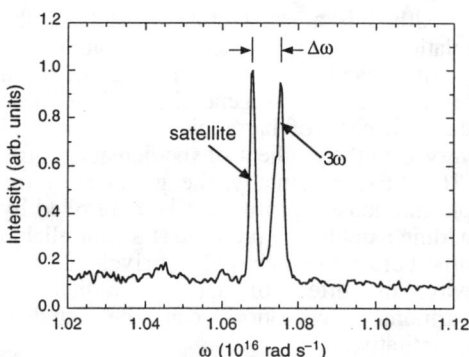

Fig. 8: The appearance of the first red-shifted satellite to 3ω from a parylene-N target (CH); I ~ 4×10^{17} W cm^{-2}, Δω = 7.4×10^{13} rad s^{-1}.

scattered laser light showed sudden line-centre depletion and broadening exactly upon the appearance of the satellite features.

We do not at present have a satisfactory explanation for these features, but we note that the frequency step between satellites corresponds approximately to the resonant electron plasma wave frequency scaled by the square root of the ratio of electron and ion masses and presumed charge-to-mass ratio, *cf.*, an associated ion plasma wave.

CONCLUSIONS

In using very high contrast pulses to which we have added back deliberate, controlled prepulses, we have partly separated the contributions of nonlinear hydrodynamics and nonlinear optics in the generation low- and mid-order laser-plasma harmonics. We have shown how previously noted spectral breakup of harmonic spectra follows from preformed plasmas caused by small prepulses, characterized the effect of prepulse in spreading harmonic emission over large solid angles, and quantified the sensitivity of harmonic generation efficiency to the gradient of preformed plasma. Finally, we have reported new observations of regular Stokes-like and anti-Stokes-like satellite features accompanying the harmonics, possibly resulting from the participation of ions near critical.

ACKNOWLEDGMENTS

The authors gratefully acknowledge the support of several agencies: RSM: the Natural Sciences and Engineering Research Council of Canada, Photonics Research Ontario, and the Fellows Program, Center for Ultrafast Optical Science, University of Michigan; DPU: grants NSF PHY 972661 and NSF STC PHY 8920108; SPL, MCD: partial support of DOE grant DE-FG03-97ER54439, Robert Welch Foundation grant F-1038, and a Faculty Research Assignment from the University of Texas.

REFERENCES

1. Burnett, N.H., Baldis, H.A., Richardson, M.C., Enright, G.D., *Appl. Phys. Lett.* **31**, 172 (1977).
2. Carman, R.L, Forslund, D.W., and Kindel, J.M., *Phys. Rev. Lett.* **46**, 29 (1981).
3. Carman, R.L., Rhodes, C.K., and Benjamin, R.F., *Phys. Rev. A* **24**, 2649 (1981).
4. Bezzerides, B., Jones, R.D., and Forslund, D.W., *Phys. Rev. Lett.* **49**, 202 (1982).
5. von der Linde, D., *et al.*, *Phys. Rev. A* **52**, R25 (1995).
6. Kohlweyer, S., *et al.*, *Opt. Commun.* **117**, 431 (1995).
7. Norreys, P.A., *et al.*, *Phys. Rev Lett.* **76**, 1832 (1996).
8. Esarey, E., *et al.*, *IEEE Trans. Plasma Sci.* **21**, 95 (1993).
9. Gibbon, P., *Phy. Rev. Lett.* **76**, 50 (1996).
10. Wilks, S.C., Kruer, W.L., and Mori, W.B., *IEEE Trans. Plasma Sci.* **21**, 120 (1993).
11. Lichters, R., Meyer-ter-Vehn, J., and Pukhov, A., *Phys. Plasmas* **3**, 3425 (1996).
12. Fedosejevs, R, *et al.*, *Phys. Rev. Lett.* **43**, 1664, (1979).
13. Marjoribanks, R.S., *et al.*, *Opt. Lett.* **18**, 361 (1993).
14. Mourou, G., and Umstadter, D., *Phys. Fluids B* **4**, 2315 (1992).
15. Chien, C.Y., *et al.*, *Opt. Lett.* **20**, 353 (1995).
16. Delettrez, J., Bonnaud, G., *LLE Review* **58**, 76 (1994).
17. Liu, X., and Umstadter, D., *Phys. Rev. Lett.* **69**, 1935 (1992).
18. Christiansen, J.P., Ashby, D.E.F.T., and Roberts, K., *Comput. Phys. Commun.* **7**, 271 (1974); Djaoui, A., and Rose, S.J., *J. Phys. B* **25**, 2745 (1992).

Interaction of TW laser pulses with high density gas jet targets near the threshold for relativistic self-focusing

G. D. Tsakiris, R. Fedosejevs[1], and X. F. Wang[2]

Max-Planck-Institut für Quantenoptik, D-85748 Garching, Germany

Abstract. We report optical investigations of the interaction of femtosecond Ti:sapphire laser pulses with underdense plasmas created from high density gas jet targets. Time-resolved shadowgraphy using a 2ω probe pulse and images of the transmitted radiation are presented for nitrogen and hydrogen. For the laser power available, the experimental results and their analysis based on a simple numerical Gaussian beam model show that ionization induced refraction dominates the interaction process for all gases except hydrogen. The numerical modeling also shows that for a given laser power there exists only a narrow density range in which self-focusing can be expected to occur. In the case of hydrogen for electron densities greater than $\sim 10^{20} cm^{-3}$, the onset of channeling expected at the critical power for relativistic self-focusing is experimentally observed.

The propagation of ultrashort high intensity laser pulses in underdense plasma is of interest for the understanding of laser interactions in this new relativistically driven plasma regime and for applications such as laser wakefield accelerators [1], x-ray lasers [2] and the fast ignitor concept for inertial confinement fusion energy [3]. The propagation of such pulses through long scale length underdense plasmas involves a competition between diffraction, ionization induced refraction and relativistic self-focusing. The threshold power for relativistic self-focusing in a uniform plasma given by the critical power of $P_c = 16.2\ n_c/n_e [GW]$, can be reduced using higher electron densities, but at the same time the accompanying refractive defocusing is increased. The present work explores this problem of propagation at relatively high gas densities both using a simple Gaussian beam propagation model and by a series of experimental measurements.

---

[1] on leave from the Department of Electrical Engineering, University of Alberta, Edmonton, Canada.
[2] on leave from the Shanghai Institute of Optics and Fine Mechanics, Chinese Academy of Sciences, Shanghai, China.

The problem of beam propagation in a medium undergoing rapid ionization right at the leading edge of a short ($\lesssim 1ps$) intense ($\gtrsim 10^{15} W/cm^2$) laser pulse can be treated using the paraxial ray equation for a Gaussian beam [4] in conjunction with the plasma refractive index which at relativistic laser intensities is given by $\eta = 1 - \frac{1}{2}\frac{\omega_p^2}{\omega_L^2}(1 + a_L^2/2)^{-1/2}$ [5,6]. Here $\omega_p = (4\pi e^2 n_e/m)^{1/2}$ is the plasma frequency, ω_L is the laser frequency, and $a_L = eA_L/mc^2$ the amplitude of the normalized vector potential.

A detailed analysis of the resulting paraxial ray equation for an ionizing plasma [7] reveals that in the power-density parameter space one can distinguish three different regions characterized by the corresponding dominant process. A specific example of such $P - N$ diagram for nitrogen is given in Fig. 1. Here $N = n_a/n_c$ denotes the normalized atomic density to the critical density. It is seen that for relative high densities refraction dominates and irrespectively of the laser power available one cannot expect to observe self-focusing. For densities lower than a limiting density N_B refraction would determine the minimum beam radius unless the laser power is high enough for the beam to self-focus. In terms of the initial beam convergence half-angle θ_0 and the maximum ionization state of the gas q_{max} this limiting density is approximately given by $N_B \approx \tan^2\theta_0/q_{max}$. For gas densities lower than N_B, self-focusing will occur if in addition the condition $P_L > P_c$ is satisfied, which in turn sets a lower density limit N_A dependent on the laser power P_L. It is easily deduced that this limit would be $N_A \sim 1/(q_{max}P_L)$. It follows that for a given laser power P_L and a specific gas target there exists a gas density range $N_A < N < N_B$ for which an ideal Gaussian beam will exhibit self-focusing (see Fig. 1). Outside this density *window* no self-focusing can occur and more specifically, for $N < N_A$ diffractive defocusing while for $N > N_B$ refractive defocusing would dominate over the self-focusing mechanism. These results although derived for the case of static-filled interaction chamber, are still useful in order to separate and study the different processes when instead a gas jet is used that limits the spatial extend of the beam interaction with the gas to a small volume around its nozzle.

The experiments were carried out using the ATLAS Ti:sapphire laser facility at Max-Planck-Institut für Quantenoptik delivering a $0.3TW$, $80mJ$, $250fs$ pulse focused at a peak vacuum intensity of $3\times 10^{17} W/cm^2$. A $500\mu m$ diameter pulsed gas jet yielding peak electron densities of up to $0.25 n_c$ was used for the interaction studies. The main optical diagnostics were: time resolved shadowgraphy using a 2ω probe and an image of the transmitted 1ω radiation at a position of $250\mu m$ past the vacuum focus position. A perspective view of the experimental layout is shown in Fig. 2a. A number of additional diagnostics not discussed here were also used, e.g. side on images of the 1ω and 2ω self-emission, Raman backscattered and forward scattered spectroscopy, and hard x-ray emission detection. A detailed report of these experimental results is given in Ref. [7].

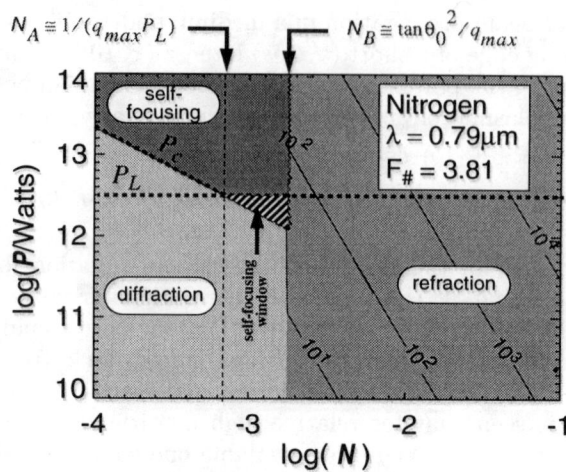

FIGURE 1. The $P - N$ diagram for the case of a Gaussian beam with the indicated parameters propagating in nitrogen. Depending on the power P of the beam and the density of the gas N, the final minimum radius would be determined by the dominant process, i.e., diffraction or refraction or self-focusing. The contours indicate the minimum achievable radius in μm in the refraction dominated regime. The diagonal dashed line represents the transition to self-focusing according to the relation $P \sim 1/(q_{max}N)$. The shaded area indicates the density *window* where self-focusing can occur for a laser power of $P_L = 3.0TW$.

A sequence of images for nitrogen gas at three different gas densities is shown in Fig. 2b (left column). The shadowgrams were taken at $9ps$ after the leading edge of the pulse has reached the vacuum focus position. Thus, the images give a snap shot of the ionized plasma produced after the interaction has finished but before any significant expansion of the plasma can occur. For a density of $N_0 = 0.014$ the outline of the ionized plasma region is clearly visible and filamentary structures can be seen inside the ionized plasma. This fine structure in the form of regular striations of the shadow has been previously observed in similar experiments and it has been attributed to the refractive splitting of intensity non-uniformities (hot spots) in the laser beam profile [8,9]. Asymmetry in the plasma structure is observed with more deflection and curvature towards the gas nozzle than away from the gas nozzle. This arises from the gradient in the background gas density which decrease in the direction away from the nozzle. Thus the upper part of the beam propagates through significantly higher gas density than the lower part of the beam. At a much higher density of $N_0 = 0.041$ the transverse size of the plasma has increased further and the asymmetry is even more pronounced than in the previous images. Whereas, at lower pressures it appears that the ionized plasma continues to the edge of the field of view, in the highest density case

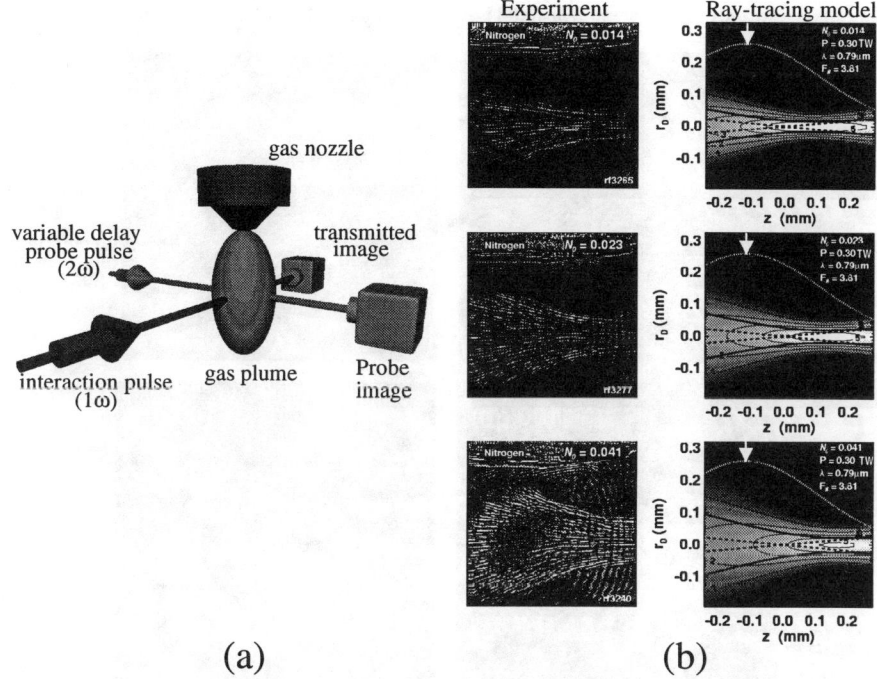

FIGURE 2. (a) Perspective view of the gas jet target and interaction region geometry. (b) Comparison of the experimentally obtained shadowgrams (left column) with the predictions of the ray tracing model (right column) for nitrogen and the indicated peak densities. The laser enters from the right and the gas jet nozzle is located at a distance of $\sim 300 \mu m$ above the laser propagation axis. In the experimental records the shadow of the gas jet orifice is visible in the upper part. In the model calculations the thick solid line represents the $1/e$ beam radius variation in the presence of the gas jet while the thick dashed line in vacuum. The ionization contours are given as labeled thin lines and the gray scale depicts the continuous variation of the ionization degree up to its maximum value of $q_{max} = 5$. The arrow gives the exact location of the peak density N_0 while the white dotted line its variation along the z axis.

the absorption and refraction is so strong that the main part of the plasma terminates at about $100 \mu m$ before the edge of the field of view. The shape of the terminating edge of the main plasma is also asymmetric occurring earlier nearer the gas jet nozzle than the side farther from the gas jet nozzle. This would agree with enhanced refraction from the higher density region of the gas jet spreading the beam more and causing the average flux to drop below the single ionization threshold of nitrogen. Also the dark region in the tail end of the plasma indicates significant absorption of the probe beam in the large, cold high density plasma.

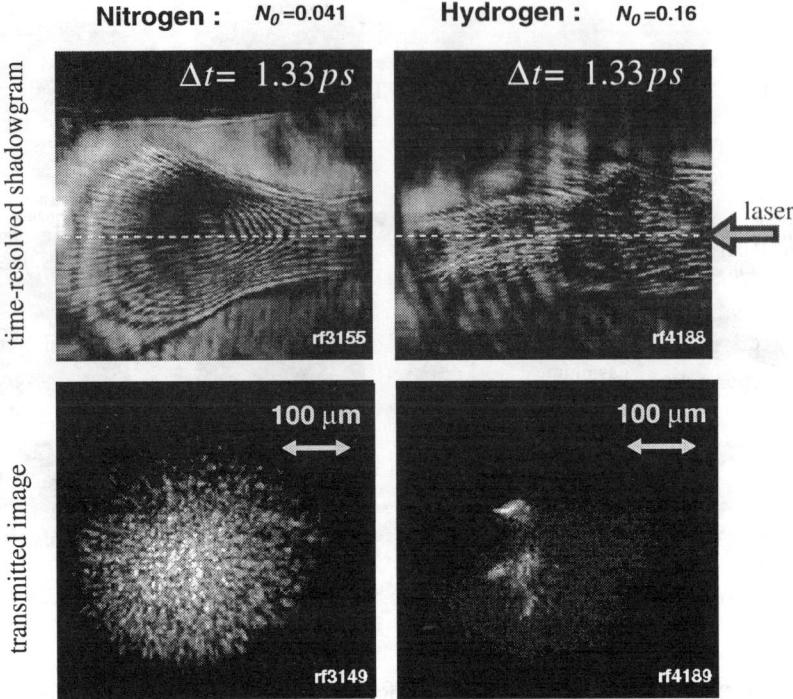

FIGURE 3. Direct comparison of the shadowgram and transmitted spot images underlining the radically different behavior of the interaction for nitrogen and hydrogen.

A direct comparison of the shadowgrams with the ray tracing model is depicted in Fig. 2b (right column) for the case of nitrogen and for the three densities in the refraction dominated region. The sharp well defined boundaries of the shadow cast by the interaction region seen in the experimental records of Fig. 2a are associated with the transition boundary between neutral gas and singly ionized nitrogen atoms. It indicates that even small intensity changes as those existing in the wings of the radial beam profile result in substantial refraction index gradients and that these gradients are produced at well defined regions where the intensity first exceeds the value required for single ionization of the nitrogen atom. In this context, it is expected that the observed shadowgrams should correspond to the outer ionization contour, $Z = 1$, predicted by the model. As can be seen the predicted refraction by the ray tracing model agrees qualitatively well with the images and it reproduces the overall behavior as the density increases.

The strikingly different behavior of the interaction for nitrogen and hydrogen is depicted in Fig. 3. For nitrogen, the time-resolved shadowgrams show an orderly progression of refraction with the beam smoothly spreading out as it propagates forward. However, for hydrogen a significantly different pattern is observed. After arriving at the vacuum focus position two prongs of light jut forward and the lateral spreading of the radiation appears constrained. Similarly, the transmitted images exhibit very different behavior. For nitrogen, the diameter increases as the transmitted spot is more and more refracted. In addition, the overall distribution of the transmitted light becomes quite smooth with a filamentary substructure similar to the filamentary appearance seen in the shadowgrams. For hydrogen on the other hand, several hot spots are observed in the transmitted radiation. This structure is indicative of large scale channeling of parts of the laser beam. The observed structures are similar from shot to shot indicating that the channeling is seeded by modulation in the laser beam profile itself.

In conclusion, a study has been carried out on the propagation of high-power Ti:sapphire laser pulses in high density gas targets. In contrast to previous investigations, we have focused our interest in the density range around $0.1 n_c$ or $n_e \sim 10^{20} cm^{-3}$ for which the threshold for relativistic self-focusing could be achieved for modest intensities of a fraction of a terawatt. While a window of relativistic self-focusing was predicted for nitrogen gas for our experimentally available power of $0.3 TW$, this was not observed in the experiments. This is attributed to the strong deviation of the beam quality from the lower order Gaussian mode on which the analysis is based. In the case of hydrogen gas for which refraction has a lesser effect due to reduced ionization threshold, clear evidence for the onset of self-focusing was observed at densities $n_e \approx 0.1 n_c$.

ACKNOWLEDGMENTS

This work was funded in part by the commission of the European Communities in the framework of the Euratom-IPP association. X. F. Wang was supported by a fellowship under the Cooperative Agreement between the Max-Planck-Gesellschaft and the Chinese Academy of Sciences.

REFERENCES

1. T. Tajima and J. M. Dawson, Phys. Rev. Lett. **43**, 267 (1979).
2. N. H. Burnett and P. B. Corkum, J. Opt. Soc. Am. B **6**, 1195 (1989).
3. M. Tabak, et al., Phys. Plasmas **1**, 1626 (1994).
4. E. E. Fill J. Opt. Soc. Am. B **11**, 2214 (1994).
5. C. E. Max, et al., Phys. Rev. Lett. **33**, 209 (1974).
6. P. Sprangle, et al., Phys. Rev. A **41**, 4463 (1990).
7. R. Fedosejevs, X. F. Wang, and G. D. Tsakiris, *Onset of relativistic self-focusing in high density gas jet targets*, accepted for publication in Phys. Rev. E, (1997).
8. A. Sullivan, et al., Opt. Lett. **19**, 1544 (1994).
9. A. J. Mckinnon, et al., Phys. Rev. Lett. **76**, 1473 (1996).

Energy Deposition and Transport Dynamics in Plasmas Produced by Intense, Short Pulse Irradiation of Atomic Clusters

T. Ditmire, R. A. Smith, E. T. Gumbrell,
J. W. G. Tisch, and M. H. R. Hutchinson

Imperial College of Science, Technology and Medicine
Blackett Laboratory
London, SW7 2BZ
United Kingdom

Abstract: We have explored the dynamics of laser energy deposition and subsequent energy transport in plasmas produced by the intense ionization of gases composed of large atomic clusters. We find that the laser energy absorption can be extremely efficient, despite the low average density of the gas. We have used time-resolved interferometery to examine the spatial and temporal evolution of the plasmas created in this manner. We find that, due to the high temperatures achieved, a supersonic ionization wave is driven by hot electron thermal energy transport.

INTRODUCTION

To date the vast majority of laser-plasma studies have been conducted by interacting an intense laser with either a solid density target or a gas of atomic monomers. In contrast to these target materials, a gas of large atomic clusters (> 1000 atoms/cluster) presents a radically different environment for laser-plasma interaction dynamics. In general, low density gases ionized by intense laser pulses are expected to exhibit very low absorption efficiency (< 1%) and remain quite cold (10 - 100 eV) (1). The presence of clusters in a gas changes this situation dramatically. Though the average atomic density is low, the local density within the cluster is near solid and, consequently, will be subject to the rapid heating experienced by a solid target from collisional inverse bremsstrahlung (2). Bright x-rays have been observed from gas target plasmas produced by intense femtosecond illumination of clusters (2,3), indicating that electron temperatures in these plasmas were quite high.

Recent studies of intense laser interactions with individual clusters have confirmed that hot electrons (up to 3 keV) are produced during the laser-cluster interaction (4), and that, furthermore, an even greater energy can be deposited in the ions when these hot, highly ionized clusters explode (5-7). These studies indicated that the clusters are rapidly heated by the laser, to a non-equilibrium, super-heated state, in large part due to the passage of the free electrons in the cluster through a Mie resonance with the laser field during the cluster expansion (8). Soon after the clusters are heated, charge separation of the hot electrons causes the clusters to explode on a sub-picosecond time scale.

We have conducted laser energy absorption measurements of intense laser pulses in gaseous media containing atomic clusters (9). We find that cluster gases are at least as efficient as solid targets in absorbing short pulse laser energy (10), resulting in a high temperature (~ 1 keV) plasma. Using short pulse laser interferometry we have further explored the dynamics of these cluster plasmas by temporally and spatially resolving the electron density profiles of the hot plasma filament produced by focusing the laser pulse through the cluster medium.

ENERGY ABSORPTION EFFICIENCY OF CLUSTERING GASES

To measure the energy absorption efficiency of an intense laser pulse in a gas of atomic clusters we used a Nd:glass laser based on chirped pulse amplification which produces 2 ps laser pulses with energy up to 0.5 J. These pulses were frequency doubled to 526 nm, and focused with an f/12 lens to a spot of 30 µm ($1/e^2$ diameter) into the output of a pulsed gas jet. This gas jet produced a gas of large clusters (>50 Å diameter) when high backing pressure is used (> 10 bar with argon). The transmitted light was collected with an eight cm diameter f/2 lens to assure that no light refracted by plasma formation at the focus was lost. The energy of the transmitted light was measured with a volume absorbing calorimeter, and the input laser energy was monitored with a photodiode. Any backscattered light was also monitored with a fast photodiode; (none was observed for any of the measurements described in this work). Imaging from the side of the plasma was also performed and showed that no significant amount of laser energy was side scattered.

The absorption fraction of the laser energy in an argon jet with 40 bar of pressure (yielding an average atomic density of 1.5×10^{19} cm^{-3} and ~ 80 Å clusters) as a function of the peak laser intensity is shown in figure 1. The absorption rises from zero at an intensity of 2×10^{13} W/cm^2 and slowly increases with intensity to 70% at 5×10^{15} W/cm^2 after which the absorption begins to drop with increasing intensity. One important feature of the data in figure 1 is the saturation in the absorption fraction followed by a drop in absorption with a rise in laser intensity. Previous modeling indicates that this feature in the data appears to be the result of the limited time over which the clusters absorb laser energy in the laser pulse (9). The rapid expansion of the clusters once they begin to be heated by the laser causes them to disassemble on a 100 - 200 fs time scale, a time scale faster then our 2 ps pulse. As a result, the leading edge of the pulse initially experiences the highest absorption as it propagates into the cluster medium. This leads to an effective pulse shortening as the pulse propagates deeper into the medium. When the pulse energy is high enough, the pulse "burns through" the cluster medium, and the trailing edge of the pulse propagates through the medium without any significant absorption.

The very high absorption observed in these measurements indicates that a large amount of laser energy can be deposited in a small volume in these clustering gas jet targets. In our experiments, the laser energy is deposited in a volume of ~ $\pi \omega_0^2 l$ ~ 5×10^{-6} cm^3 (where ω_0 is the laser focal spot radius and $l \approx 2$ mm is the plasma length .)

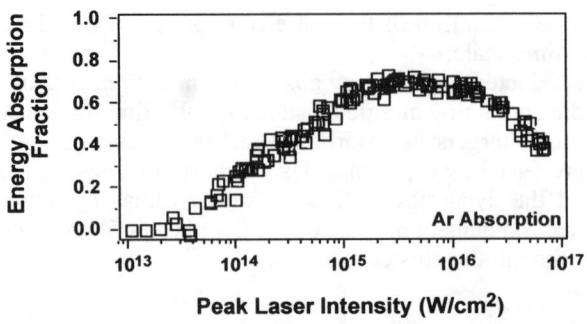

FIGURE 1. The absorption fraction of the laser energy in an argon jet as a function of the peak laser intensity. The jet was backed with 40 bar of pressure (~ 80 Å clusters).

At the highest intensity, over 0.15 J of laser energy is deposited in this small volume. With an average atomic density of 1.5×10^{19} cm^{-3} in the gas jet, this implies that up to 15 keV of energy is deposited per atom. For an argon gas with an average ionization state of 8+ (Ne-like), this absorbed energy fraction implies an initial electron temperature of ~ 0.8 - 1 keV assuming an equal distribution of energy between electrons and ions. Previous measurements of electron energies from laser heated clusters indicate that this is a reasonable value (4).

INTERFEROMETRIC PROBING OF THE LASER ENERGY DEPOSITION

To temporally and spatially probe the plasma filament we conducted short pulse interferometry on these cluster plasmas. To do this a small amount of the main 2 ps, 526 nm pulse was split off and Raman shifted in ethanol to a wavelength of 620 nm. This pulse traversed a delay leg and illuminated the plasma filament at a right angle to the propagation of the main pulse. The plasma was imaged with a telescope onto a CCD camera. Between the imaging telescope and the CCD detector the probe light passed through a Michelson interferometer, with a roof prism in one leg such that the probe light which traversed the plasma filament was interfered with reference light which passed below the plasma. The interferograms yield information on the phase shift resulting from the passage of the light through the cylindrically symmetric plasma. This shift can be Abel-inverted to yield the radial electron density profile.

An example of interferometric images taken in the cluster gas when the pulse absorption is near its maximum is shown in figure 2. Here a 30 mJ pulse (~10^{16} W/cm^2) enters the image from the left. At t = 0 the fringes are smeared in the center of the image due to ionization of the gas occurring on the same time scale as the probe pulse. At 5 ps after the main pulse has traversed the cluster medium, the fringe deviation from the plasma is clearly visible. Though the plasma channel is of a large diameter to the left of the image, where the pulse has entered, the electron density drops as the pulse propagates to the right. This is a result of the large attenuation experienced by the pulse as it travels through the cluster medium. Only a small fraction of the laser energy reaches the far end of the medium.

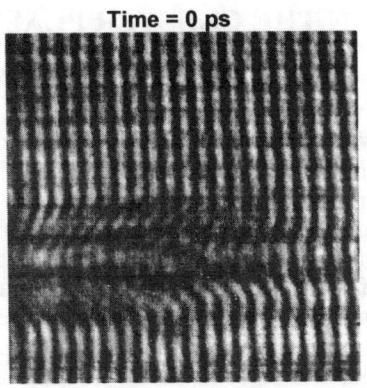 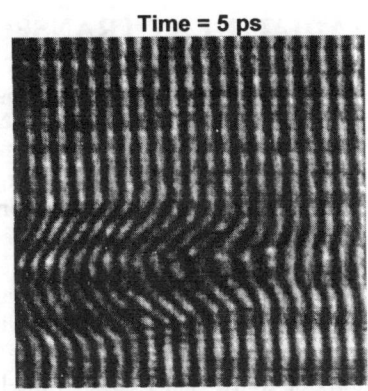

FIGURE 2. Interferometric images of a plasma created by a 2 ps 526 nm laser pulse with a peak intensity of 10^{16} W/cm^2 focused through an argon cluster gas. The laser enters from the left.

The deconvolved profile of this plama is shown in figure 3. The fall in electron density along the plasma channel is clearly evident here. This data also seems to indicate that, up to the point when the pulse energy is completely depleted, the plasma is quite uniform along the channel. The scale length for the stopping of the plasma channel at this point is only ~ 200 µm. This indicates that, though the pulse energy is absorbed as it propagates through the cluster gas, it deposits energy at a constant rate per unit length. This is consistent with the model forwarded in ref. (9) where it was hypothesized that, because the clusters expand on a fast time scale, only energy in a small temporal slice is absorbed at a given point in space. Early in the channel, the clusters absorb energy on the rising edge of the pulse and then disassemble before the remainder of the pulse arrives. This model predicts that the energy deposition will be fairly uniform until almost all of the pulse energy is depleted where the plasma channel abruptly terminates. This is confirmed by our data.

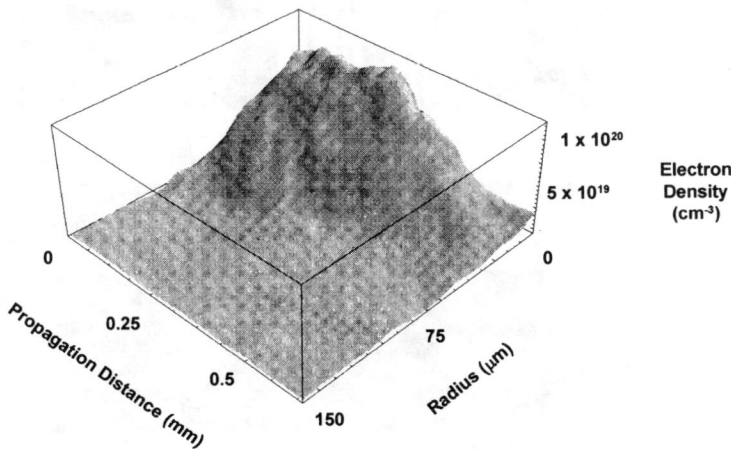

FIGURE 3. Deconvolved electron density profile from the plasma image in figure 2.

RADIAL HEAT TRANSPORT IN THE CLUSTER PLASMAS

In these laser heated clustering gases, the electron temperatures are in the vicinity of 1 keV, the radial scale lengths are of the order of 10-50 µm and the electron densities are between 10^{19} and 10^{20} cm^{-3}. We, therefore, expect that electron driven heat transport will rapidly conduct energy from the central plasma. One manifestation of this heat conduction is the evolution of an ionization wave outward from the hot core of the filament. To measure these energy transport dynamics we have used the interferometric probe to time resolve the electron density profiles.

Figure 4 shows interferometric images of the plasma filament created in an argon cluster gas at four different times with respect to the main heating pulse which contained 300 mJ of energy. 3 ps before the peak of the pulse a small amount of ionization due to the rising edge of the laser pulse is evident. At t = 0 rapid ionization by the main laser pulse in the center of the causes the fringes to smear out. From this image it is clear that the gas is initially ionized to a radius of slightly greater than 50 µm. Within 6 ps after the heating of the main pulse, heat transport from the central hot plasma has driven ionization out to a radius of over 100 µm. After 30 ps the fringe shift from the plasma is clearly visible and, though the ionization wave has slowed, heat transport continues to cause ionization at the large radius. This data indicates that

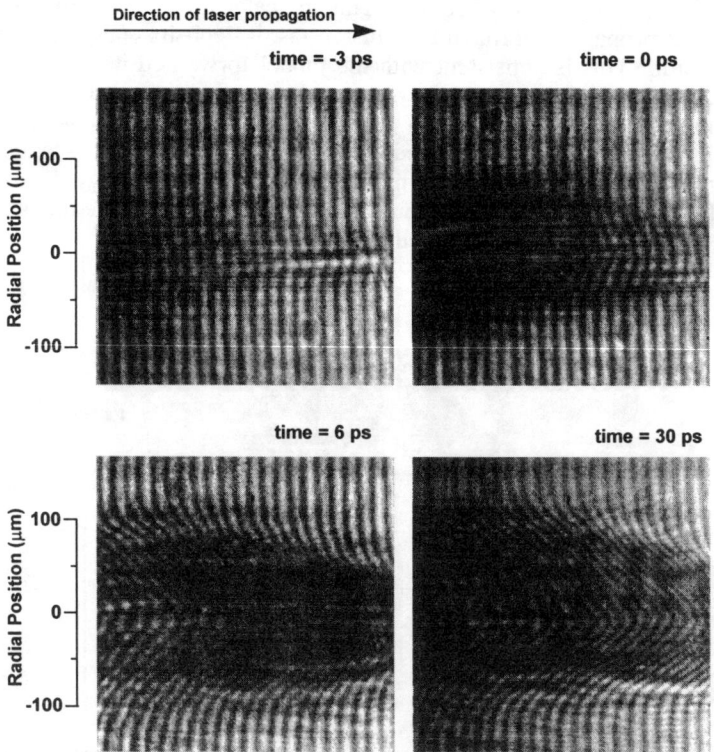

FIGURE 4. Interferograms of the hot Ar plasma at four times with respect to the laser pulse.

the hot electron ionization wave has a velocity of the order of 0.7 - 1 x 10^7 m/s. Deconvolution of the fringe images at the center of the plasma indicates that the electron density is approximately 1.3 x 10^{20} cm^{-3}, consistent with 8 times ionized Ar.

Under these conditions, where the plasma is hot and the density is not large, the mean free path of the thermal electrons in the plasma (given by $\lambda_e = (k_B T_e)^2 / 4\pi e^4 n_e (Z+1)^{1/2} \ln \Lambda$) is comparable to the radial extent of the plasma (20 - 50 µm). When this is true, standard diffusive heat transport is no longer the primary mechanism for the heat flux; instead the heat transport becomes nonlocal (11). It is dominated by hot electrons free streaming from the hot portions of the plasma to the cooler regions. In this situation an ionization wave with a velocity given roughly by $v = (k_B T_e / m_e)^{1/2}$ will result. This predicts a maximum free streaming ionization wave velocity of ~1.3 x 10^7 m/s for a 1 keV plasma. This is very close to the observed ionization wave velocity and suggests that nonlocal, free streaming electron energy transport is an important transport mechanism in these plasmas.

CONCLUSION

In conclusion we have presented a study of the energy deposition and the energy transport dynamics of intense, short pulse heated gases of atomic clusters. We find that the clusters are very efficient at absorbing the intense laser light and that very high plasma temperatures can result. We have examined the stopping of intense pulses in these gases using time resolved interferometry, and we have observed a fast ionization wave emanating from these plasmas on a picosecond time scale.

ACKNOWLEDGEMENTS

We would like acknowledge the technical assistance of A. Gregory and P. Ruthven. This work was supported by the UK EPSRC and the MOD.

REFERENCES

1. T. E. Glover, T. D. Donnelly, E. A. Lipman, A. Sullivan, and R. W. Falcone, *Phys. Rev. Lett.* **73**, 78 (1994).
2. T. Ditmire, T. Donnelly, R.W. Falcone, and M.D. Perry, *Phys. Rev. Lett.* **75**, 3122 (1995).
3. A. McPherson, B.D. Thompson, A.B. Borisov, K. Boyer, and C.K. Rhodes, *Nature (London)* **370**, 631 (1994).
4. Y.L. Shao, T. Ditmire, J.W.G. Tisch, E. Springate, J.P. Marangos, and M.H.R. Hutchinson, *Phys. Rev. Lett.* **77**, 3343 (1996).
5. J. Purnell, E.M. Snyder, S. Wei, and A.W.C. Jr., *Chem. Phys. Lett.* **229**, 333 (1994).
6. T. Ditmire, J.W.G. Tisch, E. Springate, M.B. Mason, N. Hay, R.A. Smith, J. Marangos, and M.H.R. Hutchinson, *Nature (London)* **386**, 54 (1997).
7. M. Lezius, S. Dobosz, D. Normand and M. Schmidt, *J. Phys. B: At, Mol. Opt. Phys.* **30**, L251 (1997).
8. T. Ditmire, T. Donnelly, A.M. Rubenchik, R.W. Falcone, and M.D. Perry, *Phys. Rev. A* **53**, 3379 (1996).
9. T. Ditmire, R. A. Smith, J. W. G. Tisch, and M. H. R. Hutchinson, *Phys. Rev. Lett.* **78**, 3121 (1997).
10. H.M. Milchberg, R.R. Freeman, S.C. Davey, and R.M. More, *Phys. Rev. Lett.* **61**, 2364 (1988).
11. J. F. Luciani, P. Mora, and J. Virmont, *Phys. Rev. Lett.* **51**, 1664 (1983).

Harmonic Generation during the Ionization of a Plasma by a Short Light Pulse.

Enrique Conejero Jarque and Luis Plaja

Dept. de Física Aplicada. Universidad de Salamanca.
37008 Salamanca. Spain.

Abstract.
In this contribution we study the influence of the ionization process in the radiation produced by a thin slab of a solid when it is irradiated by a short light pulse. The emitted light spectrum is calculated from the time-dependent solution of Maxwell and Lorentz equations, by means of one-dimensional particle-in-cell calculations. Ionization is described by the standard expresion for the tunnelling regime. Several new radiation modes appear in the transmited and reflected field which are identified in terms of highly non-linear wave mixing of the plasma and field frequencies due to collective effects.

INTRODUCTION

It is well-known that high-order harmonics can appear in the light reflected by a preionized plasma when the intensity of the incident field and the density of the medium are high enough. These harmonics are due to several mechanisms, such as the nonlinear Thomson scattering by the nearly-free electrons in the case of underdense plasmas. For overdense plasmas, resonance absorption, parametric instabilities, transverse density gradients or electrons which are dragged away and come back to the surface are all processes that can generate harmonics [1].

When very short pulses incide on a medium that has not been previously ionized, the radiation reflected and transmitted may be drammatically affected by the ionization process. In the limit of very high incident field intensities (above one atomic unit), the ionization process is nearly instantaneous and, therefore, has no effect in most of the interaction time. The harmonic yield in this case is equivalent to the yield from a pre-ionized plasma [2,3]. By contrast, when the field is not so intense and the ionization is produced during a few optical cycles, it cannot be ignored at all. In this case, a time-dependent ionization produces harmonic radiation as a steep spatial profile of the free electron density propagates inside the medium [4–6].

To gain insight in this problem, we have performed numerical calculations simulating the tunneling ionization of a thin plasma slab with the help of a 1D3V particle-in-cell code (one dimensional in space and three dimensional in velocity) [7,8]. The results show that the ionization process and the collective phenomena induce strong couplings between the incident field frequency and the intrinsic plasma frequency.

THE NUMERICAL MODEL

We consider a linearly polarized incident electric field $\mathbf{E_0}(x,t) = A(x,t)\sin(kx - \omega_0 t)\mathbf{e_z}$, with $A(x,t)$ a slowly-varying envelope, that incides perpendicularly to the surface of a medium whose depth is one wavelength. To study the propagation of the field we solve Maxwell equations for the electric and magnetic fields and the relativistic Lorentz equation for the movement of the particles.

The medium is ionized according the ionization rate [9]

$$W(x,t) = \frac{4}{|\mathbf{E}(x,t)|}\exp(\frac{-2}{3|\mathbf{E}(x,t)|}), \tag{1}$$

which is valid for hydrogenlike atoms in the tunneling regime. As we will use typically incident field amplitudes $E_0 \sim 0.1 a.u.$ and frequencies $\omega_0 < 0.1 a.u.$, the election of this rate is justified.

In addition to the PIC calculations, we have performed simulations of the ionizing plasma using a very simple model proposed by Brunel [4]. This model does not take into account particle dynamics, thus, reducing the problem to the solution of the set of equations [4–6]

$$\frac{\partial^2 E}{\partial x^2} - \frac{1}{c^2}\frac{\partial^2 E}{\partial t^2} = \frac{4\pi}{c^2}\frac{e^2}{m}NE, \tag{2}$$

$$\frac{\partial N}{\partial t} = W(N_0 - N), \tag{3}$$

where N_0 is the maximal density of free electrons and N the instantaneous one.

The comparison between this simple propagation model and the former PIC calculation will allow us to investigate the effect of the particle dynamics in the harmonic generation, including collective effects.

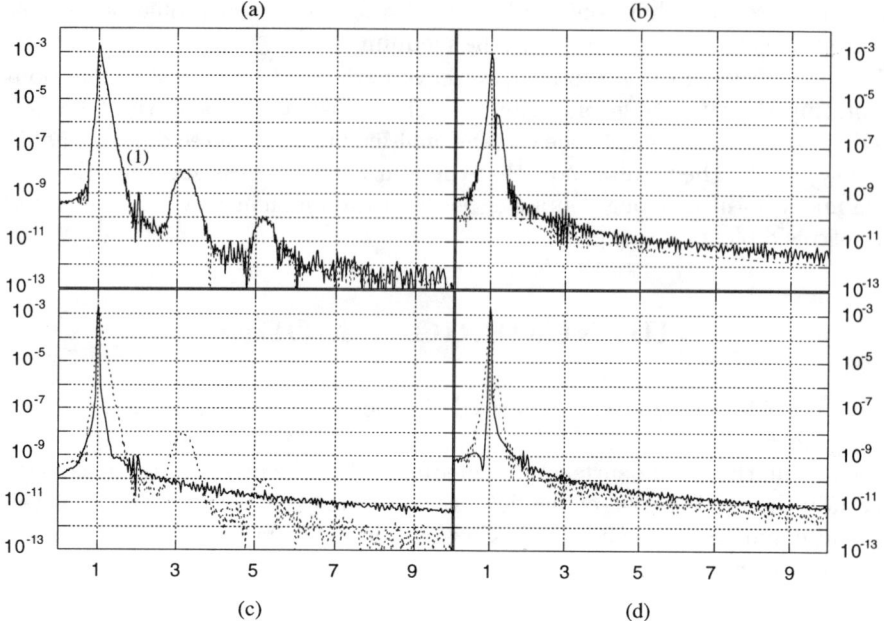

FIGURE 1. Spectra of the reflected (a) and transmitted (b) fields when $\omega_{p0} = \omega_0$. Solid line shows the result given by the PIC simulation. Dashed line shows the result given by the simple propagation model. Graphs (c) and (d) are the spectra obtained when the plasma is preionized.

RESULTS

In Fig. 1 we have depicted the harmonic spectra obtained during the propagation of a pulse of length 30 cycles with sinus square shape whose main frequency is $\omega_0 = 0.05 a.u.$ and whose maximum amplitude is $E_0 = 0.1 a.u.$. The maximum plasma frequency is in this case $\omega_{p0} = \omega_0$. Graphs 1a and 1b show the spectra of the transmitted and reflected fields, respectively. The solid line plots the spectrum resulting from the PIC calculation and the dashed line that calculated from the simple propagation model. For this parameters, the spectra obtained with both models are very similar. There is only a particular feature in the PIC calculation spectrum of the transmitted field, a small peak labelled as (1), which is not present in the simpler model calculation. The frequency of this peak coincides with $\omega_{p0} + \omega_0$, and therefore corresponds to the wave mixing between the incident field frequency and the plasma frequency, which is the signature of the collective effects.

The role of the time-dependent ionization may be singularized by repeating the PIC calculation for a pre-ionized plasma, i.e. the plasma is already ionized at $t = 0$ and no further ionization is taken into account. In figures 1c and 1d

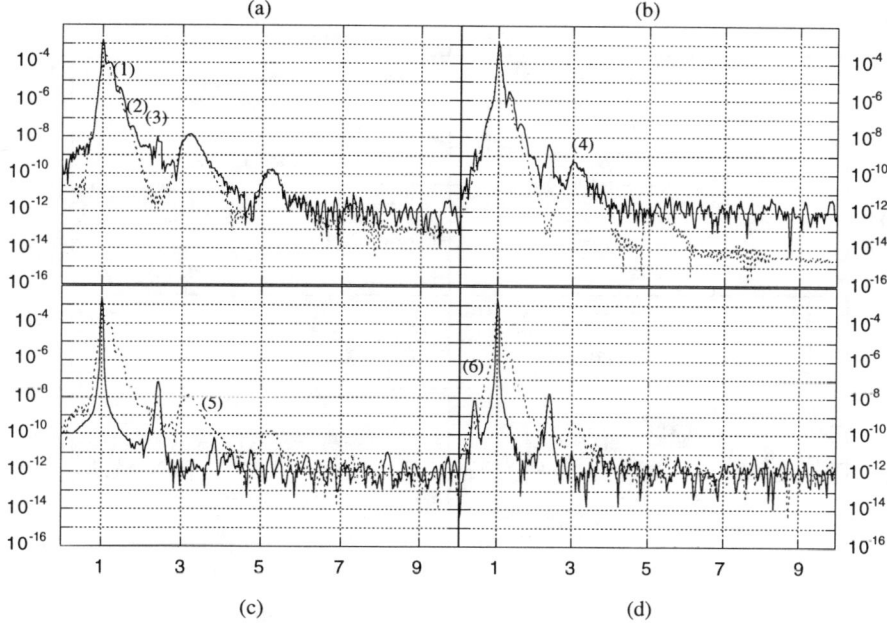

FIGURE 2. Same as previous plot when $\omega_{p0}^2 = 2\omega_0^2$.

we plot the harmonic spectrum for the pre-ionized case (solid line) against the full calculation (dashed line), shown already in Fig. 1a and 1b as a solid line. We can see that when ionization is neglected, the results are quite different, appearing only a sharp peak corresponding to the main incident frequency and traces of a response at $\omega_{p0} + \omega_0$.

We can conclude that, for underdense plasmas, the harmonic spectrum is completely dominated by the ionization process. The simple model is, therefore, a very good approximation and the inclusion of the ionization process is absolutely necessary for understanding the light emitted by the medium.

Fig. 2 shows the results obtained when the plasma frequency grows up to $\omega_{p0} = \sqrt{2}\omega_0$, remaining the rest of parameters unchanged. In this case, the differences between the PIC and the simple model begin to be important. The peaks that do not correspond to odd harmonics of the fields are clearer now and some of them can be identified in the transmitted or reflected spectra as linear combinations of the two frequencies involved in the problem: ω_{p0} (1), $2\omega_{p0} - \omega_0$ (2), $\omega_{p0} + \omega_0$ (3), $\omega_{p0} + 2\omega_0$ (4), $2\omega_{p0} + \omega_0$ (5), $\omega_{p0} - \omega_0$ (6). While some of these peaks appear also in the preionized plasma, others seem to be a result of the interference between the ionization process and the standard plasma response. They depict a very particular coupling between collective effects, whose characteristic is the plasma frequency, and ionization effects, which induce harmonics of the field frequency.

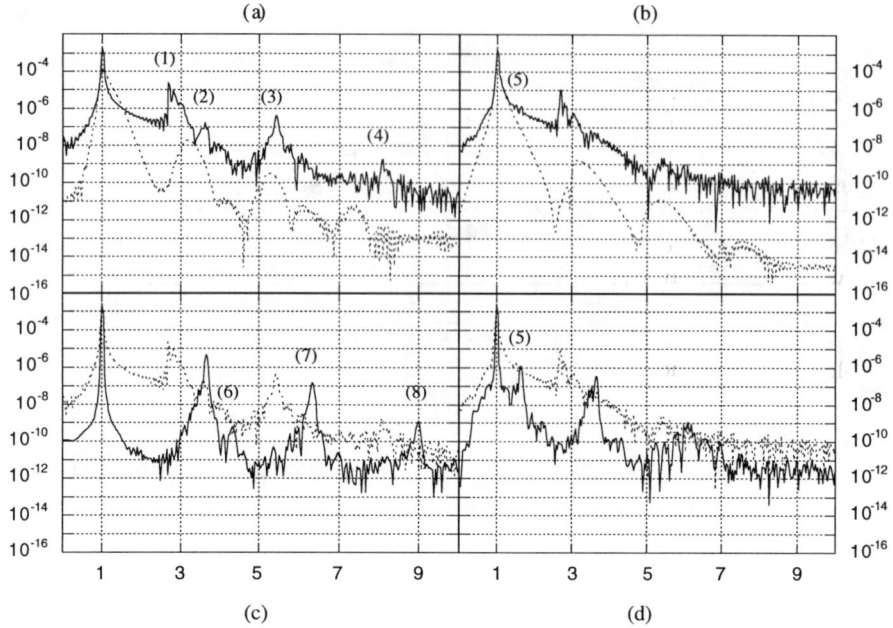

FIGURE 3. Same as previous plots for $\omega_{p0}^2 = 7\omega_0^2$.

Finally, in Fig. 3 we have increased again the plasma frequency ($\omega_{p0} = \sqrt{7}\omega_0$) and even more peaks are visible, some of them blueshifted, like the harmonics, due to propagation: ω_{p0} (1), $\omega_{p0} + \omega_0$ (2), $2\omega_{p0}$ (3), $3\omega_{p0}$ (4), $\omega_{p0} - \omega_0$ (5). Some peaks, which are present when the plasma is preformed, are inhibited due to other couplings during the ionization of the plasma: $2\omega_{p0} - \omega_0$ (6), $2\omega_{p0} + \omega_0$ (7), $3\omega_{p0} + \omega_0$ (8). In this case it is clear that the simple model does not reproduce the behavior of the medium for these parameters and the harmonics are hidden by the other frequencies.

ACKNOWLEDGEMENTS

Partial support from the Spanish Dirección General de Investigación Científica y Técnica (grant PB95-0955), from the Junta de Castilla y León (grant SA81/96) and from the European Comission Training and Mobility of Researchers Program (under contract ERBFMRXCT96-0080) is acknowledged.

REFERENCES

1. See Gibbon, P., *J. Quantum Electronics*, in print, and references therein.

2. Gibbon, P., *Phys. Rev. Lett.* **76**, 50 (1996).
3. Lichters, R., Meyer-ter-Vehn, J., and Pukhov, A., *Phys. Plasmas* **3**, 3425 (1996).
4. Brunel, F., *J. Opt. Soc. Am. B* **7**, 521 (1990).
5. Rae, S. C., and Burnett, K., *J. Phys. B* **26**, 1509 (1993).
6. Malyshev, V., Conejero Jarque, E., and Roso, L., *J. Opt. Soc. Am. B* **14**, 163 (1997). Conejero Jarque, E., Malyshev, V., and Roso, L., *J. Mod. Opt.* **44**, 563 (1997).
7. Birdsall, C. K., and Langdon, A. B., *Plasma Physics via Computer Simulation*, New York: McGraw-Hill, 1985.
8. Wallace, J. M., Forslund, D. W., Kindel, J. M., Olson, G. L., and Comly, J. C., *Phys. Fluids B* **3**, 2337 (1991).
9. Delone, N. B., and Krainov, V. P., *Multiphoton Processes in Atoms*, Berlin: Springer-Verlag, 1994.

Time-Resolved Imaging of High Harmonic Radiation Using Chirped Laser Pulses

J.W.G.Tisch[1], T.Ditmire[1], D.D.Meyerhofer[2], N.Hay[1], M.B.Mason[1] and M.H.R.Hutchinson[1]

[1]*Blackett Laboratory, Imperial College of Science Technology and Medicine, London SW7 2BZ, UK and*
[2]*Laboratory for Laser Energetics, University of Rochester, 250 East River Road, Rochester NY 14623-1299 USA*

Abstract. In this paper we describe the first spatially-resolved measurements of the time-dependence of high-order harmonic radiation. We show that the harmonic source can become annular with a radius that increases in time during the laser pulse owing primarily to plasma formation and that under appropriate conditions this effect could be used to produce ultra-short harmonic pulses.

INTRODUCTION

High order harmonic generation (HHG) [1] provides a unique source of high brightness, coherent radiation in the extreme ultraviolet (XUV) region of the spectrum. The properties of this radiation are of considerable interest from an applications viewpoint and as a probe of both the high-intensity laser-atom interaction and the phase-matching process. In particular, the temporal characterisation of femtosecond high harmonic pulses is important for applications wishing to exploit the short pulse durations and is of significance to proposed pulse-shortening schemes for harmonics [2]. However, such characterisation presents a formidable experimental challenge.

Streak cameras have been used to time-resolve relatively low-order harmonics generated with laser pulses of 50-150 ps duration [3,4], but they are inadequate for time-resolving harmonics generated with sub-picosecond pulses. Three experiments based on pump-probe type schemes have been conducted to time-resolve high harmonics with femtosecond resolution [5,6,7]. In contrast to these spatially-integrated measurements, in this paper we describe the first spatially-resolved measurements of the time evolution of high harmonic radiation (the 13th harmonic of a 780 nm laser at $\lambda_{13} = 60$ nm).

EXPERIMENT

We time-resolve the harmonic radiation by recording the spectrum of the harmonics using a conventional time-integrating XUV detector. The time-resolution comes from the linear frequency chirp on the laser pulse, $\omega_1(t) = \omega_0 + \beta t$, where β is the chirp parameter. Ignoring any additional time-dependent frequency shift of the harmonic radiation (such as ionisation blue-shifting [8]), the instantaneous harmonic frequency is simply $q\omega_1(t)$, where q is the harmonic order, so the frequency axis on the harmonic spectrum is linearly related to time by $t(\omega) = (\omega - q\omega_0)/q\beta$. A linearly chirped laser pulse can be produced in a chirped pulse amplification (CPA) laser system by changing the separation of the compressor gratings relative to the optimum-compression position [9]. If the compressed pulse duration is τ_1 (assumed to be a transform-limited Gaussian pulse with a bandwidth $\Delta\omega_1$) and the chirped pulse duration is $\tau_c \gg \tau_1$, the chirp parameter is given by $\beta \approx 2/(\tau_c \tau_1) = \Delta\omega_1/\tau_c$. The temporal resolution obtained is $\tau_{res} = q\tau_c \Delta\lambda_{spec}/\Delta\lambda_1$ where $\Delta\lambda_1 = 2\pi c \Delta\omega_1/\omega_0^2$ and $\Delta\lambda_{spec}$ is the spectral resolution (this is valid provided τ_{res} is greater than the transform limit associated with the spectral resolution, as is the case in this experiment).

We used a 10 Hz, 160 fs Ti:sapphire CPA laser [10] with $\Delta\lambda_1 \approx 65$ Å centred at a wavelength of $\lambda_1 \approx 780$ nm. The chirped pulse duration for both positive and negative chirps was set to $\tau_c \approx 1$ ps ($\beta \approx \pm 2\times 10^{13}$ s^{-1}fs^{-1}) by changing the compressor grating separation while monitoring the pulse duration using a single-shot autocorrelator and ensuring that the beam alignment and pulse spectrum remained unchanged. The laser pulse was focused by an f/17 lens (confocal parameter $b \approx 2$ mm, focal spot diameter $2w_0 \approx 50$ µm) into a Xe pulsed gas-jet of an estimated density of ~5×10^{16} cm^{-3} [11]. Measurements showed that at this low gas density, ionisation induced blue-shifting of the 13th harmonic was negligible. The jet was positioned 5 mm before the laser best-focus and 20 cm from the entrance slit of a 1-m imaging spectrometer. With the gas-jet positioned $\approx 2b$ before the best-focus position the harmonic radiation was effectively already in its far-field at the exit of the jet, so the angular distribution of the harmonic at the exit of the medium was essentially unchanged in propagating to the entrance slit of the spectrometer. Since the entrance slit was imaged onto our detector, this configuration allowed us to record the time-dependence of the harmonic intensity distribution at the exit of the jet (the harmonic source distribution).

The harmonics were detected with a microchannel plate detector producing a 2-dimensional single-shot image with wavelength dispersion in one direction and spatial resolution in the orthogonal direction. We observed the 13th harmonic spectrum in the second diffraction order of a 2400 l/mm grating, giving a temporal resolution of ~100 fs limited by the ~0.25 mm spatial resolution of our detector.

RESULTS AND SIMULATIONS

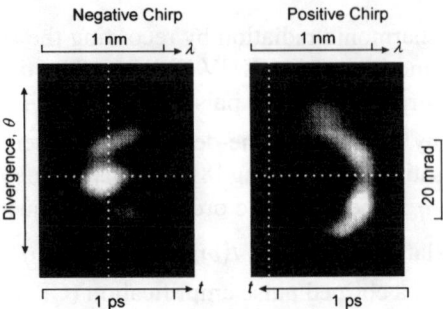

FIGURE 1. Spatio-temporal images of the 13th harmonic source for negative and positive chirped laser pulses at a peak intensity of 1.3×10^{14} Wcm^{-2}. The dotted vertical line corresponds to the temporal peak of the laser pulse ($t = 0$), while the dotted horizontal line is the laser axis ($r = 0$).

In Fig.1 we present two spatially and temporally resolved grey-scale images of the 13th harmonic source distribution for negative and positive laser chirps. The peak intensity in the Xe jet was 1.3×10^{14} Wcm^{-2} for both cases, an intensity at which significant ionisation of the xenon occurs. The vertical axis is the harmonic divergence while the horizontal axis is the pulse wavelength or time. The peak of the laser pulse is taken to be the centre of the spectrum of the 13th harmonic generated with the compressed pulse (160 fs) under conditions where ionisation induced blue-shifting was completely negligible. The crescent shapes reveal that at early time the harmonic source is initially peaked on axis and then becomes annular with a radius which increases with time during the rising edge of the laser pulse. This picture is confirmed by the fact that the crescent reverses direction when the sign of the chirp (and hence the time arrow) is reversed.

We have investigated the spatio-temporal evolution of the harmonic radiation as a function of laser intensity. Fig.2(a)-(d) shows time and space resolved images of the 13th harmonic source for four different peak intensities, 5.3, 7.3, 9.6 and 13×10^{13} Wcm^{-2}. The images in the top row are the experimental data (for $\beta > 0$), while those in the bottom row are simulations at the same peak intensity calculated from a simple model of the harmonic generation. In the plane-wave limit, with $b >> L$, where L (≈ 1 mm) is the interaction length, the harmonic intensity can be written as $I_q \propto N_0^2 I_1^p \mathrm{sinc}^2(\Delta k L/2)$, where we have used an effective power law formulation for the harmonic polarisation [1]. $N_0(r,t)$ is the neutral density which can become depleted in a space and time dependent manner owing to ionisation, $I_1(r,t)$ is the laser intensity (we assume a Gaussian laser pulse in space and time), p ($= 9$) the effective order of the process, $\Delta k(r,t) = (\pi q/\lambda_1)(n_e(r,t)/n_{crit})$ the phase-mismatch due to dispersion from free electrons created by ionisation, n_e the electron density and $n_{crit} = 1.8\times10^{21}$ cm^{-3} the critical density at $\lambda_1 = 780$ nm. Ionisation is treated using ADK tunneling ionisation rates for Xe [12].

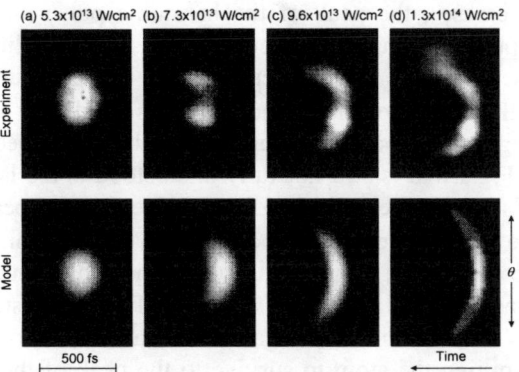

FIGURE 2. Intensity evolution of the 13th harmonic source distribution. The experimental data (100 shot averages) are shown in the top row, simulations in the bottom row. (The model does not include propagation of the harmonic from the gas-jet to the detector, so for clarity, the divergence in the simulations has been matched to the experimental data.)

The evolution of the images with increasing intensity is accurately reproduced in the simulations. The model indicates that ionisation depletion of the neutral medium is primarily responsible for the termination of the harmonic generation and the resulting crescent formation. For the lowest intensity in Fig.2, ionisation levels are quite low (~10% depletion of the neutrals on axis at the peak of the laser pulse) so the harmonic distribution in space and time is essentially the laser distribution raised to the pth power. For the higher intensities, significant depletion occurs during the laser pulse which modifies the harmonic distribution. The depleted region in the jet expands radially in time, confining the harmonic emission to an annulus limited on the outside by the nonlinear intensity dependence of the harmonic polarisation. The crescent shape observed in the spatio-temporal images is thus the time-dependence of a slice through this annular emitting region (as selected by the entrance slit of the spectrometer).

By radially integrating the harmonic signal in our 2D images at each time-step we obtain the time-dependence of the total harmonic yield. Fig.3 shows the results for images (a)-(d) in the top row of Fig.2.

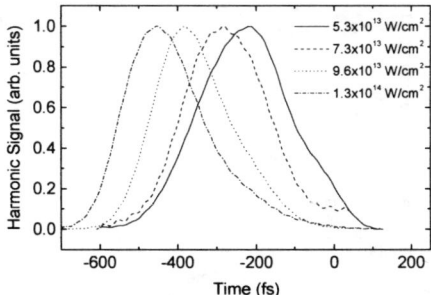

FIGURE 3. Radially integrated 13th harmonic pulses for a range of laser intensities (calculated from the experimentally measured images Fig.2(a)-(d)).

As the intensity increases, the peak of the harmonic pulse moves earlier in time relative to the laser pulse as ionisation occurs progressively earlier during the rising edge of the laser pulse (the same general trend, though less pronounced, was seen with $\beta<0$). The slower falling edges of the harmonic pulses are due to harmonic generation continuing in the wings of the focal distribution after the centre becomes depleted. An aspect of the data in Fig.3 that is still not fully understood is the large temporal offset between the peak of the positively chirped laser pulse and the peak of the harmonic pulse (~200 fs) at the lowest intensity (5.3×10^{13} Wcm^{-2}). This shift was not observed for the negative chirp case at the same intensity. A similar difference between positive and negative chirp was observed by Zhou et al. [13]. They suggested that the negative chirp might partially stabilise the atom against ionisation (compared to the positive chirp case), thus permitting the atom to survive to the peak of the pulse, whereas for the positive chirp, ionisation occurs on the rising edge. This explanation does not appear to be applicable in our case, since there is no evidence for ionisation terminating the emission in the low intensity positive chirp case, as this would manifest itself through the formation of a crescent in the 2D image, which is not observed in Fig.2(a). We are currently investigating other explanations, such as resonance effects.

At the lowest intensity in Fig.3, where ionisation effects are small, the harmonic FWHM pulse duration is $\tau_q = 250$ fs±25% (taking into account our temporal resolution). This is consistent with the expected $\tau_c/\sqrt{p}$ pulse width in the low intensity regime. The data in Fig.3 also show that the harmonic pulse duration decreases with increasing intensity owing to ionisation terminating the emission. The observed pulse shortening (about 10% for the highest intensity) is limited by the spatial integration that is normally inherent in harmonic pulse-width measurements.

Without this spatial integration, the pulse shortening arising from ionisation can be much more significant, as is demonstrated in Fig.4. Here horizontal line-outs from Fig.2(d) (top-row) at three different radial positions have been taken to obtain the time-dependence of the harmonic emission at those positions. The pulse duration of the emission on axis is ~120 fs, about a factor of 2 shorter than the spatially integrated duration for this shot, and approximately 8 times shorter than the laser pulse. This is close to our temporal resolution.

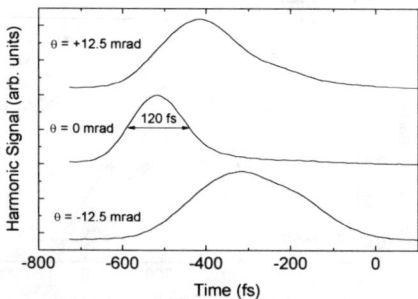

FIGURE 4. Measured angular dependence of the 13th harmonic pulse duration (corresponding to Fig.2(d)) at a peak intensity of 1.3×10^{14} Wcm^{-2}. The curves have been offset vertically for clarity.

CONCLUSION

In conclusion, we have time-resolved harmonic radiation with a resolution of ~100 fs using linearly chirped laser pulses. By positioning the gas-jet significantly before the laser focus, we found that we could measure the harmonic distribution at the exit of the nonlinear medium and thus record, for the first time, the time-evolution of the harmonic source distribution. Our data confirmed that the source can become annular with a radius that increases in time during the laser pulse and modelling indicated that ionisation depletion of the neutral medium was responsible for this behaviour. Further, for our configuration, we have shown that the harmonic pulse duration at a particular angle in the far-field is substantially shorter than the spatially integrated pulse duration. Obviously, harmonic pulse-shortening in this way can be applied to shorter laser pulses. Our modelling indicates that for a 20 fs pulse at $\sim 3 \times 10^{14}$ Wcm^{-2} the 13th harmonic pulse duration on axis would be ~3 fs. The fact that the harmonic distribution at the exit of the jet can be preserved in propagating to the detector if the jet is positioned significantly before the focus implies that a small aperture on axis some distance from the laser focus would select this ultra-short harmonic pulse.

ACKNOWLEDGEMENTS

We acknowledge financial support of the UK EPSRC. DDM acknowledges travel support from NATO contract No. CRG 930274 and the US NSF.

REFERENCES

1. L'Huillier, A., Lompré, L.A., Mainfray,G. and Manus, C. "High-Order Harmonic Generation in Rare Gases" in Atoms in Intense Laser Fields pp.139-202, Gavrila,M. ed. Boston:Academic Press (1992)
2. Antoine,P., Piraux,B., Milosevic,D.B. and Gajda,M. *Phys. Rev. A* **54** R1761 (1996)
3. Faldon,M.E., Hutchinson,M.H.R., Marangos,J.P., Muffett,J.E., Smith,R.A., Tisch,J.W.G. and Wahlström C-G. *J. Opt. Soc. Am. B* **9** 2094 (1992)
4. Starczewski,T., Larsson,J., Wahlström, C-G., Tisch,J.W.G., Smith,R.A., Muffett,J.E. and Hutchinson,M.H.R. *J. Phys. B* **27** 3291 (1994)
5. Kobayashi,Y., Yoshihara,O., Nabekawa,Y., Kondo,K. and Watanabe,S. *Opt. Lett.* **21** 417 (1996)
6. Glover,T.E., Schoenlein,R.W., Chin,A.H. and Shank,C.V. *Phys. Rev. Lett.* **76** 2468 (1996)
7. Bouhal,A., Evans,R., Grillon,G., Mysyrowicz,A., Breger,P., Agostini,P., Constantinescu,R.C., Muller,H.G. and von der Linde,D. *J. Opt. Soc. Am. B* **14** 950 (1997).
8. Wood,W.M., Siders,C.W. and Downer,M.C. *Phys. Rev. Lett.* **67** 3523 (1991)
9. Treacy,E.B. *IEEE J. Quant. Elec.* **QE5**, 454 (1969)
10. Fraser,D.J. and Hutchinson,M.H.R. *J. Mod. Opt.* **43** 1055 (1996)
11. Ditmire,T., Smith,R.A., Tisch,J.W.G. and Hutchinson,M.H.R. "Short-Pulse Interferometric Measurements of Gas Densities from a Cooled Gas Jet" submitted to *Appl. Phys. Lett.*
12. Ammosov,M.V., Delone,N.B. and Krainov,V.P. *Sov. Phys. JETP* **64** 1191 (1986)
13. Zhou,J., Peatross,J., Murnane,M.M., Kapteyn,H.C. and Christov,I.P. *Phys. Rev. Lett.* **76** 752 (1996)

Propagation in Compressed Matter of Hot Electrons created by short intense lasers

D.Batani[1], A.Bernardinello[1], V.Masella[1], F.Pisani[2], M.Koenig[2],
J.Krishnan[2], A.Benuzzi[2], S.Ellwi[3], T.Hall[3], P.Norreys[4], A.Djaoui[4],
D.Neely[4], S.Rose[4], P.Fews[5], M.Key[6]

1 Università di Milano and INFM, Italy
2 LULI, CNRS, Ecole Polytechnique, France
3 University of Essex, Colchester, UK
4 Rutherford Appleton Laboratory, UK
5 University of Bristol, UK
6 LLNL, USA

Abstract. We performed the first experimental study of propagation in compressed matter of hot electrons created by a short pulse intense laser. The experiment has been carried out with the VULCAN laser at Rutherford compressing plastic targets with two ns laser beams at an intensity $\geq 10^{14}$ W/cm^2. A CPA beam with an intensity $\geq 10^{16}$ W/cm^2 irradiated the rear side of the target and created hot electrons propagating through the compressed matter. K-α emission was used as diagnostics of hot electron penetration by putting a chloride plastic layer inside the target.

INTRODUCTION

We realised the first experimental study of propagation in compressed matter of hot electrons created by the interaction with a short pulse intense laser. The goal of the experiment was to begin to address the problems connected to the last phase of the fast ignitor approach, namely the interaction, propagation and energy deposition of hot electrons in compressed matter. To obtain this goal we used two ns laser beams to shock compress a plastic target and we used a CPA beam to generate hot electrons and study their propagation in the compressed material. As it will be seen in the following, a quite large focal spot (≥ 100 μm) was used for the CPA laser beam. This reduced the CPA intensity to a relatively low value ($\geq 10^{16}$ W/cm^2) which could seem negative for fast ignitor relevant experiments. Anyway this was a necessary requirement in our experiment since a larger CPA intensity would have meant a higher hot electron temperature T_{hot} and an increased hot electron penetration depth. But this would have required thicker targets which we would not have been able to compress in a relatively uniform way with the available ns laser energy and pulse duration. Hence we preferred to perform a cleaner experiment rather than one in which the values of different experimental parameters are just maximised.

Intermediate steps of the experiment were the characterisation of shocks and of the hot electron source (including propagation of hot electrons in the non-compressed materials). In this paper we only describe these two steps. The results describing the comparison between compressed and non-compressed targets will be the content of a following paper (1).

EXPERIMENTAL SET-UP

The experiment has been carried out with the VULCAN laser facility at the Rutherford Appleton Laboratory, where planar plastic targets (typically 20 to 100 μm thick) were irradiated on the same side with two laser beams, operating at a pulse length of the order of 2 ns with an incident intensity of $\approx 1.5 \cdot 10^{14}$ W/cm^2. The two drive beams were converted at 2ω and were delivering a total energy of ≈ 150 J used to produce a strong shock and create a high density plasma, as described in the following section. RPPs were used to produce a focal spot with FWHM ≈ 200 μm and an almost uniform irradiation.
A CPA beam with a maximum energy of ≈ 30 J, pulse duration of the order of 3 ps, focal spot size ≥ 100 μm and an intensity on target of $\approx 1.5 \cdot 10^{16}$ W/cm^2, was focused on the other side of the plastic target and created hot electrons propagating through the compressed matter. The contrast ratio with the laser pulse pedestal was $\approx 10^6$ and the beam was operated at 1.06 μm. The CPA beam was incident at 30° following Beg et al. (2) and Schnurer et al. (3).
Several diagnostics were used on each laser shots including:
- far field imaging and autocorrelator to measure the CPA focal spot and pulse duration.
- calorimeters to measure the CPA and the ns beams pulse energies.
- CR39 foils to measure the hot electron temperature with ion impact technique (2).
- X-ray active pin-hole cameras on both sides of the target to measure respectively the size of the plasmas created by the CPA and the ns laser beams.
- flat crystal Bragg mini-spectrometers to measure X-ray emission from the laser irradiated targets (we used PET crystals)

K-α emission was used as diagnostics for hot electron penetration by putting a chloride plastic layer (13.5 μm thick) in the target. The intensity of the K-α line of Cl, at ≈ 2.6 keV, was measured with the three flat crystal spectrometers placed at different angular positions with respect to the target.

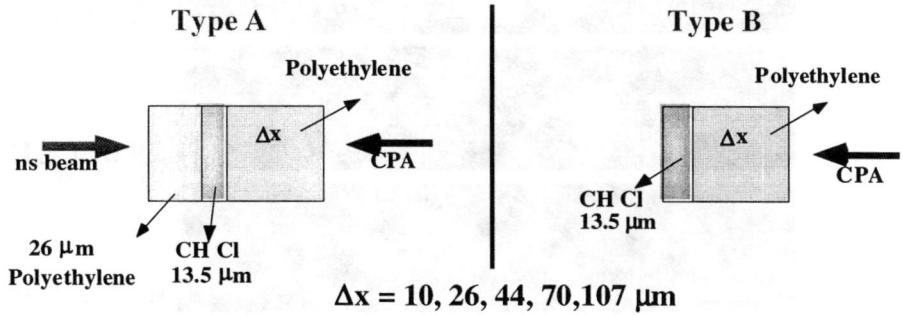

$\Delta x = 10, 26, 44, 70, 107$ μm

FIGURE 1. type of targets used in the experiment.

Two types of targets were used in the experiments (see fig. 1). Type B was used for the study of the propagation of hot electrons in cold matter and the characterisation of the hot electron source. Type A was used for the comparison of compressed/uncompressed matter. Here 26 μm of polyethylene were added on the ns beam side in order to avoid any X-ray emission from Cl due only to the ns beam alone (as experimentally checked).

SHOCK COMPRESSION

A streak camera imaged the target rear side and measured the emission of radiation produced by the emerging shock. The method is similar to what described in (4). The arrival time of the shock for different target thickness gave a velocity of about 42 µm/ns (km/s). The determination of the shock velocity was very important in our experiment for two reasons:

i) it allowed to exactly time the CPA beam. The delay between the time at which the CPA beam was fired and the time of the shock breakout was chosen so that at least a few µm of plastic were still uncompressed. Indeed we wanted to avoid the opposite case because if the CPA arrives after the shock breakout it would find a plasma expanding from the rear side and the interaction and the generation of hot electrons would be completely different.

ii) from the shock velocity, by using the Hugoniot of polyethylene (given by the Sesame tables) we could determine the other parameters of the compressed material, the pressure $P \approx 11$ Mbar, the temperature $T \approx 7 - 8$ eV, and the density $\rho \approx 3\rho_0$ where ρ_0 is the density of the cold solid plastic. We then could calculate the other parameters including the ionisation degree $Z^* \approx 1 - 2$ and the conductivity for which we used the Spitzer's formula finding $\sigma \approx 2 - 4 \; 10^4 \; (\Omega m)^{-1}$. Even if Spitzer's formula is probably not correct in our case (we are dealing with a strongly coupled and partially degenerate plasma) it is at least qualitatively correct in the sense that it gives a higher conductivity for the compressed material (due to the higher ionisation), unlikely more refined models which are really applicable only to the case of metals and give a reduced conductivity for the compressed material.

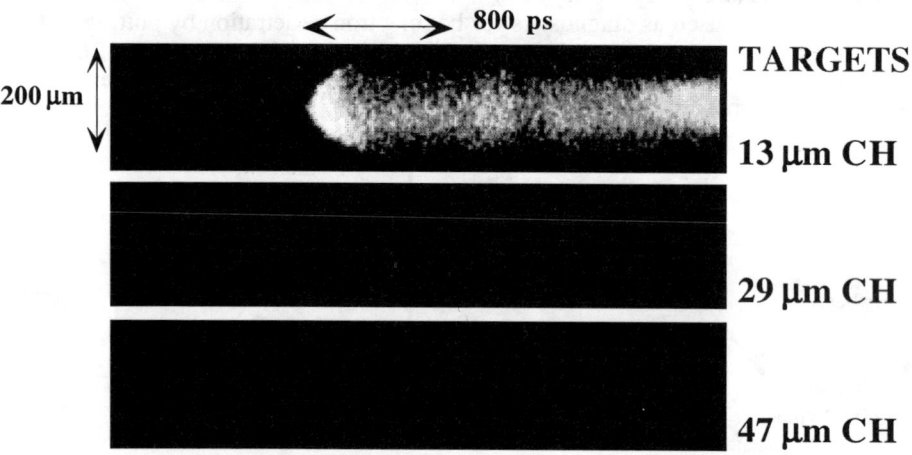

FIGURE 2. Streak camera images of shock breakouts with targets of different thickness.

CHARACTERISATION OF HOT ELECTRONS

The characterisation of hot electrons included the determination of the "hot" temperature and the study of their penetration in the uncompressed material. For the first study the most important data were those given by CR39 foils from which the energy of fast ion produced in the interaction was determined by revealing their tracks inside the plastic foils after CR39 development. These gave a temperature $T_{hot} \leq 50$ keV depending on the CPA laser energy in each shot.

Fig. 3 shows instead the results of K-α emission as a function of target thickness for type B targets (uncompressed). Their behaviour is as expected, i.e. K-α emission increases when CPA energy is increased or the target thickness is decreased.

We have interpolated such results (and similar ones obtained with uncompressed type A targets) with the formula derived by Harrach and Kidder's model (5)

$$I(x) = I_o \exp[-\beta (x/R_o)^{1/2}]$$

This model describes the propagation of hot electrons produced from a planar diffusive source (which we think is our case) into a cold solid material. It allows the mean penetration depth R_o to be defined, and also gives a formula which relates R_o to the hot electron temperature T_{hot}. For the typical density of polyethylene it reads

$$R_o(T_{hot}) = 3.8 \, 10^{-2} \, (T_{hot})^{-1.78} \, \mu m$$

using the value $\beta = 1.85$ for carbon, we found $R_o = 30 \pm 5$ µm which implies a hot electron temperature $T_{hot} \approx 40 - 50$ keV, in good agreement with what is given by CR39.

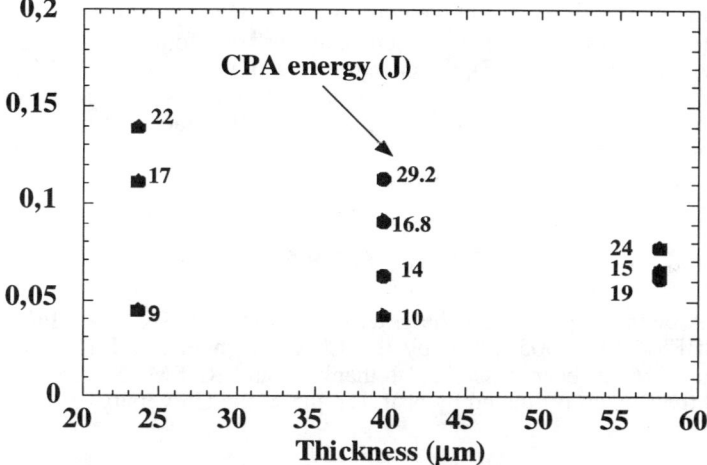

FIGURE 3. K-α yields for different CPA laser energies as a function of target thickness.

It is also interesting to note that such a value is consistent with what can be found using the scaling law for hot electron temperature given by Beg et al. (2)

$$T_{hot} \approx 100 \text{ keV } (I \lambda^2)^{1/3}$$

using our experimental value of CPA laser intensity on target. This scaling law is quite relevant to our case since it has been obtained with the same laser system, and in similar experimental conditions of the present experiment.

We have also performed an analysis of the propagation of hot electrons in the cold material based on a calculation of its stopping power with the formula

$$\frac{dE}{dx} = -\frac{4\pi e^4}{mv_0^2}(n_b L_b)$$

where e and m are charge and mass of electrons and v_o their velocity. Here n_b is the density of bound electrons and no contribution from free electrons is present in the cold material case since we are dealing with an insulator. For L_b we have used the classical expression by Bethe with an effective ionisation / excitation potential (we note that with our typical hot temperature, the electrons are not yet relativistic). We have then assumed a gaussian distribution of hot electrons with an average temperature T_{hot}, which has been determined with the scaling law by Beg et al., and divided it in 0.3 keV intervals in the energy range between 0 and 5 T_{hot}. Such monochromatic "families" of hot electrons have been propagated in the cold material which has been divided in thin slices (0.5 µm thick) in each of which the electron energy loss has been calculated using the stopping power formula.

Finally, in the chlorinated plastic layers the ionisation of Cl atoms has been calculated with the usual formula for K-shell ionisation cross section

$$\sigma \text{ (cm}^2\text{)} = 7.9 \; 10^{-14} \; \frac{\ln(U_k)}{E_k^2 U_k}$$

where $U_k = E / E_k$, and E and E_k (both in eV) are respectively the energy of the electrons and the binding energy of the K-shell electrons in Cl. This approach allows not only to find the K-α emission for each target (which again has been found to be consistent with a hot electron temperature ≤ 50 keV), but also to follow other fine details of the interaction, such as the evolution of the hot electron distribution. On the other side the limit of such approach is the straight line approximation and hence the fact that backscattering and multiple scattering of electrons are neglected. For this reasons we are currently working on the optimisation of a Monte Carlo code.

ACKNOWLDGEMENTS

This work has been supported by the E.U. Programme: TMR Laser Facility Access (contract ERBFMGEC950053) and by the LEA "High Power Laser Science". The participation of V.M has been possible also thanks to an ERASMUS grant obtained in the framework of the PIC co-ordinated by prof. L.Lanz of the University of Milan.

REFERENCES

1. T.Hall, D.Batani, M.Koenig, P.Norreys et al. sub. to Phys. Rev. Lett. (1997).
2. F.Beg et al. Phys.Plasma 4, 447 (1997).
3. M.Schnürer et al. Phys.Plasma 2, 3106 (1995).
4. D.Batani, S.Bossi, A.Benuzzi, M.Koenig, B.Faral, J.M.Boudenne, N.Grandjouan, M.Temporal, S.Atzeni, Laser and Particle Beams, 14, 211 (1995).
5. R.Harrach and R.Kidder, Phys. Rev. A, 2, 887 (1981).

Generation of Superhot Electrons by Intense Field Structures

R. R. E. Salomaa*, S. J. Karttunen[†], P. Mulser[‡],
T. J. H. Pättikangas[†], and W. Schneider[‡]

*Department of Engineering Physics and Mathematics, Helsinki University of Technology, P.O.Box 2200, FIN-02015 HUT, Finland
[†] VTT Energy, P.O.Box 1604, FIN-02044 VTT, Finland
[‡] Theoretical Quantum Electronics, Technical University of Darmstadt, D-64289 Darmstadt, Germany

Abstract. Strong, localized electrostatic fields created in laser plasma interactions act as a source of hot electrons. We have derived analytical formulas based on adiabatic invariants for explaining of the main characteristics of the electron spectra found in test particle calculations and in full wave-particle simulations. The electrons are treated relativistically. Simple models for phenomenological description of nonlinear wave damping are discussed.

INTRODUCTION

Strong electric fields appearing in plasmas can create highly superthermal particles. A common acceleration mechanism at moderate intensity levels is the resonant interaction where the particle velocities are close to the phase velocity of the field. With modern terawatt level short-pulse lasers additional acceleration effects become operational, too. Resonant interaction to which this paper focuses is encountered in various laser-plasma accelerator concepts, current generation in fusion plasmas, in stimulated scattering processes in the underdense plasma corona and in the holeboring region of the fast ignitor, and in resonance absorption at the critical density. The high-intensity accelerating field is usually confined to a limited spatial region determined by the resonance conditions of the parametric process creating the field. The acceleration may also occur as a multistage cascade as for instance when Raman backward and forward scattering are excited simultaneously [1] or in the case of phase velocity sweeping near the cut-off layer [2,3].

We have studied the electron energy spectra arising from localized electric field pulses with test particle and full PIC or Vlasov simulations in the context of current drive in fusion plasmas. The problem is quite analogous in laser-plasma interaction studies. When a test particle traverses a smooth electrostatic field pulse, typically the particle energy remains unaltered or it gets a substantial energy boost — the particle spectra exhibit clear branches. To interpret the observed features we have, as a first step, studied the case of moderately low intensities and long characteristic length and time scales. A particular feature is that as the electrons may achieve very high energies relativistic corrections have to be taken into account.

TEST PARTICLE ACCELERATION

In a test particle calculation we assume a given external electrostatic field with a finite spatial extent, i.e., the field amplitude vanishes outside the interaction region. The task is to solve the final momentum $p_{out} = p(t \to \infty)$ for various initial values $p_{in} = p(t \to -\infty)$. The plots of p_{out} versus p_{in} display several characteristic features within a random scatter of the points: branches of acceleration and of unchanged momenta, interaction cut-off velocities, etc. Two examples are given in Fig. 1 (from Ref. [1]). The field has a Gaussian envelope with a $1/e$-width of $15\omega_p/c$ where ω_p is the plasma frequency. The left hand figure represents a slow wave, with a phase velocity $v_{ph} = 0.45c$ and the right hand figure a fast wave with $v_{ph} = 0.82c$. As is evident from the dissimilarity of the figures, relativistic effects play a role.

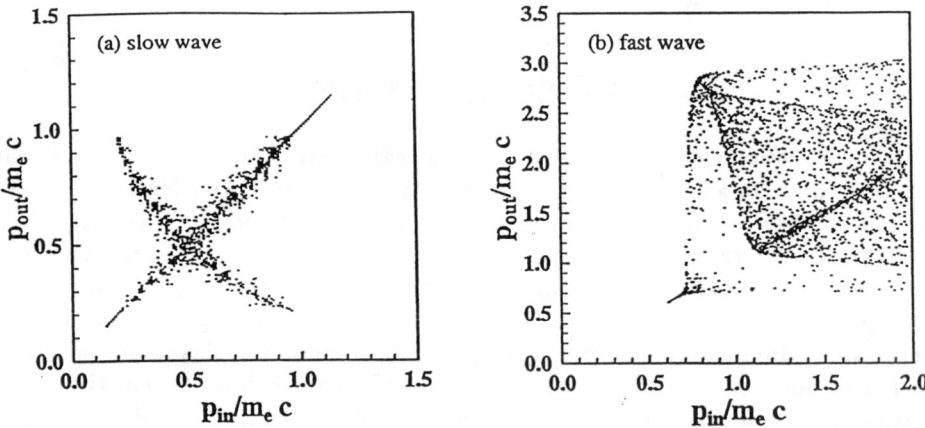

FIGURE 1. The output momentum versus the input momentum of a test electron accelerated by a slow (left hand side) and a fast (right hand side) Gaussian wave packet. Two thousand equally distributed test electrons were used. The phase of the field was randomized for each electron trajectory.

RELATIVISTIC PARTICLE TRAPPING

Let us assume an electric field which in the laboratory coordinates (z', t') has the form $E(z', t') \cos[K(z')z' - \omega t' + \phi(z', t')]$. We first make a Lorentz transformation to a coordinate system (z, t) which moves at a representative central phase velocity

$v_{\text{ph}} = \omega/K_0 < c$. As the variation of $K(z')$ can be included into $\phi(z', t')$, we put $K(z') = K_0$. We get in the moving coordinate system

$$\frac{dp}{dt} = -eE(z,t)\cos[k_0 z + \phi(z,t)] \tag{1}$$

$$\frac{dz}{dt} = \frac{pc}{(p^2 + m^2c^2)^{1/2}} \tag{2}$$

where $z = \gamma_0(z' - v_{\text{ph}}t')$, $t = \gamma_0[t' - v_{\text{ph}}z'/c^2]$, $k_0 = K_0/\gamma_0$, and $\gamma_0 = (1 - v_{\text{ph}}^2/c^2)^{-1/2}$. The velocity $dz/dt = v$ is related to the momentum by $p = mv(1 - v^2/c^2)^{-1/2}$.

For a field whose amplitude and phase does not depend explicitly on time Eqs. (1) and (2) can be integrated to give the energy conservation relation

$$(m^2c^4 + p^2c^2)^{1/2} + e\int dz\, E(z)\cos[k_0 z + \phi(z)] = W = \text{const.} \tag{3}$$

For a constant field amplitude $E(z) = E$, this can solved for v:

$$v^2/c^2 = 1 - (w - \mathcal{E}\sin\varphi)^{-2} \tag{4}$$

where $\mathcal{E} = eE/k_0 mc^2$, $\varphi = k_0 z + \phi$, and $w = W/mc^2$. Particles get trapped if $w < 1 + \mathcal{E}$. The relativistic trapping width is given by $v_{\text{tr}} = c[1 - (1 + 2\mathcal{E})^{-2}]^{1/2}$. In the laboratory coordinates the upper and lower trapping velocities are thus obtained from $v_{\pm} = (v_{\text{ph}} \pm v_{\text{tr}})/(1 \pm v_{\text{ph}}v_{\text{tr}}/c^2)$. The familiar classical formulas, $v_{\text{tr}} = 2c\mathcal{E}^{1/2} = 2(eE/k_0 m)^{1/2}$ and $v'_{\pm} = v'_{\text{ph}} \pm v_{\text{tr}}$, arise in the limit $\mathcal{E} \ll 1$.

From (4) we obtain the trajectories of non-trapped particles ($w > 1 + \mathcal{E}$)

$$ck_0(t - t_0)\mathcal{E}^{1/2}\kappa^{-1} = (1 + w + \mathcal{E})\Pi(\theta, n, \kappa) - F(\theta, \kappa) \tag{5}$$

where $F(\theta, \kappa)$ and $\Pi(\theta, n, \kappa)$ are elliptic integrals of the first and third kind, respectively [4]. The argument is $\theta = \arcsin[\sin^2\xi/(1 - n\cos^2\xi)]^{1/2}$ where $\xi = (k_0 z + \phi + \pi/2)/2$, the modulus is $\kappa = 2\mathcal{E}^{1/2}/[w^2 - (1 - \mathcal{E})^2]^{1/2}$, and the parameter is $n = -2\mathcal{E}/(w + 1 - \mathcal{E})$.

In the trapping-region two different expressions appear for the trajectories. Both of these can be obtained from (5) by making use of the transformation properties of the elliptic integrals for $\kappa^2 > 1$ or $\kappa^2 < 0$ [5]. For $1 + \mathcal{E} > w > -1 + \mathcal{E}$ we get

$$ck_0(t - t_0)\mathcal{E}^{1/2} = (1 + w + \mathcal{E})\Pi(\theta', n\kappa^{-2}, \kappa^{-1}) - F(\theta', \kappa^{-1}) \tag{6}$$

with $\theta' = \arcsin[\kappa^2\sin^2\xi/(1 - n\cos^2\xi)]^{1/2}$. In the region $-1 + \mathcal{E} > w > 1 - \mathcal{E}$ (the region $w < 1 - \mathcal{E}$ is unphysical) the trajectories are given by

$$ck_0(t - t_0)\frac{1}{2}[(1 + \mathcal{E})^2 - w^2]^{1/2} = 2\mathcal{E}\Pi(\theta', n', \kappa') + (w - \mathcal{E})F(\theta', \kappa') \tag{7}$$

where $\theta' = \arcsin(-\tan^2\xi/n')^{1/2}$, $\kappa' = [(\mathcal{E} - 1)^2 - w^2]^{1/2}[(\mathcal{E} + 1)^2 - w^2]^{-1/2}$, and $n' = (1 - w - \mathcal{E})/(1 - w + \mathcal{E})$.

In the nonrelativistic case, $\mathcal{E} \ll 1$ and $w - 1 \ll 1$, Eqs. (5) and (6) reduce to the well-known solution $\sin\xi = \text{sn}(s|\kappa)$ where $\text{sn}(s|\kappa)$ is the Jacobian elliptic function with $s = ck_0 t\mathcal{E}^{1/2}/\kappa + s_0$ (see e.g. [6]).

SPATIALLY SLOWLY VARYING FIELDS

Equations (1–2) have been solved analytically in the nonrelativistic limit for slowly varying fields, i.e. for $|(dE/dz)/kE| \ll 1$ and $|d(1/k)/dz| \ll 1$ [6,7,8]. The method of solution is based on the fact that to the lowest order in the above small quantities the action integral is an adiabatic invariant [9]. If we choose p and z as the canonical coordinates, the Hamiltonian of the system is simply that given by Eq. (3). The corresponding action integral for trapped particle trajectories

$$ J = \oint p\, dz = 2(mc/k_0) \int_{-\pi/2}^{\arcsin(w-1)/\mathcal{E}} d\varphi \sqrt{(w - \mathcal{E}\sin\varphi)^2 - 1} \tag{8} $$

is thus approximately constant. For the nontrapped particles we must extend the upper integration limit in (8) to $\pi/2$ and add a term $2\pi\sigma w(mc^2/\omega)$ where $\sigma = \text{sgn}(v' - v_{\text{ph}})$ is the sign of the particle velocity in the wave frame. Equation (8) is a generalization of the analogous non-relativistic calculation of Refs. [6,7].

Assuming that $w - 1 \ll 1$ and $\mathcal{E} \ll 1$, one obtains from Eq. (8) and the corresponding equation for trapped particle trajectories the adiabatic invariants [6,7]:

$$ 1 + y + 4\pi^{-1}\sigma(x + y)^{1/2} E(\kappa) = \text{const.} \tag{9} $$

for $\kappa^2 = 2\mathcal{E}/(w - 1 + \mathcal{E}) = 2x/(x+y) \le 1$ (nontrapped particles), and

$$ (x/2)^{1/2}[E(\kappa^{-1}) + (\kappa^{-2} - 1) K(\kappa^{-1})] = \text{const.} \tag{10} $$

for $\kappa^2 > 1$ (trapped particles). E and K are the complete elliptic integrals of the first and second kind, respectively [4]. The new notation in (9–10) is: $y = (2k^2c^2/\omega^2)(w-1)$ is the total particle energy in the wave frame in units $mv_{\text{ph}}^2/2$, and $x = (2k^2c^2/\omega^2)\mathcal{E}$ is the field strength parameter.

Particles can get an energy boost in a smoothly varying electric field pulse when their trajectory in the $(\mathcal{E}, w)$-plane (or equivalently in the (x,y)-plane) crosses the separatrix $\kappa = 1$ or when the electric field abruptly changes as for example near the critical layer in resonance absorption. The adiabatic invariant J is conserved when the separatrix is crossed. The various alternatives — particle transmission with possible energy changes, particle reflection, and trapping — have been described in detail in [7].

The use of the adiabatic invariants turns out to be very expedient, because J can be evaluated in the field free region, $J_{\text{in}} = 2\pi(p_{\text{in}}/k_0 + \sigma W_{\text{in}}/\omega)$, and because many of the characteristics of the problem involve the separatrix $\kappa = 1$ (i.e., $w = 1 + \mathcal{E}$) at which (8) yields:

$$ J = 4(mc/k_0)\{\mathcal{E}^{1/2} + (1 + \mathcal{E})\arcsin[\mathcal{E}/(1 + \mathcal{E})]^{1/2}\}. \tag{11} $$

As an excercise one can show that the equation $J_{\text{in}} = J(\kappa = 1)$ does have a solution only for a limited range of input momenta. This implies that a minimum value v_{cr} is required for v_{in} the trapping to be possible at all. In the nonrelativistic limit we have $v_{cr} = (1-8/\pi^2)^{1/2} v_{ph}$ which is expected to be a reasonably good approximation also in the relativistic case.

During the interaction the particles may be transmitted, reflected or trapped. Besides "noninteracting" electrons for which $p_{out} = p_{in}$, there appears a branch of accelerated electrons. Equation (8) defines a curve $w = w(\mathcal{E}, \mathcal{J} = \mathcal{J}_{in})$. This curve has both an upper w_{up} and a lower crossing point w_{lo} with the w-axis. The difference $w_{up} - w_{lo}$ defines the possible energy change of the particle when it traverses the field region. In the classical limit the momentum change is easily solved:

$$p_{out} = \{p_{cr}^2 + [\frac{4\sqrt{2}}{\pi} p_{ph} - (p_{in}^2 - p_{cr}^2)^{1/2}]^2\}^{1/2} \qquad (12)$$

where $p_{cr} = (1-8/\pi^2)^{1/2} p_{ph}$. Test particle simulations also manifest that electrons with $0 \leq p_{in} \leq p_{cr}$ leave the smooth field pulse region unaffected. All these features are visible in the slow wave case of Figure 1.

DISCUSSION

For short-pulse interactions the applicability of the adiabatic approximation is, of course, questionable. The test particle calculations, however, retain the characteristic features even for rather rapidly varying pulses. The main modification — as compared to the adiabatic limit — is that the scatter of the points increases, but the gross features remain unaltered. It has been shown that for pulses $E(z) = E\text{sech}^2(\alpha k_0 z)$ the adiabatic approach is a good one even for $\alpha = 0.3$ [7].

Very large amplitude plasma waves are expected to break. In cold plasmas this takes place at $eE/m\omega_p c \geq (2\gamma_0 - 2)^{1/2}$ which in our notation reads $\mathcal{E} \geq (2/\gamma_0)^{1/2}(\gamma_0 - 1)$ [10]. The wave-breaking limit in warm plasmas has recently been studied in Ref. [11]. One can conclude that values of $\mathcal{E}$ up to about unity can still be supported without wave breaking.

We have investigated in detail the electron acceleration close to the critical surface where v_{ph} diverges (see also [2]). An example is given in Fig. 2. During the preacceleration phase the assumption of slowly varying fields apply and the first energy increment can be calculated using the above results. The final energy boost takes place in a very short distance which part requires separate modelling.

According to the adiabatic model of [7,12], nonlinear wave damping takes place both at the instance of trapping or detrapping which both occur when either the boundary $\kappa = 1$ is crossed or a sudden change of the field decouples the particle-field interaction. Wave damping effects of this kind have been studied extensively, but the problem seems not yet to be exhausted [13].

Wave-damping can be described approximately by space dependent multiplicative factors, i.e. $E^2(z) = E_0^2(z) R_D(z) R_L(z)$, where $E_0^2(z)$ is the local field intensity obtained from the WKB-approximation, $R_D(z) = 1 - [S_p(z) - S_p(0)]/S_w(0)$ takes into acount trapping effects (S_p and S_w are the particle and wave energy fluxes), and $R_L(x)$ is the Landau damping decrement [3]. All dynamics is of course lost in this approach.

ACKNOWLEDGEMENTS

The financial support by the European Commission through the TMR Network SILASI (Superintense Laser Pulse - Solid Interaction) ERBFMRX-CT96-0043 and through the Association Euratom-TEKES is gratefully acknowledged.

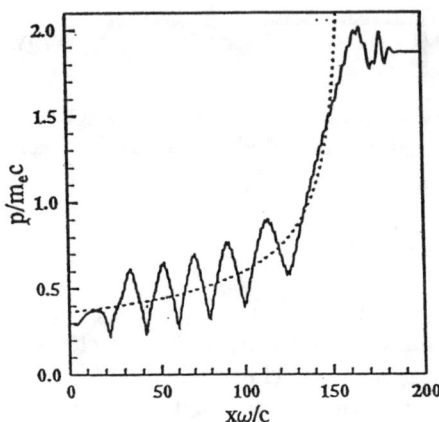
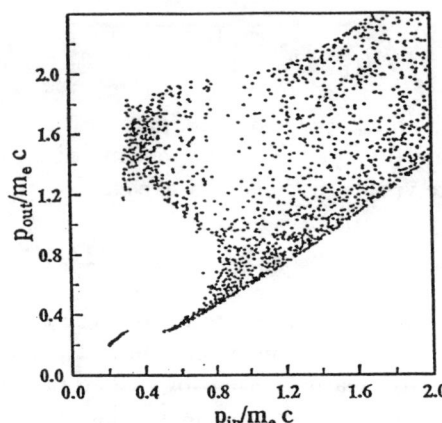

FIGURE 2. Electron acceleration close to a critical surface. A linear density ramp is assumed. Left hand figure gives a typical particle trajectory and the right hand figure the output momenta of 2000 test electrons behind the critical surface.

REFERENCES

1. Bertrand, P., Ghizzo, A., Karttunen, S. J., Pättikangas, T. J. H., Salomaa, R. R. E., and Shoucri, M., Phys. Plasmas **2**, 3115 (1995).
2. Brooks, R., and Pietrzyk, Z., Phys. Fluids **30**, 3600 (1987).
3. Karttunen, S. J., Pättikangas, T. J. H., Tala, T. J. J., and Cairns, R. A., Phys. Rev. E **56**, 56 (1997).
4. Gradshteyn, I. S., and Ryzhik, I. M., *Table of Integrals, Series, and Products*, Academic Press 1980.
5. Abramowitz, M., and Stegun, I. A., *Handbook of Mathematical Functions*, Dover, 1964.
6. Pocobelli, G., Phys. Fluids **24**, 2173 (1981).
7. Schneider, W., PhD Thesis, Technische Hochschule Darmstadt, 1984.
8. Mora, P., Phys. Fluids B **4**, 1630 (1992).
9. Kruskal, M., J. Math. Phys. **3**, 806 (1962).
10. Akhiezer, A. I., and Polovin, R. V., Sov. Phys. JETP **3**, 696 (1956).
11. Sheng, Z. M. and Meyer-ter-Vehn, J., Phys. Plasmas 4, 493 (1997).
12. Pocobelli, G., Phys. Fluids **24**, 2177 (1981).
13. Brodin, G., Phys. Rev. Lett **78**, 1263 (1997).

Hot Particle Generation in Ultrahigh Intensity Laser-Plasma Interaction

Catherine Toupin, Erik Lefebvre, and Guy Bonnaud

*Commissariat à l'Énergie Atomique, Centre d'Études de Limeil-Valenton,
94195 Villeneuve-Saint-Georges, France*

Abstract. Both one-and-a-half and two-dimensional relativistic particle-in-cell simulations of an overdense plasma impinged upon by an ultra-intense laser pulse are discussed. The results provide new features of the relativistic electron heating in relation with ion mobility and two-dimensional geometry.

I. INTRODUCTION

Although with scale much reduced compared to inertial fusion lasers, present lasers are capable of providing powers in excess of 1 terawatt in sub-picosecond pulses, within the optical-infrared region, e.g. 1.053 µm for Nd-glass lasers. After beam focusing onto a spot of roughly 10-wavelength diameter, the irradiance is currently above 10^{19} W/cm^2 (1). This ultra-high irradiance (UHI) corresponds to an electric field amplitude of 27 $I_{18}^{1/2}$ GV/cm and to an electromagnetic pressure of 300 I_{18} Mbars, where I_{18} denotes the irradiance in units of 10^{18} W/cm^2. When such a laser beam is incident upon solid matter, the atoms are strongly ionized and the free electrons get a relativistic motion: their quiver momentum, normalized to $m_e c$, is indeed written: $a_0 = 0.85 (I_{18}\lambda_0^2)^{1/2}$, where m_e denotes the electron mass, c is the velocity of light in vacuum and λ_0 is the laser wavelength in µm. When the electron density n_e exceeds the critical density $n_c = 10^{21}/\lambda_0^2$, low light irradiances are classically reflected.

These lasers have given rise to an increasing number of new kinds of experiments, which would have been impossible to conceive of earlier. Then, a large simulation effort has been made in order to give interpretation/prediction basis, with help of the new generation of vector/parallel computers. The most interesting theoretical results come from kinetic simulations based on particle-in-cell (PIC) codes: they have evidenced two dramatic effects induced by the ultra-high irradiances, namely the penetration of the incident wave in an overdense plasma (2) and absorption into relativistically-accelerated electrons (3,4).

This paper is organized as follows: Section 2 describes the model and the simulation codes used for this paper. Section 3 is concerned with the results from a one-dimensional (1-D) model where the incident pulse is assumed to stay planar whereas Section 4 deals with specific results involved in a two-dimensional (2-D) plasma by a finite-size laser wave. In Sec. 5, concluding observations are given.

II. MODEL AND SIMULATIONS

The physics we explore here is characterised by collective, relativistic and non-quantum electromagnetic effects in which the magnetic force plays a role similar to the electric field: displacements of a large number of free charges with relativistic velocities must be calculated in the presence of self-consistent electric and magnetic fields. Initially, the matter is assumed to be ionized. The lack of hierarchy among the time or space parameters makes the scenarii difficult to interpret, and pure analysis nearly untractable. Numerical computation is therefore required without resorting to any *ab initio* choice of frequencies or wavelengths.

Specific codes are required to tackle those electromagnetic mechanisms that are basically kinetic effects, involving non-Maxwellian distributions, varying strongly in space and time. The plasma can be reduced to two distributions of electrons and ions, evolving according to the Vlasov equation with self-consistent fields; these fields are computed from Maxwell equations where the charge and current densities are deduced from the particle distributions. The plasma can be considered as collisionless since the quiver motion along the laser electric field makes the electron-ion collision frequency much smaller than the electron plasma frequency. The Vlasov equation is solved with a particle-in-cell (PIC) method which needs less computing time than classical discretization in 2-D or 3-D simulations.

Both of the codes we use are fully relativistic. The 1-D code EUTERPE was designed ten years ago and has been used for both ICF and UHI contexts (5,6). The 2-D code MANET has been recently designed. It operates in 2-D plane geometry in which the beam and the plasma appear as slabs with infinite length along one transverse axis. Both codes can handle linearly polarized light. The particles are reflected on the laser-entrance side and reflected (resp. reinjected with initial thermal velocity) on the other sides when the plasma does not touch (resp. reaches) the boundary of the simulation box. Along the transverse direction, periodic conditions for the fields are used. No collisions are used here and the laser wave is normally incident on the plasma.

We use the following units: [space] = c/ω_0 = laser wavelength/2π, [time] = ω_0^{-1} = laser period/2π, with ω_0 the laser radial frequency.

III. RESULTS FROM 1-D SIMULATIONS

The illumination at normal incidence of an overdense plasma surface causes the acceleration of electrons to relativistic energies. In the opacity regime (2,4,6), the laser wave does not propagate inside the plasma. The drag and pull motion of surface electrons driven by the oscillating ponderomotive force and the charge space field causes them to exit the plasma and have an excursion in vacuum in front of the plasma surface (7-9). They undergo a strong acceleration before returning into the plasma. There, quasi no electric field exists and these electrons are nearly free-streaming in the plasma.

1-D PIC simulations, performed with fixed ion background, have shown that the fast electron temperature is dependent on both laser irradiance, plasma density and density gradient at the plasma surface (4). The results are given in Fig. (1), where we display the average energy of the electrons exiting the simulation box with energies in excess of 100 keV. If the plasma is sharp-edged, the electron heating originates from the dephasing of the electron orbits in front of the surface; this dephasing is related to the

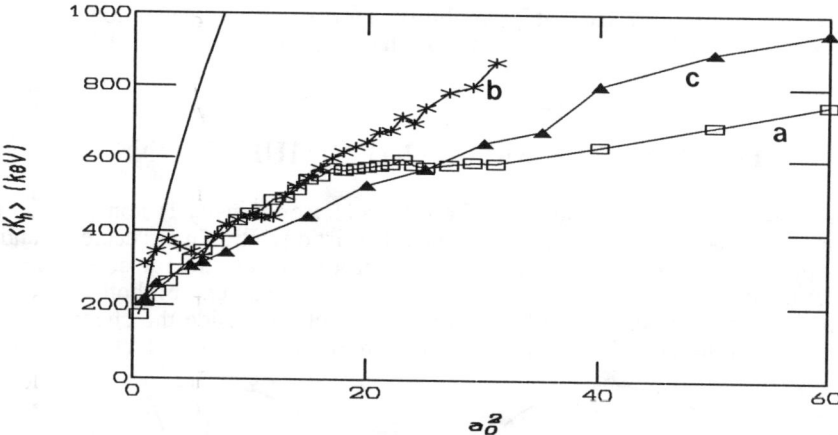

FIGURE 1. Temperature of electrons penetrating the plasma, from 1-D simulation as a function of a_0^2, for $n_e/n_c = 17$, (a) $m_i/Zm_e = \infty$, $L = 0$, (b) $m_i/Zm_e = \infty$, $L = 0.1$, (c) $m_i/Zm_e = 1836$, $L = 0$. L is the density scale-length at density n_c in units of c/ω_0. The laser pulse is always square shaped with 40 laser-period width. The continuous line corresponds to the ponderomotive potential.

time when they exit the plasma. The absorption occurs only at the plasma surface, the denser the plasma, the colder the fast electrons. For an extended plasma surface (gentle density gradient), the incident and reflected waves form a nearly standing pattern in front of the target. When the density gradient is smooth enough, the laser reflection point is located inside the plasma and the electrons that are below this turning point interact with this standing wave by oscillating stochastically with a large longitudinal amplitude (10). Their thermal energy is increased as the density-scale length L is increased and approaches the ponderomotive potential: $\Phi_p = m_e c^2 [(1+ a_0^2)^{1/2} -1]$. The electrons are accelerated inside a plasma volume. When ions are mobile, both surface and volume mechanisms are involved. Electron acceleration into the target steepens the ion density profile due to the space charge electric field. Absorption takes place in the underdense expanded plasma, at the density jump where electrons can bounce between the moving plasma surface and the propagated ion wave

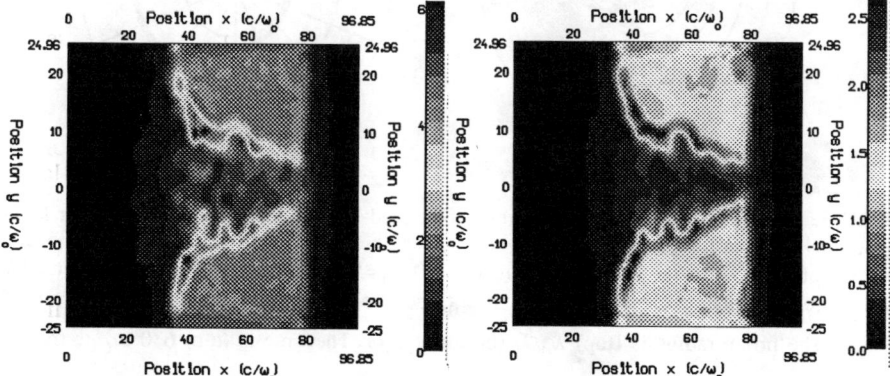

FIGURE 2. 2-D-map of (left) the electron density n_e/n_c, (right) the effective electron density $<n_e/\gamma n_c>$. 2-D simulation. $n_e/n_c = 2$, $a_0 = 3$, $m_i/Zm_e = 3672$, $T_{e0} = 5$ keV, initial plasma length 50 c/ω_0.

(shock or soliton (11)). We see that ion mobility favors electron acceleration; the typical electron energies lie between the results for sharp-edged plasma and $0.1\ c/\omega_0$-scale ramp.

IV. RESULTS FROM 2-D SIMULATIONS

The laser wave can bore a channel where the electron density is non-zero by two ways: the self-induced transparency due to the relativistic electron mass decrease and the radial expulsion due to the transverse ponderomotive force. So for conditions that would define an opacity regime in 1-D, the 2-D geometry can exhibit both opacity and transparency regimes; behind the laser front, the electrons inside the channel will be quivered by the laser and plasma electric fields. In Fig. (2), the real electron density n_e

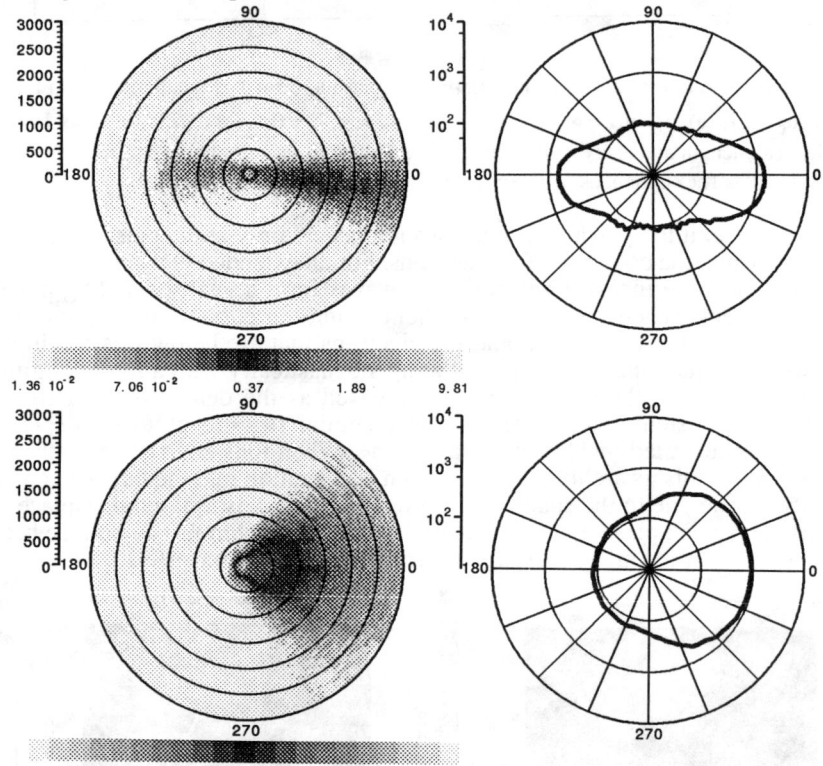

FIGURE 3. (left) Angular distribution of the accelerated electrons as a function of their kinetic energy (radially in keV), (right) temperature (keV) vs. angle. 2-D simulation with $m_i/Zm_e = 3672$ and $T_{e0} = 5$ keV. (top) $n_e/n_c = 2$, $a_0 = 3$, (bottom) $n_e/n_c = 10$, $a_0 = 5$. The electrons which are selected have crossed a circular probe disk located at (top) $16\ \lambda_0$, (bottom) $1.8\ \lambda_0$ from the plasma surface on the laser axis. The probe radius is (top) $\lambda_0/2$, (bottom) $\lambda_0/3$. The time is (top) $630\ \omega_0^{-1}$, (bottom) $100\ \omega_0^{-1}$.

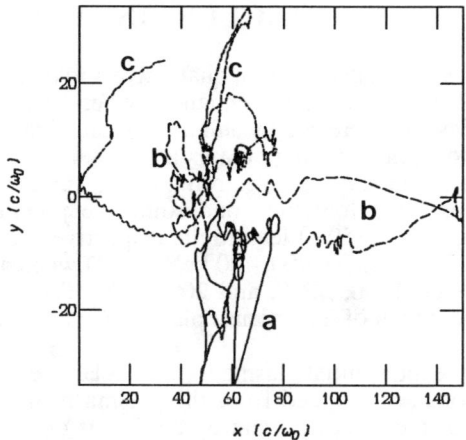

FIGURE 4. Orbits of selected electrons initially located at $x=50\ c/\omega_0$, $y=$ (a) $-22.5\ c/\omega_0$, (b) $-6.\ c/\omega_0$, (c) $0.\ c/\omega_0$. 2-D simulation. $n_e/n_c = 2$, $a_0 = 3$, $m_i/Zm_e = 3672$, $T_{e0} = 5$ keV. The plasma extends from 37 to 150 c/ω_0. The laser wave enters at time $t=0$ on the left side of the simulation box.

(the ion density, not shown here, is similar) and the optical density $\langle n_e/\gamma \rangle$ are compared: the latter has values below n_c in a horn-shaped channel surrounded by the overdense plasma. This shows that the self-induced transparency scenario, visible in 1-D runs (2) remains valid in 2-D geometry. Only one filament is visible. Both self-focusing and radial expansion initiated at different times explain its horn shape.

In Fig. (3), we have compared the electron features for two different densities $n_e/n_c = 2$ and 10. The plasma length is 19 λ_0 (resp. 2.2 λ_0). The angular distribution of the electron energies is determined by counting electrons which cross a probe disk, with size much smaller than plasma/beam sizes, over the simulation time. An average suprathermal energy is inferred from electrons with kinetic energies above 100 keV.

For $n_e/n_c = 2$, the return current is colder than the forward current, as expected. The electron distribution is centered on the laser axis and extends in a $\pm 20°$ angle lobe, which corresponds to electron emission from the channel edges; the forward energy is 1.5 MeV. The high rate of electron heating induces a large magnetic field at the edges of the transparent channel; this field tends to pinch the electron beam and then modify the beam propagation. Here the magnetic field is antisymmetric around the laser axis with one positive and one negative lobes located inside the density channel edges. The amplitude reaches $2\ 10^4$ Teslas, which is roughly two-thirds of the magnetic field of the incident laser wave. The beam propagation is bound to the hot electrons which act as a guide of the laser front into the plasma. The most spectacular effect has been revealed in 3-D simulations (12) where initial light filaments collapse into one super-intense filament.

For $n_e/n_c = 10$, the suprathermal energy which peaks at 900 keV, is lower than for $n_e/n_c = 2$; this indicates that the suprathermal energy scales with density as in 1-D (4). The angular distribution is also much wider: the aperture angle of the distribution encompasses the full beam width.

Three electron orbits are displayed to show three specific electron behaviors; they exhibit one electron that is radially expelled and does not cross the axial channel (see Fig. (4-a)), one electron which is strongly accelerated in the channel and goes back to participate to the return current (4-b) and one electron with excursion in vacuum (4-c). This orbit diagnostic allows one to track the special areas in the plasma.

V. CONCLUSION

The major conclusion can be stated on both qualitative and quantitative aspects. First, in 1-D simulations, the electron heating is enhanced when light interacts with a large volume of low density plasma, as created by an initial plasma gradient or by ion motion. Second, in 2-D simulations, hole boring creates a channel where the laser wave can self-focus; both phenomena increase the electron heating, compared to 1-D. Third, the typical electron energies lie below the ponderomotive potential: the mobile-ion simulations (n,a_0)=(2,3), (10,5), (50,5) lead to the respective average kinetic energies 1250, 700, 320 keV in 1-D and 1500, 900, 400 keV in 2-D whereas the ponderomotive potential would have given 1100, 2090, and 2090 keV. The influence of the pulse width that can control the length of underdense plasma in which the light can penetrate remains to be explored.

Since the electrons leave their initial plasma location a large electric field builds; as a consequence, the electrons are dragged from the plasma behind the laser-irradiated surface. Any limitation of the return current by collisions or by magnetic fields will cause heating limitation. Following computer studies, various experiments are trying to understand the electron heating in the MeV range (13). But a lot of work remains to be done to control the electrons useful for the fast-ignitor concept (14).

ACKNOWLEDGMENTS

The authors are indebted to A. Adolf who designed the parallelization of the 2-D PIC code on the T3D Cray computer and to J. Chave and P. Bain for the design of the diagnostics and the versatility improvement. Fruitful discussions are acknowledged with J.M. Rax.

VI. REFERENCES

1. Maine P., Strickland D., Bado P., Pessot M., and Mourou G., *IEEE J. Quantum Electron.* **24**, 398 (1988); Ferray M., Lompre L.A., Gobert O., Mainfray G., Manus C., Sanchez A., and Gomes A., *Optics Comm.* **75**, 278 (1990); Perry M.D., Patterson F.G., and Weston J., *Opt. Lett.* **15**, 381 (1990); Watteau J.P., Bonnaud G., Coutant J., Dautray R., Decoster A., Louis-Jacquet M., Ouvry J., Sauteret J., Seznec S., and Teychenne D., *Phys. Fluids B* **4**, 2217 (1992).
2. Lefebvre E. and Bonnaud G., *Phys. Rev. Lett.* **74**, 2002 (1995).
3. Wilks S.C., Kruer W.L., Tabak M., and Langdon A.B., *Phys. Rev. Lett.* **69**, 1383 (1992).
4. Lefebvre E. and Bonnaud G., *Phys. Rev. E* **55**, 1011 (1997).
5. Bonnaud G. and Reisse C. *Nuclear Fusion* May **26**, 633 (1986).
6. Lefebvre E., *PhD dissertation* (in french), Univ. of Paris (1996).
7. Brunel F., *Phys. Rev. Lett.* **59**, 52 (1987).
8. Bonnaud G., Gibbon P., Kindel J., and Williams E.A., *Laser Part. Beams* **9**, 339 (1991).
9. Gibbon P., *Phys. Rev. Lett.* **73**, 644 (1994).
10. Bauer D., Mulser P., and Steeb W.H. *Phys. Rev. Lett.* **75**, 4622 (1995).
11. Denavit J., *Phys. Rev. Lett.* **69**, 3052 (1992).
12. Pukhov A. and Meyer-ter-Vehn J., *Phys. Rev. Lett.* **76**, 3975 (1996).
13. Malka G. and Miquel J.-L., *Phys. Rev. Lett.* **77**, 75 (1996).
14. Tabak M., Hammer J., Glinsky M.E., Kruer W.L., Wilks S.C., Woodworth J., Campbell E.M., Perry M.D., and Mason R.J., *Phys. Plasmas* **1**, 1626 (1995).

Fast Electron Transport in Solid Targets

J. R. Davies and A. R. Bell

Plasma Physics, Blackett Laboratory, Imperial College, London SW7 2BZ

Abstract.
The generation of fast electrons in high intensity laser-solid interactions has been extensively studied eg [1,2]. A frequently used diagnostic for fast electrons is K_α emission from layered targets, the interpretation of such data requires a model for the transport of the fast electrons through the target [3]. Such experiments have largely been interpreted using models including only collisional effects. However recent works have shown that at high intensities electric and magnetic fields could be important [4–6]. Here we present a code which can deal with the transport of fast electrons through solid targets including both collisions and field generation. The fast electrons are represented by a relativistic Fokker-Planck equation including drag, angular scattering, electric and magnetic fields, which is solved using stochastic differential equations (SDEs). The background is represented by $\mathbf{E} = \eta \mathbf{j}_b$, where η is the resistivity and $\mathbf{j}_b$ the background current density. Changes in resistivity due to heating of the background by the fast electrons are included. Current balance is assumed allowing the electric field to be found directly from the fast electron current. Rotational symmetry is assumed. The treatment is valid for fast electron number densities much less than that of the background, fast electron energies much greater than the background temperature and time scales short enough that magnetic diffusion and thermal conduction are negligible. The neglect of ionization also limits the validity of the model.

I THE MODEL

The fast electrons are represented by a Fokker-Planck equation

$$\frac{\partial f}{\partial t} = -\frac{\partial}{\partial \mathbf{r}} \cdot (\mathbf{v} f) - \frac{\partial}{\partial \mathbf{p}} \cdot ((\mathbf{F} + <\Delta \mathbf{p}>) f) + \frac{1}{2} \frac{\partial}{\partial \mathbf{p}} \frac{\partial}{\partial \mathbf{p}} : (<\Delta \mathbf{p} \Delta \mathbf{p}> f) \quad (1)$$

$$\mathbf{F} = -e(\mathbf{E} + \mathbf{v} \times \mathbf{B})$$

$$<\Delta \mathbf{p}> = \left(<\Delta p> - \frac{p}{2} <\Delta \theta^2>\right) \frac{\mathbf{p}}{p} \quad <\Delta \mathbf{p} \Delta \mathbf{p}> = p^2 <\Delta \theta^2> \left(\mathcal{I} - \frac{\mathbf{pp}}{p^2}\right)$$

for the fast electron probability density $f(\mathbf{r},\mathbf{p},t)$, in three dimensional Cartesian geometry. Where $<\ldots>$ signifies the mean change per second due to collisions. $\mathcal{I}$ is the identity tensor, $<\Delta p>$ is the drag term, $<\Delta\theta^2>$ is the angular scattering term and other symbols have their usual meaning. The motion of the background particles has been neglected so there is no diffusion in the magnitude of the momentum. Interactions between the fast electrons and large angle scattering have also been neglected.

For electrons with energies from 10keV to a few MeV the standard result for the drag term [7] for fast electrons in solids gives

$$<\Delta p> \approx -\frac{Zne^4}{4\pi\varepsilon_0^2 mv^2}\ln\frac{K}{I_{ex}} = -\frac{Zne^4}{4\pi\varepsilon_0^2 mv^2}\ln\Lambda_l \qquad (2)$$

where Z is the atomic number, n is the background atom number density, m is the electron mass, K is the fast electron kinetic energy and I_{ex} is the mean excitation energy, which is determined by the binding of the atomic electrons. We use values of I_{ex} from ICRU Report 37 [7].

In a solid angular scattering is caused by collisions with the atoms. An approximate model for the field of an atom is an exponentially screened potential, with a screening distance (a) given by $a \approx 4\pi\varepsilon_0\hbar^2/Z^{1/3}me^2$ [8]. The de Broglie wavelength (λ_{dB}) of the fast electrons is much less than typical interatomic spacings so the atoms may be considered as independent scattering centres. Using the first Born approximation for the scattering formula gives,

$$<\Delta\theta^2> \approx \frac{Z^2ne^4}{2\pi\varepsilon_0^2}\frac{\gamma m}{p^3}\ln\frac{4\pi a}{\lambda_{dB}} = \frac{Z^2ne^4}{2\pi\varepsilon_0^2}\frac{\gamma m}{p^3}\ln\Lambda_s \qquad (3)$$

where γ is the Lorentz factor. The Born approximation requires [8] $Z\alpha\ln(\Lambda_s/2)/(v/c) < 1$ where α is the fine structure constant, so this equation is not valid at high Z. Angular scattering by electrons is neglected.

The background atom number density n and atomic number Z are assumed to be uniform and fixed. This and the neglect of collisions between the fast electrons requires the fast electrons number density to be much less than that of the background.

The only difference between these coefficients and those for a plasma is in the log terms, which are weakly varying functions of their arguments. So little adjustment is required to deal with either solids or plasmas.

The details of the fast electron generation are not dealt with, a specified distribution of fast electrons being injected into the target. For the generation of the fast electrons and for the field calculations rotational symmetry is assumed.

For the background electrons we use $\mathbf{E} = \eta\mathbf{j}_b$, as they are highly collisional compared to the fast electrons. Glinsky [5] discusses this approximation in more detail. Further assuming that the background current exactly balances the fast electron current $(\mathbf{j}_f)$ gives

$$\mathbf{E} = -\eta \mathbf{j}_f. \tag{4}$$

This is used by Glinsky [5] and Bell et al [6]. Bell et al justify this by estimating the energy in the magnetic field generated if the initial fast electron current is not cancelled, finding it to be vastly greater than the energy of the fast electrons. The currents will remain approximately in balance for a magnetic diffusion time [9]

$$t_B = \frac{\mu_0}{\eta} L^2 \tag{5}$$

where L is the radial scale length of variations in $\mathbf{j}_f$. This model is valid for time scales less than this.

The magnetic field is given by

$$\frac{\partial \mathbf{B}}{\partial t} = -\nabla \times \mathbf{E} = \nabla \times \eta \mathbf{j}_f. \tag{6}$$

This can be expressed as two separate terms $\eta \nabla \times \mathbf{j}_f$ and $\nabla \eta \times \mathbf{j}_f$.

The assumption of rotational symmetry allows non-zero $E_r(r,z)$, $E_z(r,z)$ and $B_\theta(r,z)$.

The resistivity is assumed to depend only on the background temperature. Any given function of temperature may be used. The background temperature is initially given a uniform value and is then increased by the energy lost to the background electrons (energy lost by the fast electrons minus the energy in the magnetic field) using a given specific heat capacity. Thus the resistivity can vary in space and time. In calculating the background temperature thermal conduction is neglected. The resistivity will be independent of the electric field if the drift velocity of the electrons carrying the background return current is much less than their mean speed. The requirement that the fast electron number density is much less than that of the background will usually ensure this.

There is an effective restriction on the maximum fast electron current for which the model is valid as a high current leads to rapid heating of the background (from ηj_f^2), violating the assumption that the fast electron speed is much greater than the mean speed of the background electrons. In this case the distinction between fast and background electrons rapidly vanishes and it would be better modelled by a PIC code.

The displacement current has been neglected, this is only important for very short time scales of order $\varepsilon_0 \eta$, while the return current establishes itself. The geometry used does not allow for kink instabilities. Microinstabilities, such as two stream instabilities, are not modelled in this approach. The effect of microinstabilities could be approximately accounted for by a suitable choice of the resistivity.

II NUMERICAL METHOD

To solve equation 1 we use the equivalence of Fokker-Planck equations to SDEs to give a particle, Monte-Carlo type approach [10]. This type of approach is ideal for highly localized and anisotropic distributions, which are expected in laser plasma interactions. Finite element and finite difference approaches have difficulty with such distributions. We use Ito SDEs [10] formulated in terms of a particle's position ($\mathbf{r}$), magnitude of momentum (p) and scattering angle (θ). Neglecting the force from the fields for the present they are

$$d\mathbf{r} = \mathbf{v} dt \qquad (7)$$
$$dp = <\Delta p> dt \qquad (8)$$
$$d\theta = <\Delta\theta^2>^{1/2} dW \qquad (9)$$

where dW is the increment of a Wiener process, which is the solution of a diffusion equation with a diffusion coeffcient of one. The collisional part of the code is executed in three dimensional Cartesian geometry.

We now give the routine followed in the code.

Computational particles are generated at the beginning of each time step at the target surface, placed at $z=0$, with a uniform distribution over the desired range of r and $\mathbf{p}$ using the Sobel sequence [11]. They are then assigned a number of electrons to give the desired number density. This assures good statisitics over the whole phase space. Particles are advanced by a random fraction of a time step when first generated as they are assumed to be generated at a constant rate during the time step.

The grid cells are labelled $j=1-N_r$ and $k=1-N_z$, $(j-1/2)\Delta r$, $(k-1/2)\Delta z$ giving the position of the cell centres. The boundaries $r=0$ and $z=0$ and optionally the far z boundary are reflective boundaries for the fields and particles. The fields outside the other boundaries are set to zero and the particles are allowed to escape. The fast electron current on the grid is found by weighting qv_r and qv_z from each particle, where q is the particle charge, to its four nearest grid points using area weighting [12] then dividing the totals at each grid point by the grid cell volume $(2j-1)\pi\Delta r^2 \Delta z$. This gives $\mathbf{E}^n_{j,k}$ from equation 4. The azimuthal magnetic field is then found from

$$B^n_{j,k} = B^{n-1}_{j,k} + \Delta t \left(\frac{Ez_{j+1,k} - Ez_{j-1,k}}{2\Delta r} - \frac{Er_{j,k+1} - Er_{j,k-1}}{2\Delta z} \right)^n \quad j>2 \qquad (10)$$

where the superscript gives the time step. The magnetic field at the first two radial grid points is found by a linear fit from zero at $r=0$ to the value obtained at the third radial grid point. This minimizes errors for particles near $r=0$. The fields on the particles are then found by weighting the fields from the grid points back to the particles using the same weighting scheme.

Having found the fields the momentum is advanced using an explicit first order scheme. First the rotation $eB^n \Delta t/\gamma^n m$ about the magnetic field is applied, then equations 8 and 9 are solved using

$$p^{n+1} = p^n + <\Delta p>^n \Delta t \qquad (11)$$
$$\theta = (<\Delta \theta^2>^n \Delta t)^{1/2} \Gamma^n \qquad (12)$$

where Γ is a Gaussian random variable with mean zero and variance one, a new value being generated each time step. The rotation θ is applied about an axis at a random angle perpendicular to the direction of motion. Then the acceleration from the electric field is applied. The new position is then found from

$$\mathbf{r}^{n+1} = \mathbf{r}^n + \Delta t \mathbf{v}^{n+1}. \qquad (13)$$

The energy lost by each particle in a time step is accumulated in the grid cell it is in at that time step to give the energy lost on the grid. This energy minus the energy in the magnetic field (from $B^2/2\mu_0$) gives the heating of the background. Using the given value of the specific heat capacity the temperature at each grid point is calculated and from that the resistivity using a specified function.

The various features of the code (collisons, electric field, magnetic field, variable resistivity and relativity) can be switched on or off and have all been tested independently. The purely collisional case has been extensively studied using this code [13].

III CONCLUSIONS

Given an initial distribution of fast electrons, the target material, resistivity and specific heat capacity the code described here can model the propagation of the fast electrons through the target over time scales short enough that magnetic diffusion and thermal conduction are negligible, provided that the background temperature remains significantly lower than that of the fast electrons. Such calculations have been done for semi-infinite aluminium targets irradiated by 1ps, 1μm wavelength 20μm diameter laser spots for a wide range of intensities [14]. These calculations show that field effects must be included at intensities of around 10^{17}Wcm^{-2} and higher.

Combined with a calculation of X-ray emission this code could be used to interpret K_α emission data, providing a more accurate model than the purely collisional ones currently in use.

The major drawback of the code described here as a predictive tool is the need to specify the fast electron distribution, the background resistivity and specific heat capacity, which are not well known parameters in laser solid experiments.

To be more generally applicable the model presented here must be extended to include magnetic diffusion and a more realistic treatment of the background. Work on the first of these is currently underway.

REFERENCES

1. S. J. Gitomer, R. D. Jones, F. Begay, A. W. Ehler, J. F. Kephart and R. Kristal, *Phys. Fluids* **29**, 2679 (1986).
2. P. Gibbon and E Förster, *Plas. Phys. Con. Fus.* **38**, 769 (1996).
3. P. Lee C. K., Ph.D. thesis, University of London, 1996.
4. B. Luther-Davies, A. Perry and K. A. Nugent, *Phys. Rev. A* **35**, 4306 (1987).
5. M. E. Glinsky, *Phys. Plas.* **2**, 2796 (1995).
6. A. R. Bell, J. R. Davies, S. Guerin and H. Ruhl, *Plas. Phys. Con. Fus.* **39**, 653 (1997).
7. International Committee on Radiation Units Report **37**, *Stopping Powers for Electrons and Positrons*, I.C.R.U., 1984.
8. C. J. Joachain, *Quantum Collision Theory*, 3rd Edition, North-Holland, 1987.
9. R. B. Miller, *Intense Charged Particle Beams*, Plenum Press, 1982, ch. 4.
10. see for example C. W. Gardiner *Handbook of Stochastic Methods*, 2nd Edition, Springer-Verlag, 1985, H. Risken, *The Fokker-Planck Equation*, 2nd Edition, Springer-Verlag, 1984.
11. W. H. Press, S. A. Teukolsky, W. T. Vettering and B. P. Flannery, *Numerical Recipes*, 2nd Edition, Cambridge 1992.
12. C. K. Birdsall and A. B. Langdon, *Plasma Physics Via Computer Simulation*, Adam Hilger, 1991.
13. J. R. Davies, Ph.D. thesis, University of London, 1997.
14. J. R. Davies, A. R. Bell, M. G. Haines and S. M. Guerin, submitted to *Phys. Rev. E*.

4. LASERS FOR ULTRAHIGH INTENSITY PHYSICS

Ultrahigh Intensity Laser: Present and Future

J. Nees,[†] S. Biswal,[†] F. Druon,[†] J. Faure,[†] M. Nantel,[†] G. Mourou,[†]

A. Nishimura,[•] H. Takuma,[•]

J. Itatani,[°]

J. C. Chanteloup,[*]

C. Hönninger[††]

Abstract

Over the past ten years, we have seen a revolution in the generation of ultraintense pulses, now well in the 10^{19} W/cm^2 range. If further developments in high-field lasers are to be accessible to universities and institutes, new laser materials and phase control techniques, which will result in compact, reliable systems with higher peak power, must be adopted. The choice of high-saturation-fluence gain material and the measurement and active control of temporal and spatial phase distortions for compact CPA systems of the future will be essential. Using the proper material and phase control, a focused intensity of 10^{25} W/cm^2 is theoretically possible.

[†]Univ. Michigan, Center for Ultrafast Optical Science, Room 1006 IST, 2200 Bonisteel, Ann Arbor, MI 48109-2099, USA

[•]Japan Atomic Energy Research Institute, 2-4 Shirane, Tokai-mura, Naka-gun, Ibaraki-ken 319-11, Japan

[°]Institute for Solid-State Physics, University of Tokyo, 7-22-1 Roppongi, Minato-ku, Tokyo 106, Japan

[*]Laboratories pour l'Utilisation des Lasers Intenses (LULI), CNRS UMR 100, Ecole Polytechnique, 91128 PALAISEAU CEDEX, France.

[††]Institute of Quantum Electronics, Ultrafast Laser Physics Laboratory, ETH Zurich ETH Hönggerberg, HPT D20 CH-8093 Zurich, Switzerland

CP426, *Superstrong Fields in Plasma:* First International Conference
edited by M. Lontano et al.
© 1998 The American Institute of Physics 1-56396-748-0/98/$15.00

I. INTRODUCTION

Optical nonlinearities in solids were inaccessible until lasers were introduced with focused intensities near 10^{10} W/cm^2 [1]. Likewise, the discovery of Q-switching [2, 3] opened the investigation of nonlinearities in gases near 10^{14} W/cm^2 [4]. In more recent years, relativistic nonlinearities have become accessible due to the introduction of chirped pulse amplification (CPA) [5, 6] and to the use of focused intensities above 10^{17} W/cm^2 [7].

Beginning with the first terawatt lasers built for fusion applications and proceeding to the modern tabletop terawatt lasers, a wealth of scientific discovery has been based on the interaction of strong optical fields with matter. Critical to these studies is the availability of high-focused-intensity lasers. Along with the development of femtosecond oscillators [8], the introduction of CPA has had a profound influence on the way ultrafast and high-field science is done. By bringing reliability and compactness to high-intensity lasers, these advances allow research to be done not only in national laboratories but, also, at the level of universities and industry. Scaling to higher intensity sources will require continued attention to compactness and reliability.

In order to realize of the full potential of the laser amplification process to form high focused intensity, we should use a gain material which can be pumped by simple free-running lasers or laser diodes [9]. Both the density of energy stored and the excited-state lifetime of the material must be exploited. We must also choose a gain material with sufficient bandwidth to allow short pulses to be formed. Finally, the temporal and spatial characteristics of the amplified pulses must be known and manipulated to deliver a well-behaved optical pulse to the experiment.

Following the notion that broad usage will enhance the development of applications and support, one can reason that the path to fully utilizing optical amplifiers lies in the direction of compact and efficient laser technology. As illustrated by the development of electronics, the compactness, reliability, and scalability of a technology have a strong impact on its growth. The use of the transistor, coupled with the further invention of the printed circuit board and the microchip, brought about the current phenomena of microcomputers and telecommunications. These changes were made because the new technology offered compactness, reliability, and scalability.

In a similar way, the invention of CPA brings a phenomenal reduction in scale to the production of high focused intensity. Before the introduction of CPA, high-intensity ultrafast pulses were generated by dye or excimer lasers and the use of solid-state materials with their superior energy storage capability was restricted to nanosecond-pulse amplification. Using these technologies, a terawatt laser was the size of a building. With the technique of CPA, the scale of terawatt lasers is reduced by orders of magnitude to tabletop dimensions. The requirement for pump power is also reduced by orders of magnitude. As a result, the cooling rate and the ease of cooling are improved. The size of optics and the probability of damage are reduced. The result is that terawatt lasers are widely available and used by a growing number of researchers in universities, institutes, and industry.

Where can improvements in compactness, reliability, and scalability take high-field science? This question is naturally open-ended. Immediate applications in medicine [10, 11], precision machining[12], laser film deposition[13], XUV and X-ray generation [14], and particle acceleration [15, 16, 17, 18, 19] provide some indication of practical directions for high-field lasers. Continued exploration of relativistic effects in laser-plasma and laser-particle interactions can also be projected. Though the stellar goal of exploiting vacuum nonlinearity at 10^{29} W/cm^2 [9, 20] remains hidden by several orders of magnitude in intensity, this too gives a sense of direction to the development of compact, ultra-intense light sources.

In this paper we discuss the choice of laser materials and its effect on compactness and scalability of ultraintense lasers. We also discuss the importance of both measuring and controlling the temporal and spatial pulse features in the amplification process.

II. OPTICAL GENERATION OF HIGH FIELDS

High-intensity lasers produce ultraintense pulses by concentrating a given amount of optical energy both temporally and spatially. The temporal limit is imposed by the time-bandwidth product, $\tau_{min} \cong 1/\Delta\nu$; spatially the limit of focus is $\Delta x_{min} \cong \lambda^2$. ($\tau_{min}$ is the minimal pulse duration, $\Delta\nu$ is the fluorescence bandwidth, Δx_{min} is the minimal focal diameter, and λ is the wavelength.)

A. Energy storage in laser materials

Compact and efficient storage of optical energy depends on the use of materials with high doping density. The second factor which is critical to the suitability of the gain material is its fluorescence lifetime, τ_f. This factor determines the rate at which the stored energy is depleted due to spontaneous emission, and the pump power which is needed to create a population inversion in the gain material. The pump power, P_{pump}, required to achieve the stored energy limit, is the total stored energy divided by the fluorescence lifetime of the gain material and the quantum defect:

$$P_{pump} = \frac{Nh\nu_e}{\eta\tau_f}. \qquad (1)$$

($\eta = \nu_e/\nu_{pump}$ and N is the number of active ions per unit volume.)

The fundamental relationship between the minimum pulse duration, the emission cross-section, and the upper-state lifetime in a two-level laser transition is given by [21]

$$\tau_f = \kappa \frac{\tau_{min} \lambda^2}{\sigma_e n^2}. \qquad (2)$$

(κ is a numeric constant depending on emission line shape and n is the optical index of the material.) From this relation it can be seen that materials with smaller emission cross-sections have the advantage of longer excited-state lifetimes.

B. Energy extraction at the saturation fluence of laser materials

With these factors it is also important to have a high saturation fluence in the amplifier in order to reduce the aperture size. Saturation fluence is defined [22] as U_{sat} [J/cm^2].

$$U_{sat} = \frac{h\nu_e}{\sigma_e}. \quad (3)$$

(Here h is Planck's constant and σ_e is the emission cross section at frequency ν_e.) This factor varies widely from one gain material to another.

Figure 1 shows the effect of saturation fluence on the aperture size of an amplifier system producing a given energy. The aperture required to generate 1 J of optical energy using a typical laser dye would be about 600 cm². Using solid-state materials, the same energy can be produced by an aperture of only 0.02 cm²— 30,000 times reduction in area. Dye amplifiers are also inherently difficult to pump because of their high rate of spontaneous emission. This limits the storage time of dyes to only a few nanoseconds. In contrast, solid-state materials can store energy in the excited state for 2 to 3 ms — 6 orders of magnitude longer than dyes. The long lifetimes of materials such as Yb:glass allow energy from pump lasers to be accepted over a long interval. As a result, pump sources for these materials may be simple, free-running, multimode lasers or laser diodes.

C. Theoretical peak power of laser materials

The theoretical limit of peak power, P_{max}, which can be generated from a 1-cm² area of laser material scales as the saturation fluence divided by the minimum pulse duration.

$$P_{max} \cong \frac{h\nu_e}{\sigma_e \tau_{min}} \quad (4)$$

As shown in Fig. 2, the theoretical limit for the mature technologies of Nd:glass and Ti:sapphire lies in the range of 100 TW/cm² of laser aperture. Yb:glass displays both the bandwidth and the saturation fluence necessary to exceed 1 PW/cm².

Ultimately, the focused intensity that can be obtained from a cm² aperture of gain material is limited to

$$I_{max} \cong \frac{P_{max}}{\Delta x_{min}} \cong \frac{U_{sat}}{\lambda^2 \tau_{min}} \quad (5)$$

Producing peak power at the theoretical limit from any laser material is dependent on extracting the stored energy from the material with the necessary bandwidth to form a short, intense pulse.

D. Avoiding nonlinear phase accumulation

Before the introduction of CPA, nonlinear phase accumulation, as described by the B integral,

$$B = \frac{2\pi}{\lambda} \int n_2(z)\, I(z)\, dz, \quad (6)$$

set the upper limit on the peak power of amplified nanosecond signals to the order of a few GW/cm², or mJ/cm² for picosecond pulses. Such an operating level is several orders of magnitude above saturation fluence in solid-state amplifiers and leads to gross inefficiency. Consequently, solid-state lasers were operated far below their potential for peak power generation.

Following the most widely used recipe for CPA, a short, broadband pulse is generated by a mode-locked oscillator. A single pulse is selected from a 10^8-Hz pulse train and stretched in a grating-based stretcher, the degree of stretching being determined by the limit of nonlinear phase accumulation. The stretched pulse is then used to seed a regenerative or multi-pass amplifier. After passing through one or more stages of amplification, the pulse is compressed by a grating pair to form a short, highly intense pulse.

During this process, the peak power of the pulse in the amplifying medium is reduced by the ratio of stretched pulse duration, $\tau_{stretched}$, to the minimum pulse duration, τ_{min}. This ratio can be in excess of 10^5. The resulting nonlinear phase accumulation is also reduced in proportion.

The product, $\sigma_e N$, defines the small signal gain coefficient, g_0, and the length of material, L, necessary to achieve the desired overall gain, G_0, can be calculated by

$$L = \frac{\ln(G_0)}{g_0} = \frac{\ln(G_0)}{\sigma_e N}. \tag{7}$$

Note that the length of material which contributes to nonlinear phase accumulation may be reduced through the use of highly doped (high N) gain material.

One of the challenges in working with high-saturation-fluence materials is avoiding damage. The threshold for damage to reflective optics at 3-ns, near the limit of present stretching technology, is 20 J/cm^2. When a laser material cannot be used at saturation fluence, it is possible to efficiently extract stored energy below the saturation fluence [23]. This increases flexibility in the choice of materials and allows the extraction of energy to be maximized. The development of new techniques to allow extended stretching into the 10–100-ns regime, or improvement of damage threshold can lead to better use of high-saturation-fluence materials. Maintaining active control of beam wave front can reduce the margin required for damage-free operation of CPA systems. Finally, because the aperture of the gain material will be smaller than that offered by low-saturation-fluence materials, the probability for damage is reduced.

E. Promising new storage materials

From the foregoing statements it can be seen that an alternative to Nd:glass systems, which maintains diode pumpability but also supports shorter pulses, would allow higher peak power to be developed. A diode-pumpable alternative to Ti:sapphire with higher saturation fluence, which can also support short pulses, is also needed. Candidates should have high saturation fluence, broad bandwidth emission, long excited-state lifetime, and high doping concentration. They should also have absorption in a wavelength range where high-average-power diode lasers are available.

Properties of Yb-doped materials. A range of materials which incorporate Yb^{3+} as an active ion, including Yb:YAG [24, 25, 26] and other crystals [27], Yb:SFAP (apatite) [28, 29], Yb:KGW (tungstate) [30, 31] and Yb:glass [32, 33, 34] has recently attracted the attention of many investigators. In all of these materials, Yb^{3+} displays a simple energy level structure. Since energy levels above the excited-state manifold do not exist, there is virtually no excited-state absorption. The lack of intermediate levels and the large separation between the excited-state and ground-state manifolds reduces the number of nonradiative paths for de-excitation. All display a broad pump band in the 900-nm range with a long upper-state lifetime, which is ideal for laser diode pumping. The spectral proximity of the emission wavelength to the pump band results in a quantum defect less than 10%, reducing the amount of heat left in these materials by excess pump energy.

Each of the two levels of the laser transition in Yb^{3+} is split by local fields into manifolds. In many hosts, the energy levels within these manifolds overlap to form continuous spectra due to broadening of the transitions, though in crystalline hosts the individual transitions tend to be more pronounced. With the wide variety of Yb-doped glasses and crystals available, the choice among the foregoing laser parameters is quite flexible.

Of these materials, those most capable of high-peak-power generation due to high saturation fluence and broad bandwidth are the glasses. These materials also offer greater flexibility in fabrication. The curve in Fig. 3 shows the absorption and emission cross sections for Kigre QX:Yb glass. It has a saturation fluence of 40 J/cm^2 at the peak lasing wavelength of 1.01 µm and a 20-fs minimum pulse duration. The 2-ms fluorescence lifetime of QX:Yb sets the requirement for pump power to the order of 40 kW/cm^2. This opens the path for simple pump lasers and laser diode pumping. The density of Yb^{3+} ions which can be incorporated into this material is as high as $2 \times 10^{21}/cm^3$ [35]. Thus, a petawatt laser could be made from a piece of glass the size of a dime (1 mm thick and 1.7 cm in diameter). Applying the theoretical peak intensity formula (5) to this material yields more than 10^{23} W/cm^2 for each cm^2 of gain aperture. Ultimately, a CPA system with a 10-cm-diameter amplifier could generate up to 10^{25} W/cm^2 focused intensity.

Experiments involving Yb:glass. Numerous experiments have been done to demonstrate the utility of Yb:glass for a variety of applications. CPA was demonstrated by Walton *et al.* [36] in a continuous-wave, Ti:sapphire-pumped, Yb:germano-silicate fiber producing fluence up to 15J/cm^2 from the fiber core. The utility of multi-pulse depletion of gain was also demonstrated in this system. Researchers at Lucent Technologies [37] and Polaroid [38] have demonstrated high-power, continuous-wave operation of Yb-doped fiber lasers. Under diode pumping, Hönninger was able to achieve 100 mW at 60-fs pulse duration, mode locking with a semiconductor saturable absorber mirror [39]. Using a similar cavity and pump configuration, Hönninger and collaborators at the Center for Ultrafast Optical Science (CUOS) in Michigan used CPA in a regenerative amplifier to produce 50-µJ pulses at several hundred hertz.

Using a free-running, flashlamp-pumped, Ti:sapphire laser as a pump source, QX:Yb glass was gain switched with >50% slope efficiency and >30% efficiency with respect to absorbed pump energy [40]. With the addition of a Pockels cell and polarizer, CPA has been performed at an efficiency of nearly 10% with respect to absorbed pump power [41]. Further optimization of the fluence in the regenerative amplifier is the subject of a paper submitted by Biswal for publication in this issue. Experiments are planned at CUOS that will employ flashlamp-pumped Ti:sapphire [42] and Cr:LiSAF [43] lasers to simulate diode pumping to the level of 10 to 20 J.

III. MANIPULATION AND CONTROL OF OPTICAL FIELDS FOR HIGH FIELD GENERATION

A. Measurement and management of phase distortions

As experiments probe to higher intensities, the light which strays from the focal interaction either temporally (along the beam axis) or spatially (in the transverse dimensions) provides an increasing level of unwanted signal. In some cases, the generation of excitation prior to the desired interaction may be so great as to screen or eliminate the desired process [44]. Hence, for the purpose of maintaining a well-behaved beam, increased attention to the measurement and correction of temporal and spatial distortions will be necessary.

The initial spark that launched solid-state materials into the femtosecond domain was the introduction of phase control to lasers [5, 6, 8, 45]. Within the oscillators, the degree of stretching and compression is typically small. In CPA the stretcher coupled with matched compressor so dramatically modifies the phase of a short pulse that the time-bandwidth product may be changed by a factor exceeding 10^5 [46]. The nearly quadratic (as a function of time) phase function used in the CPA system is an analog of propagation after a lens in the spatial domain [47]. In either case, the presence of phase errors produces unwanted broadening of the minimal pulse features. In addition to phase distortions in the dispersive optics of the system, the problem of background spontaneous emission from the amplifiers also poses a disturbance to which some experiments are sensitive [45].

B. Measurement and management of temporal distortions

The problems of phase measurement have already attracted considerable interest, as was recently demonstrated by the appearance of several papers detailing methods of frequency-resolved optical gating and related techniques for deriving the functional form of ultrashort-pulse electric fields [48].

The problem of correcting for phase distortions has also been addressed in the field of ultrafast optics for some time now. Beginning with the first paper on conjugate stretching and compression for amplification [49], it has been recognized that the dispersive properties of the materials used in CPA would make exact recompression difficult. As the bandwidth of CPA systems continues to grow, the compensation of higher orders of phase error is becoming increasingly important.

The usual approach taken in producing a bandwidth-limited output from a CPA system is to balance positive and negative dispersion in each order of phase, beginning with quadratic and proceeding through quartic or quintic [50, 51, 52, 53]. By analyzing the effect of different materials and optical elements on phase, certain elements can be added or adjusted to affect primarily one order. This process is linked with the measurement of the electric field to determine the final outcome. The use of fixed optics in this way to produce a desired phase function has been successful in producing pulses as short as 18 fs at the millijoule level [47].

A more flexible approach involves the introduction of an addressable phase element into a zero-dispersion stretcher to impress an adaptable phase on the pulse [54, 55]. Such a system could in principle be used to compensate for a limited amount of nonlinear phase distortion. The availability of new liquid-crystal devices and deformable mirrors [56] makes this approach attractive for systems requiring flexibility in pulse form or automated pulse-to-pulse adjustment.

Through the use of high-order phase control and pulse cleaning, the loss of energy from the peak of the intense laser pulse can be minimized. As a result, the efficiency of conversion from energy stored in the amplifying medium to energy in the short pulse will be increased. These means of obtaining higher focused intensities do not require the addition of large pump lasers, the increase of the laser aperture, or larger gratings. It is a potentially inexpensive way to increase focused intensity.

For experiments involving solid-density targets, the measurement and control of amplified spontaneous emission (ASE) is especially important. However, it cannot be addressed by the foregoing techniques. Light that is spontaneously emitted in the amplifier during the first pass of the signal overlaps the signal temporally, spatially, and spectrally. This poses a problem both for measurement and for control. Through the amplification system, ASE maintains its initial ratio of intensity with the signal, yet it lacks the coherence necessary to undergo compression. The result is that the compressed pulse is accompanied by a pedestal of energy nearly equal to its own energy in some cases. The duration of the background signal is on the order of a few nanoseconds and lies between 4 and 6 orders of magnitude below the signal in intensity. This dynamic range is sufficient to frustrate most measurement techniques, yet the power is sufficient to pre-ionize experimental targets and disable many solid-density plasma interactions.

In a paper submitted to *Selected Topics in Quantum Electronics* [44], Nantel *et al.* discuss the problem of temporal contrast in Ti:sapphire lasers. They detail the use of a high-dynamic-range correlation and a plasma-shuttered streak camera to analyze the pulse contrast. They also discuss the suppression of pre-pulse energy through the use of a short-pulse preamplifier and a saturable absorber. Contrast improvements of two orders of magnitude are demonstrated experimentally.

C. Measurement and management of spatial distortions

In a similar way, the spatial quality of the beams throughout the CPA system should be preserved or corrected, because these corrections can significantly enhance the intensity of light at focus. Considering the problem of concentrating light at focus in the presence of wave-front distortions, it is common to refer to the Rayleigh criterion when choosing optics for beam focusing [57]. This criterion states that the presence of a $\lambda/4$ purely spherical aberration decreases the maximum focal intensity by 20% from the theoretical limit. It may be argued, however, that the optical system of a CPA laser offers far more complexity than the simple case of spherical aberration. CPA systems often include Gaussian thermal lensing, uneven intensity, aperturing due to non-uniform pump intensity, and slight nonlinear distortion. The Maréchal criterion is more general. It guarantees 80% of the theoretical maximum intensity at focus, provided that distortions are less than $\lambda/14$. To maintain such a tolerance through the entire optical system to the compression gratings would be quite expensive. Spatial filtering and wave-front correction are alternatives. In contrast to spatial filtering, where stray energy is discarded, the use of deformable mirrors for wave-front correction brings stray energy back into the focal spot. Consequently, the use of deformable mirrors to correct for wave-front distortions will become a decided advantage.

Before applying a figure to a deformable mirror to correct distortion, it is necessary to measure the wave front. One particularly useful interferometric wave-front measurement device was recently introduced by Primot [58, 59]. This device uses a hexagonal phase grating to produce three-beam sheering interference. By resolving the three-beam interference pattern into orthogonal directions, two-dimensional wave-front distortions may be recorded. Like many temporal phase retrieval techniques, the achromatic three-wave lateral shearing interferometer (ATWLSI) produces redundant information which may be used to check the accuracy of the phase measurement. This technique has the advantage over the more prominent Shack-Hartman wave-front sensor in that it has an adjustable sensitivity from levels greater than a wave of distortion to as low as one hundredth of a wave. Furthermore, the diffractive nature of the beam splitter in the ATWLSI makes it inherently achromatic and broadband.

Applying this device to the single-shot measurement of beam distortions as a function of output energy in a 50-fs CPA laser has already given valuable insight into the degree of whole-beam self-focusing arising in materials at the 45-mJ level [60]. In order to separate thermal distortions in the amplifier chain from the nonlinear self-focusing, a piece of glass was placed both in the compressed beam and in its weak reflection. By taking the difference in the wave front between these cases, the degree of nonlinear effect in the glass can be measured. Operating at a pulse intensity of 45 mJ in a 2-cm elliptical beam, the peak nonlinear effect produced a $\lambda/2$ wave-front distortion, as shown in Fig. 4.

By controlling the phase of the optical pulse approaching the experimental volume, a more accurate determination of the conditions in the experiment may be made. Furthermore, the wave front may be modified to enhance the experimental conditions. To first approximation, this means correcting for wave-front distortions to achieve a diffraction-limited focus. But, by using focal profiles other than Gaussian, it would be possible to reduce the degree of unwanted background signal. Just as reduction of pre-pulse energy reduces the interference of pre-plasma, a reduction of stray light would reduce low-intensity-induced background signals.

D. The spatio-temporal mix

It has been noted that the distortions that take place in the compressor are still more complex because they mix the spatial and temporal domains. As a result, restoration to an undistorted wave front is an intractable problem. This can be avoided to some extent by correcting the wave front before the gratings and using low-distortion gratings. Another approach would be to partially pre-compensate for compressor distortions, using temporal and spatial adaptive optics and an iterative procedure to minimize the total system distortions. The optimization of second harmonic generation signal to minimize temporal pulse duration from a "zero-dispersion stretcher" [56] bears a relation to this problem.

IV. SUMMARY AND CONCLUSION

The rapid and sustained growth of High-Field Science is due in part to the widespread use of the ultraintense lasers employing CPA technology. Continued growth in this field will require further attention to compactness and reliability, features which tend to make CPA systems more accessible to university- and institute-level research. We have outlined the theoretical limits of CPA lasers in the range of 10^{23} W/cm^2 for each cm^2 of gain aperture and noted the importance of Yb-doped materials, particularly glasses, in moving toward those limits.

In addition, the advantages of exercising spatial and temporal control over the phase of high-intensity pulses have been noted. The building blocks for fine measurement and control of both spatial and temporal phase exist. Frequency-resolved optical gating techniques for the characterization of ultrashort pulses are well developed and new high-contrast measurement techniques can quantify pre-pulse energy to seven orders of magnitude. Static correction of phase in the stretcher is well established. This capability will be further refined by adaptable optics. Improvement of pulse contrast by pre-amplification of short pulses provides needed pulse contrast of 10^7 to 10^8. Spatial phase distortions can be measured in a single shot with resolution variable from several λ to $\lambda/100$. Potential for improving focal intensity and correcting for mild nonlinearities will make the use of deformable mirrors a key factor in tailoring ultraintense pulses to the demands of various experiments.

With the use these materials and techniques, the move to the next generation of high-field lasers will not involve a move to a larger laser facility. On the contrary, the next compact CPA lasers will produce important results in the broad scientific community and find applications in medicine and materials processing and other fields. Such lasers will also bring us one step closer to the limits of focused intensity.

V. ACKNOWLEDGMENTS

The authors would like to thank Michael Myers of Kigre Inc. for providing samples of various Yb-doped glass for our experiments. We are grateful to O. Peterson at Los Alamos National Laboratory and M. Perry at Lawrence Livermore National Laboratory for providing Cr:LiSAF lasers for the continuation of this work. Discussions and collaboration with Gleb Vdovin of the Technical University of Delft have opened our eyes to the possibility of inexpensive deformable mirror technology. This work is sponsored by the National Science Foundation through the Center for Ultrafast Optical Science (CUOS) (contract No. STC-PHY 8920108). Marc Nantel was supported in part by the FCAR fund. J.-C. Chanteloup was supported in part by a collaboration between LULI and CUOS-NSF-CNRS grant #94N92/0043).

REFERENCES

1. P. A. Franken, A. E. Hill, C. W. Peters, and G. Weinreich, "Generation of optical harmonics," *Phys. Rev. Lett.* **7**, 334 (1961).
2. R. W. Hellwarth, "Q switched modulation of lasers," *Lasers* 1, A. K. Levine, ed. (Merckel Dekker, Inc., New York, 1966), p. 253.
3. R. W. Hellwarth, "Control of fluorescent pulsations," *Advanced Quantum Electronics*, J. R. Singer, ed. (Columbia University Press, New York, 1961).
4. C. K. Rhodes, A. Szoke, "Transmission of coherent optical pulses in gaseous SF_6," *Phys. Rev.* **184**, 25–37 (1969).
5. D. Strickland and G. Mourou, *Opt. Commun.* **56**, 219 (1985).
6. P. Maine, D. Strickland, P. Bado, M. Pessot, and G. Mourou, "Generation of ultrahigh peak power pulses by chirped pulse amplification," *IEEE J. Quantum Electron.* **24**, 398–403 (1988).
7. B. C. Stuart, M. D. Perry, J. Miller, G. Tietbohl, S. Sherman, J. A. Britten, C. Brown, D. Pennington, V. Yanovski, and K. Whartman, "125-TW Ti:sapphire Nd:glass laser system," *Opt. Lett.* **22**, 242 (1997).
8. D. E. Spence, P. N. Kean, W. Sibbett, "60-fsec pulse generation from a self-mode-locked Ti:sapphire laser," *Opt. Lett.* **16**, 42 (1991).
9. G. Mourou, "The ultrahigh peak power laser: present and future," *Appl. Phys. B* **65**, 205–211 (1997).
10. F. H. Loesel, M. H. Niemz, J. F. Bille, T. Juhasz, "Laser-induced optical breakdown on hard and soft tissues and its dependence on the pulse duration: experiment and model," *IEEE J. Quantum Electron.* **32**, 1717–1722 (1996).
11. R. M. Kurtz, X. Liu, T. Juhasz, V. M. Elner, "Tissue ablation as a function of laser pulsewidth," *Digest IEEE/LEOS 1996 Summer Topical Meeting*. Adv. App. of Lasers in Materials and Proc. 5–9 Aug., Keystone, CO, USA.
12. P. P. Pronko, S. K. Dutta, J. Squier, J. V. Rudd, D. Du, and G. Mourou, "Machining of submicron holes using a femtosecond laser at 800 nm," *Opt. Commun.* **114**, 106–110 (1995).
13. F. Qian, R. K. Singh, S. K. Dutta, and P. P. Pronko, "Laser deposition of diamondlike carbon films at high intensities," *Appl. Phys. Lett.* **67**, 3120 (1995).
14. J. Workman, A. Maksimchuk, X. Liu, U. Ellenberger, J. S. Coe, C. Y. Chen, and D. Umstadter, "Control of bright picosecond x-ray emission from intense subpicosecond laser-plasma interactions," *Phys. Rev. Lett.* **75**, 2324 (1995).
15. D. Umstadter, S.-Y. Chen, A. Maksimchuk, G. Mourou, R. Wagner, "Nonlinear optics in relativistic plasmas and laser wakefield acceleration of electrons," *Science* **273**, 472–475 (1996).
16. A. Modena, Z. Najmudin, A. E. Dangor, C. E. Clayton, K. A. Marsh, C. Joshi, V. Malka, C. B. Darrow, C. Danson, D. Neely, F. N. Walsh, "Electron acceleration from the breaking of relativistic plasma waves," *Nature* **377**, 606–608 (1995).
17. T. Ditmire, J. W. G. Tisoh, E. Springate, M. B. Mason, N. Hay, R. A. Smith, J. Marangos, M. H. R. Hutchinson, "High-energy ions produced in explosions of superheated atomic clusters," *Nature* **386**, 54–56 (1997).
18. P. A. Van Rompay, M. Nantel, P. Pronko, "Pulse contrast effects on energy distributions of C^{1+} to C^{4+} ions for high-intensity 100-fs laser ablation plasmas," accepted for publication in *Applied Surface Science*.
19. C. P. J. Barty, C. L. Gordon-III, S. E. Harris, B. E. Lemoff, F. Raksi, C. Rose-Petruck, K. R. Wilson, V. V. Yakovlev, K. Yamakawa, G. Y. Yin, "Multi-terawatt femtosecond lasers for high field physics," Future of Accelerator Physics. The Tamura Symposium. Nov. 1994, Austin, TX, USA, *AIP-Conference-Proceedings* **356**, 310–321 (1996).
20. A. I. Nikishov, V. I. Ritus, *Sov. Phys. JEPT* **19**, 1191 (1964) and **20**, 757 (1965).

21. W. Koechner, *Solid-State Laser Engineering*, (Springer-Verlag, Berlin, Heidelberg, New York, 1996), p. 15. Spontaneous emission rate A_{21} is replaced by $1/\tau_f$ and τ_{min} is taken to be $1/\Delta\nu$. The remaining numerical factors are taken up in κ.

22. A. E. Siegman, *Lasers* (University Science Books, Mill Valley, CA, USA, 1986). Each laser text uses different notation for saturation fluence. This notation is modified from Siegman.

23. S. Biswal, J. Itatani, J. Nees, and G. Mourou, "Efficient Extraction below the saturation fluence in a low-gain low-loss regenerative amplifier," submitted for publication in *Selected Topics in Quantum Electronics*.

24. P. Lacovara, H. K. Choi, C. A. Wang, R. L. Aggarwal, T. Y. Fan, "Room-temperature diode-pumped Yb:YAG laser," *Opt. Lett.* **16**, 1089–1091 (1991).

25. U. Brauch, A. Giesen, M. Karszewski, C. Stewen, and A. Voss, "Multi-watt diode-pumped Yb:YAG thin disk laser continuously tunable between 1018 and 1053 nm," *Opt. Lett.* **28**, 713–715 (1995).

26. D. C. Brown, "Ultrahigh-average-power diode-pumped Nd:YAG and Yb:YAG lasers," *IEEE J. Quantum Electron.* **33**, 861–873 (1997).

27. L. D. DeLoach, S. A. Payne, L. L. Chase, L. K. Smith, W. L. Kway, W. F. Krupke, "Evaluation of absorption and emission properties of Yb3+ doped crystals for laser applications," *IEEE J. Quantum Electron.* **29**, 1179–1191 (1993).

28. C. D. Marshall, L. K. Smith, R. J. Beach, M. A. Emanuel, K. I. Schaffers, J. Skidmore, S. A. Payne, B. H. T. Chai, "Diode-pumped ytterbium-doped $Sr_5(PO_4)_3F$," *IEEE J. Quantum Electron.* **32**, 650–656 (1996).

29. C. D. Orth, S. A. Payne, W. F. Krupke, "A diode pumped solid state laser driver for inertial fusion energy," *Nuclear Fusion* **36**, 75–116 (1996).

30. N. V. Kuleshov, A. A. Lagatski, A. V. Podlipenski, and V. P. Mikhailov, "Pulsed laser operation of Yb-doped $KY(WO_4)_2$ and $KGd(WO_4)_2$," *Opt. Lett.* **22**, 1317–1319 (1997).

31. N. V. Kuleshov, A. A. Lagatsky, V. G. Shcherbitsky, V. P. Mikhailov, E. Heumann, T. Jensen, A. Diening, G, Huber, "CW laser performance of Yb and Er, Yb-doped tungstates," *Appl. Phys. B* **B64**, 409–413 (1997).

32. M. J. Weber, J. E. Lynch, D. H. Blackburn, and D. J. Cronin, "Dependence of the stimulated emission cross section of Yb^{3+} on host glass composition," *IEEE J. Quantum Electron.* **19**, 1600–1608 (1983).

33. H. M. Pask, R. J. Carman, D. C. Hana, A. C. Tropper, C, J. Mackenchie, P. R. Barber, J. M. Dawes, "Ytterbium-doped silica fiber lasers: versatile sources for the 1–1.2 μm region," *IEEE Selected Topics in Quantum Electron.* **1**, 2–13 (1995).

34. X. Zou, H. Toratani, "Evaluation of spectroscopic properties of Yb3+ -doped glasses," *Phys. Rev. B* **52**, 15889–15897 (1995).

35. Private communication with Michael J. Myres of Kigre Inc.

36. D. T. Walton, J. Nees, and G. Mourou, "Broad-bandwidth pulse amplification to the 10-μJ level in ytterbium doped germanosilicate fiber," *Opt. Lett.* **21**, 1061–1063 (1996).

37. D. Inniss, D. J. DiGiovani, T. A. Strasser, A. Hale, C. Headly, A. J. Stentz, R. Pedrazzani, D. Tipton, S. G. Kosinski, D. L. Brownlow, K. W. Quoi, K. S. Kranz, R. G. Huff, R. Espindola, J. D. LeGrange, G. Jacobawitz-Veselka, D. Baggavarapu, X. He, D. Caffey, S. Gupta, S. Srinivasan, K. McEuen, and R. Patel, "Ultrahigh-power single-mode fiber lasers from 1.065 to 1.472 using Yb-doped cladding-pumped and cascaded Raman lasers," Conference on Lasers and Electro-Optics (May 18–23, Baltimore, MD, 1997), CDP31-2.

38. M. Muendel, B. Engstrom, D. Kea, B. Laliberte, R. Minns, R. Robinson, B. Rockney, Y. Zhang, R. Collins, P. Gavrilovic, A. Rowley, "35-W cw single-mode ytterbium fiber laser at 1.1 μm," Conference on Lasers and Electro-Optics (May 18–23, Baltimore, MD, 1997), CDP30-2.

39. C. Hönninger, F. Morier-Genoud, M. Moser, U. Keller, L. R. Brovelli, and C. Harder, "Efficient and tunable diode-pumped femtosecond Yb:glass lasers," accepted for publication in *Optics Letters*.
40. S. Biswal, J. Nees, G. Mourou, A. Nishimura, "Efficient gain-switched operation of a highly-doped Yb:phosphate glass laser," Advanced Solid State Lasers, C. Pollack and W. Bosenberg, eds., (OSA TOPS Vol. 10, 1997), pp. 119–121.
41. S. Biswal, F. Druon, J. Nees, G. Mourou, A. Nishimura, "Ytterbium:glass CPA regenerative amplifier pumped by a free-running laser," Conference on Lasers and Electro-Optics (May 18–23, Baltimore, MD, 1997), CThC1.
42. E. Erickson, S. Owada, S. Satou, A. Nishimura, A. Ohzu, A. Sugiyama Y. Maruyama, T. Arisawa, H. Takuma, J. Nees, S. Biswal, G. Mourou, "12 J Flashlamp Pumped Ti-sapphire Laser," submitted to Advanced Solid State Lasers (Coeur d'Alene, ID, 2–4 February, 1998).
43. T. Ditmire and M. D. Perry, "Amplification of femtosecond pulses to above 1 J with large aperture Cr:LiSAF amplifiers," Generation, Amplification, and Measurement of Ultrashort Laser Pulses II, *Proc. SPIE Opt. Eng.* **2377**; 301–310 (1995). Experiments are planned using the final amplifier head from this system as a free-running laser to simulate diode pumping at 920–950 nm.
44. M. Nantel, J. Itatani, A. C. Tien, J. Faure, D. Kaplan, M. Bouvier, T. Buma.,P. A. VanRompay, J. Nees, P. P. Pronko, D. Umstadter, and G. Mourou, "Temporal contrast in Ti:sapphire lasers: characterization and control," submitted for publication in *Selected Topics in Quantum Electronics*.
45. R. L. Fork, C. V. Shank, R. Yen, C. A. Hillman, "Femtosecond optical pulses," IEEE J. Quantum Electron. **19**, 500–506 (1983). Refers to the original work in dye lasers.
46. C. P. J. Barty, T. Guo, C. Le-Blanc, F. Raksi, C. Rose-Petruck, J. Squier, K. R. Wilson, V. V. Yakovlev, K. Yamakawa "Generation of 18-fs, multi-terawatt pulses by regenerative pulse shaping and chirped-pulse amplification," *Opt. Lett.* **21**, 668–670 (1996).
47. B. H. Kolner, M. Nazarathy,"Temporal imaging with a time lens," *Opt. Lett.* **14**, 630–632 (1989).
48. G. Taft, A. Rundquist, M. M. Murnane, I. P. Christov, H. C. Kapteyn, K. W. DeLong, D. N. Fittinghoff, M. A. Krumbugel, J. N. Sweetser, R. Trebino, "Measurement of 10-fs laser pulses," *IEEE Selected Topics Quantum Electron.* **2**, 575–585 (1996). See, also, included references.
49. M. Pessot, P. Maine, G. Mourou, "1000 times expansion/compression of optical pulses for chirped pulse amplification," *Opt. Commun.* **62**, 419–421 (1987).
50. B. E. Lemoff and C. P. Barty, "Quintic-phase-limited, spatially uniform expansion and recompression of ultrashort optical pulses," *Opt. Lett.* **18**, 1651 (1993).
51. W. E. White, F. G. Patterson, R. L. Combs, D. F. Price, and R. L. Shepherd, "Compensation of higher-order frequency-dependent phase terms in chirped-pulse amplification systems," *Opt. Lett.* **18**, 1343 (1993).
52. S. Backus, C. G. Durfee III, G. Mourou, H. C. Kapteyn, and M. M. Murnane, "0.2 TW laser system at 1 kHz," *Opt. Lett.* **22**, 1256 (1997).
53. S. Backus, C. G. Durfee III, H. C. Kapteyn, and M. M. Murnane, accepted for publication in *Review of Scientific Instruments* (1997).
54. A. Efimov, D. H. Reitze, "Spectral adaptive optics: phase compensation for ultrashort chirped pulse amplifier systems," Generation, Amplification, and Measurement of Ultrashort Laser Pulses III (28–30 Jan. 1996, San Jose, CA, USA) *SPIE* **2701**, pp. 190–197.
55. "Free optimization of a compressor using second harmonic generation and an addressable phase-amplitude modulator," presented as a post-deadline paper at Ultrafast Optics (4–7 August 1997, Monterey, CA).

56. G. V. Vdovin, "Spatial light modulator based on the control of the wavefront curvature'," *Opt. Commun.* **115**, 170–178 (1995).
57. M. Born and E. Wolf, *Principles of Optics* (Pergman Press, Oxford, 1992), chpt. 10.
58. J. Primot, "Three-wave lateral shearing interferometer," *Appl. Opt.* **32**, 6242–6247 (1993).
59. J. Primot, L. Sogno, B. Fracasso, K. Haggarty, Opt. Eng. **36**, 901–904 (1997).
60. J-C. Chanteloup, F. Druon, M. Nantel, G. Mourou, "Single shot wavefront measurement using a three wave interferometer," Conference on Lasers and Electro-Optics (18–23 May 1997, Baltimore, MD), post-deadline CPD9.

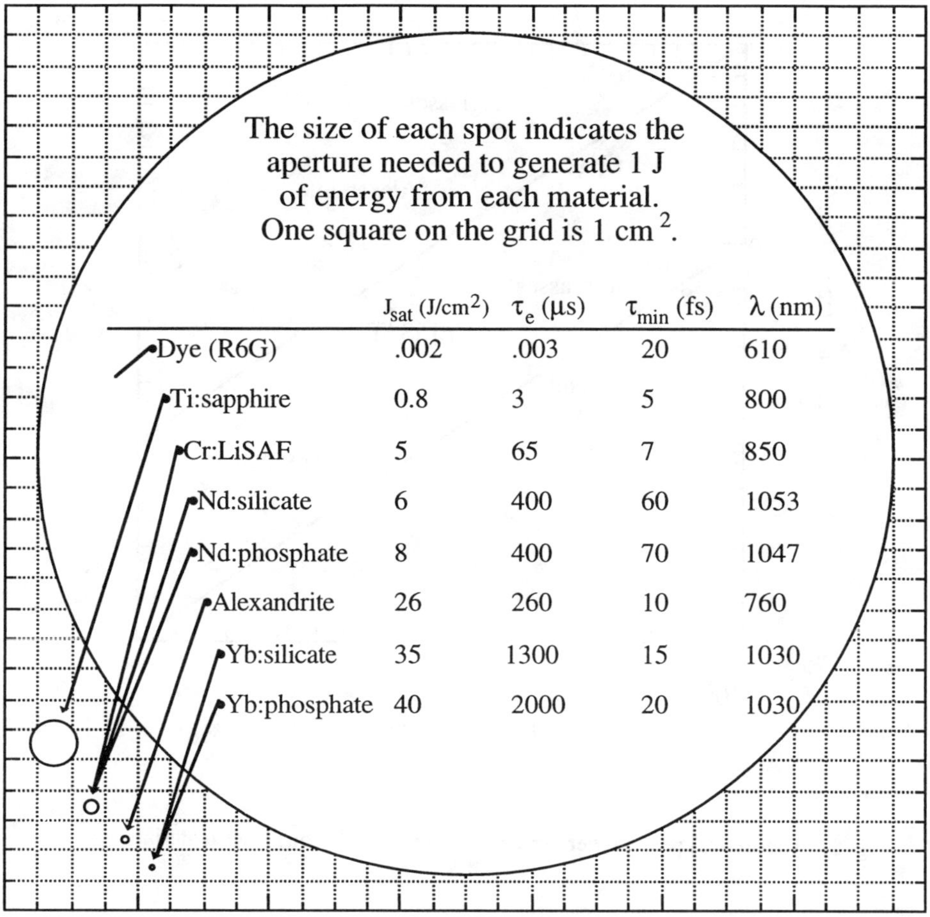

	J_{sat} (J/cm^2)	τ_e (μs)	τ_{min} (fs)	λ (nm)
Dye (R6G)	.002	.003	20	610
Ti:sapphire	0.8	3	5	800
Cr:LiSAF	5	65	7	850
Nd:silicate	6	400	60	1053
Nd:phosphate	8	400	70	1047
Alexandrite	26	260	10	760
Yb:silicate	35	1300	15	1030
Yb:phosphate	40	2000	20	1030

The size of each spot indicates the aperture needed to generate 1 J of energy from each material. One square on the grid is 1 cm^2.

Fig. 1. Emission cross section, lifetime, and minimum pulse duration of various laser materials. Materials with larger areas require larger apertures to generate a given energy.

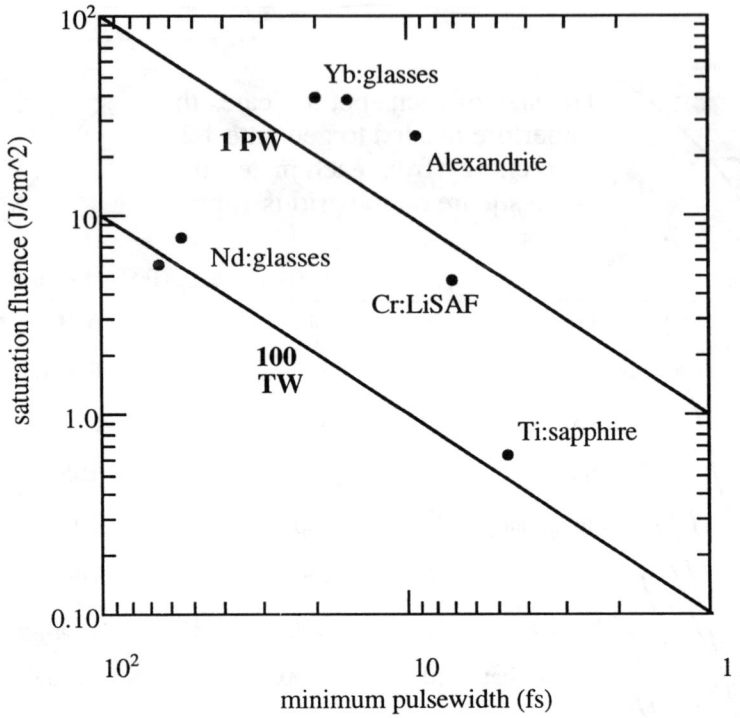

Fig. 2. Theoretical peak power per cm^2 aperture for various laser materials.

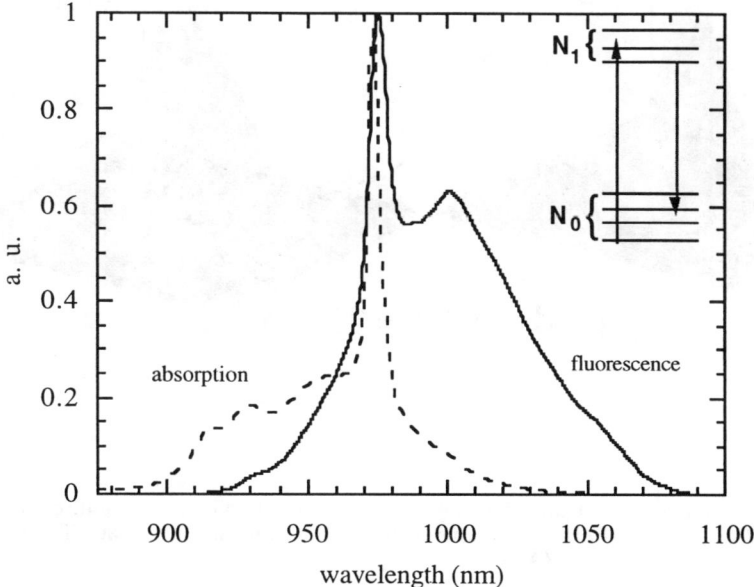

Fig. 3. Emission and absorption curves for Kigre QX:Yb glass. Its low emission cross section, long fluorescence lifetime, and broad bandwidth make it suitable for super-intense pulse generation.

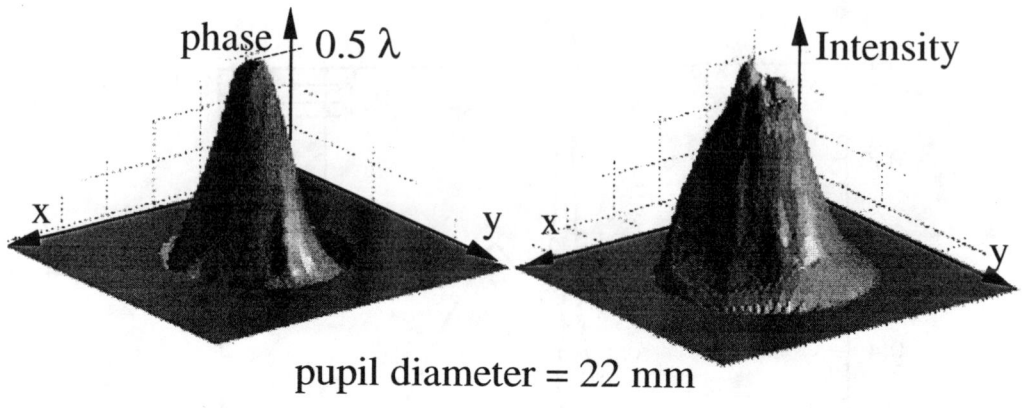

Fig. 4. Nonlinear Phase Distortion from a 50-fs, 45-mJ, 2x3-cm² beam propagating through 1 cm of BK-7 glass. Wave fronts are measured single-shot using an Achromatic Three-Wave Sheering Interferometer (ATWLSI).

Terawatt Picosecond CO$_2$ Laser Technology for Strong Field Physics Applications

I.V. Pogorelsky

Accelerator Test Facility, BNL, Upton, NY 11973, USA

Abstract. The first terawatt picosecond (TWps) CO$_2$ laser is under construction at the BNL Accelerator Test Facility (ATF). TWps-CO$_2$ lasers, having an order of magnitude longer wavelength than solid state lasers, offer new opportunities for strong-field physics research. For laser wakefield accelerators (LWFA) the advantage of the new class of lasers is due to a gain of two orders of magnitude in the ponderomotive potential. The demonstrated large average power of CO$_2$ lasers is important for the generation of hard radiation through Compton back-scattering of the laser off energetic electron beams. We discuss applications of TWps-CO$_2$ lasers for LWFA modules of a tentative electron-positron collider, for a γ-γ (or γ-lepton) collider, for a possible "table-top" source of high-intensity x-rays and gamma rays, and the generation of polarized positron beams.

INTRODUCTION

Terawatt picosecond lasers are known as the sources of the most intense electromagnetic radiation and strongest electric and magnetic fields available for laboratory research. Their development stimulated a number of strong-field physics disciplines to emerge during the last decade. One of such applications is laser-driven high-gradient particle accelerators. According to equation

$$|E_L|^2 = 2I_L / \varepsilon_0 c \tag{1}$$

or
$$E_L[TV/m] = 15 P_L^{1/2}[TW] / r_L[\mu m], \tag{1a}$$

where E_L, I_L, P_L, r_L are correspondingly the laser electric field, intensity, power, and focal spot radius, focusing of a terawatt laser beam to r_L=10 μm results in an intensity of $I_L \approx 10^{18}$ W/cm^2 and, associated with it, an electric field of $E_L \approx 1$ TV/m. This exceeds by four orders of magnitude accelerating fields available in conventional particle accelerators. Various methods to convert at least a portion of such an enormous transverse fast-oscillating field into a longitudinal accelerating electric field have been proposed. Using the LWFA technique [1,2], about 30 GeV/m electron acceleration gradients have been demonstrated [3,4]. In these experiments solid state terawatt lasers based on mode-locked picosecond pulse generation and pulse chirped amplification [5] and operating at wavelength near $\lambda \approx 1$ μm have been utilized.

Emerging TWps-CO$_2$ lasers having wavelength $\lambda \approx 10$ μm may provide an additional boost to laser accelerator development towards practical devices. For the most part, this is based on the fact that the energy of oscillatory motion acquired by the electron from an electromagnetic wave, called ponderomotive potential

$$W_{osc} = e^2 E_L^2 / 2m\omega^2 \qquad (2)$$

(where e and m are correspondingly the electron charge and mass and ω is the laser frequency $\omega = 2\pi c/\lambda$) is quadratically proportional to λ. Thus, any process, where the field-induced electron oscillation is paramount, is dramatically enhanced. The examples of such processes are: avalanche and tunneling ionization, plasma wave ponderomotive excitation, and relativistic self-focusing, which are especially relevant for electron acceleration in plasma, such as in the LWFA method.

The same physical effect is responsible for the λ-proportional increase in Compton scattering cross-section thus opening additional prospects for relatively compact high-brightness x-ray and gamma sources [6].

The approach to utilize the long-wavelength laser radiation for particle acceleration and x-ray generation is pursued at the Brookhaven Accelerator Test Facility (ATF) where the first TWps-CO_2 laser is under construction [7].

In the present paper we review the opportunities provided by TWps-CO_2 laser technology for high-energy physics. This includes novel compact particle accelerators and monochromatic x-ray sources and prospects for their future development towards the electron-positron (e^--e^+) and gamma (γ) colliders of the TeV energy range.

TWps-CO_2 LASERS

An important physical parameter that enables generation and amplification of picosecond laser pulses is the gain spectral bandwidth. In solid state lasers, radiation transitions in outer electron shells of active ions are broadened to 5-50 THz due to the perturbative action by the host matrix. Such a broad gain spectrum makes possible the generation and amplification of picosecond, and even femtosecond, laser pulses by the mode locking technique. Unlike the solid state, the spectral gain in the molecular gas discharge is periodically modulated by the rotational structure. Due to the discrete spectrum, and for other physical and technical reasons, mode-locking techniques do not work for CO_2 lasers as well as for solid state lasers.

However, alternative methods to produce picosecond and sub-picosecond CO_2 laser pulses have been developed. One of them is semiconductor optical switching [8]. Using this method, subpicosecond slices out of a multi-nanosecond CO_2 laser output may be produced using a conventional mode-locked solid-state laser.

Pressure broadening of the CO_2 gain spectrum at ~10 atm into a 1 THz wide quasi-continuum permits direct amplification of multi-terawatt λ=10 µm laser beams.

For a τ_L=1 ps pulse propagating in a 10-atm amplifier, the estimated small-signal gain is 3-4%/cm and the extractable specific energy is ~20 mJ/cm^3. Taking into account that the total discharge volume may exceed 10 l, the possibility of extraction of as high as 100 J of energy in a picosecond pulse from a reasonably compact CO_2 laser amplifier looks realistic. However, the limiting factor to high energy extraction may be the damage threshold of the output window that is at the level of 0.5 J/cm^2. For an optical window of the ~100 cm^2 size, the extractable energy is 30-50 J which still permits ~30 TW peak power at a 1-ps laser pulse duration.

Thus, to attain terawatt peak power, a ~10-atm, ~10-l CO_2 amplifier is required. To maintain a uniform discharge under such conditions, the following requirements should be satisfied: a) strong penetrating preionization, b) ~1 MV voltage applied to the discharge, and c) the energy load of several kilojoules deposited in a relatively short, ≤300 ns, time interval. The first laser with such parameters is under construction at the ATF.

The ATF laser is intended as a test bench for proof-of principle evaluation of TWps-CO_2 technology for such strong-field physics applications as high-gradient laser accelerators and high-intensity Compton x-ray sources. For these purposes, the ATF also provides a high-brightness 50-MeV electron beam from a photocathode RF linac synchronized within subpicoseconds to the CO_2 laser pulse.

Fig.1 shows the principal optical diagram of the ATF TWps CO_2 laser system. The 1 MW, 100 ns pulse produced by a 1-atm CO_2 laser oscillator is sliced at a semiconductor switch controlled by the picosecond Nd:YAG laser. The high power will be attained via regenerative amplification and four additional passes through the preamplifier followed by three passes in the 10-atm, 10-l final amplifier with the beam expansion to its full 10-cm aperture.

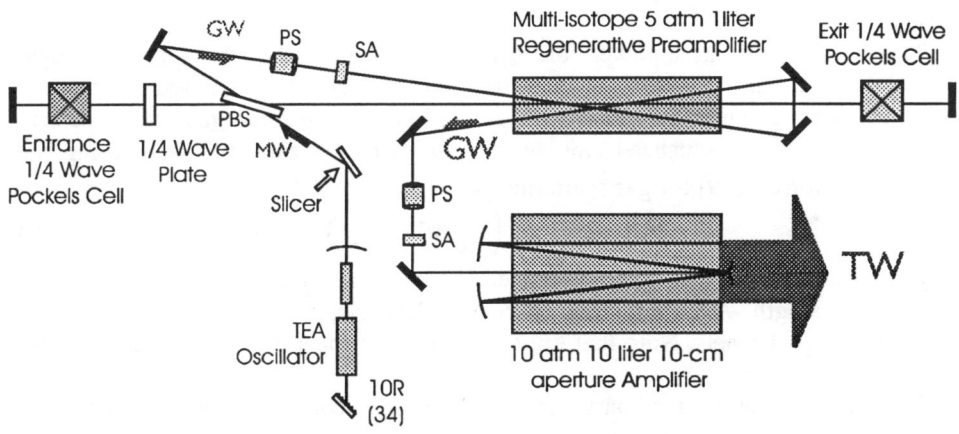

FIGURE 1. Optical diagram of the ATF TWps-CO_2 laser;
PS - plasma shutter; SA - saturable absorber; PBS - polarizing beam splitter

LASER WAKEFIELD ACCELERATORS AND e$^-$-e$^+$ COLLIDERS

Progress in the exploration of particle interactions relies upon the development of a new-generation of accelerators on a TeV energy scale. One of the prospective approaches is a linear e$^-$-e$^+$ collider based on high-gradient laser acceleration.

Among the known laser acceleration techniques, the LWFA [1] is considered as the most reliable approach. The LWFA method is based on the ponderomotive charge separation and a relativistic wake formation when a short laser pulse propagates in underdense plasma. The amplitude of the accelerating field, E_a, due to the charge separation in a plasma wave is

$$E_a[V/cm] = \left(a^2/\sqrt{1+a^2}\right)\sqrt{n_e[cm^{-3}]}, \tag{3}$$

where n_e is electron density in plasma, and a is the dimensionless laser vector-potential

$$a = eE_L/mc\omega = 0.3E_L[TV/m]\lambda[\mu m]. \tag{4}$$

From Eqs.(3) and (4) we see that $E_a \propto \lambda^2$ for $a \ll 1$ and $E_a \propto \lambda$ for $a \gg 1$. Thus, a 10-µm CO_2 laser is capable of producing an accelerating gradient at least 10 times higher than the 1-µm laser of the same intensity. This is due to the stronger ponderomotive potential of plasma electrons oscillating in a lower-frequency electromagnetic field.

Two options to build the 2.5 TeV multi-stage plasma-channeled LWFA linac using CO_2 or solid state lasers are illustrated by Table 1. Both design options are aimed to attain a luminosity $\Lambda = 10^{35}$ cm^{-2}s^{-1} [8] defined as

$$\Lambda = N_e^2 f \zeta / 4\pi\sigma_\perp^2, \tag{5}$$

where N_e is the number of particles per bunch, ζ is a number of bunches per train, f is the laser repetition rate, and $\sigma_\perp$ is the e-beam cross-section at the interaction point.

The parameters entering Table 1 are chosen according to the following prime considerations:

1. The 50 TW peak laser power foreseeable with state-of-the art laser technology, and the laser pulse duration - close to the experimentally demonstrated minimum.
2. The plasma channels are filled with 100% ionized hydrogen gas at the density equal to $0.5n_e$. Normalized emittance of the particle beam calculated using the Highland formula [9] for gas scattering is

$$\varepsilon_n[m] \approx 3 \times 10^{-15} \left(n_e[cm^{-3}]/E_a^2[MeV/m]\right). \tag{6}$$

3. Plasma channel length for every accelerator stage is ~$30z_0$, where $z_0 = \pi r_L^2/\lambda$ is the Rayleigh length. A 20 cm long evacuated dead space is assumed between the accelerating channels. Note that optics of the same focal length are used for both lasers.
4. The maximum number of particles per bunch is calculated by the condition that space charge field of the electron bunch does not effect the wakefield structure: $N_e \leq n_e(c/\omega_p)^3$.

We see that both design approaches illustrated in Table 1 demonstrate the LWFA capabilities to attain the desired 2.5 TeV electron energy in a compact multi-stage accelerator. However, the calculated requirements of the laser driver for two cases are essentially different.

With the 1-µm laser, the short τ_L results in a proportionally small λ_p and N_e. Then, in order to satisfy the high luminosity requirements, the laser repetititon rate and, hence, the average output power should increase quadratically. This becomes orders of magnitude beyond any reasonable expectation for picosecond solid state laser technology and does not fit into the anticipated wall-plug power limits.

TABLE 1. Prospective Comparative Characteristics for Standard LWFA Driven with 1-μm and 10-μm Lasers

Laser Parameters		
Laser wavelength, λ [μm]	10	1
Energy [J]	50	5
Pulse length, τ_L [ps]	1	0.1
Power, P [TW]	50	50
Focal spot radius, r_L [μm]	300	30
Laser field, E_L [TV/m]	0.4	4
Dimensionless laser strength, a	1.3	1.3
Repetition rate. f [kHz]	0.2	20
Average power [kW]	10	100
Wakefield Parameters		
Plasma density, n_e [cm^{-3}]	3×10^{15}	3×10^{17}
Plasma wavelength, λ_p [μm]	600	60
Acceleration gradient, E_a [GeV/m]	4.5	45
Pump depletion length [cm]	280	28
Phase detuning length [cm]	230	23
Assumed channel length [cm]	100	10
Energy gain per stage [GeV]	4.5	4.5
Collider Parameters		
Electrons/bunch. N_e	3×10^9	3×10^8
Number of bunches per pulse, ζ	3	3
ε_n due to gas scattering [m]	4×10^{-7}	4×10^{-7}
$\sigma_\perp$ at β^*=5 μm focus, [Å]	7	7
Luminosity, Λ [cm^{-2}s^{-1}]	10^{35}	10^{35}
Bunch repetition rate, ζf [kHz]	0.6	60
Number of stages	555	555
Total length [m]	666	166

On the contrary, the parameters specified for the 10-μm laser look feasible for TWps-CO$_2$ laser technology.

Another potential advantage of using a longer period plasma wave is the ease in producing the seed electron bunch which is short enough to fit into the small portion of the wake period thus ensuring the good beam quality (small energy spread and emittance). For example, at τ_L=1 ps and the resonance plasma wavelength λ_p=600 μm the desirable electron bunch duration is $\tau_b \leq 200$ fs. Contemporary photocathode RF guns tend to approach these requirements. In particular τ_b=370 fs electron bunches of 2.5×10^8 electrons, $\Delta p/p = 0.15\%$, and ε_n=0.5 mm.mrad have been demonstrated with the ATF photocathode RF gun [10].

The above comparative analysis of two alternative approaches to the 2.5 TeV collider serves to illustrate the problems related to laser wavelength scaling and should not be considered as the collider design proposal. For example, we did not address such important aspects as stage coupling, other sources of the emittance degradation, etc. Still, we believe that the presented approach may be helpful in designing the criteria for choosing the appropriate laser driver for compact GeV accelerators and for future linear colliders.

X-RAY AND GAMMA SOURCES BY COMPTON SCATTERING OF CO_2 LASER BEAMS

Table-Top Laser Synchrotron Source

Synchrotrons equipped with wiggler magnets are the sources of x-ray fluxes at the level of 10^{18} photon/sec. According to another approach to a relatively compact high-brightness x-ray generator called laser synchrotron source (LSS), the laser beam acts on relativistic electrons as an electromagnetic wiggler with a period 10^4-10^5 times shorter than the magnetic undulator. Thus, LSS permits significant downsizing of the electron accelerator, or produces proportionally heavier photons than a conventional synchrotron source operating at the same e-beam energy.

A combination of a high-gradient LWFA with LSS may open a route to table-top wakefield LSS operating in x-ray and gamma regions. Proof-of-principle table-top LSS may be realized at the ATF using the 5-TW CO_2 laser and a 5 MeV photocathode electron gun. As shown in Fig.2 the CO_2 laser beam is split into two beams which are focused by parabolic mirrors at the entrance and exit of a plasma channel to drive both LWFA and LSS.

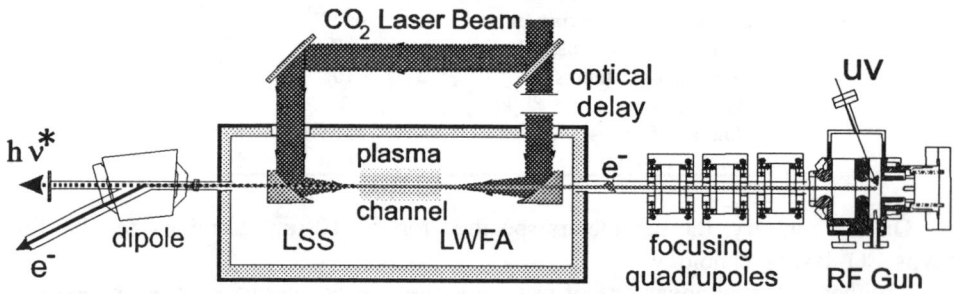

FIGURE 2. Diagram of the table-top laser wakefield LSS

PIC simulations predict 250 MeV acceleration when a 4 TW CO_2 laser beam is focused into the waveguide with parameters shown in Table 2 [11]. At the exit of the waveguide, electrons accelerated in the plasma wake field interact with the second CO_2 laser beam, generating x-ray fluxes which can be orders of magnitude above numbers obtained with conventional synchrotron light sources.

TABLE 2. Design Parameters for Table-Top LSS

LWFA	
Electron Energy [MeV]	5
Bunch Charge [nC]	0.1
Bunch Duration FWHM [fs]	300
Laser Peak Power [TW]	4
Laser Pulse Duration [ps]	3
Plasma Density [cm^{-3}]	3.5×10^{16}
Channel Radius [μm]	60
Channel Length [cm]	4
Acceleration gradient [GV/m]	6
Energy Gain [MeV]	250
LSS	
Laser Peak Power [TW]	1
Laser Pulse Duration [ps]	3
Laser Focus Radius [μm]	30
X-ray Photon Energy [keV]	470
X-ray Pulse Duration [fs]	300
X-ray Photons per Pulse	3×10^9
X-ray Peak Flux [photons/s]	10^{22}

Gamma and Positron Sources for Linear Colliders

By Compton backscattering of the laser photons from the TeV electron beam, a high brightness TeV photon beam can be created. It opens an additional opportunity to study a variety of interaction processes by colliding e^-, e^+ and γ beams in any combination and at independently controlled polarizations.

The expression for the maximum gamma photon energy for linear (single photon) Compton backscattering is

$$\hbar\omega_\gamma = (x/x+1)E_e, \qquad (7)$$

where E_e is the electron energy, and $x = 4E_e\hbar\omega/m^2c^4$. At $x \gg 1$, the Compton photon energy approaches the electron energy, $\hbar\omega_\gamma \approx E_e$. For CO_2 laser, $x=1$ at $E_e=0.5$ TeV. Thus, the long wavelength of the CO_2 laser used for the $e^\pm \Rightarrow \gamma$ converter at $E_e=2.5$ TeV does not significantly degrade ω_γ.

Another strong requirement to the laser wavelength is set by rescattering of gamma photons on the laser beam into pairs through the reaction $\gamma+\lambda \Rightarrow e^-+e^+$. This occurs when $\omega\omega_\gamma > m^2c^4/\hbar^2$. Based on this condition and using Eq.(7), the optimum laser wavelength is derived:

$$\lambda[\mu m] = 4.2 E_e[\text{TeV}]. \qquad (8)$$

Then, for the 2.5 TeV collider the laser with $\lambda=10.5$ μm is the optimum choice.

For $\tau_L=1$ ps, the probability of $e^\pm \Rightarrow \gamma$ conversion,

$$\chi = \sigma_C\, \mathbf{E}_L / \hbar\, \tau_L c^2, \qquad (9)$$

where $\sigma_C=1.9\times10^{-25}$ cm^2 is the Compton scattering cross-section, reaches unity at the laser pulse energy $E_L \approx 1$ J.

The laser pulse repetition rate should match that of the e$^-$-e$^+$ collider. The TWps-CO$_2$ laser technology is envisioned to provide the laser source delivering picosecond pulses of a 1 J energy at several kW of average power to satisfy the requirements of the γ-γ collider. Relatively compact ~10 l discharge, high-pressure, fast-flow CO$_2$ lasers operating at a ~100 Hz repetition rate may serve this purpose when the energy stored in the laser medium is extracted by a train of a hundred pulses of a 1 ps length each following at a ~1 ns period. Such a regime looks not just feasible but also quite efficient, permitting extraction of a good portion of the stored CO$_2$ laser energy. Overall electric efficiency of the laser may approach 10%.

Lasers may be used also in polarized positron sources for e$^-$-e$^+$ colliders. Here, the backward Compton scattering serves as an intermediate process followed by pair production on a target or via two-photon rescattering. Polarization of the produced particles is controlled by the input laser beam. Capable of high average power and delivering ten times more photons than solid state lasers of the similar energy, picosecond CO$_2$ lasers become the optimum choice for this application as well. The projected positron source for the Japan Future Collider [12] employs a hundred of 1.5 kW CO$_2$ lasers of 150 Hz repetition rate and 50 ps pulse duration. This project appears likely to become the biggest utilization of CO$_2$ lasers in fundamental science.

ACKNOWLEDGEMENTS

The author wishes to thank I. Ben-Zvi and T. Tajima for helpful discussions. The work is supported by the US Department of Energy.

References

1. Tajima T. and Dawson J.M., *Phys. Rev. Lett.*, **43**, 267 (1979)
2. Sprangle P., Esarey E., Ting A., and Joice G., *Appl. Phys. Lett.*, **53**, 2146 (1988)
3. Madena A., Najmudin Z., Dangor A.E., et al., *Nature*, **377**, 606 (1995)
4. Nakajima K., Fisher D., Kawakubo T., et al., *Phys. Rev. Lett.*, **74**, 4428 (1995)
5. Strickland D. and Mourou G., *Opt. Commun.*, **56**, 219 (1985)
6. Sprangle P., Ting A., Esarey E., and Fisher A., *J. Appl. Phys.*, **72**, 5032 (1992)
7. Pogorelsky I.V. et al., "The First Terawatt Picosecond CO$_2$ Laser for Advanced Accelerator Study at the Brookhaven ATF", *Proceedings of 7th Advanced Accelerator Concepts Workshop*, Lake Tahoe, CA, October 12-18, 1996, to be published
8. Corkum P.B., *IEEE J. Quant. Electron.*, **QE-21**, 216 (1985)
9. Highland V.L., *Nucl. Instr. & Methods in Phys. Res.*, **129**, 497 (1975)
10. Wang X.J., Qui X., and Ben-Zvi I., *Phys. Rev.*, *E* **54**, R3121 (1996)
11. Bulanov S.V., et al., *IEEE Trans. on Plasma Sci.*, **24**, 393 (1996)
12. Okugi T., et al., *Jpn. J. Appl. Phys.* **35**, 3677 (1996)

How To Measure Femtosecond Pulses

M.A. Franco, J.-F. Ripoche, H.R. Lange, J.P. Chambaret, P. Rousseau,
B.S. Prade and A. Mysyrowicz

Laboratoire d'Optique Appliquée
Centre National de la Recherche Scientifique URA 1406
ENSTA-Ecole Polytechnique
91761 Palaiseau Cedex, France

Abstract. We discuss techniques allowing to measure the duration of ultrashort laser pulses. In particular, a new method is described, which recovers the phase, amplitude and absolute intensity of femtosecond pulses, allowing to reconstruct the temporal pulse profile with femtosecond resolution. It is based on the spectral analysis, at different time delays, of a probe pulse experiencing cross-phase modulation induced by a pump pulse.

1. INTRODUCTION

The spectacular recent advance in ultra-short laser pulse technology has brought with it a need for accurate techniques of measuring ultrashort pulse durations. There is no electronic detector responding on a subpicosecond time scale; one must resort to nonlinear optical techniques in order to time-resolve such short optical pulses. The basic idea is to convert the measurement from the time domain into the space domain.

Since the speed of light is 0.3 µm per femtosecond in free space, femtosecond accuracy can be then achieved using submicron mechanical accuracy, well within the specifications of precision translation stages.

The most common method for the evaluation of the duration of subpicosecond pulses is the background-free second harmonic autocorrelation technique. Its principle is shown in figure 1. The optical pulse to be measured is divided in two pulses of equal intensity, one of which passes through a variable delay line. Both pulses are then recombined in a KDP crystal under a small angle. Second harmonic light at frequency 2ω is created in the direction of propagation of each beam, but also along the bisector if the crystal is properly oriented for phase matching conditions. Phase matching insures that both beams at ω and 2ω propagate at the same phase velocity inside the nonlinear crystal, so that energy transfer occurs always constructively from beam at ω to beam at 2ω.

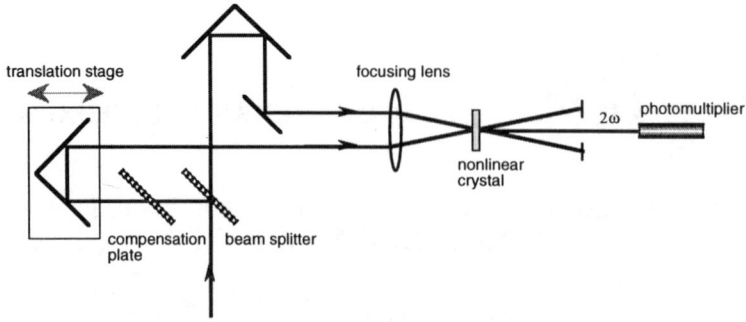

FIGURE 1. Second harmonic autocorrelator for pulse width measurement.

The signal at 2ω, time integrated by a slow photodetector, is proportional to the intensity autocorrelation function

$$\Gamma(\tau) = \int_{-\infty}^{+\infty} I(t)I(t-\tau)dt \qquad (1)$$

An example of an autocorrelation trace from a 30 fs pulse emitted by a Ti:Sapphire laser is shown in figure 2.
For very short pulses, below 50 fs, one often resorts to an interferometric autocorrelator. The beam configuration is that of a Michelson interferometer with both beams propagating collinearly inside the doubling crystal after the delay line. The signal recorded by the detector corresponds to the fourth power of the sum of the field amplitudes (one quadrature from the harmonic generation process, another quadrature from the detector which measures the square of the resulting field).

$$\Gamma(\tau) = \int_{-\infty}^{+\infty} [E_1(t) + E_2(t+\tau)]^4 dt \qquad (2)$$

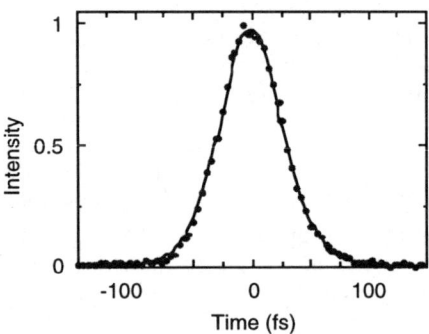

FIGURE 2. Autocorrelation trace of a 40 fs pulse (dots) together with a theoretical fit (line) assuming a hyperbolic sech pulse shape.

It therefore displays fringes corresponding to successive constructive and destructive interferences. In order to extract the pulse duration, one counts the number of fringes under the interference pattern envelope, knowing that the distance between two successive fringes corresponds to one wavelength of incident light. For a properly adjusted interferometer with $E_1=E_2$, the signal amplitude at 2ω varies between a value $2 I_o^2$ independent of time delay (for large delays) and a value $16 I_o^2$ at the maximum of the highest fringe, where I_o is the intensity of the fundamental pulse at ω. An example of such an autocorrelation trace for a pulse of 4.5 fs duration, recently obtained by Nisoli et al.(1), is shown in figure 3. It represents the shortest laser pulse produced so far.

For many experiments performed at very high intensities, it is imperative to measure the amount of laser energy preceding the main pulse over a very large dynamic scale. Indeed, when dealing with multitcrawatt pulses focused on a target, a prepulse laser intensity even 10^{-7} below the peak value still corresponds to an intensity of 10^{12} W/cm^2 or more, sufficient to ionize any material, modifying thereby the target conditions at the instant of arrival of the main pulse. In order to detect the eventual presence of prepulse energy on a large dynamic scale, one can resort to third-order correlation. Here the fundamental pulse at ω is mixed with the harmonic of the same pulse at 2ω to generate a pulse at 3ω by sum frequency generation via the second-order nonlinearity of a KDP crystal. By varying the time delay between pulses at ω and 2ω, one can explore the front of an intense pulse over a large dynamic range, reaching 10 orders of magnitude. Figures 4 and 5 show an experimental set-up for that purpose together with corresponding results (2).

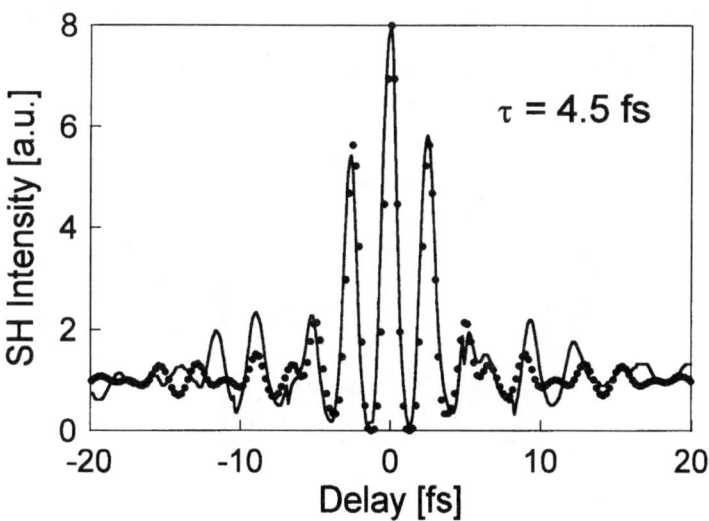

FIGURE 3. Characterization of a pulse by interferometric autocorrelation from Nisoli et al. (1). Full points : experimental datas. Continuous line : theoretical fit.

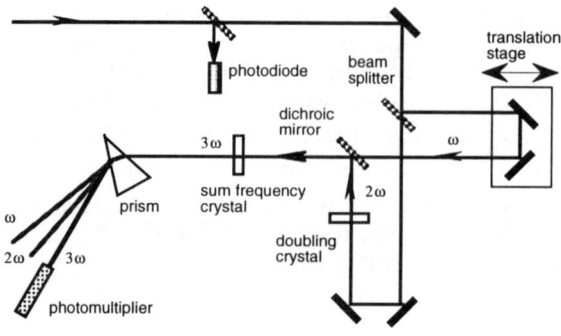

FIGURE 4. Third-order correlator. See ref (2)

Although practical and in wide use, autocorrelators have serious limitations. They only give the pulse duration if the pulse temporal shape is known a priori. A standard pulse time profile, such as a $\text{sech}^2(t)$ pulse is generally assumed. However, pulses are often far from ideal. In particular, a chirp (a time-varying shift of the carrier frequency) can be easily imprinted to a pulse during its propagation through dispersive or nonlinear optical elements inside as well as outside the laser oscillator-amplifier system. Autocorrelation traces, especially interferometric traces are affected by the presence of phase distortion and can therefore give misleading results on the duration of the pulse energy envelope. More generally, the outcome of many experiments depends on the eventual presence of chirp, an information not obtained from autocorrelation traces.

New techniques have been recently developed, aimed at reconstructing the exact temporal lineshape of ultrashort pulses (3,4). This requires a knowledge of both the amplitude and the phase of the complex radiation field. One such method developed during the last few years and currently tested in several laboratories, is the so-called FROG technique, an acronym for Frequency-Resolved Optical Gating (4). Its principle is shown in figure 6.

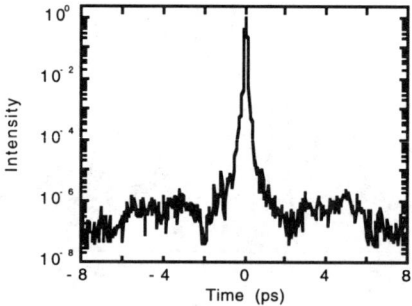

FIGURE 5. Third-order correlation of a 50 fs pulse. See ref (2)

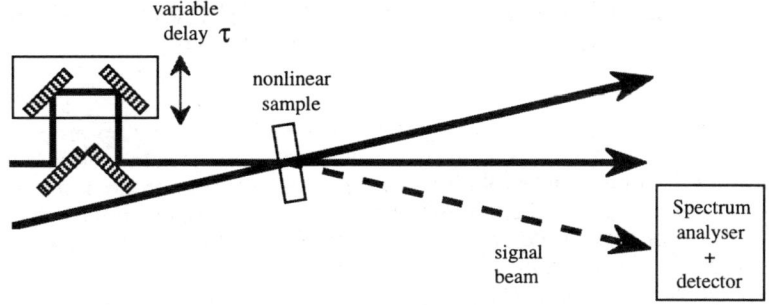

FIGURE 6. Experimental set-up for FROG.

A coherent signal beam is created in a transparent nonlinear medium by the overlap of the incident and gating beam. The spectrum of the signal is recorded as a function of time delay between the incident pulses. Depending on the situation, a different physical effect is used to generate the signal (second harmonic or third harmonic generation, polarisation rotation or self-diffraction). An algorithm extracts the information from the set of N spectra by an iterative process which minimises the difference between the measured and the calculated signal. For details about this method, the reader is referred to the literature (4).

We describe here another more recent approach (5,6), the principle of which is shown in figure 7. Instead of generating a new signal proportional to a gating function, one records the change of the probe spectrum induced by a pump pulse. This change of spectrum is due to cross-phase modulation and thus affects the phase of the pulse only, but not its modulus. It can be viewed therefore as a FM phase retrieval method in contrast to FROG which is a AM technique in the sense that a gating function modulates the amplitude of the generated signal.

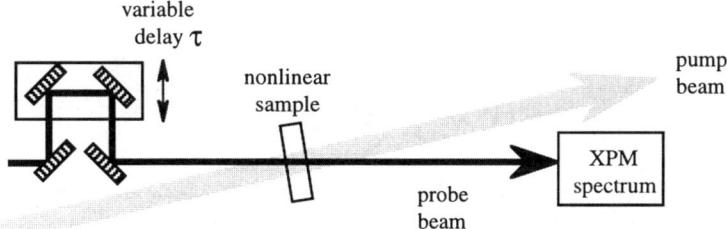

FIGURE 7. Experimental set up of the pump-probe scheme.

2. GENERAL DESCRIPTION OF THE CROSS-PHASE MODULATION (XPM) METHOD

The method is based on the acquisition of N probe spectra in a pump-probe experiment. Both pump and probe beam are derived from the same primary beam and have therefore the same time evolution. A reference spectrum of the probe beam is first taken in the absence of the pump. Then, N-1 cross-phase-modulated spectra are taken after transmission of the probe beam through a plate of glass which is illuminated by the pump beam, at different delays between pump and probe. From the set of cross-phase-modulated spectra, the modulus of the pulse and its phase are extracted with an inverse algorithm described below. If diffraction and dispersion effects are negligible, a condition satisfied by using a thin plate and a large diameter beam, the intensity profile I(t) of the probe pulse is conserved. Consequently the equation describing the propagation of the probe pulse envelope in the nonlinear medium reduces to a multiplication of the complex field with a nonlinear phase factor in the time domain :

$$E^z(t,\tau) = E^0(t) e^{i q_{med} I_p(t-\tau)} \tag{3}$$

where $E^0(t)$ denotes the normalized electric field of probe beam at the entrance of the medium and τ is the delay between pump and probe pulses. The coefficient q_{med} is related to experimental parameters by :

$$q_{med} = \frac{2\pi}{\lambda_0} n_2 I_{max} z \tag{4}$$

where λ_0 is the central laser wavelength, n_2 the optical Kerr coefficient, z the thickness of the nonlinear sample and I_{max} gives the peak laser intensity, a quantity to be determined. Physically q_{med} describes the maximum value of nonlinear phase modulation accumulated during propagation.

The power spectrum $F^z(\nu, t)$ of the probe pulse, the experimentally accessible quantity, can be expressed as the square modulus of the Fourier transform of the temporal pulse:

$$F^z(\nu,\tau) = \left| \int_{-\infty}^{+\infty} E^z(t,\tau) e^{2i\pi\nu t} dt \right|^2 \tag{5}$$

A two-dimensional map of the difference between the probe spectrum measured at time delay τ and the reference spectrum can be formed. The corresponding spectrograms yield an intuitive guide for pulse diagnostics. As an example, three numerical cases corresponding to a pulse with a negative or positive chirp as well as a bandwidth limited pulse are shown in figure 8. The presence of a chirp and its sign can be readily recognised by inspection of these images. Another example is given in figure 9a for a pulse which has undergone selfphase modulation and in figure 9b for a sequence of two pulses. We show now how information on pulse shape can be extracted in a quantitative way.

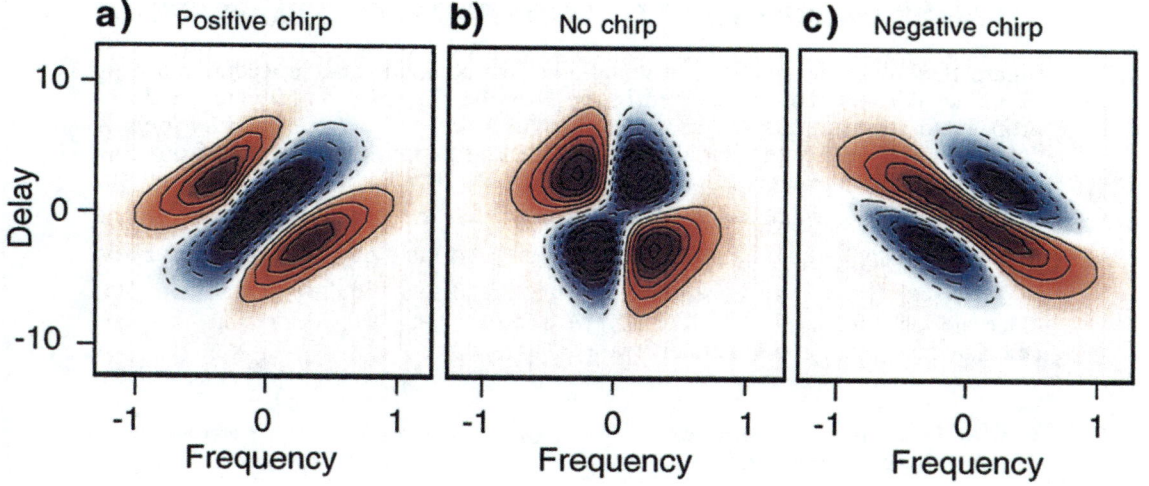

FIGURE 8. Numerically generated 2D image for a) negative, b) zero, c) positive quadratic chirp.

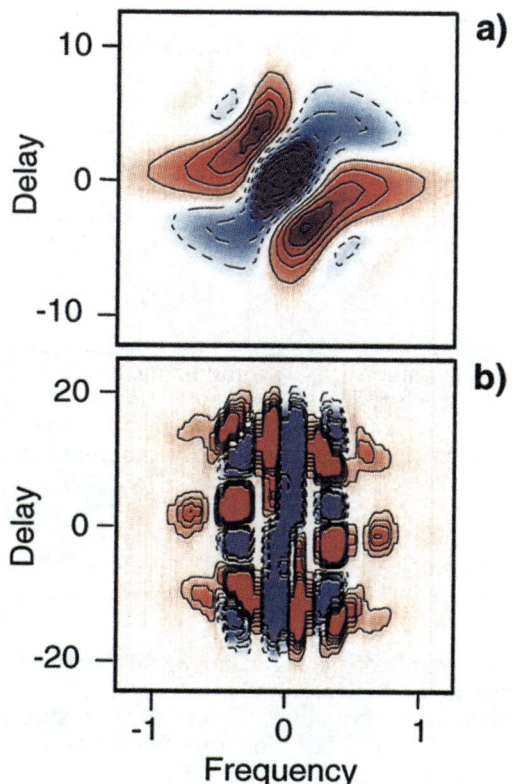

FIGURE 9. Numerically generated 2D image for a) self-phase modulated pulse and b) double pulse.

3. THE PHASE-RETRIEVAL ALGORITHM FOR XPM SPECTRA

Figure 10 shows a schematic representation of the computational procedure allowing to extract the information encoded in the N probe spectra. One reference and one propagated spectrum are used as experimental inputs for a single internal loop. A description of the substitution algorithm extracting the phase of the field from both spectra is found in reference 6. The phase calculated by applying the algorithm to the first set of spectra with delay τ_1 is then introduced as the initial guess for the second internal loop applied on the reference spectrum $F^0(\nu)$ and the XPM spectrum $F^z(\nu, \tau_2)$. N-1 such two spectra substitution schemes are attached in the external loop. This external loop is repeated over all N-1 XPM spectra as long as the discrepancy between calculated and measured spectra decreases.

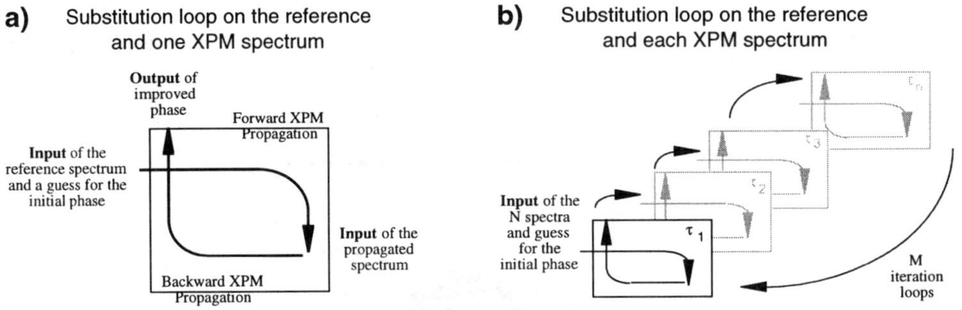

FIGURE 10. Scheme of the algorithm: (a) the internal loop with a substitution scheme for the reference spectrum and one XPM spectrum, (b) the external loop over the N-1 XPM spectra which is iterated M times until convergence is achieved.

The algorithm yields, in addition to the phase, the absolute value of the experimental intensity: one retains the value of q_{med} minimizing the RMS distance between the estimated and the correct value. The q_{med} value yields the maximum intensity of the pulse and, after integration in the time domain, the absolute energy density of the pulse. The comparison with the experimentally measured pulse energy allows a verification of the pulse characterization.

4. PERFORMANCE OF THE METHOD FOR NUMERICAL CASES

We have tested the robustness of the algorithm against noise. Three sources of noise have been introduced in numerical simulations: a) random noise on the spectral amplitude to simulate detector shot noise; b) random fluctuations in the value of q_{med} among the N-1 spectra to simulate laser intensity fluctuations in multi-shot operation; c) a systematic error simulating an experimental inaccuracy in the time delay τ. The random noise is generated independently in each spectrum.

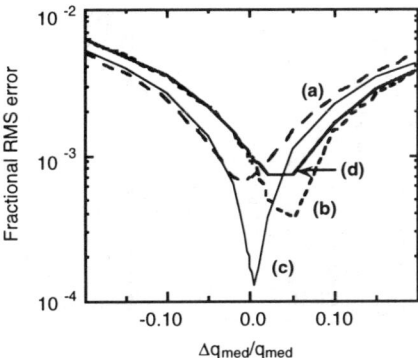

FIGURE 11. Influence of noise on the algorithm: the error D_F as a function of the relative error in q_{med} for a pulse with satellites and higher phase distortion.

The curves of figure 11 show the impact of (a) 0.1% spectral noise, (b) fluctuations of 10% in the value of laser intensity, (c) an inaccuracy in the time delay of 1% and (d) the combined effect of all three error sources. The simulation indicates that, all three sources of error combined, it is reasonable to expect convergence close to 10^{-3} in the experimental case, after one hundred iterations.

Figure 12 illustrates the effect of another type of phase fluctuations introduced in the form of random phase jumps. A series of 22 spectra, calculated at successive delays was given as an input to the algorithm. The information extracted by running the algorithm shows excellent agreement between the initial and reconstructed pulse.

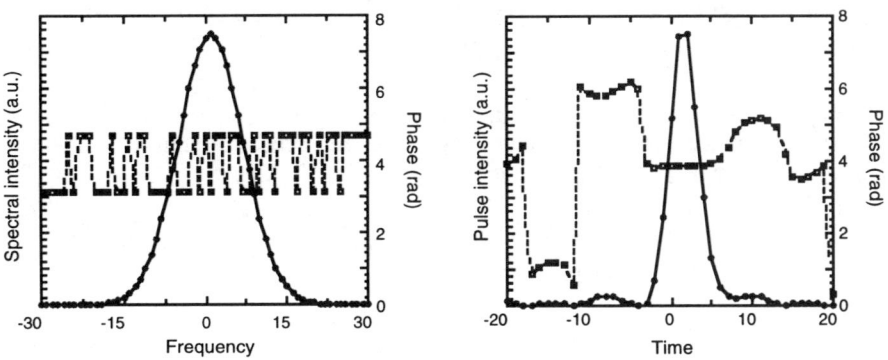

FIGURE 12. Illustration of the robustness of the algorithm.
Left hand side : initial (lines) and reconstructed (dots) spectrum intensity and phase.
Right hand side : initial (lines) and reconstructed (dots) pulse intensity and phase.

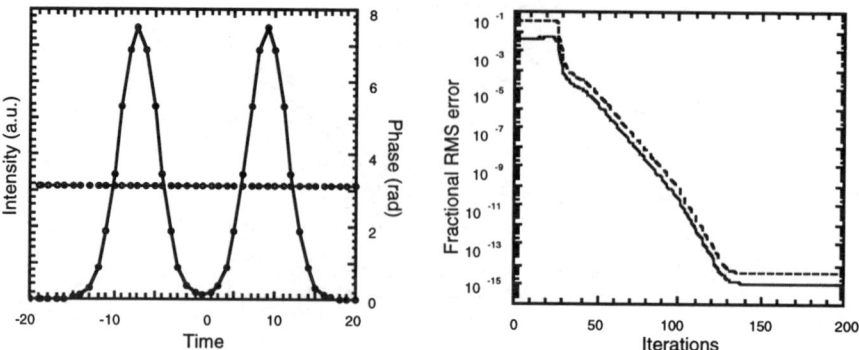

FIGURE 13. Double pulse:
-Left hand side : initial (lines) and reconstructed (dots) pulses intensities and phase.
-Right hand side : error on the spectrum (solid) and on the pulse (dashed) as a function of the number of iterations.

The numerical study of computer generated pulses allows to test also the algorithm in cases recognized as difficult in other phase retrieval methods. Convergence of the algorithm to the round off error of the computer was found in all encountered cases. By introducing an appropriate number of spectra, even very complex pulse shapes could be solved.

As an example, figure 13 shows the numerical reconstruction of a train of 2 pulses, obtained with 32 spectra. The error D_F on the spectrum and the error D_I on the pulse are shown with a solid and dashed line respectively. This case indicates that even highly modulated amplitudes in the time domain can be reconstructed with the algorithm.

5. ALGORITHM-FREE DETERMINATION OF THE PULSE DURATION

By simply plotting the first spectral moment of the spectra, one can evaluate the pulse duration without the need of any algorithm. The first order frequency moment for a given time delay τ is defined as :

$$\langle v(\tau) \rangle = \frac{\int_{-\infty}^{+\infty} v F^z(v, \tau) dv}{\int_{-\infty}^{+\infty} F^z(v, \tau) dv} \tag{6}$$

As shown in reference (7), this quantity corresponds to the derivative of the intensity autocorrelation function

$$\langle v(\tau) \rangle \propto q_{med} \frac{d}{d\tau} \int_{-\infty}^{+\infty} I(t) I(t - \tau) dt \tag{7}$$

Thus, the experimental values of $<v(\tau)>$ serve as a precious consistency test for the validity of the method. The first derivative of $<v(\tau)>$ being an antisymmetric function, an asymmetry of the curve denotes the presence of either misalignment or higher-order nonlinear response of the sample, such as multiphoton absorption or a retarded Kerr effect. In that case, the XPM method does not apply.

6. EXPERIMENTAL RESULTS

The method has been implemented on two Ti:Sapphire CPA laser systems. The first one, which used lenses in the pulse stretching stage, delivers pulses 130 fs in duration. The second laser which uses specially designed reflective optics in the stretcher delivers pulses of 30 fs duration.

Three examples of the performance of the method obtained with the first laser are shown in Figure 14.

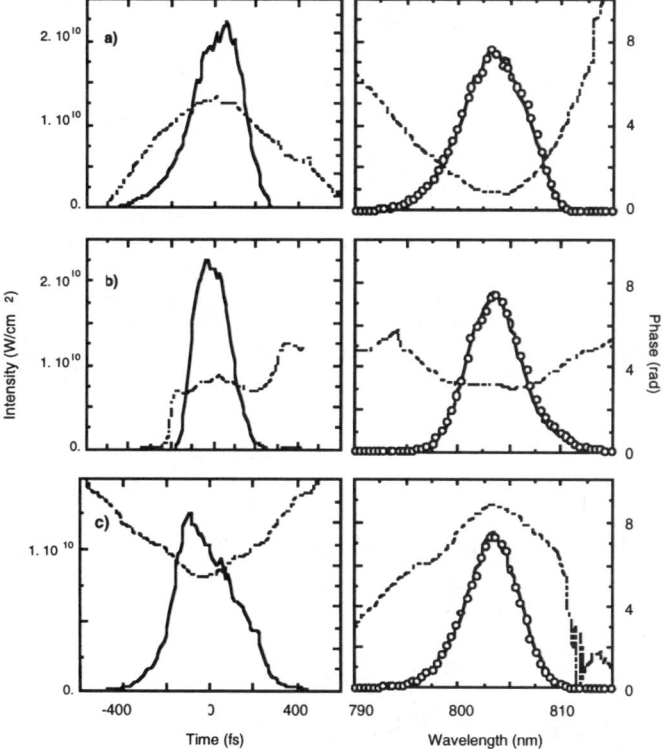

FIGURE 14. Retrieval of three experimental pulses at different positions of the compressor grating: (a) $\Delta l=-1.0$ mm, (b) $\Delta l=0$ mm and (c) $\Delta l=1.5$ mm. The left column displays the retrieved temporal pulse profile and phase (dashed). The right column shows the square modulus (measured: solid, retrieved: circles) and phase (dashed) of the spectrum.

Figure 14(a) shows the effect of a positive chirp, obtained by a reduction of the distance between the gratings in the compression stage by 1mm from the optimal value (b), and in (c) the effect of a negative chirp (grating separation increased by 1.5mm with respect to the position (b). The pulses display the expected broadening to $T_{FWHM} = 280$ fs in case (a) and to $T_{FWHM} = 310$ fs in case (c). In case (b) the duration is $T_{FWHM} = 200$ fs. In addition to pulse broadening, one can notice a change of pulse shape with a steeper rising edge in (a) and a steeper falling edge in (c). The energy density determined from the algorithm gives 6.6 mJ/cm^2 for case (a), 4.7 mJ/cm^2 for case (b) and 4.1 mJ/cm^2 for case (c), in reasonable agreement with the experimentally measured energies. The obtained error on the spectra varies between $D_F = 10^{-2}$ and $D_F = 5 \times 10^{-3}$ after 80 iterations and lies in the order of the minimum convergence predicted in the simulation of the experimental errors.

For typical application to the characterization of a CPA laser chain the method requires about 20 spectra, 200 iterations and a non-linearity corresponding to a value $q_{med} \approx 1$. This number of 20 spectra could be reduced without loss of accuracy but is the minimum necessary to extract a good autocorrelation trace from the curve of first spectral moment, providing an independent check of internal consistency.
In figure 15, pulses reconstructed with the 30 fs laser-amplifier system are shown. In order to explore the pulse shape over a large dynamic range, 60 spectra were recorded. One sees that it is possible to reconstruct the pulse over at least 4 decades of intensity.
In figure 16 we show experimental and calculated spectrograms corresponding to cases 14a and 15.

The method has also been applied to the characterization of a femtosecond UV pulse obtained by second harmonic generation in a KDP or BBO doubling crystal of a pulse comparable to that of figure 15 (see figure 17). The pedestal just before the main peak is reduced, because of the quadratic power law in the doubling process which increases the contrast ratio of the pulse. For the same reason one would expect also a pulse shortening. However, the UV pulse duration is not reduced, but rather increased. This is due to pulse broadening in the doubling crystal, as checked by measurements with thinner doubling crystals, which show reduction of the UV pulse duration.

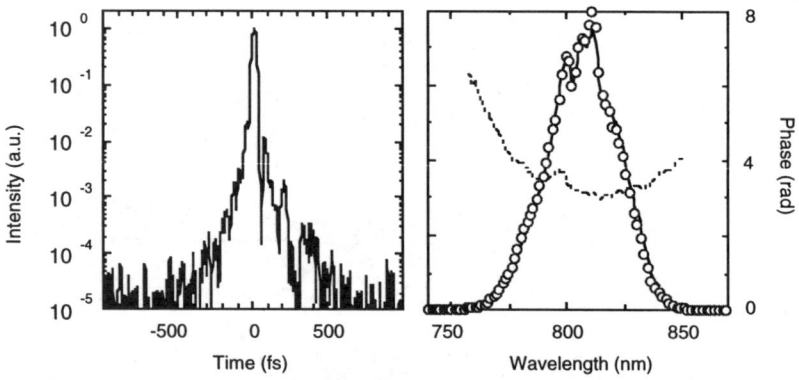

FIGURE 15. Reconstructed pulse profile (left) and spectrum (right) at 800 nm. The FWHM of the pulse is 30 fs.

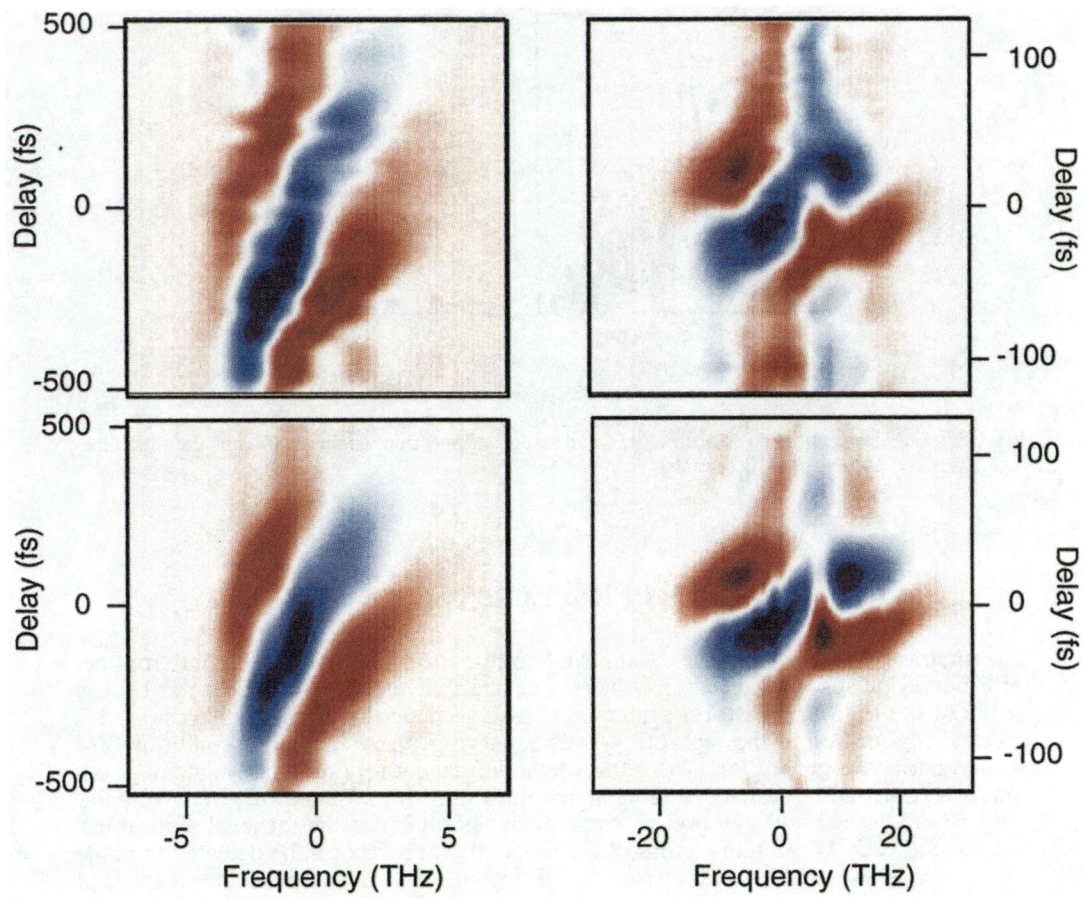

FIGURE 16.
Left hand side :
- top : measured spectrogram of a 280 fs pulse with positive chirp;
- bottom : reconstructed spectrogram from XPM algorithm (see also figure 14a).

Right hand side :
- top : measured spectrogram of a 30 fs pulse close to its transform limit;
- bottom : reconstructed spectrogram from XPM algorithm (see also figure 15).

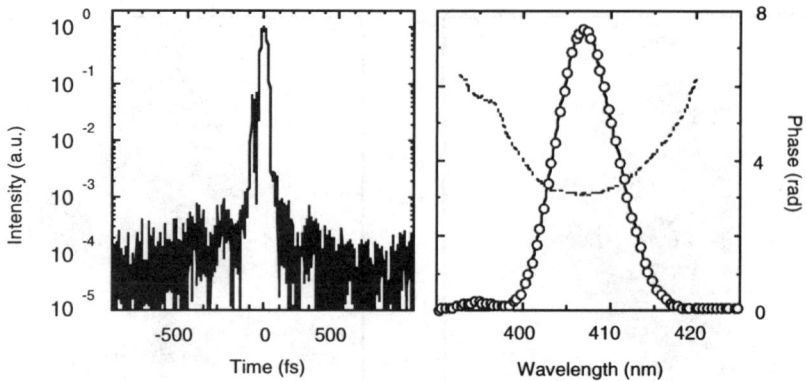

FIGURE 17. Measured pulse profile (left) and associated spectrum (right) at 400 nm after doubling in a 200 µm BBO crystal.

7. PERSPECTIVES

The method described above could find applications in the attosecond regime. Attosecond pulses will necessarily have a central wavelength shifted to the UV by virtue of the Fourier transform principle. Classical autocorrelators are useless in this spectral region due to the lack of crystals satisfying phase-matching conditions for second harmonic generation. The XPM method does not rely on phase-matching, but requires transparency of the nonlinear medium and the absence of group velocity dispersion (at least with the present stage of development of the retrieval algorithm). We have already verified that cross-phase modulation spectra can be detected in noble gases such as Ar, which are transparent well into the UV. This opens the prospect of measuring the duration of high-order harmonics generated in gases or at the interface between a gas and a solid (see the contributions of Wahlstrom and von der Linde in this volume).

ACKNOWLEDGMENTS

We would like to thank Y.-B. André for the expert assistance to the acquisition system. H. R. Lange gratefully acknowledges the support by a Marie-Curie-Fellowship of the European Community (grant ERBFMBICT950065).

REFERENCES

1. Nisoli, M., De Silvestri, S., Svelto, O., *Optics Let.* **22**, 522-524 (1997).
2. Antonetti, A., Blasco, F. , Chambaret, J.P., Chériaux, G. , Darpentigny, G. , Le Blanc, C., Rousseau, P., Ranc, S., Rey, G. , Salin, F., *Appl. Phys. B* **65**, 197-204 (1997).
3. Chilla, J.L.A., Martinez, O.E., *Optics Let.* **16**, 39-41 (1991).
Paye, J., *IEEE J. Quantum Electr.* **30**, 2693-2697 (1994).
4. Kane, D.J., Trebino, R., *IEEE J. Quantum Electr.* **29**, 571-579 (1993).
Trebino, R., DeLong, K.W., Fittinghoff, D.N., Sweetser, J.N., Kane, D.J., *Rev. Sci. Instrum* **68,** 3277-3295 (1997).
5. Prade, B.S., Schins, J.M., Nibbering, E.T.J., Franco, M.A , Mysyrowicz, A., *Optics Comm.* **113**, 79-84 (1994).
6. Franco, M.A., Lange, H.R., Ripoche, J.F., Prade, B.S., Mysyrowicz, A., *Optics Comm.* **140**, 331-340 (1997).
7. Ripoche, J.F., Prade, B. , Franco, M. , Grillon, G., Lange, R., Mysyrowicz, A., *Optics Comm.* **134**, 165-170 (1997).

Methods for the Shaping High-Power Picosecond Laser Pulses with a High-Contrast Ratio

V. A. Malinov, A. V. Charukchev, V. N. Chernov, N.V.Nikitin,
S.L.Potapov, V. M. Efanov* and P. M. Yarin*

*Scientific Research Institute for Complex Testing "S.I.Vavilov State Optical Institute",
Sosnovy Bor, Leningrad Region, 188537 Russia*
*A. F. Ioffe Physical-Technical Institute, St. Petersburg, 194021, , Russia

Abstract. We present the performance of the electrooptical system based on four Pockels cells with 10 and 20 mm diameters, each of them is driving by its own drift step recovery diode pulse generator. We are developing electro-optic deflector system for CPA laser using two identical deflectors (diverging and converging) and three spatial filters. The results of numerical modeling of the time - dependent distributions of the intensity in the beam are presented. A peak-to-background intensity ratio more than five orders is achieved by this technique. We have developed a new pulse generator based on single drift step recovery diode producing two identical electrical pulses with output voltage up to 15 kV, FWHM of 1.5 ns, rise time of 0.7 ns and jitter of 100 ps at a 100 Hz repetition rate to electro-optic deflectors.

1. INTRODUCTION

A high contrast between the main ultrashort laser pulse and the prepulse is required for many laser-plasma interactions, in particular to the generation of ultrafast x-ray pulses (1,2). In this work we report the methods of the pedestal energy suppression on "Progress-P" 30 TW CPA picosecond phosphate Nd:glass laser facility (3).

2. ELECTROOPTIC SYSTEM OF CPA LASER "PROGRESS-P"

The electrooptic and pulse synchronization/driver system of the 30 TW CPA laser system "Progress-P" is shown in Figure 1. This system provides the passage of the single pulse from the continuous wave mode-locked Nd:YLF oscillator that produces 60 ps pulses at a 100 MHz repetition rate to the output of the laser when the pump level of rod amplifiers reaches the maximum. Besides, it launches the laser and plasma diagnostic devices.

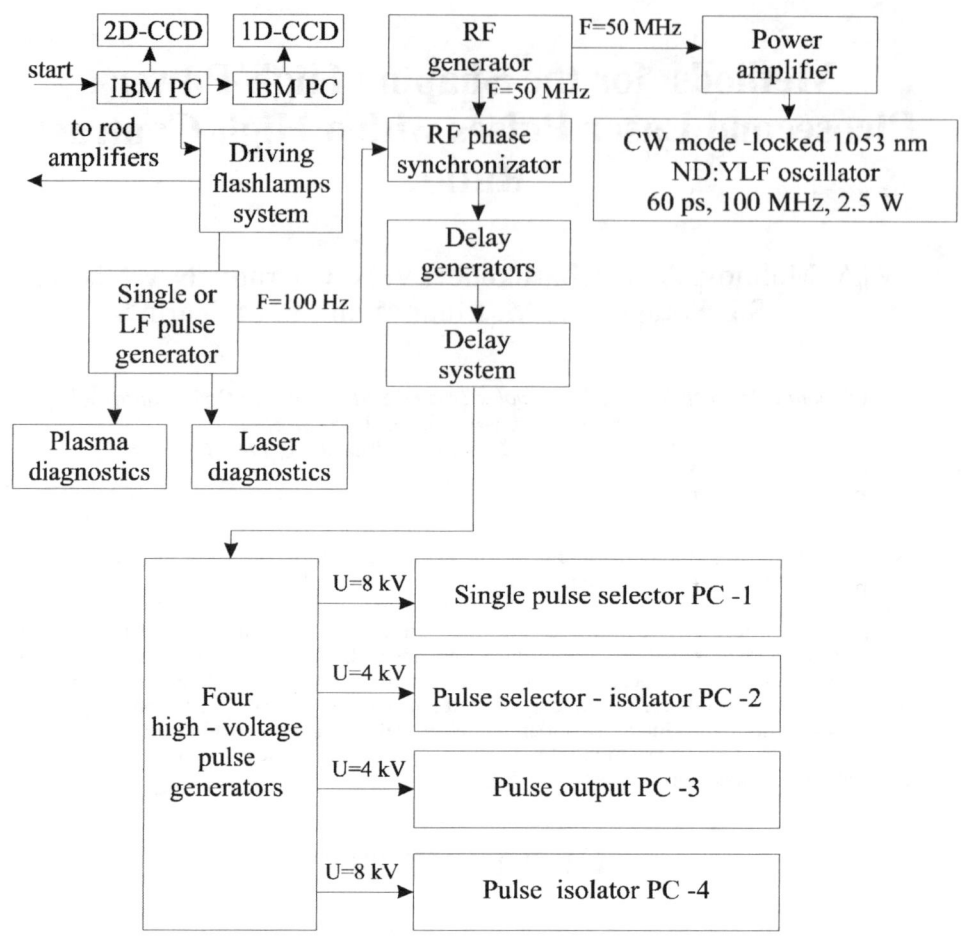

FIGURE 1. Block diagram of the pulse synchronization/driver system of the 30 TW CPA laser system «Progress-P»

The key elements of this system are four crystal KD*P Pockels cells (PC) driven electronically and connected in series. Polarizers in the PC's are dielectric polarizers at Brewster's angle. The high-voltage pulse generators based on the drift step recovery diodes (DSDR) (4) capable of producing 4-8 kV pulses up to 1 kHz with a FWHM of 5 ns and rise time of 1.5 ns are used to drive the PC's. The first PC used as a selector with a contrast of at least 700:1 is used to select a single pulse from the pulse train; the next two PC's used as selector and output ejector (20 mm diameter, quarter-wave voltage) are ensured the passage through the multipass amplifier and rejection from it. The latter, fourth isolator PC (20 mm diameter, half-wave voltage 8 kV ,) is located between the 20 and 45 mm diameter amplifiers. The optical transmission of the PC's driven by DSDR generators was sampled for different time delays between the arrival of the optical and electrical pulses at the Pockels cells. The optical 0.1-0.9 rise times are 1.2-1.5 ns with a jitter 100 ps and some long-term (approximately one week) delay-drift. The timing synchronization between the single laser pulse and firing the rod amplifiers is accomplished by use of a rf phase synchronizator and low frequency or single pulse generator with subnanosecond accuracy.

As a result, the amplified spontaneous emission which passed through grating compressor was determined by the fast photo-diode and calorimeter and occurred to be less than 50 µJ energy at 5 ns duration. The energy contrast ratio was more than 10^5:1 of laser pulse.

3. DRIFT STEP RECOVERY DIODE PULSE GENERATOR

The high pulsed voltage to Pockels cells is produced by drift step recovery diode pulse generators based on solid state plasma opening switches with inductive storage system. The peak load voltage may be many times higher than the voltage at which the energy has been stored. Being semiconductor devices, DSRD has small life time and low jitter. Maximum repetition rate is limited mainly by heat and may be as high as megahertz. Generator is based on well known symmetrical circuit with two arms shown in Figure 2.

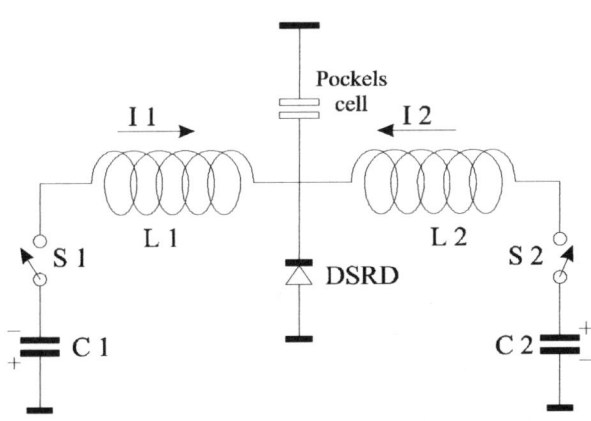

FIGURE 2. Schematic of Pockels cell circuit.

The p+nn+ structure is made by deep diffusion. Initially, energy storage capacitors C1, C2 are charged and switches S1, S2 are opened. When the switch S1 is closed, capacitor C1 discharges via inductor L1 and the diode DSRD. The discharge current is a forward current for DSDR, the resistance of DSDR is low and the current I1 in circuit C1, S1, L1, DSRD oscillates. Minority carriers lifetime in DSRD is large - ten and more microseconds. The total amount of electron-hole parts injected in DSRD during the fist forward half-period of current oscillation is equal to the charge passed through the diode. When the current changes it's direction the diode remains in high conducting state due to stored electron-hole pairs. If, at the moment when current I1 crosses the zero-level, the second switch S2 closes, the C2 discharge current I2 is added to the current of L1, C1 circuit. The total current trough DSRD is doubled. At this moment all energy stored in C1 and C2 are accumulated in L1, L2 inductors and their currents reach maximum values. During the time of DSRD switch off process, it's current is being switched into Pockels cell which can be considered as capacitor.

4. DEFLECTOR SYSTEM

Characteristics of the deflector system are very promising for use in CPA lasers in order to achieve high contrast of the laser pulses and to protect diffraction grating in stretcher from damage because of back-reflections. Earlier we successfully used deflector system in six-beam Nd-glass laser «Progress» in laser fusion experiments (5). In these experiments pulses with a power contrast up to 10^8 1 ns before the main pulse arrival were generated.

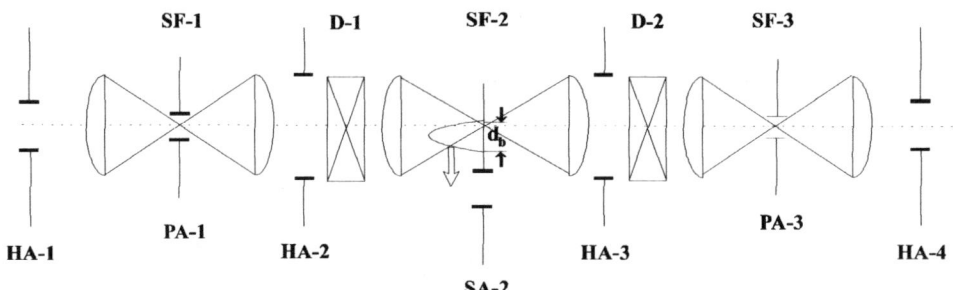

FIGURE 3. Scheme of deflector system.

The optical scheme of deflector system (Fig. 3) consists of two identical deflectors (diverging D-1 and converging D-2) , three spatial filters (SF 1-3) and four hard apertures (HA 1-4). The spatial filters consist of spherical lenses and focal pinholes (PA-1, SA-2 and PA-3). The filters relay consecutively the image of the input aperture HA-1 to output aperture HA-4. After passing the deflector system the contrast of pulse increases as described below . The diverging deflector deflects a focused beam in the focal plane of the «contrast» aperture SA-1 and the converging deflector compensates fully for this beam deflection. The pulse contrast is being formed while the beam scans the «contrast» aperture up to peak power. The axial propagation of the beam sets its peak power so that the beam structure in space and time is entirely determined by the clipping and diffraction of the beam by the hard apertures and the shape of the «contrast» aperture. The off-axial propagation determines the pulse contrast. Pulses with sharp heading edges can be generated if we suppress the long wings of the focal spot distribution of the laser beam in the plane SA-2. In order to suppress these wings one needs to remove the diffraction scattering of the beam because of the clipping by hard apertures and the bulk scattering in the crystals. The calculations have shown that the diffraction scattering can be neglected if HA-1 diameter is 0.7 times as much as HA-2 diameter.

The contrast of the pulse was investigated by calculating the distributions of the intensity in a beam and losses experienced at each component of the system while the beam scans the SA-2. The beam propagation was described using the scalar diffraction theory in the aberration-free approximation. The field in the output of the system is a result of the subsequent operations representing filtering of the input field in the spatial domain, Fourier transform, filtering in the frequency domain and inverse Fourier transform of each spatial filter.

In the calculations we used our experimental results involving beam scattering in the optical scheme of the deflector system with $LiNbO_3$ crystals (6) with the next parameters: diameter of $LiNbO_3$ crystal of 4.5 mm , focal length of all lenses of 1000 mm, divergence of the beam of 0.43 mrad, velocity of scanning of 3.75 mrad/ns and diameter of focal spot of 0.85 mm. The calculated power contrast ratio of the laser pulses in this deflector system is shown in Figure 4. The intensity peak-to-background ratio is better than 10^5 230 ps before the main pulse.

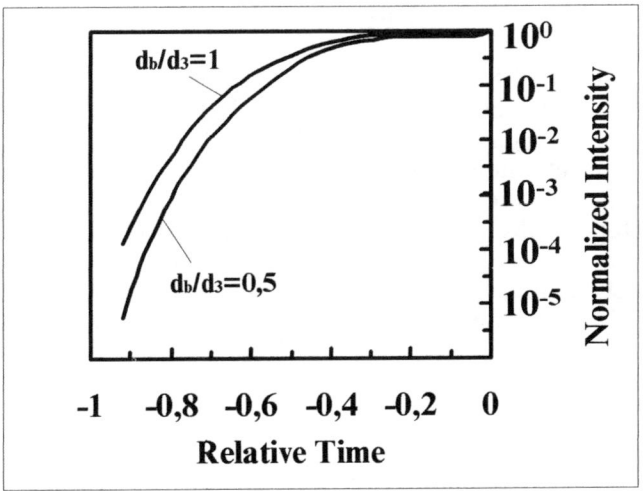

FIGURE 4. Power contrast ratio of the laser pulses in a deflector system. Relative time is t/t_0, where $t_0 = d_b/v = 230$ ps, d_b - diameter of focal spot in SA-2, d_3 - diameter of the pinhole of the output spatial filter, v - velocity of scanning.

As shown in Fig. 4, with the decrease of diameter of the pinhole of the output spatial filter (harder filtration of the beam) the contrast is increasing. For this system we have developed a new pulse generator based on single DSDR producing two identical pulses with output voltage up to 15 kV, FWHM of 1.5 ns, rise time of 0.7 ns and jitter of 100 ps. We plan to install this deflector system between the stretcher and multipass amplifier of the 30 TW CPA laser facility "Progress-P" and considerably increase contrast ratio of laser pulses.

5. CONCLUSIONS

We have described the electrooptic system based on four Pockels cells with 10 and 20 mm diameters driving by drift step recovery diodes which suppressed the amplified spontaneous emission to the level of less than 50 µJ energy at 5 ns duration on the laser facility "Progress-P". The energy contrast ratio was more than 10^5:1. We have optimized the parameters of the deflector system and shown that the intensity peak-to-background ratio is better than 10^5 230 ps before the main pulse. In the future we intend to set up the described system on the laser facility "Progress-P".

ACKNOWLEDGMENTS

This work was supported by the International Scientific and Technology Center under Project ISTC #107-94.

REFERENCES

1. Kieffer, J.C., et al., Phys. Fluids **B 5,** 2676-2681 (1993).
2. Perry M.D., Mourou G., Science **264**, 917-924 (1994).
3. Borodin, E.G., et al., «The «Progress-P» 30 TW picosecond Nd:glass facility», presented at the 13th International Conference on Laser Interactions and Related Plasma Phenomena, Monterey, USA, April 13-18, 1997.
4. Grehov I.V., Efanov V.M.., Kardo-Susoev A.F., Shenderey Sov. Pisma v GTF, **9**, 435 -437 (1983).
5. Andreev A.A., et al., Sov. J. Kvantovay Electronika, **17,** 1306-1310 (1990).
6. Kruchanovski V.I., et al., Sov. J. Kvantovay Electronika, **12**, 372-375 (1985).

30 TW Laser Facility "Progress-P"

V. G. Borodin, A. V. Charukchev, V. N. Chernov, V. M. Komarov,
S. V. Krasov, V. A. Malinov, V. M. Migel, N. V. Nikitin,
V. S. Popov, S. L. Potapov

Scientific Research Institute faor Complex Testing
"S.I.Vavilov State Optical Institute"
Sosnovy Bor, Leningrad Region, 188537 Russia

Abstract. Chirped pulse amplification was implemented in one of six amplifier chains "Progress" phosphate Nd:glass laser system. Laser system configuration and performance are presented. Formation of 300ps chirped pulse at 1053 nm with energy up to 1 J is made by using developed starting laser which consists of Nd:YLF oscillator, optical fiber, stretcher and three amplifiers with output aperture 20 mm. The large amplifier chain of the laser system includes three rod amplifiers with the aperture of output rod of 85mm. Preliminary experiments have been carried out yielding output chirped pulses of up to 45 J and compression them to 1.5 ps by grating compressor.

1. INTRODUCTION

The use of chirped pulse amplification (CPA) [1] has resulted in the generation of high power ultrashort laser pulses for study of high-density plasma physics and ultrafast x-ray emission [2]. Here we describe the design and performance of developed on base of six-beam phosphate Nd:glass «Progress» [3] CPA laser system capable of producing picosecond laser pulses with peak power about 30 TW.

2. STARTING LASER SYSTEM

A scheme of CPA laser system is shown in Figure 1. It consists of four parts: the starting laser system, large amplifier chain, compression stage and focusing stage. Starting laser produces a chirped pulses with energy up to ~1 J and duration 300ps at wavelength 1053 nm. This pulses are used both to further amplification in large amplifier chain and to compression to ~1.5 ps duration by using of pair of two 1700 lines/mm gold -coated holographic gratings for laser-plasma interaction experiments under average

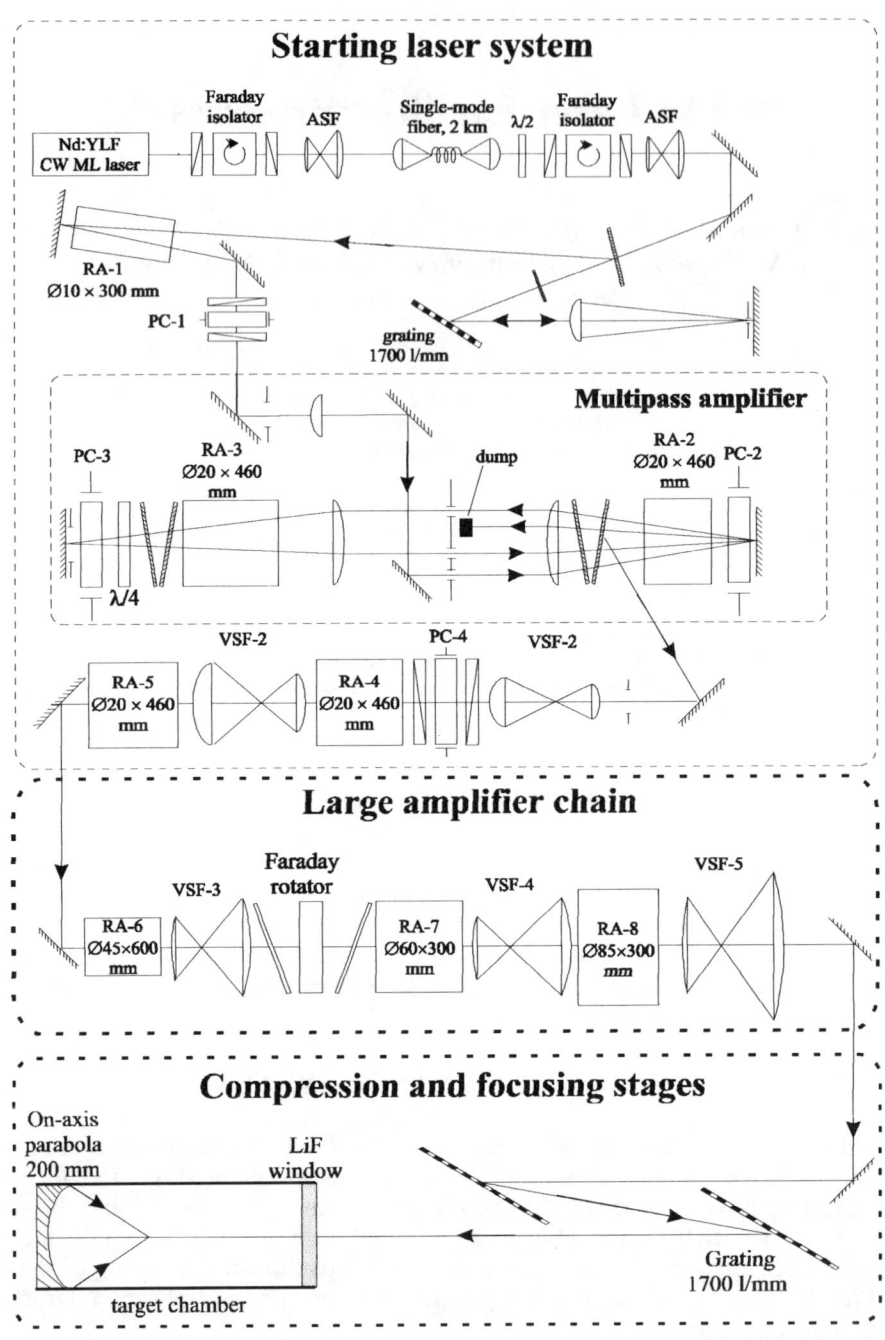

FIGURE 1. Scheme of laser system «Progress-P».

intensity of $10^{17} W/cm^2$. The chirped pulses are obtained by a cw mode-locked Nd:YLF oscillator generating a 100-MHz train of 60 ps pulses with average power ~2.5 W, 2 km single-mode optical fiber and four-pass stretcher with single 1700 lines/mm gold-coated holographic grating. As a result of combined effects of group velocity dispersion and self-phase modulation, the pulse on output of fiber has been broadened to a 150 ps duration with spectrum of 2.4 nm FWHM. The stretcher has been passed only central part of spectrum ~1.7 nm FWHM as shown in Figure2 for elimination of nonlinear chirp [4] and expansion of laser pulse to 300 ps duration.

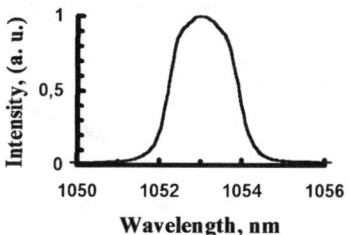

FIGURE 2. Spectrum of pulse after stretcher.

Amplification of pulse from 0.3 nJ to ~J level is performed in three rod amplifiers: two-pass, multipass and single-pass with output aperture of 20 mm. All the amplifiers contain phosphate Nd:glasses GLS22 or KGSS0180 with maximum gain near 1053 nm. A single laser pulse is selected and controlled through laser system by four Pockels cells which driving by high-voltage pulse generator based on the drift step recovery diodes [5]. The generators are produced 4-8 kV pulses with a FWHM of 5 ns and rise time ~1.5 ns. The main amplification of pulse (from 50 nJ up to 200 mJ) is carried out in developed multipass amplifier with net gain $\sim \cdot 10^6$. Amplifier consists of two $\varnothing 20 \times 460$ mm Nd:glass rods with small signal gain up to ~50. The optical scheme of multipass amplifier is disaligned conjugated resonator with relaying of end flat mirrors by use the 1.1 m focal length positive lens pair. In common focus of lenses in air is positioned three-pinholes system with 4-mm spacing between pinholes. The pulse is seeded into amplifier through one of pinholes and removed by Pockels cell. Optimization was made the pinhole diameter for producing the high-quality beam and reducing amplified spontaneous emission pedestal. The diameters of two front pinholes are chosen equal diffraction limited spot size ~ 0.5 mm for 6-mm amlifying beam and the last pinhole diameter is chosen equal 0.35 mm. Further amplification in single-pass amplifiers $\varnothing 20 \times 460$ mm produces in excess of 0.8 J energy with nearly diffraction limited divergence of radiation. The amplified spontaneous emission which passed through grating compressor as determined by the fast photo-diode and calorimeter are less then 10 µJ energy and 5ns duration and makes the energy contrast ratio ~10^5:1 of laser pulse.

3. LARGE AMPLIFIER CHAIN

To obtain the pulses with the energy up to 45 J we used a three-stage relay-imaged amplifier chain, which is a part of the amplifier chain of the "Progress" facility. In the chain, rod stages from GLS22 glass with $\varnothing 4.5\times 60$ cm, $\varnothing 6\times 30$ cm and $\varnothing 8.5\times 30$ cm dimensions are used with small signal gains of 25, 7.5, and 5.5, respectively. Calculations were made of amplification of the beam with the fill factor equal to 0.8, in the geometrical approximation, with account for saturation of the amplifier by means of the Frantz-Nodvick equation. The calculated and measured output energy as a function of input energy to large amplifier chain is shown in Figure 3.

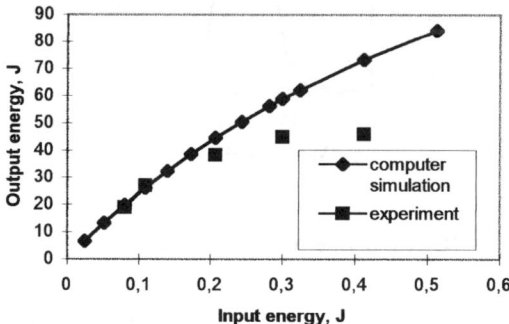

FIGURE 3. Measured and calculated output energy of «Progress-P» laser system as a function of input energy to large amplifier chain for 300 ps pulse duration.

This results show that the beam with the energy ~0.3 J obtained at the output of the starting laser should be applied to the input of the amplifier, to obtain the output beam with the required energy of about 45 J (with consideration for the losses in the compressor). For this regime the B-integral accumulating in section with aperture of 85 mm is equal ~1.3. At the output, the beam is expanded by the vacuum spatial filter up to the diameter of ~200 mm to reduce the energy density in the compressor down to the value by 4 to 5 times lower than the damage threshold equal to ~250 mJ/cm^2. The divergence of beam is twice as much as diffraction limit.

4. COMPRESSION AND FOCUSING STAGES

The pulse is compressed in the single-pass compressor on two gold-coated holographic diffraction gratings with dimensions of 420 mm × 210 mm and 1700 lines/mm. The compressor efficiency is over 80 %. The duration of the compressed pulse for 80 mm beam was measured using a second-order single-shot autocorrelator. Measured autocorrelation trace and spectrum of the compressed pulse are shown in Figure 4. The FWHM of laser pulse is 1.5 ps and width of spectrum is 1.5 nm.

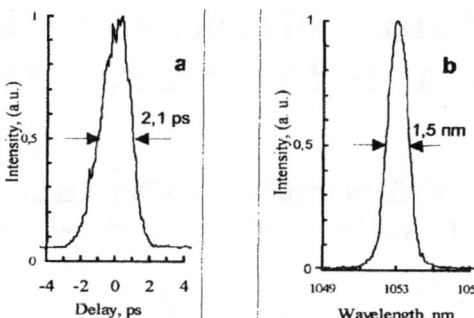

FIGURE 4. Measured single-shot autocorrelation trace (a) and spectrum (b) of the compressed pulse. The FWHM is 1,5 ps assuming Gaussian pulse profile.

The beam is focused to the target by means of the on-axis mirror parabola with focal length of 200 mm,. The beam is injected in the target chamber through the high optical quality LiF-window with low nonlinear refractive index coefficient.

5. CONCLUSION

In conclusion, we have developed the CPA laser system capable of producing ~30 J in 1.5 ps pulses. Preliminary laser experiments are carried out. As a result, the experiments will be performed on exposure of targets at the intensity over $10^{19} W/cm^2$.

This work was supported by International Scientific and Technology Center under project #107.

REFERENCES

1. Main P., Strickland D., Bado P., Pessot M., Mourou G. IEEE J. Quantum Electron., **24**, 398 (1988).
2. Perry M.D., Mourou G. Science, **264**, 917 (1994).
3. Alekseev V.N. et al. Izv. AN SSSR . Ser. Fizich., **48**, 1447 (1984).
4. Heritage J.P.,Thurston R.N.,Tomlinson W.J., Weiner A.M., Stolen R.H. Appl.Phys.Lett., **47**
5. Grehov I.V., Efanov V.M.., Kardo-Susoev A.F., Shenderey S.V. Pisma v GTF, **9**, 435 (1983).

Methods and means of superstrong field formation on multiterawatt laser facility "Progress-P"

V. N. Chernov, V. G. Borodin, A. V. Charukchev, V. A. Malinov,
V. M. Migel, N. V. Nikitin, R. R. Gerke* and I. Yu. Yusupov*

*Scientific Research Institute for Complex Testing "S. I. Vavilov State Optical Institute",
Sosnovy Bor, Leningrad Region, 188537 Russia*
**S. I. Vavilov State Optical Institute, Birzhevaya Liniya, 12, St. Petersburg, Russia*

Abstract. We present the parameters of compressor and focusing systems of 30-TW Nd:glass laser facility "Progress-P" and results of experimental investigation of beam propagation through this systems. Near diffraction-limited beam quality is obtained at output laser amplifier chain by use the low-thermal phosphate Nd:glass and high quality optical elements. Output 180 mm beam is compressed using two holographic gratings with dimensions 420x210 mm and focused on targets by means on-axis parabolic mirror with focal length of 200 mm. About 50% energy is obtained in 8 μm focal spot in target chamber for low power beam. Investigations of the focal spot characteristics for high-power beam are under way.

INTRODUCTION

The important problems of superstrong field formation by multiterawatt picosecond laser systems with chirped pulse amplification are to obtain near diffraction-limited beam quality on output laser amplifier, to achieve high damage threshold and effectiveness of diffraction grating compressor, to prevent the nonlinear breakup of compressed high-intensity pulse under propagation to vacuum target chamber and to produce high optical quality focusing optics with supersmall focal spot. In this work we present the parameters of compressor and focusing systems of 30-TW Nd:glass laser facility "Progress-P" (1) with chirped pulse amplification and results of experimental investigation of low power beam propagation through this systems.

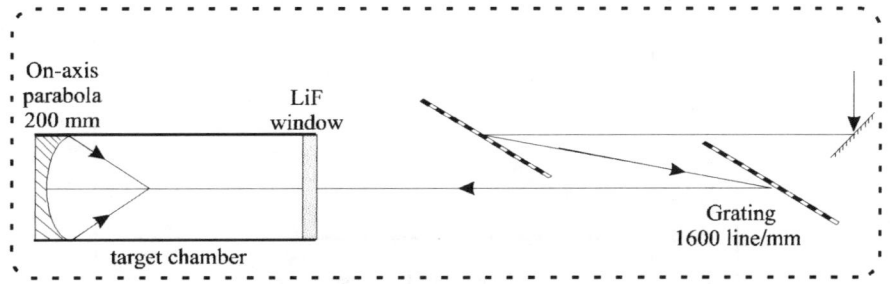

FIGURE 1. Scheme of compression and focusing systems.

SCHEME OF COMPRESSION AND FOCUSING SYSTEMS

Scheme of compression and focusing systems is shown in Figure 1. The pulse after propagation through rod amplifier chain with output 85-mm rod and expansion to 180 mm by output vacuum spatial filter is compressed in a single-pass air compressor on two gold-coated holographic diffraction gratings with dimensions of 420 mm×210 mm and 1600 lines/mm. The gratings have damage threshold of 260 mJ/cm^2 for 1-ps pulses and high optical quality as shown from interferogram of grating substrate on Figure 2. The grating separation is equal 2.5 m and compressor efficiency is over 80 %. To prevent the nonlinear breakup of compressed high-intensity pulse under propagation to vacuum target chamber we minimized path of compressed pulse in air and directed the beam in chamber through high optical quality 2-cm-thick LiF-window with low nonlinear refractive index coefficient $n_2 = 0.35 \cdot 10^{-13}$ esu. For 30-TW laser beam the estimated B-integral accumulated in air and window after compression is ~3.

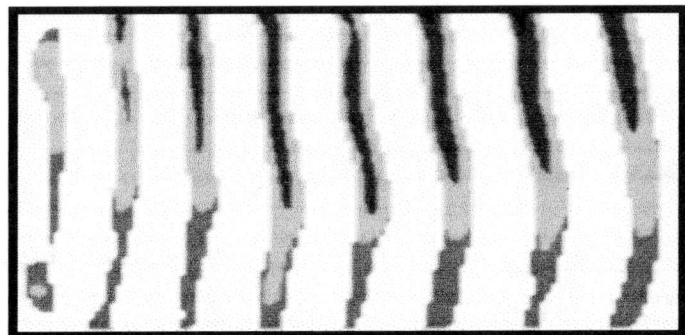

FIGURE 2. Interferogram at wavelength 0,63 μm of difraction grating substrate with 420mm×210mm dimensions.

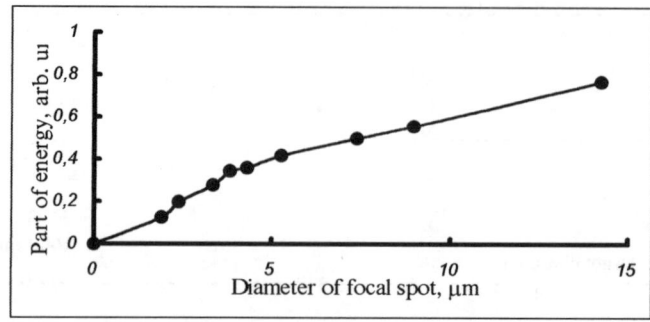

FIGURE 3. Measured energy distribution of unaberrated 200 mm diameter laser beam (λ=1053 nm) in focal plane of on-axis parabola with focal length 200 mm.

The beam is focused to the target by means of the axis mirror parabola with focal length of 200 mm. We measured focusing characteristics of parabola in itself for unaberrated beam. As shown in Figure 3 the parabola focuses ~50% of the energy of 200-mm unaberrated beam of cw Nd:YLF laser in the 8 μm diameter spot (it exceeds the diffraction limited spot by 3 times).

LASER BEAM PROPAGATION EXPERIMENTS

Laser pulse with energy 20-45 J and duration of 300 ps was obtained at output of amplifier chain. This pulse was compressed to 1.5 ps with spectral width of 1.5 nm. We measured far field intensity distribution on the input of compressor using optical system with equivalent focal length of 200 m. Target chamber parabola microspot intensity distribution was recorded using optical system with magnification of 260 and film or CCD-camera as a detectors.

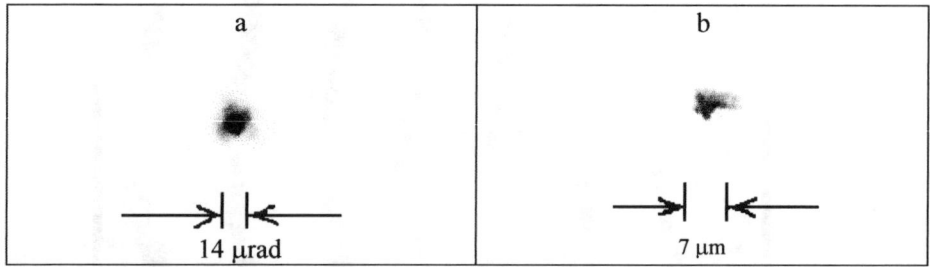

FIGURE 4. Photo of chain amplifier output beam in far field (a) and laser microspot in focal plane of parabola in target chamber (b).

At present, we investigated this beam spatial characteristics only for low energy output laser beam (less than 100 mJ). Near diffraction-limited beam divergence

(14 μrad) is obtained for this regime at output laser amplifier chain by use the low-thermal phosphate Nd:glasses GLS22 and KGSS0180/18 and high quality optical elements (Fig. 4, a). As shown at Figure 4, b, the significant fraction of beam energy is focused in target chamber to 2-times diffraction-limited ~ 5 μm spot.

CONCLUSION

In conclusion, we developed the compression and focusing systems of laser facility capable of producing up to 30 TW in 1.5 ps pulses which can be focused significant fraction of energy (50%) in ~3-times diffraction-limited spot in absence of nonlinear aberrations in beam. Investigations of the focal spot characteristics for high-power beam are under way.

This work was supported by International Scientific and Technology Center under project #107.

REFERENCES

1. Borodin, V. G. et al. "The "Progress-P" 30 TW picosecond Nd:glass facility", presented at the Thirteenth International Conference on Laser Interactions and Related Plasma Phenomena, Monterey, USA, April 13-18, 1997.

High Intensity Ultraviolet Laser and Next Generation Sources

Fiorenzo G. Omenetto*,⁑, Keith Boyer*, , James W. Longworth*,
Tom Nelson*, W. Andreas Schroeder* and Charles K. Rhodes*,⁂

*Laboratory for Atomic, Molecular and Radiation Physics, rm.2136, University of Illinois at Chicago, 845 W. Taylor Street, Chicago, Illinois 60607-7059.
⁑Università di Pavia, Dipartimento di Elettronica, Via Ferrata 1, 27100 PAVIA, Italy and INFM, sezione di Pavia.
⁂TARA, University of Tsukuba, Tennodai 1-1-1, Tsukuba, Ibaraki Japan

Abstract. Performance of a Ti:Sapphire/KrF* laser system leading to the generation of what could prove to be the first multi-terawatt pulses reported in this spectral region will be presented. An attractive alternative to hybrid excimer lasers is offered by direct UV generation from laser crystals that can support generation and amplification of ultrashort pulses. The design of a high-intensity ultraviolet laser system based on Cerium-doped colquiirites is examined.

INTRODUCTION

The advances in ultrashort pulse generation and the refinement of chirped pulse amplification (CPA) techniques have been the driving force in the development of laser systems with peak powers at and beyond the Terawatt (10^{12} W) level (1-7). Most high power systems developed to date, work in a spectral region between 0.8 and 1.1 microns, and Terawatt and Petawatt (10^{15} W) class lasers have been demonstrated in Ti:Sapphire, Nd:glass and Cr:LISAF based systems. These systems are moving towards shorter pulse durations, now commonly of the order of tens of femtoseconds, and design goals are heading towards high repetition rate (kHz) (8) multiterawatt systems. The progress in this field has certainly been inspired by the improvements in solid-state laser materials for ultrashort pulse generation. An analogous statement does not generally hold, however, for the ultraviolet wavelengths.

Efforts directed towards the development of ultrahigh peak power systems in this spectral region can be justified by the expectation of observing a distinct set of physical phenomena caused by the different properties of the stimulating radiation, among which we mention, *inter alia* (a) the greater energy per photon (b) the tighter focus obtainable and (c) the shorter period of the stimulating electric field. A means to obtain TW-class lasers in the ultraviolet is offered by the combination of excimer lasers and ultrafast oscillators. Efforts in this direction have been undertaken by a number of groups (9,10). In the present paper, the laser system operating at the University of Illinois at Chicago will be outlined, followed by an overview of promising new solid-state laser materials directly emitting in the UV, and the first tests performed on them.

The Laser System

The laser currently operational is a hybrid $TiAl_2O_3/KrF^*$ system. The initial pulses are obtained with a Ar-ion pumped KLM Ti:Sapphire oscillator optimized for stable operation at $\lambda=745$ nm with a single-plate birefringent filter. The average mode-locked power is 350 mW ($P_{pump}=6.7$ W) for a pulse duration of $\tau = 85$ fs. Amplification in the red is performed through a conventional CPA technique. Gold-replica gratings (groove density of 1200 lines/mm) are used in a single grating stretcher/single grating compressor combination (stretch factor ~1000). The previously employed dye preamplifier/triple pass Ti:Sapphire combination (11) has been substituted by a home built regenerative amplifier (12) also illustrated in Fig. 1. Pumping is achieved by means of a Continuum Q-switched Nd:YAG, furnishing 25 mJ of energy to the amplifier crystal. The combination of the prism pair and curved end mirror allows continuous spectral tunability of the amplifier cavity which for our purposes is tuned for operation at 745 nm. The SF10 prisms also provide approximate compensation for the dispersion of the other optical elements in the cavity. Typical energy outputs are of 1.3 mJ (with fluctuations of ±5% caused primarily by instabilities in the pump energy). This gives us, as expected, considerable improvement over the dye-cell arrangement in terms of energy stability, beam profile and gain. The repetition rate of the amplified pulses is kept to 2 Hz for convenience, given the limitations on the repetition rate of the final excimer amplifier stage. Frequency conversion takes place at this point, after recompression of the pulses from the regenerative amplifier in two KDP crystals of thickness 3 mm and 1 mm, respectively, for doubling and mixing of the 372 nm radiation with the residual 745 nm fundamental. Seeding energies of at least 10 µJ from the tripler are desirable, to guarantee a good contrast ratio between the signal and the amplified spontaneous emission (ASE) in the KrF^* amplifier. Average output energies from the tripler are of 18 µJ at 248 nm (for a 0.7 mJ 745 nm input), which are sent to seed the excimer preamplifier. To ensure good beam quality, the beam is spatially filtered in vacuum after the tripler by focusing it into a 35 µm pinhole which reduces the seed energy to 12 µJ. The first UV amplification stage is performed in a KrF^* excimer (Lambda Physik 201 MSC EMG) arranged in a double-pass off-axis geometry (12). The device is operated with a 22 kV discharge voltage and a pressure of 2000 mbar (100 mbar F_2, 120 mbar Kr, 1800 mbar Ne). The off-axis angles in the two paths are designed to achieve optimum amplification over the whole beam dimension. The output energies are measured to be an average of 1.5 mJ. The duration of the 248 nm pulse, after the double-pass excimer, obtained through a two-photon fluorescence (TPF) measurement gives a value of 160 fs (13). The beam is recollimated after amplification and the contrast ratio (signal to ASE) at this point is measured to be in excess of 10^3. The beam is then directed through a telescope beam expander before entering the final amplification stage which consists in a large aperture (10 cm) excimer amplifier (14). This device is operated at relatively low pressure and low gain in order to reduce wave-front distorsions and ASE. The pulses, produced at a repetition rate of 0.4 Hz exhibit a 10-shot average energy above 0.25 J, with peak recorded shots of 0.75 J. Characterization and optimization of the spectral and temporal features of the final pulse are presently underway. From the previous performance of the system, there is reason to believe that there will not be considerable broadening of the pulse through the final amplification stage furnishing evidence of a laser system that can reach the 2-3 Terawatt power level in the UV region of the spectrum (248 nm). Preliminary observations have shown improved experimental conditions, with a more homogeneous beam profile and near diffraction limited focal spot (measured to be < 3 µm in diameter in the target chamber) indicating improved efficiency in the delivery of the stimulating radiation for high-intensity physics experiments.

Solid-State Ultraviolet Lasers

Rare-earth doped colquiirites, $LiSrAlF_6$ (LiSAF), $LiCaAlF_6$ (LiCAF), and $LiSrGaAlF_6$ (LiSGAF), are solid-state materials of particular interest for UV generation. Ultrahigh peak-power laser systems operating in the IR adopting a colquiirite crystal host (with Chromium doping) have been demonstrated (5) and the possibility of extending this technology to the ultraviolet is justified by the demonstration of laser action in the UV using Cerium doped LiCAF (15,16) and Cerium doped LiSAF. For the sake of this discussion, we will examine the latter, but the considerations hold true, in general, for the other crystal host as well.

Ce:LiSAF has an emission bandwidth located between 285 and 295nm (17), hence theoretically capable of supporting pulses as short as 10 fs. A number of technical issues, inherent to operation at shorter wavelengths, and to the performance of the host and other materials (especially in the amplification stages) under high ultraviolet fluences, remain to be verified experimentally. The advantages, however, would be considerable and would entail, among other things, reduced size of the laser system, phase control of the generated pulses, and the possibility of supporting shorter pulse duration in comparison to the hybrid system described above.

The design of the system is based upon the application of Kerr Lens Modelocking (KLM) to a Ce:LiSAF crystal. This well established technique requires *(i)* a gain bandwidth sufficient to support a short optical pulse, *(ii)* no significant non-linear absorption (i.e. two-photon absorption) at the oscillating wavelength and *(iii)* a suitable non-linear refractive index n_2. By examining the properties of the various components in the melt, an estimate of the two-photon absorption edge for LiSAF can be inferred: LiF, SrF_2, and AlF_3 are materials which all have two-photon absorption edges falling in the region from 105 to 130 nm. Assuming the influence of the dopant to be negligible, the two-photon absorption edge, is conservatively estimated to be at about 125 nm, or 9.9 eV, indicating that there will be no significant two-photon absorption. Furthermore, from this value for the band gap, we can estimate the non-linear index of refraction (n_2) of LiSAF at 290 nm using the theoretical method developed by M. Sheik-Bahae et al. (18). A value $n_2 \sim 1.4 \times 10^{-20}$ m²/W (positive) is obtained (comparable to n_2 for Ti:Sapphire at 800 nm, i.e. $\sim 3 \times 10^{-20}$ m²/W (19)).

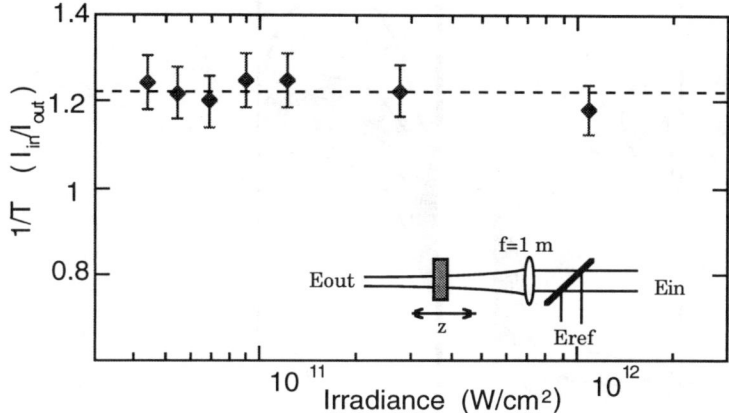

FIGURE 1. Inverse transmission data for a 1 mm Ce:LiSAF sample. Each data point is the average of 100 shots to account for energy fluctuations in the beam. The dashed line represents the best fit to the data (slope = 0)

In order to confirm these assumptions, a series of tests was performed on a 1 mm-thick sample of Ce:LiSAF (1% CeF$_3$, Lightning Optical Co.) to verify the non-linear absorption properties of the material. A conventional transmission measurement was performed by using 248 nm, 160 fs pulses, as illustrated in fig. 1. The best fit to the inverse transmission data yields β=0, where β is the two photon absorption coefficient, as described by Taylor et al. (20), in support of the predicted value for the material's band-gap. Furthermore, inspecting the sample after irradiation, no damage is detectable.

The design of the cavity can be based on the common geometries employed in the infra-red (i.e. Ti:sapphire oscillators), with one of the main points being the appropriate choice of the suitable pump laser. A first order solution could be the use of the fourth harmonic of a CW mode locked Nd:YLF (263nm), which is a close match to the absorption profile of the active Ce^{3+} ion and provides, with proper cavity length matching, gain modulation that can assist the KLM mechanism (synchronous pumping). On the other hand, the progress in diode pumped solid-state IR sources, such as the ones based on Nd-doped YAG, and YVO$_4$, offer a number of alternatives which can deliver, after frequency conversion, more pump power in aid of the non-linear modelocking process as well. By comparison of the parameters estimated for Ce:LiSAF with those of a typical hard-aperture KLM Ti:Sapphire laser, we note that the critical power for self-focusing, P_c, is significantly lower (P_c prop. λ^2/n_2) thus indicating the possibility of operation at lower pump power levels. The projected energy output, assuming a 100 fs pulse duration (a conservative estimate) yields an average power output of ~30mW for KLM operation with ~ 0.5nJ/pulse (21,22). Modelling of the cavity, using an ABCD matrix code similar to the one first employed by Magni et al. (23) has been performed and is shown in fig.2. The simulation shows, indeed, favorable regions for KLM operation.

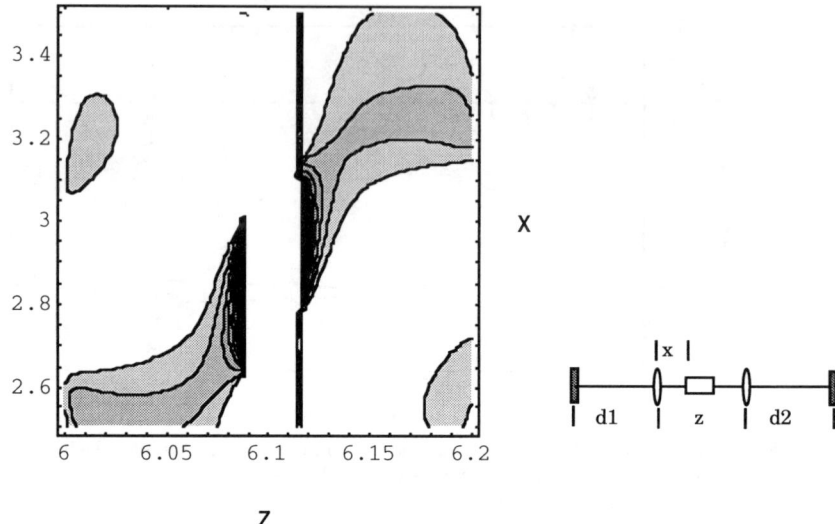

FIGURE 2. Plot of the Kerr-lens sensitivity parameter $\delta = Pc/P (w_K/w_{CW} - 1)$, for the case $f_1=f_2=3$ cm, $d_1=80$ cm, $d_2=100$ cm. Darker areas represent increasing negative values of δ, indicating more discrimination between the CW and Kerr spot sizes, thus favorable regions for modelocking.

Once these pulses have been generated, amplification would be possible by the application of standard Chirped Pulse Amplification (CPA) techniques. It is important to note that all the optics and gratings necessary to treat ultraviolet radiation are commercially available. Multipass amplifiers are an option for these (and possibly further) stages. Regenerative amplification is an attractive means to treat the seed pulses coming from the oscillator. Once again, important considerations have to be made in the design of the device. For instance, the short upper state lifetime (28ns) of Ce:LiSAF implies a short cavity to allow a significant number of round trips during that lifetime (a cavity length of 0.5 meters, for instance, and a single pass gain of 2 would prove sufficient to yield several microjoules of energy per pulse). Also, a relatively long pulse would be required to pump this amplifier. Among many options, a frequency-quadrupled Q-switched Nd:YAG laser, for instance, would provide 266 nm radiation in a several-nanosecond pulse (22). The only serious limitations would appear to lie in the material properties of the Ce:LiSAF gain medium itself, i.e. the saturation fluence of the material and the thermal effects. The gain cross sections for Ce:LiSAF (17) suggest a maximum saturation fluence of ~220mJ/cm^2. This, along with possible thermal limits, establishes the requirement of a large beam aperture in this last stage. Recently, Q-switched operation of Ce:LiCAF has been demonstrated to yield 14 mJ pulses at 290 nm, indicating the possibility of pushing performances of these materials towards TW-power levels, in conjunction with appropriate pulse compression techniques.

It should be pointed out that the output of every stage of this system is readily applicable to numerous areas of scientific activity, aside from the specific use of the system as a whole for high-intensity experiments, among which the generation of X-rays from atomic rare gas clusters (24). The pulses generated by the oscillator alone or with the first amplifier stage, would open the door to coherent control experiments with tuneable UV pulses. Fluorescence spectroscopy of proteins and nucleic acids, applications to photocatalysis related research and manufacture of pharmaceuticals are well suited to ultrafast optical pulses in the ultraviolet. The system as a whole would also provide a compact, high powered UV source for use in many other applications such as specialized forms of lithography or microfabrication. It seems plausible to expect the same trend that has occurred in the recent years for IR sources, to extend to the ultraviolet region of the spectrum. The progress on these materials is consistent (25), opening the path to the availability of new versatile sources at shorter wavelengths and to new powerful tabletop systems.

ACKNOWLEDGEMENTS

The authors would like to thank G. Quarels (Lightning Optical Co.) for providing the Ce:LiSAF sample. Support for this research was provided under contracts with SDI/NRL (N00014-93-K-2004), ARO (DAAH04-94-G-0089) and the University of California/Lawrence Livermore National Laboratory (B328353).

REFERENCES

1. Ditmire, T. and Perry M., Optics Letters, **18** (6), 426 (1993)
2. Barty, C. P. J., Guo, T., LeBlanc, C., Raksi, F., Rose-Petruck, C., Squier, J., Wilson, K. R. , Yakolev, V. V. and Yamakawa, K., Optics Letters, **21** (9), 668-670 (1996)
3. Zhou, J., Huang, C. P., Shi, C., Murnane, M. M. and Kapteyn, H. C., Optics Letters, **19** (2), 126-128(1994)
4. Sullivan, A., Bonlie J., . Price D. F and White, W. E., Optics Letters, **21** (8), 603 (1996)
5. Beaud, P., Richardson, M., Miesak, E. J. andChai, B. H. T., Optics Letters, **18** (18), 1550-1552 (1993)
6. Antonetti, A., Chambaret, J. P., Cheriaux, G, Curley, P. F., Darpentigny, G., LeBlanc, C. and Salin, F. : *Ultrafast Phenomena 8*, , (OSA, Washington D.C. 1996) p.160
7. Y. Nabekawa, K. Kondo, N. Sarukura, K. Sajiki and S. Watanabe, Optics Letters, **18** (22), 1922-1924 (1993)
8. Backus, S., Durfee, C., Mourou, G., Kapteyn, H., and Murnane, M., Optics Letters, **22** (16), 1256-1258 (1997)
9. Ross, I. N., Damerell, A. R., Divall, E. J., Evans, J., Hirst, G. J., Hooker,C. J., Houliston, J. R., Key, M. H.,Lister, J. M. D., Osvay, K., Shaw,. M. J., Optics Comm., **109**, 288-297 (1994)
10. Szatmári, S., Schäfer, F. P., Müller-Horsche, E. and Mückenheim, W, Optics Comm., 63, 305 (1987)
11. Bouma, B., Luk,T. S., Boyer, K.and Rhodes, C. K., JOSA B, **10** (7), 1180-1186 (1993)
12. Nelson,T., Omenetto, F. G., Longworth, J. W., Schroeder, W. A., and Rhodes, C. K., Applied Optics, in press

13. Omenetto, F. G., Boyer, K., Longworth, J. W., McPherson, A., Schroeder, W. A., Rhodes, C. K.," Applied Optics,**36** (15), 3421-3424, 1997
14. Luk, T. S., McPherson, A., Gibson, G., Boyer, K. and Rhodes, C. K., Optics Letters, **14**, 1113-1115 (1989)
15. Dubinskii, M. A. et al., *J. Mod. Opt*, **40**, 1, (1993)
16. Dubinskii, M. A. et al., Optics Letters, **22** (18), 994-996, (1997)
17. Marshall, C. D., Speth, J. A., Payne, S. A., Krupke, W. F., Quarles, G. J., Castillo, V., and Chai, B. H. T., J. Opt. Soc. Am. B **11**, 2054 - 2065, (1994)
18. Sheik-Bahae, M., Hutchings, D. C., Hagan, David J., and Van Stryland, E., IEEE J. Quant.Elect. **QE-27**, 1296 - 1309 (1991)
19. Salin, F., Squier, J., and Piché, Opt. Lett. **16**, 1674 1676 (1991)
20. Taylor, A. J., Gibson, R. B., and Roberts, J. P., Optics Letters **13** (10), 814-816, 1988
21. Omenetto, F. G., Nelson,T., Longworth, J. W., Schroeder, W. A., and Rhodes, C. K., *"Design and Analysis of a New Solid-state Deep UV Laser based on Ce:LiSAF,"* in Laser Spectroscopy, Ed. M. Inguscio, M. Allegrini, A.. Sasso, World Scientific, Singapore, p. 360-361, 1995
22. Nelson,T., Omenetto, F. G., Longworth, J. W., Schroeder, W. A., and Rhodes, C. K., , *"Design and Analysis of an all Solid-State UV Laser based on $Ce^{3+}LiSrAlF_6$,"* in "High Power Lasers", NATO-ASI series, Ed. R. Kossowsky, M. Jelinek, R. F. Walter, Kluwer Ac., London, p. 177-,184 1995.
23. Magni,V., Cerullo, V. and De Silvestri, S., Optics Comm., **101**, 195 (1993)
24. McPherson, A., Thompson, B. D., Borisov, A. B., Boyer, K., and Rhodes, C. K. (1994) Multiphoton-induced X-ray emission at 4-5 keV fromXe atoms with multiple core vacancies., Nature **370**, 631 - 634.
25. Dubinskii, M. A., "Recent Developments in Ce-activated Tunable Solid-State UV Lasers," in *Conference on Lasers and Electro-Optics, Vol.10 of Technical Digest Series* (Optical Society of America, Washington D.C., 1997), CThN1 p. 404

Development of a High-Peak and High-Average Power Ti:Sapphire Laser System

K. Yamakawa, M. Aoyama, S. Matsuoka, Y. Akahane, H. Takuma,
D. Fittinghoff* and C. P. J. Barty*

Advanced Photon Research Center, KANSAI Research Establishment,
Japan Atomic Energy Research Institute, Tokai, Ibaraki 319-11, Japan
*University of California, San Diego, Urey Hall, La Jolla, California 92093-0339, USA

Abstract. We have developed a two-stage Ti:sapphire amplifier system, which is capable of producing 16 fs, 10 TW pulses at a 10 Hz repetition rate. We also describe an extension of this system to a peak power of 100 TW and an average power of 20 W level.

1. INTRODUCTION

The application of the chirped-pulse amplification (CPA) technique (1) to broadband solid-state materials makes possible the development of table-top terawatt femtosecond lasers. Recently this technique was extended to produce multiterawatt pulses of less than 30 fs in duration (2, 3). Such ultrashort pulses are useful for a variety of high-field applications such as the generation of ultrafast x-ray radiation and ultrahigh-order harmonic generation. We describe a compact Ti:sapphire amplifier system which produces 16 fs 10 TW laser pulses at a 10-Hz repetition rate. To our knowledge, these results represent the shortest-duration terawatt level pulses yet produced and the highest amplifier efficiency used in any Ti:sapphire CPA system. This result demonstrates that laser pumping of Ti:sapphire can be a very efficient means of producing both high peak and high average power pulses. We also report the present status of our Ti:sapphire

CPA laser system which is designed to produce a peak power of 100 TW for a pulse duration of 20 fs and an average power of 20 W at a 10 Hz repetition rate.

2. DESIGN AND PERFORMANCE OF THE LASER SYSTEM

A schematic of the laser system is shown in Fig. 1. The system consists of a Ti:sapphire oscillator, a pulse expander, a regenerative amplifier, a 4-pass amplifier, and a pulse compressor. Seed pulses were generated with an all-solid-state mirror-dispersion-controlled Ti:sapphire oscillator (4) capable of producing ~ 10 fs pulses with an average power of 230 mW. A typical interferometric autocorrelation of the pulses is shown in Fig. 2. Before amplification, the pulses from the oscillator were stretched by a factor of 100,000 in an all-reflective, cylindrical-mirror-based pulse expander (5). The bandbass of this expander is roughly 140 nm. The FWHM duration of the output of the expander is ~ 1.1 ns after 4 passing the expander. Such large stretching ratios (~ 110,000) enable the amplifiers to be operated above the saturation fluence of Ti:sapphire (~ 1 J/cm^2) without intensity-dependent optical damage to optical components in order to extract the energy efficiently.

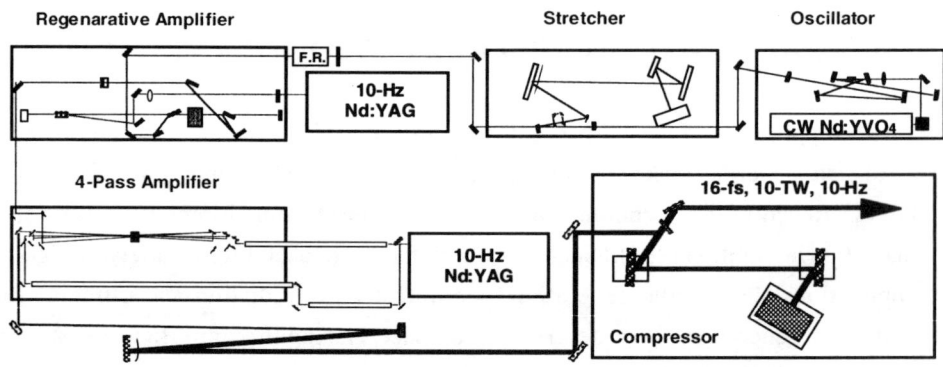

FIGURE 1. A schematic of a 10 TW, 16 fs, 10 Hz Ti:sapphire laser system.

The stretched pulses are then first amplified in a regenerative amplifier (6). The regenerative amplifier is 1.8 m long and uses two cavity mirrors.

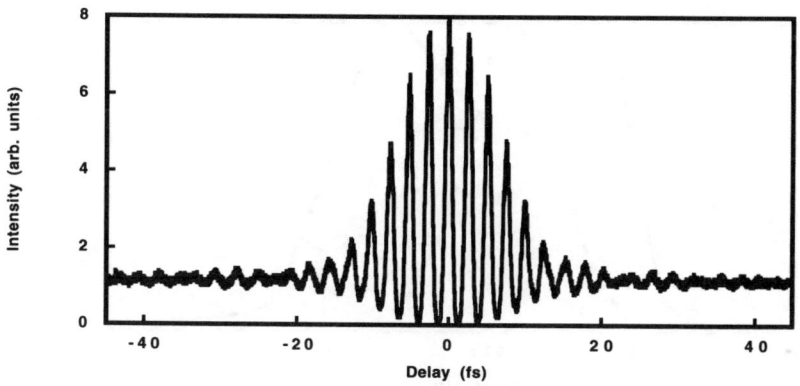

FIGURE 2. Measured interferometric autocorrelation of the mode-locked pulses.

The Ti:sapphire crystal is end pumped with a 50 mJ frequency-doubled Q-switched Nd:YAG laser which produces 7-ns pulses at a 10 Hz repetition rate. This regenerative amplifier provides a net gain of approximately 10^7 for twelve cavity round trips, which leads to gain narrowing during amplification. We have investigated angle-tuned thin etalons to broaden the amplified spectrum in the regenerative amplifier (7). By selectively amplifying the wings of the spectrum with two 3 micron thick etalons in the resonator, the spectrum of the amplified pulse was broadened to 82 nm FWHM as shown in Fig. 3.

Output from the regenerative amplifier was up collimated to an ~ 6 mm diameter with a Galilean telescope and then introduced into the four-pass amplifier. This amplifier utilizes a 20 mm diameter 15 mm long Ti:sapphire crystal (0.15% doping) that is pumped with a 690 mJ frequency-doubled Q-switched Nd:YAG laser at a 10 Hz repetition rate. For high efficiency the pulse fluence on the last pass in the amplifier was designed to be ~ 1.6 J/cm^2. Under this condition, this amplifier has demonstrated > 90% of the theoretical maximum conversion efficiency of 790-nm radiation to 532-nm pump light. This result well agrees with a Frantz-Nodvik simulation. At this high fluence, however, an amplified pulse spectrum is red-shifted due to saturated amplification and is resultant a reshaped spectrum FWHM of ~ 60 nm. One of the thin etalons in the regenerative amplifier can be tuned to pre-compensate for spectral shifting which occurs during 4-pass amplification. As a result, the amplified spectrum is increased to 72 nm FWHM.

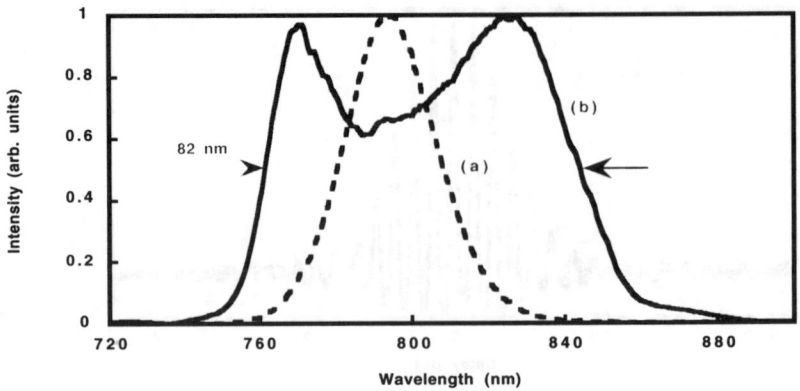

FIGURE 3. The spectra for amplified pulses (a) with no intracavity filter and (b) with the pulse shaper consisting of two 3 µm etalons.

The transmission of the compressor, including the gold-coated turning optics, was ~ 50%, yielding a compressed output pulse energy of ~ 160 mJ. A typical single-shot autocorrelation trace of the compressed pulse is shown in Fig. 4. The FWHM of the measured autocorrelation differs from that of the transform limited pulse shape by less than 1.4 fs. The duration of the transform limit, as calculated from the measured, amplified spectrum after the compressor is 15 fs. The high degree of agreement suggests that the compressed pulses are very nearly transform limited.

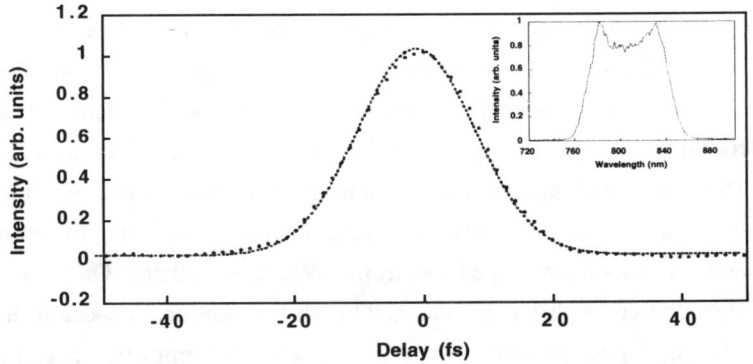

FIGURE 4. Measured (doted line) and calculated (broken line) autocorrelation. Inset: measured amplified spectrum after compression.

The spatial beam quality was determined by focusing the attenuated output with a 2 m focal-length spherical mirror and measuring the spot size at the focus with a CCD camera. At a 6 TW power level, the near field beam profile closely remains Gaussian and the spatial quality of the beam is near diffraction limited in both s- and p-planes. Under the full power conditions (~ 10 TW), the near field profile of the beam closely resembles that of the pump beam (near flat-top) and the beam quality of the p-plane is decreased to about 2.5 times that of the flat-topped beam. The decreased beam quality in this plane may be due to different beam divergence in both s- and p-planes of the Nd:YAG pump laser beam.

3. TOWARDS A 100 TW

In the present system, the output of the two-stage Ti:sapphire amplifiers is introduced into a final double-pass amplifier which is designed to produce > 3 J of radiation. A layout of a 100 TW, 20 fs, 10 Hz Ti:sapphire laser system is shown in Fig. 5. The double-pass amplifier is pumped by a frequency-doubled output from a custom built Nd:YAG laser which is capable of producing ~ 7 J of 532 nm radiation at 10 Hz. The final vacuum compressor consists of four 1200 grooves gold coated gratings and should have overall efficiency greater than 60%. Thus the energy and duration of the compressed pulse are expected to be approximately 2 J and 20 fs, respectively, approaching a peak power of 100 TW.

4. CONCLUSION

In conclusion, we have demonstrated the generation of 10 TW near transform-limited 16 fs duration pulses at a 10 Hz repetition rate from a compact two-stage Ti:sapphire amplifier system. The laser system is currently being upgraded to produce laser pulses of 20-fs duration with approaching peak and average powers of 100 TW and 20 W level.

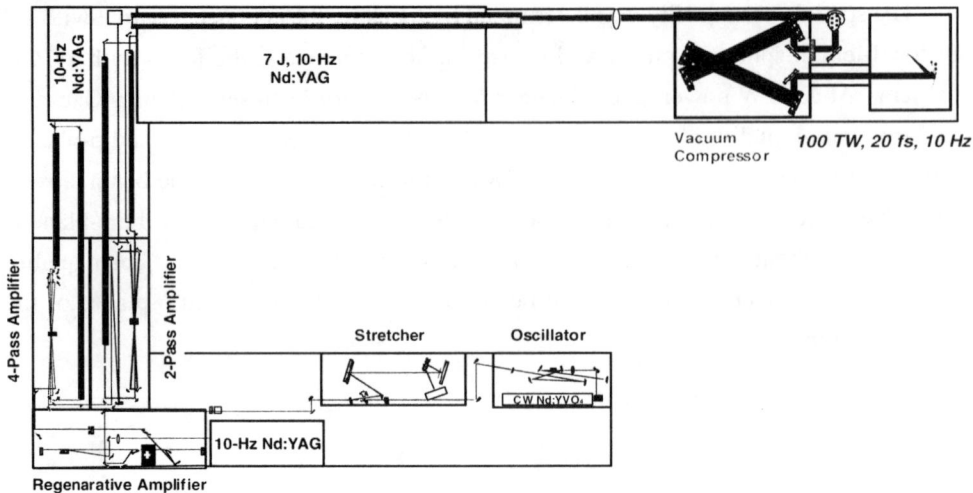

FIGURE 5. An optical layout of a 100 TW, 20 fs, 10 Hz Ti:sapphire laser system.

ACKNOWLEDGMENT

The authors acknowledge T. Kase for his technical assistance and T. Matoba for his stimulating support.

REFERENCES

1. Strickland, D. and Mourou, G., Opt. Comm. **56**, 219-221 (1985).
2. Barty, C. P. J., Guo, T., Le Blanc, C., Raksi, F., Rose-Petruck, C., Squier, J., Wilson, K. R., Yakovlev, V. V., and Yamakawa, K., Opt. Lett. **21**, 668-670 (1996).
3. Zhou, J., Huang, C.-P., Murnane, M. M., and Kapteyn, H. C., Opt. Lett. **20**, 64-66 (1995).
4. Stingl, A., Spielmann, Ch., Krausz, F., and Szipocs, R., Opt. Lett. **19**, 204-206 (1994).
5. Lemoff, B. E. and Barty, C. P. J., Opt. Lett. **18**, 1651-1653 (1993).
6. Yamakawa, K., Chiu, P. H., Magana, A., and Kmetec, J. D., IEEE J. Quantum Electron. **QE-30**, 2698-2706 (1994).
7. Barty, C. P. J., Korn, G., Raksi, F., Rose-Petruck, C., Squier, J., Tien, A.-C., Wilson, K. R., Yakovlev, V. V., and Yamakawa, K., Opt. Lett. **21**, 219-221 (1996).

Design Considerations for the Stretcher of a 35-fs Chirped Pulse Amplification Laser System

M.B. Mason and M.H.R. Hutchinson

Imperial College of Science, Technology and Medicine, Blackett Laboratory, Prince Consort Road, London SW7 2BZ, United Kingdom

Abstract. We have developed a code which will model different configurations and designs of stretcher and compressor systems. The code allows us to compare the performance of systems in terms of pulse duration, pulse contrast and alignment sensitivity. The results of the code have enabled us to design a chirped pulse amplification system that will stretch and recompress 35 fs pulses.

INTRODUCTION

As the peak intensity available from chirped pulse amplification (CPA) lasers increases, so does the pulse contrast requirements for many high intensity experiments. For example, a focused intensity of 10^{18} Wcm^{-2} demands a contrast ratio of about eight orders of magnitude to prevent the preformation of a plasma on a solid target by a laser prepulse. In addition, the very large bandwidth associated with an ultra-short laser pulse (e.g. 30 nm for a 30 fs gaussian profile pulse) means a great deal of care has to be taken to limit high-order phase errors in the CPA process. These usually insignificant errors become important for pulse durations below 50 fs and can dramatically reduce the contrast as well as temporally broaden the compressed pulse.

Currently, we are upgrading our Ti:Sapphire system which delivers 150 fs pulses with an energy of up to 30 mJ (1). The upgrade involves replacing the short-pulse oscillator with an ultra-short pulse oscillator, purchased from Kapteyn-Murnane Labs (2), running at 35 fs as well as increasing the amount of amplification. The stretcher and compressor pair have to be redesigned to give an increased stretch of approximately 10000 times the input pulse duration and to accommodate the extra bandwidth associated with 35-fs pulses.

In the design of an ultra-short pulse CPA system, it is critical to be able to compare various stretcher configurations numerically, not only in terms of the final pulse duration and contrast ratio, but also with respect to the sensitivity to alignment and optical quality of components. We have developed a ray-tracing code which we use to model different stretcher and compressor configurations.

MODEL DESCRIPTION

The code works on three levels that operate with different degrees of accuracy. The first is a quick overview of the system in order to make an outline of a design. This is achieved by using approximate equations and geometrically traced rays, to describe the stretcher and compressor. The code provides us with system statistics such as the duration of the stretched pulse and the sizes of the optical components including, most importantly, the size of the diffraction gratings.

The second level code traces a small bundle of rays through the stretcher and compressor and obtains their phase contributions as a function of frequency. From this, up to the fifth order of dispersion can be accurately calculated, the second order being the amount of temporal chirp in the system. The phase in the system can be calculated from a simple formula (Eqn. 1) taken from Treacy's pulse compression paper (3).

$$\phi(\omega) = \frac{path \times \omega}{c} + R(\omega) \qquad (1)$$

In this equation, ω is the angular frequency of the ray being traced, *path* is the optical length of the ray through the system, c is the speed of light and $R(\omega)$ is a constant which takes into account the apparently non-intuitive phase imparted by the diffraction gratings. For example, for first order diffraction in a compressor, it is given by an addition of -2π for every groove on the grating from a position perpendicularly opposite the incident ray on the first grating. The Taylor expansion of equation 1 about the central frequency, ω_o, gives phase orders as a function of angular frequency:

$$\phi(\omega) = \phi_o + \beta'(\omega - \omega_o) + \tfrac{1}{2}\beta''(\omega - \omega_o) + \tfrac{1}{6}\beta'''(\omega - \omega_o) + \ldots \qquad (2)$$

All terms (ϕ_o, β', β'', etc.) can be calculated from the physical parameters of the stretcher and compressor. Of these, the terms ϕ_o and β' are relatively unimportant constants and represent the central frequency phase and time average time delay respectively. The third term, β'', is the most significant and represents the temporal chirp, which has a negative value for a compressor and a positive value for a stretcher. The higher order, non-linear terms give departures from an ideally matched stretcher-compressor system and become increasingly significant for shorter pulses. Using the information from the medium level code, the compressor can be matched so as to completely cancel the temporal chirp added by the stretcher and, as far as possible, the higher order terms as well.

The third level code traces several thousand rays through the system giving us information about the duration and contrast of the compressed pulse. The code defines a particular pulse shape (gaussian or sech squared) and Fourier transforms it to obtain the associated frequency components. The stretcher is then geometrically ray-traced

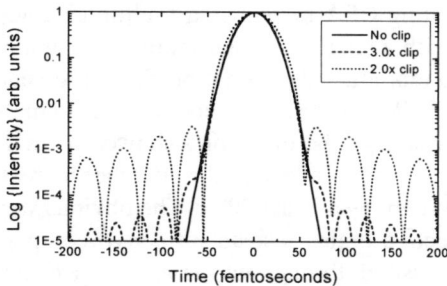

FIGURE 1. Output pulse profiles showing the effects of hard spectral clipping. The solid line is the original pulse profile, the dashed line shows clipping at three times the FWHM bandwidth of the pulse and the dotted line shows clipping at two times the FWHM bandwidth. The broadening of the pulse can clearly be seen for the two times clip even though this is plotted on a log scale.

within these frequencies and a phase function is calculated. This phase is added onto the phase component of the pulse that arises from the Fourier transform.

The modelling of the amplifier chain is basic, the only things that are included are spectral clipping and the material dispersion of the optical components within it. Most of the significant spectral clipping comes from the stretcher and compressor and is caused by the limited sizes of the components. Clipping in the amplifier chain comes from components such as polarisers and the gain medium, but this is 'soft' clipping as the frequency components are not abruptly cut as with 'hard' clipping in the stretcher and compressor.

The compressor is then geometrically ray-traced and the corresponding phase function added to the pulse. Finally the pulse is inverse Fourier transformed to give the temporal profile of the stretched and recompressed pulse. We can see from these profiles that soft clipping serves to temporally broaden the pulse as the bandwidth of the pulse is reduced, whereas hard clipping introduces 'wings' onto the output pulse as well as broadening it (see Fig. 1). This is due to the fact that high spectral frequency components such as those associated with sharp spectral features correspond to high frequency components in the output pulse profile.

With this final tool small adjustments can be made to the design of the system in order to achieve the shortest compressed pulse together with the highest possible contrast. As well as optimising the performance of the CPA system, the sizes of expensive components such as the diffraction gratings have to be kept to a minimum, and has to be technically feasible, robust and as easy to align as possible.

STRETCHER DESIGNS

In the literature there are many examples of stretcher design, ranging from the very first two lens, two grating type (4) to a simple reflective systems with only one grating (5) to a more complicated system comprising two reflective optics (6). The stretcher used in our existing CPA system is a traditional folded, unit magnification telescope configuration with a single lens and grating. A first check on the reliability of the CPA Model was to reproduce the output profiles of the current laser system with a refractive stretcher optic. We were able to model this extremely well.

One of the first results of the modelling showed us that because of chromatic and spherical aberration, the use of a lens in the stretcher increases the oscillator pulse duration, in sub 100-fs system, by at least 50%. The obvious way to remove chromatic aberration is to use reflective optics instead refractive components (a rule also best adhered to, as much as possible, throughout the amplifier chain). For this reason the systems that we considered were a simple folded stretcher arrangement with the lens replaced by a curved mirror (5) and a system based on the Öffner triplet containing two curved mirrors which act in a similar way to a reflective doublet (6).

MODELLING RESULTS

The first system to be modelled was the simpler of the two reflective systems. Using the first and second levels of the code, a system was designed which would stretch 35 fs pulses to 300 ps (Fig. 2a). The input angles into the stretcher and compressor were chosen to be as close as possible to Littrow configuration so as to give the highest possible diffraction efficiency. Fine adjustments were made to the design using the third level program; this involved changing the compressor grating separation by a small amount to completely cancel the second order phase from the stretcher. The initial results from this part of the optimisation is a plot of pulse duration and pulse contrast (defined as the ratio of the height of the second highest peak on the pulse profile to the height of the main peak) as a function of compressor grating separation (Fig. 3). From this plot a compromise can be established between

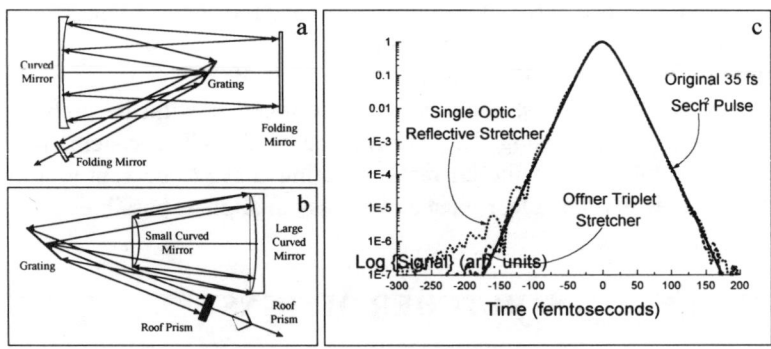

FIGURE 2. Ray diagrams for (a), the single reflective optic system and (b), the Öffner triplet based system. The output pulse profiles of the two systems are shown in part (c).

the shortest recompressed pulse duration and the best pulse contrast. The shape of the pulse duration curve gives an indication of the sensitivity of the compressor to alignment. The more shallow the curve the less sensitive to alignment it is, as the pulse duration does not change as rapidly with grating separation. The sharp structure in the contrast curve appears at first to be unphysical, but when compared with figure 4, which shows the pulse profile as a function of compressor grating separation, we can see that as the peak of a pre or post pulse becomes less sharp, i.e. changes from a second order curve to third order, the place where the contrast is measured 'slides' down to the next lowest peak giving a sharp drop in the apparent pulse contrast. Although there is no longer a pre or post pulse peak at that point, there is still a large structure at the base of the pulse which is important and would be able to pre-ionise a solid target. The output pulse profile for this stretcher for the optimum compressor grating is shown in figure 2c.

The second, more complicated, system was then modelled and provided additional verification of the model, as the pulse profiles produced matched the results contained within (6). The Öffner system is shown in figure 2b and the output pulse profile is shown in figure 2c, to compare with the output of the simple reflective system. This system performs very well and recompresses an initial 35 fs pulse to 35.5 fs (compared to 36.5 fs for the single optic system) with a contrast of 1×10^{-6} (compared to 5×10^{-5}). The plot shown in figure 4 for pulse duration and contrast as a function of compressor grating separation for the Öffner system looks very similar to the single optic system showing almost equal sensitivity to compressor grating alignment. From these results the two systems compare very well, with only small advantages in compressed pulse duration and contrast from the Öffner system. The simplicity of the single optic stretcher is attractive but we need more information to be able to choose between them.

The next step was to look at the tolerance of the stretcher and compressor to alignment. This was done by displacing important components and seeing how much the compressor had to be adjusted to re-optimise the pulse duration. First the folding

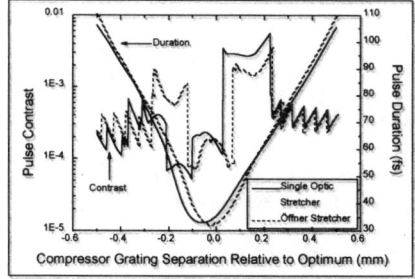

FIGURE 3. Pulse contrast and duration as a function of compressor grating separation for the single optic and Öffner stretchers

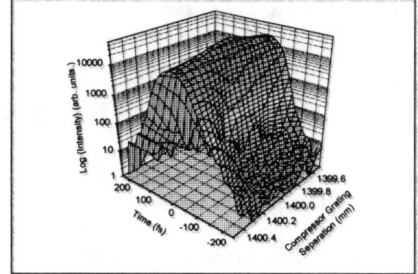

FIGURE 4. Pulse profile as a function of compressor grating separation for the single optic stretcher.

mirror in the stretcher was longitudinally displaced by ±0.1 mm. The Öffner system required a compressor grating movement of 0.5 mm to optimise the pulse duration to 37.9 fs whereas the single optic system needed only 0.1 mm of adjustment optimising the duration also to 37.9 fs. Then a mismatch between the input angle of the stretcher and compressor of ±0.1° was introduced. The two systems performed equally, requiring a compressor grating shift of 1.0 mm for the Öffner system and 0.9 mm for the single optic system, both optimising the pulse duration to 40.1 fs. This is obviously a very critical parameter to match. Finally the curved optics in the stretcher were laterally shifted by ±0.1 mm. In the Öffner system, there are two optics, the small optic required a compressor grating shift of 0.2 mm optimising to 37.5 fs and the large optic required a shift of 0.3 mm optimising to 36.8 fs. The single optic system required a compressor grating shift of 7.8 µm, in other words, this stretcher is very insensitive to the alignment of the optic.

Overall, the single optic stretcher performed better than the Öffner system being less sensitive to misalignments of the optics. More importantly the misalignment of the simpler system resulted in less of an increase in pulse duration than with misalignments of the Öffner system.

CONCLUSION

A number of different stretcher designs have been modelled, investigating their performance with respect to output pulse duration, pulse contrast and alignment sensitivity. We have chosen to use a simple single reflective optic, single diffraction grating configuration as this provides us with the best compromise of all the above factors.

ACKNOWLEDGMENTS

Financial support by the Engineering and Physical Sciences Research Council is gratefully acknowledged.

REFERENCES

1. Fraser,D.J. and Hutchinson,M.R.H., *Journal of Modern Optics* **43**, 1055-1062 (1996).
2. Asaki,M.T., Huang,C.P., Garvey,D., Zhou,J.P., Kapteyn,H.C., Murnane,M.M., *Optics Letters* **18**, 977-979 (1993)
3. Treacy,E.B., *IEEE Journal of Quantum Electronics* **5**, 454-458 (1969)
4. Martinez,O.E. *Journal of the Optical Society of America B* **3**, 929-934 (1986)
5. Zhou,J., Huang,C.-P., Murnane,M.M. and Kapteyn,H.C., *Optics Letters* **20**, 64-66 (1995)
6. Cheriaux,G., Rousseau,P., Salin,F., Chamberet,J.P., Walker,B. and Dimauro,L.F., *Optics Letters* **21**, 414-416 (1996)

Implementation of a CPA Line-focus Travelling-wave for Highly Efficient Saturated Lasing of Ne-like Ti and Ge

C Danson, PV Nickles[1], R Allott, A Behjat[2], J Collier, A Demir[2],
MP Kalachnikov[1], M Key, C Lewis[3], D Neely, DA Pepler,
GJ Pert[4], M Schnurer[1], W Sandner[1], VN Shlyaptsev[1], G Tallents[2],
J Warwick[3], E Wolfrum[5] and J Zhang[5].

Rutherford Appleton Laboratory, Chilton, Didcot, Oxon, OX11 0QX, UK.
Telephone: (0)1235 821900 Fax: (0)1235 445888 e-mail: c.danson@rl.ac.uk
[1]Max Born Institut, Berlin, Germany
[2]University of Essex, Colchester, CO4 3SQ, UK
[3]Queen's University of Belfast, Belfast, BT7 1NN, UK
[4]University of York, Heslington, York, YO1 5DD, UK
[5]Clarendon Laboratory, University of Oxford, Oxford, OX1 3PU, UK

Abstract. The development of X-ray lasers is marked by the progression toward shorter wavelengths using less drive energy. A recent X-ray laser experiment at the Rutherford Appleton Laboratory (RAL) demonstrated saturated X-ray laser operation in the Ne-like Ti and Ge X-ray laser schemes at 32.6 and 19.6 nm respectively, with a drive energy of only a few joules. In this paper we describe the laser development to generate the laser drive pulses and travelling wave line-focus target irradiation geometry required for this experiment.

INTRODUCTION

The X-ray laser experiment was carried out on the VULCAN high power laser system, delivering synchronised nanosecond and subpicosecond pulses in a multi-beam configuration. Short pulse (ps) generation at ultra-high intensities is achieved using the technique of chirped pulse amplification (CPA) (1). For this experiment both pulses were configured in line focus geometry and overlapped on target. The short pulse was synchronised to the falling edge of the nanosecond long pulse allowing optimised plasma conditions to be formed. The travelling wave was optimised by inserting a gold coated diffraction grating and associated mirrors, at the output of the rod amplifier stage of the VULCAN laser chain. The first order diffraction imposes a tilt on the laser wave front in one dimension due to the path length difference across the beam. The near-field is imaged through the system and the path difference is preserved through the focusing system and onto target generating a travelling line focus.

THE LINE FOCUS

The creation of both the short and long pulse line foci relies on the optical aberration that is introduced by spherical mirrors operated at an angle relative to the optic axis (2). A schematic of the target chamber optical configuration is shown in Figure 1. For the ultra-short CPA pulse the incoming laser beam is focused to a point (surrogate focus) using an off-axis parabolic (OAP) mirror. The spherical mirror then images this point source to the target plane. The angled nature of the spherical mirror introduces a large astigmatism into the beam producing two, one dimensional 'images' which are line foci. The second of the line foci which lies in the horizontal plane is the one that is used. The line focus always points toward the surrogate focus as can be seen from the geometry shown in Figure 1. The long pulse pre-ionising beam is delivered to target in a very similar way to the short pulse. The beam is brought to a surrogate focus using a 275 mm focal length lens and the line focus generated using a 315 mm focal length spherical mirror used off-axis.

The geometry leads to an inherent optical path difference between rays arriving on one side of the line focus with respect to the other. This optical path difference can be significant for the ultra-short pulses used. This means that the incident laser pulse will propagate from one side of the target to the other with a finite velocity that is normally several times the speed of light. This propagation velocity increases rapidly with reduced incidence angle onto the spherical mirror.

The target lengths are typically 10 mm which means that the transit time for an

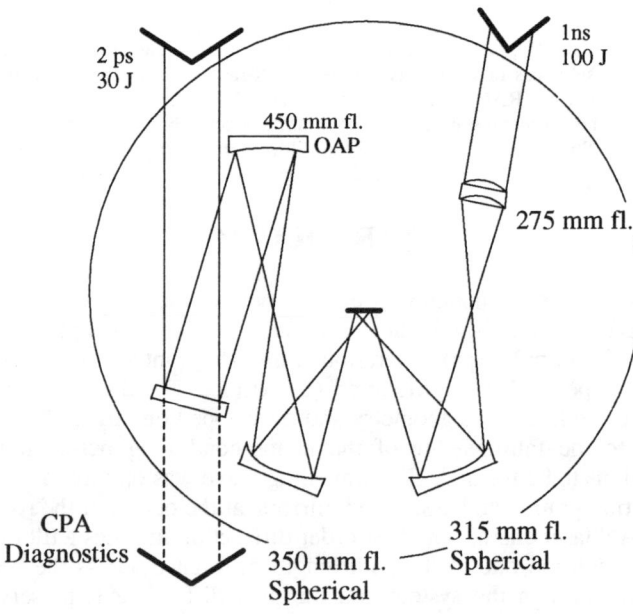

FIGURE 1. Optical layout of the target interaction chamber for the X-ray laser experiment.

X-ray laser photon from one end of the line focus to the other is typically tens of picoseconds. The X-ray laser upper state lifetime (gain) only lasts for a few picoseconds so if the plasma were to be formed at the same time everywhere along the line length then the gain at the end of the line will have dissipated by the time the X-ray photons arrive. The travelling wave effect can therefore significantly assist the X-ray laser process. One consequence is that the X-ray laser becomes directional with significantly higher output in the plasma propagation direction. The typical phase velocities encountered in this scheme are of the order 2-3 times the speed of light (c). Ideally, the phase velocity should be reduced to exactly c to ensure that that the X-ray photons always see the peak of the available gain throughout the whole length of the line.

The input beam dimensions to the target interaction chamber were 88 mm horizontally and 130 mm vertically. The off-axis parabola had a focal length of 450 mm producing a F5.1 surrogate cone. The spherical mirror, located 350 mm from the surrogate had a 350 mm focal length and was operated at an incidence angle of 15.5°. This gives a distance of 195 mm from the surrogate to line focus centre. This focusing system was modelled by the optical design program ZEEMAX (3) which gave a propagation time of 15.9 ps over a 12.0 mm length line focus which gives a phase velocity of 2.5 c.

OPTIMISATION OF THE TRAVELLING-WAVE

The phase velocity can be optimised by inserting a diffraction grating into the near-field of the incoming laser beam (4). The dispersive nature of a diffraction grating means that an incident beam will experience a lateral time shear. The insertion of a diffraction grating into the beam will add to the inherent optical path difference that the beam experiences at the line. A schematic of the grating configuration is shown in Figure 2. The correct choice of diffraction grating will then ensure that the plasma formation rate is equal to the speed of light.

The grating is inserted into the laser chain, at a point between the end of the rod amplifier system and the start of the disk amplifier system. This point was chosen because the beam flux is at a level which is below the threshold for laser damage of the grating and the beam aperture is <50 mm diameter allowing inexpensive commercial gratings to be used. The VULCAN laser system is image relayed using a series of spatial filters between amplifier stages. The gating plane is therefore approximately imaged into the target interaction area. This imaging, coupled with the overall system magnification means that the pulse lengthening due to the dispersion induced by the grating is minimised.

TABLE 1. Effect on the phase velocity of changing the angle of incidence onto the spherical mirror

Incidence Angle	Line focus length (mm)	Phase velocity
14.5°	10.24	0.95 c
15.5°	12.00	1.02 c
16.5°	13.87	1.08 c

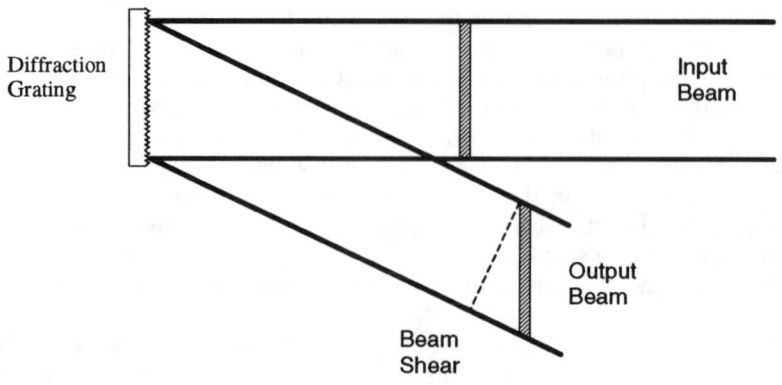

FIGURE 2. Schematic layout of the grating inserted to optimise the travelling line focus.

For matching to 'c' a propagation time of 40 ps is required which equates to an extra 7.2 mm optical path length difference across the 88 mm input beam. However, account must be taken of the beam expansion encountered from the grating insertion point to the target area. The beam size is expanded by a factor of 3.8 so at the grating insertion point the additional path length required is 7.2 mm across a 23.3 mm beam. The diffraction grating line density required is 294 lines mm^{-1} when operated at normal incidence, the diffracted angle would be 18.4°. A commercial 300 lines mm^{-1} is used at the expense of introducing a 2 % increase in the path difference. This is however over compensated by the 5% change in beam size as a result of the diffraction. These factors result in a path difference of 7.0 mm using the commercial grating which gives a nominal overall path difference of 11.77 mm and a phase velocity of 1.02 c.

It is possible to fine tune the phase velocity by changing the incidence angle onto the spherical mirror. Table 1 illustrates the possible tuning range.

CPA Pulsewidth Optimisation

Insertion of a diffraction grating to optimise the travelling wave geometry introduces a divergence into the beam via the spectral dispersion of the additional grating such that the compressor (or stretcher) gratings have to be re-tuned in order to achieve a minimum pulsewidth. The act of re-tuning is simply to adjust the grating separation and measure the pulsewidth at each point thus generating a tuning curve. In order to perform such a measurement part of the CPA beam must be split into a diagnostics line. The final turning mirror in the CPA beamline allows 2% of the beam to be directed onto the CPA diagnostics tables. The beam is spatially compressed by a factor of x7 before passing onto the diagnostics line. A fraction of the beam is allowed to pass

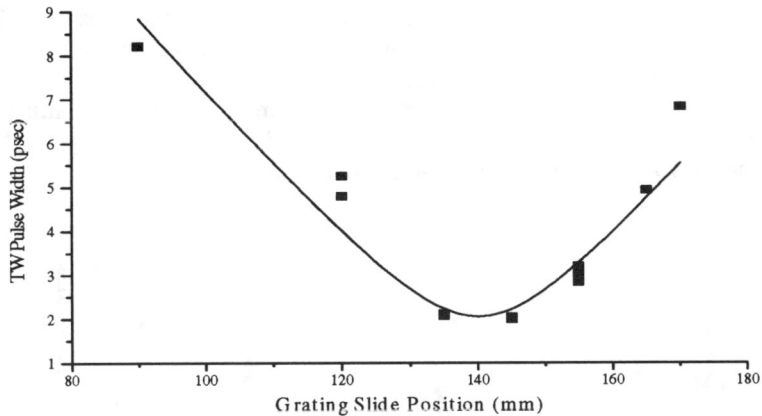

FIGURE 3. Compression grating tuning curve with the travelling wave grating in the laser system.

through a single shot autocorrelator which measures the pulse duration, the remaining fraction passes to a spectrometer allowing the bandwidth of the pulse to measured for each shot. Pulsewidths were measured using low energy rod amplifier shots from the VULCAN laser system. Varying the stretcher gratings whilst keeping the compressor gratings fixed ensures that the beamline does not require realignment between shots. The data obtained is presented in the tuning curve in Figure 3 where the solid line represents a polynomial fit. The data indicates a minimum pulsewidth of 2 psec for the travelling wave geometry. The insertion of the grating will impose a minimum compressed pulse duration because the divergence imposed on the beam will not allow the pulse to be completely recompressed.

X-RAY LASER RESULTS

Titanium and Germanium stripes on glass with a thickness between 0.6 and 1.0 µm and a width of about 200-300 µm as well as massive slab targets were used. The amplification length was varied with the target length. The targets were centred in the line focus. The primary diagnostic along the target axis was a common flat field grazing incidence XUV-spectrometer with a 1200 line/mm aperiodically ruled grating. Focusing perpendicular to the dispersion direction was provided by a cylindrical mirror. The time integrated spectrum was recorded with a back-thinned, cooled, 16 bit CCD-camera. At higher X-ray laser signals Al-filters were used in front of the spectrometer to prevent saturation of the X- CCD camera. The corresponding absorption values were taken into account to estimate the intensity of the lasing signal.

The applicability of the efficient transient excitation scheme for higher z-material to reach shorter wavelength lasing was convincingly demonstrated for neon-like Ge. Due to the large energy range delivered by the VULCAN CPA-beam it was also possible to investigate the amplification behaviour of this Ge XRL for target lengths up to 10 mm. The recorded lasing signal of the 3p-3s transition at 19.6 nm is displayed in Figure 4, showing both a high small signal gain value of about 30 cm^{-1} for targets with ≤ 5 mm, and clear evidence of saturation-like behaviour for ≥ 6 mm. It is worth mentioning that strong lasing signals have been recorded with only 2.1 J mm^{-1} pump energy in the short CPA beam and 3.3 J mm^{-1} in the long pulse pump beam.

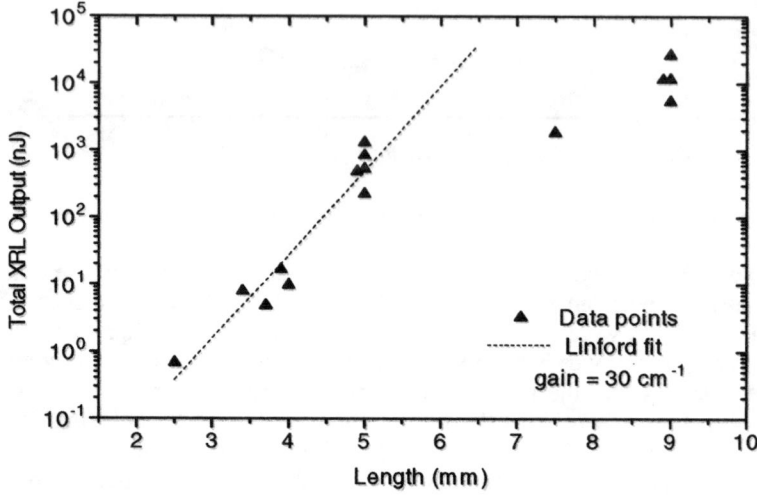

FIGURE 4. The integrated energy output of the Ge, 19.6 nm X-ray laser as a function of target length.

It was shown for the first time that the new transient gain excitation scheme can be brought into saturation on the 3p-3s line in Ti at 32.6 mm as well as on the 3p-3s line in Ge at 19.6 nm with only several Joules pump energy. This represents 1-2 orders of magnitude reduction in pump energy as compared with common quasistationary XRL systems.

ACKNOWLEDGEMENTS

This experiment was funded under contract ERBFMGECT 950053 of the EU Training and Mobility of Researchers (TMR) Large Facility Access Programme.

REFERENCES

1. Strickland D., and Mourou G., Opts Comm, **56**, 219 (1985)
2. Ross, I.N. et al., Applied Optics **76**, p 1584 (1987)
3. ZEEMAX, Focus Software Inc., Tucson, Arizona.
4. Moreno, J. C., Nilsen, J., and Da Silva L. B., Opts Comm, **110**, 585 (1994)

A multi-channel soft X-ray flat-field spectrometer

D Neely, D. Chambers[2], C Danson, P. Norreys, S. Preston[2],
F. Quinn[3], M. Roper[3], J. Wark[2] and M. Zepf[2]

Rutherford Appleton Laboratory, Chilton, Didcot, Oxon, OX11 0QX, UK.
Telephone: (0)1235 821900 Fax: (0)1235 445888 e-mail: d.neely@rl.ac.uk
[2]Clarendon Laboratory, University of Oxford, Oxford, OX1 3PU, UK
[3]Daresbury Laboratory, Warrington, Cheshire, WA4 4AD

Abstract. The development of a three channel soft X-ray spectrometer which was specifically designed to resolve the angular intensity distribution of laser generated high harmonics is reported. The spectrometer uses a novel design of three tapered glancing angle mirrors to subtend angles of 0 ± 13, 46 ± 21 and 108 ± 35 mrad in the horizontal plane from the source. These angles were selected to characterise emission within an f5 angular cone, assuming radial symmetry. The soft X-ray mirrors are tapered in width to form an elliptical profile under the application of a bending moment. Design studies demonstrated that by the application of different bending moments, a similar shaped tapered mirror can be used for imaging in each of the three channels. The total subtended solid angle of the instrument in the three channels is 1 millisteradian. The measured resolution of the system is reported as well as the absolute throughput efficiencies in the spectral range 4 - 40 nm.

INTRODUCTION

Significant advances have been made recently in the production of coherent soft X-ray sources from laser matter interactions (1, 2). At the Vulcan and Titania laser systems of the Rutherford Appleton Laboratory work has concentrated on high harmonic production from solid (3) and gaseous (4) interactions and soft X-ray lasers. Grazing incidence flat-field spectrometers have been the primary soft X-ray diagnostic. To increase sensitivity and provide a high signal to noise ratio a collection mirror is used to image the source onto the detector plane. Measuring the brightness and output energy of each source requires angular resolution and a calibrated throughput. This report details the design and testing of an instrument with a total collection angle of 1 millisteradian, throughput efficiencies of ~20 % at 10 nm and spectral resolving powers in the 300 range, with a total wavelength coverage of 4 - 100 nm.

SPECTROMETER DESIGN

The spectrometer is based around a Hitachi aperiodic 1200 l/mm flat-field concave grating (5). The spectral focusing properties of this grating were investigated using a ray tracing program. This program followed a bundle of rays from a point source which uniformly intersected the 50 mm long 5649 mm radius of curvature concave grating surface and located the optimum spectral resolution plane for each wavelength. Diffraction from the grating surface was dealt with using the equation $\text{Sin}^{-1}(\underline{n} \wedge \underline{s} - m\lambda\rho) = \theta_{\text{Out}}$ where ρ is the groove density (6), λ is the wavelength being examined, n the grating substrate unit normal, m the order, θ_{Out} the output angle and s the unit vector along the ray being examined.

To obtain maximum detection sensitivity it is necessary to simultaneously obtain optimum spatial and spectral resolution. Spatial resolution was obtained using a set of collection mirrors located between the source and the grating. Each imaging mirror's surface is orthogonal to the grating surface, thus, the spatial imaging properties of the mirrors and spectral focusing/ dispersion of the grating can be considered separately. To combine them for optimum sensitivity, the solution is a circular image plane $I(\lambda)$ centred about the flat-field grating, of radius r = |MI| - |MG| where M is the mirror and G is the grating. However, the soft X-ray sensitive CCD detector (a Princeton Instruments back-thinned 1024 x 1024 pixel chip of size 25x25 mm) is planar. When this chip is placed on the circular detector plane (r ~ 0.24 m) no point on the chip is further than 160 µm from the ideal plane. In the wavelength range 5 - 35 - 90 nm the effective f number subtended by the grating onto the image plane varies from f/80 - f/20 - f/12. In the wavelength range of interest 5 - 35 nm his corresponds to an equivalent spectral broadening of less than one third of a pixel and is therefore acceptable. The final instrument essentially consists of four parts, a collection and imaging system, a diffraction grating, a back thinned CCD detector (which can be moved to intercept the circular image plane) and a light tight housing and associated filters see in figure 1.

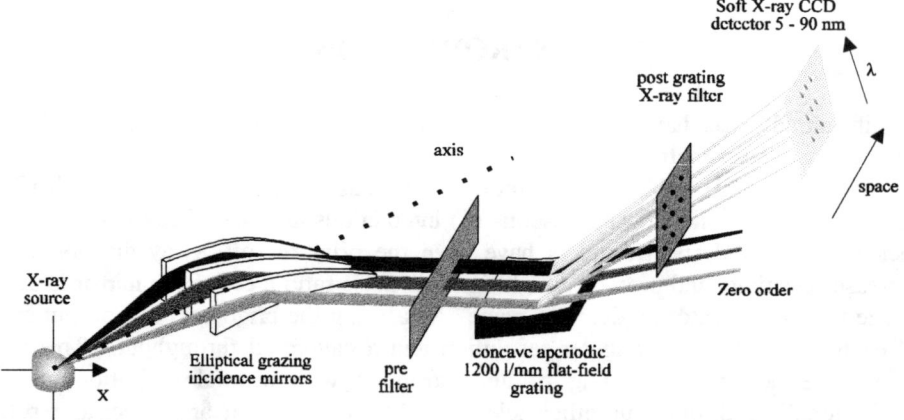

FIGURE 1. Schematic layout showing the triple mirror collection optics and flat-field spectrometer design

COLLECTION MIRROR DESIGN

The imaging collection optics set, consists of three 300 nm gold overcoated 1 mm thick, 160 mm long BK7 glass mirrors of RMS surface roughness < 1 nm, orientated to focus in a direction perpendicular to the dispersion direction of the grating. Their effect is to produce three rows of dispersed images of the source on the detector plane, each corresponding to a different angular channel. Considerations demonstrated that a tapered mirror which can be bent into an elliptical profile through the application of a bending moment (7) was the optimum design to give maximum flexibility. The profile of the tapered mirror was calculated by first fixing the location and curvature of the three elliptical mirrors which would subtend the desired angular section from the source. The taper width of each mirror was then calculated such that the application of a given bending moment at the end of the mirror would produce the desired curvature along its length. Figure 2 shows the calculated taper width as a function of position along the mirror. Calculations clearly demonstrated that to within the manufacturing tolerances (± 0.2 mm width) the application of a different force on the tapered mirror profile shown in Figure 2 would produce an acceptable elliptical profile. Table 1 shows the semi-major and minor elliptical profiles selected for the imaging optics, the average glancing angle, total angle subtended and average curvature of each of the mirrors.

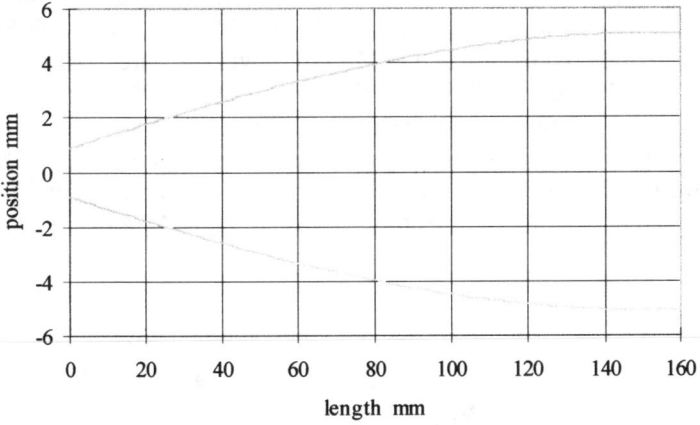

FIGURE 2. Tapered mirror width as a function of length along the mirror from the point where the bending force is applied.

TABLE 1. Design details for the three elliptically bent imaging mirrors.

Average glancing angle mrad	Semi-major axis length mm	Semi-minor axis length mm	Angle subtended mrad	Average curvature rads / m
40	350.3	13	26	.13
70	350.9	23.5	42	.22
120	352.4	40	70	.37

THROUGHPUT EFFICIENCY

The diffraction efficiency of the flat-field grating and reflection efficiencies of the imaging mirrors were calibrated absolutely using beam line 6.1 of the Daresbury synchrotron facility (8). This beamline provides a sub mrad collimated source of $\approx 1\%$ bandwidth tuneable over the spectral range 4 - 50 nm. Each test grating was inserted into a diffractometer and then it's entrance slits adjusted to only illuminate the central 80 % of the grating surface, thus avoiding over-fill. An AXUV-100 soft X-ray sensitive photo-diode connected through a charge sensitive amplifier was scanned to intercept the incident, reflected and diffracted signals from the grating at each glancing angle under study. The current from a grid placed before the diffractometer was also recorded and used to eliminate the effect of the slow decay in beam intensity during a scan. By varying the charge accumulation time at each output sample angle and adjusting attenuation filters, a dynamic range of $> 10^4$ was easily achievable. Large angular scans carried out using a grating in a dispersive mode showed that the incident beam onto the diffractometer was not monochromatic but had some contamination present, $\approx 2\%$ between 4 - 20 nm and $\approx 20\%$ between 20 - 50 nm. Using a combination of suitable filters and spectral deconvolution, the effect of this was subtracted from the raw data to extract the test efficiencies under investigation. Figure 3 shows the measured efficiency in 1st, 2nd and 3rd order for grating 38907-1-1 (a ~2 year old grating) at grazing incidence angle of 70 mrad. In a similar fashion the specular reflectivity of the gold overcoated BK7 mirrors was measured and found to be in good agreement with the calculated values using the Henke data (9).

The Hitachi grating was operated at a grazing angle of 70 mrad in the vertical plane and subtended an angle of 7.5 mrad as seen from the laser focus. The throughput for each mirror channel which is the product of the reflectivity, grating efficiency, and the solid angle collected by each mirror and the grating is shown in figure 4. As can be seen, the

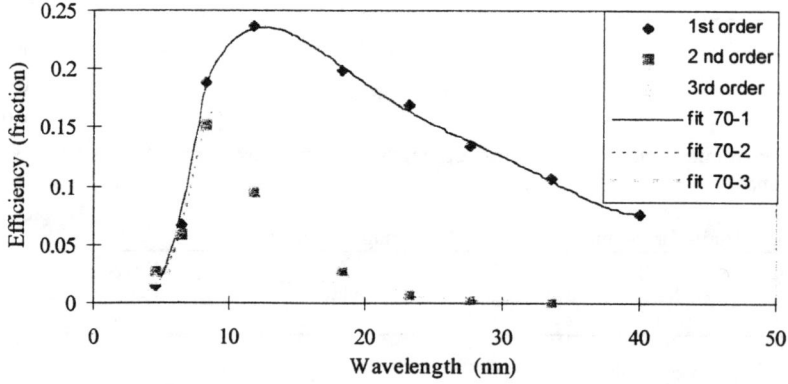

FIGURE 3. First, Second and Third diffraction efficiencies for a 1200 l/mm aperiodic grating at 70 mrad grazing angle.

inner mirror had the smallest grazing angle (and highest reflectivity at all wavelengths), but its smaller solid angle resulted in its total collection efficiency being significantly less than the central mirror. The same argument applies to the total collection efficiency of the central mirror with respect to the outer mirror. The relative throughput of the mirrors was verified experimentally by measuring the intensities of the incoherent spectral line emission. Al filters were placed before and after the focusing mirrors to eliminate laser light and to attenuate the harmonic radiation. The addition of a limited aperture filter between the grating and detector efficiently discriminates against fluorescent and scatter giving significantly reduced background.

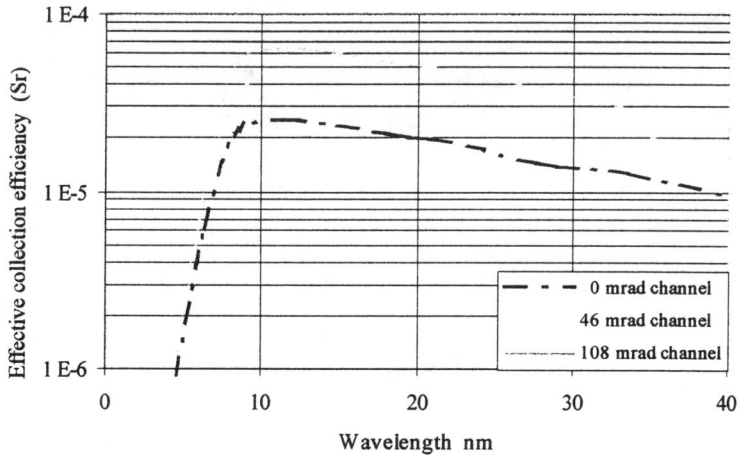

FIGURE 4. Graph showing the effective collection efficiency for each channel of the triple mirror spectrometer.

EXPERIMENTAL

In an experiment (10) conducted using the Titania 248.6 nm KrF laser up to 200 mJ was delivered to target at a pulse length of 350 fs. The laser beam was focused above a solenoid-valve gas jet using an f/12.5 off-axis parabolic mirror to a 10-30 times diffraction limited spot. The variation of 7th to 13th harmonic was studied as a function of the gas jet backing pressure and of the axial position of the laser focus see ref (10). Harmonic energies from neon targets of up to 1 µJ were measured. Results showed (see figure 5) that negligible harmonic emission was recorded in the outer channel (in the absence of refraction in the gas target) which lies outside the f/12.5 cone of the drive laser. The measured spatial resolution of the system was ~ 0.1 mm. This is attributable to poor manufacturing tolerance and problems with applying the desired bending forces. The recombination line widths at low gas jet backing pressures shows that the instrument spread function of the spectrometer has a FWHM of 4 pixels corresponding to a spectral resolution of about 0.1 nm at 35 nm. This resolution was sufficient to observe Stark broadening and line reversal effects in He and blue shifting and spectral broadening of harmonics. Subsequent refocussing reduced the spectral width to single pixel, which corresponds to ~ 25 pm, the theoretical maximum.

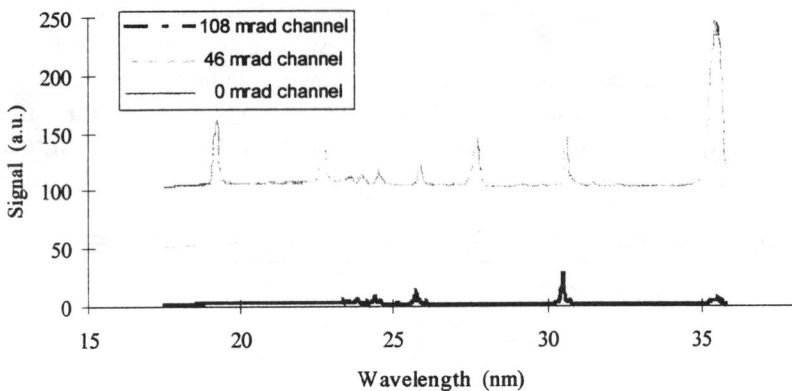

FIGURE 5. Graph showing typical time integrated CCD signals on the three angular channels of the spectrometer of emission from neon. Note the absence of the harmonic emission on the 108 mrad channel which lies outside the drive laser beam cone angle.

CONCLUSION

The use of a triple mirror collection system combined with a calibrated flat-field grating has enabled the accurate measurement of absolute harmonic conversion efficiencies. Finite element analysis is presently underway which will considerably improve the deliverable spatial resolution and sensitivity of the spectrometer.

REFERENCES

1. Salieres P, Ditmire T, Perry M, L'Huillier A and Lewenstein M, J. Phys. B **29**, 4771 (1996).
2. Zhou J, Peatross J, Murnane M M and Kapteyn H C, Phys. Rev. Lett. **76** 752 (1996).
3. Norreys P A et al, Phys. Rev. Lett, Vol 76, No 11, p1832-1835 (1996)
 Chambers D M, Preston S G, Zepf M, Castro-Colin M, Key M H, Wark J S, Dangor A E, Dyson A, Neely D and Norreys P A, J. App. Phys. **81** 2055 (1997).
4. Preston S G, Sanpera A, Zepf M, Blyth W J, Smith C G, Burnett K, Key M H, Wark J S, Neely D and Offenberger A A, Phys. Rev. A **53** R31 (1996).
5 Hitachi Part No. 001-0437 blank size30x50x10 mm Blaze angle 87 degrees.
6. Kitta T, Harada T, Nakano N and Kuroda H. Applied Optics, **22**, 4 512-513 (1983).
7. Underwood J H and Turner D, SPIE **106**, 125-135 (1977)

8 Neely D, Chambers D, Quinn F and Roper M, RAL technical report TR-96-074
9. Henke B L, Lee P, Tanata T J, Shimabukuio R L and Fujikona B K, Atomic and Nuclear data tables, **27** 1-144 (1982)
10. Zepf M et al, These proceedings

200TW Upgrade of the Vulcan Nd:glass Laser Facility

C.B. Edwards, C.N. Danson, M.H.R. Hutchinson, D. Neely and B. Wyborn

Central Laser Facility, Rutherford Appleton Laboratory, Chilton, Didcot, Oxon, OX11 0QX, UK.

Abstract. We describe progress on the upgrade of the Central Laser Facility's Nd:glass laser facility, Vulcan, to the 200 TW, 10^{20} W.cm^{-2} regime using Chirped Pulse Amplification (CPA) (1). The system uses a commercial Titanium Sapphire oscillator followed by a grating pulse stretcher and a phosphate glass rod and disc amplifier chain which boosts the pulse energy to 120 J at 150 mm aperture. The pulse is compressed using large aperture diffraction gratings (420 x 210 mm) in single pass and then focused on target using reflective optics.

INTRODUCTION

The current Vulcan upgrade will provide a 200 TW sub picosecond CPA beam to target for ultra high intensity (10^{20} Wcm^{-2}) laser plasma interactions. The upgrade specification includes the provision of the main CPA beam synchronised to the six 108 mm diameter long pulse beams (80 ps - 3 ns) in multiple geometries. To accommodate this requirement a new interaction chamber has been designed and commissioned.

The Vulcan upgrade comprises four main parts:

- Pulse generation and stretching systems
- Amplification to the 120 J level
- Pulse compression
- Provision of a new target chamber to enable a wide range of laser matter interaction experiments.

PULSE GENERATION AND STRETCHING

The VULCAN high power Nd:glass laser is configured with Master Oscillator Power Amplifier (MOPA) architecture. Two amplification chains, each fed by a choice of synchronised long or short pulse oscillators give the system multiple pulse length capabilities. At the output of the system one amplifier chain delivers two 150 mm diameter beams to target; the other drives six 108 mm beams.

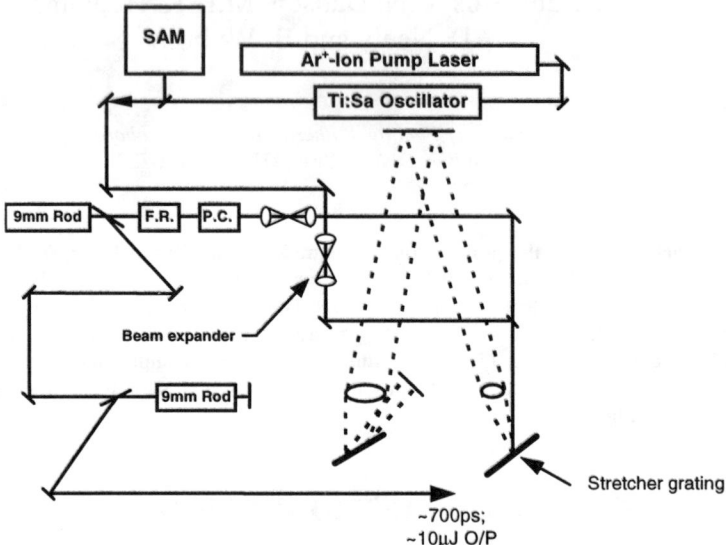

FIGURE 1. Optical schematic of the Vulcan front end

A schematic of the oscillator, pulse stretcher and pre-amplification system is shown in figure 1. A choice of two oscillators is available to provide ultra short pulses for CPA operations. One, a Spectra Physics Tsunami, 11 W Argon ion pumped Titanium Sapphire system operating at 1.053 µ produces pulses of 120 fs duration. The other, a Time-Bandwidth Products diode pumped Nd:phosphate glass laser, uses the solid-state Saturable Absorber Mirror (SAM) technique to produce high contrast pulses of 170 fs duration.

The train of pulses from the oscillator is temporally stretched in a two grating (1740 lines mm^{-1}) double passed stretcher, producing pulses of ~1 ns duration. The pulse train is then injected into a pre-amplification stage of two double passed 9 mm rod amplifiers where the single pulse energy is boosted to the 10 µJ level.

A single pulse is selected using a series of Pockels cells and then injected into one of the main Vulcan amplifier chains. A Faraday rotator provides optical isolation between the oscillator and pre-amplification stages.

LASER AMPLIFICATION

The main VULCAN amplifier system consists of two rod amplifier chains, each using a double passed 9 mm rod and single pass 16, 25, and 45 mm rod amplifiers. The three larger rods have typical gains of 30, 17 and 10 respectively.

Pockels cells or Faraday isolators are used between each stage of amplification to limit the growth of amplified spontaneous emission (ASE) and to prevent self-lasing. Spatial filters are used to control the beam size and to control the growth and propagation of high spatial frequency intensity modulations.

FIGURE 2. Disc amplifier stage of the Vulcan Nd:glass laser facility

After amplification in the rod chains, the beams are switched into the "six beam" 108 mm disc amplifier line and the 150 mm "backlighting" line. In each case the beam is split and then amplified in double passed 108 mm disc amplifiers. The two beams on the "six beam" system are then divided into three and amplified in single pass 108 mm disc amplifiers. The two beams of the "backlighting" line pass through an expanding vacuum spatial filter followed by amplification in a 150 mm amplifier. The output array of disc amplifiers is shown in figure 2.

The Vulcan facility delivers a maximum output of 2.5 kJ to target areas. Each of the six 108 mm amplifier lines gives 300 J in 1 ns, or 50 J in 100 ps; the two backlighting amplifiers deliver 450 J each in 1 ns, or 100 J in 100 ps. This output can be frequency converted to 527 nm with ~60% efficiency.

Two target areas are available for experiments using the full eight beam capability of Vulcan, supporting spherical, cylindrical and cluster irradiation geometry. A third target area is used for two beam experiments

PULSE COMPRESSION

The maximum CPA beam energy delivered to target is determined by the laser induced damage threshold of the compression gratings. New gratings with a working aperture of 190 mm are being installed during the current upgrade. The diffraction efficiency into first order is 89%. Combined with a peak damage threshold of 500 mJcm^{-2} this gives a safe operating regime of 80 J to target.

The new compressor follows the 35 TW design already proven on the Vulcan CPA beam line (2). The first grating is in air, with the second housed in a purpose built vacuum vessel. An anti-reflection coated window located close to the first grating provides the vacuum interface. To maintain alignment during pump down and let-up of the grating chamber, the second grating is held on a mounting fixed directly to the floor but mechanically isolated from the chamber by bellows.

After leaving the final grating the beam is transported to the interaction chamber in vacuum to minimise B-integral, thereby preserving beam quality. The beam is steered with a 98% reflectivity 45° mirror and focused using either off-axis or on-axis parabolic mirrors. The 2% leakage from the final mirror is directed to the diagnostics table where near-field and far-field intensity distribution, spectrum and bandwidth are measured.

NEW TARGET CHAMBER

The Vulcan facility supports a broad and diverse community of users. Versatility and ease of use are therefore essential to maintaining the productivity of the scientific programme. This has been a particularly important consideration in the design of the new target chamber for ultra-high intensity interactions.

The new interaction chamber is shown in figure 3. It takes the form of a cylinder 1.8 m in diameter, 0.95 m long, constructed from nickel coated mild steel. The main influence on the design of the chamber was requirement for six-beam interaction capability, in various geometries, synchronised with the ultra-high intensity CPA beam.

The chamber incorporates 12 rectangular ports of 600 x 300 mm aperture, centred at 0°, ±25°, ±50° and ±90° from the horizontal. Eight 300 mm diameter circular ports are located in each end flange and additional circular ports are located at the top and bottom of the chamber. Two large hinged doors at each end give good internal access to the chamber. These doors can be replaced by six beam cluster or line focus assemblies as required.

Beam train optics, diagnostics and alignment systems are mounted on a frame which is mechanically isolated from the chamber using a bellows system identical to the final grating mount. Tests show that the angular deflection of this frame on chamber pumpdown is less than 0.05 mrad.

During experiments the target chamber must be cycled between atmospheric pressure and vacuum at the 10^{-4} mbar level within the 20 minute duty cycle of the Vulcan laser. This is achieved using a combination of roughing, booster and turbo-molecular pumps. Water vapour is removed using a liquid nitrogen cooled 'cold finger'. To achieve the necessary pumping speed the grating chamber is normally maintained at 10^{-4} mbar with only the interaction chamber volume cycled with dry nitrogen. This procedure also minimizes the possibility of grating contamination by migration of oil or target debris from the target chamber.

TARGET CHAMBER CONFIGURATIONS

The new target chamber has been designed to accommodate four main irradiation geometries as described below. The high irradiance CPA interaction beam and lower energy CPA probes are available synchronised to within 50 ps of the six 108 mm long pulse (80 ps - 1.5 ns) beams.

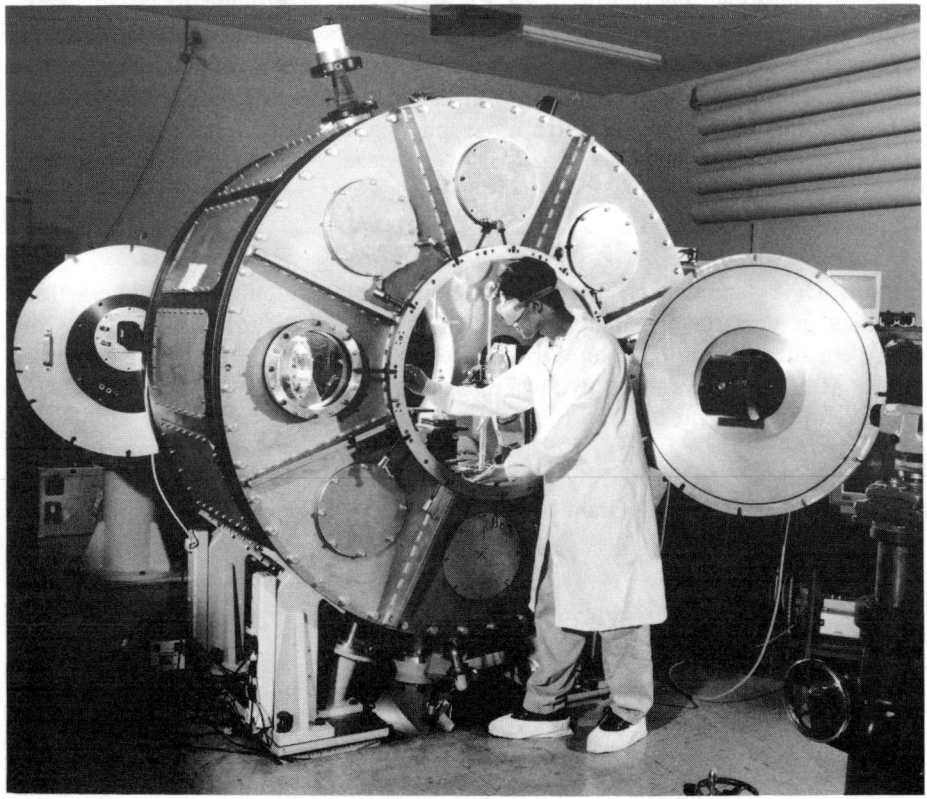

FIGURE 3. The new 200TW target chamber

Cylindrical 6 beam geometry. Five of the 108 mm beams are brought through the rectangular ports in the chamber, four at ± 30° and one at + 90° from the horizontal. The remaining beam is brought through an end window in one of the end flanges, vertically below chamber centre. F10 focusing optics are available in this configuration.

Spherical 6 beam geometry. Four of the 108 mm beams are brought through rectangular ports in the side wall. The remaining two beams through the end flanges. The f10 focusing optics are located externally to the chamber.

Cluster geometry. The f10 lenses are attached to one of the chamber end flanges and the six 108 mm beams brought to a focus at chamber centre. Alternatively, two special flanges can be attached to either end of the CPA chamber and three beams brought in on each.

Line focus. Line focus geometry is achieved using a set of six f2.5 aspheric doublets and concave mirrors mounted inside the chamber. Line foci from 0.5 - 40 mm in length are available.

In all the above geometries the CPA beam can be brought to target using various focusing configurations. Point foci are generated using either a 0.225 m On Axis Parabola (ONAP), a 0.6 m focal length Off Axis Parabola (OAP), a 1.5 m OAP or a 5.0 m focal length spherical mirror. A line focus of length 5 to 20 mm can be generated using an OAP and a tilted spherical mirror of 0.2-0.3 m ROC.

CONCLUSION

New CPA facilities on Vulcan are giving access to ultra-high interaction intensities in versatile irradiation geometries. When the current programme of facility enhancement is complete, irradiances on target in the 10^{20} Wcm^{-2} regime will be available from the CPA beam, synchronised to the 1.5 kJ, six beam capability of the facility.

This work was supported by a grant from the UK Engineering and Physical Sciences Research Council (Grant No. GR/K74180).

REFERENCES

1. Strickland D., and Mourou G., Opts Comm, **56**, 219 (1985)
2. CN Danson et al.,Well characterised 10^{19}Wcm^{-2} operation of an ultra-high power Nd:glass laser, J.Modern Optics (In Press)

Saturation in Transient Gain Scheme of Collisionally Pumped Germanium X-Ray Laser

K.A. Janulewicz[*], P.V. Nickles[*], M.P. Kalachnikov[*], M. Schnürer[*],
W. Sandner[*], S.B. Healy[§], G.J. Pert[§], P.J. Warwick[#], C.L.S. Lewis[#],
C. Danson[†], D. Neely[†], E. Wolfrum[†], A. Behjat[¶],
A. Demir[¶], G. Tallents[¶]

[*] *Max Born Institute, Rudower Chaussee 6, D-12489 Berlin, Germany*
[§] *Department of Physics, University of York, Heslington, York YO1 5DD, UK*
[#] *Department of Physics, Queen's University of Belfast, Belfast BT7 1NN, UK*
[†] *Ratherford Appleton Laboratory, Chilton, Didcot, Oxfordshire OX11 0QX, UK*
[¶] *Department of Physics, University of Essex, Colchester CO4 3SQ, UK*

Abstract: Simulation are presented for the collisionally pumped germanium X-ray laser working in the transient gain scheme. The results obtained are compared to the experimental observables registered during the joint european experiment conducted at RAL. Strong refraction influence on the output has been confirmed by a 3-D raytracing. Reliability of the hydrodynamics code and preliminary results of the pump conditions optimisation are discussed as well.

INTRODUCTION

The collisionally pumped X-ray laser has proved, since its first demonstration in 1984 (1), to be the most reliable scheme for gain creation in the XUV region of the spectrum. Pumping process occurs if sufficient amount of energy is delivered to strip an atom to neon- or nickel-like electron configuration and to excite effectively 3p (3d) levels manifolds by collisions. The strongest transitions occur between 3p and 3s levels, although recently some work, indicating significant gain on the 3d-3p transitions, has been done (2). However, this effect is strong for the elements with lower Z and is neglected here
This scheme in its classical form involves long, higly energetic single pumping pulse and succeded in maximal gain coefficients of 5-8 cm^{-1}, depending on the pump variant used (2-5). Due to the long interaction times these lasers were working in a quasi-static regime and as a consequence a significant part of the energy delivered was lost for the inversion creation process.

Recently a high transient gain coefficient of ~19 cm^{-1}, with duration of tens picoseconds has been reported at 326 Å with Ne-like titanium ions (6). This experiment was motivated by the theoretical work (7) indicating that transient high gain could be produced by rapid rise of the electron temperature in the regions with high density of the Ne-like ions. This rapid rise of the temperature requires a very short and strong pulse irradiating a preformed plasma. It should be noted that a similar scheme based on two pulse (long/short) configuration was proposed independently a few years ago in (8) but motivation of this work on Ni-like tantalum foil was essentially different. The transient gain scheme for the selenium laser was recently analysed theoretically in (9).

Such a break-through in the gain values available in the XUV region of spectrum has attracted more attention to the transient gain scheme and fruited by much theoretical (2, 9-12) and experimental _(6, 13-15) work. The most comprehensive analysis of the lasing in this regime performed in terms of the experimental parameters is presented in (9). This work was a base for a joint experiment (MBI Berlin, QU Belfast, Rutherford-Appleton Laboratory and Essex University) conducted at Rutherford-Appleton Laboratory. This experiment demonstrated for the first time saturation in Ne-like titanium (13) and germanium (14) x-ray lasers pumped by relatively low energy beams. The estimated gain coefficients were equal to 35 cm^{-1} and 30 cm^{-1}, respectively. It should be noted that recently also the LLNL group successfully repeated the experiment on Ne-like Ti (15). New interesting prospects for the transient gain scheme are analysed in (10). It is shown by numerical simulations, that applying the combination long/short pulse accompanied by a very limited increase of the pump energy the "water window" could be reached for Ne-like yttrium and the wavelength of 62 Å for Ne-like germanium.

The main advantage of the transient gain scheme is significant reduction in the pump energy requirements. The first succesful experiment on titanium required only 4-6 J in the prepulse and ~2 J in the main pulse. The saturation of the Ti-X-ray laser output during the last joint experiment required only few joules delivered to the target and for the germanium this value was a little bit higher. All experiments mentioned above were conducted with arbitrary pump pulses and limited control over their parameters. The simulations presented in this paper are to model the experiment conducted, in order to test our code reliability and determine the possible ways of the pump conditions optimisation. The germanium collisionally pumped X-ray laser is subject to analysis due to its well known theoretical and experimental characteristics.

MODEL

The computational model is described in detail in [10]. Simulations are performed with a 1.5-D hydrodynamics/atomic physics code EHYBRID and the output data (gain coefficient and the plasma density distribution) is post-processed by a 3-D raytracing. This code determines the real volume of the x-ray source, beam divergence and deflection independently in both direction, i.e. perpendicular (radial) and parallel (transverse)to the target, intensity distribution in the near and far fields, as well as

creats an "averaged cell". The latter includes averaged gain coefficient, gain-length product, saturation intensity and line parameters. Thus, by the weighting process the effective parameters corresponding to an uniform active medium are determined. Both codes confirmed their reliability for the longer pumping pulses, although it was found that under some conditions the hydrodynamics code overestimates the energy absorbed. It should be stressed that the raytracing has a stationary character and corresponds to the travelling wave pumping scheme.

RESULTS

The results refer to germanium slab target of thickness 10^{-4} cm, width 10^{-2} cm and length 0.5 - 1.0 cm, irradiated with a combination of two pulses (long/short) at 1.06 μm. A reduced Gaussian, characterised by a rise time and a full width at the half-maximum (FWHM) is used in the simulations of the experiment. As the rise time was not known in the experiment this parameter is taken to have an arbitrary value. It could be a source of some discrepencies between theory and experiment as the simulation show, that the plasma hydrodynamics is dependent to some extent on the rise time. A symmetric, trapezoidal shape with known lengths of the ramp and the flat part is used in other calculations. The experimental pulses are assumed to have the energy of 40 J (long prepulse) and 24 J (main pulse) in the case of the 1.0 cm target. These values are, for the halved target, reduced to 34 and 21 J ($\sim 10^{15}$ W/cm^2), respectively (14). The preforming pulse has width of 600 ps (FWHM) and the same value is taken for the rise time. The main pulse of length 7 ps (FWHM) is assumed to have the rise time of 5 ps.

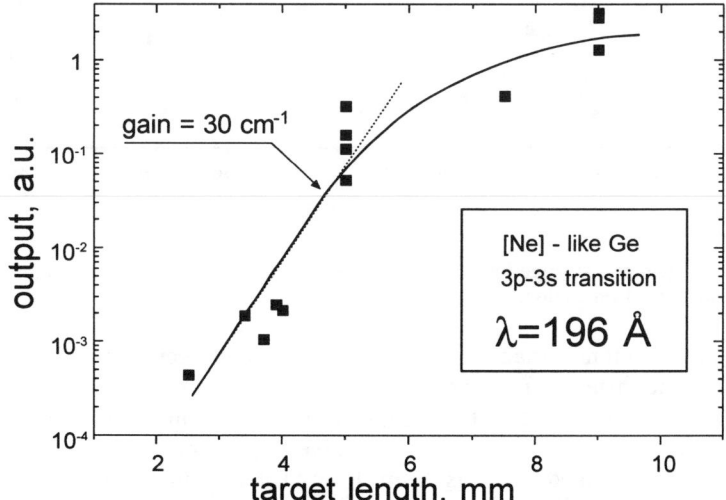

Figure 1. Output signal at 196 Å vs. the target length for germaniumin in the joint experiment

The measured X-ray laser output at 196 Å as a function of the target length is plotted in Fig. 1. The estimate of the gain coefficient was done by applying the least square approximation method together with the Linford's formula. It has been found the gain coefficient of 30 cm^{-1} and gain-length product of ~15 which corresponds to the saturation threshold.

The numerical simulations performed for two lengths of the target gave the local maximal gain coefficient and averaged ion stage radial distributions which are shown in Fig.2. Surprisingly, the gain coefficient for the shorter target is strongly reduced to the value of ~65 cm^{-1}, while the same parameter for the 1 cm target is equal to 116.5 cm^{-1}. The reason for this is clear from the same figure. The average ion stage for the shorter target is in a broad region higher than 22 and this is equivalent to the reduced aboundence of the Ne-like ions. It means, the code may overestimate the energy absorbed also for the shorter pulses. Due to this uncertainty we shall concentrate in our further analysis on the 1 cm-target.

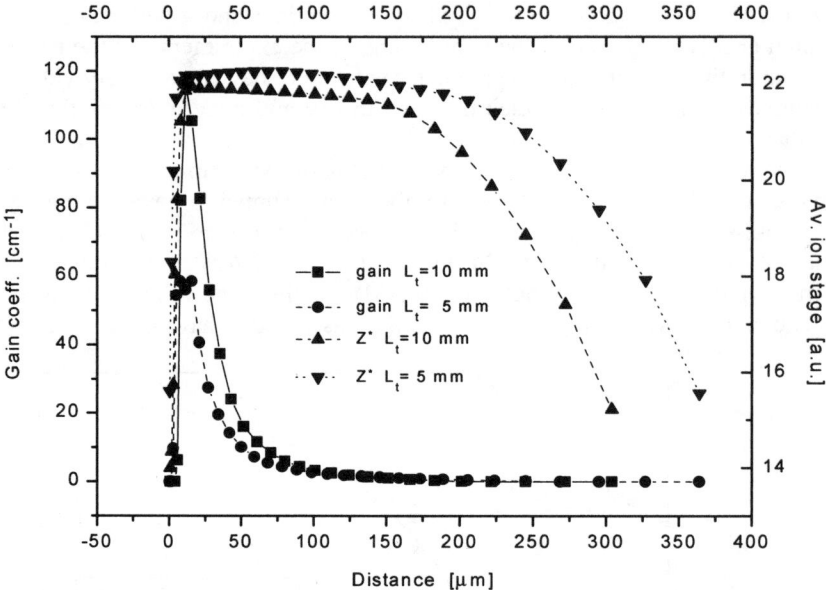

Figure 2. Local gain and average ion stage in the germanium plasma for two different target lengths. Result of the numerical simulation.

The results of different variants of the raytracing are collected in Table I. All results correspond to the distance from the target where the local gain coefficient is maximal. The outputs refer to the targets of the length 1 cm, 0.5 cm and 1 cm both latter traced 2 ps after the local gain maximum. The available experimental estimates (14) are also given for some parameters. It is seen from Table I that some of experimental parameters compare favourably to the theoretical estimates. The last column refer to the hydrodynamic data obtained for 1 cm target and then raytraced as if it was 5 mm target. Such a treatment could be considered as justified if the expansion along the target is very limited. The small signal gain coefficients, saturation intensity, output

TABLE I. Output of the raytracing

Parameter	Experiment	Target of 1 cm length	Target of 1 cm length and 2 ps delay	Target of 0.5 cm length and 2ps delay
Gain coefficient [cm^{-1}]	30	20.1	28.65	73.74
Gain-length product	15	47.3	56.7	36.13
Spontaneous emission [W/cm^2]		3.68×10^2	4.06×10^2	6.68×10^3
Saturation intensity [W/cm^2]		8.93×10^9	7.5×10^9	1.44×10^{10}
Output intensity [W/cm^2]	6.5×10^9	1.04×10^{11}	1.20×10^{11}	9.84×10^9
Output power [W]	0.65×10^6 [a]	2.95×10^6 [b]	3.94×10^6 [b]	1.95×10^6 [b]
Far field deflect. (radial/transv.) [mrad]	9.8±0.6 (1 cm) 9.9±0.3 (5 mm)	7.05 / 9.71	8.9 / -6.4	5.66 / 4.11
Far field diverg. (radial/transv) [mrad]	7.4±0.2 (1 cm) 8.7±0.2 (5 mm)	2.13 / 10.3	2.97 / 6.84	1.04 / 4.50

[a] 100x100 μm plasma face
[b] 300x300 μm plasma face

power and the x-ray beam deflection show reasonable agreement. Strong influence of the delay at the raytracing moment suggests that the beam refraction is a dominating factor in the output creation. The reduction of 15% in the local small signal gain coefficient value is compensated due to the plasma expansion and creation a larger volume with the favourable conditions for the lasing. Effective gain coefficient increases by 40%. At the same time the divergence and deflection, both in the transverse direction are strongly modified. Both parameters increase in the radial direction whilst strong reduced in the transverse one.

DISCUSSION

The results presented in the previous section referred to the simulations of the lasing in the transient gain regime under the experimental conditions which gave the saturation effect with the limited amount of the pump energy. The calculations input data were determined in the experiment and partly taken arbitrarily. The relation between the experimental anf theoretical results is under this circumstances reasonable, especially as far as the general trends in the behaviour of plasma and and x-ray radiation are concerned. However, the problem of the energy absorption overestimation and its potential consequences must be checked very carefully also for the shorter laser pulses. If it is the case the pump energy requirements must be multiplied by factor 2. On the other hand the raytracing is usually done at the moment and for the place where the local (not averaged along the optical path) small signal gain approaches ist maximal value. Our simulation show this is not the optimal choice and some delay (here not optimised), accompanied by the decrease of the local gain coefficient, can lead to the significant increase in its effective value (averaged over the optical path) (see Table I). As this dramatic change occurs on very short time scale one must analyse very carefully the theoretical and experimental data. The x-ray beam divergence and deflection change as well and it means the quasistatic raytracing is not able to give full information about the output. More importantly, the travelling wave pumping with phase velocity equal to c can appear to be not optimal.

REFERENCES

1. D.L. Matthews, P.L. Hagelstein, M.D. Rosen, M.J. Eckart, N.M. Ceglio, A.U. Hazi, H. Medecki, B.J. McGowan, J.E. Trebes, B.L. Whitten, E.M. Campbell, C.W. Hatcher, A.M. Hawryluk, R.L. Kaufman, L.D. Pleasance, G. Rambach, J.H. Scofield, G. Stone and T.A. Weaver, *Phys. Rev. Lett.* **54**, 110 (1985)
2. J. Nielsen, *Phys. Rev. A* **55**, 3271-3274 (1997)
3. J. Nielsen and J.C. Moreno, *Phys. Rev. Lett.* **74**, 3376-3379 (1995)
4. R. Kodama, D. Neely, D.M, Y. Kato, H. Daido, K. Murai, G. Yuan, A. MacPhee, C.L.S. Lewis, *Phys. Rev. Lett.* **73**, 3215 (1994)
5. B.J. McGowan, S. Maxon, L.B. Da Silva, D.J. Fields, C.J. Keane, D.L. Matthews, A.L Osterheld, J.H. Scofield, G. Shimkaveg, G.F. Stone, *Phys. Rev. Lett.* **65**, 420 (1990)
6. P.V. Nickles, V.N. Shlyaptsev, M. Kalachnikov, M. Schnürer, I. Will, W. Sandner, *Phys. Rev. Lett.* **78**, 2748-2751 (1997)
7. Yu.V. Afanas'ev, V.N. Shlyaptsev, *Sov. J.Quantum Electron.* **19**, 1606 (1989)
8. S. Maxon, K.G. Estabrook, M.K. Prasad, A.L. Osterheld, R.A. London, D.C. Eder, *Phys. Rev. Lett.* **70**, 2285 (1993)
9. K.G. Whitney, A. Dasgupta, and P.E. Pulsifer, *Phys. Rev. E* **50**, 468-473 (1994)
10. S.B. Healy, K.A. Janulewicz, J.A. Plowes, G.J. Pert, *Opt. Commun.* **132**, 442-448 (1996)
11. S.B. Healy, K.A. Janulewicz, G.J. Pert, accepted for publication in *Opt. Commun.*
12. J. Nielsen, *JOSA B* **14**, 1511-1514 (1997)
13. M.P. Kalachnikov, P.V. Nickles, M. Schnürer, W. Sandner, V.N. Shlyaptsev, C. Danson, D. Neely, E Wolfrum, C.L.S. Lewis, P.J. Warwick, A. Behjat, A.Demir, G. Tallents, submitted to *Phys. Rev. Lett.*
14. P.V. Nickles, M. Schnürer, M.P Kalachnikov, W. Sandner, V.N. Shlyaptsev, C. Danson, D. Neely, E Wolfrum, A. Behjat, A.Demir, G. Tallents, P.J. Warwick, C.L.S. Lewis, in *Proceedings of SPIE* **3156**, (1997)
15. A.L. Osterheld, J. Dunn, B.K.F. Young, R.L. Shepherd, D.F. Price, W.E. White, D.C. Eder, S.J. Moon, V.N. Shlyaptsev, in *Proceedings of SPIE* **3156**, (1997)

5. APPLICATIONS OF ULTRA-INTENSE FIELDS

Recent progress in coherent XUV generation at RAL

M. Zepf[b], J. Zhang[a], D. Chambers[a], A.E. Dangor[b], A.G. MacPhee[c], J. Lin[g], E. Wolfrum[a], J. Nilsen[d], T.W. Barbee, Jr.[d], C. Danson[e], M.H. Key[d], C.L.S. Lewis[c], D. Neely[e], P.A. Norreys[e], S.G. Preston[a], R.M.N. O'Rourke[c], G.J. Pert[f], R. Smith[g], G.J. Tallents[g], I.F. Watts[b], J.S. Wark[a]

[a] Department of Physics, Clarendon Laboratory, University of Oxford, Oxford, OX1 3PU, UK
[b] Blackett Laboratory, Imperial College of Science, Technology and Medicine, London, SW7 2BZ, UK
[c] School of Mathematics and Physics, Queen's University of Belfast, Belfast, BT7 1NN, UK
[d] Lawrence Livermore National Laboratory, Livermore, CA 94550, USA
[e] Central Laser Facility, Rutherford Appleton Laboratory, Chilton, Oxon, OX11 0QX, UK
[f] Department of Physics, University of York, York, YO1 5DD, UK
[g] Department of Physics, University of Essex, Colchester, CO4 3SQ, UK

Abstract. The research into coherent XUV generation has made considerable progress since the first X-ray laser was demonstrated over a decade ago. Progress has been made in three diverse areas: XUV lasers, harmonics from gases and, more recently, harmonics from solid targets. We report the first demonstration of saturation in nickel-like X-ray lasers with conversion efficiencies of $>10^{-6}$. We will also report on the generation of high yield, high order harmonics from gases and solids. In gases conversion efficiencies of $>10^{-6}$ and harmonics as short as 6.7 nm were observed. Solid harmonics were demonstrated at very high yields ($10^{-4} - 10^{-6}$) and very high orders (up to the 75th) using irradiances on target of up to 10^{19} Wcm^{-2}µm^2.

INTRODUCTION

The generation of coherent XUV radiation has commanded a large world wide research effort, involving groups from many countries and several different schemes (1-9). These schemes vary from attempting to generate lasing action in highly ionised atoms to the production of very high order harmonics from high power short pulse lasers. Additionally to these laser generated sources, large facilities such as DESY in Hamburg are planning XUV free electron lasers to produce high brightness, short pulse radiation (10). The effort in the United Kingdom has concentrated in university research programs at the Rutherford Appleton Laboratory (RAL) with X-ray lasers being covered by research groups from Queen's University, University of York, University of Essex and the University of Oxford. Harmonic generation from gaseous media is being investigated by the University of Oxford and harmonics from solid targets are investigated by Imperial College in a collaborative effort with the University of Oxford and RAL. This paper will attempt to provide an overview over the recent work of these groups and compare the performance of the various sources of XUV radiation.

An important objective in the development of XUV sources is to deliver a coherent, high power output at successively shorter wavelengths toward the "Water Window"(1). Such XUV sources are required for holography and microscopy of biological specimens and for deflectometry, interferometry and radiography of dense plasmas relevant to inertial confinement fusion and laboratory astrophysics and many other applications (11-14). One way of achieving such a source are saturated X-ray lasers. Saturated operation is very important because it means that the maximum power possible for a given volume of excited plasma is extracted by the stimulated emission. Saturated X-ray lasers also ensure an output energy sufficient for most applications and tend to produce a consistent output with little variation from shot to shot. Lasers are characterised by the product of the laser gain coefficient and the length of the laser gain region (gain-length product). Saturation requires a gain-length product exceeding 15, which means millions fold amplification of the spontaneous emission by the stimulated emission.

Saturation has been observed in neon-like (Ne-like) X-ray lasers at wavelengths longer than 15 nm on the J = 2-1 transitions in plasmas of Ge, Se and Y and on the J = 0-1 transition in Zn, Ar, Ge and Ti plasmas when driven by the largest optical laser facilities in the world. However, these X-ray lasers are difficult to scale to the shorter wavelengths, with the currently available laser driver energy. Nickel-like (Ni-like) X-ray lasers, in principle, have a more favourable scaling of laser wavelength with drive-laser energy but have so far had difficulties in providing a saturated output. Much effort has been devoted toward developing Ni-like X-ray lasers, but the resulting gain-length product and efficiency have been low (2,3). Earlier experiments have shown that the intensity and the efficiency of Ni-like X-ray lasers can be greatly enhanced by use of multiple equal intensity short pulses (15). We will report experiments which demonstrate the saturation of Ni-like lasers between 14 nm (Ag) and 7.3 nm (Sm) using a suitable combination of driving pulses at conversion efficiencies between $6 \cdot 10^{-7}$ and $9 \cdot 10^{-6}$.

In addition to the X-ray laser technology, significant advances have been made over the past few years in the production of coherent XUV radiation by generating ultra-high harmonics of the driving field (4-9). Two methods have been explored to date. Firstly the interaction of intense laser pulses with gaseous targets and secondly by the interaction of an extremely intense pulses (10^{17} - 10^{19} Wcm^{-2}µm^2) with a solid target. While the total yields and source brightness of these sources is still second to the output of saturated X-ray lasers, this approach has several advantages. Firstly, it is conceivable that small laser systems, so called table-top terawatt lasers, will be capable of producing harmonic emission both efficiently and at high repetition rates. Harmonics are also intrinsically a short pulse source since the harmonic pulse length follows that of the generating pulse, and they are, given a suitable driving laser, tuneable.

The mechanisms responsible for the generation of the harmonic radiation are quite different in the two cases considered here. In the case of gaseous targets a strong laser field, comparable to the Coulomb field of the atom, enables electrons to tunnel through the Coulomb barrier and therefore give rise to large non-linear susceptibilities. The symmetry of the atomic potential dictates that only odd and even harmonics are generated (16,17). In the other case the harmonics are generated by the non-linear motion of the electrons across the density step at the critical density surface. This density step can either be the solid-vacuum interface of the target (for sufficiently short, say < 100 fs pulses with high contrast) or the density step produced by ponderomotively steepening the critical density surface of the preplasma, formed by a prepulse or by the leading edge of the main pulse(18,19).

SATURATED NI-LIKE X-RAY LASERS

Relaxing the density gradients in collisionally excited X-ray lasers through multiple pulse technique has proven to be very effective in enhancing drive efficiency(20,21). In the multi-pulse technique, the first pulse heats and ionises the plasma but the density gradients are too steep for laser propagation. The plasma then expands, before the arrival of the following pulses, creating a larger scale length plasma. During the following laser pulses, a larger, more uniform gain region is produced which allows for good laser propagation (20, 21). Aiming at improving efficiency and enhancing the gain-length product, a specially designed pulse configuration was implemented to drive Ni-like X-ray lasers in our experiments presented here. Since it is not necessary to have a preplasma in an ionisation stage as high as Ni-like, a low intensity (10% ~ 30% of total energy) laser pulse (prepulse) was used to create a preplasma with a lower ionisation. The preplasma was then allowed to cool down for a much longer time (> 2 ns), than those used in other experiments. This results in preferential absorption of the main pulse energy in the plasma region where laser gain is generated. The long delay also provides an excited plasma column with much reduced density gradients to enable the X-ray laser beam to propagate a longer distance at higher density and therefore obtain sufficient amplification for saturation to occur(20). The preplasma was then rapidly heated by a short main pulse while maintaining a low ion temperature and avoiding over ionisation. The combined effects of these factors brought about stable regions of Ni-like ions and resulted in the saturated operation of Ni-like X-ray lasers with much higher efficiency.

The experimental set-up is similar to that described in Ref. 20. Three beams of the VULCAN Nd:glass laser with a 75 ps duration at 1.05 μm were used in a standard off-axis focus geometry, which provides a line focus with 25 mm length and 100 μm width, giving an irradiance of up to 4×10^{13} Wcm^{-2} on targets. Deploying another three beams 180° opposed in a second line focus on a second target produced a plasma with an opposed density gradient which helps compensate for the refraction of the X-ray laser beam from the first plasma.

Flat slab targets used in the experiment were typically 18 mm long with 100 μm wide medium-Z metal (Ag, In, Sn and Sm) stripes coated on the glass substrates. The targets were aligned so that they were parallel with an adjustable separation (in the direction perpendicular to the target surfaces) between the surface planes and an axial separation of 500 μm between the two targets. Since the X-ray laser pulse duration is comparable to the propagation time, travelling wave excitation for the two successive targets is desirable to achieve maximum amplification. To approximate this condition, the three drive beams for the first slab target were therefore timed 60 ps earlier than the three beams for the second target.

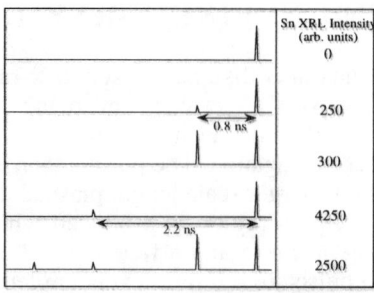

Figure 1. Schematic diagram of drive pulse configuration and the relative peak output intensity observed on the 12 nm laser line from a single 18 mm long Sn target. No lasing action was observed in the single pulse configuration.

As mentioned before, an optimised drive pulse configuration is very critical for both relaxing the density gradients and enhancing absorption efficiency in order to achieve saturated operation in Ni-like X-ray lasers at shorter wavelengths. To find an optimal drive pulse configuration, we carried out a systematic study on the Sn laser output intensity from a simple standard 18 mm long target irradiated by varied pulse configurations at similar focused intensity of 2×10^{13} Wcm^{-2}. The drive pulse configurations included: a single 75 ps pulse; a single main pulse preceded at 800 ps and 2.2 ns by a single prepulse (10~30%); two equal intensity pulses separated by 800 ps; two equal main pulses separated by 800 ps with double prepulses 2.2 ns early. This is summarised schematically in Fig. 1.

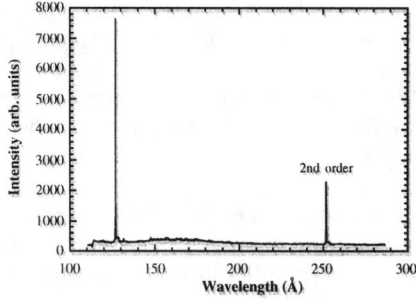

Figure 2. Axial spectrum from a 18 mm long single flat In target showing that the 12.6 nm laser line and its second order completely dominate the spectrum. The laser line has been attenuated by a factor of 300 to avoid saturation of the CCD detector.

In the single pulse drive configuration, we could not observe any laser signal with a single 75 ps pulse at 4×10^{13} Wcm^{-2}. In the 2.2 ns prepulse drive configuration, which was the standard configuration used in whole of the experiments presenting here, the Sn laser intensity did not vary much, when we changed the prepulse level between 5% and 50% while remaining the same separation (2.2 ns) and the same intensity (2×10^{13} Wcm^{-2}) in the main pulse. This demonstrates the importance of preplasmas and also the trivial contribution of the ionisation stage in preplasmas. The effect on the Sn laser is dramatic as the pulse separation is changed for the prepulse configuration. The output intensity of the Sn laser driven the 2.2 ns prepulse configuration is about 20X stronger than that driven by 800 ps prepulse configuration and about 15X stronger than that driven by two equal intensity pulses. The Sn laser driven by the more complicated double prepulse plus double main pulse drive configuration (Fig. 1) produced a weaker signal.

Under the optimised drive pulse configuration, the Ni-like X-ray lasers were found to be very monochromatic X-ray sources, with the laser lines completely dominating the output spectra as shown in figure 2. Figure 2 shows the time integrated output of an In target. The outputs are completely dominated by the Ni-like J = 0-1 laser transition at 12.6 nm and its second order. The intensity of the laser line was attenuated by a combination of Si and C filters to avoid saturation of the CCD detector.

The axial spectra of the other Ni-like elements investigated are similar to the above spectrum with the saturated laser line completely dominating the spectrum. We are not going to show them here for the sake of space. We performed a series of experiments to characterise the Ni-like X-ray lasers as we optimised the target and drive pulse configurations. The transverse separation between two slab lasers was found to be optimised for a separation between 150 µm and 175 µm for the elements investigated. The temporal behaviour of the Ni-like X-ray lasers was characterised using an X-ray streak camera with 10 ps time resolution. The pulse duration was found to be approximately 34 ps for a single flat target and 53 ps for a double target, due to the contributions of the two individual targets.

This makes the X-ray lasers generated by this configuration one of the shortest pulse X-ray lasers yet produced. The divergence of the laser was measured directly using the spatial dimension of the streak camera. The laser emission from the single target has a divergence of 1.8 mrad and peaks at 2.5 mrad off-axis.

These small divergence and deflection angles are evidence for a much improved density gradient due to our special drive pulse configuration. For comparison, the laser emission from a coupled double target shows a strong coupling effects: Its divergence is only 1.2 mrad, implying an ~ 5 µm spatial coherence length at the exit plane of the Sm X-ray laser (22).

The Sm X-ray laser beam from the double target is almost parallel to the target surface with a deflection angle as small as ~1 mrad. This is certainly an advantage to applications. The angle of the peak intensity and divergence are influenced by refraction of the beam in the plasma. Both deflection and divergence angles of the Ni-like X-ray lasers were improved by the refraction compensating double target geometry.

Table 1. Main characteristics of the Ni-like X-ray lasers developed at RAL

	Ag X-ray laser	In X-ray laser	Sn X-ray laser	Sm X-ray laser
Wavelength	14.0 nm	12.6 nm	12.0 nm	7.3 nm
Gain Coefficient	7.2 cm^{-1}	9.0 cm^{-1}	11.5 cm^{-1}	8.4 cm^{-1}
Divergence	2.1 mrad	2.2 mrad	2.7 mrad	1.2 mrad
Deflection	4.5 mrad	0.9 mrad	1.2 mrad	1.0 mrad
Output Energy	90 µJ	300 µJ	690 µJ	313 µJ
Output Intensity	6.9×10^{10} Wcm^{-2}	8.0×10^{10} Wcm^{-2}	9.7×10^{10} Wcm^{-2}	2.0×10^{11} Wcm^{-2}
Conversion Efficiency	6.0×10^{-7}	2.0×10^{-6}	9.0×10^{-6}	2.1×10^{-6}

The output intensity was measured by imaging the near field at the laser output onto a back thinned CCD camera. The output intensity can be estimated by integrating the total photons emitted from the output aperture and using an absolute calibration for the CCD detector, the calibrations for the X-ray multilayer mirrors, filters and the measured angularly integrated X-ray laser pulse duration. The output energy of the lasers for an optimised separation between the two targets varied between ~90 µJ (Ag) and ~690 µJ (Sn), corresponding to a conversion efficiencies of 6×10^{-7} (Ag) to 9×10^{-6} (Sn), which is the highest efficiency achieved for Ni-like X-ray lasers. The maximum intensities were about 6.9×10^{10} (Ag) and 2×10^{11} Wcm^{-2} (Sm). The estimated uncertainty in this measurement is a factor of 1.5 and is due predominantly to uncertainty in the filter attenuation. Table 1 summarises the results for the various Ni-like lasers. The parameters of the Ni-like X-ray lasers are summarised in table 1.

HARMONICS FROM GASEOUS TARGETS

Harmonics from gaseous targets have received much attention over the past decade(7-9). They can be generated by short pulse lasers (several ps to a few fs) with relatively modest intensity requirements (10^{13} to 10^{17} Wcm^{-2}) attainable by many table top short pulse lasers around the world. The use of gas targets allows straightforward operation at very high repetition rates using a gas jet, as the target is not destroyed in the process. Much effort has gone into the optimisation of the harmonic generation process, both in terms of the conversion efficiency and the shortest attainable wavelength. The shortest harmonics wavelength is limited by the well known formula (16,17)

$$E_{max} = I_p + 3.2 U_p \tag{1}$$

This is an expression for the highest energy trajectory, which will lead to a recollision with the atomic core, where U_p is the ponderomotive potential and I_p is the ionisation potential. From this we can see that to generate very short wavelength harmonics we need the highest possible intensity and a long wavelength. The highest useful intensity is limited by the depletion of the medium (atoms or ions) through ionisation during the pulse. Thus we require ultra-short, high intensity pulses to optimise the highest order obtainable. For this reason most research has centred on using Ti:Sapphire lasers at around 800 nm. On the other hand we can consider using a medium with a high ionisation potential, which can withstand high intensities. However, this is only feasible using UV lasers for because of larger wavepacket spreading and phase considerations in the single atom response (22).

Experiments performed at the Sprite and Titania KrF lasers (248 nm) have investigated harmonic response using short wavelength drivers. The results suggest that the harmonics up to the cut-off wavelength for the neutral noble gases can be generated at very high conversion efficiencies (> 10^{-6} at 35 nm). Also the shortest wavelength harmonics to date have been observed from the noble gas ions (up to the 37th at 6.7 nm) (9), albeit at significantly lower conversion efficiencies. Figure 3 shows the conversion efficiencies for a range of harmonics generated using a KrF laser. These conversion efficiencies are significantly higher than achievable with, say, Ti:Sapphire or Nd:glass lasers (8,9). Also the highest observable harmonics is comparable in wavelength and conversion efficiency to those produced by long wavelength lasers. This indicates that for the wavelength regime > 20 nm, UV drive lasers are preferable to infra-red or visible lasers.

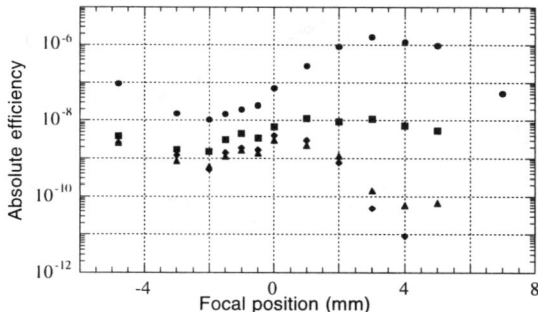

Figure 3. Variation of the conversion efficiency of the 7th (35.5 nm, circles), 9th (27.6 nm, squares), 11th (22.6 nm, triangles), 13th (19.1 nm, diamonds) harmonic with focal position. The conversion efficiencies are optimised for different focal positions where the intensity is optimised for their generation. The harmonics were recently generated using the Titania KrF laser at 248 nm.

HARMONICS FROM SOLID TARGETS

Harmonics from solid targets were first observed in nanosecond experiments using CO_2 lasers with irradiances between 10^{17} and 10^{18} Wcm^{-2}μm^2 in the late 1970's (23,24). The advent of CPA lasers has resulted in such irradiances now being available from short-pulse, optical lasers. This has lead to a renewed interest, both theoretical (18,19) and experimental (4-6), in this method of generating coherent, short wavelength radiation. Particle-in-Cell simulations have suggested that very high harmonic orders (>60) can be generated at very high efficiencies (>10^{-6}) (18,19). This has been confirmed experimentally at RAL (6) using 1 μm light at irradiances of 10^{19} Wcm^{-2}μm^2 with an efficiency of 10^{-6}. Figure 4 shows the conversion efficiencies obtained for a range of intensities.

This source of XUV radiation has a very high potential for Laboratory Applications as it can be generated using table-top terawatt lasers (4,5). Lasers such as ATLAS at the MPQ should soon be capable of exploring the regime demonstrated at RAL. These harmonics also display a very favourable scaling to shorter wavelengths: The conversion efficiency for a given harmonic order remains constant for the same value of $I\lambda^2$ irrespective of driver wavelength. Experimental evidence for this scaling can be found by comparing the results of Carman et al. (24) to the experimental results obtained at RAL. This implies that using a frequency tripled Ti:Sapphire harmonics at 3.5 nm can be generated at $I\lambda^2 = 10^{19}$ Wcm$^{-2}\mu$m^2 at 10^{-6} conversion efficiency. This exciting prospect remains to be verified.

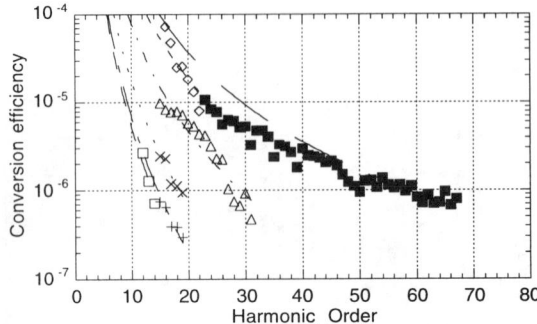

Figure 4. Conversion efficiency of harmonics from solid targets for a range of irradiances. They data points vary from $I\lambda^2 = 5.5 \ 10^{17}W\mu$m^{2}cm^{-2} (open squares) over $2.9 \ 10^{18}$Wμm^{2}cm^{-2}, $3.6 \ 10^{18}$ Wμm^{2}cm^{-2}, $6.5 \ 10^{18}$Wμm^{2}cm^{-2} to $1.1 \ 10^{19}$Wμm^{2}cm^{-2} (diamonds & solid squares). Data taken from (6).

Several decisive challenges remain however to develop harmonics from solid targets into a source that is ready for most possible applications. Currently the source brightness is limited by a large bandwidth (25), probably caused by self phase modulation in the underdense coronal plasma and a large anglular distribution caused by instabilities at the critical density surface (6,26,27). It may be possible to overcome these limiting factors by using ultra-high contrast, < 100 fs pulses, thus keeping the extent of the coronal plasma to a minimum and reducing growth time and rates for the surface instabilities. Preliminary research on this topic will be published in this conference proceeding (M. Zepf et al.), which indicates that the limitations can be overcome, or at least suppressed to a substantial degree.

COMPARISON OF XUV SOURCES

The three laser generated XUV sources have quite different characteristics and are at quite different stages of their development. Nevertheless it is of interest to compare their relative performance. A suitable figure of merit will usually be spectral brightness - i.e. the power per unit area, per unit solid angle, per unit frequency interval. These comparisons are sometimes difficult, because not all parameters are known with sufficient accuracy. The brightest source, especially at short wavelengths, are currently the saturated Ne-like and Ni-like X-ray lasers. A comparison of the sources discussed in this paper compared to other current sources can be seen in Figure 5. The spectral brightness is the lower bound in each case. Harmonics from gases trail somewhat in this comparison, but does not necessarily mean, that they are less useful. As pointed out, they are by far the easiest source to generate and are easily produced at very high repetition rates. If single pulse operation is not an imperative, this can increase photon flux per second by a factor $>10^3$ using kHz repetition rate lasers.

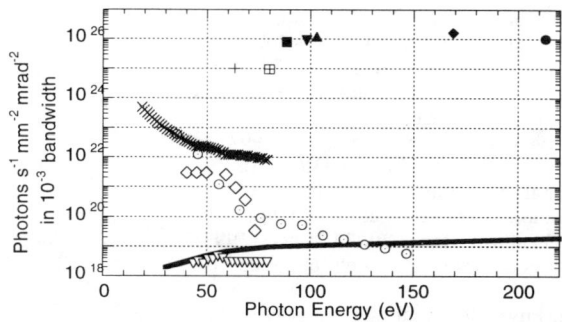

Figure 5. Comparison of the spectral brightness of XUV sources. The solid symbols represent the Ni-like X-ray lasers from Ag (14 nm) to Gd (5.7 nm) demonstrated at RAL. The vertical cross is the Ne-like Ge-laser and the crossed box the Ne-like Y laser. The diagonal crosses represent the solid harmonics from (6) the open diamonds and the open triangles represent the gas harmonic results from (8) the open circles the KrF gas harmonics (this paper) and the solid line is the ALS UV bending magnet.

A very interesting observation is the good performance of the harmonics from solid targets, despite the large fractional bandwidth (currently $\Delta\lambda/\lambda \sim 10^{-2}$) and the isotropic angular distribution. Controlling these parameters would improve the current performance by over 3 orders of magnitude. Using short wavelength drive lasers to produce light at a given wavelength could improve this figure by another 2-3 orders of magnitude. This potential has yet to proven experimentally, but makes further research very timely.

ACKNOWLEDGEMENTS

The authors acknowledge the assistance of the Rutherford Appleton Laboratory laser and other staff. The work has been supported by several grants from the United Kingdom Engineering and Physical Sciences Research Council. The work of the LLNL authors was performed under the auspices of the U.S. Department of Energy by the Lawrence Livermore National Laboratory under Contract No. W-7405-Eng-48. One of us (D.C.) would like to acknowledge support from AWE plc.

REFERENCES

1. R.C. Elton, X-ray Lasers (Academic Press, New York, 1990).
2. C.L.S. Lewis et al., X-ray Lasers 1992, E.E. Fill, Editor, IOP Publishing, Conf. Series No. **125**, 23 (1992).
3. B.J. MacGowan et al., Phys. Fluids B**4**, 2326 (1992).
4. S. Kohlweyer et al., Opt Comm., **117**, 431, (1995).
5. D. von der Linde er al., Phys. Rev. A, **52**, R25 (1995).
6. P.A. Norreys et al., Phys. Rev. Lett, **76**, 1832 (1996).

7. A. L'Huillier et al., Phys. Rev. Lett, **66**, 2200, (1991).
8. T. Ditmire et al., Phys. Rev. A., **51**, R902, (1995).
9. S.G. Preston et al., Phys. Rev. A., **53**, R31 (1996).
10. Rossbach, J., Nucl. Inst.& Methods In Phys. Res. A, **375**, 269 (1996).
11. J.E. Trebes et al., Science, **238**, 517 (1987).
12. L.B. Da Silva el al., Science, **258**, 269 (1992).
13. R. Cauble et al., Science, **273**, 1093 (1996).
14. D.H. Kalantar et al., Phys. Rev. Lett. **76**, 3574 (1996).
15. H. Daido et al., Phys. Rev. Lett. **75**, 1074 (1995).
16. P.B. Corkum, Phys. Rev. Lett, **71**, 1994, (1993).
17. J.L. Krause et al., Phys. Rev. Lett, **68**, 3535 (1992).

18. P. Gibbon, Phys. Rev. Lett., **76**, 50 (1996).
19. R. Lichters et al., Phys. Plasmas, **3**, 3425, (1996).
20. J. Zhang et al., Phys. Rev. A**54**, R4653 (1996).
21. J. Nilsen and J.C. Moreno. Phys. Rev. Lett. **74**, 337 (1995).
22. J.B. Watson et al. Phys. Rev. A., **51**, 1458 (1995).
22. J. Zhang et al., Science **276**, 1097 (1997).
23. N.H. Burnett et al., Appl. Phys. Lett, **31**, 172 (1977).
24. R.L. Carman et al., Phys. Rev. Lett, **46**, 29 (1981).
25. J. Zhang et al., Phys. Rev. A., **54**, 1597 (1996).
26. M. Zepf et al., Phys. Plasmas, **3**, 3242 (1996).
27. S.C. Wilks et al., Phys. Rev. Lett, **69**, 1383 (1992).

The Scope and Present Status of JAERI "Advanced Photon Research" Program

Hiroshi Takuma

JAERI Kansai Research Establishment at Tokai
Tokai, Naka-Gun, Ibaraki 319-11 Japan

Abstract: In consideration of revolutionary importance of CPA lasers and their potential applications in a broad range of science and technology, a research program called "Advanced Photon Research" has been launched at JAERI by the Science and Technology Agency. The program includes the development of repetitive ultra-short pulse Ti:sapphire laser with the highest peak power and the shortest pulse duration and also high repetition rate Ti:sapphire laser systems. Presently 10 Hz generation of 9.6 TW peak power in 16.6 fs pulse duration was demonstrated, and it is expected to be up-graded to 100 TW, 20 fs, 10 Hz system before the end of 1997. Those laser systems will be used in the researches on the x-ray lasers, and also in the basic study of laser acceleration. Other applications of CPA lasers and x-ray lasers in various field of science and technology are also included in the scope of the present program.

INTRODUCTION

Recent development of ultra-short ultra-high intensity laser technology based on CPA concept(1) brought such a system that can generating a train of pulses with 100 TW peak power and 10 fs duration in a compact size that can be installed in a laboratory room of normal size (2).

There are two important features in such a system. One is the compactness of the system which brought high field phenomena, which used to be a subject needing a building sized laser system which only existed in a fusion laboratory, to a laboratory of normal size available for faculty level researches. Another feature, which may be even more important than the former in many cases, is that an ultra-high peak intensity of electromagnetic radiation can be produced in a time duration of 10 fs.

Irradiation intensity obtained by the use of a modern CPA system can be high enough for us to readily observe high field effect such as above threshold ionization of atoms, and for even higher intensity, relativistic effect such as self-focusing and transmission of laser pulse in the over dense plasma.

The energy of ultra-short laser pulse should be absorbed only by electrons when the pulse duration is on the order of or less than 10 fs. Various new scheme of absorption of super-intense electromagnetic wave have been investigated theoretically and experimentally. After the absorption of laser energy by electron system, the thermalization of the electrons, energy transfer from the electron system to the ion system occurs successively, giving us opportunity to investigate various relaxation phenomena in variety of material phases. Interaction of the electromagnetic waves with matter should show different behavior following such relaxation processes: for example, when the laser pulse has a steep rise front, the quivering motion of the electrons is prominent before the laser energy is absorbed by the electron system, and the emission of the short wavelength electromagnetic radiation due to bremsstrahlung in the non-relativistic region and relativistic Larmor radiation in deep relativistic region may be observed. After the electrons are excited to the ionized states, the mechanism of the radiation-matter interaction may depends on the state of the matter, providing very interesting new subjects to be investigated.

The new field of science opened by modern CPA lasers is extremely interesting not only as the subject of basic research, but also for a broad range of applications. For example, the fact that the laser energy is primarily given to the electron system gives the cutting processes by CPA lasers a non-thermal feature such as seen in the process using UV lasers. Such a feature appears in the cutting of all kinds of materials, regardless the state of the cut material: metal, semiconductor, insulator or even living tissue can be treated in a same manner. This gives us a very promising future of high precision cutting in various fields including medical surgery.

X-ray generation is another application of CPA lasers which is expected to have a broad range of applications. The mechanisms of x-ray generation by a CPA laser provide new subjects to be investigated, extending in a broad area including emission of x-rays from electrons making quivering motion in the electromagnetic waves including high order harmonic generation, amplified stimulated emission in a recombination plasma (recombination type x-ray laser) or inverted population produced by the electron excitation in a plasma including highly ionized atoms having stable electron configuration as the major component (x-ray laser by electron collisional excitation), the black body radiation from high temperature high density plasma, and Thomson and Compton scattering with an electron beam.

Laser acceleration is another new field having a potential capability of developing an electron accelerator with 10^4 times of acceleration rate per unit length compared with the present systems based on acceleration in the radiofrequency field. Although only proof of principle type experiments are going on at present and a lot of component developments are needed before designing a test bench laser accelerator, it should well worth to spend a good effort for the realization of laser accelerators, because the merit we can expect on the laser accelerator is quite large.

In consideration of the present sate of art and future prospect of super-strong field science, a program called "Advanced Photon Research" was launched at the Science and Technology Agency of Japan, and a new institution called Kansai Research Establishment was established in October 1995. The program may include research activities belonging to five categories: laser developments, x-ray lasers in a broad sense, laser acceleration, other application researches, and basic technology developments. The following part of the present article is dedicated to describe the prospect and the present status of this program.

LASER DEVELOPMENTS

In order to keep research activities having sufficient impact in the frontiers of the super-strong field sciences, a great deal of effort should be dedicated for the development of laser systems used in the present program, and naturally laser development consist the major part of the program, in order to keep the laser sources compatible with the most advanced systems. In consideration of our needs, which may be described as (a) needs to generate the highest irradiation intensity available in the state of art, (b) needs of reasonably high irradiation intensity in a large volume and (c) needs of moderately high peak power and high repetition rate, we develop laser source of two types: (i) a repetitive ultra-short ultra-high intensity laser system in order to take care of the first two types of needs and (ii) high repetition rate laser systems to take care of the last type of needs.

Repetitive Ultra-High Intensity Ti:Sapphire Laser System

In order to give a feature which is distinctly different from other competitive systems existing in the world, we try to make the pulse duration of our system the shortest one. Setting the purpose of our first stage development as 1 PW, 20 fs system with a repetition rate on the order of a Hz, we presently have a system generating 9.6 TW nearly diffraction limited output in a pulse duration of 16.6 fs with a repetition rate of 10 Hz. This system is designed as a part of a 100 TW, 20 fs system with a repetition rate of 10 Hz, the assembling work of which being already completed, expecting the first 100 TW operation within 1997.

The detail of the design of our 100 TW system and the performance of 9.6 TW operation is described elsewhere(3). The system is composed of Kerr lens mode-locked oscillator using chirped mirrors generating a train of pulses having less than 10 fs duration, followed by a pulse stretcher which stretches the pulse to about 1 ns duration, band-width controlled regenerative amplifier, four-path amplifier and two-path amplifier. The highest care is taken to compensate the effect of higher order (up to the fourth order at present) nonlinear indices of refraction in designing the system, and the bandwidth reduction by saturated amplification is compensated by inserting double etalons in the regenerative amplifier. The far-field pattern of 9.6 TW output, which is obtained by operating up to the four-path amplifier, shows diffraction limited near Gaussian intensity distribution.

The four-path amplifier shows the pump (532 nm) to output (320 mJ at 800 nm) conversion efficiency of 53 percent, with a good agreement with the theoretical value. The two-path amplifier is designed to give 800 nm amplified pulse with 55 percent efficiency by >7 J of green pump. The vacuum compressor with 1200 groove/mm gold plated gratings will generate 2J pulse with 20 fs compressed pulse duration with 60 percent efficiency.

The design work for 1 PW system will follow the generation of 100 TW, together with development of the green pump source for it. One very important requirement to a laser system generating such extremely high peak power is to have very high contrast. We are keen in developing a system with high contrast ratio especially because generation of the highest peak power with the shortest pulse duration is our propaganda. Besides the use of saturable absorber and Faraday isolator, we also investigate new types of saturable absorbers to cot off the leading tail. One interesting example is to use a very thin layers of near critical density plasma. Our PIC code analysis predicts that such a system cuts off the leading edge of the incident pulse and, as the result, makes the pulse duration much shorter than the input(4).

High Repetition Rate Ti:Sapphire System

For many of the applications of TW lasers the most appropriate peak power is on the order of a TW or less, and high repetition rate is required instead. The required pulse width is different from one application to the other. For applications in cutting various matters, there shouldn't be much need of the effort to obtain extremely short pulse, and harmonic generation of shorter wavelength soft x-rays, for example those within or near the "water window", by high order harmonic generation is known to need pulse duration of 25 fs or shorter. The design work of a high repetition rate Ti:sapphire laser will be made in consideration of its application.

One of the technical difficulties is to obtain sufficiently high intensity pumping pulse. We started this kind of development by working on pumping lasers, and developed all solid state Nd:YAG zigzag slab laser MOPA system generating 105 W average power in green.

Development of Yb:Glass Systems

Although Ti:sapphire is the most suitable material for CPA laser system for the shortest pulse duration up to 1 PW peak power, the required size of the crystal needed to generate over 1 PW level is too large to obtain optically uniform one by the crystal grawth technology in the present state of art. Cooling a large crystal is another technical difficulty we must encounter in order to develop repetitive system generating repetitive pulses of more than several tens of joules.

In order to open possibility of going beyond 1 PW repetitive CPA laser system, and also to develop small scale CPA lasers for applications needing modestly short pulse duration of the order of 100 fs, we have started joint research on Yb:glass CPA laser system pumped by free running lasers in collaboration with CUOS of the University of Michigan. The proof of principle experiments as the first step toward ultra-high intensity Yb:glass CPA laser development was successful (5). The main problem at present is the pump source. Because the Yb system has an energy level structure of quasi three level, the lower state of the laser transition is overlapping with the ground state, absorption at the laser wavelength is extemely high unless the medium is pumped intensely by the pumping source having sufficiently high brightness. This is the penalty we must pay in order to enjoy its advantage of having extremely high quantum efficiency.

On the other hand, Yb system has potential advantage of being able to pump directly by the fundamental output of semiconductor laser diode arrays. Possibility of high doping with which the laser medium can have a high energy storage of 30 times of Nd system is another important advantage. Low radiative cross section is advantageous from the stand point of seeking a compact system, although it may be regarded as disadvantage of having extremely low gain per length in designing the front end. At this moment until high power diode array as a high brightness pumping source at 940 nm, we have developed flash-lamp pumped Ti:sapphire rod laser generating pulses of more than 5 joules at this wavelength to simulate the pumping with a high brightness LD array.

X-RAY LASER RESEARCHES

Generation of x-ray is the application having our first priority, partly because of the responsibility of JAERI as a national laboratory to keep up the frontiers of radiation sciences, and partly because of the importance of the potential application of the x-ray sources generated by ultra-high intensity lasers in various field of science and technology. We summarize such a source as "x-ray lasers" in a broad sense.

X-Ray Lasers Excited by CPA Lasers

The most interesting subject to be investigated as an application of CPA lasers is the x-ray lasers in its true sense of the word: x-ray generation by stimulated emission in a medium having inverted population between the x-ray transition of atoms. There have been a significant effort to develop x-ray lasers using highly ionized atoms as the working media by excitation of high energy lasers developed for fusion researches. By the use of compact CPA laser system, x-ray laser research has been brought into normal scale laboratories, and several pioneering works have been carried out with a good success (6).

However, before the experimental works are carried out, we make careful numerical examination in order to know how and what extent the efficiency of the x-ray lasers may be improved by the use of CPA lasers with a reasonably good accuracy. Actually, there have been no systematic investigations to evaluate the impact of CPA lasers in x-ray laser researches so far, as far as the knowledge of the present author is concerned. It is also important to introduce new schemes such as the use of multiple pulses of various combinations to prepare a medium with which efficient laser performance can be expected by the excitation of a CPA laser pulse.

In order to carry on such researches, it is our essential need to have reliable atomic codes and data base for a broad range of atoms in combination with PIC or hydrodynamic codes to deal with a laser excited plasmas. The most of works to develop such atomic codes and data base have been carried out in combination with experiments done in much slower time scale. By the use of CPA lasers in such experiments, we may opportunities to improve the accuracy of the codes and data taking advantage of observing extremely fast processes. Such investigations on atomic processes are also included in our research program, not only using those codes and data in the x-ray laser researches.

One example of our x-ray laser experiments is a recombination type x-ray lasers using Ne diluted in hydrogen gas. Hydrogen works as the donor of low energy electrons to improve the recombination rate in the x-ray laser of Li type Ne. Preliminary analysis by PIC code predicts gain of 16 cm^{-1} for a Ne density of 7×10^{17} cm^{-3}, with a mixing ratio of 1 neon to 100 hydrogen, under the irradiation of 10^{17} W/cm^2 irradiation of 800 nm radiation. Another example is the electron collisional excitation type x-ray lasers using a plasma prepared by the irradiation of a prepulse on a thin metal foil, although detailed analysis will not be given here.

Generation of gain in the medium is not only a needed process in x-ray lasers. We also need the knowledge of propagation of CPA laser beam in plasma, time evolution of the laser produced plasma, and ray trace of the x-ray radiation in the excited media are essential subject to be understood before designing the optimized x-ray lasers.

X-Ray Emission by CPA Laser Irradiation

The emission of x-ray radiation from various materials is also very interesting subject to be investigated by the use of CPA lasers. When a CPA laser output with ultra-high peak power and ultra-fast rise leading edge is focused on a surface of material having solid-state density, the first phenomenon we expect is the quivering motion of the electrons in the material. Nonlinearity due to the relativistic motion together with the Coulomb interaction with nuclei cause the quivering electrons to emit x-ray radiation.

Then the excitation of the electrons with ATI and various heating processes of the electrons (such as resonant absorption) follow, bringing in high degree of incoherence in the quivering electron motion. Such situation may makes a very peculiar state of matter where the ions still remain at the same position as the solid, and many ionized and heated electrons are spread over the material. As the energy is transferred to the ions gradually, and the material start to expand. Such a complicated process strongly depends on how long is the duration of the laser pulse. The emission of x-ray radiation should be related strongly with such a short term evolution of the states of the matter, and detailed investigation of such processes is needed to understand the x-ray emission from the matter completely. The experimental probing techniques of such phenomena having sufficiently high time resolution are needed to investigate such processes with high accuracy.

LASER ACCELERATION

Another interesting application of CPA lasers is the acceleration of electrons by the super-strong electric field generated by the laser pulse. Several possibilities have been proposed, and our interest is concentrated in the acceleration in the wake-field generated by an intense ultra-short laser pulse propagating in low density plasma. The scheme of laser wake-field acceleration was proposed a long time ago (7), and the existence of wake-field has been experimentally shown recently.

Our scenario in laser acceleration research is (i) to observe the wake-field by interferometric measurement, (ii) to show the experimental evidense of the wake-field acceleration, (iii) secure sufficient working distance (for example, by self channeling of the laser beam), (iv) develop injection system for a good number of electrons at the right phase (temporary and spacially) of the wake-field, and (v) finally to improve the quality and efficiency as an accelerator.

So far acceleration of the electrons injected by a linear accelerator of around 100 MeV was demonstrated. Experimental observation of the wake-field was also successful. Interesting behavior and long channeling in low density plasma were also studied.(8). Those are only preliminary works, and further works are planned along above shown scenario.

OTHER APPLICATIONS

The broad range of applications of CPA lasers and x-ray laser are expected to follow the researches on the line described above. A part of the applications such as x-ray microscope observation of living tissues are already in progress. As one of the government institutions, JAERI Kansai is supposed to operate as a national center of application researches of CPA lasers and x-ray lasers including physical, chemical, industrial and medical applications. We also expect that the collaboration will expand internationally.

Another types of researches we expect to commit in near future is the generation of high energy x-rays by the interaction of CPA laser output with high energy electron beams. Advantage of JAERI Kansai is that there is another division for application researches of the orbital radiation of a high energy electron storage ring called Spring-8, which has just started normal operation. Beside the main ring, synchrotron injector and/or 1 GeV LINAC pre-injector may give us opportunity to try such kind of new trials.

REFERENCES

1. D. Strickland and G. Mourou, Opt. Commun. **56**, 219 (1985).
2. C.P.J. Barty et al., Opt. Lett., **21**, 668 (1996).
3. K. Ymakawa et al., P-M-6, SUPERSTRONG FIELD IN PLASMAS, Varenna, 1997.
4. K. Nagashima et al., O-F-2, SUPERSTRONG FIELD IN PLASMAS, Varenna, 1997.
5. J. Nees et al., in the Proceedings of JAERI-ICFA Conference on Advanced Accelerators, 1997.
6. B.E. Lemoff et al., Phys. Rev. Lett., **74**, 1574 (1995).
7. T. Tajima and J.M. Dawson, Phys. Rev. Lett., **43**, 267 (1979).
8. K. Nakajima et al., Phys. Rev. Lett. **74**, 4428 (1995).

High Energy Electron Acceleration by Laser Wakefields

K. Nakajima, H. Nakanishi and A. Ogata
High Energy Accelerator Research Organization,
Tsukuba, Japan

M. Kando
Institute for Chemical Research, Kyoto University, Uji, Japan

H. Ahn, H. Dewa, S. Kondo, H. Kotaki and H. Sakai
Japan Atomic Energy Research Institute, Tokai, Japan

T. Ueda, M. Uesaka, T. Watanabe and K. Yoshii
Nuclear Engineering Research Laboratory,
The University of Tokyo, Tokai, Japan

Abstract. The laser wakefield electron acceleration over 100 MeV has been demonstrated in an underdense plasma driven by a 2 TW, 90 fs laser pulse synchronized with 17 MeV RF linac electron injector at 10 Hz. The electron acceleration was enhanced at a pressure corresponding to the plasma density higher than the resonant density due to gas ionization and self-channeling effects. The wakefield excitation has been confirmed by measuring the electron density oscillation of the plasma wave with the frequency domain interferometer.

I INTRODUCTION

Laser-plasma accelerators have been conceived to be the next-generation particle accelerators, promising ultrahigh field particle acceleration and a compact size compared with conventional accelerators [1]. Among a number of laser accelerator concepts, the laser wakefield accelerator [2] has been revived because of a simple mechanism similar to the plasma wakefield accelerator [3] and availability of high peak power, ultrashort pulse lasers. It has been experimentally demonstrated that laser wakefield acceleration has great potential to produce ultra-high-field gradients of plasma waves excited by intense ultrashort laser pulses [4,5]. Recently a wakefield of the order of 10 GeV/m in a plasma has been directly observed by the use of a compact terawatt laser system so called T^3 lasers [6]. In a homogeneous plasma, however, diffraction of the laser propagation limits the laser-plasma interaction distance to the extent of the vacuum Rayleigh length. This effect deducts the advantage of ultrahigh gradient acceleration from laser-driven accelerators. It would be of importance for practical application of the laser wakefield accelerator concept to be able to generate a high energy gain as well as high gradient acceleration. Therefore it is essential for laser wakefield acceleration to achieve a long interaction of an intense ultrashort laser pulse with an underdense plasma in order to increase the energy gain from tens of MeV to the order of GeV. Although a self-modulated laser wakefield can generate the accelerating field exceeding 100 GeV/m, this mechanism could produce at most the energy gain of ~ 100 MeV because of a short acceleration length limited by dephasing of accelerated electrons or depletion of pump pulses [5]. In order to exceed a limit of the acceleration length restricted by diffraction, optical guiding has been proposed as a promising way of propagating a high-power laser pulse over many Rayleigh lengths in a plasma [7]. A laser beam may be guided through a plasma of which the refractive index along the optical axis is sufficiently so high as to compensate diffraction. The relativistic self-guiding in a homogeneous plasma has been predicted to occur above the critical power, given by $P_c = 17(\omega_0^2/\omega_p^2)$ GW where ω_0 is the laser frequency and ω_p is the plasma frequency [8]. Recent experiments using intense femtosecond laser pulses report that a long distance self-channeling has been observed under the critical power of the relativistic self-focusing [9]. We have also observed self-channeling of a 2 TW laser pulse with 90 fs duration occurred over a few cm even under the relativistic critical power [10]. We made an attempt to demonstrate laser wakefield acceleration by intense ultrashort laser pulses propagating owing to self-channeling in a plasma. Here we report the experimental results of laser wakefield acceleration of an externally injected electron beam synchronized with laser pulses.

II LASER WAKEFIELD ACCELERATION

The laser wakefield results from the ponderomotive force exciting the density oscillation of a plasma with the frequency $\omega_p = \sqrt{4\pi e^2 n_e/m_e}$ for the ambient electron density n_e in a plasma. Assuming a Gaussian laser pulse of a temporal $1/e$ half-width σ_z with the peak power P, a peak amplitude of the accelerating wakefield is

$$eE_z = \frac{\Omega_0 P}{\sqrt{\pi} m_e c^2}\left(\frac{\lambda_0}{\lambda_p}\right)\left(\frac{k_p \sigma_z}{2 Z_R}\right)\exp\left(-\frac{k_p^2 \sigma_z^2}{4}\right), \qquad (1)$$

where Ω_0 is the vacuum resistivity (377Ω), λ_0 is the laser wavelength, λ_p is the plasma wavelength, $k_p = 2\pi/\lambda_p$ and Z_R is the vacuum Rayleigh length, i.e. $Z_R = \pi R_0^2/\lambda_0$, where R_0 is the spot radius at the focus. The maximum amplitude is achieved at $\lambda_p = \pi \sigma_z$. The maximum energy gain of relativistic electrons is given by $\Delta W = eE_z L_{ac}$ with the acceleration length L_{ac}. Diffraction limits the acceleration length to $L_{ac} \simeq \pi Z_{eff}$, where Z_{eff} is the effective Rayleigh length for the laser propagation in a plasma. For the resonant plasma density, $n_e = 1/(\pi r_e \sigma_z^2)$ where r_e is the classical electron radius,

$$\Delta W_{\max}[\text{MeV}] \simeq 850 P[\text{TW}] \lambda_0[\mu\text{m}]/\tau_0[\text{fs}] \times (Z_{eff}/Z_R), \quad (2)$$

where τ_0 is the FWHM pulse duration, $c\tau_0 = 2\sqrt{\ln 2}\sigma_z$. Note that the maximum energy gain is independent of a focal spot size of laser pulses for Gaussian propagation in a homogeneous plasma where $Z_{eff} = Z_R$.

III LWFA TEST FACILITY

We have constructed the LWFA test facility consisting of the T^3 laser system and the electron beam injector [11]. The Ti:sapphire T^3 laser system based on the chirped-pulse amplification at $\lambda_0 = 790$ nm produces output pulses compressed by a grating compressor to 90 fs with an energy of $>$ 200 mJ corresponding to a peak power of $>$ 2 TW at the repetition rate of 10 Hz. It is necessary to inject an electron beam with an appropriate initial energy so that electrons can be trapped and accelerated by wakefields. We used the 2856 MHz RF linac as an electron injector to produce a 17 MeV single bunch beam with a 10 ps FWHM pulse duration containing $\sim$ 1 nC at the repetition rate of 10 Hz.

The setup for acceleration experiments is shown in Fig. 1. Focusing optics and the injection electron beamline were installed in the vacuum chamber filled with He gas. Laser pulses were focused with f/10 off-axis parabolic (OAP) mirror with a focal length of 480 mm. The measured focal spot radius was

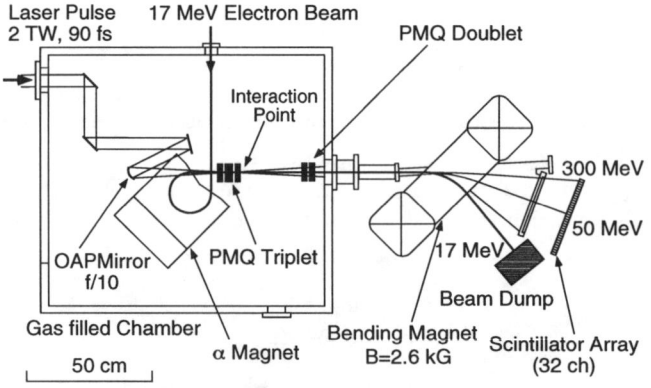

FIGURE 1. Schematic of the experimental setup.

13μm. Since the focusing force of the radial wakefield exists at $r < R_0/2$,

electrons injected to the diameter less than a half laser spot size would be trapped and accelerated by the wakefield. An electron beam from the injector is brought to a focus in the chamber with the FWHM beam size of 0.8 mm through a beamline consisting of a triple focusing magnet and a permanent magnet quadrupole (PMQ) triplet. The RF linac and the beamline were separated with a $50 \mu m$ thick titanium window from the interaction chamber to maintain ultrahigh vacuum in the electron injector. Since this window caused emittance blow-up due to multiple scattering of electrons, the collimator slit was installed at the downstream of the window to reduce the beam emittance. Beam collimation reduced an electron charge to ~ 100 pC per pulse.

An electron pulse was synchronized to laser pulses with the phase locked control of the mode-locked oscillator. The phase locked loop maintained synchronization of the oscillator repetition period (79.33 MHz) with every 36th RF period of the linac (2856 MHz). We measured a timing jitter between the laser pulse and Cherenkov radiation from the electron beam with the streak camera with a time resolution of 200 fs. Synchronization between two pulses was achieved within the rms jitter of 3.7 ps.

Inprovements of the injection electron beam will be accomplished by the use of a photocathode RF gun and a bunch compression chicane. The photocathode RF gun has been developed to produce a good quality elctron beam with the normalized emittance of $\sim 1\pi$ mm-mrad and the pulse duration of ~ 1 ps. The electron pulse delivered from the photoinjector is further compressed to ~ 100 fs through the magnetic bunch compressor after acceleration in the 17 MeV RF linac in order to achieve a complete phase matching of the electron beam with the laser wakefield.

IV ELECTRON ACCELERATION EXPERIMENT

The energy of accelerated electrons was measured with the magnetic spectrometer consisting of a dipole magnet and an array of 32 scintillation detectors. The pulse heights of the detectors were recorded with ADC triggered by a gate signal synchronized with the linac electron pulse. Injected electrons undergoing no acceleration were swept out of the detectors by the spectrometer magnet following a PMQ doublet. The spectrometer covered the energy range of $10 \sim 300$ MeV. The energy calibration of the spectrometer was made by varying the magnetic field to measure a 17 MeV electron beam from the RF linac.

The timing between laser and electron pulses was adjusted by changing a phase delay of the reference RF to the phase locked loop so that streak images of two pulses were overlapped. After a He gas was filled in the acceleration chamber, fine adjustment of overlapping two spots of laser and electron beams was carried out within 50 μm. Two sets of pulse height data of the scintillator array were taken with pump laser pulses and without them as a background.

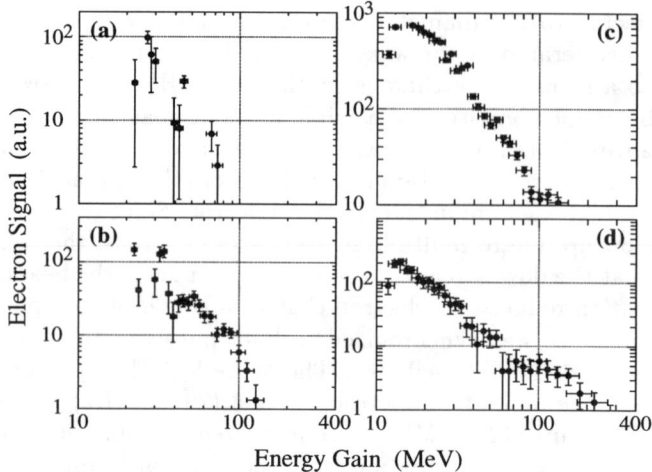

FIGURE 2. Energy gain spectra of accelerated electrons for a) 3.4 Torr, P=0.9 TW, b) 20 Torr, P=0.9 TW, c) 2 Torr, P=1.8 TW, and d) 20 Torr, P=1.8 TW.

The pulse height was averaged over 500 to 1000 shots to reduce a signal fluctuation. A net pulse height proportional to the number of electrons accelerated was obtained from subtracting the data without the pump pulses from the data with them. The number of electrons was estimated to be ranged from 2 to 4 per ADC count for all detectors.

In the acceleration experiment the gas pressure of He was scanned from 0 Torr to 300 Torr. Figure 2 shows energy gain spectra of electrons accelerated in the wakefield pumped by the laser peak power of 0.9 TW and 1.8 TW. The maximum energy gain up to 300 MeV was obtained from these data. In the vacuum pressure of 4.3 mTorr we confirmed no electron acceleration occurred when an electron pulse was injected with a proper delay. In order to make a proof of wakefield acceleration, we have measured off-timing interaction of electrons with respect to the pump laser pulse. When the electron pulse preceded the pump laser pulse, no acceleration of electrons was observed. Accelerated electrons visibly appeared as the electron pulse was delayed. No accelerated electron was observed at a delay of 1 ns. We have investigated the self-trapping of plasma electrons due to wakefields without the electron beam injection as the gas pressure was increased up to 760 Torr. We observed no electron accelerated to higher energies than ~ 1 MeV. It implies that the large amplitude wakefield excitation caused by the stimulated Raman instability may be suppressed for ultrashort laser pulses of 100 fs in contrast with the self-modulated laser wakefield.

The side scattered laser light from the plasma region were imaged onto a CCD camera synchronized with a 10 Hz laser trigger through a 10 nm FWHM

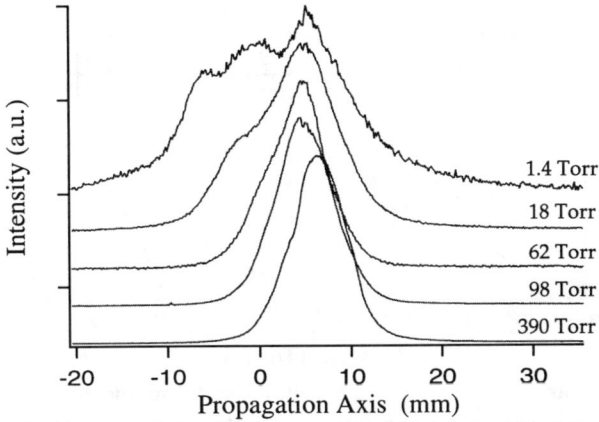

FIGURE 3. The lineouts of the side scattered laser light at various He gas pressures for P=2.5 TW. The laser pulse propagates from the right to the left of the figure.

interferential filter to measure the laser intensity distribution. Figure 3 shows an intensity profile of the scattered light projected onto the propagation axis for various gas pressures. The laser pulse was focused at the position of 0 mm on the propagation axis in vacuum. Since the wakefield amplitude is proportional to the laser intensity, the lineout width of the scattered light proportional to the laser intensity gives a good estimate of the acceleration length.

The acceleration field gradient was obtained from the measured maximum energy gain and the FWHM length of the scattered light lineout equal to $2Z_{eff}$. Figure 4 shows the peak accelerating gradient for the laser peak powers of 0.9 TW and 1.8 TW as a function of the gas pressure of He. The data indicate a good agreement with theoretical prediction based on the linear fluid model below 10 Torr, assuming the focal spot radius of 13 μm. It is found that acceleration occurs even at much higher pressures than a plasma wake resonance. Recent particle-in-cell simulations elucidate that a large amplitude wakefield can be excited by self-modulation of the laser pulse induced in rapidly ionizing plasmas at the pulse front [12]. In Fig. 4 the acceleration gradient jumps to $\sim$ 15 GeV/m above 20 Torr for 1.8 TW. It has been observed that the lineout of the scattered laser light was turned out to be a constant length of $\sim$ 1 cm above 20 Torr.

V LASER PROPAGATION MEASUREMENT

We have investigated the propagation of ultrashort terawatt-power laser pulses in gases. In a gas a plasma is produced via tunneling ionization due to

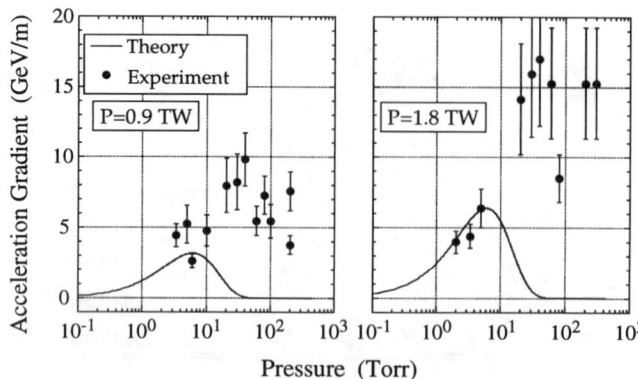

FIGURE 4. The peak accelerating field gradient deduced from the maximum energy gain and the acceleration length. The solid curve shows theoretical expectation for the spot radius $R_0 = 13\mu m$.

an intense laser field higher than the threshold intensity of 9×10^{15} W/cm^2 for He^{2+} ionization. In the vacuum chamber filled with helium (He) gas, the forward scattered laser light was imaged onto a charge-coupled-device (CCD) camera coupled to the microscope objective through a 10 nm FWHM interferential filter to measure a focused spot size along the propagation axis. The spot size was measured at each location along the propagation axis as shown in Fig. 5 (a) for 20 Torr. The axial Intensity distribution calculated from the measured spot size is shown in Fig. 5 (b). The effective Rayleigh length defined as a half width at the half maximum of the intensity distribution is much longer than the vacuum Rayleigh length corresponding to the measured spot size at the best focus. This self-guiding takes place at the lower power than the critical power for the relativistic self-guiding. It is implied that a mechanism of this self-channeling may be attributed to leaking mode self-effect due to field-induced saturable ionization recently presented as a new mechanism of self-channeling of an ultrashort laser pulse in underdense plasmas [13]. When a plasma filament with a sharply bounded cross-section is created via tunneling ionization owing to strong dependence of the field ionization on the field intensity, a laser pulse is guided in the form of a leaking mode with small losses over many Rayleigh lengths by a strong reflection of an electromagnetic wave on a sharp plasma boundary. Assuming an uniform electron density to be saturated inside the plasma waveguide with the radius of r_p, the ratio of the leakage distance Z^* to the vacuum Rayleigh length Z_R is given by $Z^*/Z_R = k_p r_p^3 / r_0^2$, where r_0 is the spot size of a laser pulse focused on a gas. For $k_p r_p \geq 1$, a long distance self-channeling, $Z_{eff} \approx Z^* \gg Z_R$, takes place through the plasma waveguide with the radius, $r_p \gg r_0$.

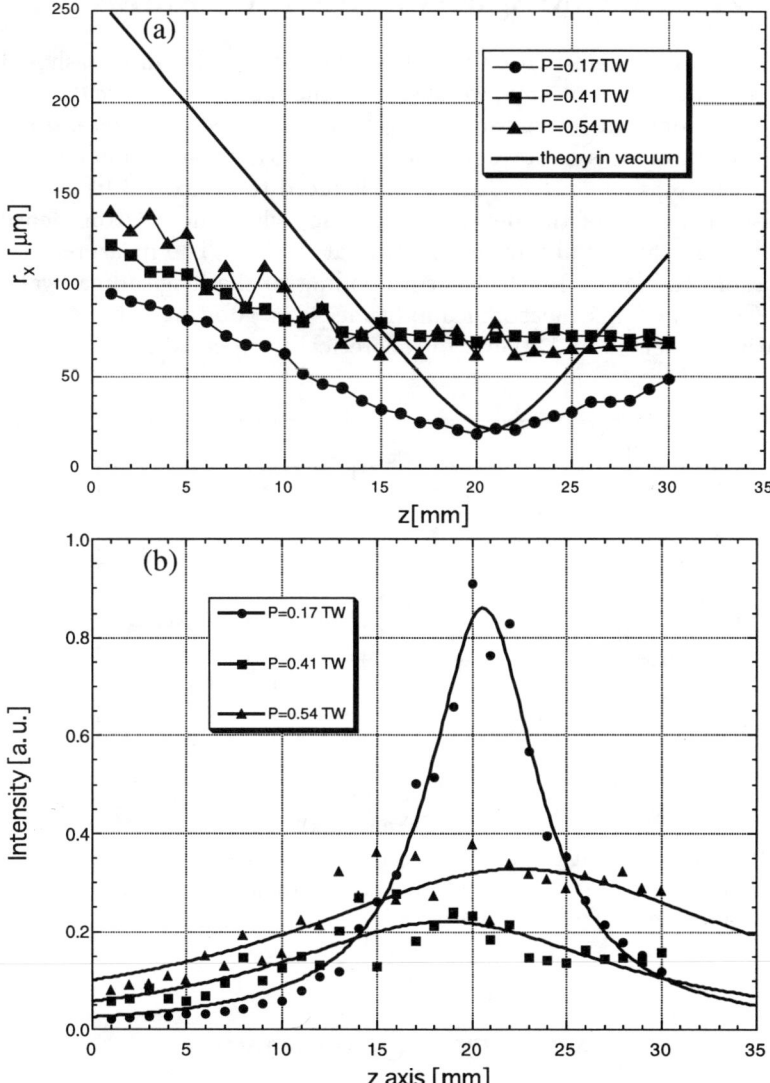

FIGURE 5. (a) Spot radius measured along the axis and (b) Intensity distribution as a function of propagation distance in a He gas of 20 Torr.

VI PLASMA WAVE MEASUREMENT

In order to make confirmation of wakefield excitation by an ultrashort laser pulse in an underdense plasma, we measured the plasma wave oscillation with the frequency domain interferometer [6] to be reported elsewhere. Figure 6 shows the electron plasma wave measured at 2 Torr for the pump peak power of 1 TW. The plasma electron density is deduced to be 9.6×10^{16} cm^{-3} from the measured period of the density oscillation, while the electron density is 1.4×10^{17} cm^{-3} for a fully ionized plasma at 2 Torr. The measured density perturbation was $\langle \delta n/n_e \rangle \sim 15\%$ corresponding to the longitudinal wakefield of ~ 2 GeV/m. This measured amplitude is in good agreement with the accelerating wakefield of 2.2 GeV/m theoretically expected for 1 TW.

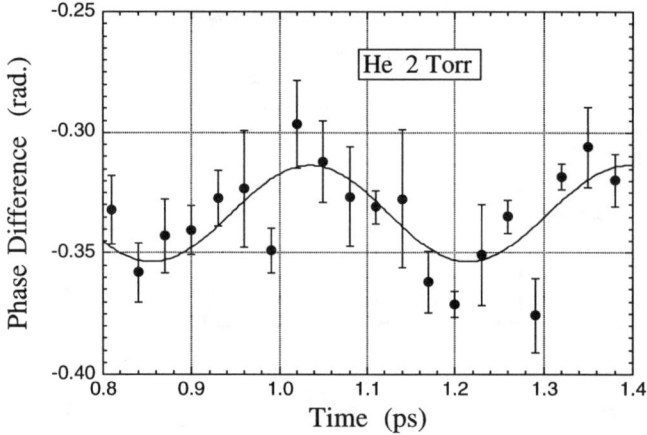

FIGURE 6. Measurement of the plasma density oscillation excited by a 1 TW pump power in a He gas of 2 Torr. The solid curve shows a fit of the plasma wave with oscillation period of 360 fs.

VII CONCLUSIONS

In conclusion we have carried out electron acceleration by wakefields excited by intense ultrashort laser pulses delivered from a 2 TW, 90 fs T^3 laser system synchronized with the 17 MeV RF linac electron beam injector at the repetition rate of 10 Hz. We have observed high energy electrons accelerated over 100 MeV up to 300 MeV by the wakefield of ~ 15 GeV/m excited over a few cm long underdense plasma. Acceleration enhancement at higher pressures of He gas than the resonant pressure may be elucidated from the pulse self-modulation due to ionization and self-channeling of the laser pulse in a plasma. These effects help to achieve an efficient electron acceleration up to the higher energy than 1 GeV. A direct measurement of the electron density oscillation by the frequency domain interferometry verified wakefield excitation consistent with the results of acceleration experiments.

REFERENCES

1. T. Tajima and J. M. Dawson, Phys. Rev. Lett., **43**, 267 (1979).
2. P. Sprangle et al., Appl. Phys. Lett. **53**, 2146 (1988).
3. P. Chen, J. M. Dawson, R. W. Huff, and T. Katsouleas, Phys. Rev. Lett., **54**, 693 (1985).
4. K. Nakajima et al., Phys. Rev. Lett., **74**, 4428 (1995).
5. A. Modena et al., Nature (London) **377**, 606 (1995); D. Umstadter et al., Science **273**, 472 (1996); S. P. Le Blanc et al., Phys. Rev. Lett. **77**, 5381 (1996).
6. J.R. Marquès et al., Phys. Rev. Lett., **76**, 3566 (1996); C.W. Siders et al., Phys. Rev. Lett., **76**, 3570 (1996).
7. E. Esarey, P. Sprangle, J. Krall, A. Ting, and G. Joyce, Phys. Fluids B, **5**, 2690 (1993).
8. G. Z. Sun, E. Ott, Y. C. Lee, and P. Guzdar, Phys. Fluids **30**, 526 (1987); D. C. Barnes, T. Kurki-Suonio, and T. Tajima, IEEE Trans. Plasma Sci. **PS-15**, 154 (1987); P. Sprangle, C. M. Tang, and E. Esarey, IEEE Trans. Plasma Sci. **PS-15**, 145 (1987).
9. A. Braun et al., Opt. Lett., **20**, 73 (1995); E. T. J. Nibbering, et al., Opt. Lett., **21**, 62 (1996).
10. M. Kando et al., Advanced Accelerator Concepts 1996, to be published.
11. K. Nakajima, Phys. Plasmas, **3**, 2169 (1996).
12. D. L. Fisher and T. Tajima, Phys. Rev. E **53**, 1844 (1996); S. Kato, Y. Kishimoto, and J. K. Koga, Proc. of the 1996 Int. Conf. on Plasma Physics (ICPP 96), **2**, 1710 (1997) (The Soc. of Plasma Science and Nuclear Fusion Research, Nagoya).
13. A. M. Sergeev, M. Lontano, and A. V. Kim, OSA 1997 Technical Digest Series, **7**, 118 (1997).

Measurement of Ultra High Field Propagation and The Excited Wakefield

H. Dewa*, H. Ahn*, M. Kando[†], H. Kotaki*,
K. Nakajima[††] and A. Ogata[††],

*Japan Atomic Energy Research Institute
Shirakata Shirane 2-4, Tokai, Ibaraki, 319-11 Japan
[†] Kyoto University
Gokanosho, Uji, Kyoto, 611 Japan
[††] High Energy Accelerator Research Organization (KEK)
Oho 1-1, Tsukuba, Ibaraki, 305 Japan

Abstract. Ultra high field propagation of a intense laser was measured. The self-focusing of the laser was observed and it depends on the laser power and the gas pressure. The wakefield measured using frequency-domain interferometry was consistent with a theoretical prediction.

INTRODUCTION

The laser wakefield acceleration (1,2) is expected to be one of the next-generation accelerators because of its high electric field gradient and compact size. The laser wakefield accelerator accelerates electrons by the strong wakefield in the gas plasma induced by a high intensity short pulse laser. The acceleration gain depends on the acceleration length and the acceleration gradient of the wakefield. The acceleration length is mainly limited by the propagation length of the focused laser in a gas.

The relativistic self-channeling has been predicted to occur above the critical power, given by $Pc = 17(\omega_0^2/\omega_p^2)$ GW where ω_0 is the laser frequency and ω_p is the plasma frequency. Recently a long distance channeling has been observed under the critical power (3,4), which enabled us to obtain the acceleration gain higher than that expected without self-channeling effect. In our experiments of the wakefield acceleration exceeding 100 MeV energy gain was attained at the laser power of 1.8 TW and the helium pressure more than 20 Torr. Such a high energy gain might be realized in term of the self-channeling effect and the field gradient enhancement by its gas ionization.

Owing to the self-channeling effect of a laser pulse, the propagation length of the laser in a gas reaches many vacuum Rayleigh length. We have measured the propagation of a intense laser at different gas pressures and laser intensities, and report the results in this paper.
In order to study the wakefield excitation in a plasma, the plasma oscillation induced by a intense pulse laser has been measured by the frequency-domain interferometry. This measurement is important because the acceleration gradient in the wakefield can

be estimated by the measurement and is compared with results of the acceleration experiments.

LASER PROPAGATION

In a vacuum, the waist size of a gaussian laser beam is given by the following equation,

$$2w_0 = \frac{4\lambda}{\pi}\frac{S}{D} \tag{1}$$

here λ is the wave length of the laser, S is the focal length of the off-axis parabolic mirror, and D is the diameter of the laser on the mirror. The propagation length in a vacuum is calculated by the Rayleigh length

$$z_d = \frac{\pi w_0^2}{\lambda} \tag{2}$$

We assumed the intensity distribution as a gaussian distribution

$$I(x,y,z) = I_0(z)\exp(-\frac{x^2}{2\sigma_x^2})\exp(-\frac{y^2}{2\sigma_y^2}). \tag{3}$$

The energy density $I_0(z)$ on the beam axis is expressed by the following functions of z

$$I_0(z) \cong \frac{I_{0max}}{1+\left(\frac{z}{z_d}\right)^2} \tag{4}$$

where $I_0(0) = I_{0max}$ is the laser intensity density on the beam axis at the beam waist.

Due to the self-channeling effect, the propagation of the laser in gas is different from the predictions of the equation (4). In order to estimate the increase of the propagation length in a gas, the effective Rayleigh length z_{eff} is introduced as

$$I_0(z) = \frac{I_{0max}}{1+\left(\frac{z}{z_{eff}}\right)^2} \tag{5}$$

MEASUREMENT OF LASER PROPAGATION

In order to generate the ultra high field, we have prepared the Ti:sapphire laser system and the optical focusing system. The nominal peak power of the Ti-Sa laser is 2 TW and the time duration is 90 fs. Then the laser go into a chamber filled with the helium gas. The laser is focused by an off-axis parabolic mirror located in the chamber.

The propagation of the high intensity laser was measured by the setup shown in Fig 1. In our measurement, λ is 790 nm, S is 480 mm, and D is 40 mm. The waist size $2w_0$ and the vacuum Rayleigh length are calculated as 12 μm and 0.15 mm.

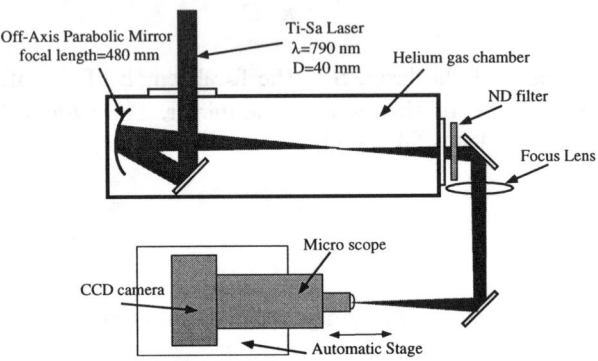

Figure 1 Setup for the laser propagation measurement

The images of the laser transverse profile were taken by the CCD camera connected to the microscope placed on the automatic stage. The propagation of the laser along the beam axis was measured by moving the stage longitudinally. These profile images were taken under the different conditions of the helium gas pressures and the laser power.

The gas pressure dependence of the measured peak energy density $I_0(z)$ at 20 Torr gas pressure is shown in Fig. 2.

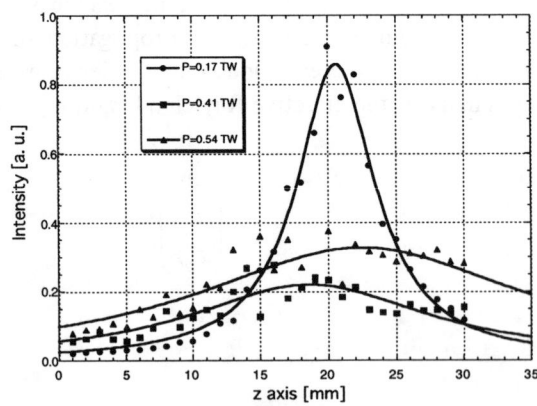

Figure 2 Intensity distribution of the laser on the beam axis at 20 Torr

Here we have obtained $I_0(z)$ by using the next equation, where P is the peak power of the laser pulse,

$$I_0(z) = \frac{P}{2\pi\sigma_x\sigma_y}$$ (6)

When the power of the laser is low, the peak value of the laser intensity is high and the propagation length is short. We calculated the effective Rayleigh length by using the equation (4).

The gas pressure dependence of the effective Rayleigh length is shown in Fig. 3. The propagation length in a gas is much longer than vacuum Rayleigh length, and it depends on the both the gas pressure and the laser power. At the low laser power of 0.17 TW, the effective Rayleigh length is almost constant to the gas pressure change. But when the laser power is higher, the effective Rayleigh length increase in the higher gas pressure region. It is clear that the effective Rayleigh length strongly depends on the laser power.

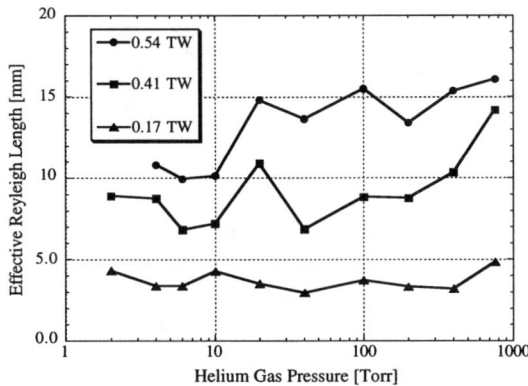

Figure 3 Pressure dependence of the laser propagation length

WAKEFIELD MEASUREMENT

In order to estimate the wakefield in a gas plasma, we used a frequency-domain interferometry, by which we can obtain the plasma density oscillation. The schematic of the frequency-domain interferometry measurement is shown in Fig. 4. The two probe lasers of the second harmonic at interval times of $T = 3\pi / \omega_p$, and the pump laser are aligned on the same path in the gas chamber. We used another probe lasers reflected on the backside of the mirror as the reference. We obtain a interference fringe in the wavelength measured by the monochrometor. The peak of fringe have a shift of the wavelength according to the difference of the plasma densities when two probes pass the plasma. The plasma density modulation can be calculated by the following equation

$$\delta n = \frac{\delta \Phi \lambda' n_e}{\pi L_p} \qquad (9)$$

where $\delta\Phi$ is a fringe shift in phase, λ' is a wave length of the probe laser, L_p is a interaction length. When the automatic stage of the pump laser was moved, the time relation between pump laser and the two probe laser change and the time dependence of the density modulation can be measured.

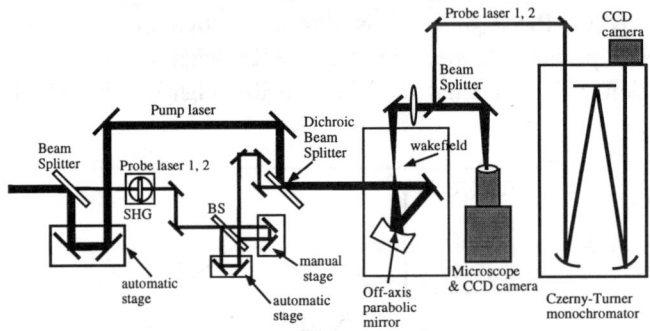

Figure 4 Schematic view of the wakefield measurement

The plasma density oscillation measured at 2 Torr for the pump of 1 TW is shown in Fig. 5. If we assume the full ionization of the helium gas, the plasma density perturbation $\delta n/n_e \sim 0.15$ corresponds to the longitudinal field gradient of ~ 3 GeV/m. This result is in good agreement with the accelerating wakefield theoretically expected.

Two dimensional distribution of the wakefield was obtained for the laser peak power of 1.4 TW. The contour plot of the wakefield is shown in Fig. 6. The oscillation of the plasma density is clearly measured. The longitudinal wakefield is dominant in the 2–D plasma distribution. Therefore the longitudinal electric field generated by the intense laser is higher than the transverse field, which is very important for the acceleration of electrons.

CONCLUSION

We measured the propagation of the intense laser in a gas. Owing to the self-channeling effect, the propagation length measured was longer than that expected in the vacuum. The self-channeling depends on gas pressure and laser power. We also measured the plasma density oscillation excited by a intense laser. The acceleration gradient estimated from the result was consistent with the acceleration experiment. The 2-D plasma distribution indicates the longitudinal field was dominant in the excited plasma.

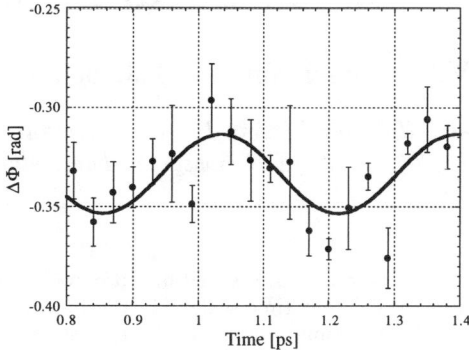

Figure 5 The wakefield oscillation excited by a 1 TW pump power in a He gas of 2 Torr. The solid curve show a fit of the plasma wave with oscillation period of 360 fs.

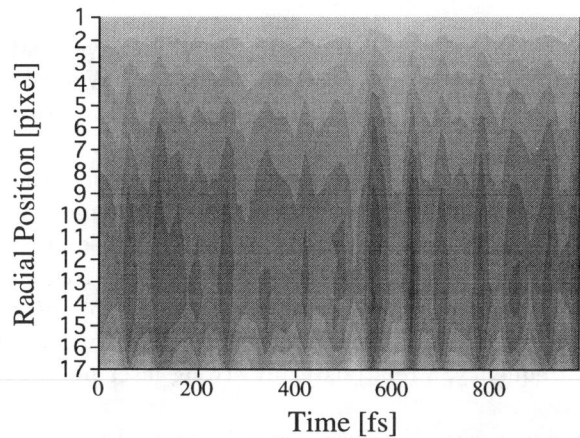

Figure 6 Two dimensional wakefield distribution measured at 6 Torr gas pressure and 1.4 TW laser power

REFERENCES

1. T. Tajima and J. M. Dawson, *Phys. Rev. Lett.*, **43**, 267 (1979)
2. P. Sprangle et al., *Appl. Phys. Lett.* **53**, 2146 (1988)
3. A. Braun et al., *Opt. Lett.*, **20**, 73 |(1995)
4. E. T. J. Nibbering, et al., *Opt. Lett.*, **21**, 62 (1996)
5. J. R. Marqués, et al., *Phys. Rev. Lett.*, **76**, 3566 (1996)
6 C. W. Siders et al., *Phys. Rev. Lett.*, **76**, 3570 (1996)

Mega-Ampere Ion Currents and Nuclear Reactions in the Focal Laser Spot

V.V. Goloviznin and T.J. Schep

*FOM-Institute for Plasma Physics "Rijnhuizen",
P.O. Box 1207, 3430 BE Nieuwegein, The Netherlands*

Abstract. The plasma response to a picosecond multiterawatt laser-pulse is characterized by several distinctly different time scales. In the present paper we consider the plasma evolution on the slow time scale typical for excitations of the ionic component. A relativistically strong laser pulse (10^{18}–10^{19} W/cm^2) is shown to drive, under proper conditions, mega-ampere ion currents with typical ion energies of 50–150 keV. This feature of the ultra-intense laser-plasma interaction may be of interest for nuclear fusion since the ion energy involved is close to the resonance peak of the cross-section for the D–T reaction.

The terawatt power domain is currently under active investigation, both theoretical and experimental. At an irradiance of about 10^{18} W/cm^2 and higher, the quiver velocity of electrons in the laser field becomes relativistic, which leads to considerable changes in plasma optics. In contrast to the low intensity propagation regime, such a pulse has been shown not to diverge in plasma due to diffraction, but to converge forming multiple foci which eventually merge into a single channel [1, 2, 3]. In addition to this self-channeling, the intense laser pulse becomes self-modulated [4] with a period determined by the plasma frequency, and excites ultrastrong (up to 100 GV/m) longitudinal electric fields [5] and multimegagauss magnetic fields in its wake [6, 7]. Finally, one must note that due to the Lorentz shift of the plasma frequency, a relativistically strong laser pulse can propagate in overdense plasmas.

All the above phenomena, however, are related to the fast response of the medium, which is associated with the excitation of its electronic component while the ion motion is neglected. The ion response, in general, did not receive much attention in the current literature. In the present report we would like to accentuate that the low-frequency domain may be of considerable interest as well. Namely, we find that a properly shaped terawatt laser pulse can drive mega-ampere ion current with a typical ion energy of 50–150 keV, which covers the region of the resonance peak in the D–T cross-section. The conversion efficiency of the pulse energy into ion kinetic energy turns out to be noticeably high, up to 10–12 %.

Although the ion motion as such has been noticed by several authors [8], the above mentioned magnitude of the ion current and the corresponding ion energy seem to be surprisingly high. One must note, however, that certain conditions should be fulfilled to reach this regime. In the present paper, we discuss the necessary conditions and assess their accessibility at today's technological level.

Our consideration is based on a simplified model of ion current generation. It takes into account neither laser pulse evolution nor the kinetic response of the plasma electrons. We assume the laser pulse to have already reached the (nearly) steady-state regime of self-channeling, when it just propagates as a "rigid" object of a given longitudinal and transverse extent. This approximation is justified by the results of recent experiments [2, 3] and numerical simulations [7]. As for the plasma electrons, we neglect their fast oscillations that take place on a typical time-scale of ω_p^{-1}, where the plasma frequency ω_p is given by $\omega_p^2 = 4\pi e^2 n/m$ with n being the plasma density and m the electron mass. In fact, the leading edge of the pulse will generate electron density oscillations in the bulk of the pulse. Even if the pulse profile is initially smooth (on the length scale determined by the plasma wavelength), relativistic self-focusing [1] will inevitably steepen the leading edge, so that the magnitude of the excited wake-field will be substantial. However, these fast electronic oscillations are not important for the ion dynamics under consideration, because their influence will be effectively averaged over many periods due to ion inertia. Therefore only the slow shift of the equilibrium point due to the presence of the laser field is relevant for the problem. Effectively, the situation looks like if the plasma electrons just smoothly follow the ponderomotive potential of the pulse; one may call this an "adiabatically slow" excitation of plasma waves. As we will see, the adiabatic approximation considerably simplifies the calculation of the ion response (note that in the following we always refer to time-averaged rather than instantaneous characteristics of the process).

An important observation consists in the fact that in the adiabatic approximation ions experience practically the same net influence from the laser pulse as the electrons, even though the direct ponderomotive forces acting upon ions and electrons differ by orders of magnitude. Indeed, a charged particle placed in a non-uniform optical field feels an effective (averaged over an optical period) ponderomotive force equal to (here and further $\hbar = c = 1$)

$$\vec{F}_{PM} = -\frac{1}{4M}\vec{\nabla}\left(\frac{eE_0}{\omega_0}\right)^2, \qquad (1)$$

where M is the mass of the particle, e is its charge, ω_0 is the laser frequency, and E_0 is the (position-dependent) amplitude of the optical electric field. Since an ion is about three orders of magnitude heavier than an electron, one may safely neglect the direct ponderomotive force acting upon ions, as compared to that acting upon electrons. But on the other hand, electrons are easily movable, and as soon as an external perturbation exists, they will move until the external force is compensated by the induced charge separation. This induced electrostatic field, in turn, leads to an additional force acting upon ions. In the case of hydrogen (deuterium, tritium) plasma, the latter is equal in magnitude but opposite in sign with respect to the electrostatic force acting upon electrons. In the adiabatic approximation, when the net force experienced by electrons is assumed to always remain close to zero, one immediately arrives at the conclusion that the *net ion force* practically coincides with the *ponderomotive force acting upon electrons*: $\vec{F}_{i\ net} \simeq \vec{F}_{e\ PM}$. For convenience, we re-write the expression for this force in the following form:

$$\vec{F}_{i\ net} \simeq -\frac{m}{4}\vec{\nabla}\eta^2 \, , \qquad (2)$$

where $\eta = eE_0/m\omega_0$ is the dimensionless amplitude of the laser field.

Adiabaticity thus provides that (i) the ponderomotive force $\vec{F}_{e\ PM}$ drives ions rather than electrons (the latter, of course, are also moved but with just a low velocity typical for the ion motion), and (ii) the driving force is determined by the laser pulse parameters and is essentially independent of the plasma parameters (provided the plasma density is still sufficiently high to balance the external ponderomotive force). In a sense, the situation is unique, since quasistatic external forces in plasma are usually screened by electrons and hardly drive the plasma ions. In contrast, a terawatt laser pulse, as we will see below, is able to drive a considerable ion current inside the plasma.

A very simple model may then be adopted to describe ion current generation. As was already said, the laser pulse parameters (in the co-moving frame of reference) are assumed to be fixed. As the pulse passes by, it expells charged particles outwards; the corresponding ion force is given by Eq. 2. We will see below that a typical length of the pulse is much larger that its radius, therefore $\vec{F}_{i\ net}$ practically has only radial components, and so does the ion velocity. The initial ion distribution is uniform in space, and the initial particle velocity is just zero because any realistic thermal velocity of an ion is much lower that its final velocity induced by the interaction. By solving the Newton equation

$$\ddot{r} = -\frac{m}{4M_i}\frac{\partial\eta^2(r)}{\partial r} \qquad (3)$$

for different initial radial positions, one easily finds the particle distribution over final velocities upon passage of the laser pulse. In the above, M_i is the ion mass.

From Eq. 3 one can easily see that the laser pulse with an on-axis intensity of $\eta(r=0) = \eta_0$ can accelerate ions up to a maximum kinetic energy of $\varepsilon_{max} = m\eta_0^2/4$. In practical terms, it means $\varepsilon_{max} \simeq 130$ keV for $\eta_0 = 1$. As for the time scale involved, a rough estimate of the ion acceleration is given by $\ddot{r} \sim (m\eta_0^2/4M_i a)$, where a is the radius of the laser pulse; by equating the radial shift of an ion to the radius of the pulse, one finds the following estimate for the total pulse length necessary to reach the final particle energy of the order of 50–100 keV:

$$\frac{L}{a} \sim \frac{1}{\eta_0}\left(\frac{8M_i}{m}\right)^{1/2} \qquad (4)$$

The laser pulse should thus be properly shaped, since the optimal condition for the ion current generation is provided by a certain length-to-radius ratio. For $\eta_0 \sim 1$ and $M_i/m \sim 4000$, this gives $L/a \sim 150$–200.

The numerical simulations we discuss below have been performed for a D–T mixture (with equal amounts of both atoms). To be specific, we have choosen the plasma density to be $n = 3 \times 10^{20}$ cm^{-3}, which is three times lower than the critical density for a laser wavelength of $\lambda \simeq 1$ μm. For simplicity, the laser-pulse shape was assumed to be rectangular in the longitudinal direction and Gaussian in the transverse direction. As for the size of the pulse, recent measurements [2, 3] and numerical simulations [7] of the laser-pulse dynamics under conditions close to our ones show formation of a channel with a transverse size of about 5 μm. We assume therefore the radius of the pulse in the steady-state regime to be $a = 5 \mu$m. The on-axis laser intensity has been taken to be $\eta_0 = 1$. For a circularly polarized laser pulse, such choice of parameters implies an instantaneous power of $P \simeq 1.1$ TW, well within the limits of modern laser technology. The results of the simulations are shown in Figs. 1–2.

Figure 1 shows the power P_i carried by the accelerated ions as a function of the laser-pulse length-to-radius ratio. The power is seen to rapidly increase until the pulse becomes quite long, with the length-to-radius ratio of about 150–200. Then the ion power saturates because there are no more ions available in the rear end of the pulse. The saturated ion power is about 0.08 TW which is nearly 8 % of the input laser power.

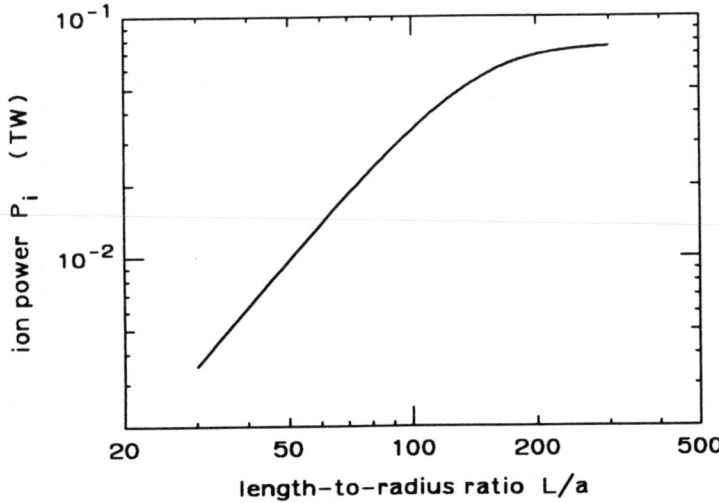

FIGURE 1. Power P_i carried by the accelerated ions as a function of the laser-pulse length-to-radius ratio. The radius of the (Gaussian) laser pulse is assumed to be $a = 5$ μm, the on-axis intensity is $\eta_0 = 1$.

The total energy deposited in plasma depends on the length of the plasma channel. Self-channeled propagation of a terawatt laser pulse through up to 850 μm of background plasma has been reported recently [2]; the length of the channel was not limited by physical mechanisms but simply by the width of the gas jet. If one takes 1 mm as an estimate of the length of the plasma channel, then for a typical pulse energy of about 2.5 J (see below) the energy conversion efficiency appears to be be rather high, about 10–12 %. It can probably be further optimized by proper choice of the background plasma density and the input laser power.

Particle distribution over the final velocity appears to be essentially non-Gaussian, with a broad plateau and two pronounced peaks corresponding to the maximal velocities of the deuterium and tritium ions. For the conditions considered, the mean kinetic energy at which ions escape from the region occupied by the laser pulse is about 40 keV. Taking into account that for our conditions the saturated ion power is about 0.08 TW, one can easily estimate the scale of the ion current involved. With an average ion energy of about 40 keV, this amounts to the injection of about 2 MA (!) of ion current from the path of the pulse into plasma.

As soon as we discuss ion current generation in the context of nuclear fusion, the reaction yield is an important parameter. The yield Y is defined as

$$Y = \int \sigma_{DT}(v)\, v\, \frac{dN}{dv}\, dv , \qquad (5)$$

where $\sigma_{DT}(v)$ is the cross-section of the reaction as a function of the ion velocity. Y is proportional to the number of neutrons produced by the accelerated ions per

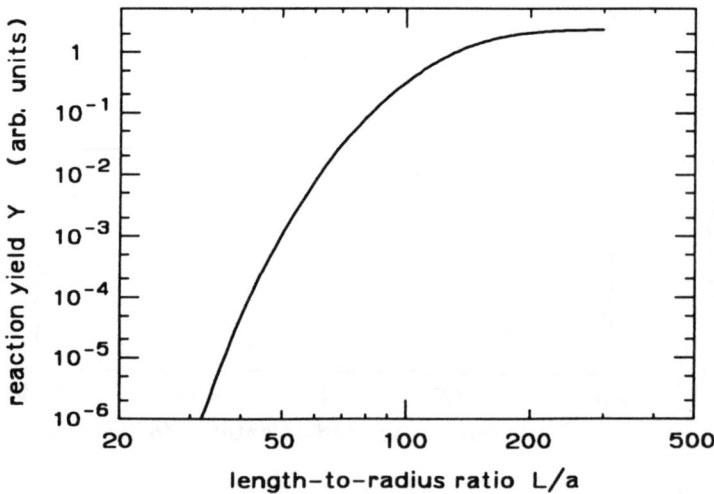

FIGURE 2. Nuclear reaction yield Y as a function of the pulse length-to-raduis ratio.

unit time. Its dependence on the length-to-radius ratio is shown in Fig. 2. The reaction yield is seen to strongly depend on the shape of the pulse: it increases by six orders of magnitude when L/a increases by three times from about 30 to 100. This is connected with the sharp resonance peak in the cross-section at a deuteron energy of about 100 keV. For $L/a < 50$–60, the yield is negligibly small because of the absence of particles with an energy close to the resonant one.

One may conclude from Fig. 2 that the optimal length-to-radius ratio lies between 150 and 200, in agreement with our earlier estimates. A shorter pulse does not produce ions energetic enough, while usage of a longer pulse is redundant because the ion power is already saturated (one should note, however, that the optimum value depends on the on-axis power; the above result corresponds to $\eta_0 = 1$). If we assume the steady-state radius of the pulse to be about 5 μm, then the optimal pulse duration is 2.5–3 ps. For a pulse power of about 1 TW, the total energy carried by the pulse is 2.5–3 J, which is achievable with modern T^3 lasers.

To conclude, we have shown that a properly shaped terawatt laser pulse can drive mega-ampere ion currents inside plasma. For a length-to-radius ratio of about 150–200, the ion energy reaches 50–150 keV — the range that may be of interest for nuclear fusion since it covers the resonance peak in the D–T cross-section. Experimental observation of this phenomenon seems to be feasible with modern T^3 lasers, which would result in spectacular appearance of direct nuclear neutrons from the focal laser spot.

This work was performed as part of the research programme of the Stichting voor Fundamenteel Onderzoek der Materie (FOM) and was supported by the Nederlandse Organizatie voor Wetenschappelijk Onderzoek (NWO).

REFERENCES

1. Max,C., Arons,J., and Langdon,A.B., *Phys. Rev. Lett.* **33**, 209 (1974); Sun,G.Z. et al., *Phys. Fluids* **30**, 526 (1987); Sprangle,P., Tang,C.M., and Esarey,E., *IEEE Trans. Plasma Sci.* **PS-15**, 145 (1987).

2. Wagner,R., et al, *Phys. Rev. Lett.* **78**, 3125 (1997).

3. Borghesi,M., et al, *Phys. Rev. Lett.* **78**, 879 (1997).

4. Sprangle,P., Esarey,E., Krall,J., and Joyce,G., *Phys. Rev. Lett.* **69**, 2200 (1992); Andreev,N.E., et al, *Sov. JETP Letters*, **55**, 571 (1992).

5. Modena,A., et al, *Nature* **377**, 606 (1995).

6. Bulanov,S.V., Pegoraro,F., and Pukhov,A.M., *Phys. Rev. Lett.* **74**, 710 (1995).

7. Pukhov,A., and Meyer-ter-Vehn,J., *Phys. Rev. Lett.* **76**, 3975 (1996).

8. Young,P.E., et al, *Phys. Rev. Lett.* **75**, 1082 (1995); Beg,F.N., et al, *Phys. Plasmas* **4**, 447 (1997); Lawson,W.S., Rambo,P.W., and Larson,D.J., *Phys. Plasmas* **4**, 788 (1997); Ditmire,T., et al., *Phys. Rev. Lett.* **78**, 3121 (1997).

6. ASTROPHYSICS APPLICATIONS OF ULTRAINTENSE LASER PULSES

Ultradense Hydrogen in Astrophysics, High-Pressure Metal Physics and Fusion Studies

Setsuo Ichimaru[*] and Hikaru Kitamura[†]

[*] *Max-Planck-Institut für Quantenoptik*
Hans-Kopfermann-Str. 1, 85748 Garching, Germany
[*] *Department of Physics, University of Tokyo, Hongo, Tokyo 113, Japan*
[†] *Institute for Solid State Physics, University of Tokyo, Roppongi, Tokyo 106, Japan*

Abstract. Phase diagrams of hydrogen are constructed through first-principles calculations of the equations of state for metallic and insulator phases. On the bases of these theories of the equations of state and the electric resistivity, it is shown that the results of recent shock-metallization experiments can be consistently interpreted in terms of *first-order* metal-insulator transitions, involving discontinuous changes in density, entropy and enthalpy. The first-order transitions then predict a discontinuous distribution of density and resistivity near the Jovian surface, with a large magnetic Reynolds number enough to sustain prominent magnetic activities. A phase diagram for freezing and ferromagnetic transitions provides a basic account of strong magnetization observed in magnetic white dwarfs. Feasibility of a novel scheme of fusion studies in ultradense metallic hydrogen is examined in light of these experimental and theoretical developments.

INTRODUCTION

Ultradense hydrogen under high pressures as found in stellar interiors may be expected to undergo transformation between phases, such as *metallization, crystallization* and *magnetization*. All of these phase transitions are not only of great interest in condensed-matter physics but, since hydrogen is the most abundant chemical element in the Universe, their nature crucially affects fundamental issues in astrophysics, such as the generation of energy and magnetism in the interiors of stars and planets and their transport to the stellar and planetary surfaces. Thus, the physics of hydrogen constitutes a vital element in our insight into the formation, structure and evolution of these astronomical objects.

The recent experiments (1,2) on shock metallization in dense hydrogen have led to predict discontinuous structures in density and resistivity distributions inside the planet Jupiter (3). These structures can help resolve outstanding issues such as its prominent magnetic activities (4) and the release of excess energy in infrared (5). A phase diagram of ultradense hydrogen pertaining to freezing and ferromagnetic transitions (6) is relevant to elucidation of those systems of fermions expected in the outer layer of a hydrogen-rich white dwarf.

The rates of nuclear fusion processes, such as the thermonuclear and pycnonuclear reactions, are influenced strongly by the state that a dense matter may assume (7). A huge enhancement of the reaction rate, arising from dense-matter effects, albeit ineffective for the solar nuclear reactions or for inertial-confinement-fusion (ICF) experiments (a "sun on the Earth"), provides a physical mechanism vital to supernovae, a final stage of stellar evolution. Recent adcances in laser technology as well as in ultrahigh-pressure metal physics may possibly make a "supernova on the Earth" scheme utilizing the enhanced pycnonuclear processes in ultradense metallic hydrogen (8) an attractive and possibly even realizable prospect for fusion studies.

STATES OF HYDROGEN MATTER

A hydrogen atom is a bound state between an electron and a proton. The orbital radius of a bound electron in the ground state is known as the Bohr radius, $a_B \approx 0.53$ Å. The binding energy of such a hydrogen atom constitutes a unit of energy called the Rydberg with the value of approximately 13.6 eV. These quantities are representative of typical length and energy scales in the atomic physics of hydrogen.

A hydrogen molecule is a bound state between two hydrogen atoms. In its ground state, the average inter-proton spacing is 0.742 Å. It may dissociate into two hydrogen atoms or be ionized by releasing an electron. The dissociation energy and the ionization potential of a hydrogen molecule are 4.477 eV and 15.43 eV, respectively; the dissociation energy of a molecular ion, H_2^+, is 2.467 eV. In dense hydrogen, processes of dissociation are followed immediately by those of ionization.

If the number density of protons in hydrogen matter is n, a sphere with the characteristic volume, $1/n$, will have a radius a given by

$$a = \left(\frac{3}{4\pi n}\right)^{1/3} .$$

This is called the Wigner-Seitz radius (9). If the density is high so that $a < a_B$ (corresponding to $n > 1.6 \times 10^{24}$ cm^{-3}), wave functions of the orbital electrons in neighboring hydrogen atoms or molecules overlap each other considerably so that a given electron can no longer be considered as attached to a given proton: something akin to ionization has set in, the electrons become conduction electrons, and such a process is called pressure ionization.

E. Wigner and H. B. Huntington in 1935 were the first to predict the possibility of such a metallic modification of hydrogen at extreme pressure (10). They calculated the energy of a body-centered lattice of hydrogen as a function of the lattice constant and, by comparison of the result with the energy of the molecular form, arrived at the conclusion that hydrogen is expected to undergo a *first-order*, metal-insulator (MI) transition at an ultrahigh density or in a pressure range of megabars.

The thermodynamics of the MI transitions may be summarized as follows: In the coexistence conditions between the metal and insulator phases, all the thermodynamic quantities except for the pressure, the temperature and the chemical potential are discontinuous; such discontinuities are the characteristics of what is known as a first-order transition. In the MI transition, the discontinuous quantities—density, entropy and enthalpy—are greater in the metallic phase than in the insulator phase; these discontinuities vanish at the critical point, C_{MI}. The insulator phase is either a fluid or a solid, consisting of molecular hydrogen. The metal phase consists of protons and electrons, the protons forming either a fluid or a Coulomb solid known as a Wigner crystal.

An insulator-to-metal transition (i.e., metallization) proceeds in the direction of decreasing the chemical potential, accompanied by an increase of the temperature and/or decrease of the pressure. It is an *endothermic* process (3), analogous to our familiar vaporization and melting transitions: the entropy and the enthalpy increase. Recent shock-compression experiments (1,2) have elucidated the nature of these metallization transitions.

An electron liquid is a one-component plasma of electrons immersed in a uniform compensating background of positive charges (9). Electrons are fermions, with spin 1/2, obeying Fermi statistics. What this means is that wave functions of two identical fermions with parallel spins are antisymmetric; in other words, they change their signs when the positions of the two fermions are interchanged. In particular, the value of the wave function vanishes when two identical fermions occupy the same position; interpreted physically, identical fermions with parallel spins simply cannot occupy the same position and thus they repel each other. This explains the origin of the spin-discriminating (repulsive) exchange forces between such identical fermions. Protons, too, are fermions and thus these exchange forces and the ordinary Coulomb forces, both repulsive, are effective between protons as well as between electrons.

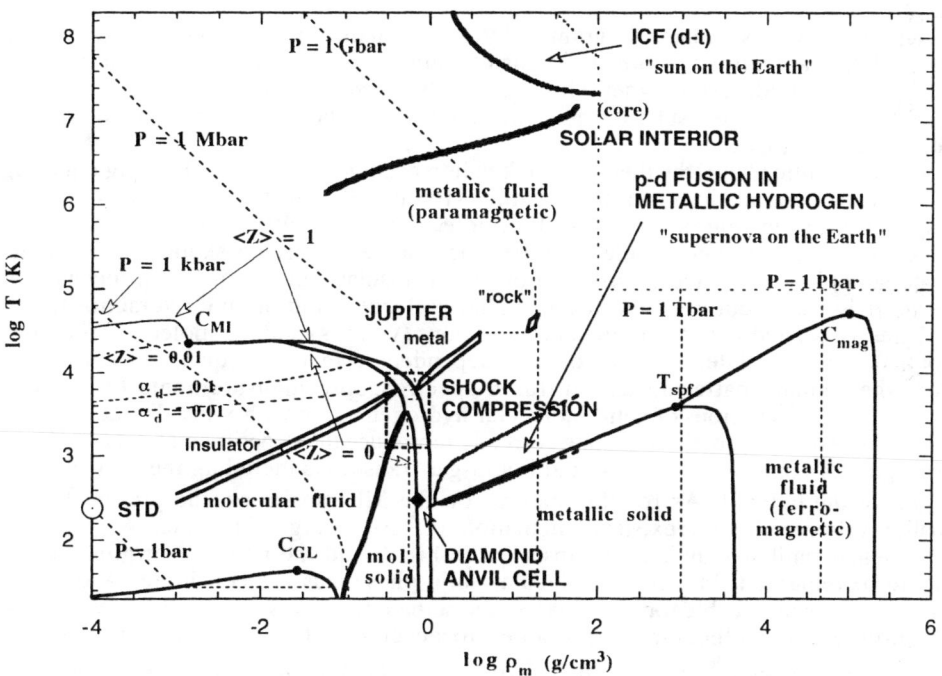

FIGURE 1. Phase diagram of hydrogen: The dotted curves are isobars. $\langle Z \rangle$ denotes the degree of ionization; α_d, the degree of atomic dissociation.

It has been expected that an electron liquid may undergo a magnetic transition (9), from a spin-non-aligned, paramagnetic phase to a spin-aligned, ferromagnetic phase, near the conditions for Wigner crystallization, a phase transition of dilute electrons into a crystalline state at low temperatures. A magnetic transition takes place basically through competition between the spin-orientation-dependent exchange processes which favor a ferromagnetic state and the contributions of kinetic energy which favor a paramagnetic state. Analogous situations exist in the case of a freezing transition, where the repulsive Coulomb forces favor an inhomogeneous distribution such as one predicted in a Wigner crystal, while the kinetic processes favor a uniform distribution characteristic of a fluid state.

We may thus convert these calculations of magnetic and freezing transitions for the electron liquids and thereby derive the phase diagram of these transitions for dense metallic hydrogen (6), a system of fermions relevant to the outer layer of a hydrogen-rich white dwarf. The strength of magnetization arising from the nuclear spin alignment of protons in the resultant ferromagnetic hydrogen is expected at $\sim 10^7$ gauss, high enough to account for the origin of the observed strong magnetization in so-called magnetic white dwarfs (11).

PHASE DIAGRAMS

Figure 1 exhibits a phase diagram of hydrogen matter. It was constructed through the explicit calculations of the equations of state for hydrogen in the metallic (solid, paramagnetic fluid, ferromagnetic fluid, partially ionized atomic and molecular fluid) phases as well as in the insulator (molecular solid, molecular fluid, atomic and molecular fluid) phases (3,6,9).

In the metallic phase, the theory has taken into account the effects of strong coupling between electrons and ions near the conditions for the MI transitions as well as those of *lowering* or *elimination* of atomic and/or molecular levels in the bound states, due to screening by (dense) metallic electrons (9); these screening effects should *accelerate* the processes of ionization and/or dissociation. In the insulator phase, quantum-mechanical Lennard-Jones potentials were used for the molecular and atomic interactions; free energies associated with the molecular vibrations (vibrons) and the molecular rotations (rotons) were calculated with the density-dependent, effective frequencies determined from the Raman-scattering and infrared-spectroscopic measurements (12). The pressure-density relations of solid molecular hydrogen agree with x-ray diffraction data (12). The resultant equations of state correctly reproduce the experimental data on dense hydrogen for solid-liquid (13,14) and liquid-gas transitions including the critical point (C_{GL}) for the latter.[1] As for the features on the MI transitions, the present theory qualitatively corroborates existing calculations (15,16), as Fig. 2 illustrates.

Salient quantitative differences observed in Fig. 2 and the physical origins thereof should be remarked: In (15), a molecular-solid phase was not considered and so the theory was not applicable for $T < 5000$ K; the screening effects of metallic electrons on the atomic and molecular bound states to accelerate the metallization were not appropriately taken into account in (15,16), so that (15) predicts $<Z> = 0.23$ at $\rho_m = 0.65$ g/cm^3 and $T = 10^4$ K, while the present theory predicts $<Z> = 1.0$, a fully metal-

---

[1] The work (13) observed a fluid-solid coexistence pressure of 9.9 kbar at 100 K, while the present theory predicts 10.8 kbar. The Brillouin measurement (14) in a diamond-anvil cell observed a coexistence pressure of 55 kbar at 300 K, while the present theory predicts 50 kbar.

lized state. The critical point (C_{MI}) in (15,16) remained high (in pressure) because of the neglect of the electronic screening effects for acceleration of ionization; the number of Monte Carlo particles used in the simulations in (16) (32 electrons and 32 protons) was too small to be able to account for such an effect. Separately (3), we have calculated the values of P and $<Z>$ at $\rho_m = 0.005$ g/cm^3 in two ways, *with* and *without* these electronic screening effects taken into account, as functions of the temperature. In the former calculations, P and $<Z>$ exhibited discontinuities at $T = 2.245 \times 10^4$ K ($<Z>$ from 0 to 1), while in the latter both varied continuously.

Approximate parameter domains for the hydrogen matter appropriate to the solar interior, the Jovian interior (17), ICF researches, ultrahigh-pressure metal experiments (shock compression and diamond-anvil-cell compression), proton-deuteron (*p-d*) fusion in ultradense metallic hydrogen and hydrogen under standard (STD) conditions are likewise illustrated in the figure.

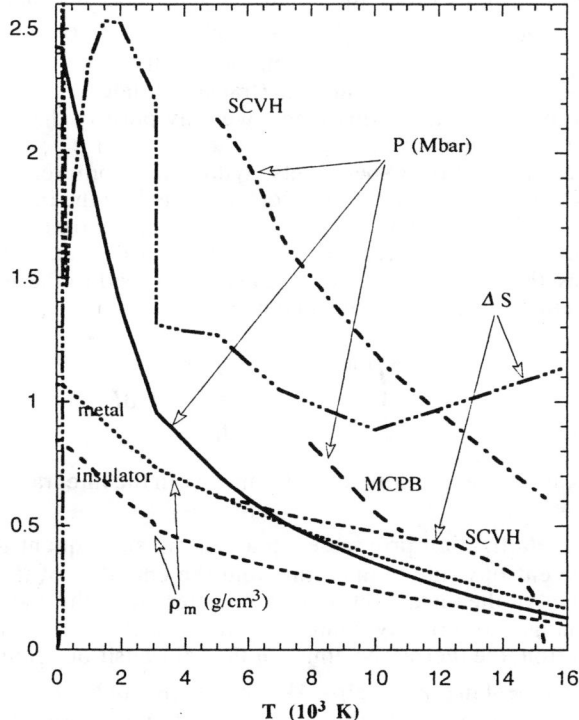

FIGURE 2. The characteristics of the metal-insulator transitions: $\Delta S \equiv S_{\text{metal}} - S_{\text{insulator}}$ denotes the discontinuous change of the entropy per a proton mass in units of the Boltzmann constant. SCVH (15); MCPB (16).

SHOCK-COMPRESSION EXPERIMENTS

Metallization of molecular hydrogen, though elusive in the diamond-anvil-cell experiments (12,18), was recently demonstrated by compression experiments through shock wave reverberating between electrically insulating sapphire (Al_2O_3) anvils (1). The pressure and the electric resistivity of hydrogen, attained in the experiment, were accurately measured and were interpreted consistently with the theoretical predictions of the MI transitions and the electric resistivity (19).

In analyzing the experimental data, we examine the relevant time and length scales. Let the axial width of the compressed hydrogen be x, which typically assume values on the order of 100 μm. A hydrodynamic or compressional time τ_H, estimated as x divided by a sound velocity, may take on a value on the order of 10 ns. An electronic or ionization time τ_E, estimated as x divided by the Fermi velocity, may take on value of approximately 60 ps. Since $\tau_E \ll \tau_H$, we find that metallization develops *instantaneously* in a manner analogous to electric breakdown.

When the metallization is partial, the heavy (metal) component ρ_{pl}, deposited uniformly in the volume, should in principle phase separate from the light (insulator) component ρ_{ins} by gravity. Since a maximally possible gravitational displacement may be estimated at its free-fall value, $g\tau_H^2 \approx 1$ fm, where $g \approx 980$ cm/s^2, we find that the phase separation predicted in Figs. 1 and 2 cannot actually take place in the experiments; the hydrogen matter may remain in a uniform (transient) state.

In light of these numerical comparisons, we may portray the shock-metallization experiments (1) in two sequential stages:

Compression: In each run of the experiments, hydrogen is compressed from the state ($P \approx 1$ bar, $T \approx 20$ K) through reverberating shock imparted by the impactor with a kinetic energy on the order of 50 kJ. The hydrogen pressure takes on its *highest* value in the run when the state of hydrogen reaches the insulator side (i.e., $<Z> = 0$) of the MI coexistence curves; the pressures measured in (1) correspond to these maximum values. The enthalpy increment in the hydrogen, given by

$$\Delta W = \int_{\text{initial}}^{\text{final}} T\,dS + \int_{\text{initial}}^{\text{final}} V\,dP \ ,$$

has typically increased by approximately 6 kJ, mostly in the integral of VdP, during this stage.

Metallization: As metallization progresses in a time τ_E subsequent to that maximum-pressure state, the entropy, the temperature and the enthalpy of the hydrogen matter increase. Efficiency of the metallization may be determined through delicate matching between these changes and the dynamic conditions of the compression cell. For an efficient metallization, the thermodynamics of the MI transitions summarized in Fig. 2 implies that the increment in the enthalpy ΔW be imparted in hydrogen in a thermal form (i.e., the integral of TdS, with the change in the entropy dS stemming from redistribution in the microscopic electronic states) rather than in a hydrodynamic form (i.e., the integral of VdP). A shock (and a diamond-anvil) compression by itself does not give optimum conditions; the enthalpy has increased only by ~0.3 kJ through the metallization experiments (1). We remark that a shock analysis of the Rankine-Hugoniot type is not applicable to such a non-hydrodynamic process in a short time-scale of τ_E.

To achieve an efficient metallization, an additional physical mechanism such as injection of an intense, ultra-short laser pulse into the compressed hydrogen would have to be considered.

The resistivity measured in a given compression experiment provides a measure of metallization attained. The (partial) mass density ρ_{pl} of the metallized hydrogen may then be calculated with the aid of the resistivity formula (14a) in (19). This density is connected to the pressure in the final state through the consideration that the (negative) Coulombic pressure (9) in dense hydrogen is the major source of the pressure decrease in the metallized state (3). In Fig. 3, we plot the corresponding values of the density ($\rho_m = \rho_{pl} + \rho_{ins}$) and the temperature between the initial and final states of metallization in the seven experimental runs of the compression experiments for hydrogen (1); the results are seen consistent with the phase diagram.

STELLAR AND PLANETARY MAGNETISM

The magnetic Reynolds number, R_m, in magnetohydrodynamics is a number representing the ratio between the convective effect of a conductive fluid dragging and stretching the magnetic lines of force and the dissipative effect due to decay of electric currents by the resistivity. The larger R_m, the more efficiently are the magnetic activities and field strengths sustained and amplified by the fluid motion.

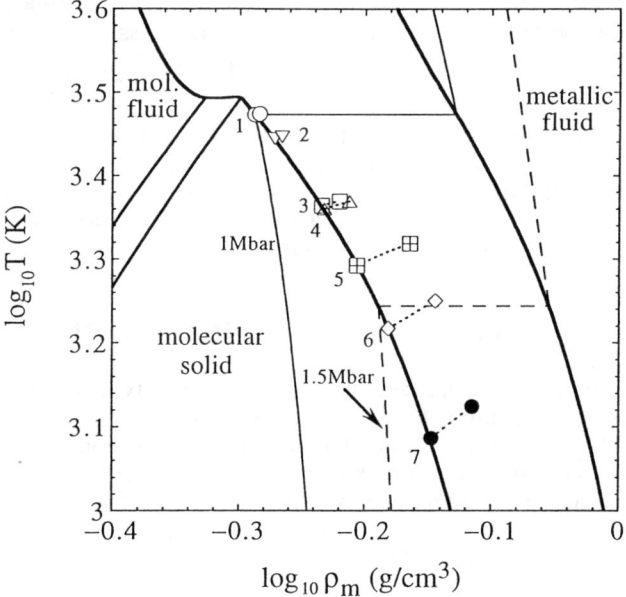

FIGURE 3. The initial (insulator) and final (metallized) states of metallization assessed for the seven runs of the experiments (1). The diagram describes the MI coexistence curves; the molecular (insulator) fluid-solid coexistence curves; and isobars at 1.0 and 1.5 Mbar.

The first-order MI transitions in hydrogen lead us to predict the existence of a boundary layer near the surface of Jupiter across which the mass density and the resistivity change discontinuously. Assuming the temperature of the boundary layer at 6.5×10^3 K, we calculate the density in the outer insulator side to be 0.34 g/cm^3 and that of the inner metallic side, 0.54 g/cm^3, where the resistivity ρ_E is computed at 1.34×10^{-4} $\Omega \cdot$cm, only about 100 times greater than that of copper. The magnetic Reynolds number associated with this configuration may then take on a magnitude, $4\pi R_J^2 \omega_J / \rho_E c^2 \approx 0.86 \times 10^{12}$, where R_J (= 7.14×10^4 km) is the radius of Jupiter and ω_J (= 15.2 day^{-1}) is the angular velocity of Jupiter's self rotation. Comparing this with a corresponding estimate $\sim 1.0 \times 10^8$ for the solar magnetic activities (20), we find that prominent magnetic activities may be amply sustained near Jupiter by the presence of a highly conductive metallic hydrogen at the MI boundary near the surface.

Physical mechanisms for the origin of strong magnetization in the magnetic white dwarfs may likewise be elucidated by the magnetic phase diagram (Fig. 1) of metallic hydrogen (6). The hydrogen density at the magnetic critical point, C_{mag}, on the order of 10^5 g/cm^3 may be expected in an outer shell of a hydrogen-rich white dwarf, located presumably within ~5% of the radius from the stellar surface. As for the strengths of magnetic field, we have detected in (6) that some of the observed field strengths (21) are greater by 1~2 orders of magnitude than those (~10^7 gauss) which can be produced by the nuclear ferromagnetism alone. For those white dwarfs with super-strong magnetic fields, a separate amplification mechanism, such as a "dynamo" (11), is called for, where the nuclear ferromagnetism may provide the "seed" fields for such amplification.

To examine the efficiency of such amplification, we again evaluate the magnetic Reynolds number associated with a rotating white dwarf:[2] Assuming the radius and the angular velocity of a white dwarf to be ~5000 km and ~2π/day, respectively, we find $R_m \approx 6.2 \times 10^{14}$ using the value, 3.75×10^{-10} $\Omega \cdot$cm, for ρ_E calculated at the critical conditions, C_{mag}. Comparing this evaluation with the corresponding estimates above, ~0.86×10^{12} and ~1.0×10^8, for Jovian and solar magnetic activities (where the ferromagnetic seed fields are not available), we find that considerable strengths of the magnetic fields may be sustained around a white dwarf by the presence of highly conductive, ferromagnetic hydrogen near the surface.

A "SUPERNOVA ON THE EARTH"

In the Sun, the nuclear reactions take place most vigorously near the core, where the mass density and the temperature of the metallic hydrogen are 56.2 g/cm^3 and 1.55×10^7 K, respectively. The ICF researchers attempt to realize the conditions analogous to those near the solar core by the compression of deuterium-tritium fuel pellets to mass densities and temperatures on the order of 3~60 g/cm^3 and ~10^8 K; hence, a "sun on the Earth." At such a high temperature, required for the thermonuclear reactions, one anticipates that there is an assortment of dynamic instabilities to overcome in the compression of hydrogen matter.

In condensed matter, of which metallic hydrogen is an example, electrons act to weaken or *screen* the Coulomb repulsion between the atomic nuclei. In a metallic sub-

---

[2] In Fig. 2 of (11), correlations between observed white-dwarf magnetisms, on the one hand, and periods of rotation and H-rich atmospheres in white dwarfs, on the other, have been investigated.

stance under ultrahigh pressures, effects of such screening become so conspicuous that the rates of nuclear reactions at relatively low temperatures take on values independent of the temperature (7). These processes, called the *pycnonuclear reactions* (22), are thought to be applicable in a white-dwarf progenitor of supernova.

In addition to these screening effects by electrons, strong spatial correlation between atomic nuclei in dense matter due to their Coulomb repulsion, the very correlation effect manifested in freezing, tends to enhance the reaction rates through an effective reduction of the overall internuclear repulsion (7). The enhancement factor in a dense fluid increases steeply as the temperature is lowered, while that in a solid sharply increases with the temperature since the reaction rates in solids depend sensitively on amplitudes of the atomic vibrations. Hence a maximum enhancement may be attained near the conditions of freezing transition. In a white-dwarf supernova progenitor, enhancement by a factor of 30 to 40 orders of magnitude may be anticipated. Virtually no enhancement is expected, however, in solar interior or in an ICF plasma.

In Fig. 4, we compare the *p d* pycnonuclear contact probabilities (7) with the deuteron-triton (*d-t*) thermonuclear contact probabilities (7), as functions of the center-of-mass energy, E, in temperature units. Assumed in the calculations are: *d-t* thermonuclear reactions at 10^8 K; *p-d* pycnonuclear reactions at 800 K, with number density of the screening electrons at 1.2×10^{24} cm^{-3}, appropriate to liquid-metallic hydrogen at 10 Mbar. The "Gamow peak" for the thermonuclear contact probabilities is seen at about 3×10^8 K. We observe that the *enhanced* contact probabilities for the *p-d* pycnonuclear reactions can be comparable in magnitude to the contact probabilities for the *d-t* thermonuclear reactions.

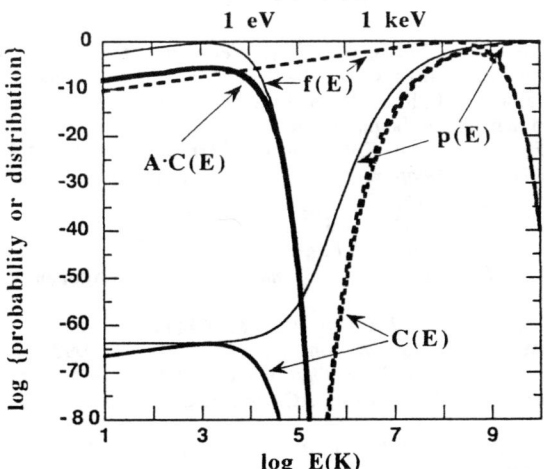

FIGURE 4. Thermonuclear (dashed curves) vs. pycnonuclear (solid curves) reactions: p(E) denotes a penetration probability through the Coulomb barrier between a pair of reacting nuclei with a center-of-mass energy, E(K), in temperature units; f(E), the Boltzmann distribution; C(E) = p(E)·f(E), a contact probability between a pair of reacting nuclei; A·C(E), enhanced contact probability with A denoting the enhancement factor (7).

Possibilities of combined utilization of the pycnonuclear *p-d* reactions at lower temperatures and their enhancement due to the strong Coulomb correlation, both applicable in ultradense metallic hydrogen near the freezing conditions, lead to a proposal of a "supernova on the Earth" scheme (7,8) for nuclear fusion researches: A reverberating shock imparted by a low-speed impactor may be used for a low-entropy compression of a *p-d* mixture with a total mass of ~0.3 mg, say, from a molecular-*solid* (insulator) state at ~1 bar and ~10 K to a state on the insulator side of the MI transitions at ~2 Mbar and ~300 K. The concept would be *to inject at this instant super-intense, ultra-short laser pulses, made possible by very recent advances in laser technology, into the compressed hydrogen to ensure efficient metallization.* The compression exerted continuously by the impactor may bring the resultant *metallic* hydrogen finally into a further compressed state at ~10 Mbar and ~800 K, say, near the freezing conditions, when an outburst of enhanced pycnonuclear *p-d* reactions would be expected. Recent experimental and theoretical progress in ultrahigh-pressure metal physics makes such a scheme an attractive and possibly even realizable prospect for the fusion studies.

ACKNOWLEDGMENTS

One of the authors (S. I.) acknowledges support of Alexander von Humboldt-Stiftung for his reseach at Max-PlanckInstitut für Quantenoptik as a *Humboldt-Forschungspreisträger*. Work of H. K. was supported in part by Ministry of Education of Japan through Grant No. 3455.

REFERENCES

1. Weir, S.T., Mitchell, A.C., and Nellis, W.J., *Phys. Rev. Lett.* **76**, 1860 (1996).
2. Da Silva, L.B., Celliers, P., Collins, G.W., Budil, K.S., Holms, N.C., Barbee, T.W. Jr., Hammel, B.A., Kilkenny, J.D., Wallace, R.J., Ross, M., and Cauble, R., *Phys. Rev. Lett.* **78**, 483 (1997).
3. Kitamura, H., and Ichimaru, S., to be published.
4. See e.g., Kennel, C.F., and Coroniti, F.V., *Ann. Rev. Astron. Astrophys.* **15**, 389 (1977).
5. See e.g., Hubbard, W.B., *Rev. Geophys. Space Phys.* **18**, 1 (1980).
6. S. Ichimaru, S., *Phys. Lett. A* **235**, 83 (1997).
7. See e.g., Ichimaru, S., *Rev. Mod. Phys.* **65**, 255 (1993).
8. Ichimaru, S., *J. Phys. Soc. Jpn* **60**, 1437 (1991).
9. Ichimaru, S., *Statistical Plasma Physics II: Condensed Plasmas* (Addison-Wesley, Reading, MA, 1994).
10. Wigner, E., and Huntington, H.B., *J. Chem. Phys.* **3**, 764 (1935).
11. See e.g., Chanmugam, G., *Ann. Rev. Astron. Astrophys.* **30**, 173 (1992).
12. Mao, H.K., and Hemley, R.J., *Rev. Mod. Phys.* **66**, 671 (1994).
13. Mills, R.L., Liebenberg, D.H., Bronson, J.C., and L.C. Schmidt, *J. Chem. Phys.* **66**, 3076 (1977).
14. Shimizu, H., Brody, E.M., Mao, H.-k., and Bell, P.M., *Phys. Rev. Lett.* **47**, 128 (1981).
15. Saumon, D., and Chabrier, G., and Van Horn, H.M., *Astrophys. J. Suppl.* **99**, 713 (1995).
16. Magro, W.R., Ceperley, D.M., Pierleoni, C., and Bernu, B., *Phys. Rev. Lett.* **76**, 1240 (1996).
17. Van Horn, H.M., *Science* **252**, 384 (1991).
18. Mao, H.-k., Hemley, R.J., and Hanfland, M., *Phys. Rev. Lett.* **65**, 484 (1990); Hemley, R.J., Mao, H.-k., Goncharov, A.F., Hanfland, M., and Struzhkin, V. *Phys. Rev. Lett.* **76**, 1667 (1996).
19. Kitamura, H., and Ichimaru, S., *Phys. Rev. E* **51**, 6004 (1995).
20. See, e.g., Ichimaru, S., *Phys. Plasmas* **3**, 233 (1996).
21. Weisheit, J., in *Elementary Processes in Dense Plasmas: Proc. Oji International Seminar*, eds. S. Ichimaru and S. Ogata (Addison-Wesley, Reading, MA, 1995) p. 61.
22. Cameron, A.G.W., *Astrophys. J.* **130**, 916 (1959).

Supernova Hydrodynamics Experiments on Nova

B. A. Remington, S. G. Glendinning, K. Estabrook, R. J. Wallace, R. London, R. A. Managan, A. Rubenchik, D. Ryutov, and K. S. Budil

Lawrence Livermore National Laboratory, P. O. Box 808, Livermore, CA 94550

J. Kane and D. Arnett

Stewart Observatory, University of Arizona, Tucson, AZ 85721

R. P. Drake

University of Michigan, Ann Arbor, MI 48109-2143

R. McCray

University of Colorado-Boulder, JILA, Boulder, CO 80309-0440

E. Liang

Rice University, Houston, Texas 77251

Abstract.
We are developing experiments using the Nova laser to investigate (1) compressible nonlinear hydrodynamic mixing relevant to the first few hours of the supernova (SN) explosion and (2) ejecta-ambient plasma interactions relevant to the early SN remnant phase. The experiments and astrophysical implications are discussed. We discuss additional experiments possible with ultra-high-intensity lasers.

Two phases of core-collapse supernova (SN) evolution, the core hydrodynamic mixing in the first few hours and colliding plasmas during early SN remnant formation,(Arnett et al. 1989; McCray 1993) are areas rich with possibilities for supporting laboratory experiments. We report here on two experiments utilizing the Nova laser(Kilkenny 1992) to create the relevant plasma environment. We conclude with a discussion of a new class of astrophysical experiment – radiative fireball relevant to gamma-ray bursts (GRB) – potentially possible with ultra-intense, short-pulse lasers.

We start with a progenitor for SN1987A similar to that shown in Fig. 1a from Arnett (1996), and calculate the hydrodynamic evolution using the SN hydrodynamics code PROMETHEUS (Fryxell et al. 1991; Müller et al. 1991). We will focus on the instabilities at the He-H interface. To economize on computing time, we model only $M_r/M_o \geq 5$, depositing the explosion energy, $E = 1.5 \times 10^{51}$ ergs at $M_r/M_o = 5$. This launches a strong radial shock that reaches the He-H interface ($M_r/M_o = 6$) after a transit time of about 100 sec. We show

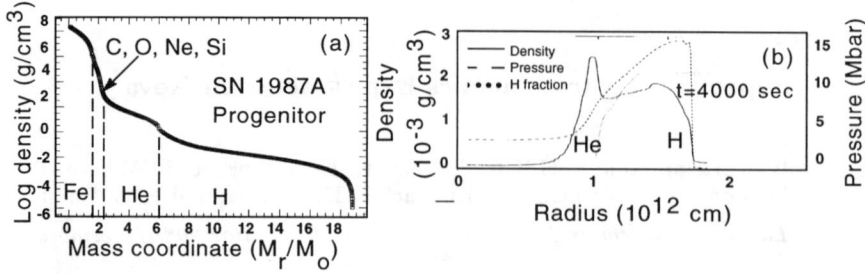

Figure 1. Supernova simulations in 1D of SN1987A.

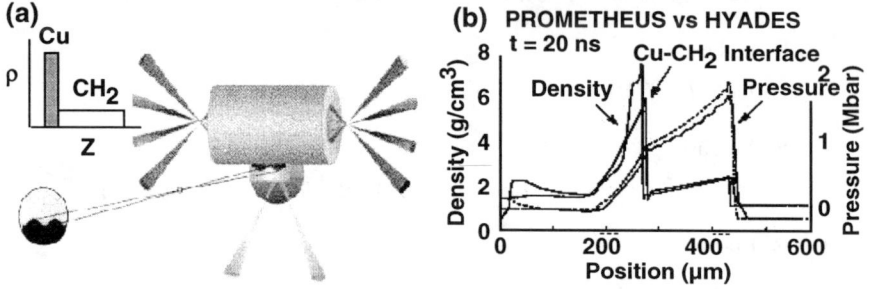

Figure 2. Supernova hydrodynamics experiment.

the density and pressure profiles at a time of 4000 sec in Fig. 1b. Note that at the He-H interface (R ≈ 1.0×10^{12} cm), the pressure and density gradients are crossed, that is, $\nabla \rho \cdot \nabla P < 0$, such that the He layer is being decelerated by the lower density H layer. This situation is unstable to the Rayleigh-Taylor (RT) instability and perturbations at the interface grow in time. (Fryxell et al. 1991; Müller et al. 1991)

The experimental configuration adopted for these laser experiments is illustrated in Fig. 2a and is described in more depth elsewhere (Remington et al. 1995; Kane et al. 1997; Peyser et al. 1995). Eight of the ten Nova laser beams at a duration of 1 ns and total energy of 12 kJ are focused into a 3.0 mm long, 1.6 mm diameter Au hohlraum (cylindrical radiation cavity) converting to a ~ 190 eV thermal x-ray drive. The experimental package is planar, a 85 μm Cu (ρ =8.9 g/cm^3) foil backed by 500 μm of CH$_2$ (ρ = 0.95 g/cm^3). A λ = 200 μm wavelength, η_o = 20 μm amplitude sinusoidal ripple is imposed at this embedded interface. The package is mounted across a hole in the hohlraum wall, so that the inner (smooth) side of the Cu sees the x-ray drive. Diagnosis of the interface is through side-on x-ray radiography, using the remaining two Nova beams focused onto an Fe backlighter disk to generate a 5 ns pulse of He-α x-rays at 6.7 keV. In this side-on view, the opaque Cu appears as a shadow, and the CH$_2$ is essentially transparent.

In Fig. 2b we show the results of modeling in 1D with HYADES,(Larsen & Lane 1994) and PROMETHEUS. (Fryxell et al. 1991; Müller et al. 1991) HYADES is a 1D Lagrangian code with multigroup radiation transport and tabular equation of state (EOS), and PROMETHEUS is a 3D Eulerian Piecewise Parabolic Method (PPM) code using (here) an ideal gas EOS. We use a measured radiation temperature, $T_r(t)$, as the source input to HYADES, and achieve an impulsive shock acceleration, followed by a protracted deceleration, qualitatively similar to the He-H interface in the SN. We do a high resolution HYADES run, including multigroup radiation transport, for the first 2.45 ns, at which time the shock is approaching the Cu-CH_2 interface. We then map the results to 1D or 2D PROMETHEUS. We compare the results for pressure and density at 20 ns from a continuous 1D HYADES run including radiation transport versus that from PROMETHEUS (Fig. 2b). The mapping works well. Note, the pressures for the Nova experiment, 1-2 Mbar, are not too different from those of the SN (10-15 Mbar), as shown in Fig. 1b, though the SN densities are lower by a factor of about 10^3.

We consider the difference of scales between the SN and the Nova experiment. If we assume that the mixing is dominated by the RT instability, then in the nonlinear regime, the fluid flow can be characterized by a spatial scale of order the perturbation wavelength λ and velocity of order the perturbation terminal bubble velocity $v_B \propto (g\lambda)^{1/2}$. Here g corresponds to the acceleration and we have assumed constant Atwood number. A hydrodynamic time scale is then given by $\tau = \lambda/v_B \propto (\lambda/g)^{1/2}$, and the hydrodynamics equations are invariant under the scale transformation (Hecht et al. 1994; Alon et al. 1995) $\lambda \to a_1 \lambda$, $g \to a_2 g$, and $\tau \to (a_1/a_2)^{1/2}\tau$. We illustrate this transformation, using characteristic scales taken from the simulations shown in Figs. 1 and 2. At 4000 sec for the SN, the deceleration of the He-H interface is $g_{SN} = -1.5 \times 10^4 cm/s^2$, the density gradient scale length is $L_{SN} = \rho/\nabla\rho = 8 \times 10^{10}$ cm, and the dominant perturbation wavelength is approximated to be $\lambda_{SN} \approx 10 L_{SN} = 8 \times 10^{11}$ cm. For the Nova experiment at 20 ns we have $g_{Nova} = -2.5 \times 10^{13} cm/s^2$, $\lambda_{Nova} = 2 \times 10^{-2}$ cm, and a characteristic time interval of $\tau_{Nova} = 5$ ns. The scale transformation is given by $a_1 = \lambda_{SN}/\lambda_{Nova} = 4 \times 10^{13}$, and $a_2 = g_{SN}/g_{Nova} = 6 \times 10^{-10}$. The corresponding hydrodynamically equivalent time interval for the SN is then given by $\tau_{SN} = (a_1/a_2)^{1/2}\tau_{Nova} = 1.3 \times 10^3$ sec, which is a reasonable time scale for the SN instability evolution that we are investigating. This scale transformation is not precisely correct, because we have not taken into account the effects of decompression, finite layer thickness, and shocks. Nevertheless, the Nova experiment appears to address the nonlinear compressible hydrodynamics similar to that at the He-H interface of a Type II SN during the first few hours.

In Fig. 3 we show a 2D image from the experiment at 33 ns (Fig. 3a) compared with results from the 2D simulations at 30 ns (Fig. 3b). The experimentally observed perturbation has evolved into the classic nonlinear RT bubble-and-spike shape with peak-to-valley amplitude $\eta_{P/V} \approx 1$, and there are faint indications of Kelvin-Helmholtz roll-ups at the tip of the spike and along its sides. For the simulations, we use the same mapping scheme in 2D as we did in 1D, only now the Cu-CH_2 interface has a $\lambda = 200$ μm wavelength, $\eta_o = 20$ μm amplitude sinusoidal ripple. By 30 ns the perturbation has grown with the opposite phase to an overall peak-to-valley amplitude of $\eta_{P/V} \approx 180 \mu m \approx \lambda$,

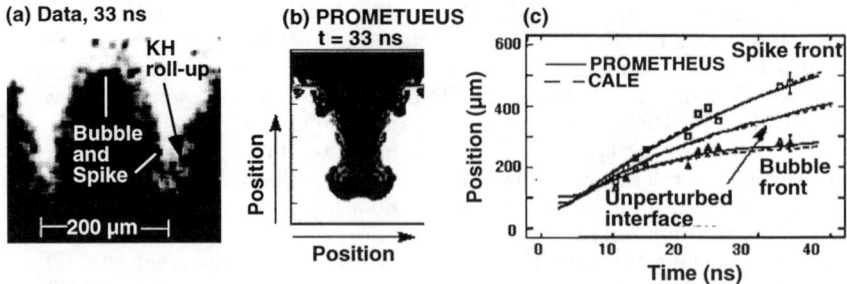

Figure 3. Comparison of data with the simulations.

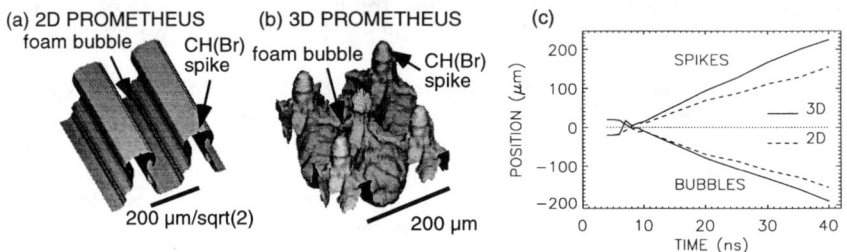

Figure 4. Simulation of RT growth in the nonlinear regime of a 2D versus 3D single mode perturbation. (2D=dotted, 3D=solid).

as shown in Fig. 3b. The shape of the perturbation has changed from sinusoidal to bubble-and-spike, indicating that the interface has evolved well into the nonlinear regime. In Fig. 3c we show the evolution of the spike and bubble fronts, compared with the predictions from both PROMETHEUS and CALE, a 2D ALE code (Tipton 1996). The locations of the 2D bubble front and spike front are reproduced very well by both hydrodynamics codes.

We have very recently designed an experiment to illustrate the differences in single-mode RT growth between 2D and 3D, as illustrated in Fig. 4. For target fabrication reasons, the target materials used were changed to CH(Br) for the dense material ($\rho = 1.53$ g/cm^3) and $CH_{1.1}O_{0.33}$ foam for the lower density material ($\rho = 0.07$ g/cm^3). These densities were chosen to maintain the same inflight Atwood number (~ 0.6). The 2D perturbation was a single sinusoid, $\eta = \eta_o \cos(kx)$, with k=$2\pi/\lambda$, and $\lambda = 200$ μm/$\sqrt{2} = 141$ μm and $\eta_o = 20\mu$m. The 3D perturbation corresponded to crossed sinusoids, $\eta = \eta_o \cos(k_x x)\cos(k_y y)$, with $k_{x,y} = 2\pi/\lambda_{x,y}$, where $\lambda_x = \lambda_y = 200\mu$m and $\eta_o = 20\mu$m. Results from the simulations in the nonlinear regime (t = 30 ns) for the 2D case are shown in Fig. 4a and for the 3D case in Fig. 4b. The spike and bubble front trajectories are juxtaposed in Fig. 4c, showing that the perturbation spike grows $\sim 50\%$ more in 3D. The effect for the bubble is less, but again the bubble grows more in 3D. Similar effects have been observed by a number of other groups doing 3D simulations (Tryggvason & Unverdi 1990; Dahlburg et al. 1993, 1995; Hecht et

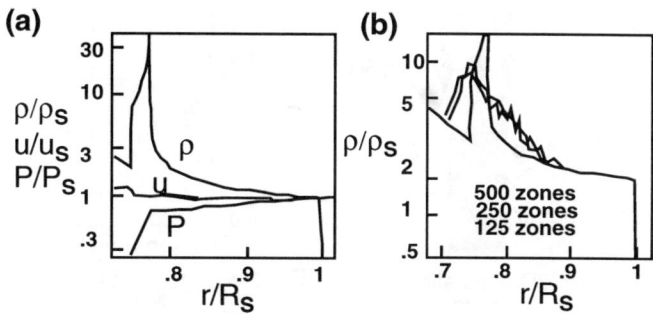

Figure 5. The density, velocity, and pressure profiles for a supernova ejecta-stellar wind interaction. (Reproduced from Chevalier et al. 1992)

al. 1995). Experiments will be conducted on the Nova laser in the near future to test this prediction. We are now doing simulations to investigate whether a similar effect occurs in the supernova.

Supernovae remnant formation is one of the classic problems of astrophysics. With SN1987A, we have for the first time the opportunity to watch the time-dependent dynamics of the early stages of SN remnant as it evolves. High-velocity supernova ejecta sweep up the surrounding ambient plasma, left over from the stellar wind of the SN progenitor. At the contact discontinuity (the place where the two plasmas meet), shocks are launched forward into the ambient plasma ("forward shock") and backward into the SN ejecta ("reverse shock"), as illustrated with the 1D profiles of density, velocity, and pressure shown in Fig. 5a (from Chevalier et al. 1992). At the contact discontinuity (at $r/R_S = 0.77$ in Fig. 5a), the pressure and density gradients have opposite signs, that is, $\nabla P \cdot \nabla \rho < 0$. Consequently, the shocked circumstellar plasma (of lower density but higher pressure) acts to decelerate the shocked SN ejecta (of higher density but lower pressure). Such a situation is hydrodynamically unstable due to the Rayleigh-Taylor instability, with strong RT growth at the contact discontinuity predicted from 2D simulations (Chevalier et al. 1992; Chevalier & Blondin 1995; Borkowski et al. 1997). This mixing at the contact discontinuity smears out the density profile, as illustrated in Fig. 5b (reproduced from Chevalier et al. 1992). The details of what to expect when the SN1987A ejecta impacts the ring nebula will depend on the structure of both the projectile assembly and the ring. It would be highly beneficial to be able to test these models experimentally prior to the awaited collision.

Hence, our second experiment is focused on testing our understanding of the colliding plasma dynamics in a situation qualitatively similar to that of the ejecta of SN1987A. Our goal is to develop the experiment and model it with the astrophysics codes used to make the predictions of the upcoming ejecta-circumstellar ring collision expected for shortly after the year 2000.

Our initial approach to experimentally simulate the ejecta-wind interaction hydrodynamics is shown in Figure 6a. We use about 20 kJ of laser energy at 0.35 μm laser wavelength, in a 1 ns pulse, to heat a 3 mm long by 1.6 mm

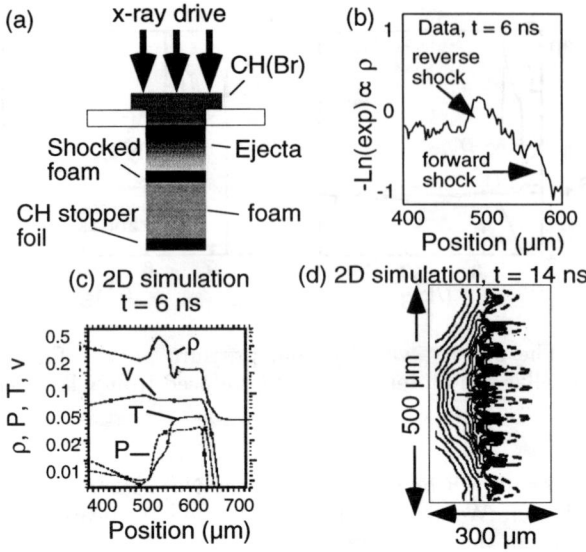

Figure 6. (a) A schematic of the laser experiment. (b) A lineout at 6 ns from the data showing a 1D density profiles at 6 ns. (c) Profiles from the 1D LASNEX simulation at 6 ns showing density (g/cm^3), pressure/100 (Mbar), ion temperature (keV), and velocity ($\times 10^8$ cm/s). (d) Isodensity contours at 14 ns from a 2D LASNEX simulation. The ejecta is flowing into the foam from left to right.

diameter cylindrical gold cavity (a hohlraum) to a temperature of about 220 eV. The x-ray flux ablates a CH plug (doped with Br to reduce the transmission of higher-energy x-rays) which is mounted in a 700 μm diameter hole in the hohlraum. The ablation drives a very strong ($\sim$ 50 Mbar) shock through the CH(Br), ejecting plasma at about 30 eV from the rear of the plug. This plasma (the ejecta) expands and cools. The leading edge of the expansion is a high-Mach-number plasma flow (about Mach 10), although it is at well below solid density. The ejecta impacts a 700 μm diameter cylinder of SiO$_2$ aerogel foam located 150 μm away and having a density of 40 mg/cm^3. In response, the flowing ejecta stagnates and a shock is driven forward into the foam (forward shock), as well as back into the ejecta (reverse shock).

We diagnose these experiments by x-ray backlighting at 4.3 keV (Sc He$_\alpha$) to obtain radiographs of the shocked matter. We show a profile of -ln(exposure) $\propto$ density from the data in Fig. 6b and from a LASNEX (Zimmerman & Kruer 1975) simulation in Fig. 6c, both at t = 6 ns. In both the data and simulation, we observe a clear forward shock in the foam, a reverse shock in the ejecta, and a contact discontinuity in between. From the simulations, we see that the shock breaks out of the CH(Br) at about 2 ns, at which time the back edge of the CH(Br) is at a density of about 2 g/cm^3 (compression of $\sim$ 2), pressure of 45 Mbar, and temperature of 30 eV. The foam is impacted by the ejecta about 1

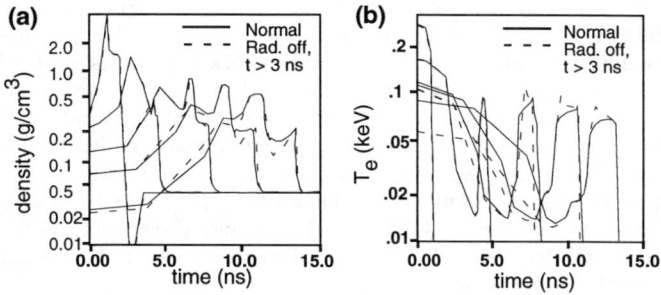

Figure 7. Look-ahead simulations of a similar ejecta-ambient plasma interaction experiment at a higher drive of $T_r = 300$ eV.

ns later, suggesting that the high velocity tail of the ejecta is moving at ~ 150 μm/ns = 150 km/s. We also show in Figure 6c the pressure, temperature, and velocity of the ejecta-foam assembly from the LASNEX simulation at 6 ns, that is, about 3 ns after the ejecta first starts sweeping up the foam. The contact discontinuity is located at a position of about 560 μm in both the data and the simulation, and the peak densities from the simulation on either side of the contact discontinuity in the ejecta (foam) are 0.65 g/cm^3 (0.25 g/cm^3). The pressure is continuous across the contact discontinuity at a peak value of 3.5 Mbar, the peak temperature is about 50 eV, and the velocity of the projectile assembly is about 1×10^7 cm/sec.

The region near the contact surface at the front of the ejecta is RT unstable (Chevalier et al. 1992; Chevalier & Blondin 1995; Borkowski et al 1997). This is illustrated in Fig. 6d for a 2D simulation of the laser experiment. In the latter, a seed perturbation of wavelength $\lambda = 50$ μm and initial amplitude $\eta_o = 1\mu$m was imposed on the surface of the foam. By 14 ns, strong RT growth of the perturbation well into the nonlinear regime is visible, due to the $\nabla P \cdot \nabla \rho < 0$ configuration at the contact discontinuity (which is indicated by the dashed curve). Comparing Fig. 6d with Fig. 10 from Chevalier et al. (1992), we conclude that both the Nova experiment and the early SN remnant evolution access roughly similar levels of nonlinear RT hydrodynamics. A similar conclusion can be arrived at by comparing linear regime RT growth rates, σ_{linear} (e.g. Eq. 6 of Müller et al. 1991), with total duration Δt_{tot} of the growth: $1/\sigma_{linear} \ll \Delta t_{tot}$ for both cases. We intend to use this experiment to test the theories and models being used to predict the behavior of SN 1987A, well in advance of the upcoming SN ejecta-ring nebula impact.

As a "look-ahead" exercise, we have done simulations where the drive radiation temperature was increased from 220 eV to $T_r = 300$ eV. Such a drive should be routine on the National Ignition Facility (NIF) laser currently being constructed at LLNL (Paisner et al. 1994; Lindl 1995). Density profiles are shown at 2 ns intervals in Fig. 7a for a nominal simulation (solid) compared to a simulation where radiation is turned off after 3 ns (dashed curve). The corresponding temperature profiles are shown in Fig. 7b. The radiation serves to distribute the heat more evenly behind the forward shock, smoothing out the density valley at the contact discontinuity. The effects of the radiation on the ensuing hydrodynamics will become more prominent at higher drive temperatures, lower target densities, and upon the multiple shock reflections that will occur when the forward shock impacts a surrogate higher density ring. Experiments where the hydrodynamics depends on correct treatment of radiation transport should be possible on the larger NIF laser currently being built.

As impressive as they seem, supernovae no longer occupy the status of most enigmatic phenomena of the universe. That title almost surely belongs to the mysterious gamma-ray bursts (GRB) (Fishman and Hartmann, 1997; Tavani, 1997; Meszaros and Rees, 1997). As described by Fishman and Hartmann, on the average of about 3 times a day, powerful pulses of gamma rays can be observed, each corresponding to the emission in the span of seconds or minutes more energy our sun will emit in its entire lifetime. The phenomenon has resisted study, since the flashes come from random directions in space and seemingly vanish wihtou a trace. This all changed on Feb. 28, 1997. The Italian-Dutch Beppo-SAX stellite detected GRB 970228, which lasted some 80 seconds. Within 8 hrs, the same spacecraft trained an x-ray telescope to that location in the sky and found a source of x-rays that was fading fast. Thereafter, ground based, and eventually the Hubble Space Telescope turned to this location, finding a rapidly fading source. If GRBs are extragalactic, as most currently believe, the represent the most powerful explosions in the univserse, eclipsing even supernovae. Not surprisingly, GRBs are the hottest topic in the astrophysics community.

Data is sparse, and theories and models abound (Tavani, 1997). Possibilities include the coalescence of compact stars, neutron stars, or black holes; failed supernovae; newborn spinars; or special jet phenomena in distant galaxies. But at the heart of essentially all models is a rapidly expanding relativistic, fireball dissipating its energy in internal (reverse) and external (forward) radiative shocks. The GRBs are initially observed with the bulk of the radiation in the 0.1 - 1 MeV range, and cool subsequently through the x-ray into the UV and optical regimes. Such high initial photon energies carrying the bulk of the radiation energy is most easily explain by a small, exceedingly hot source, $T \sim$ 0.1-1 MeV. Lasers such as the PetaWatt laser (Perry reference) are observed to produce such temperatures by focusing 0.5 kJ down to a $\sim$ 30 μm spot in a 1/2 ps pulse. Simulations of the interaction lead to a strong hemispherical blast wave being lauched in the first 100 ps after the laser pulse that propagates radially outwards. Such a blast wave may be observable in current experiments, and bear the generic characteristics of the external and internal shock models proposed to explain gamma-ray bursts. This presents an exciting opportunity to employ state-of-the-art laser facilities such as the PetaWatt laser to investigate one of most enigmatic events of current high energy astrophysics, gamma-ray bursts.

In conclusion, we are developing experiments to investigate (1) hydrodynamic instabilities relevant to core collapse supernovae in the first few hours, and (2) ejecta-ambient plasma interaction experiments relevant to the early stages of SN remnant formation. Initial results from both experiments look promising.

References

Alon, U., Hecht, J., Ofer, D., & Shvarts, D. 1995, Phys.Rev.Lett, 74, 534
Arnett, W. D. et al. 1989, ARA&A, 27, 629
Arnett, David 1994, ApJ, 427, 932
Arnett, D. 1996, in Supernovae and Nucleosynthesis, Princeton University Press, Princeton, NJ, 1996
Borkowski, K. J., Blondin, J. M., & McCray, R. 1997, ApJ, 476, L31; 1997, ApJ, in press
Chevalier, R. A., Blondin, J. M., & Emmering, R. T. 1992, ApJ, 392, 118
Chevalier, R. A. & Blondin, J. M. 1995, ApJ, 444, 312
Dahlburg, J. P. et al.1993, Phys. Fluids B, 5, 571; 1995, Phys. Plasmas, 2, 2453
Fryxell, B., Müller, W., & Arnett, D. 1991, ApJ, 367, 619
Hecht, Jacob, Alon, Uri, & Shvarts, Dov 1994, Phys. Fluids, 6, 4019
Hecht, J. et al.. 1995, Laser Part. Beams, 13, 423
Kane, J. et al. 1997, ApJ, in press
Kilkenny, J. D. 1992, Rev. Sci. Instrum., 63, 4688
Larsen, J. T. & Lane, S. M. 1994, J. Quant. Spect. Rad. Trans., 51, 179
Lindl, John 1995, Phys. Plasmas, 2, 3933
McCray R. 1993, ARA&A, 31, 175
Müller, W., Fryxell, B., & Arnett, D. 1991, A&A, 251, 505
Paisner, J. A., Campbell, E. M., & Hogan, W. J. 1994, Fusion Tech., 26, 755
Peyser, T. et al. 1995, Phys.Rev.Lett, 75, 2332
Remington, B. A. et al. 1995, Phys. Plasmas, 2, 241
Tipton, R. 1996, private communication
Tryggvason, G. & Unverdi, S. O. 1990, Phys. Fluids A, 2, 656
Zimmerman, G. B. & Kruer, W. L. 1975, Comments Plasma Phys. Controlled Fusion, 2, 51
Fishman, G. J. and Hartmann, D. H., Sc. Am. (July 1997), pg. 46.
Meszaros, P. and Rees, M. J., ApJ, 476, 232 (1997).
Tavani, M., ApJ, 483, L87 (1997).
Perry, M. D. and Moruou, G., Science, 264, 917 (1994).

Laboratory Astrophysics with Intense and Ultra-intense Lasers

Hideaki Takabe

Institute of Laser Engineering, Osaka University
Yamada-oka 2-6, Suita, Osaka 565 Japan

Abstract. Since the first laser generation by T. H. Maiman in July 1960, the laser technology has made a big progress. With the energetic lasers, the plasmas characterized by the temperature and density comparable to the interior of the Sun have been produced in laser fusion experiments. It will be possible to study complex hydrodynamic and radiation phenomena relating to, for example, the physics of evolution of stars by use of the intense lasers. We try to clarify a possible research scenario according to which a part of the critical issues to be studied in the astrophysics will be solved in model experiments with the intense and ultra-intense lasers. The paper is mainly focused on the topics related to hydrodynamic and radiation phenomena.

INTRODUCTION

Astrophysics has based on a traditional style such as "looking up the sky " (observation) and "imaging what's happening far in the sky" (theory). Recently, a rapid progress in observational technologies such as HST(Hubble Space Telescope), ASCA(X-ray satellite), COBE(Cosmic Microwave Background Explorer) etc is opening a new world in astrophysics, and the check and reconstruction of theories have been in progress intensively by comparing them to the observational data. If, in addition, it becomes possible to carry out model experiments of astrophysics in laboratories, a new evolution will come to the astrophysics, and the astrophysics will grow as a precise science.

The astrophysics aims at clarifying the evolution scenario from the Big Bang to the present structure of the universe. Therefore, if ambiguities in the physical processes in galaxy formation, star formation, increase of heavy elements due to nuclear synthesis in supernovae, and so on can be clearer with model experiments, the entitled research will grow as a new research field for astrophysics. It is meaningless just to say that the laser-produced plasmas are resemble to those observed in the universe. We have to look for a research scenario with which a part of the critical issues for astrophysics can be solved with lasers.

We can divide the basic physics which controls the evolution of the universe after a few minutes from the Big Bang roughly into four; namely, (1)hydrodynamics and magneto hydrodynamics, (2)atomic physics and radiation transport, (3)nuclear reactions, and (4)gravitational interaction. It is difficult to design some model experiments of item (4) in laboratories, while a variety of model experiments can be designed concerning the items (1), (2), and (3) with intense and ultra-intense lasers.

Based on a scaling transformation, the important physics such as hydrodynamic instabilities in supernova explosion and interaction between high Mach number shock wave and stellar matters can be modeled in a laboratory, and the controlling physics can be clarified. Not only the hydrodynamics but also so-called radiation hydrodynamics should be studied. Although heavy elements such as the irons are small abundance in the universe, they are not necessarily fully ionized and the many bound electrons contribute to complex structure of line emission and absorption. By producing such high temperature and high density plasmas by laser heating, the spectral opacity can be studied.

Line X-rays and γ-rays are good objects for new astrophysics research. For example, observational data with X-ray satellite "ASCA" are used to estimate abundance in supernova remnant (SNR). The line X-rays are emitted from a SNR, which is still in the ionizing phase even after 10^4 years from a supernova (SN) explosion. In order to calculate the material abundance of Si, Mg, and so on from the data, non-LTE atomic physics code is necessary. The code should be checked experimentally and the laser produced ionizing plasma can be a good candidate. In this case, the temperature is almost same as SNR plasma, while the density and the time scale are very different. However, it will be a reasonable model experiment by adjusting the ionization parameter defined by the product of the density and time.

In the present paper, we show a possibility of model experiments by focusing on 3 topics. The first one is blast wave phenomena of high Mach number, which relate to star burst and supernova explosion. The second one is hydrodynamic instability and convection, which relate to the mixing in SN explosion as infered for the case of SN1987A. The third one is opacity and non-LTE physics, which relate to nova light curve etc.

Let us clarify the importance and necessity for promoting the Laboratory Astrophysics with intense lasers. In the astrophysics, especially relating to evolution of stars, complex hydrodynamic phenomena should be studied from the first principle. There is no other candidate to produce plasmas in the laboratories whose temperature and density are comparable to those of stars. In addition to the observational technology such as HST, by carrying out model experiments, the astrophysics will newly evolve as a precise science.

In laser fusion research, on the other hand, hydrodynamic instability and mixing issues obstract our pass as a critical issue to be solved. There is no promise that the complex hydrodynamic phenomena are surely controlled, even if we improve laser uniformity and target sphericity technically. To promote the entitled research will enrich the laser fusion research from the point of academia. In addition, the laser fusion research will qualitatively change toward a fruitful direction by getting astrophysicists around the foot of the "laser fusion" mountain.

WHY LASERS FOR ASTROPHYSICS

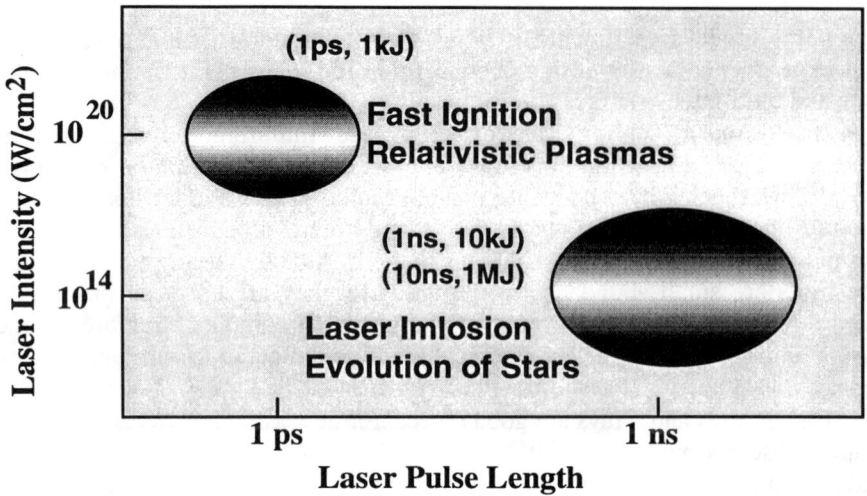

FIGURE 1. Physics of laser fusion is related to that of astrophysics in two regimes

Relating to laser fusion, the intensity of lasers ranges from 10^{14} W/cm^2 for driving implosion hydrodynamics in the central ignition scheme to 10^{20} W/cm^2 for igniting the edge of compressed core in the fast ignition scheme as shown schematically in Fig. 1. When the laser light of intensity of 10^{14}-10^{15} W/cm^2 is irradiated on a solid target, the plasma temperature increases up to keV and an extremely high pressure in the range of 100 Mbar is generated. With use of such high pressures, a strong blast wave, high density and high temperature plasmas seen near stars can be generated in laboratories.

We can enumerate another ways, such as shock tube and high explosive to generate high pressure. However, the typical pressure generated by the former is several tens of atmosphere and the Reynolds number is low. In case of the latter, the pressure reaches a few hundreds of kbar, while ionization of high dense matter can not be expected. With intense laser the matter with density of about 1 g/cm^3 can be pressurized up to 100 Mbar. These papameters are typical ones of, for example, SN1987A explosion around hundreds of seconds after explosion. Since the hydrodynamic instability became drastic after this time, model experiments with lasers can be reasonable for studying this matter.

With ultra-intense lasers generated by CPA (Chirped Pulse Amplification), on the other hands, the solid matter is heated without hydrodynamic motion and relativistic electron plasma with energy of several MeV can be confined in dense matter. Then, it will be possible to generate electron-positron plasmas and it may become possible to study relativistic plasma physics seen near Black Holes and neutron stars.

It is very important to point out a similarity in research style between laser fusion and astrophysics. Laser fusion research is promoted by balanced coupling of four elements as shown in Fig. 2. In experiment, a model experiment is necessary to study a certain element of physics and a theoretical model should be

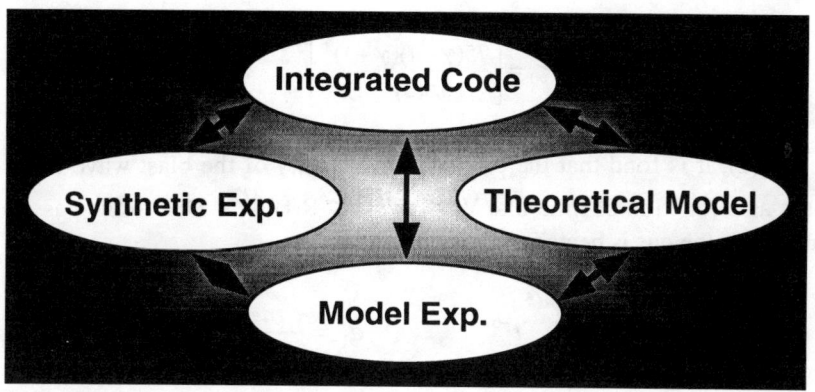

FIGURE 2. Four critical elements for studying complex physics like fusion and astrophysics

checked through comparison with the model experiment. After this process for a variety of theoretical models, they are installed into an integrated implosion code which simulates all scenario from laser irradiation to fusion burn. The integrated code is essential to analyze and predict complex physics such as laser implosion experiment, where many of different physics couple at the same time.

In astrophysics, the observation is like a synthetic experiment and the same reseach style as shown in Fig. 2 are required. However, the model experiment especially relating to the evolution of stars has not been available. The intense and ultra-intense lasers are a strong candidate to fill up the thirstiness of astrophysics.

HIGH MACH NUMBER BLAST WAVES

It has been widely observed that strong shock waves propagate in the interstellar medium and galaxy, when an extremely large amount of energy is released in a short time at a localized point due to supernova explosion or star burst. These phenomena are important relating to star birth and origin of heavy elements and resultant natural abundance. The starting point to discuss the blast wave is well-known Sedov (Taylor-Sedov-von-Neuman) self-similar solution. However, the molecular gas in the universe is heated up to the temperature of keV to be ionized. The blast wave, therefore, is not an ideal one, but complex phenomena coupled with radiation transport could be seen. We can model such phenomena in a chamber filled with a gas, by irradiating intense laser on a target.

Provided that the energy E is released at point r = 0 at time t = 0, the radius of the blast wave is calculated to be

$$R(t) = \xi_0 \left(\frac{E}{\rho_0}\right)^{1/5} t^{2/5} \tag{1}$$

where ρ_0 is the density of gas and ξ_0 is non-dimensional constant determined only with the specific heat ratio of the gas γ and give by Chernyi, for example, to be[1]

$$\xi_0 = \left[\frac{75(\gamma-1)(\gamma+1)^2}{16\pi(3\gamma-1)}\right]^{1/5} \quad (2)$$

From Eq. (1), it is fond that the propagation velocity of the blast wave front (V) decreases as a function of time like $V \propto t^{-3/5} \propto R(t)^{-3/2}$. Due to the passage of the blast wave, the gas is heated to the temperature

$$T(t) = \xi_1 \left(\frac{E}{\rho_0}\right)^{2/5} t^{-6/5} \quad , \quad \xi_1 = \frac{m}{\gamma+1}\left(\frac{2}{5}\xi_0\right)^2 \quad (3)$$

where m is the mass of a gas particle.

Since the radiative properties depend upon the temperature, let us evaluate how we can generate a scale-transformed blast wave of supernova explosion with laser irradiation. We assume the following parameters for a typical SN blast wave. The density $\rho_0 = mn_0$, where m is a proton mass and $n_0 = 1$ cm^{-3}. The energy E = 10^{51} (erg) and t = 10^4 years (3×10^{11}s). Then, the radius of Eq. (1) is 40 light years (ly) and the temperature of Eq. (3) is 300 eV. These are reasonable value comparing to Vela SNR[2]. For laser produced blast wave, we take the parameters E = 1 kJ and t = 10 ns. Requiring the temperature of Eq.(3) to be 300 eV, we obtain the gas density 3×10^{17}cm^{-3}, namely about 1% of the atmosphere. The radius of Eq. (1) is 6 mm. These are appropriate vales as laboratory experiment.

An example of laser-produced blast wave is shown in Fig. 3[3]. A plane aluminum target is set in nitrogen gas chamber and laser is irradiated from the left on the target surface. As the result, the ejecting velocity of aluminum plasma becomes 700 km/s and a high Mach number blast wave is generated on the left side. On the right side, in contrast, a turbulent structure is seen. This is thought to

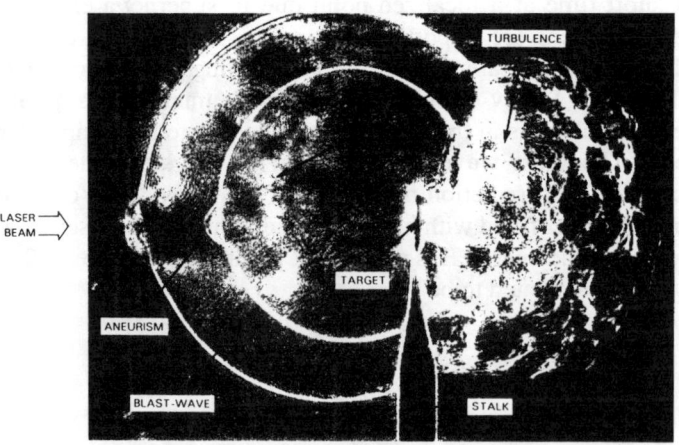

FIGURE 3. Blast wave and turbulence generatedby laser irradiation[3]

be due to hydrodynamic instability of accelerated aluminum target.

For example, by using one more beam to irradiate another target located on the right of Fig. 3, we can study physics of collision of blast wave with turbulence[4]. In the space, density in not necessarily uniform and it is interesting to see what happens when a shock wave collides a bumpy cloud. So-called baloclinic term ($\nabla \rho \times \nabla P$) causes vortex generation and anomalous energy dissipation of the shock wave will be expected[5].

In the space, there is a weak background magnetic field of about 3×10^{-5} Gauss[6]. In some case, the effect of the magnetic field becomes important in blast wave propagation. A model experiment by Ripin et al[7] indicates that a filament structure can be seen in the later stage of blast wave expansion because of Rayleigh-Taylor like instability by magnetic pressure. Such filament formation possibly explains the origin of the filament seen in the Crab nebula.

The blast wave seen, for example, in Fig. 3 seems to be stable (spherical). The blast wave become unstable, however, when the ionization and radiation cooling becomes drastic near the front[8]. When the blast wave is high-Mach number and the gas is ionized, a simple Sedov solution is not applicable to explain the complex phenomena. By doing a variety of model experiments, we may see more diverseness in the phenomena relating to supernova explosions and star bursts.

HYDRODYNAMIC INSTABILITY AND CONVECTION

It can be expressed that the explosion of SN1987A was a magnificent "experiment" in the universe. This event has pushed researchers to study hydrodynamic instabilities in the explosion scenario. Then, technologies such as computer simulation have made a big progress. Even after ten years from the explosion, we recognize that SN1987A is a very important object in clarifying astrophysics. The recent news on observation of collision of ejecting material with a stellar ring is picking up the interests of many people.

It is no exaggeration to say that the history of laser fusion research is a history of fight with hydrodynamic instability. The present status and future prospects on the hydrodynamic instability is given recently in a review paper[9]. As hydrodynamic instability at the density surface, which is a contact surface, density jump surface, or even a continuous region with a density gradient, we can enumerate Richtomyer-Meshkov(R-M) and Rayleigh-Taylor(R-T) instabilities.

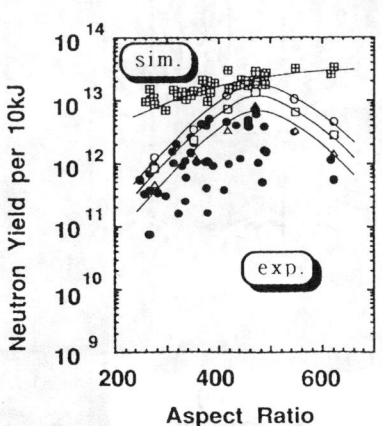

FIGURE 4. Neutron yields as a function of glass shell aspect ratio. Turbulent mixing model can reproduce the experimental yields

The R-T instability is an instability driven by the mechanism that an excess potential energy of fluids is released as fluid kinetic energy, when a heavy fluid is posed on a light fluid in a gravitational or inertial forces. It is well-known that the instability is an absolute instability localized around the density surface and the linear growth rate is $(kg)^{1/2}$, where k is the wave number of the perturbation and g the gravitation.

In laser fusion experiments, the fuel is heated and compressed by imploding a fuel pellet of plastic or glass capsule with multi-beam laser irradiation. This physical process is called "implosion". The experimental DT neutron yield are plotted in Fig. 4[10]. The horizontal axis is the aspect ratio of the glass shell defined to be a ratio of the radius to the thickness. The neutron yields are values per 10 kJ green laser irradiation. The corresponding one-dimensional neutron yields are plotted with ⊞ marks. It is clear from Fig. 4 that the experimental values are one or two orders of magnitude lower than the computational ones. It has been concluded that the degradation is due to the hydrodynamic instability.

When a capsule is accelerated toward the target center, the outer surface of the capsule becomes R-T unstable. The ablative flow across the unstable surface, however, modifies the growth rate of the instability to the form:

$$\gamma = \alpha\sqrt{kg} - \beta k V_a \qquad (4)$$

where $\alpha=0.9$ and $\beta = 3-4$ and V_a is the ablation velocity at the density peak. Thanks for this ablative stabilization mechanism, shorter wavelength modes are stabilized[11].

Around the final compression after the acceleration phase, however, the contact surface between fuel and the capsule becomes R-T unstable because of the

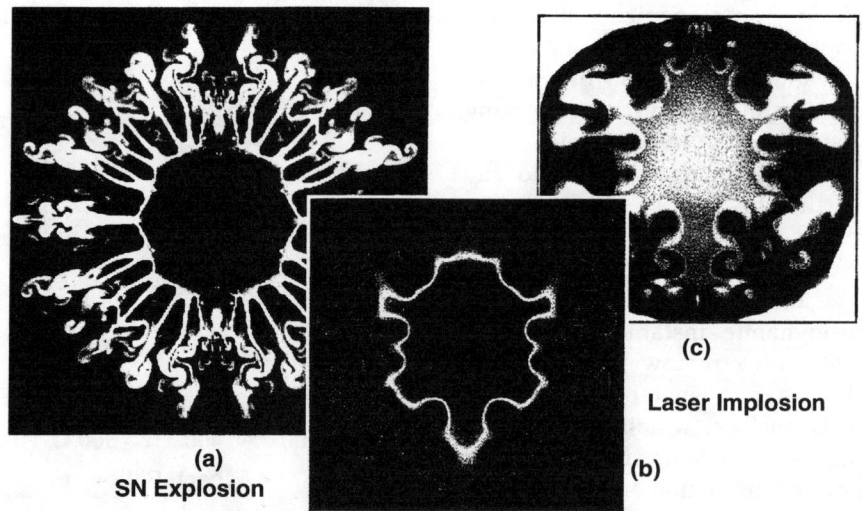

FIGURE 5. Two dimensional simulations of supernova explosion (size:10^8 km) and laser implosion (size:0.1mm).

deceleration of capsule material by the fuel compressed and heated at the target center. This is called R-T instability in the stagnation phase. This is seen in Fig. 5, where the capsule material penetrates into the fuel as seen in Fig. 5-(c). Figures 5-(b) and 5-(c) show a two-dimensional simulation results for the temperature and density, respectively. Because of such hydrodynamic instability, the experimental neutron yield are reduced compared to 1-D simulational ones as shown in Fig. 4

In order to study the discrepancy in Fig. 4, we have developed a mixing model with which penetration of glass material into the fuel region due to the instability and the resultant fuel cooling are modeled. The resultant neutron yields are plotted with □ marks in Fig. 4. It is found that the 5-10 % of the directed kinetic energy of the implosion should be converted into the turbulence energy through the acceleration phase in order to explain the experimental results.

When SN1987A exploded, the He/H contact surface is at first accelerated outward by a shock wave and then decelerated by a pressure gradient formed by a reflected shock wave[12]. In this deceleration phase, the contact surface becomes R-T unstable. As the result, relatively heavy helium penetrates outward with fingure-like structure[13]. Due to material mixing driven by such R-T instability in the supernova with mass more than 20 solar mass, the ^{56}Ni which is nuclear synthesized in the interior are pushed toward the outer region. As the result, the γ-rays emission stemming from the nuclear decay of radioactive element (^{56}Ni -> ^{56}Co -> ^{56}Fe) occurs in the outer region. This is the mechanism for why the x-rays which are produced through Compton scattering process of the γ-rays have been observed much earlier than predicted with a one-dimensional SN explosion code[14].

SN1987A is Type II supernova, whose explosion is due to gravitational collapse of a single massive star. As the result of the explosion, relatively light elements O, Mg, Si etc are scattered into the space. Relatively heavy elements like Fe, on the other hand, mainly originate Type Ia supernovae. In this case, the explosion is triggered by the propagation of nuclear deflagration wave ignited near the central core of C/O rich White Dwarfs. The turbulent burning wave physics is now a hot topics and many studies are on going[15].

In this case, the turbulence is generated by the R-T instability of the nuclear deflagration front and the physics is resemble to the deflagration wave of ablation surface driven by laser irradiation. Inserting parameters of Type Ia supernova into Eq. (4), for example, we can obtain the shortest wavelength of the unstable mode ℓ_{min} = 5 (km) [16]. By taking the size of a White Dwarfs for the longest wavelength of the unstable modes, we obtain $\Gamma = \ell_{max}/\ell_{min} = 10^3$. This indicates that vortices whose sizes are different over 3 orders of magnitude co-exist in the turbulence. It should be noted that in case of high gain laser fusion targets, most unstable mode has the spherical wave number of a few hundreds and the instability of deflagaration front of $\Gamma = 10^2$-10^3 should be also treated.

The physics of convection in the interior of stars is very important to determine the ages of a stars, which is strongly related to the estimation of the age of the universe. If we can model a slowly growing hydrodynamic phenomena in laboratory, an energy transport through the convection of hot-dense matter will be clarified. Consequently, evaluation of the ages of stars will be modified.

Since the explosion of SN1987A, many of multi-dimensional simulations have been carried out[17]. It is concluded that in order to explain the time history of hard x-ray observed by x-ray satellite "Ginga", it is necessary to assume about 5 % of velocity nonuniformity at the beginning[18]. However, the reason why such level of nonuniformity was generated or existed is still open question and the origin of this nonuniformity has been studied with multi-dimensional codes[19]. Hydrodynamic phenomena of the early stage of supernova explosion are of complex physics where neutrino transport, shock wave, and nuclear synthesis couple, while it attracts theoretical scientists as a gland challenge with super-computer[6]. The laser implosion is also a complex physics where the energy transports of x-ray radiation, fusion-product particles, and electrons couple with hydrodynamic phenomena. The both stand on the same base from the point of physics models and numerical schemes.

Recently, a model experiment of hydrodynamic instability in supernova explosion has been carried out with NOVA laser[20]. This experiment aims at study of evolution of spike and mushroom structure in the similar condition to the supernovae. The experimental results have been compared to two-dimensional simulations with laser-fusion oriented code and supernova oriented code. The nonlinear evolution of bubble and spike are well reproduced with both codes, but the structures, for example, are different and more detail studies are required.

OPACITY AND NON-LTE ATOMIC PHYSICS

Two physical mechanisms control the transport of energy generated in the central region of stars. One is the material convection stemming from the hydrodynamic instability explained in the previous section and the other is the radiation transport. So-called radiation hydrodynamics is one of the most important subjects in astrophysics.

The radiation transport itself is governed by a simple equation. By defining I_ν the radiation energy flux per unit energy interval (keV), the equation for I_ν (erg/cm^2 s keV rad) is given to be

$$\frac{1}{c}\frac{\partial}{\partial t}I_\nu + \Omega \cdot \nabla I_\nu = \eta_\nu - \kappa_\nu I_\nu \tag{5}$$

where Ω is the unit vector indicating the direction of radiation propagation. The propagation velocity is assumed to be the speed of light c. In Eq.(5), η_ν and κ_ν are spectral emissivity and opacity.

Equation (5) is easy to understand, while it is very hard to solve directly because of multi-dimensionality including angular distribution of photon[21]. In addition, it is a hard job to determine the spectral opacity and emissivity of partially ionized hot plasmas[22]. Although it is hard even in assuming local-thermodynamic equilibrium (LTE) for plasma atoms, it becomes a grand challenge for the case of non-LTE plasmas. In many cases in laser and astrophysical plasmas, non-LTE problems appear to be essential.

Recently, it has become possible to calculate the complicated spectral structure of line groups based on statistical models called UTA(Unresolved Transition Array)[23] and STA(Super Transition Array)[24]. In addition, material opacities can be calculated from the first principle with much CPU time of CRAY and have been compared to data obtained in opacity experiment with intense lasers[25]. The fine structure stemming from line transition group has been well reproduced with, for example, OPAL code[26]. Such code has been used to explain the observation data of Petersen diagram of variable Cepheids[26], the light curve of Cygni 1978 Nova[28] and so on. There are many related topics concerning the precise opacity. For example, the cooling rate of white dwarfs(WD) is controlled by radiation transport and, for example, is important to determine the age of a galaxy. If the cooling rate is much faster than expected, the WD's can also be a candidate of Dark Matter.

Not only the opacity but also non-LTE atomic process is an essential subject to analyze observed data. Recently, encouraged by a success of, for example, ASCA x-ray satellite, astrophysics with line x-ray and γ-ray are pursued[29]. One of the key subjects is study of the origin of heavy element in the universe. The SNR is a target and line emission from the SNR has been compared to a simulation based on time dependent ionization model. In this case, the ionization parameter τ (= tn) defined by the product of time and density is a controlling parameter. For example, SNR of Tycho is evaluated to have t of 400 years and n of roughly 1 cm^{-3}, and therefore $\tau = 10^{10} cm^{-3} s$. In laser produced plasma, such value of ionization parameter can be easily realized with 1 ns pulse and the plasma density of $10^{19} cm^{-3}$. So, the laser plasma can be a good test bed to check the non-LTE atomic process code for astrophysical purpose.

Due to the limitation of page, I left more detail explanation and more fruitful aspect regarding radiation and atomic process issues in laboratory astrophysics in another opportunity.

SUMMARY

The compressible hydrodynamics coupled with atomic physics should be studied to understand a variety of phenomena observed in the universe. In supernova explosion and star burst, high-Mach number blast wave plays important role and it is required to clarify the physics of hydrodynamics coupled with atomic process. In addition, radiation process is a key element to transfer energy inside stars. It appears to be important in many cases through the coupling with hydrodynamics, and is called radiation-hydrodynamics. Such physics can be studied with intense and ultra-intense lasers.

The physics of laser fusion is governed by compressible hydrodynamics and radiation physics. In laser implosion, the matter has been compressed to the temperature and density comparable to those of stars. For studying laser fusion, theoretical works on hydrodynamics and radiation are essential. Since these topics are directly related to those in astrophysics, the study of laser fusion can give its progress through the theoretical and computational researches. In addition, we can carry out a study of astrophysics in laboratories, provided that model experiments are planed to be done with intense and ultra-intense lasers.

ACKNOWLEDGMENTS

The author would like to thank Prof. K. Mima and other staffs of ILE, Osaka University for valuable discussion. He also thanks Prof. K. Nomoto of University of Tokyo for valuables discussions on astrophysics and encouragement.

REFERENCES

[1] Ya. B. Zel'dovich and Yu. P. Raizer, "Physics of Shock Waves and High-Temperature Hydrodynamic Phenomena" (Academic, 1966) Chap.I, p. 99.
[2] B. Aschenbach et al, Nature **373**, 587(1985)
[3] B. H. Ripin et al., in "Laser Interaction and Related Plasma Phenomena" Vol.7 (Plenum, 1986), p.837.
[4] B. H. Ripin et al, "Laser Interaction with Matter" eds. G. Velarde et al, (World Scientific, 1989), p. 196.
[5] D. Rotman, Phys. Fluids A**3**, 1792 (1991).
[6] M.L. Norman, Phys. Today, Oct. (1996), p.42.
[7] B. H. Ripin et al, Phys. Fluids B**5**, 3491 (1993).
[8] J. Grun et al., Phys. Rev. Lett. **66**, 2738 (1991).
[9] H. Takabe, "Hydrodynamic instability in laser implosion and supernova explosion", to be published.
[10] Y. Isayama et al, J. Plasma and Fusion Res. **70**, 756 (1994).
[11] H. Takabe et. al., Phys. Fluids **28**, 3676 (1985)
[12] T. Ebisuzaki et al, Ap J. **344**, L65 (1989).
[13] I. Hachisu et al, Ap J. **358**, L57 (1990): Ap J. **390**, 230 (1992).
[14] S. Kumagai et al, Ap J. **345**, 412 (1989).
[15] P. Puiz-Lapuente et al., ed., "Thermonuclear Supernovae", NATO ASI Ser.(C) Vol. 486 (Kluwer Acad. Pub., Dordrecht, 1977) Chaps. VII and VIII.
[16] E. Livne and D. Arnett, Astrophys. J. **415**, L107 (1993).
[17] E. Muller et al, A &A **251**, 505 (1991).
[18] I. Hachiso, "Nagare", **16**, 40 (1997) in Japanese.
[19] K. Sato et al., in "Elementary Process in Dense Plasma" ed. S. Ichimanu (Addison-Wesley, Mass, 1995) p.35.
[20] J. Kane et al., Astrophys. J. **478**, L76 (1997).
[21] J. M. Stone et al, Ap J. Suppl. **80**, 819 (1992).
[22] H. Takabe and T. Nishikawa, J. Quant. Spectrosc. Radiat. Transfer, **51**,379 (1994).
[23] J. Bauche et al., Phys. Rev. A **20**, 2424 (1979).
[24] A. Bar-Shalom, et.al, Phys. Rev. A. **40**, 3183 (1989).
[25] L. B. DaSilva et al., Phys. Rev. Lett. **69**, 438 (1992).
[26] F. J. Rogers and C. A. Iglesias, in "Elementary Process in Dense Plasma" ed. S. Ichimanu (Addison-Wesley, Mass, 1995) p.181.
[27] P. Moskalik et al., Astrophys. J. **385**, 685 (1992).
[28] M. Kato and I. Hachisu, Astrophys. J. **437**, 802 (1994).
[29] H. Tsunemi et al., in "Thermonuclear Supernovae", NATO ASI Ser.(C) Vol. 486 (Kluwer Acad. Pub., Dordrecht, 1977) p.561.

AUTHOR INDEX

A

Adam, J. C., 32
Ahn, H., 516, 526
Akahane, Y., 461
Allott, R., 473
Anderson, D., 79
Andiel, U., 264
Andreev, A. A., 61
Aoyama, M., 461
Arnett, D., 551

B

Backus, S., 331
Barbee, Jr., T. W., 499
Barbini, A., 253
Barty, C. P. J., 461
Batani, D., 288, 372
Bauer, D., 201
Bayanov, V. I., 61
Behjat, A., 473, 491
Bell, A. R., 147, 389
Benuzzi, A., 288, 372
Bernardinello, A., 288, 372
Biswal, S., 397
Bonnaud, G., 383
Borisov, A. B., 322
Borodin, V. G., 445, 450
Bossi, S., 288
Boyd, T. J. M., 164
Boyer, K., 322, 454
Budil, K. S., 551
Budnik, F. W., 342
Bulanov, S. V., 113, 123

C

Califano, F., 113, 123, 282
Chambaret, J. P., 423
Chambers, D. M., 264, 479, 499
Chang, Z., 296
Chanteloup, J. C., 397
Charukchev, A. V., 439, 445, 450
Chen, S.-Y., 103

Cheng, Z., 304
Chernobrovtseva, M. D., 79, 85
Chernov, V. N., 439, 445, 450
Chessa, P., 253
Chizhov, S. A., 61
Christov, I., 296
Collier, J., 473
Conejero Jarque, E., 360
Cornolti, F., 55, 201, 282

D

Dangor, A. E., 264, 499
Danson, C. N., 473, 479, 485, 491, 499
Davies, J. R., 389
Demir, A., 473, 491
De Silvestri, S., 304
Dewa, H., 516, 526
Ditmire, T., 314, 354, 366
Djaoui, A., 372
Downer, M. C., 342
Drake, R. P., 551
Druon, F., 397
Durfee, C. G., 331
Dzhidzhoev, M. S., 241

E

Edwards, C. B., 485
Efanov, V. M., 439
Eidmann, K., 270
Ellwi, S., 372
Ermolaev, A. E., 129
Esirkepov, T. Zh., 113
Estabrook, K., 551

F

Faral, B., 288
Faure, J., 397
Fedosejevs, R., 348
Feurer, T., 213
Fews, P., 372
Fill, E. E., 336

Fittinghoff, D., 461
Franco, M. A., 423

G

Gallant, P., 231
Gerke, R. R., 450
Gildenburg, V. B., 15
Giulietti, A., 253
Giulietti, D., 253
Gizzi, L. A., 253
Glendinning, S. G., 551
Goloviznin, V. V., 532
Gordienko, V. M., 241
Guérin, S., 32
Gumbrell, E. T., 354

H

Hain, S., 201
Hall, T. A., 288, 372
Häßner, R., 213
Hay, N., 366
Healy, S. B., 491
Héron, A., 32
Higaki, H., 342
Hönninger, C., 397
Hutchinson, M. H. R., 314, 354, 366, 467, 485

I

Ichimaru, S., 541
Itatani, J., 397

J

Janulewicz, K. A., 491
Jarque, E. C., 55
Jiang, Z., 231
Joachain, C. J., 3, 129
Joukov, M. A., 241

K

Kalachnikov, M. P., 473, 491
Kando, M., 516, 526
Kane, J., 551
Kapteyn, H. C., 296, 331
Karttunen, S. J., 377
Key, M. H., 372, 473, 499
Kieffer, J. C., 231
Kim, A. V., 15, 79, 85
Kingham, R. J., 147
Kishimoto, Y., 192
Kitamura, H., 541
Koenig, M., 288, 372
Komarov, V. M., 445
Kondo, S., 516
Kotaki, H., 516, 526
Kozlov, A. A., 61
Krasov, S. V., 445
Krausz, F., 304
Krishnan, J., 372
Kulcsár, G., 342
Kurnin, I. V., 61
Kylstra, N. J., 129

L

Lange, H. R., 423
Laval, G., 32
Le Blanc, S. P., 342
Lefebvre, E., 383
Lenzner, M., 304
Lewenstein, M., 276
Lewis, C. L. S., 473, 491, 499
Liang, E., 551
Lichters, R., 41, 141
Lin, J., 499
Lisak, M., 79
London, R., 551
Longworth, J. W., 454
Lontano, M., 15, 85, 113
Lopes, N., 135
Löwer, Th., 288

M

Macchi, A., 55, 282
MacPhee, A. G., 499

Magnitskii, S. A., 73
Maksimchuk, A., 103, 342
Malinov, V. A., 439, 445, 450
Managan, R. A., 551
Marangos, J. P., 314
Marjoribanks, R. S., 342
Masella, V., 372
Mason, M. B., 314, 366, 467
Matsuoka, S., 461
McCray, R., 551
McPherson, A., 322
Mendonça, J. T., 49, 135
Meyerhofer, D. D., 366
Meyer-ter-Vehn, J., 41, 93, 141, 153, 159
Michelmann, K., 213
Migel, V. M., 445, 450
Mima, K., 176, 182
Miquel, J. L., 231
Mora, P., 32
Mourou, G., 397
Muller, L., 288
Mulser, P., 201, 377
Murnane, M. M., 296, 331
Mysyrowicz, A., 423

N

Nagashima, K., 192
Nakajima, K., 516, 526
Nakanishi, H., 516
Nantel, M., 397
Naumova, N. M., 113
Neely, D., 372, 473, 479, 485, 491, 499
Nees, J., 397
Nelson, T., 322, 454
Nickles, P. V., 473, 491
Niedermeier, S., 213
Nikitin, N. V., 439, 445, 450
Nilsen, J., 499
Nishikawa, K., 176, 182, 186
Nishimura, A., 397
Nisoli, M., 304
Norreys, P. A., 264, 372, 479, 499

O

Ogata, A., 516, 526
Oliveira e Silva, L., 135
Omenetto, F. G., 322, 454

O'Rourke, R. M. N., 499

P

Pättikangas, T. J. H., 377
Pegoraro, F., 113, 123
Pépin, H., 231
Pepler, D. A., 473
Pert, G. J., 473, 491, 499
Peyrusse, O., 231
Pfund, R. E. W., 141
Pisani, F., 372
Plaja, L., 67, 276, 360
Platonenko, V. T., 73
Platonov, K. Y., 61
Pogorelsky, I. V., 415
Popov, V. S., 445
Potapov, S. L., 439, 445
Prade, B. S., 423
Prandi, R., 123
Preston, S. G., 479, 499
Pretzler, G., 264, 336
Pukhov, A. M., 41, 93, 113, 159

Q

Quesnel, B., 32
Quinn, F., 479
Quiroga-Teixeiro, M., 15

R

Remington, B. A., 551
Rhodes, C. K., 322, 454
Ripoche, J.-F., 423
Roper, M., 479
Rosa, C. C., 135
Rose, S., 372
Roso, L., 67, 276
Rousseau, P., 423
Rubenchik, A., 551
Ruhl, H., 282
Rundquist, A., 296
Ryutov, D., 551
Rzążewski, K., 276

S

Saemann, A., 270
Sakai, H., 516
Salomaa, R. R. E., 377
Sandner, W., 473, 491
Sarkisov, G. S., 103
Sartania, S., 304
Sauerbrey, R., 213
Savel'ev, A. B., 241
Schep, T. J., 532
Schillinger, H., 213
Schneider, W., 377
Schnürer, M., 473, 491
Schroeder, W. A., 322, 454
Sergeev, A. M., 15, 85
Shashkov, A. A., 241
Sheng, Z.-M., 153, 159
Shlyaptsev, V. N., 473
Smith, R., 499
Smith, R. A., 314, 354
Solovyev, N. A., 61
Spielmann, Ch., 304
Springate, E., 314
Stagira, S., 304
Svelto, O., 304

T

Takabe, H., 560
Takuma, H., 192, 397, 461, 509
Tallents, G. J., 473, 491, 499
Tarasishin, A. V., 73
Tatarinov, A. V., 164
Teychenne', D., 253
Theobald, W., 213
Thompson, B. D., 322
Tisch, J. W. G., 314, 354, 366
Toupin, C., 383
Tsakiris, G. D., 264, 348
Tsintsadze, L. N., 170, 176, 182, 186
Tsintsadze, N. L., 49, 170

U

Ueda, T., 516
Uesaka, M., 516
Umstadter, D., 103, 342

V

Vankov, A. B., 61
Vitcu, A., 342
Volkov, R. V., 241
von der Linde, D., 221
Vshivkov, V. A., 113

W

Wagner, R., 103, 342
Wallace, R. J., 551
Wang, H., 296
Wang, X. F., 348
Wark, J. S., 264, 479, 499
Warwick, P. J., 473, 491
Watanabe, T., 516
Watts, I. F., 264, 499
Wolfrum, E., 473, 491, 499
Wyborn, B., 485

Y

Yamakawa, K., 461
Yarin, P. M., 439
Yashin, V. E., 61
Yoshii, K., 516
Yusupov, I. Yu., 450

Z

Zepf, M., 264, 479, 499
Zhang, J., 473, 499
Zhao, L., 342